# Essential Concepts in Molecular Pathology

# Essential Concepts in Molecular Pathology

## Second Edition

Edited by

**William B. Coleman**
American Society for Investigative Pathology, Rockville, MD, United States

**Gregory J. Tsongalis**
The Audrey and Theodor Geisel School of Medicine at Dartmouth, Hanover, NH, United States
Laboratory for Clinical Genomics and Advanced Technology (CGAT), Department of Pathology
and Laboratory Medicine, Dartmouth Hitchcock Health System, Lebanon, NH, United States

ACADEMIC PRESS
An imprint of Elsevier

Academic Press is an imprint of Elsevier
125 London Wall, London EC2Y 5AS, United Kingdom
525 B Street, Suite 1650, San Diego, CA 92101, United States
50 Hampshire Street, 5th Floor, Cambridge, MA 02139, United States
The Boulevard, Langford Lane, Kidlington, Oxford OX5 1GB, United Kingdom

**British Library Cataloguing-in-Publication Data**
A catalogue record for this book is available from the British Library

**Library of Congress Cataloging-in-Publication Data**
A catalog record for this book is available from the Library of Congress

ISBN: 978-0-12-813257-9

For Information on all Academic Press publications
visit our website at https://www.elsevier.com/books-and-journals

*Publisher:* Stacy Masucci
*Acquisition Editor:* Tari K. Broderick
*Editorial Project Manager:* Sam W. Young
*Production Project Manager:* Stalin Viswanathan
*Cover Designer:* Matthew Limbert

Typeset by MPS Limited, Chennai, India

Working together
to grow libraries in
developing countries

www.elsevier.com • www.bookaid.org

# Dedication

This textbook contains a concise presentation of essential concepts related to the molecular pathogenesis of human disease. Despite the succinct form of this material, this textbook represents the state-of-the-art and contains a wealth of information representing the culmination of innumerable small successes that emerged from the ceaseless pursuit of new knowledge by countless experimental pathologists working around the world on all aspects of human disease. Their ingenuity and hard work have dramatically advanced the field of molecular pathology over time and in particular in the last several decades. This book is a tribute to the dedication, diligence, and perseverance of the individuals who have contributed to the advancement of our understanding of the molecular basis of human disease. We dedicate *Essential Concepts in Molecular Pathology, Second Edition* to our colleagues in the field of experimental pathology and to the many pioneers in our field whose work continues to serve as the solid foundation for new discoveries related to human disease. In dedicating this book to our fellow experimental pathologists, we especially recognize the contributions of the graduate students, laboratory technicians, and postdoctoral fellows, whose efforts are so frequently taken for granted, whose accomplishments are so often unrecognized, and whose contributions are so quickly forgotten.

*Essential Concepts in Molecular Pathology, Second Edition* is dedicated to the memory of Jennifer K. Denning, who passed away on June 3, 2018 a few months short of her 26th birthday due to complications associated with her treatment for lymphoma. Jennifer was a vivacious, interesting, adventurous, and energetic young woman that spent her life advocating for children's literacy, women's rights, and other worthwhile issues. While she accomplished much in her life, she did not complete everything that she intended to accomplish due to her diagnosis with lymphoma in the Fall of 2017. Following her diagnosis, Jennifer underwent a rigorous schedule of treatment with chemotherapy. Her therapy was successful, but she succumbed to complications brought on by the treatments she received. Jennifer's story is important in that it shows that for many diseases, we continue to rely upon crude drugs that have many unintended consequences for the patient. Jennifer reminds us that we still have far to go and much to learn before sophisticated, personalized, and effective treatments can be implemented for all diseases, maximizing therapeutic efficacy while minimizing adverse effects of treatment. Inspired by her memory, we will endeavor to work harder and longer in pursuit of new advances in biomedical science that will translate into tangible benefits for patients in the clinic.

We also dedicate *Essential Concepts in Molecular Pathology, Second Edition* to the many people that have played crucial roles in our successes. We thank our many scientific colleagues, past and present, for their camaraderie, collegiality, and support. We especially thank our scientific mentors for their example of research excellence. We are truly thankful for the positive working relationships and friendships that we have with our faculty colleagues. We also thank our students for teaching us more than we may have taught them. We thank our parents for believing in higher education, for encouragement through the years, and for helping our dreams into reality. We thank our brothers and sisters, and extended families, for the many years of love, friendship, and tolerance. We thank our wives, Monty and Nancy, for their unqualified love, unselfish support of our endeavors, understanding of our work ethic, and appreciation for what we do. Lastly, we give a special thanks to our children, Tess, Sophie, Pete, and Zoe. Their achievements and successes are a greater source of pride for us than our own accomplishments. As young adults they are not just our children, but also our friends. Although they are now grown up, our children continue to provide an unwavering bright spot in our lives, to exhibit unbridled enthusiasm and boundless energy, and to give us a million reasons to take an occasional day off from work just to have fun.

**William B. Coleman and Gregory J. Tsongalis**

# Contents

## 16. Molecular basis of lymphoid and myeloid diseases   247

*Joseph R. Biggs and Dong-Er Zhang*

## 17. Molecular basis of diseases of immunity   271

*David O. Beenhouwer*

## 18. Molecular basis of pulmonary disease   285

*Dani S. Zander and Carol F. Farver*

## 19. Molecular basis of diseases of the gastrointestinal tract   323

*Antonia R. Sepulveda and
Armando J. Del Portillo*

# List of contributors

**Philippe Aftimos, MD,** Institut Jules Bordet, Université Libre de Bruxelles, Brussels, Belgium

**M. Rabie Al-Turkmani, PhD,** Laboratory for Clinical Genomics and Advanced Technology (CGAT), Department of Pathology and Laboratory Medicine, Dartmouth Hitchcock Medical Center, Lebanon, NH, United States

**Hatem A. Azim, Jr., MD, PhD,** Institut Jules Bordet, Université Libre de Bruxelles, Brussels, Belgium

**Sheldon I. Bastacky, MD,** Department of Pathology, University of Pittsburgh School of Medicine, Pittsburgh, PA, United States

**David O. Beenhouwer, MD,** Department of Medicine, David Geffen School of Medicine at University of California, Los Angeles, CA, United States; Division of Infectious Diseases, Veterans Affairs Greater Los Angeles Healthcare System, Los Angeles, CA, United States

**Jaideep Behari, MD, PhD,** Division of Gastroenterology, Hepatology, and Nutrition, Department of Medicine, Pittsburgh Liver Research Center, School of Medicine, University of Pittsburgh, Pittsburgh, PA, United States

**María Berdasco, MD,** Josep Carreras Leukaemia Research Institute (IJC), Barcelona, Spain

**Joseph R. Biggs, PhD,** Department of Pathology and Division of Biological Sciences, University of California San Diego, La Jolla, CA, United States

**Mariana Brandão, MD,** Institut Jules Bordet, Université Libre de Bruxelles, Brussels, Belgium

**Sheldon Campbell, MD, PhD, FCAP,** Department of Laboratory Medicine, Yale School of Medicine, VA Connecticut Healthcare System, New Haven, CT, United States

**Wai-Yee Chan, PhD,** School of Biomedical Sciences, Faculty of Medicine, The Chinese University of Hong Kong, Sha Tin, Hong Kong

**William B. Coleman, PhD,** American Society for Investigative Pathology, Rockville, MD, United States

**Massimiliano M. Corsi Romanelli, MD, PhD,** Department of Biomedical Sciences for Health, Università degli Studi di Milano, Milan, Italy; U.O.C SMEL-1 Patologia Clinica IRCCS Policlinico San Donato, Milan, Italy

**Robin D. Couch, PhD,** Department of Chemistry and Biochemistry, George Mason University, Manassas, VA, United States

**Justin B. Davis, PhD,** Center for Applied Proteomics and Molecular Medicine, George Mason University, Manassas, VA, United States

**Sophie J. Deharvengt, PhD,** Laboratory for Clinical Genomics and Advanced Technology (CGAT), Department of Pathology and Laboratory Medicine, Dartmouth Hitchcock Medical Center, Lebanon, NH, United States

**Armando J. Del Portillo, MD, PhD,** Department of Pathology and Cell Biology, Columbia University College of Physicians and Surgeons, New York, NY, United States

**Virginia Espina, PhD, MT(ASCP),** Center for Applied Proteomics and Molecular Medicine, George Mason University, Manassas, VA, United States

**Manel Esteller, PhD,** Josep Carreras Leukaemia Research Institute (IJC), Barcelona, Spain; Physiological Sciences Department, School of Medicine and Health Sciences, University of Barcelona (UB), Barcelona, Spain; Institució Catalana de Recerca i Estudis Avançats (ICREA), Barcelona, Spain

**Carol F. Farver, MD,** Director of Division of Education, Department of Pathology, University of Michigan Medical School, Ann Arbor, MI, United States

**Michael D. Feldman, MD,** Professor of Pathology, The Perelman School of Medicine, University of Pennsylvania, Philadelphia, PA, United States

**Susan L. Fink, MD, PhD,** Department of Laboratory Medicine, University of Washington, Seattle, WA, United States

**Margaret Flanagan, MD,** Department of Pathology, Stanford University, Palo Alto, CA, United States

**Claudia Fredolini, PhD,** SciLifeLab, School of Biotechnology, KTH – Royal Institute of Technology, Solna, Sweden

**William K. Funkhouser, MD, PhD,** Department of Pathology and Lab Medicine, UNC School of Medicine, Chapel Hill, NC, United States

**Matthias E. Futschik, PhD,** School of Biomedical Sciences, Faculty of Medicine and Dentistry, Institute of Translational and Stratified Medicine (ITSMED), University of Plymouth, Plymouth, United Kingdom

**Emanuela Galliera, PhD,** Department of Biomedical, Surgical and Oral Sciences, Università degli Studi di Milano, Milan, Italy; IRCCS Galeazzi Orthopedic Institute, Milan, Italy

**Avrum I. Gotlieb, MDCM, FRCPC,** Department of Laboratory Medicine and Pathobiology, Faculty of Medicine, University of Toronto, Laboratory Medicine Program, University Health Network, Toronto, ON, Canada

**Robert F. Hevner, PhD, MD,** Department of Neurological Surgery, Seattle Children's Hospital Research Institute, Seattle, WA, United States

**W. Edward Highsmith, Jr., PhD,*** Department of Laboratory Medicine and Pathology, Mayo Clinic, Rochester, MN, United States

**Christopher Dirk Keene, MD, PhD,** Department of Pathology, University of Washington, Seattle, WA, United States

**Nigel S. Key, MD,** Department of Medicine, Division of Hematology/Oncology, University of North Carolina, Chapel Hill, NC, United States

**Christine M. Koellner, MS, CGC,** Department of Laboratory Medicine and Pathology, Mayo Clinic, Rochester, MN, United States

**Joel A. Lefferts, PhD,** Laboratory for Clinical Genomics and Advanced Technology (CGAT), Department of Pathology and Laboratory Medicine, Dartmouth Hitchcock Medical Center, Lebanon, NH, United States

**John J. Lemasters, MD, PhD,** Departments of Drug Discovery & Pharmaceutical Sciences and Biochemistry & Molecular Biology, Medical University of South Carolina, Charleston, SC, United States

**Markus M. Lerch, MD,** Department of Medicine A, University Medicine Greifswald, Greifswald, Germany

**Lance A. Liotta, PhD,** Center for Applied Proteomics and Molecular Medicine, George Mason University, Manassas, VA, United States

**Youhua Liu, PhD,** Department of Pathology, University of Pittsburgh School of Medicine, Pittsburgh, PA, United States

**Karen H. Lu, MD,** Department of Gynecologic Oncology, University of Texas MD Anderson Cancer Center, Houston, TX, United States

**Nicholas W. Lukacs, PhD,** Department of Pathology, University of Michigan Medical School, Ann Arbor, MI, United States

**Alice D. Ma, MD,** Department of Medicine, Division of Hematology/Oncology, University of North Carolina, Chapel Hill, NC, United States

**Karlyn Martin, MD,** Department of Medicine, Division of Hematology/Oncology, University of North Carolina, Chapel Hill, NC, United States

**Julia Mayerle, MD,** Department of Medicine II, Ludwigs-Maximilian-University Munich, Munich, Germany

*Deceased

**Kara A. Mensink, MS, GCG,** Department of Laboratory Medicine and Pathology, Mayo Clinic, Rochester, MN, United States

**Samuel C. Mok, PhD,** Department of Gynecologic Oncology, University of Texas MD Anderson Cancer Center, Houston, TX, United States

**Satdarshan P.S. Monga, MD,** Division of Experimental Pathology, Department of Pathology, Pittsburgh Liver Research Center, School of Medicine, University of Pittsburgh, Pittsburgh, PA, United States; Division of Gastroenterology, Hepatology, and Nutrition, Department of Medicine, Pittsburgh Liver Research Center, School of Medicine, University of Pittsburgh, Pittsburgh, PA, United States

**Thomas J. Montine, MD, PhD,** Department of Pathology, Stanford University, Palo Alto, CA, United States

**Jason H. Moore, PhD,** Division of Informatics, Department of Biostatistics and Epidemiology, Institute for Biomedical Informatics, The Perelman School of Medicine, University of Pennsylvania, Philadelphia, PA, United States

**Markus Morkel, PhD,** Laboratory of Molecular Tumor Pathology and Tumor Systems Biology, Charité — Universitätsmedizin Berlin, Berlin, Germany

**Karl Munger, PhD,** Department of Developmental, Molecular and Chemical Biology, Tufts University School of Medicine, Boston, MA, United States

**Zoltan Nagymanyoki, MD, PhD,** Department of Pathology, West Pacific Medical Laboratory, Santa Fe Springs, CA, United States

**Robert D. Nerenz, PhD,** Assistant Professor of Pathology and Laboratory Medicine, Dartmouth-Hitchcock Medical Center, Lebanon, NH, United States

**Alan L.-Y. Pang, PhD,** TGD Life Company Limited, Hong Kong Science Park, Sha Tin, Hong Kong

**Emanuel Petricoin, III, PhD,** Center for Applied Proteomics and Molecular Medicine, George Mason University, Manassas, VA, United States

**Catherine Ptaschinski, PhD,** Department of Pathology, University of Michigan Medical School, Ann Arbor, MI, United States

**Reinhold Schäfer, PhD,** Charité Comprehensive Cancer Center, Charité — Universitätsmedizin Berlin, Berlin, Germany; German Cancer Consortium (DKTK), German Cancer Research Center, Heidelberg, Germany

**Matthias Sendler, MD,** Department of Medicine A, University Medicine Greifswald, Greifswald, Germany

**Antonia R. Sepulveda, MD, PhD,** Department of Pathology, George Washington University School of Medicine and Health Sciences, Washington, DC, United States

**Christine Sers, PhD,** Laboratory of Molecular Tumor Pathology and Tumor Systems Biology, Charité — Universitätsmedizin Berlin, Berlin, Germany

**Lawrence M. Silverman, PhD,** Department of Pathology, University of Virginia Health System, Charlottesville, VA, United States

**Joshua A. Sonnen, PhD,** Department of Pathology, University of Utah, Salt Lake City, UT, United States

**Christos Sotiriou, MD, PhD,** Institut Jules Bordet, Université Libre de Bruxelles, Brussels, Belgium

**Roderick J. Tan, MD, PhD,** Renal-Electrolyte Division, Department of Medicine, University of Pittsburgh School of Medicine, Pittsburgh, PA, United States

**Gregory J. Tsongalis, PhD, HCLD,** Laboratory for Clinical Genomics and Advanced Technology (CGAT), Department of Pathology and Laboratory Medicine, Dartmouth Hitchcock Medical Center, Lebanon, NH, United States; The Audrey and Theodor Geisel School of Medicine at Dartmouth, Hanover, NH, United States

**Vesarat Wessagowit, MD, PhD,** The Institute of Dermatology, Rajvithi Phyathai, Bangkok, Thailand

**Eli S. Williams, PhD,** Department of Pathology, University of Virginia Health System, Charlottesville, VA, United States

**Kwong-Kwok Wong, PhD,** Department of Gynecologic Oncology, University of Texas MD Anderson Cancer Center, Houston, TX, United States

**Dani S. Zander, MD,** Department of Pathology and Laboratory Medicine, University of Cincinnati, Cincinnati, OH, United States

**Dong-Er Zhang, PhD,** Department of Pathology and Division of Biological Sciences, University of California San Diego, La Jolla, CA, United States

**Weidong Zhou, MD, PhD,** Center for Applied Proteomics and Molecular Medicine, George Mason University, Manassas, VA, United States

# About the editors

William B. Coleman, Ph.D. is the Executive Officer for the American Society for Investigative Pathology (Rockville, MD). Prior to taking this position with the American Society for Investigative Pathology, Dr. Coleman spent 28 years at the University of North Carolina School of Medicine (Chapel Hill, NC), first as a postdoctoral fellow (1990−95) and then as a faculty member (1995−2018) in the Department of Pathology and Laboratory Medicine. During his time at the UNC School of Medicine, Dr. Coleman served as director of Graduate Studies for the Molecular and Cellular Pathology Ph.D. Program (2006−12; now the Pathobiology and Translational Medicine Ph.D. Program), was a co-founder of the UNC Program in Translational Medicine and served as its co-director (2006−15) and then its director (2015−18), was affiliated with the Curriculum in Toxicology, the Cancer Biology Training Program, and was a member of the UNC Lineberger Comprehensive Cancer Center. Dr. Coleman was active in teaching biomedical graduate students and is a four-time recipient of the Joe W. Grisham Award for Excellence in Graduate Student Teaching from the Molecular and Cellular Pathology graduate students at the UNC School of Medicine. Prior to becoming an employee, Dr. Coleman was active in the leadership of the American Society for Investigative Pathology, serving in various roles including President (2015−16). Dr. Coleman was honored with the ASIP Outstanding Investigator Award in 2013 from the American Society for Investigative Pathology. He is also a long-time member of the American Association for Cancer Research. He serves as Senior Associate Editor for *The American Journal of Pathology*, and is an Associate Editor for *BMC Cancer*, and *PLoS One*, and serves on the editorial boards of *Clinica Chimica Acta*, *Experimental and Molecular Pathology*, *Archives of Pathology and Laboratory Medicine*, *Laboratory Investigation*, and *Current Pathobiology Reports*, and has served as an ad hoc reviewer for 99 other journals. Dr. Coleman's major research interests are in the molecular pathogenesis of human cancers, with a specific interest in breast cancer epigenetics, liver carcinogenesis, and lung cancer biology. His research was funded by the NIH/NCI, The Susan G. Komen Breast Cancer Foundation, Friends for an Earlier Breast Cancer Test, and the UNC Lineberger Comprehensive Cancer Center. Dr. Coleman is the author of over 140 original research articles, reviews, and book chapters. In addition, Dr. Coleman has co-edited or co-authored ten books on topics related to molecular pathology, molecular diagnostics, and the molecular pathogenesis of human cancer.

Gregory Tsongalis, Ph.D., H.C.L.D., is the Vice Chair for Research and the Director of the Laboratory for Clinical Genomics and Advanced Technology (CGAT) in the Department of Pathology and Laboratory Medicine at the Dartmouth-Hitchcock Medical Center and Norris Cotton Cancer Center (NCCC) in Lebanon, NH. He is a professor of Pathology and Laboratory Medicine at the Audrey and Theodor Geisel School of Medicine at Dartmouth in Hanover, NH and a member of the NCCC Molecular Therapeutics Program and the gastrointestinal and breast cancer clinical oncology groups. In 2016 he became a member of Dartmouth College's Program in Experimental and Molecular Medicine (PEMM), and he has served on the advisory board of the Health Care Genetics Professional Science Master's Degree Program and Diagnostic Genetic Sciences Program at the University of Connecticut (Storrs, CT). His area of expertise is in the development and implementation of clinical molecular diagnostic technologies. His research interests are in the pathogenesis of human cancers, personalized medicine and disruptive technologies. He has authored/edited twelve textbooks in the field of molecular pathology, published more than 230 peer reviewed manuscripts, and has been an invited speaker at both national and international meetings. He has served on numerous committees of the American Association for Clinical Chemistry, the American Society for Investigative Pathology, the Federation for American Societies for Experimental Biology, and the Association for Molecular Pathology (where he is a Past President). He is active in the Alliance for Clinical Trials in Oncology, the Association for Molecular Pathology, the American Association for Clinical Chemistry, the American Association of Bioanalysts, and the American Society for

Investigative Pathology. He serves on the editorial boards of 8 journals including *Clinical Chemistry*, *Experimental and Molecular Pathology*, and the *Journal of Molecular Diagnostics*. In 2016, Dr. Tsongalis received the Norris Cotton Cancer Center Award for Excellence, in 2017 the Association for Molecular Pathology (AMP) Jeffrey A. Kant Leadership Award, and in 2019 the American Society for Investigative Pathology Robbins Distinguished Educator Award. He also serves on numerous corporate scientific advisory boards.

# Preface

Pathology is the scientific study of the nature of disease and its causes, processes, development, and consequences. The field of pathology emerged from the application of the scientific method to the study of human disease. Thus, pathology as a discipline represents the complimentary intersection of medicine and basic science. Early pathologists were typically practicing physicians who described the various diseases that they treated and made observations related to factors that contribute to the development of these diseases. The description of disease evolved over time from gross observation to structural and ultrastructural inspection of diseased tissues based upon light and electron microscopy. As hospital-based and community-based registries of disease were developed, the ability of investigators to identify factors that cause disease and assign risk to specific types of exposures expanded to increase our knowledge of the epidemiology of disease. While descriptive pathology can be dated to the earliest written histories of medicine and the modern practice of diagnostic pathology dates back perhaps 200 years, the elucidation of mechanisms of disease and linkage of disease pathogenesis to specific causative factors occurred more recently from studies in experimental pathology. The field of experimental pathology embodies the conceptual foundation of early pathology - the application of the scientific method to the study of disease - and applies modern investigational tools of cell and molecular biology to advanced animal model systems and studies of human subjects. Whereas the molecular era of biological science began over 60 years ago, recent advances in our knowledge of molecular mechanisms of disease have propelled the field of molecular pathology. These advances were facilitated by significant improvements and new developments associated with the techniques and methodologies available to pose questions related to the molecular biology of normal and diseased states affecting cells, tissues, and organisms. Today, molecular pathology encompasses the investigation of the molecular mechanisms of disease and interfaces with translational medicine where new basic science discoveries form the basis for the development of new therapeutic approaches and targeted therapies for the new strategies for prevention, and treatment of disease.

With the remarkable pace of scientific discovery in the field of molecular pathology, basic scientists, clinical scientists, and physicians have a need for a source of information on the current state-of-the-art of our understanding of the molecular basis of human disease. More importantly, the complete and effective training of today's graduate students, medical students, postdoctoral fellows, medical residents, allied health students, and others, for careers related to the investigation and treatment of human disease requires textbooks that have been designed to reflect our current knowledge of the molecular mechanisms of disease pathogenesis, as well as emerging concepts related to translational medicine. Most pathology textbooks provide information related to diseases and disease processes from the perspective of description (what does it look like and what are its characteristics), risk factors, disease-causing agents, and to some extent, cellular mechanisms. However, most of these textbooks lack in-depth coverage of the molecular mechanisms of disease. The reason for this is primarily historical — most major forms of disease have been known for a long time, but the molecular basis of these diseases are not always known or have been elucidated only very recently. However, with rapid progress over time and improved understanding of the molecular basis of human disease the need emerged for new textbooks on the topic of molecular pathology, where molecular mechanisms represent the focus.

In *Essential Concepts in Molecular Pathology, Second Edition* we have assembled a group of experts to discuss the molecular basis and mechanisms of major human diseases and disease processes, presented in the context of traditional pathology, with implications for translational molecular medicine. *Essential Concepts in Molecular Pathology, Second Edition* is an abbreviated version of *Molecular Pathology: The Molecular Basis of Human Disease, Second Edition*, that contains several distinct features. Each chapter focuses on essential concepts related to a specific disease or disease process, rather than providing comprehensive coverage of the topic. Each chapter contains key concepts, which capture the essence of the topic covered. In place of long lists of references to the primary literature, each chapter provides a list of suggested readings, which include pertinent reviews and/or primary literature references that are deemed to be most important to the reader. This volume is intended to serve as a multi-use textbook that would be appropriate as a classroom teaching tool for medical students, biomedical graduate students, allied health students, advanced undergraduate students, and others. We anticipate that this book will be most useful for teaching students in courses where the full textbook is not needed, but the concepts included are integral to the course of study. This book might also be useful for students that are enrolled in courses that utilize a traditional pathology textbook as the primary text, but need the complementary concepts related to molecular pathogenesis of disease. Further, this textbook will be valuable for pathology residents and other postdoctoral fellows that desire to advance their understanding of molecular mechanisms of disease beyond what they learned in medical/graduate school, and as a reference book and self-teaching guide for practicing basic scientists and physician scientists that need to understand the molecular concepts, but do not require comprehensive coverage or complete detail. To be sure, our understanding of the many causes and molecular mechanisms that govern the development of human diseases is far from complete. Nevertheless, the amount of information related to these molecular mechanisms has increased tremendously in recent years and areas of thematic and conceptual consensus have emerged. We hope that *Essential Concepts in Molecular Pathology, Second Edition* will accomplish its purpose of providing students and researchers with a broad coverage of the essential concepts related to the molecular basis of major human diseases in the context of traditional pathology so as to stimulate new research aimed at furthering our understanding of these molecular mechanisms of human disease and advancing the theory and practice of molecular medicine.

**William B. Coleman and Gregory J. Tsongalis**

# Acknowledgments

The editors would like to acknowledge the significant contributions of a number of people to the successful production of *Essential Concepts in Molecular Pathology, Second Edition*.

We would like to thank the individuals who contributed to the content of this volume. The remarkable coverage of the state of the art in the molecular pathology of human disease would not have been possible without the hard work and diligent efforts of the 70 authors of the individual chapters. Many of these contributors are our long-time colleagues, collaborators, and friends, and they have contributed to other projects that we have directed, and we sincerely appreciate their willingness to contribute once again to a project that we found worthy. We especially thank the contributors to this volume who were willing to work with us for the first time. This group also includes some of our long-time friends and colleagues, as well as some new friends. We look forward to working with all of these authors again in the future. Each of these contributors provided us with an excellent treatment of their topic, and we hope that they will be proud of their individual contributions to the textbook. Collectively, we can all be proud of this volume, as it is proof that the whole can be greater than the sum of its parts.

Special thanks to Ms. Mara Conner (Senior Acquisitions Editor, Academic Press - Elsevier) who worked with us on the first edition of this textbook. She embraced the concept of this textbook when our ideas were not yet fully developed and encouraged us to pursue the project. She was receptive to the model for this textbook that we envisioned and worked closely with us to evolve the project into its final form. Without Mara's early support, the first edition of this textbook would not have been so successful and this second edition would not have been possible.

We would also like to thank the many people who work for Academic Press - Elsevier that made this project possible. We have not met and do not know many of these people, but we appreciate their efforts to bring this textbook to its completed form. Special thanks goes to three key people who made significant contributions to this project on the publishing side and proved to be exceptionally competent and capable. Ms. Tari Broderick (Senior Acquisitions Editor, Bioscience and Translational Medicine, Academic Press - Elsevier) provided excellent oversight (and optimistic patience) during the construction and editing of this edition of the textbook and has become our valued colleague as we develop new projects. Mr. Sam Young (Editorial Project Manager, Elsevier) provided excellent support to us throughout this project. As we interacted with our contributing authors, collected and edited manuscripts, and began production of the textbook, Sam assisted us greatly by being a constant reminder of deadlines, helping us with communication with the contributors, and generally providing support for details small and large, all of which proved to be critical. Mr. Stalin Viswanathan (Publishing Services Manager, Elsevier) worked with us closely to ensure the integrity of the content of the textbook as it moved from the edited manuscripts into their final form. We thank him for his direct involvement with the production and also for directing his excellent production team. It was a pleasure to work with Tari, Sam, and Stalin on this project. We hope that they enjoyed it as much as we did, and we look forward to working with them again soon.

**William B. Coleman and Gregory J. Tsongalis**

# Chapter 1

# Molecular mechanisms of cell death

**John J. Lemasters**
*Departments of Drug Discovery & Pharmaceutical Sciences and Biochemistry & Molecular Biology, Medical University of South Carolina, Charleston, SC, United States*

**Summary**

Although many stimuli cause death of cells, the mode of cell death typically follows one of two patterns. The first is necrosis, or oncosis. Oncotic necrosis is most often the result of profound metabolic disruption and is characterized by cellular swelling leading to plasma membrane rupture with release of intracellular contents. The second pattern is apoptosis, a form of programmed cell death. Apoptosis causes the orderly resorption of individual cells initiated by well-defined pathways involving activation of proteases called caspases. In contrast to necrotic cell death, which typically occurs from adenosine triphosphate (ATP) depletion, apoptosis is an ATP-requiring process. However, in both modes of cell death, mitochondrial permeabilization and dysfunction typically develop. In some instances, apoptosis and necrosis share signaling pathways as extreme endpoints on a phenotypic continuum of lost cell viability.

## Introduction

A common theme in disease is death of cells. In diseases ranging from stroke to congestive heart disease to alcoholic cirrhosis of the liver, death of individual cells leads to irreversible functional loss in whole organs and ultimately mortality. For such diseases, prevention of cell death becomes a basic therapeutic goal. By contrast in neoplasia, the purpose of chemotherapy is to kill proliferating cancer cells. For either therapeutic goal, understanding the mechanisms of cell death becomes paramount.

## Modes of cell death

Although many stresses and stimuli cause cell death, the mode of cell death typically follows one of two patterns. The first is necrosis, a pathological term referring to areas of dead cells within a tissue or organ. Necrosis is typically the result of an acute and usually profound metabolic disruption, such as ischemia (loss of blood supply). Since necrosis is an outcome rather than a process, the term oncosis has been introduced to describe the process leading to necrotic cell death, but the term has yet to be widely adopted. Here, the terms oncosis, oncotic necrosis, and necrotic cell death will be used synonymously to refer both to the outcome of cell death and the pathogenic events precipitating cell killing.

The second pattern is programmed cell death, most commonly manifested as apoptosis, a term derived from ancient Greek for the falling of leaves in autumn. In apoptosis, specific stimuli initiate execution of well-defined pathways leading to orderly resorption of individual cells with minimal leakage of cellular components into the extracellular space and little inflammation. Whereas necrotic cell death occurs with abrupt onset after adenosine triphosphate (ATP) depletion, apoptosis may take hours to go to completion and is an ATP-requiring process without a clearly distinguished point of no return. Although apoptosis and necrosis were initially considered separate and independent phenomena, an alternate view is emerging that apoptosis and necrosis can share initiating factors and signaling pathways to become extremes on a phenotypic continuum.

Essential Concepts in Molecular Pathology. DOI: https://doi.org/10.1016/B978-0-12-813257-9.00001-2

# Structural features of necrosis and apoptosis

## Oncotic necrosis

Cellular changes leading up to onset of necrotic cell death include formation of plasma membrane protrusions called blebs, mitochondrial swelling, dilatation of cisternae of the endoplasmic reticulum (ER), dissociation of polysomes, and cellular swelling leading to rupture with release of intracellular contents (Table 1.1, Fig. 1.1). After necrotic cell death, characteristic histological features of loss of cellular architecture, vacuolization, karyolysis (dissolution of chromatin), and increased eosinophilia soon become evident (Fig. 1.2). Cell lysis evokes an inflammatory response, attracting neutrophils and monocytes to the dead tissue to dispose of the necrotic debris by phagocytosis and defend against infection (Fig. 1.3). In organs like heart and brain with little regenerative capacity, healing occurs with scar formation, namely replacement of necrotic regions with fibroblasts, collagen and other connective tissue components. In organs like the liver that have robust regenerative capacity, cell proliferation can replace areas of necrosis with completely normal tissue within a few days. The healed liver tissue shows little or no residua of the necrotic event, but if regeneration fails, collagen deposition and fibrosis will occur instead to cause cirrhosis.

**TABLE 1.1**  Comparison of necrosis and apoptosis.

| Necrosis | Apoptosis |
| --- | --- |
| Accidental cell death | Controlled cell deletion |
| Contiguous regions of cells | Single cells separating from neighbors |
| Cell swelling | Cell shrinkage |
| Plasmalemmal blebs without organelles | Zeiotic blebs containing large organelles |
| Small chromatin aggregates | Nuclear condensation and lobulation |
| Random DNA degradation | Internucleosomal DNA degradation |
| Cell lysis with release of intracellular contents | Fragmentation into apoptotic bodies |
| Inflammation and scarring | Absence of inflammation and scarring |
| Mitochondrial swelling and dysfunction | Mitochondrial permeabilization |
| Phospholipase and protease activation | Caspase activation |
| ATP depletion and metabolic disruption | ATP and protein synthesis sustained |
| Cell death precipitated by plasma membrane rupture | Intact plasma membrane |

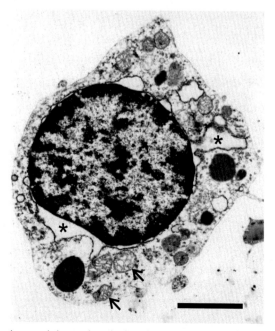

**FIGURE 1.1**   Electron microscopy of oncotic necrosis in a rat hepatic sinusoidal endothelial cell after ischemia/reperfusion. Note cell rounding, mitochondrial swelling (arrows), rarefaction of cytosol, dilatation of the ER and the space between the nuclear membranes (*), chromatin condensation, and discontinuities in the plasma membrane. Bar is 2 μm.

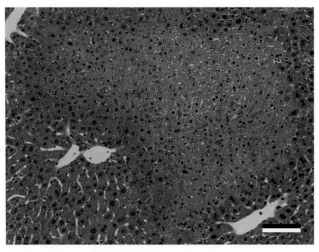

FIGURE 1.2   Histology of necrosis after hepatic ischemia/reperfusion in a mouse. Note increased eosinophilia, loss of cellular architecture, and nuclear pyknosis and karyolysis. Contrast to lower left and right areas that are non-necrotic. Bar is 50 μm.

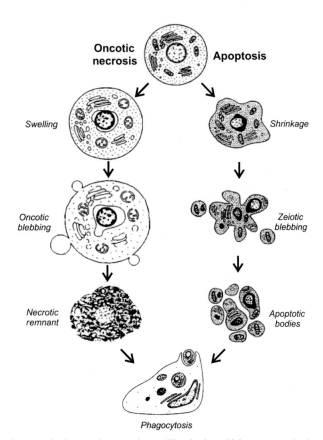

FIGURE 1.3   Scheme of necrosis and apoptosis. In oncotic necrosis, swelling leads to bleb rupture and release of intracellular constituents which attract macrophages that clear the necrotic debris by phagocytosis. In apoptosis, cells shrink and form small zeiotic blebs that are shed as membrane-bound apoptotic bodies. Apoptotic bodies are phagocytosed by macrophages and adjacent cells. *Adapted with permission from Van CS, Van Den BW. Morphological and biochemical aspects of apoptosis, oncosis and necrosis. Anat Histol Embryol 2002;31:214−23.*

## Apoptosis

Unlike necrosis, which often occurs in response to an imposed unphysiological stress, apoptosis is a process of physiological cell deletion that has an opposite role to mitosis in the regulation of cell populations. In apoptosis, cell death occurs with little release of intracellular contents, inflammation, and scar formation. Individual cells undergoing apoptosis separate from their neighbors and shrink rather than swell. Distinctive nuclear and cytoplasmic changes also occur, including chromatin condensation, nuclear lobulation and fragmentation, formation of numerous small cell surface blebs (zeiotic blebbing), and shedding of these blebs as apoptotic bodies that are phagocytosed by adjacent cells and macrophages for lysosomal degradation (Table 1.1, Fig. 1.3). Characteristic biochemical changes also occur, typically activation of a cascade of cysteine-aspartate proteases, called caspases, leakage of pro-apoptotic proteins like cytochrome *c* from mitochondria into the cytosol, internucleosomal deoxyribonucleic acid (DNA) degradation, degradation of poly(ADP-ribose) polymerase (PARP), and movement of phosphatidyl serine to the exterior leaflet of the plasmalemmal lipid bilayer. Thus, apoptosis manifests a very different pattern of cell death than oncotic necrosis (Table 1.1, Fig. 1.3).

## Cellular and molecular mechanisms underlying necrotic cell death

### Metastable state preceding necrotic cell death

Cellular events culminating in necrotic cell death are somewhat variable from one cell type to another, but certain events occur regularly. As implied by the term oncosis, cellular swelling is a prominent feature of oncotic necrosis. In many cell types, swelling of 30—50% occurs early after ATP depletion associated with formation of blebs on the cell surface (Fig. 1.4). These blebs contain cytosol and ER but exclude larger organelles. Bleb formation is likely due to cytoskeletal alterations after ATP depletion, whereas swelling arises from disruption of cellular ion transport. Mitochondrial swelling and dilatation of cisternae of ER and nuclear membranes accompany bleb formation (see Fig. 1.1). After longer times, a metastable state develops, which is characterized by mitochondrial depolarization, lysosomal breakdown, ion dysregulation, and accelerated bleb formation with more rapid swelling. The metastable state lasts only a few minutes and culminates in rupture of a plasma membrane bleb (Fig. 1.4). At onset of the metastable state, nonspecific pores appear to open, permitting uptake of electrolytes (principally sodium and chloride) and initiating rapid swelling driven by colloid osmotic (oncotic) forces (Fig. 1.5). Bleb rupture leads to loss of metabolic intermediates such as those that reduce tetrazolium dyes, leakage of cytosolic enzymes like lactate dehydrogenase, uptake of dyes like trypan blue, and collapse of all electrical and ion gradients. This all-or-nothing breakdown of the plasma membrane permeability barrier is long-lasting, irreversible, and incompatible with continued life of the cell.

### Mitochondrial dysfunction and ATP depletion

Ischemia as occurs in strokes and heart attacks is perhaps the most common cause of necrotic cell killing. In ischemia, oxygen deprivation prevents ATP formation by mitochondrial oxidative phosphorylation, a process providing up to 95% of ATP utilized by highly aerobic tissues. The role of mitochondrial dysfunction in necrotic killing can be assessed experimentally by the ability of glycolytic substrates to rescue cells from lethal cell injury (Fig. 1.6). As an alternative source of ATP, glycolysis partially replaces ATP production lost after mitochondrial dysfunction. Maintenance of as little as 15% or 20% of normal ATP then rescues cells from necrotic death. Glycolysis also protects against toxicity from

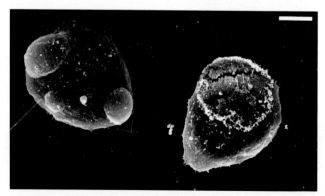

**FIGURE 1.4** Bleb rupture at onset of necrotic cell death. After metabolic inhibition with cyanide and iodoacetate, inhibitors of respiration and glycolysis, respectively, a surface bleb of the cultured rat hepatocyte on the right has just burst. Note the discontinuity of the plasma membrane surface in the scanning electron micrograph. The hepatocyte on the left is also blebbed, but the plasma membrane is still intact, and viability has not yet been lost. Bar is 5 μm. *Adapted with permission from Herman B, Nieminen AL, Gores GJ, Lemasters JJ. Irreversible injury in anoxic hepatocytes precipitated by an abrupt increase in plasma membrane permeability. FASEB J 1988;2:146—51.*

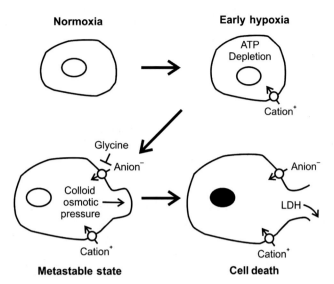

**FIGURE 1.5**   Plasma membrane permeabilization leading to necrotic cell death. Early after hypoxia and other metabolic stresses, ATP depletion leads to inhibition of the Na,K-ATPase and opening of monovalent cation channels causing cation gradients ($Na^+$ and $K^+$) to collapse. Swelling is limited by impermeability to anions. Later, glycine and strychnine-sensitive anion channels open to initiate anion entry and accelerate bleb formation and swelling. Swelling continues until a bleb ruptures. With abrupt and complete loss of the plasma membrane permeability barrier, viability is lost. Supravital dyes like trypan blue and propidium iodide enter the cell to stain the nucleus, and cytosolic enzymes like lactate dehydrogenase (LDH) leak out. *With permission from Lemasters JJ, Qian T, He L, et al. Role of mitochondrial inner membrane permeabilization in necrotic cell death, apoptosis, and autophagy. Antioxid Redox Signal 2002;4:769–81.*

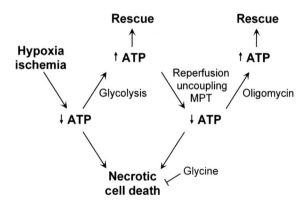

**FIGURE 1.6**   Progression of mitochondrial injury. Respiratory inhibition inhibits oxidative phosphorylation and leads to ATP depletion and necrotic cell death. Glycine blocks plasma membrane permeabilization causing necrotic cell death downstream of ATP depletion. Glycolysis restores ATP and prevents cell killing. Mitochondrial uncoupling as occurs after reperfusion due to the mitochondrial permeability transition (MPT) activates the mitochondrial ATPase to futilely hydrolyze glycolytic ATP, and protection against necrotic cell death is lost. By inhibiting the mitochondrial ATPase, oligomycin prevents ATP depletion and rescues cells from necrotic cell death if glycolytic substrate is present. *With permission from Lemasters JJ, Qian T, He L, et al. Role of mitochondrial inner membrane permeabilization in necrotic cell death, apoptosis, and autophagy. Antioxid Redox Signal 2002;4:769–81.*

oxidant chemicals, suggesting that mitochondria are also a primary target of cytotoxicity in oxidative stress. However, in pathological settings like ischemia, glycolytic substrates are rapidly exhausted.

## Mitochondrial uncoupling in necrotic cell killing

Mitochondrial injury and dysfunction are progressive (Fig. 1.6). Respiratory inhibition as occurs in anoxia causes ATP depletion and ultimately necrotic cell death. Glycolysis can replace this ATP supply, although only partially in highly aerobic cells, to rescue cells from necrotic killing. However, when mitochondrial injury progresses to uncoupling (inner membrane permeability to hydrogen ions), accelerated ATP hydrolysis occurs that is catalyzed by the mitochondrial ATP synthase working in reverse. Since glycolytic ATP production cannot keep pace, ATP levels fall profoundly, and necrotic cell death ensues. In the progression from respiratory inhibition to uncoupling, mitochondria become active agents promoting ATP depletion and cell death.

## Mitochondrial permeability transition

### Inner membrane permeability

In oxidative phosphorylation, respiration drives translocation of protons out of mitochondria to create an electrochemical proton gradient composed of a negative inside $\Delta\Psi$ and an alkaline inside pH gradient ($\Delta$pH). ATP synthesis is then linked to protons returning down this electrochemical gradient through the mitochondrial ATP synthase. This chemiosmotic proton circuit requires the mitochondrial inner membrane to be impermeable to ions and charged metabolites.

In some pathophysiological settings, however, the mitochondrial inner membrane abruptly becomes non-selectively permeable to solutes of molecular weight up to about 1500 Da. $Ca^{2+}$, oxidative stress, and numerous reactive chemicals induce this mitochondrial permeability transition (MPT), whereas cyclosporine A and pH less than 7 inhibit. The MPT causes mitochondrial depolarization, uncoupling, and large amplitude mitochondrial swelling driven by colloid osmotic forces. Opening of highly conductive permeability transition (PT) pores in the mitochondrial inner membrane underlies the MPT. Conductance is so great that opening of a single PT pore may be sufficient to cause mitochondrial depolarization and swelling.

The composition of PT pores is uncertain. In one model, PT pores are formed by the adenine nucleotide transporter (ANT) from the inner membrane, the voltage dependent anion channel (VDAC) from the outer membrane, the cyclosporine A binding protein cyclophilin D (CypD) from the matrix, and possibly other proteins (Fig. 1.7A). Although once widely accepted, the validity of this model has been challenged by genetic knockout studies showing that the MPT still occurs in mitochondria that are deficient in ANT, VDAC and CypD. More recently, PT pores are proposed to form in association with the $F_1F_O$-ATP synthase (Fig. 1.7B); with spastic paraplegia 7 (SPG7), a mitochondrial AAA-type membrane protease; or with the inorganic phosphate carrier of the inner membrane. An alternative model for the PT pore is that oxidative and other stresses damage membrane proteins that then misfold and aggregate to form PT pores in association with CypD and other molecular chaperones (Fig. 1.7C).

### pH-dependent ischemia/reperfusion injury

Ischemia is an interruption of blood flow and hence oxygen supply. In ischemic tissue, anaerobic metabolism causes tissue pH to decrease by a unit or more. The naturally occurring acidosis protects against necrotic cell death during ischemia and also after various toxic stresses. After reperfusion, the protection of acidotic pH is lost, and onset of necrotic cell death occurs. Much of reperfusion injury is attributable to recovery of pH, since reoxygenation at low pH prevents cell killing entirely, whereas restoration of normal pH without reoxygenation produces similar cell killing as restoration of pH with reoxygenation, a so-called pH paradox (Fig. 1.8). Cell killing in the pH paradox is linked specifically to intracellular pH and occurs independently of changes of cytosolic and extracellular free $Na^+$ and $Ca^{2+}$.

### Role of the mitochondrial permeability transition in pH-dependent reperfusion injury

pH below 7 inhibits PT pores, and recovery of intracellular pH to 7 or greater after reperfusion induces the MPT (Fig. 1.8). Mitochondria depolarize during ischemia, but after reperfusion at normal pH, mitochondria repolarize initially and in parallel with recovery of intracellular pH to neutrality, the MPT occurs (Fig. 1.8 and Fig. 1.9). ATP depletion and necrotic cell death then follow. Reperfusion in the presence of PT pore blockers (*e.g.*, cyclosporine A and its derivatives) prevents mitochondrial inner membrane permeabilization, depolarization and cell killing. Notably, cyclosporine A protects when added only during the reperfusion phase. Thus, the MPT is the proximate cause of pH-dependent cell killing in ischemia/reperfusion injury.

### Oxidative stress

Reactive oxygen species (ROS) and reactive nitrogen species (RNS), including superoxide, hydrogen peroxide, hydroxyl radical, and peroxynitrite, have long been implicated in cell injury leading to necrosis (Fig. 1.10). Reperfusion after ischemia stimulates intramitochondrial ROS formation, MPT onset and cell death (Fig. 1.9). In neurons, excitotoxic stress with glutamate and N-methyl-D-aspartate (NMDA) receptor agonists also stimulates mitochondrial ROS formation, leading to the MPT and excitotoxic injury.

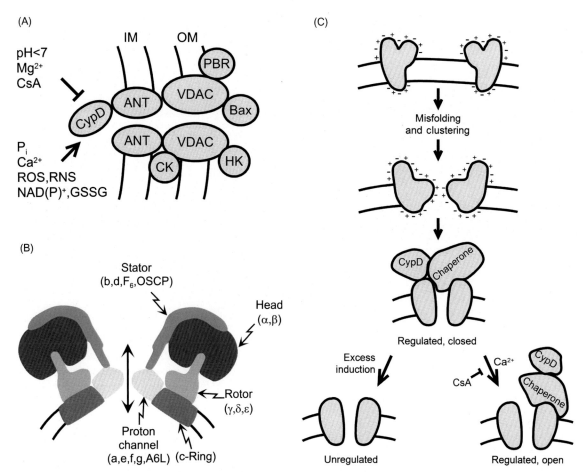

**FIGURE 1.7** Models of mitochondrial permeability transition pores. In one model (A), PT pores are composed of the adenine nucleotide translocator (ANT) from the inner membrane (IM), cyclophilin D (CypD) from the matrix and the voltage-dependent anion channel (VDAC) from the outer membrane (OM). Other proteins, such as the peripheral benzodiazepine receptor (PBR), hexokinase (HK), creatine kinase (CK), and Bax may also contribute. PT pore openers include $Ca^{2+}$, inorganic phosphate (Pi), reactive oxygen and nitrogen species (ROS, RNS), and oxidized pyridine nucleotides $(NAD(P)^+)$ and glutathione (GSSG). A newer model (B) has PT pores forming in $F_1F_O$-ATP synthase dimers at the interface between monomers (or possibly in association with c-rings). OSCP (oligomycin sensitivity-conferring protein), a, b, c, d, e, f, g, $\alpha$, $\beta$, $\gamma$, $\delta$, $\varepsilon$, A6L and F8 are subunits of the synthase. An alternative proposal (C) suggests that oxidative and other damage to integral inner membrane proteins leads to misfolding. These misfolded proteins aggregate at hydrophilic surfaces facing the hydrophobic bilayer to form aqueous channels. CypD and other chaperones block conductance of solutes through these nascent PT pores. High matrix $Ca^{2+}$ acting through CypD leads to PT pore opening, an effect blocked by cyclosporine A (CsA). As misfolded protein clusters exceed the number of chaperones to regulate them, constitutively open channels form. Such unregulated PT pores are not dependent on $Ca^{2+}$ for opening and are not inhibited by CsA. *Adapted with permission from Kim JS, He L, Qian T, Lemasters JJ. Role of the mitochondrial permeability transition in apoptotic and necrotic death after ischemia/reperfusion injury to hepatocytes. Curr Mol Med 2003;3:527−35.*

Iron potentiates injury in a variety of diseases and is an important catalyst for hydroxyl radical formation from superoxide and hydrogen peroxide (Fig. 1.10). During oxidative stress, acetaminophen hepatotoxicity and hypoxia/ ischemia, lysosomes release chelatable (loosely bound) iron with consequent pro-oxidant cell damage. This iron is taken up into mitochondria by the mitochondrial calcium uniporter and helps catalyze mitochondrial ROS generation. Iron chelation with desferal prevents mitochondrial ROS formation and decreases cell death.

## Other stress mechanisms inducing necrotic cell death

### Poly (ADP-ribose) polymerase

Single strand breaks induced by ultraviolet (UV) light, ionizing radiation, and ROS (particularly hydroxyl radical and peroxynitrite) activate PARP. PARP transfers ADP-ribose from $NAD^+$ to the strand breaks and elongates ADP-ribose polymers attached to the DNA. Excess consumption of $NAD^+$ in this fashion leads to $NAD^+$ depletion, disruption of

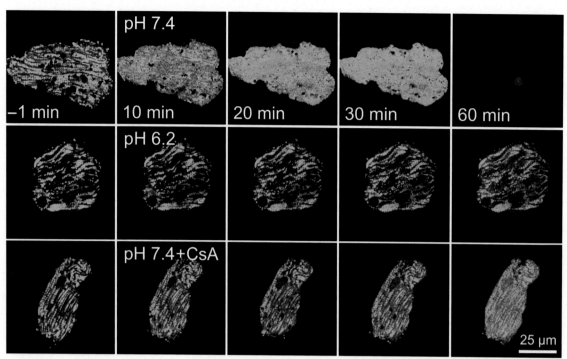

**FIGURE 1.8** Mitochondrial inner membrane permeabilization in adult rat cardiac myocytes after ischemia and reperfusion. After loading mitochondria of cardiac myocytes with calcein, cells were subjected to 3 h of anoxia at pH 6.2 (ischemia) followed by reoxygenation at pH 7.4 (A), pH 6.2 (B), or pH 7.4 with 1 μM CsA (C). Red-fluorescing propidium iodide was present to detect loss of cell viability. Note that green calcein fluorescence was retained by mitochondria at the end of ischemia (1 min before reperfusion), indicating that PT pores had not opened. After reperfusion at pH 7.4, mitochondria progressively released calcein over 30 min at which time calcein was nearly evenly distributed throughout cytosol. After 60 min, all cellular calcein was lost, and the nucleus stained with PI, indicating loss of viability. After reperfusion at pH 6.2 (B) or at pH 7.4 in the presence of CsA (C), calcein was retained and cell death did not occur. Thus, reperfusion at pH 7.4 induced onset of the MPT and necrotic cell death that were blocked with CsA and acidotic pH. *Adapted with permission from Kim JS, Jin Y, Lemasters JJ. Reactive oxygen species, but not Ca²⁺ overloading, trigger pH- and mitochondrial permeability transition-dependent death of adult rat myocytes after ischemia-reperfusion. Am J Physiol Heart Circ Physiol 2006;290:H2024–34.*

ATP-generation, and ATP depletion-dependent cell death. PARP-dependent necrosis is an example of programmed necrosis, since PARP actively promotes a cell death-inducing pathway that otherwise would not occur. Necrotic cell death also frequently occurs when apoptosis is interrupted, as by caspase inhibition. Such caspase-independent cell death is the consequence of mitochondrial dysfunction or other metabolic disturbance.

### Plasma membrane injury

An intact plasma membrane is essential for cell viability. Detergents and pore-forming agents like mastoparan from wasp venom defeat the barrier function of the plasma membrane and cause immediate cell death. Immune-mediated cell killing can act similarly. In particular, complement mediates formation of a membrane attack complex that in conjunction with antibody lyses cells. Complement component 9, an amphipathic molecule, inserts through the cell membrane, polymerizes, and forms a tubular channel visible by electron microscopy. Indeed, a single membrane attack complex is sufficient to cause lysis of an individual erythrocyte.

# Pathways to apoptosis

## Roles of apoptosis in biology

Apoptosis is an essential event in both the normal life of organisms and in pathobiology. In development, apoptosis sculpts and remodels tissues and organs, for example, by creating clefts in limb buds to form fingers and toes. Apoptosis is also responsible for reversion of hypertrophy to atrophy and immune surveillance-induced killing of pre-neoplastic and virally infected cells. Each of several organelles can give rise to signals initiating apoptosis. Often these signals converge on mitochondria as a common pathway to apoptotic cell death. In most apoptotic signaling, activation of caspases 3 or 7

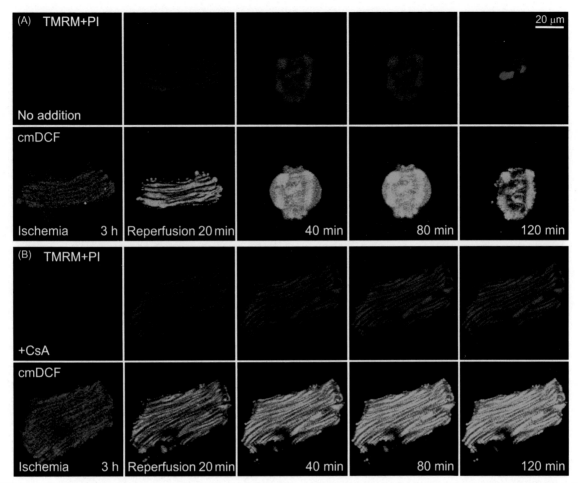

FIGURE 1.9 Mitochondrial ROS formation after reperfusion. Myocytes were co-loaded with red-fluorescing tetramethylrhodamine methyester (TMRM) and green-fluorescing chloromethyldichlorofluorescin (cmDCF) to monitor mitochondrial membrane potential and ROS formation, respectively. At the end of 3 h of ischemia, mitochondria were depolarized (lack of red TMRM fluorescence). After 20 min of reperfusion, mitochondria took up TMRM, indicating repolarization, and cmDCF fluorescence increased progressively inside mitochondria (A). Subsequently, hypercontraction and depolarization occurred after 40 min, and viability was lost within 120 min, as indicated by nuclear labeling with red-fluorescing propidium iodide. When cyclosporine A was added at reperfusion (B), mitochondria underwent sustained repolarization, and hypercontracture and cell death did not occur. Nonetheless, mitochondrial cmDCF fluorescence still increased. By contrast, reperfusion with antioxidants prevented ROS generation and MPT onset with subsequent cell death (data not shown). Thus, mitochondrial ROS generation induces the MPT and cell death after ischemia/reperfusion. *Adapted with permission from Kim JS, Jin Y, Lemasters JJ. Reactive oxygen species, but not Ca²⁺ overloading, trigger pH- and mitochondrial permeability transition-dependent death of adult rat myocytes after ischemia-reperfusion. Am J Physiol Heart Circ Physiol 2006;290:H2024−34.*

from a family of caspases (Table 1.2) begins execution of the final and committed phase of apoptotic cell death. Caspase 3/7 has many targets. Degradation of the nuclear lamina and cytokeratins contributes to nuclear remodeling, chromatin condensation, and cell rounding. Endonuclease activation leads to internucleosomal DNA cleavage. The resulting DNA fragments have lengths in multiples of 190 base pairs, the nucleosome to nucleosome repeat distance. In starch gel electrophoresis, these fragments produce a characteristic ladder pattern. DNA strand breaks are also recognized in tissue sections by the terminal deoxynucleotidyl transferase-mediated dUTP nick-end labeling (TUNEL) assay. Additionally, caspase activation leads to cell shrinkage, phosphatidyl serine externalization on the plasma membrane, and formation of numerous small surface blebs (zeiosis). Unlike necrotic blebs, zeiotic blebs contain membranous organelles. However, not all apoptotic changes depend on caspase 3/7 activation. For example, release of apoptosis-inducing factor (AIF) from mitochondria and its translocation to the nucleus promotes DNA degradation in a caspase 3-independent fashion.

Pathways leading to activation of caspase 3 and related effector caspases like caspase 7 are complex and variable between cells and specific apoptosis-instigating stimuli, and each major cellular structure can originate its own set of unique signals to induce apoptosis (Fig. 1.11). Pro-apoptotic signals are often associated with specific damage or perturbation to the organelle involved. Consequently, cells choose death by apoptosis rather than life with organelle damage.

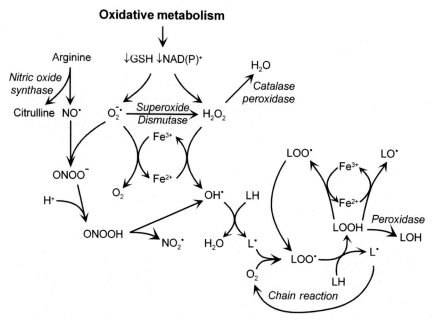

**FIGURE 1.10** Iron-catalyzed free radical generation. Oxidative stress causes oxidation of GSH and NAD(P)H, important reductants in antioxidant defenses, promoting increased net formation of superoxide ($O_2^{-\bullet}$) and hydrogen peroxide ($H_2O_2$). Superoxide dismutase converts superoxide to hydrogen peroxide, which is further detoxified to water by catalase and peroxidases. In the iron-catalyzed Haber Weiss reaction (or Fenton reaction), superoxide reduces ferric iron ($Fe^{3+}$) to ferrous iron ($Fe^{2+}$), which reacts with hydrogen peroxide to form the highly reactive hydroxyl radical ($OH^\bullet$). Hydroxyl radical reacts with lipids to form alkyl radicals ($L^\bullet$) that initiate an oxygen-dependent chain reaction generating peroxyl radicals ($LOO^\bullet$) and lipid peroxides (LOOH). Iron also catalyzes a chain reaction generating alkoxyl radicals ($LO^\bullet$) and more peroxyl radicals. Nitric oxide synthase catalyzes formation of nitric oxide ($NO^\bullet$) from arginine. Nitric oxide reacts rapidly with superoxide to form unstable peroxynitrite anion ($ONOO^-$), which decomposes to nitrogen dioxide and hydroxyl radicals. In addition to attacking lipids, these radicals also attack proteins and nucleic acids.

## Plasma membrane

The plasma membrane is the target of many receptor-mediated signals. In particular, death ligands (*e.g.*, tumor necrosis factor α, or TNFα; Fas ligand or FasL; tumor necrosis factor-related apoptosis-inducing ligand, or TRAIL) acting through their corresponding receptors (TNF receptor 1, or TNFR1; Fas; death receptor 4 and 5, or DR4/5) initiate activation of apoptotic pathways. Binding of ligands like TNFα leads to receptor trimerization and formation of a complex with adapter proteins (*e.g.*, TNF receptor-associated death domain protein, or TRADD). After receptor dissociation, a death-inducing signaling complex (DISC) forms through association with Fas-associated protein with death domain (FADD) and pro-caspase 8, which are internalized. Pro-caspase 8 becomes activated and in turn proteolytically activates other downstream effectors (Fig. 1.12). In Type I signaling, caspase 8 activates caspase 3 directly, whereas in Type II signaling, caspase 3 cleaves Bid (novel BH3 domain-only death agonist) to truncated Bid (*t*Bid) to activate a mitochondrial pathway to apoptosis. Similar signaling occurs after association of FasL with Fas (also called CD95) and TRAIL with DR4/5.

Many events modulate death receptor signaling in the plasma membrane. For example, the extent of gene and surface expression of death receptors is an important determinant in cellular sensitivity to death ligands. Stimuli like hydrophobic bile acids can recruit death receptors to the cell surface and sensitize cells to death-inducing stimuli. Surface recruitment of death receptors may also lead to self-activation even in the absence of ligand. Death receptors localize to lipid rafts containing cholesterol and sphingomyelin. After death receptor activation, ceramide forms from sphingomyelin hydrolysis, which promotes raft coalescence and formation of molecular platforms that cluster components of DISC. Glycosphingolipids, such as ganglioside GD3, also integrate into DISCs to promote apoptosis.

## Mitochondria

### Cytochrome *c* release

Bid is a BH3 only domain member of the B-cell lymphoma-2 (Bcl2) family that includes both pro- and anti-apoptotic proteins (Fig. 1.13). *t*Bid formed after caspase 8 activation translocates to mitochondria where it interacts with either

**TABLE 1.2** Mammalian caspases.

| Initiator caspases | Molecular weight of proenzyme (kDa) | Active subunits (kDa) | Prodomain | Amino acid target sequence for proteolysis |
|---|---|---|---|---|
| Caspase 2 | 51 | 19/12 | Long with CARD | VDVAD |
| Caspase-8 | 55 | 18/11 | Long with two DED | (L/V/D)E(T/V/I)D |
| Caspase-9 | 45 | 17/10 | Long with CARD | (L/V/I)EHD |
| Caspase-10 | 55 | 17/12 | Long with two DED | (I/V/L)EXD |
| Caspase-12 | 50 | 20/10 | Long with CARD | ATAD |
| **Effector caspases** | | | | |
| Caspase-3 | 32 | 17/12 | Short | DE(V/I)D |
| Caspase-6 | 34 | 18/11 | Short | (T/V/I)E(H/V/I)D |
| Caspase-7 | 35 | 20/12 | Short | DE(V/I)D |
| **Inflammatory caspases** | | | | |
| Caspase 1 | 45 | 20/10 | Long with CARD | (W/Y/F)EHD |
| Caspase 4 | 43 | 20/10 | Long with CARD | (W/L)EHD |
| Caspase 5 | 48 | 20/10 | Long with CARD | (W/L/F)EHD |
| Caspase 11 | 42 | 20/10 | Long with CARD | (V/I/P/L)EHD |
| **Other caspases** | | | | |
| Caspase 14 | 42 | 20/10 | Short | (W/I)E(T/H)D |

Caspases are evolutionarily conserved aspartate specific cysteine-dependent proteases that function in apoptotic and inflammatory signaling. Initiator caspase are involved in the initiation and propagation of apoptotic signaling, whereas effector caspases act on a wide variety of proteolytic substrates to induce the final and committed phase of apoptosis. Initiator and inflammatory caspases have large prodomains containing oligomerization motifs such as the caspase recruitment domain (CARD) and the death effector domain. Effector caspases have short pro-domains and are proteolytically activated by large pro-domain caspases and other proteases. Proteolytic cleavage of pro-caspase precursors forms separate large and small subunits that assemble into active enzymes consisting of two large and two small subunits. Caspase activation occurs in multimeric complexes that typically consist of a platform protein that recruits pro-caspases either directly or by means of adaptors. Such caspase complexes include the apoptosome and the death-inducing signaling complex (DISC). Caspase 14 plays a role in terminal keratinocyte differentiation in cornified epithelium.

Bak (Bcl2 homologous antagonist/killer) or Bax (a conserved homolog that heterodimerizes with Bcl2), two other pro-apoptotic Bcl2 family members, to induce cytochrome $c$ release through the outer membrane into the cytosol. Cytochrome $c$ in the cytosol interacts with apoptotic protease activating factor-1 (Apaf-1) and pro-caspase 9 to assemble haptomeric apoptosomes and an ATP (or deoxyadenosine triphosphate, or dATP)-dependent cascade of caspase 9 and caspase 3 activation.

Cytochrome $c$ release from the space between the mitochondrial inner and outer membranes appears to occur via formation of specific pores in the mitochondrial outer membrane. Except for the requirement for either Bak or Bax, the molecular composition and properties of cytochrome $c$ release channels remain incompletely understood. Alternatively, cytochrome c release can occur as a consequence of the MPT due to mitochondrial swelling and rupture of the outer membrane.

After the MPT, progression to apoptosis or necrosis depends on other factors. If the MPT occurs rapidly and affects most mitochondria of a cell, as happens after severe oxidative stress and ischemia/reperfusion, a precipitous fall of ATP (and dATP) will occur that actually blocks apoptotic signaling by inhibiting (d)ATP-requiring caspase 9/3 activation. With ATP depletion, oncotic necrosis ensues. However, when alternative sources for ATP generation are present (*e.g.*, glycolysis), then necrosis is prevented and caspase 9/3 becomes activated, and caspase-dependent apoptosis occurs instead (Fig. 1.14). Crosstalk between apoptosis and necrosis also occurs in other ways. For example, after TNFα binding to TNRR1, recruitment of receptor-interacting serine/threonine-protein kinase 1 (RIPK1) can activate NADPH oxidase leading to superoxide generation and sustained activation of c-Jun nuclear kinase (JNK), resulting in oncotic necrosis rather than apoptosis.

## Regulation of the mitochondrial pathway to apoptosis

Mitochondrial pathways to apoptosis vary depending on expression of pro-caspases, Apaf-1, and other proteins. Some terminally differentiated cells, particularly neurons, do not respond to cytochrome $c$ with caspase activation and

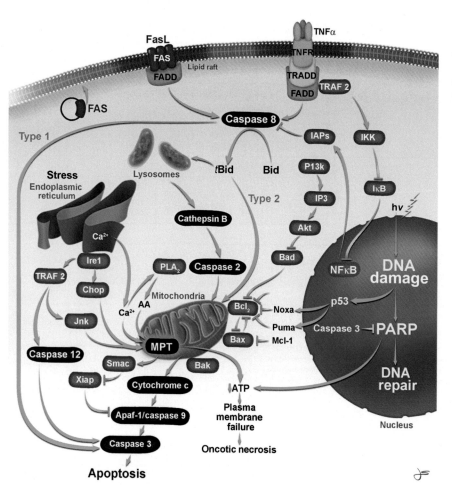

**FIGURE 1.11** Scheme of apoptotic signaling from organelles. See text for details. *Adapted with permission from Lemasters JJ. Dying a thousand deaths: redundant pathways from different organelles to apoptosis and necrosis.* Gastroenterology *2005;129:351–60.*

apoptosis, which may be linked to lack of Apaf-1 expression. Anti-apoptotic Bcl2 proteins, like Bcl2, Bcl extra long (Bcl-xL), and myeloid cell leukemia sequence 1 (Mcl-1), block apoptosis and are frequently overexpressed in cancer cells (Fig. 1.13). Anti-apoptotic Bcl2 family members form heterodimers with pro-apoptotic family members like Bax and Bak, to prevent the latter from oligomerizing into cytochrome *c* release channels.

Inhibitor of apoptosis proteins (IAPs), including X-linked inhibitor of apoptosis protein (XIAP), cellular IAP1 and 2 (cIAP1/2), and survivin, oppose apoptotic signaling by inhibiting caspase activation. Many IAPs recruit E2 ubiquitin-conjugating enzymes to promote ubiquitination of target proteins and subsequent proteosomal degradation. Some IAPs inhibit apoptotic pathways upstream of mitochondria at caspase 8, whereas others like XIAP inhibit caspase 9/3 activation downstream of mitochondrial cytochrome *c* release. Additional proteins like Smac suppress the action of IAPs, providing an "inhibitor of the inhibitor" effect promoting apoptosis. Smac is a mitochondrial intermembrane protein that is released with cytochrome *c*. Smac inhibits XIAP and promotes apoptotic signaling after mitochondrial signaling. Thus, high Smac to XIAP ratios favor caspase 3 activation after cytochrome *c* release. Other pro-apoptotic proteins released from the mitochondrial intermembrane space during apoptotic signaling include AIF, endonuclease G, and high temperature requirement A2 (HtrA2/Omi, a serine protease that degrades IAPs).

Disruption of mitochondrial function induces fragmentation of larger filamentous mitochondria into smaller more spherical structures. Such changes are also often prominent in apoptosis. Mitochondrial fission is mediated by dynamin-like protein type 1 (Drp1), a large cytosolic GTPase mechanoenzyme, and fission-1 (Fis1) in the outer membrane. Drp1 forms complexes with pro-apoptotic Bcl2 family members like Bax to promote cytochrome *c* release during apoptosis. Mitochondrial fusion depends on optic atrophy-1 (Opa1) in the inner membrane, which is mutated in dominant optic atrophy, and mitofusin 1 and 2 (Mfn1/2), two proteins in the outer membrane. Fission events in mitochondria seem to promote apoptotic signaling, since dynamin-like protein type 1 (Drp-1) overexpression promotes apoptosis, whereas Mfn1/2 overexpression retards apoptosis.

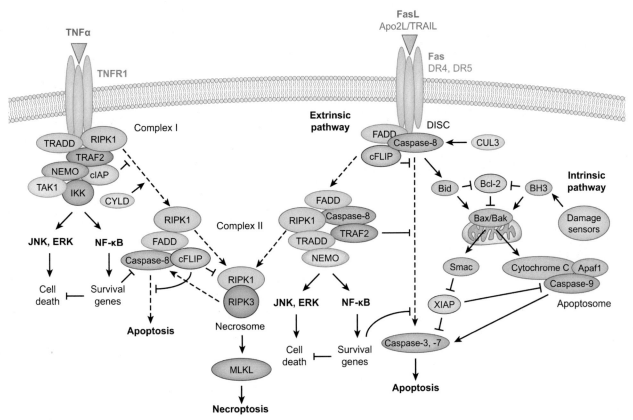

**FIGURE 1.12** Apoptotic and necroptotic signaling from death receptors. On the right, ligand (FasL, TRAIL) binding to death receptors (Fas, DR4, or DR5) causes association of FADD, FLICE-like inhibitory protein (cFLIP) and pro-caspase-8 to form a primary death-inducing signaling complex (DISC) at the plasma membrane, which activates caspase-8 and triggers apoptosis. Caspase-8 directly activates the executioner proteases, caspase-3 and -7 or proteolytic processes the BH3-only protein BID, which then activates BAX and BAK to induce mitochondrial outer membrane permeabilization and release of cytochrome *c* and second mitochondria-derived activator of caspases (Smac) from mitochondria. In association with Apaf1 and pro-caspase-9, cytochrome *c* forms an apoptosome that activates caspase-9, which in turn stimulates caspase-3 and -7. Smac augments apoptotic signaling by preventing XIAP (X-linked inhibitor of apoptosis protein) from inhibiting caspase -3, -7, and -9. A secondary cytoplasmic complex (Complex II) can form subsequently, which mediates prosurvival and other cell functions via NF-κB, JNK and ERK. On the left, ligation of TNFα to TNFR1 leads to formation of primary complex (Complex I) at the plasma membrane with the additional proteins TRADD, RIPK1, TRAF2, NF-kappa-B essential modulator (NEMO), cIAP, transforming growth factor beta-activated kinase-1 (TAK1) and IKK. Complex I mediates NF-κB, JNK, and ERK-dependent cell survival signaling and other non-death functions. Deubiquitination of RIPK1 by CYLD (a deubiquitinase) enables formation of a secondary cytoplasmic complex (Complex II) that activates caspase-8 homodimers and triggers apoptosis. Caspase-8 through heterodimeric association with cFLIP suppresses necroptosis by cleaving RIPK1. Thus, caspase-8 inhibition by cFLIP or other inhibitor protein permits RIPK1 to recruit RIPK3 and form a third complex called the necrosome. RIPK1 phosphorylates RIPK3, driving phosphorylation and oligomerization of MLKL. MLKL translocates to the plasma membrane to trigger permeabilization and necroptosis, a form of programmed necrosis. By contrast to the death receptor-induced extrinsic pathway, the intrinsic apoptotic pathway is controlled by the Bcl-2 protein family and is activated by a variety of cellular stresses, including DNA damage and metabolic stress. Damage sensors, including p53 and AKT, activate specific BH3 proteins by inducing their gene expression or post-translational modification. Activated BH3 proteins stimulate pro-apoptotic BAX and BAK directly or by abrogating their inhibition by anti-apoptotic Bcl-2 family members, typically Bcl-2 or Bcl-XL. BAX and BAK permeabilize the mitochondrial outer membrane, which releases cytochrome *c* and other pro-apoptotic proteins from the mitochondrial intermembrane space to activate caspase-3 and -7 and apoptotic cell death. *Adapted with permission from Ashkenazi A, Salvesen G. Regulated cell death: signaling and mechanisms. Annu Rev Cell Dev Biol 2014;30:337—56.*

## Anti-apoptotic survival pathways

Ligand binding to death receptors can also activate anti-apoptotic signaling to prevent activation of apoptotic death programs. Binding of adapter proteins like TNFR-associated factor 1 or 2 (TRAF1/2) and TRADD to death receptors leads to recruitment of cIAP1/2 whose E3 ubiquitin ligase activity causes polyubiquination of RIPK1, recruitment of additional proteins, and activation of IκB kinase (IKK). IKK in turn phosphorylates IκB, an endogenous inhibitor of nuclear factor κB (NFκB), leading to proteosomal IκB degradation. IκB degradation relieves inhibition of NFκB and allows NFκB to activate expression anti-apoptotic genes, including IAPs, Bcl-xL, inducible nitric oxide synthase (iNOS), and other survival factors. Nitric oxide from iNOS produces cGMP-dependent suppression of the MPT, as well as

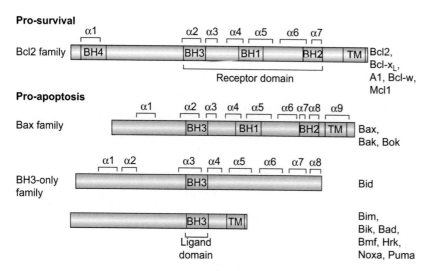

**FIGURE 1.13** Bcl2 family proteins. BH1–4 are highly conserved domains among the Bcl2 family members. Also shown are α-helical regions. Except for A1 and BH3 only proteins, Bcl2 family members have carboxy-terminal hydrophobic domains to aid association with intracellular membranes. *With permission from Cory S, Adams JM. The Bcl2 family: regulators of the cellular life-or-death switch. Nat Rev Cancer 2002;2:647–56.*

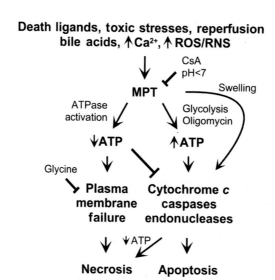

**FIGURE 1.14** Shared pathways to apoptosis and necrosis.

S-nitrosation and inhibition of caspases. In many models, apoptosis after death receptor ligation occurs only when NFκB signaling is blocked.

The phosphoinositide 3-kinase (PI3) kinase/proto-oncogene product of the viral oncogene v-akt (Akt) pathway is another source of anti-apoptotic signaling. When phosphoinositide 3-kinase (PI3 kinase) is activated by binding of insulin, insulin-like growth factor (IGF) and various other growth factors to their receptors, phosphatidylinositol trisphosphate (PIP3) is formed that activates Akt/protein kinase B, a serine/threonine protein kinase. One consequence is the phosphorylation and inactivation of Bad, a pro-apoptotic Bcl2 family member, but other anti-apoptotic targets of PI3 kinase/Akt signaling also exist. In cell lines, withdrawal of serum or specific growth factors typically induces apoptosis due to suppression of the PI3 kinase/Akt survival pathway.

# Nucleus

In the extrinsic pathway, death receptors initiate apoptosis by either a Type I (nonmitochondrial) or Type II (mitochondrial) caspase activation sequence. In the intrinsic pathway, by contrast, events in the nucleus activate apoptotic signaling. For example, ultraviolet or ionizing irradiation causes DNA damage leading to activation of the p53 nuclear transcription factor and consequent expression of genes for apoptosis and/or cell-cycle arrest, especially pro-apoptotic Bcl2 family members like p53 upregulated modulator of apoptosis (PUMA), NOXA and Bax, and the cell cycle arrest protein, 21 kDa promoter (p21) (Fig. 1.11). PUMA, NOXA, and Bax translocate to mitochondria to induce cytochrome $c$ release by similar mechanisms as in the extrinsic pathway. To escape p53-dependent induction of apoptosis, many tumors, especially those from the gastrointestinal tract, have loss of function mutations for p53.

DNA damage also activates PARP. With moderate activation, PARP helps mend DNA strand breaks, but with strong activation PARP depletes $NAD^+$ and compromises ATP generation to induce necrotic cell death (see above). Caspase 3 proteolytically degrades PARP to prevent this pathway to necrosis. Thus, DNA damage can lead to either necrosis or apoptosis depending on which occurs more quickly—PARP activation and ATP depletion, or caspase 3 activation and PARP degradation.

## Endoplasmic reticulum

The ER also gives rise to pro-apoptotic signals. Oxidative stress and other perturbations can inhibit ER calcium pumps to induce calcium release into the cytosol. Uptake of this calcium into mitochondria may then induce a $Ca^{2+}$-dependent MPT and subsequent apoptotic or necrotic cell killing (Fig. 1.11). Mitochondrial uptake of ER calcium is further facilitated by specific physical contacts between mitochondrial and ER membranes. ER calcium release into the cytosol can also activate phospholipase A2 and the formation of arachidonic acid, another promoter of the MPT.

ER calcium depletion also disturbs folding of newly synthesized proteins inside ER cisternae to cause ER stress and the unfolded protein response (UPR). Blockers of glycosylation, inhibitors of ER protein processing and secretion, various toxicants, and synthesis of mutant proteins can also cause ER stress. Calcium-binding chaperones, including glucose-regulated protein-78 (GRP78) and glucose-regulated protein-94 (GRP94), mediate detection of unfolded and misfolded proteins. In the absence of unfolded/misfolded proteins, GRP78 inhibits specific sensors of ER stress, but in the presence of unfolded proteins GRP78 translocates from the sensors to the unfolded proteins to cause sensor activation by disinhibition. The main sensors of ER stress are RNA-activated protein kinase (PKR), PKR like ER kinase (PERK), type 1 ER transmembrane protein kinase (IRE1), and activating transcription factor 6 (ATF6). PKR and PERK are protein kinases whose activation leads to phosphorylation of eukaryotic initiation factor-2a (eIF-2α). Phosphorylation of eIF-2α suppresses ER protein synthesis, a negative feedback that can relieve the unfolding stress. Ire1 is both a protein kinase and a riboendonuclease that initiates splicing of a preformed mRNA encoding X-box-binding protein 1 (XBP) into an active form. ATF6 is another transcription factor that translocates to the Golgi after ER stress where proteases process ATF6 to an amino-terminal fragment that is taken up into the nucleus. Together Ire1 and ATF6 increase gene expression of chaperones and other proteins to alleviate unfolding stress.

A strong and persistent UPR induces Ire1- and ATF6-dependent expression of C/EBP homologous protein (CHOP) and continued activation of Ire1 to initiate apoptotic signaling (Fig. 1.11). Association of TRAF2 with activated IRE1 leads to activation of caspase 12 and JNK. Caspase 12 activates caspase 3 directly, whereas JNK and CHOP promote mitochondrial cytochrome $c$ release as a pathway to caspase 3 activation.

## Lysosomes

Lysosomes and the associated process of autophagy (self-digestion) are another source of cell death signals. Autophagic cell death is characterized by an abundance of autophagic vacuoles in dying cells and is especially prominent in involuting tissues, such as post-lactation mammary gland. In autophagy, isolation membranes (also called phagophores) envelop and then sequester portions of cytoplasm to form double membrane autophagosomes. Autophagosomes fuse with lysosomes and late endosomes to form autolysosomes. The process of autophagy acts to remove and degrade cellular constituents, an appropriate action for a tissue undergoing involution. Originally considered to be random, autophagy can be selective for specific organelles, especially if they are damaged. For example, stresses inducing the MPT seem to signal autophagy of mitochondria.

Depending on the specific setting, autophagy both promotes or prevents cell death. In some circumstances, suppression of certain autophagy genes decreases cell death, whereas under other conditions, autophagy protects against cell death. When autophagic processing and lysosomal degradation are disrupted, cathepsins and other lysosomal hydrolases can be released to initiate mitochondrial permeabilization and caspase activation. In addition, lysosomal extracts cleave Bid to $t$Bid, and cathepsin D, another lysosomal protease, activates Bax. By contrast, caspases may cleave autophagy-related proteins, suppressing the execution of autophagic processes. Overall, physiological autophagy in response to nutrient deprivation and stress/damage to organelles (mitochondria, endoplasmic reticulum, *etc.*) is protective, whereas excess dysregulated autophagy promotes cytolethality. The varied and seemingly contradictory responses to autophagy again illustrate how cell death pathways are in general cell type and context-dependent.

## Shared pathways to necrosis and apoptosis

In many instances of apoptosis, mitochondrial permeabilization with release of cytochrome $c$ is a common pathway leading to a final and committed phase of cell death. At higher levels of stimulation, the same factors that induce apoptosis frequently also cause ATP depletion and a necrotic mode of cell death. Such necrotic cell killing is a consequence of mitochondrial dysfunction. In general, apoptosis is a better outcome for the organism, since apoptosis promotes orderly resorption of dying cells, whereas necrotic cell death releases cellular constituents to induce an inflammatory response that can extend tissue injury. Because of shared pathways, an admixture of necrosis and apoptosis occurs in many pathophysiological settings.

For stimuli inducing the MPT, a graded response seems to occur (Fig. 1.15). When limited to a few mitochondria, the MPT stimulates mitochondrial autophagy (mitophagy) and elimination of the damaged organelles—a repair mechanism. With greater stimulation, more mitochondria undergo the MPT, and apoptosis begins to occur due to cytochrome $c$ release from swollen mitochondria leading to caspase activation. Cathepsin leakage from an overstimulated autophagic apparatus likely also promotes apoptotic signaling. As the majority of mitochondria undergo the MPT, oxidative phosphorylation fails and ATP plummets, which precipitates necrotic cell death while simultaneously suppressing ATP-requiring apoptotic signaling.

## Programmed necrosis

Some authors subdivide cell death into three major categories: apoptosis (Type I), autophagic cell death (Type II), and necrosis (Type III). Additionally in the last several years, several forms of caspase-independent programmed necrosis have also been characterized, including ferroptosis, pyroptosis and necroptosis, as well as autophagic cell death and PARP-dependent cell death discussed above. Even in ischemia-reperfusion-induced necrosis, the enzyme cyclophilin D, a cis-trans peptidyl prolyl isomerase, enables induction of the MPT, and thus necrotic cell death after ischemia-reperfusion cannot be considered fully unprogrammed.

## Ferroptosis

Ferroptosis is iron-dependent, non-apoptotic cytolethality first described in certain cancer cell lines in response to the small molecule, erastin, and is driven by oxidative stress and subsequent lipid peroxidation. Recent evidence indicates that erastin induces opening of voltage dependent anion channels in mitochondrial outer membranes, leading to enhanced mitochondrial metabolite exchange and a sequence of mitochondrial hyperpolarization, iron-dependent reactive oxygen species generation, lipid peroxidation, glutathione depletion and stress kinase activation with ensuing mitochondrial dysfunction and ultimately cell death. Thus, ferroptosis may be a special case of iron-dependent necrotic cell death occurring after oxidative stress, ischemia-reperfusion and drug-induced hepatotoxicity.

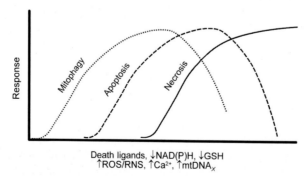

**FIGURE 1.15** Progression of mitophagy, apoptosis, and necrosis. Stimuli that induce the MPT produce a graded cellular response. Low levels of stimulation induce autophagy as a repair mechanism. With more stimulation, apoptosis begins to occur in addition due to cytochrome $c$ release after mitochondrial swelling. Necrosis becomes evident after even stronger stimulation as ATP becomes depleted. With highest stimulation, autophagy and apoptosis as ATP-requiring processes become inhibited, and only oncotic necrosis occurs. MPT inducers include death ligands, oxidation of NAD(P)H and GSH, formation of reactive oxygen species (ROS) and reactive nitrogen species (RNS), and mutation of mitochondrial DNA (mtDNA$_X$) which causes synthesis of abnormal mitochondrial membrane proteins.

## Pyroptosis

In response to intracellular pathogens, inflammasomes form comprised of a sensor like nucleotide-binding domain—like receptor (NLR), an adapter called ASC (apoptosis-associated speck-like protein containing a caspase activation and recruitment domain) and pro-caspase-1 or 11, which lead to caspase activation, proteolytic activation of proinflammatory interleukin-1β and interleukin-18, gasdermin D cleavage and a lytic cell death called pyroptosis. Cell death by pyroptosis removes the replication niche of intracellular pathogens, thereby promoting microbial killing by secondary phagocytes. Aberrant activation of pyroptosis may also contribute to sepsis and other diseases, whereas many pathogens have evolved to evade pyroptosis.

## Necroptosis

Death receptors and TLRs can also drive a nonapoptotic form of cell death called necroptosis. In necroptosis, receptor interacting serine/threonine protein kinase 3 (RIPK3) phosphorylates mixed lineage kinase domain-like (MLKL) pseudokinase, which causes cell lysis by forming pores in the plasma membrane (Fig. 1.12). A related protein, receptor interacting protein-1 (RIP1), regulates entry into programmed cell death or activation of pro-survival of NFκB signaling. Unlike apoptosis, necroptosis (or any other form of necrosis) induces inflammation and a robust immune response to help eliminate tumor-causing mutations and viruses.

## Extracellular vesicles and Exosomes

Cell lysis also releases a variety of damage-associated molecular pattern (DAMP) molecules, such as high-mobility group box 1 protein (HMGB1), mitochondrial DNA, ATP, N-formyl peptides, cardiolipin, and ATP, that promote inflammation principally through activation of toll-like receptors (TLR). Even prior to cell death, dying and stressed cells release extracellular vesicles (EVs), which include exosomes (<150 nm in diameter), as well as microvesicles/shedding particles and apoptotic bodies (both >100 nm) Exosomes are formed by exocytosis of late endosomes and multivesicular bodies, a type of lysosome, with plasma membranes. Exosomes act as delivery vehicles for DAMPs, various other proteins and both messenger and small interfering RNA (mRNA and siRNA). Microvesicles/shedding particles arise from the budding off of cell surface blebs, whereas apoptotic bodies are budded off zeiotic blebs and the fragmented remains of apoptotic cells. These larger EVs also communicate ongoing cellular stress to adjacent tissue and to the organism as a whole.

## Concluding remark

The various forms of apoptosis and necrosis are prominent events in pathogenesis. An understanding of cell death mechanisms forms the basis for effective interventions to either prevent cell death as a cause of disease or promote cell death in cancer chemotherapy.

## Acknowledgments

This work was supported, in part, by Grants AA022815, AA021191, DK073336 and CA184456 from the National Institutes of Health. Imaging facilities were supported, in part, by P30 CA138313 and S10 OD018113.

## Key concepts

- A common theme in disease is the life and death of cells. In diseases like stroke and heart attacks, death of individual cells leads to irreversible functional loss, whereas in cancer the goal of chemotherapy is to kill proliferating tumor cells. The mode of cell death typically follows one of two patterns: necrosis and apoptosis.
- Necrosis is the consequence of metabolic disruption with ATP depletion and is characterized by cellular swelling leading to plasma membrane rupture with release of intracellular contents. Apoptosis is a form of programmed cell death that causes orderly resorption of individual cells initiated by well-defined ATP-requiring pathways involving activation of proteases called caspases.
- In some pathophysiological settings, the mitochondrial inner membrane abruptly becomes permeable to solutes up to 1500 Da. This mitochondrial permeability transition causes uncoupling of oxidative phosphorylation, ATP depletion, mitochondrial swelling, and pro-apoptotic cytochrome c release that can lead to both necrosis and apoptosis.

- Each of several organelles gives rise to signals initiating apoptotic cell killing. Often these signals converge on mitochondria to cause cytochrome c release and Apaf-1-dependent caspase 9 and 3 activation as a final common pathway to apoptotic cell death.
- Death ligands like TNFa and Fas ligand activate their corresponding receptors in the plasma membrane to initiate caspase signaling cascades and the mitochondrial pathway to cell death. Inhibitor of apoptosis proteins (IAPs) oppose apoptotic signaling by inhibiting caspase activation.
- DNA damage activates p53, a nuclear transcription factor, and expression of pro-apoptotic Bcl2 family members like PUMA, NOXA, and Bax that translocate to mitochondria to induce cytochrome c release. Many tumors have loss of function mutations for p53 to escape p53-dependent apoptosis.
- Accumulation of unfolded/misfolded proteins in the ER causes ER stress. Initially, ER stress increases expression of molecular chaperones with inhibition of other protein synthesis to alleviate the unfolding stress. With prolonged ER stress, apoptotic pathways are activated. Lysosomes and the associated process of autophagy (self-digestion) are yet another source of pro-apoptotic signals. Some consider autophagic cell death as a separate category of programmed cell death.
- Dying cells release exosomes and extracellular vesicles (EVs) containing damage-associated molecular patterns (DAMPs) and both messenger and small interfering RNA (mRNA and siRNA) to communicate ongoing cellular stress to adjacent tissue and promote inflammation through toll-like receptors (TLR).
- Apoptosis and necrosis can share common signaling pathways to be extreme end points on a phenotypic continuum.

## Suggested readings

[1] Bayly-Jones C, Bubeck D, Dunstone MA. The mystery behind membrane insertion: a review of the complement membrane attack complex. Philos Trans R Soc Lond B Biol Sci 2017;372.
[2] Brenner D, Blaser H, Mak TW. Regulation of tumour necrosis factor signalling: live or let die. Nat Rev Immunol 2015;15:362−74.
[3] Carraro M, Checchetto V, Szabo I, Bernardi P. F-ATP synthase and the permeability transition pore: fewer doubts, more certainties. FEBS Lett 2019;.
[4] Cosentino K, Garcia-Saez AJ. Bax and bak pores: are we closing the circle? Trends Cell Biol 2017;27:266−75.
[5] Dorstyn L, Akey CW, Kumar S. New insights into apoptosome structure and function. Cell Death Differ 2018;25:1194−208.
[6] Green DR. The coming decade of cell death research: five Iddles. Cell 2019;177:1094−107.
[7] He L, Lemasters JJ. Regulated and unregulated mitochondrial permeability transition pores: a new paradigm of pore structure and function? FEBS Lett 2002;512:1−7.
[8] Hernandez C, Huebener P, Schwabe RF. Damage-associated molecular patterns in cancer: a double-edged sword. Oncogene 2016;35:5931−41.
[9] Hetz C, Papa FR. The unfolded protein response and cell fate control. Mol Cell 2018;69:169−81.
[10] Hirschhorn T, Stockwell BR. The development of the concept of ferroptosis. Free Radic Biol Med 2019;133:130−43.
[11] Hirsova P, Ibrahim SH, Verma VK, Morton LA, Shah VH, LaRusso NF, et al. Extracellular vesicles in liver pathobiology: small particles with big impact. Hepatology 2016;64:2219−33.
[12] Jaeschke H, Lemasters JJ. Apoptosis versus oncotic necrosis in hepatic ischemia/reperfusion injury. Gastroenterology 2003;125:1246−57.
[13] Lemasters JJ. Evolution of voltage-dependent anion channel function: from molecular sieve to governator to actuator of ferroptosis. Front Oncol 2017;10:1−4.
[14] Ogretmen B. Sphingolipid metabolism in cancer signalling and therapy. Nat Rev Cancer 2018;18:33−50.
[15] Orrenius S, Gogvadze V, Zhivotovsky B. Calcium and mitochondria in the regulation of cell death. Biochem Biophys Res Commun 2015;460:72−81.
[16] Pernas L, Scorrano L. Mito-morphosis: mitochondrial fusion, fission, and cristae remodeling as key mediators of cellular function. Annu Rev Physiol 2016;78:505−31.
[17] Ramirez MLG, Salvesen GS. A primer on caspase mechanisms. Semin Cell Dev Biol 2018;82:79−85.
[18] Stahl PD, Raposo G. Extracellular vesicles: exosomes and microvesicles, integrators of homeostasis. Physiology (Bethesda, MD) 2019;34:169−77.
[19] Tang D, Kang R, Berghe TV, Vandenabeele P, Kroemer G. The molecular machinery of regulated cell death. Cell Res 2019;29:347−64.
[20] Terman A, Kurz T. Lysosomal iron, iron chelation, and cell death. Antioxid Redox Signal 2013;18:888−98.
[21] Vida A, Marton J, Miko E, Bai P. Metabolic roles of poly(ADP-ribose) polymerases. Semin Cell Dev Biol 2017;63:135−43.
[22] Weerasinghe P, Buja LM. Oncosis: an important non-apoptotic mode of cell death. Exp Mol Pathol 2012;93:302−8.
[23] Win S, Than TA, Zhang J, Oo C, Min RWM, Kaplowitz N. New insights into the role and mechanism of c-Jun-N-terminal kinase signaling in the pathobiology of liver diseases. Hepatology 2018;67:2013−24.
[24] Zhang X, Lemasters JJ. Translocation of iron from lysosomes to mitochondria during ischemia predisposes to injury after reperfusion in rat hepatocytes. Free Radic Biol Med 2013;63:243−53.
[25] Zhong Z, Lemasters JJ. A unifying hypothesis linking hepatic adaptations for ethanol metabolism to the Proinflammatory and Profibrotic events of alcoholic liver disease. Alcohol Clin Exp Res 2018;42:2072−89.

Chapter 2

# Acute and chronic inflammation induces disease pathogenesis

Catherine Ptaschinski and Nicholas W. Lukacs

*Department of Pathology, University of Michigan Medical School, Ann Arbor, MI, United States*

**Summary**

The recognition of pathogenic insults can be accomplished by a number of mechanisms that function to initiate inflammatory responses and mediate clearance of invading pathogens. This initial response, when functioning optimally, will lead to minimal leukocyte accumulation and activation for the clearance of the inciting agent and have little effect on homeostatic function. However, often the inciting agent elicits a very strong inflammatory response, either due to host recognition systems or due to the agent's ability to damage host tissue. Thus, the host innate immune system mediates the damage and tissue destruction in an attempt to clear the inciting agent from the system. These initial acute responses can have long-term and even irreversible effects on tissue function. If the initial responses are not sufficient to facilitate the clearance of the foreign pathogen or material, the response shifts toward a more complex and efficient process mediated by lymphocyte populations that respond to specific residues displayed by the foreign material. Normally, these responses are coordinated and only minimally alter physiologic function of the tissue. However, in unregulated responses the initial reaction can become acutely catastrophic, leading to local or even systemic damage to the tissue or organs, resulting in degradation of normal physiologic function. Alternatively, the failure to regulate the response or clear the inciting agent could lead to chronic and progressively more pathogenic responses. Each of these potentially devastating responses has specific and often overlapping mechanisms that have been identified and lead to the damage within tissue spaces. A series of events take place during both acute and chronic inflammation that lead to the accumulation of leukocytes and damage to the local environment.

## Introduction

The recognition of pathogenic insults can be accomplished by a number of mechanisms that function to initiate inflammatory responses and mediate clearance of invading pathogens. This initial response, when functioning optimally, will lead to minimal leukocyte accumulation and activation for the clearance of the inciting agent and have little effect on homeostatic function. However, often the inciting agent elicits a very strong inflammatory response, either due to host recognition systems or due to the agent's ability to damage host tissue. Thus, the host innate immune system mediates the damage and tissue destruction in an attempt to clear the inciting agent from the system. These initial acute responses can have long-term and even irreversible effects on tissue function. If the initial responses are not sufficient to facilitate the clearance of the foreign pathogen or material, the response shifts toward a more complex and efficient process mediated by lymphocyte populations that respond to specific residues displayed by the foreign material. The failure to regulate the response or clear the inciting agent could lead to chronic and progressively more pathogenic responses.

## Leukocyte adhesion, migration, and activation

### Endothelial cell expression of adhesion molecules

The initial phase of the inflammatory response is characterized by a rapid leukocyte migration into the affected tissue. Upon activation of the endothelium by inflammatory mediators, upregulation of a series of adhesion molecules is initiated that leads to the reversible binding of leukocytes to the activated endothelium. The initial adhesion is mediated by L-, E-, and P-selectins that facilitate slowing of leukocytes from circulatory flow by mediating rolling of the leukocytes on the activated endothelium. The selectin-mediated interaction with the activated endothelium potentiates the

*Essential Concepts in Molecular Pathology. DOI: https://doi.org/10.1016/B978-0-12-813257-9.00002-4*

likelihood of the leukocyte to be further activated by endothelial-expressed chemokines, which mediate G protein−coupled receptor−induced activation. If the rolling leukocytes encounter a chemokine signal and an additional set of adhesion molecules is also expressed, such as intracellular adhesion molecule 1 and vascular cell adhesion molecule 1 (VCAM-1), the leukocytes firmly adhere to the activated endothelium. The mechanism of chemokine-induced adhesion of the leukocyte is dependent on actin reorganization and a conformational change of the β-integrins on the surface of the leukocytes. Subsequently, the firm adhesion allows leukocytes to spread along the endothelium and to begin the process of extravasating into the inflamed tissue following chemoattractant gradients that guide the leukocyte to the site of inflammation.

The initial binding of the leukocytes to E and P selectins is mediated by interaction with glycosylated ligands expressed on the leukocytes, most commonly P-selectin glycoprotein ligand 1 (PSGL1). Once the leukocyte is tethered by the selectin molecules on the vessel wall, a series of binding and release events allow the leukocyte to roll along the activated endothelial cell surface. The expression of phosphoinositide 3-kinase-γ (PI3Kγ) in activated endothelium appears to be critical for securing the tethered leukocyte to the vessel wall, whereas spleen tyrosine kinase signals downstream of PSGL1 in the bound leukocyte and regulates the rolling process.

In addition to selectin-mediated leukocyte rolling, there are a number of instances in which specific integrins can also participate in leukocyte rolling. The CNS, intestinal track, and lung are three examples where this has been specifically observed. Using in vitro flow analysis, cells expressing α4β7 integrins can roll on immobilized mucosal vascular addressin cell adhesion molecule 1 that is highly expressed in the intestinal tract. Likewise, very late antigen 4 (VLA4, the α4β1 integrin) supports monocyte and lymphocyte rolling in vitro and lymphocyte rolling in the central nervous system (CNS). In many cases, the cooperation of selectin-mediated and integrin-mediated rolling and adhesion is required for leukocyte rolling.

The transition from leukocyte rolling to firm adhesion depends on several distinct events to occur in the rolling leukocyte. First, the integrin needs to be modified through a G protein−mediated signaling event enabling a conformational change that exposes the binding site for the specific adhesion molecule. Second, the density of adhesion molecule expression needs to be high enough to allow the leukocyte to spread along the activated endothelium and appropriate integrin clustering on the leukocyte surface. Finally, it appears that a phenomenon known as outside-in signaling is also necessary for strengthening the adhesive interactions through several important signaling events that include FGR and HCL, two SRC-like protein tyrosine kinases. Together, these coordinated events facilitate preparation of leukocytes for extravasation through the endothelium into the inflamed tissue.

Transendothelial cell migration of leukocytes requires that numerous potential obstacles be managed. After firmly adhering to the activated endothelium, leukocytes appear to spread and crawl along the border until they reach an endothelial cell junction that has been appropriately *opened* by the inflamed environment. A number of molecules have been implicated in this route of migration, but PECAM-1 (also called CD31) has been the most thoroughly studied and appears to be functionally required for the process with targeted expression at the endothelial cell junction region. Another protein, junction adhesion molecule-A (JAM-A), has also been shown to be associated with migration of cells through the tight junctions of endothelial layer of vessels and is found on the surface of several leukocyte populations including PMNs. It appears that PECAM1 and JAM-A are utilized in a sequential manner to allow movement through the endothelial barrier.

The final obstacle for the leukocyte to traverse prior to entering into the tissue from the vessel is the basement membrane. The model that has been proposed over the years suggests that metalloproteinases (MMPs) are activated to degrade the basement membrane extracellular matrix (ECM), enabling leukocytes to penetrate toward the site of inflammation. More recently, studies have shown that leukocytes utilize low-expression regions (LERs), which have reduced expression of basement membrane components. MMPs and neutrophil elastase have been implicated in the remodeling of LERs.

## Chemoattractants

Over the past several years, researchers have identified multiple families of chemoattractants that can participate in the extravasation of leukocytes. Perhaps the most readily accessible mediator class during inflammation is the complement system. These proteins are found in circulation or can be generated de novo upon cellular stimulation. Upon activation, bacterial products or immune complexes through the alternative or classical pathways mediate cleavage of C3 and/or C5 into C3a and C5a that can provide an immediate and effective chemoattractant to induce neutrophil and monocyte activation. The role of C3a as an anaphylactic agent illustrates the importance of this early activation event on mast cell biology. In addition, C5a stimulates neutrophil oxidative metabolism, granule discharge, and adhesiveness to vascular endothelium. Altogether, these functions of C3a and C5a indicate that they are potent inflammatory mediators.

A second mediator system that is involved in early and immediate leukocyte migration is the leukotrienes, a class of lipid mediators that are preformed in mast cells or are quickly generated through the efficient arachadonic acid pathway induced by 5-LO. In particular, leukotriene B4 (LTB4) has especially been implicated in the early induction of neutrophil migration, but also can generate long-term problems during inflammation. LTB4 can be rapidly synthesized by phagocytic cells (PMNs and macrophages) following stimulation with pathogen products. Furthermore, the LTB4 receptor has been implicated in recruitment of T-lymphocytes that mediate chronic inflammatory diseases. In particular, the LTB4 receptor BLT1 has been implicated in preferential recruitment of Th2 type T-lymphocytes during allergic responses. In addition to LTB4, the cysteinyl leukotrienes, LTC4, D4, and E4, also appear to have some chemotactic activity, suggesting that targeting the conversion of arachodonic acid into the leukotriene pathway may be generally beneficial.

Chemokines represent a large and well-characterized family of chemoattractants composed of over 50 polypeptide molecules that are expressed in numerous acute and chronic immune responses. Chemokines fall under four classes, based on the N-terminal cysteine residues. The majority of chemokines are classified as CC or CXC, with a few described as XC or CX3C. The determination of what leukocyte populations are recruited to a particular tissue during a response is dictated by the chemokine ligands that are induced and the specific receptors that are displayed on subsets of leukocytes. This latter aspect can be best observed during acute inflammatory responses, such as in bacterial infections, when the cellular infiltrate is primarily neutrophils and it is the production of CXCR-binding chemokines that mediate the process. Likewise, when the acute inflammatory mechanisms are not able to control the infectious process, immune cytokines, such as interferon (IFN)-$\gamma$ and interleukin (IL)-4, tend to drive the production of chemokines that facilitate the recruitment of mononuclear cells, macrophages, and lymphocytes to the site of infection. In addition to their ability to bind to cellular receptors, chemokines are also able to bind to glycosaminoglycans. This provides several advantages with respect to the maintenance of chemoattractant gradients in inflamed tissue over long periods of time, with the most intense signals maintained at the site of inflammation. Chemokines are important at the endothelial border where they mediate firm adhesion of leukocytes undergoing selectin-associated rolling to activate their $\beta$-integrins to the activated endothelial cells and subsequently direct migration of these cells to the site of inflammation. Finally, at higher concentrations (such as those found at the site of inflammation), chemokines induce leukocyte activation for effector function (for instance, degranulation). Thus, the progressive movement of leukocytes from the endothelial cell border in activated vessels through their arrival at the site of inflammation relies on the coordinated expression and interaction of chemokines and adhesion molecules.

The activation of leukocytes by chemoattractants is mediated by a common signaling cascade via G protein−coupled receptor pathways. These pathways can be monitored by calcium mobilization and cellular migration mediated by the G$\alpha$ and G$\beta$/$\gamma$ subunits, respectively. Signaling through these subunits results in the activation of important downstream pathways, including PI3K, MAPK, and FAK. The PI3K pathway has been the most thoroughly studied after a chemotactic signal and depends on the P110 subunits $\alpha$, $\beta$, and $\gamma$. The importance of PI3K was best demonstrated in studies using cells deficient in PI3K P110$\gamma$, as well as in localization studies that identified activated P110$\gamma$ at the leading edge of migrating cells. Once activated, PI3K appears to couple with the p85 adaptor protein that mediates activation of other pathways, including rac, rho, and Cdc42 GTPases leading to actin polymerization. These PI3K-induced responses can be actively regulated by PTEN (phosphatase and tensin homology deleted on chromosome ten protein) and involved in directional sensing. This system regulates directional leukocyte movement by differential cellular localization of key signaling molecules that regulate actin and other structural proteins.

## Acute inflammation and disease pathogenesis

The initiation of a rapid innate immune response to invading pathogens is essential to inhibit the colonization of microorganisms or to sequester toxic and noxious substances. The primary mechanism is activation of edema and local fluid release to flood the affected tissue, along with early activation of the complement system in response to bacterial components, resulting in cleavage of C3 and C5. The subsequently recruited phagocytic cells engulf invading pathogenic bacteria and quickly activate to begin producing LTB4 as well as early response cytokines (such as the cytokines IL-1 and TNF) that enhance phagocytosis and killing. The early response cytokines subsequently activate resident cell populations to produce other important mediators of inflammation, such as IL-6 and IL-8 (CXCL8), and promote cytokine cascades that lead to continued leukocyte migration and activation. This multipronged approach to the activation of the inflammatory response and inhibition of the pathogen expansion is normally tightly regulated. However, in situations in which the inflammatory stimuli are intense, such as in a bolus dose of bacteria, severe trauma, or in burn victims, the acute inflammatory response can become dangerously unregulated. The host/patient can quickly become subjected to a systemic inflammatory response even though the initial insult may be quite localized. This is a result of mediator production (especially TNF and IL-1) that is systemically delivered to multiple organs, creating an overproduction of leukocyte chemoattractants in distal organs and inducing inflammatory cell influx. In this form of septic response, the overwhelming PMN recruitment and activation to multiple organs can lead to tissue damage and organ dysfunction.

While these events can affect any tissue of the organism, the liver and lung appear to be primary targets due to their relatively high numbers of resident macrophage populations that can quickly respond to the inflammatory cytokine signals. In the lung, the development of acute respiratory distress syndrome (ARDS) is often observed in patients experiencing a septic insult. Although early research focused on TNF and IL-1 as lead targets to combat these responses, clinical trials using specific inhibitors, along with more clinically relevant animal models, have not shown any benefit to blocking these central inflammatory mediators in acute diseases, such as sepsis. These failures are likely due to the fact that the early response cytokines are produced and cleared prior to the induction of the most severe aspects of disease. More recent targets include complement receptors (C3aR and C5aR) as well as numerous chemokines and their receptors to attempt to block the recruitment and activation of the cell populations during the responses. In mouse models, blocking complement signaling significantly alters the outcome of severely unregulated septic responses. However, this has yet to be confirmed during human disease.

In the case of viral infections, the system must deal with clearance in a different manner since the ability to recognize and phagocytize virus particles is not reasonable due to their size. One of the most effective early means of blocking the spread is through the immediate production of type I IFN. This class of mediators facilitates the blockade of spread by both altering the metabolism of infected cells and by promoting the production of additional anti-viral factors in uninfected cells to reduce the chance of successful viral assembly and further spread. While the anti-viral effects of type I IFNs was initially identified many years ago, researchers continue to attempt to fully understand the mechanisms that are initiated by this class of mediators.

## Pattern recognition receptors and inflammatory responses

### Toll-like receptors

The initiation of acute inflammation and the progression of chronic disease are often fueled by infectious agents that provide strong stimuli to the host. These responses evolved to be beneficial for the rapid recognition of pathogenic motifs that are not normally present in the host during homeostatic circumstances. While there are now a number of diverse families of pattern recognition systems, the best-characterized family is the Toll-like receptors (TLR). The Toll system was first discovered in *Drosophila* as a crucial part of the anti-fungal defense for the organism. These molecules primarily recognize pattern-associated molecular patterns (PAMPs) from invading pathogens. The TLR family in mammalian species consists of transmembrane receptors that reside either on the cell surface or within the endosome and that characteristically consist of leucine rich repeats (LRR) for motif recognition and an intracytoplasmic region for signal transduction (Fig. 2.1). While TLRs are most notably identified on immune/inflammatory cell populations, they can also be expressed on non-immune structural cells, such as epithelial cells and provide an activation signal for the initiation of inflammatory mediator production. Cellular activation signals are transmitted by TLRs via cytoplasmic adapter molecules that initiate a cascade of now well-defined activation pathways including NF-κB, IRF3, IRF7, as well as a link to MAPK pathways. These activation pathways provide strong stimuli that alert the host with *danger signals* that allow effective immune cell activation.

One of the first molecules in this family that was identified was TLR4, which primarily recognizes lipopolysaccharide (LPS; also known as endotoxin), a component of the cell wall of gram-negative bacteria. The TLR4 activation pathway is unique among TLRs as it signals via multiple adaptor proteins, including MyD88, TRIF, and MD-2, making it the most dynamic TLR within the family. In addition to recognizing LPS, RSV F protein, and other complex carbohydrate and lipid molecules from pathogens, TLR4 has also been demonstrated to recognize free fatty acids from host adipose tissue and may contribute to the inflammatory syndrome observed during obesity and in type 2 diabetes.

Other TLR family members recognize distinct and now better-defined factors that allow the immediate activation of the innate immune system and subsequent signaling of the adaptive immune responses. TLR2 appears to have the most diverse range of molecules that are recognized directly including peptidoglycan, mycoplasm lipopeptide, a number of fungal antigens, as well as a growing number of carbohydrate residues on parasitic, fungal, and bacterial moieties. In addition, TLR2 can heterodimerize with TLR1 and TLR6 to further expand its recognition capabilities. TLR5

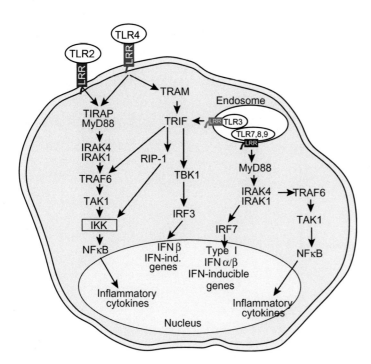

FIGURE 2.1 TLR activation leads to induction of diverse inflammatory, chemotactic, and activating cytokine production.

specifically recognizes flagellin and is therefore important for recognizing both Gram-negative and Gram-positive bacteria. While the previously described TLRs are expressed on the cell membrane, a number of TLRs are predominantly expressed in endosomal membrane compartments of innate immune cells, including TLR3, TLR7, TLR8, and TLR9. These pathogen recognition receptors (PRR) are involved in recognition of nucleic acid motifs including dsRNA (TLR3), ssRNA (TLR7 and TLR8), and unmethylated CpG DNA (TLR9). Together, these TLRs function in the recognition of viral and bacterial pathogens that enter the cell via receptor-mediated endocytosis or that are actively phagocytized. Thus, these pathways are important for the initiation of innate cytokines, including TNF, IL-12, and type I IFN, as outlined in Fig. 2.1. However, the activation of antigen-presenting cells via the TLR pathways is also extremely important for integrating acute inflammatory events with acquired immunity.

## Cytoplasmic sensors of pathogens

While the TLR proteins have been the best characterized, it is now evident that they are not the only molecules that are important for recognition of pathogenic insults. If pathogens infect directly into the cytoplasm or escape endosomal degradation pathways, the host cell must have the ability to recognize and deal with the cytoplasmic insult. Similar to the TLR system, NOD-like receptors (NLR; also known as CATERPILLAR proteins) are able to recognize specific pathogen patterns that are distinct from host sequences. NOD1 and NOD2 sense bacterial molecules produced by the synthesis and/or degradation of peptidoglycans (PGNs) (Fig. 2.2). Specifically, NOD1 recognizes PGNs that contain meso-diaminopimelic acid produced by Gram-negative bacteria and some Gram-positive bacteria, while NOD2 recognizes muramyl dipeptide that is found in nearly all PGNs. Other members of this family include NALPs (NACHT-, LRR-, and pyrin domain—containing proteins), IPAF (ICE-protease activating factor), and NAIPs (neuronal apoptosis inhibitor proteins). An interesting aspect of NLRs is that they contain a conserved caspase associated receptor domain (CARD) that was initially related to proteins involved in programmed cell death or apoptosis. The central protein of the inflammasome is caspase 1, which is bound by the CARD of the NLRs. A simplified model of the NLR activation pathway suggests that binding to caspase 1 leads to processing of pro-IL-1b and pro-IL-18 to their active and released forms. More recently, studies have shown that the activation of the inflammasome is triggered by cell stress pathways, such as endoplasmic reticulum stress, and if not controlled leads to NLRP3-mediated activation, caspase 1, and uncontrolled inflammation. In addition to the inflammasome-mediated activation of pro-inflammatory cytokine release, NOD proteins have also been shown to induce NFkB and MAPK activation through an RICK/RIP2 signaling pathway. This opens up a number of additional gene activation events via this bacterial-induced pathway.

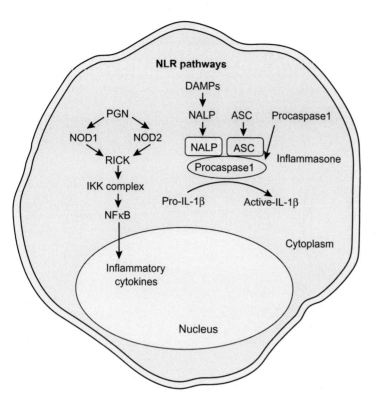

**FIGURE 2.2** Cytoplasmic pathogen receptors allow activation of cytokines responsible for the upregulation of inflammatory processes.

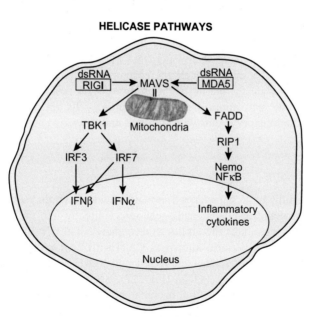

**FIGURE 2.3** Helicase proteins induce early response and activating cytokines by recognition of dsRNA.

Several years ago it was recognized that PKR proteins had the ability to recognize dsRNA. Interestingly, one of the functions of PKR is that it phosphorylates the alpha-subunit of eukaryotic translation initiation factor 2, which causes inhibition of cellular and viral protein synthesis. This latter function works in concert with type I IFN that is also a product of PKR activation for creating an anti-viral environment. PKR has a long history of investigation related to anti-viral effects of infected cells and appears to provide an important aspect to the response against a productive virus infection.

More recent investigations have identified two helicase proteins that have the ability to recognize dsRNA, RIG-I (retinoic acid-inducible gene), and MDA5 (melanoma differentiation-associated gene). The activation of the protein products of either of these genes in the cytoplasm leads to an immediate activation of type I IFNs (Fig. 2.3).

The signaling pathway that RIG-I utilizes was initially surprising since it also utilizes its CARD to interact with a mitochondria-associated protein, MAVS. This interaction leads to a scaffold involving TRAF-3, TBK-1, and IRF-3 activation. Overall, the number of proteins that have the ability to recognize dsRNA not only demonstrates redundancy in the system but also indicates the importance of being able to detect this specific PAMP that is a clear sign of a productive viral infection.

## Pattern recognition and pathologic consequences

One of the potential pathogenic side effects of activation of cells by TLRs, NLRs, and other pathogen-sensing receptor systems is the initiation of a wide range of inflammatory molecules that, if not properly controlled, could lead to detrimental pathogenic responses. Two important activation systems that are strongly upregulated by PRRs are the type I IFN (IRF-mediated) and early response cytokine (NF-κB−mediated) cascades. Both of these systems have a strong impact on the intensity and direction of the inflammatory responses through their ability to drive chemokine production. This response is greatly enhanced and perpetuated in more persistent pathogenic insults or with non-pathogenic stimuli that cannot be cleared, such is the case with silica.

In addition to PAMPs, the immune system also recognized danger-associated molecular patterns, or DAMPs. These signals can come from endogenous agents, such as signals from damaged tissue during injury. While DAMP activation of TLRs drives tissue repair, it has also be implicated in inflammatory disease, such as rheumatoid arthritis (RA) and atherosclerosis. DAMPs known to activated TLRs such as heat-shock proteins, high-mobility group box protein 1 (HMGB1), host DNA, and fibrinogen have been found in the synovia of patients with RA, while high HMGB1 levels are found in lesions of patients with multiple sclerosis and was indicative of inflammation. These studies are supported by work in TLR-deficient mice that demonstrate a role for TLRs in DAMP recognition and chronic inflammation, and injection of DAMPs results in increased inflammation. Together, this indicates an important role for endogenous activation of TLRs in inflammatory disease.

## Innate lymphoid cells

Over the past several years a novel classification of innate immune cells has emerged known as innate lymphoid cells (ILCs) that play an essential role in the initiation and regulation of inflammation. These cells are derived from lymphoid progenitors but have no lineage markers (lin-) and do not rearrange their antigen receptors. They include natural killer cells (ILC1) that produce IFN-γ, Group 2 ILCs (ILC2) that produce IL-5 and IL-13, and group 3 ILCs (ILC3) that produce IL-17 and IL-22. These cells can respond quickly to innate signals produced during inflammation and produce their cytokines to appropriately impact the immune environment during pathogenic insult. The ILC1 cells (natural killer [NK] cells), which produce IFN-γ in response to IL-12 and IL-18, appear to be especially important in early responses to viruses and help to keep the virus at bay until the acquired immune response can terminal clear the virus through cytotoxic T-cell responses and/or antibody responses. The ILC2 cells respond to epithelial-derived signals, such as IL-25, IL-33 and TSLP, and produce IL-5 and IL-13 that are necessary, at least in the gut, to clear parasitic infections through the induction of mucus and eosinophil accumulation. Finally, the most recently identified ILC3 cells may be important for anti-bacterial responses (IL-17) as well as stabilization of the epithelial barrier function (IL-22). Together, these early response cells likely help to dictate the local immune environment to instruct the acquired immune responses. Numerous studies are currently underway to better understand the role of these cells in homeostasis and disease.

## Regulation of acute inflammatory responses

A number of well-described regulators may be suited for management of acute inflammatory responses. Several anti-inflammatory mediators have been investigated, including IL-10, transforming growth factor beta (TGFβ), IL-1 receptor antagonist (IL-1ra), as well as IL-4 and IL-13. Perhaps the most attractive anti-inflammatory cytokine with broad-spectrum activity is IL-10. The function of IL-10 appears to be important during normal physiologic events, as IL-10 gene knockout mice develop lethal inflammatory bowel dysfunction. The importance of this cytokine has been demonstrated in models of endotoxemia and sepsis that neutralized IL-10 during the acute phase and led to increased lethality. In addition, administration of IL-10 to mice protects them from a lethal endotoxin challenge. The downregulation of the inflammatory cytokine mediators by IL-10 therapy suggest an extremely potent anti-inflammatory agent that might be used for intervention of inflammation-induced injury. IL-10 may also have a role in promoting end-stage disease if not properly regulated. Thus, its role as a therapeutic has been questioned.

Other cytokines also function to suppress the inflammatory response. TGFβ has anti-proliferative effects in macrophages and lymphocytes, downregulates cytokine production, and inhibits inflammation. In TGFβ gene knockout mice, a progressive inflammatory response was observed early in neonatal development (day 14 after birth) throughout the body including heart, lung, salivary glands, and in virtually every organ at later time points. Thus, the use of this cytokine as a therapeutic must be viewed very cautiously as TGFβ may promote fibrogenic outcomes and can skew the T-cell response toward a Treg or Th17 phenotype. Another cytokine that has demonstrated anti-inflammatory functions is IL-4. IL-4 is a member of the Th2-type cytokine family and has the ability to downregulate the production of inflammatory cytokines from macrophages. However, like TGFβ, IL-4 has a number of alternate functions that should be viewed carefully, including upregulation of VCAM-1, IgE antibody isotype switching, Th2 cell skewing, and fibroblast activation. IL-4 induces progression of pulmonary granulomatous responses, allergic airway eosinophilia, and proliferation and collagen-gene expression in pulmonary fibroblast populations. The use of either TGFβ or IL-4 as a therapeutic anti-inflammatory agent will likely not be considered.

## Chronic inflammation and acquired immune responses

Perhaps one of the most difficult and important aspects of disease pathogenesis to regulate is when and how to turn off an immune/inflammatory response. Uncontrolled or inefficient immune responses can lead to continual inflammatory cell recruitment and tissue damage that, if persistent and unregulated, can result in organ dysfunction. There are numerous pathogen-related and non-pathogen-related diseases that have been classically regarded as being caused by chronic inflammation, including RA, chronic obstructive pulmonary disease, asthma, mycobacterial diseases, multiple sclerosis (MS), and viral hepatitis, among others. However, additional diseases associated with chronic inflammation have now been recognized to have defects in regulation of inflammation including atherosclerosis, obesity-related diseases, as well as cancer. The persistence of the inflammatory response during chronic inflammatory disorders can shift in the cellular composition of the leukocyte populations that accumulate. The presence of activated T-lymphocytes and B-lymphocytes likely indicates the presence of a persistent antigen that induces cell-mediated and humoral immune responses. It is critical to regulate T-lymphocytes since they are central to the activation and regulation of the acquired immune response, as well as the intensity of PMN and macrophage activation.

### T-lymphocyte regulation of chronic inflammation

The nature, duration, and intensity of episodes of chronic inflammatory events are largely determined by the presence and persistence of antigen that is recognized and cleared by acquired immune responses. Thus, the regulation of T-cells is central to the outcome of the inflammatory/immune responses and is mediated through a combination of cytokine environment and transcription factor regulation (Fig. 2.4). When a pathogenic insult is encountered, the most effective immune response is a cell-mediated Th1-type response, which is induced by IL-12 with STAT4 activation and characterized by IFN production along with T-bet transcription factor expression. The immune response must be modulated and begin to shift to a less harmful response for the tissue, which in T-cells are regulated by IL-4 production and STAT6 activation leading to GATA3 transcription factor expression. One of the aspects of the shifted response that can

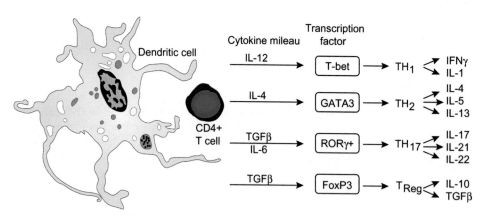

FIGURE 2.4 The cytokine environment during antigen presentation controls T-cell differentiation by regulation of specific transcription factor activation.

be detrimental is the shift toward tissue remodeling designed to promote both restoration of function and host protection.

An area that has held a significant level of interest has been the differentiation of T-regulatory (Treg) cells during the development of chronic responses. This cell subset has been divided into several subpopulations, including natural Treg cells and inducible Treg cells (iTreg). Natural Treg cells develop in the thymus and are essential for control of autoimmune diseases, whereas inducible iTreg cells develop following an antigen-specific activation event and appear to function to modulate an ongoing response. In addition, Treg cells can also be subdivided based on the mechanism of inhibition that they use, such as production of IL-10 and/or TGFβ or use of CTLA-4. Some common ground has been forged in these cell populations based on the expression of Foxp3 transcription factor, although apparently not all Treg populations express this protein. Support for the importance of Foxp3+ Treg cells comes from studies with mice missing this factor and in humans who have mutations in FOXP3, both of which develop multi-organ autoimmune diseases. Thus, this cell population appears to be centrally important for the regulation of immune responses, and defects in this pathway may lead to chronic disease phenotypes.

Our understanding of the role of T-cell subsets in chronic disease has led to an explosion of data that has described subsets of T-cells (Th17) that characteristically produce IL-17. The role of this subset of T-cells during chronic inflammatory diseases is becoming better understood. IL-17–producing cells were first described as an important component of antibacterial immunity. Subsequently, the Th17 subset has been identified as having a central role in the severity of autoimmune responses, in cancer, in transplantation immunology, as well as in infections. Interestingly, the critical aspect of whether T-cells will differentiate into a Th17 cell depends on the expression RORγt transcription factor. Similar to the Treg cells, the differentiation of these cells depends on exposure to TGFβ, but is additionally dependent on IL-6 or other STAT3 signals along with RORγt. The differentiation of Th17 cells also appears to be enhanced by IL-23, an IL-12 family cytokine that is upregulated in APC populations upon TLR signaling. While it is not yet clear, the relationship between Treg and Th17 development and regulation appears to be closely controlled, perhaps dependent on whether IL-6 is present in the inflammatory environment along with TGFβ. As outlined in Fig. 2.4, T-cell activation and cytokine phenotypes depend on the distinct transcription factor expressed for the activation of the particular T-cell subset. Clearly, these subsets and the cytokines that they produce dictate the outcome of a chronic response not only based on the cascade of mediators that they induce but also by the leukocyte subsets that are used as end-stage effectors during the responses.

## B-lymphocyte and antibody responses

The pathogenic role of the humoral immune system has been implicated in a number of chronic disease phenotypes, including allergic responses, autoimmune diseases, arthritis, vasculitis, and any other disease where immune complexes are deposited into tissues. Antibodies produced by B-lymphocytes are a primary goal of the acquired immune response to combat infectious organisms at mucosal surfaces, in the circulation, and within tissues of the host. To be effective, antibodies need to have the ability to bind to specific antigens on the surface of pathogen and through their Fc portion to facilitate phagocytosis by macrophages and PMNs for clearance and complement fixation for targeted killing of the microorganism by the lytic pathway. One of the more pathogenic side effects of antibody effector function is the inappropriate activation of PMNs leading to release of their granular products and destroying host tissue. This is often a problem in autoimmune diseases such as systemic lupus (SLE) and RA, where a wide array of antibodies directed against self-antigens is formed. Autoantibodies directed against tissue antigens can also induce damage due to FcR cross-linking on phagocytic cells including PMN, macrophages, and NK cells and a central mechanism for initiating local inflammation and damage within autoimmune responses.

The induction of allergic responses in developed nations has been steadily increasing for the past three decades. Incidence of food, airborne, and industrial allergies has a significant impact on the development of chronic diseases in the skin, lung, and gut, including atopic dermatitis, IBD, and asthma. The production of immunoglobulin E (IgE) leading to mast cell and basophil activation is the central mechanism that regulates the induction of these diseases. As the antibody isotype produced by the B-cell is governed by the T-lymphocyte response and production of specific cytokines, determining mechanisms that regulate T-cells during allergic diseases has been central to research in these fields. In particular, IL-4 production from T-lymphocytes is key to isotype switching to IgE in B-cells. More recently, pharmaceutical focus on inhibiting IgE-mediated responses has centered on directly clearing IgE from the host using an anti-IgE antibody as a therapeutic, or in blocking the IgE receptors on the cells. This treatment appears to be especially efficacious in patients with severe food allergies who are at risk for anaphylactic responses.

## Exacerbation of chronic diseases

While much of the research in this field has centered on understanding the factors involved and defining targets for therapy of chronic disease, less research has focused on what exacerbates and/or extends the severity of these diseases. It appears that a common initiating factor for the exacerbation is an infectious stimulus, bacterial or viral. In fact, in diseases such as MS or SLE, a number of viruses and bacteria have been implicated as the causative agents for the initiation and/or exacerbation of the responses. This is most often manifested in diseases where an antigenic response is the underlying cause of the chronic disease and the antigen is environmental or host available, such as allergic asthma or autoimmunity. The strong activation of immune cells locally within the affected tissue would provide the reactivation of a well-regulated immune response. The mediators that are upregulated, IL-12 and/or IL-23, in dendritic cells might dictate the type of effector response that is initiated, such as Th1 and/or Th17, respectively.

A common activation pathway for these responses is the use of molecules that can quickly and effectively recognize pathogens, such as the TLR family members. While these molecules are clearly expressed on immune cell populations and facilitate an effective host response, it may be their inappropriate expression on non-immune cells, such as epithelial cells and fibroblasts, which presents the host with the most detrimental response. While it has not been clearly established, a number of studies have indicated that TLR expression on non-immune cells in chronic lesions is upregulated and would presumably predispose these tissues to hyperstimulation during infectious insults. The continual reactivation of tissue inflammation with infectious insults, such as in the lung and gut, could provide the mechanism for tissue damage and potentially remodeling that over time could lead to gradual but continuous organ dysfunction.

## Tissue remodeling during acute and chronic inflammatory disease

Repair of damaged tissues is a fundamental feature of biological systems and, properly regulated, has little harmful effect on normal organ function. Damage to tissues can result from various acute or chronic stimuli, including infections, autoimmune reactions, or mechanical injury. In some cases acute inflammatory reactions, such as ARDS in septic patients, can result in a rapid and devastating disorder that is complicated by significant lung fibrosis and eventual dysfunction. However, more common chronic inflammatory disorders of organ systems (including pulmonary fibrosis, systemic sclerosis, liver cirrhosis, cardiovascular disease, progressive kidney disease) and the joints (such as RA and osteoarthritis) are a major cause of morbidity and mortality and enormous burden on healthcare systems. A common feature of these diseases is the destruction and remodeling of ECM that has a significant effect on tissue structure and function. Chronic inflammation, tissue necrosis, and infection lead to persistent myofibroblast activation and excessive deposition of ECM, including collagen type I, collagen type III, fibronectin, elastin, proteoglycans, and lamin, which promote formation of a permanent fibrotic scar. Most chronic fibrotic disorders have a persistent irritant that stimulates production of proteolytic enzymes, growth factors, fibrogenic cytokines, and chemokines. Together, they orchestrate excessive deposition of connective tissue and a progressive destruction of normal tissue organization and function (Fig. 2.5).

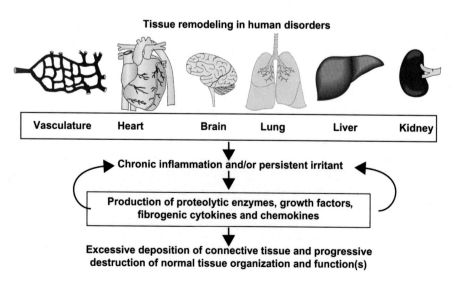

**FIGURE 2.5** The persistent production of cytokines, fibrogenic growth factors, and proteolytic enzymes can result in tissue remodeling and eventual organ dysfunction.

A mechanism that counteracts deposition of ECM and formation of fibrotic foci is activation of matrix metalloproteinases (MMPs), which represent a class of catalytic enzymes that degrade various components of ECM. All MMPs are composed of shared molecules but have different primary structures. Some MMPs are constitutive or homeostatic (MMP-2) and expressed in most cells under normal conditions. Others are inducible (MMP-9) or inflammatory. Activities of MMPs are always dependent on a balance between proteinases and natural inhibitors (tissue inhibitors of metalloproteinases or TIMPs). Overall, the maintenance of ECM deposition may depend on the active removal and proper arrangement of the components allowing proper restructuring of the tissue and basement membrane.

## Pro-fibrogenic cytokines and growth factors involved in fibrotic tissue remodeling

Alterations in the balance of cytokines can lead to pathological changes, abnormal tissue repair, and tissue fibrosis. The most well-studied cytokines involved in these processes include TGFβ, TNFα, platelet-derived growth factor, basic fibroblasts growth factor, monocyte chemoattractant protein 1, macrophage inflammatory protein-1α, and IL-1, IL-13, and IL-8.

## TGFβ

TGFβ is one of the most well-studied pro-fibrotic cytokines. Upregulation of TGFβ1 has been associated with pathological fibrotic processes in many organs (Table 2.1), such as lung fibrotic diseases, cataract formation, systemic sclerosis, renal fibrosis, heart failure, and many others. The TGFβ family represents a group of multifunctional cytokines that includes at least five known isoforms, three of which are expressed by mammalian cells (TGFβ1–3). TGFβ isoforms are known to induce the expression of ECM proteins in mesenchymal cells and to stimulate the production of protease inhibitors that prevent enzymatic breakdown of the ECM. In addition, some ECM proteins (such as fibronectin) are known to be chemoattractants for fibroblasts and are released in increased amounts in response to TGFβ. Thus, while TGFβ is necessary for normal repair processes, its overexpression plays a pivotal role in deposition of ECM and end-stage disease.

**TABLE 2.1** TGFβ contributes to fibrosis of these diseases by excessive matrix accumulation.

| Organ | Disease |
|---|---|
| Eye | Graves ophthalmopathy |
| | Conjunctival cicatrization |
| Lung | Pulmonary fibrosis |
| | Pulmonary sarcoids |
| Heart | Cardiac fibrosis, cardiomyopathy |
| Liver | Cirrhosis |
| | Primary biliary cirrhosis |
| Kidney | Glomerulosclerosis |
| | Interstitial fibrosis |
| Pancreas | Chronic or fibrosing pancreatitis |
| Skin | Hypertrophic scar |
| | Keloids |
| | Scleroderma |
| Subcutaneous Tissue | Dupuytren contracture |
| Endometrium | Endometriosis |
| Peritoneum | Sclerosing peritonitis |
| | Postsurgical adhesion |
| Retroperitoneum | Retroperitoneal fibrosis |
| Bone | Renal osteodystrophy |
| Muscle | Polymyositis, dermatomyositis |
| | Muscular dystrophy |
| | Eosinophilia-myalgia |
| Bone Marrow | Myelofibrosis |
| Neuroendocrine | Carcinoid |

**TABLE 2.2** Targeting of TNFα in chronic inflammatory disorders.

| Approved for use | Pilot studies and mixed results | Clinical failures |
|---|---|---|
| Rheumatoid arthritis | Vasculitis (small and large vessel) | Congestive heart failure |
| Juvenile idiopathic arthritis | Glomerulonephritis | Multiple sclerosis |
| Crohn disease | Joint prosthesis loosening | Chronic obstructive pulmonary disease |
| Psoriatic arthritis | Polymyositis | Sjogren syndrome |
| Ankylosing spondylitis | Systemic sclerosis | Wegener granulomatosis |
| Psoriasis | Amyloidosis | Hepatitis |
| Ulcerative colitis | Sarcoidosis | Systemic lupus erythematosus |
| Behçet disease | Ovarian cancer | |
| | Steroid-resistant asthma | |
| | Refractory uveitis | |

## TNFα

TNFα is a well-known early response cytokine that is key in initiating responses and is rapidly expressed in response to many kinds of stress (mechanical injury, burns, irradiation, viruses, bacteria). TNFα is a pro-inflammatory mediator that is involved in the ECM network, as shown in the healing infarct and collagen synthesis by cardiac fibroblasts. The pathologic events in asthma also correlate with increased TNFα production both in vivo and in vitro in cellular isolates from asthmatic patients. A direct role of TNFα in fibrogenesis was demonstrated using human epithelioid dermal microvascular endothelial cell cultures.

Rheumatoid arthritis (RA) represents a chronic disease that highlights the importance of TNFα. Using a collagen-induced human RA model in DBA/J mice, researchers showed that anti-TNF antibodies ameliorate arthritis and reduce joint damage. This led to the use of TNF blockade in human RA. With the success of anti-TNF treatment in RA, this approach was tested in a number of other chronic disorders, including inflammatory bowel disease, asthma, and graft versus host during bone marrow transplantation. Successful targeting of TNF has been observed in a number of chronic diseases where anti-TNF therapy has been approved for use and is now also being examined in numerous additional inflammatory, infectious, and neoplastic diseases (Table 2.2). However, anti-TNF therapy is not appropriate in all diseases, as there have been a number of failed clinical trials, including those using anti-TNF treatment for MS and congestive heart failure. In addition, a potential side effect of using anti-TNF therapy is the increased susceptibility to infectious organisms.

## Key concepts

- Acute and chronic inflammation are coordinated responses that rely on early cytokine production, adhesion molecule expression, and chemoattractant directed migration of leukocytes.
- Pathogen recognition is facilitated by distinct families of molecules designed to "alert" the immune system and respond to the insult prior to colonization and mediate clearance.
- The nature of the innate immune response early in disease dictates the characteristics and severity of chronic inflammatory response.
- The activation and phenotype of the CD4+ T helper lymphocyte has a primary role in dictating the development and features of chronic inflammation.
- Exacerbation of chronic disease can most often be mediated by infectious organisms that initiate strong innate immune responses that subsequently trigger the acquired immune system.
- Chronic inflammation often leads to tissue remodeling that can cause long-term dysfunction and increased disease severity.

## Suggested readings

[1]  Forlow SB, White EJ, Barlow SC, et al. Severe inflammatory defect and reduced viability in CD18 and E-selectin double-mutant mice. J Clin Invest 2000;106:1457−66.

[2]  Dunne JL, Ballantyne CM, Beaudet AL, Ley K. Control of leukocyte rolling velocity in TNF-alpha-induced inflammation by LFA-1 and Mac-1. Blood 2002;99:336–41.

[3]  Alon R, Dustin ML. Force as a facilitator of integrin conformational changes during leukocyte arrest on blood vessels and antigen-presenting cells. Immunity 2007;26:17–27.

[4]  Woodfin A, Voisin MB, Imhof BA, et al. Endothelial cell activation leads to neutrophil transmigration as supported by the sequential roles of ICAM-2, JAM-A, and PECAM-1. Blood 2009;113:6246–57.

[5]  Islam SA, Thomas SY, Hess C, et al. The leukotriene B4 lipid chemoattractant receptor BLT1 defines antigen-primed T cells in humans. Blood 2006;107:444–53.

[6]  Oyoshi MK, He R, Li Y, et al. Leukotriene B4-driven neutrophil recruitment to the skin is essential for allergic skin inflammation. Immunity 2012;37:747–58.

[7]  Rot A, von Andrian UH. Chemokines in innate and adaptive host defense: basic chemokinese grammar for immune cells. Annu Rev Immunol 2004;22:891–928.

[8]  Abreu MT. Toll-like receptor signalling in the intestinal epithelium: how bacterial recognition shapes intestinal function. Nat Rev Immunol 2010;10:131–44.

[9]  Lee MS, Kim YJ. Signaling pathways downstream of pattern-recognition receptors and their cross talk. Annu Rev Biochem 2007;76:447–80.

[10]  Caruso R, Warner N, Inohara N, Nunez G. NOD1 and NOD2: signaling, host defense, and inflammatory disease. Immunity 2014;41:898–908.

[11]  Bronner DN, Abuaita BH, Chen X, et al. Endoplasmic reticulum stress activates the inflammasome via NLRP3- and caspase-2-driven mitochondrial damage. Immunity 2015;43:451–62.

[12]  Lebeaupin C, Proics E, de Bieville CH, et al. ER stress induces NLRP3 inflammasome activation and hepatocyte death. Cell Death Dis 2015;6: e1879.

[13]  Tian J, Avalos AM, Mao SY, et al. Toll-like receptor 9-dependent activation by DNA-containing immune complexes is mediated by HMGB1 and RAGE. Nat Immunol 2007;8:487–96.

[14]  Walker JA, McKenzie AN. Development and function of group 2 innate lymphoid cells. Curr Opin Immunol 2013;25:148–55.

[15]  Flavell RA, Sanjabi S, Wrzesinski SH, Licona-Limon P. The polarization of immune cells in the tumour environment by TGFbeta. Nat Rev Immunol 2010;10:554–67.

[16]  Josefowicz SZ, Lu LF, Rudensky AY. Regulatory T cells: mechanisms of differentiation and function. Annu Rev Immunol 2012;30:531–64.

[17]  Bacchetta R, Passerini L, Gambineri E, et al. Defective regulatory and effector T cell functions in patients with FOXP3 mutations. J Clin Invest 2006;116:1713–22.

[18]  Patel DD, Kuchroo VK. Th17 cell pathway in human immunity: lessons from genetics and therapeutic interventions. Immunity 2015;43:1040–51.

[19]  Harvima IT, Levi-Schaffer F, Draber P, et al. Molecular targets on mast cells and basophils for novel therapies. J Allergy Clin Immunol 2014;134:530–44.

[20]  Feldmann M. Development of anti-TNF therapy for rheumatoid arthritis. Nat Rev Immunol 2002;2:364–71.

# Chapter 3

# Infection and host response

Susan L. Fink[1] and Sheldon Campbell[2]

[1]*Department of Laboratory Medicine, University of Washington, Seattle, WA, United States,* [2]*Department of Laboratory Medicine, Yale School of Medicine, VA Connecticut Healthcare System, New Haven, CT, United States*

**Summary**

The mammalian immune system deploys complex mechanisms to identify and defeat infections. Innate immunity recognizes molecular patterns associated both with entire classes of pathogens and with cell damage and distress. Adaptive immunity develops target-specific recognition and attack molecules aimed at specific and unique molecular elements of pathogens. A panoply of effector mechanisms directed by the innate and adaptive immune systems delay, damage, expel, contain, or destroy invading microbes. Unfortunately, most pathogenic microbes multiply hundreds of times as rapidly as we; so for every complex and elegant mechanism of immunity, microbes have found ways to subvert or evade. This chapter provides an outline of the components of mammalian immunity and, using five model organisms, explore a few of the astonishing array of mechanisms pathogenic microbes use to maintain themselves and proliferate inside the human body. Additionally, we attempt to describe the regulatory subtlety and discretion of both the immune system and the pathogen; the former to avoid overreaction and self-damage, the latter to utilize the resources of the host for efficient maintenance and transmission. Human infections are the outcome of these intricate, edged molecular conversations between host and pathogen.

## Introduction

The mammalian immune system deploys complex mechanisms to identify and defeat infections. However, most pathogenic microbes multiply hundreds of times as rapidly as people, so for every elegant mechanism of immunity, some microbe has found a way to subvert or evade it. This chapter provides an outline of the components of mammalian immunity and, using the model organisms *Staphylococcus aureus*, *Mycobacterium tuberculosis*, herpes simplex virus (HSV), *Trypanosoma brucei*, and the human immunodeficiency virus (HIV), explore a few of the mechanisms pathogenic microbes use to maintain themselves and proliferate inside the human body, despite all the immune system might do to prevent them. Human infections are the outcome of these intricate molecular conversations between host and pathogen.

## Microbes and hosts—balance of power?

Infectious disease is one of the major driving forces of evolution. Humans have a generation time of roughly 20 years, and even small mammals reproduce in weeks to months. In contrast, microbial generation times range from minutes to days. Thus, microbes evolve hundreds to thousands of times more rapidly than their vertebrate hosts. In this context, it is hardly remarkable that microbes have found numerous ways to exploit multicellular creatures for their own ends.

Large multicellular creatures represent concentrated, extremely rich nutrient sources for microbes. Therefore, the survival of multicellular creatures requires that they have sufficient defenses to prevent easy invasion and consumption. Recent advances in basic immunology illustrate the breadth and depth of the adaptations that have evolved to protect multicellular organisms from microbial invasion. However, the wondrous complexity and power of mammalian host defenses serve only as a backdrop to the even more astonishing complexity of microbial strategies for evading them. In the great scheme of things, humans (and other multicellular organisms) survive only because the microbes find us more useful alive.

Essential Concepts in Molecular Pathology. DOI: https://doi.org/10.1016/B978-0-12-813257-9.00003-6

Many aspects of host physiology have a role in preventing infection, in addition to the elements normally thought of as the immune system. Physical and chemical barriers such as skin, mucosal surfaces, and gastric acidity prevent invasion by microbes. Behavioral adaptations, such as avoidance of decayed food and slapping at insects, are most likely driven by preventing exposure to microbial threats.

It is also important to remember that not all microbes are pathogenic and cause disease. In fact, a multitude of symbiotic microorganisms inhabit discrete anatomical niches of the human body, in totality referred to as the microbiome. Mutually beneficial interactions, that are essential for health, occur between humans and their microbiome. Therefore, the immune system must be carefully controlled to support and manage the presence of beneficial microbes, while preventing infection by pathogens.

## The structure of the immune response

The response to invading microbes consists of three major arms: (1) the innate immune system, which recognizes pathogens and cellular damage and induces rapid protective responses, (2) adaptive immunity, which mounts a slower pathogen-specific response, and (3) effector mechanisms, directed by both innate and adaptive immunity, which inactivate pathogens.

Innate immunity constitutes the first line of host defense and provides early recognition and removal of pathogens and subsequent triggering of inflammation and an adaptive immune response. Preformed molecules in blood, extracellular fluids, and leukocyte granules provide an immediate response to invading microbes and tissue injury. Vascular damage triggers activation of the coagulation and kinin pathways, which mitigate blood loss and induce inflammation. Antimicrobial peptides are produced by specialized epithelial cells of barrier surfaces and also stored within leukocyte granules. Antimicrobial enzymes such as lysozyme digest bacterial cell walls. The complement system of plasma proteins directly lyse pathogens and also mark pathogens for uptake by phagocytic cells.

The innate immune system detects infection through germline-encoded sensors, called pattern recognition receptors (PRRs). Two major categories of molecules are recognized by PRRs: (1) essential conserved microbial components and (2) markers of tissue damage or death. The combination of microbial markers and markers of cellular distress or injury are the first signals that alert the host to the presence of possible infection. PRRs include membrane-bound receptors on the cell surface and endocytic compartments and cytosolic intracellular receptors. Several families of PRRs have been identified, with Toll-like receptors (TLRs) being the most studied. Activation of PRRs by ligand binding triggers intracellular signaling pathways that result in production of cytokines, chemokines, and cell-surface molecules. Depending on the tissue site, types of microbial structures, and category of cellular distress recognized, cells of the adaptive immune system and a broad range of effector mechanisms are recruited.

Microbial antigens are processed and presented to lymphocytes, which then orchestrate an adaptive immune response. The adaptive immune response is embodied in (1) T-cells, which regulate immune responses and invoke powerful effector mechanisms, and (2) B-cells, which produce antibodies. CD4$^+$ helper T-cells recognize peptides presented by major histocompatibility complex (MHC) class II molecules and differentiate into subsets, each with different functions. CD8$^+$ T-cells recognize peptides presented by MHC class I molecules and differentiate into cytotoxic effector T-cells that kill pathogen-infected cells. Antibodies are the secreted form of the B-cell receptor and directly neutralize some organisms. Antibodies also opsonize microbes to direct their ingestion by phagocytes and by initiate complement activity.

The essence of the adaptive immune response is somatic genetic variation, which produces diverse, antigen-specific molecules (antibodies and T-cell receptors). Each lymphocyte produces only a single receptor or antibody, generated by chromosomal gene rearrangements and controlled introduction of mutations. Lymphocytes are elaborately selected to eliminate self-reactive receptors and to favor cells with receptors to pathogens. While this process is complex, time-consuming, and wasteful, the specificity of the adaptive immune response makes it a central component of the mammalian defense system. Furthermore, specific antigens clonally expand and prime antigen-specific lymphocytes, which persist after the infection is cleared. These cells are then ready to more rapidly and effectively respond to subsequent pathogen encounters, and thereby provide immunological memory.

Immune effector mechanisms include phagocytic cells (such as neutrophils and activated macrophages) and soluble factors (such as complement). These effectors are often directed and controlled by the antigen-specific immune response, but have the potential to cause significant damage to host tissues. A huge range of effector mechanisms limit and eliminate infection by direct antimicrobial activity, or by creating physical or chemical barriers to microbial proliferation and spread. A partial list of effectors is provided in Table 3.1.

**TABLE 3.1** Host effector mechanisms.

| Name | Properties | Effector mechanisms |
|---|---|---|
| **Soluble effectors** | | |
| Complement system | Proteolytic cascade, activated by antibody, directly by microbial components, or via PRRs. | Direct destruction of pathogens via pore-formation. Recruit inflammatory cells. Enhance phagocytosis and killing. |
| Coagulation system | Proteolytic cascade, activated by tissue and vascular damage. | Prevents blood loss. Bars access to bloodstream. Proinflammatory. |
| Kinin system | Proteolytic cascade triggered by tissue damage. | Proinflammatory. Causes pain response. Increases vascular permeability to allow increased access by plasma proteins. |
| Antibodies | Antigen-specific proteins produced by B-cells. Recognize a broad range of antigens. | Directly neutralize pathogens. Activate complement. Opsonize pathogens to enhance phagocytosis and killing. |
| **Cellular effectors** | | |
| Monocyte/macrophage | Have PRRs to recognize pathogens; activated by specific T-cells and chemokines. | Phagocytosis and microbial killing via multiple mechanisms. Antigen presentation. |
| Dendritic cell | Ingest large amounts of extracellular fluid; migrate to lymph node to present antigen to naïve T-lymphocytes. | Antigen uptake, transport, and presentation to T-lymphocytes. Initiate adaptive immune response. |
| Neutrophil | Have PRRs to recognize pathogens, activated antibody, and complement. | Phagocytosis and microbial killing via multiple mechanisms. |
| Eosinophil | Recognize antibody-coated parasites. | Killing of multicellular pathogens. |
| Basophil/mast cell | Associated with IgE-mediated responses. | Release of granules containing histamine and other mediators of anaphylaxis. |
| NK-cell | Lymphocyte lacking antigen-specific reactivity; recognize PAMPs of intracellular pathogens, activated by chemokines and by membrane proteins of infected cells. | Induce death of infected cells via membrane pores and induced apoptosis. |
| B-lymphocyte | Recognize antigens presented by APCs; regulated by T-cells and chemokines. | Produce antibody. |
| T-lymphocyte | Recognize antigens presented by APCs; regulate major portions of both adaptive and innate immunity. | Directly kill infected cells via membrane pores and induced apoptosis. Activate macrophages. Many other functions. |

The distinction between innate immunity, adaptive immunity, and effector mechanisms is rather arbitrary, and many responses involve all three. For example, the proteins of the coagulation cascade recognize tissue damage and microbial components, recruit inflammatory cells, and by blocking blood flow to infected tissues, limit spread of infections. T-cells carry PRRs, which participate in activation of pathogen-reactive cells, express T-cell receptors (which recognize specific pathogen antigens), and participate directly in cell-dependent cytotoxicity to eliminate infected cells.

## Regulation of immunity

Because inflammatory responses are metabolically costly and capable of causing enormous damage to tissues, the immune system is tightly regulated. Soluble effector systems, such as the complement and coagulation cascades, have soluble inhibitors that usually confine these responses to the area where the initiating stimulus occurs. Cellular effectors are activated and inhibited via both chemokines and direct signaling via adhesion molecules and receptors. Adaptive immune responses undergo elaborate screening to eliminate self-reactive cells. After clearance or control of infection, resolution of the immune response avoids tissue damage and chronic inflammation.

## Pathogen strategies

To evade the flexible and powerful system of host defenses, successful pathogens have evolved complex strategies. As examples, we have selected five pathogens: (1) *S. aureus*, which evades, and in some cases overloads, innate and

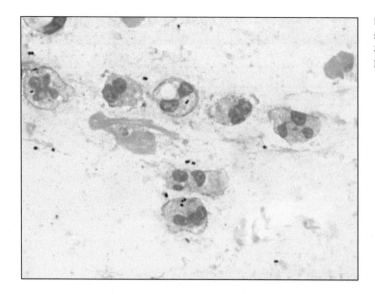

FIGURE 3.1 *Staphylococcus aureus* and neutrophils. A gram-stained smear from a patient with *S. aureus* pneumonia. Staphylococci are located both extracellularly and within neutrophils (Gram stain, 1000×).

adaptive immune responses; (2) *M. tuberculosis*, an intracellular bacterium, which proliferates inside host macrophages; (3) HSV, a complex DNA virus, which successfully disrupts intracellular mechanisms of viral control; (4) the African trypanosomes (*T. brucei* species), bloodstream-dwelling protists that evade antibody via a remarkable strategy for generating antigenic diversity; and (5) HIV, a small RNA virus, which turns the immune system on itself and generates enormous molecular diversity during infection. These five organisms serve for discussion of the major aspects of immune function, in the context of meeting infectious challenges.

## *Staphylococcus aureus*: the extracellular battleground

*S. aureus* is a gram-positive extracellular bacterium that is part of our commensal microbiota (Fig. 3.1). It is a versatile pathogen causing infections that range from superficial infections of skin and soft tissue to life-threatening systemic disease. The first lines of defense against *S. aureus* are the recognition molecules and effector cells of the innate immune system; but *S. aureus* engages mechanisms to subvert the innate immune response of the host.

### The innate immune system: recognition of pathogens

After *S. aureus* breaches intact skin or mucosal lining, which constitutes the border between the outside world and the human body, it encounters the proteins and cellular elements of host innate immune defense. One such multifunctional defense is the family of plasma proteins called complement which was initially recognized for its ability to enhance the antibacterial powers of antibodies. Hence, this activity that *complemented* the activity of antibodies became known as complement.

Three major activation mechanisms invoke complement. The complement system can be activated by antibodies, which is termed the *classical pathway*. The *lectin pathway* is initiated by host receptors that bind carbohydrate ligands on microbial intruders. The third, or *alternative pathway*, is stimulated by spontaneous hydrolysis of complement component C3, which is negatively regulated by host factors and stimulated by foreign surfaces. All three pathways converge at the formation of the C3 convertases, which cleave C3 and generate its active fragments C3a and C3b. C3b covalently binds microbial surfaces, tagging them for more efficient phagocytosis since phagocytic cells have complement receptors. Accumulation of C3b on the bacterial cell surface also allows assembly of C5 convertases, which initiate formation of the lytic membrane-attack complex consisting of C5b–C9. Gram-negative bacteria, but not gram-positive bacteria such as *S. aureus*, are successfully lysed by the pore-forming membrane attack complex.

Early in *S. aureus* infection, the bacterium encounters resident macrophages. Macrophages express several receptors that identify bacterial constituents, such as TLRs (Table 3.2), mannose receptors, complement receptor C3R, glucan receptors, and scavenger receptors. After bacteria or bacterial constituents bind to their receptors, the macrophage engulfs them. The phagosome fuses with the lysosome to form the phagolysosome, degradative enzymes and

**TABLE 3.2** Recognition of microbial products through toll-like receptors.

| Receptor | Ligands | Microorganisms recognized | Notes |
|---|---|---|---|
| TLR-2 (TLR-1, TLR-6) heterodimers | Peptidoglycan, bacterial lipoprotein and lipopeptide, porins, yeast mannan, lipoarabinomannan, glycophosphatidyl-inositol anchors | Gram-positive bacteria, mycobacteria, *Neisseria*, yeast, trypanosomes | Carried on macrophages |
| TLR-3 homodimer | Double-stranded RNA | Viral RNAs | |
| TLR-4 homodimer | Lipopolysaccharide | Gram-negative bacteria | Carried on macrophages |
| TLR-5 homodimer | Flagellin | Gram-negative bacteria | Carried on intestinal epithelium; interacts directly with ligand |
| TLR-9 homodimer | DNA with unmethylated CpG motifs | Bacteria | Intracellular receptor |

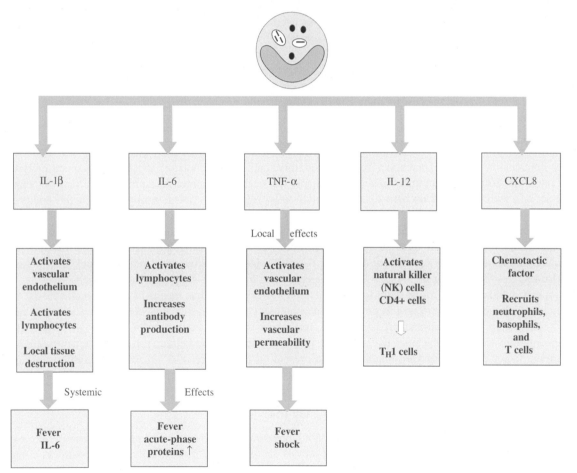

**FIGURE 3.2** Chemokines secreted by macrophages in response to bacterial challenge. Chemokines secreted by macrophages have both local and systemic effects, which mobilize defenses to infection, but may have unfortunate consequences as well.

antimicrobial substances are released, and the contents of the phagolysosome are digested, then peptide fragments are presented to the adaptive immune system via MHC class II.

TLR activation kicks off an intracellular signal transduction pathway that ultimately leads to expression of inflammatory cytokines. Important cytokines that are secreted by macrophages in response to bacterial products include IL-1, IL-6, tumor necrosis factor α (TNFα), the chemokine CXCL8, and IL-12. These molecules have powerful effects and start off the local inflammatory response (Fig. 3.2). A critical task of macrophages is to recruit neutrophils to the site of infection in an attempt to keep the infection localized.

As sheer killers of bacteria, macrophages pale in comparison with polymorphonuclear leukocytes, (PMNs or neutrophils), which circulate in the blood, awaiting a call to enter infected tissue. At sites of inflammation, endothelial cells expose molecules called selectins, which capture circulating neutrophils and stimulate them to migrate into the underlying tissue. Once in the tissues, neutrophil receptors recognize inflammatory signals including complement fragments C3a and C5a, chemokines produced by activated macrophages, and molecules from the bacteria themselves. These signals activate intracellular signaling cascades, resulting in neutrophil activation and migration toward the site of infection.

Once in the infectious battleground, neutrophils unleash microbial killing programs of degranulation, phagocytosis, and neutrophil extracellular trap (NET) formation. Neutrophils rapidly phagocytose pathogens marked by complement and/or antibody opsonization for intracellular killing. An extensive array of anti-microbial compounds are stored in ready-to-use neutrophil granules, whose deployment is carefully regulated to not damage the host. These antimicrobial molecules include cationic peptides in the defensin and cathelicidin families. Neutrophil granules fuse with phagosomes to contribute to intracellular killing, or fuse with the plasma membrane, releasing their contents into the tissue. Activated neutrophils undergo an oxidative burst, which produces toxic reactive oxygen species. In addition, neutrophils can expel their nuclear contents with cytosolic and granular proteins, producing NETs, which trap and prevent pathogen dissemination.

Most infections are cleared by these ubiquitous and induced responses of innate immunity, often so swiftly and efficiently that the host scarcely notices. Once an infectious agent escapes innate mechanisms and spreads from the point of entry, it faces an adaptive immune response characterized by an extensive process in the draining lymph node in which pathogen-specific lymphocyte clones are selected, expanded, and differentiated.

*S. aureus* is such a successful pathogen because it expresses a multitude of virulence genes (Table 3.3) that act together to evade and subvert three critical elements of the innate immune response: (1) recruitment and actions of inflammatory cells, (2) complement activation, and (3) antimicrobial mechanisms.

## Inhibition of inflammatory cell recruitment and phagocytosis

*S. aureus* contains an arsenal of antiadhesive, anti-migratory, and anti-chemotactic proteins that specifically interfere with host inflammatory cell recruitment (Table 3.3). Staphylococcal superantigen-like protein-5 (SSL-5) blocks neutrophil interactions with P-selectin on activated endothelial cells. Neutrophil transmigration through the endothelium is mediated in part by the endothelial cell adhesion molecule ICAM-1. *S. aureus* answers with production of Extracellular Adherence Protein that binds to ICAM-1, thereby interfering with extravasation of neutrophils at the site of infection. Once in the tissue, neutrophils are directed by a gradient of chemokines and other signals toward the invading pathogen. SSL-5 and chemotaxis inhibitory protein of *S. aureus* (CHIPS) bind chemokine receptors and block their activation. *S. aureus* also secretes the enzyme staphopain A, which cleaves and inactivates chemokine receptors.

Once a neutrophil manages to get close to *S. aureus*, the bacterium still has means to evade phagocytosis. *S. aureus* expresses surface-associated anti-opsonic proteins and a polysaccharide capsule that compromise efficient neutrophil phagocytosis. Protein A is a wall-anchored *S. aureus* protein that binds the Fc portion of immunoglobulin G (IgG) and coats the bacterium with IgG molecules that are in the incorrect orientation to be recognized by the neutrophil Fc receptor. Clumping factor A is a fibrinogen-binding protein present on the surface of *S. aureus* that binds to fibrinogen and coats the surface of the bacterial cells with fibrinogen molecules, additionally complicating the recognition process.

## Inhibition of complement activation: you can't tag me!

*S. aureus* has also evolved an array of complement-modulating strategies. The secreted protease, aureolysin cleaves and inactivates the central complement component C3. *S. aureus* also secretes a protein called *Staphylococcus* complement inhibitor that stabilizes bound C3 convertases and renders them less active. Additionally, extracellular fibrinogen-binding protein and extracellular complement-binding protein bind C3 and block its deposition on the bacterial cell surface (Table 3.3). *S. aureus* not only prevents complement deposition, but also eliminates bound C3b and IgG. Host plasminogen that is attached to the bacterial cell surface is activated by the *S. aureus* enzyme staphylokinase to plasmin, which then cleaves surface-bound C3b and IgG (Table 3.3). The effect of these molecules is reduced opsonization, and hence reduced phagocytosis.

## Inactivation of antimicrobial mechanisms

*S. aureus* not only prevents inflammatory cell recruitment and phagocytosis, but has mechanisms to protect itself from the deadly efforts of these cells. Secreted staphylococcal nuclease enzymatically destroys neutrophil NETs. If *S. aureus*

**TABLE 3.3** Examples of virulence factors responsible for immune evasion by *Staphylococcus aureus*.

| Name of factor | Abbreviation | Function | Interference with host response |
|---|---|---|---|
| Anti-inflammatory peptides | | | |
| Chemotaxis inhibitory protein of *S. aureus* | CHIPS | Binds to C5aR and formylated protein receptor | Blocks chemotaxis |
| Staphylococcal complement inhibitor | SCIN | Stabilizes C2a-C4b and Bb-C3b convertases | Inhibits complement |
| Toxins | | | |
| Staphylococcal superantigen-like protein-5 | SSL-5 | Binds to P-selectin glycoprotein ligand-1 | Inhibits neutrophil recruitment |
| Staphylococcal superantigen-like protein-7 | SSL-7 | Binds to complement C5; binds to IgA | Inhibits complement |
| β-hemolysin | Hlb | Lysis of cytokine-containing cells | Cytotoxicity |
| γ-hemolysin | Hlg | Lysis of erythrocytes and leukocytes | Cytotoxicity |
| Panton-Valentine leukocidin | PVL | Stimulates and lyses neutrophils and macrophages | Change in gene expression of staphylococcal proteins; important in necrotizing pneumonia |
| Leukocidins D, E, M | LukD, LukE, LukM | Lysis of erythrocytes and leukocytes | Cytotoxicity |
| Exotoxins with superantigen activity, enterotoxins Toxic shock syndrome toxin-1 | Se TSST-1 | Food poisoning when ingested; septic shock when systemic | Bridge MHC-II–TCR without antigen presentation; confer nonspecific T-cell activation and/or T-cell energy; downregulate chemokine receptors |
| Secreted expanded repertoire adhesive molecules | | | |
| Coagulase | Coa | Activates prothrombin and binds fibrin | Antiphagocytic |
| Extracellular adherence protein | Eap | Binds to endothelial cell membrane molecules, binds to ICAM-1 and T-cell receptors | Blocks neutrophil and T-cell recruitment; inhibits T-cell proliferation |
| Extracellular fibrinogen-binding protein | Efb | Binds to fibrinogen; binds to complement factors C3 and inhibits its deposition on the bacterial cell surface | Inhibits complement activation beyond C3b, thereby blocking opsonophagocytosis; binds to platelets and blocks fibrinogen-induced platelet aggregation |
| Microbial surface components recognizing adhesive matrix molecules | | | |
| Clumping factor A | ClfA | Binds to fibrinogen | Antiphagocytic |
| *S. aureus* protein A | Spa | Binds to Fc portion of IgG and TNF receptor 1 | Antiopsonic, antiphagocytic; modulates TNF signaling |
| Extracellular Enzymes | | | |
| Catalase | CatA | Inactivates free hydrogen peroxide | Required for survival, persistence, and nasal colonization |
| Staphylokinase | Sak | Plasminogen activator | Antidefensin; cleaves IgG and complement factors |
| Capsular polysaccharides | | | |
| Capsular polysaccharide type 1, 5, and 8 | CPS 1, CPS 5, CPS 8 | Masks complement C3 deposition | Antiphagocytic effect |

Modified after Chavakis T, Preissner KT, Herrmann M. The anti-inflammatory activities of Staphylococcus aureus. Trends Immunol 2007;28:408–18.

is engulfed by a neutrophil, it utilizes protective mechanisms to help it survive in the phagosome. *S. aureus* prevents fusion of phagosomes with toxic granules. Two superoxide dismutase enzymes help *S. aureus* to avoid the lethal effects of reactive oxygen species that are formed during the respiratory burst. Modifications to the cell wall teichoic acid and other cell wall components change the cell-surface charge such that the affinity of cationic, antimicrobial defensin peptides is reduced. Staphylokinase binds defensin peptides, and the extracellular metalloprotease aureolysin cleaves and inactivates certain defensin peptides.

*S. aureus* turns from defense to offense and protects itself by directly killing its attackers. One of the cardinal features of *S. aureus* is its ability to secrete several classes of cytolytic toxins (hemolysins, leukocidins, and cytolytic peptides; Table 3.3) that damage host cell membranes. They contribute to the development of abscesses with pus formation by direct neutrophil killing.

### Staphylococcal toxins and superantigens: turning the inflammatory response on the host

One of the most serious, life-threatening infections with *S. aureus* is toxic shock syndrome. It is caused by secreted exoenzymes and exotoxins (Table 3.3). Enterotoxins cause a fairly benign gastroenteritis (food poisoning) when ingested, but act as a superantigen in systemic infections. Superantigens bind MHC class II on antigen-presenting cells and link it to T-cell receptors on CD4 T helper cells. Binding occurs without the need of an antigenic peptide and allows superantigens to polyclonally activate T-cells, leading to T-cell proliferation and a massive release of cytokines. This systemic cytokine storm causes a systemic inflammatory response with reduced blood pressure and intravascular coagulation causing multiorgan failure and death. It is ironic that in toxic shock most of the deleterious effects on the host result from the exaggerated host immune response. After all, it is rarely in the interest of *S. aureus* to kill its host.

## *Mycobacterium tuberculosis* and the macrophage

*M. tuberculosis* is one of the most important causes of worldwide morbidity and premature death. Mycobacteria are also known as acid-fast bacilli; the staining property of acid-fastness results from the thick, lipid-rich cell wall of the organisms. These cell wall lipids serve a structural function and influence interactions with host cells.

Spread by the aerosol route, organisms are deposited in the lungs, where they encounter, and enter, their first immunological barrier, the alveolar macrophages. *M. tuberculosis* circumvents the macrophage killing machinery and utilizes these host cells as a niche for replication and persistence. Many pathogens utilize the intracellular compartment to evade host responses. While the intracellular lifestyle avoids some host defenses, such as complement and neutralizing antibody, other mechanisms of immunity operate inside host cells. Successful intracellular pathogens must therefore possess strategies to manipulate their host cell for their own benefit. These include directing entry into a permissive intracellular compartment, blocking killing mechanisms and limiting exposure to the cell-mediated immune system.

### Mycobacterium and macrophage: the pathogen chooses its destiny

While tubercle bacilli are capable of extracellular growth, in the early stages of infection this does not occur for long, since alveolar macrophages rapidly phagocytose them (Fig. 3.3). In phagocytic cells, different receptors drive different processing of the phagosome. Thus, pathogens can control their intracellular fate by choosing the receptor that recognizes them. In the case of mycobacteria, multiple receptors may be involved, and the details of their interaction in vivo are not yet known. However, one of the major receptors is complement receptor type 3 (CR3), a pathway that limits macrophage activation.

Once internalized, several mycobacterial molecules actively interfere with phagosome-lysosome fusion. For example, the mycobacterial cell wall glycolipid, lipoarabinomannan (LAM), blocks cytosolic calcium increases and contributes to inhibition of phagosome-lysosome fusion. Phosphatidylinositol-3-phosphate (PI3P), a molecule synthesized by host phosphatidylinositol-3-kinase (PI3K), serves to direct proteins involved in the fusion process to the endosomal membrane. *M. tuberculosis* inhibits PI3K and secretes a lipid phosphatase (SapM) that hydrolyzes PI3P. Additionally, phagosomes containing living mycobacteria recruit the host protein coronin 1. A coronin-dependent calcium influx activates calcineurin, which blocks lysosome fusion.

During the course of normal phagosome maturation, the macrophage vacuolar (V)-ATPase subunits are trafficked to the phagosome. The V-ATPase pump transports protons across the phagosomal membrane to cause phagosome acidification, which activates degradative enzymes that require a low pH. The *M. tuberculosis* protein PtpA binds a component of the macrophage V-ATPase and blocks pump assembly and phagosome acidification.

Despite all these mechanisms, *M. tuberculosis* has very little ability to block phagosome-lysosome fusion in *activated* macrophages, stimulated via cytokines and TLRs. Probably the most important cytokines in activating macrophages to kill mycobacteria are interferon-γ and TNFα. Both in animal models and in humans, suppression of these pathways by drugs or mutations results in vulnerability to tuberculosis.

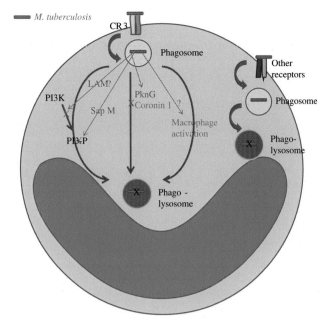

**FIGURE 3.3** *Mycobacterium tuberculosis* and the macrophage. *Gray arrows* are host endosome processing pathways; the *pink lines* are mycobacterial mechanisms. Fusion of mycobacteria into mature phagolysosomes usually leads to death of the organism, so mycobacteria select their endocytotic pathway and interfere with mechanisms designed to result in phagosome-lysosome fusion.

## The adaptive response to *M. tuberculosis*: containment and the granuloma

Antigens from intracellular pathogens located in the endosomal compartment are primarily presented on class II MHC molecules and recognized by CD4 T-cells. This is not an exclusive arrangement, since some antigen from the endosomal compartment is transported to the cytosol and presented to CD8 T-cells via MHC class I, in a process termed cross-presentation. The ability of *M. tuberculosis* to prevent phagosome-lysosome fusion likely limits the ability of macrophages to proteolytically digest *M. tuberculosis* proteins and subsequently load processed peptides in the context of MHC class II molecules.

However, *M. tuberculosis* evasion of T-cell activation is not complete and antigen-specific T-cells are generated during infection. CD4 T-cells differentiate into specialized types of helper T-cells. The best-known division is into $Th_1$ and $Th_2$ type cells. $Th_2$-type CD4 T-cells express cytokines that activate and induce class-switching in B-cells, resulting in an immune response centered on antibody production. In contrast, the major activity of $Th_1$-type cells is to activate macrophages via interferon-$\gamma$, IL-2, TNF$\alpha$, and other cytokines. Differentiation of CD4 T-cells into the $Th_1$ and $Th_2$ lineages is controlled by the cytokines they encounter during the early stages of activation. Interferon-$\gamma$, secreted by macrophages and dendritic cells, induces differentiation toward the $Th_1$ phenotype. Since interferon-$\gamma$ is a major cytokine produced by $Th_1$-type cells, the $Th_1$-type immune response is self-enforcing and tends to be stable unless other influences perturb the balance.

Activated macrophages are capable of killing ingested mycobacteria. The lesion that results from the effective immune response to *M. tuberculosis* infection is the granuloma, shown in Fig. 3.4. Structurally, a tuberculous granuloma consists of a central area of necrosis, surrounded by macrophages (both activated and nonactivated), then a mixture of lymphocytes, macrophages, tissue cells, and fibroblasts. The lymphocytes are mainly $Th_1$-type CD4 cells, though CD8 cells are also present.

The granuloma is effective in containing, but typically not in eradicating, the mycobacteria. The tubercle bacillus may exploit the granuloma to maintain chronic infection—bacteria in granulomas escape apoptotic macrophages to be phagocytosed by newly recruited ones. The bacillus mediates this apoptosis—recruitment—reinvasion cycle in part by secreting an effector molecule known as ESAT6. ESAT6 both induces macrophage apoptosis (releasing mycobacteria before they can be killed) and recruits new victim-macrophages.

Persisting bacteria may remain viable for the life of the host, either within the necrotic center of the granuloma or in dynamic equilibrium within the inflammatory region of the lesion, or both. Under conditions of waning or suppressed immunity, and driven by mycobacterial manipulation of TNF-related pathways, the persisting mycobacteria can proliferate, spread, cause disease, and also escape, typically via the airborne coughed-out route, to a new host.

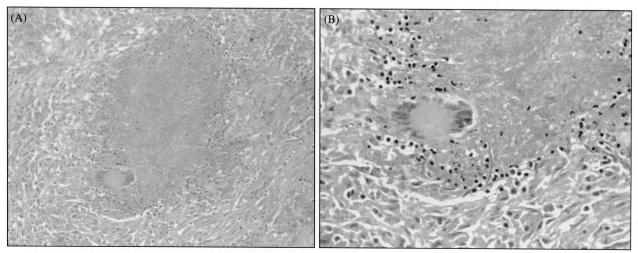

**FIGURE 3.4**   Mycobacterial granuloma. Hematoxylin and eosin−stained sections of a mycobacterial granuloma. Central necrosis and an inflammatory response consisting of macrophages, lymphocytes, and fibroblasts are apparent. A large multinucleated giant cell, characteristic of the granulomatous reaction, is also present. (A) low magnification, (B) high magnification.

*M. tuberculosis* exploits the intracellular compartment to maintain itself in the host. Antibody responses to tuberculosis infection are weak, not protective, and possibly harmful. Instead, cell-mediated immune responses leading to macrophage activation, directed primarily by CD4 T-cells, lead to an at least partially protective response. A rather delicate balance of factors works in the interest of the pathogen. It is able to maintain itself for prolonged periods of time in a single host, awaiting an opportunity for transmission.

# Herpes simplex virus: taking over

HSV types 1 and 2 are DNA viruses that cause disease from painful oral and genital lesions to life-threatening brain and systemic infections. HSV is responsible for both acute and reactivation disease.

## Defense against viruses: subversion and sacrifice

Viruses contain nucleic-acid genomes packaged with proteins. Enveloped viruses are surrounded by a lipid membrane from the host cell, whereas nonenveloped viruses lack a membrane. Viruses begin the process of infection by attachment to cell-surface receptors, then use either the endocytic pathway or nonendocytic routes to enter cells. They then utilize the protein-synthetic and other machinery to assemble new viral particles. Mechanisms by which viruses damage cells are illustrated in Fig. 3.5.

Because viruses hijack the host machinery for much of their metabolism and replication, the immune system must identify molecular features unique to viruses and absent from host cells. The best-characterized viral PAMPs are the nucleic acids of viral genomes or viral replication intermediates. Endosomal TLRs sense incoming viral nucleic acids; TLR-3 recognizes double-stranded RNA, TLR-7 recognizes single-stranded RNA, and TLR-9 recognizes unmethylated CpG-containing DNA. Human susceptibility to HSV encephalitis is associated with genetic deficiencies in TLR3 and its signaling pathway. The cytoplasmic pathogen recognition proteins RIG-1 and MDA-5 sense viral double-stranded RNA, while cytosolic viral DNA is detected by the receptors IFI16 and cGAS. Viral recognition by any of these receptors activate the interferon-regulatory factors IRF3 and IRF7 and regulatory factor NF-κB, which induce a range of cytokines, most importantly antiviral interferons (IFN) α and β.

IFNα/β induce primary defense against viral infections, binding to surface receptors and initiating a signaling cascade through the Janus-family kinases. This leads to the transcription of hundreds of IFN-stimulated genes, which together yield an antiviral state effective against many different types of viruses. Some IFN-stimulated genes include (1) $2' - 5'$ oligoadenylate synthetase (OAS), an enzyme that activates ribonuclease L to digest viral RNAs and (2) dsRNA-dependent protein kinase R (PKR), which modifies eukaryotic initiation factor 2α, leading to arrest of translation. Both OAS and PKR are activated by detection of foreign double stranded RNA. Other interferon actions include

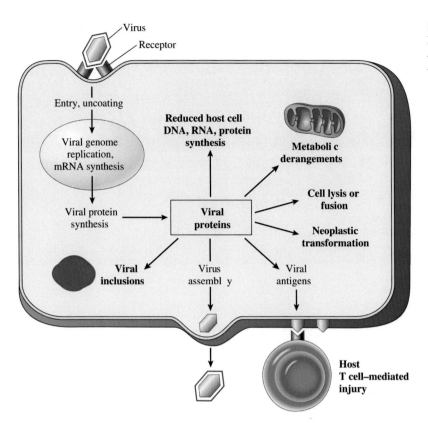

**FIGURE 3.5** How viruses damage cells. Mechanisms by which viruses cause injury to cells. *Reprinted by permission from Elsevier Saunders. Robbins and Cotran pathologic basis of disease. 7th ed. p. 357, Copyright 2004.*

upregulation of IFNα/β synthesis (a positive feedback loop) and activation of macrophages, dendritic cells, and natural killer cells.

IFN also plays a role in the adaptive immune response. Cytoplasmic antigens, including viral proteins, are presented by MHC class I to activate virus-specific CD8$^+$ cytotoxic T-cells. This process begins with protein degradation by the proteasome, which produces peptides that are loaded onto MHC class I molecules in the ER lumen, and transported through the secretory pathway to the plasma membrane. IFN alters the proteasome to favor production of peptides for MHC class I presentation and upregulates the proteins involved in transport of the complex to the cell surface.

The final defense of virally infected cells is apoptosis, programmed cell death either through induction by cytotoxic T-cells (with appropriate co-stimulation by dendritic cells and CD4 T-cells) or other mechanisms triggered by viral infection. Two major signaling pathways trigger apoptosis: (1) via extrinsic death receptors, such as the TNFα receptor and (2) via intracellular signals. In each case, a series of proteases called caspases are activated and destroy the critical infrastructure of the cell.

## Herpes simplex virus on the high wire: a delicate balancing act

HSV has a large genome for a virus, consisting of ~150 kb, with at least 74 genes. An image of an HSV infection is shown in Fig. 3.6. While fewer than half the genes are required for replication in cell cultures, viruses isolated from human hosts almost always have the full complement; genes not required for growth in vitro are mainly involved in evading, inhibiting, or subverting host responses.

HSV encodes an RNAse known as *v*iral *h*ost *s*hutoff (vhs) that degrades mRNA and globally inhibits response gene expression. By destabilizing mRNA, vhs dampens the IFN response, blocks activation of dendritic cells, and reduces production of proinflammatory cytokines and chemokines. A number of additional HSV proteins inhibit components of the IFNα/β response system (Table 3.4). The viral RNA-binding protein US11 inhibits PKR and $2' - 5'$ oligoadenylate synthetase, preventing it from shutting off translation. US11 is actually trafficked intercellularly and acts on adjacent uninfected but interferon-stimulated cells, priming them for infection. Meanwhile, ICP 34.5 activates host protein phosphorylase 1α, which reactivates initiation factor 2α after inactivation by PKR. ICP 34.5 also inhibits one of the key signaling molecules activated by PRR recognition to prevent both IRF3 and NF-κB signaling. HSV protein ICP0 acts in

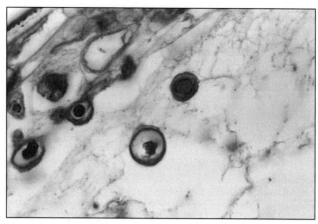

**FIGURE 3.6** Histopathology of a herpes simplex lesion. Herpes virus blister in mucosa. High-power view shows host cells with glassy intranuclear herpes simplex inclusion bodies. *Reprinted by permission from Elsevier Saunders. Robbins and Cotran pathologic basis of disease. 7th ed. p. 366, Copyright 2004.*

**TABLE 3.4 Interferon actions and HSV reactions.**

| Mechanism | Effect | HSV response |
|---|---|---|
| Activities that inhibit viral gene expression | | |
| Activation of ribonuclease L ds-RNA–dependent phosphorylation of ribosomal initiation factor | Digest viral RNAs Arrest protein synthesis | ICP0 inhibits ribonuclease L Block the kinase responsible for phosphorylation; increase activity of phosphorylase, which restores activity |
| Activities that enhance inflammatory responses | | |
| Alteration of proteasome to favor production of peptides for class I MHC Upregulation of MHC class I and associate mechanisms Activation of antigen-presenting and effector cells | Increase presentation of antigen to adaptive immune system Increase presentation of antigen to adaptive immune system Accelerate antibody and cell-mediated immune responses; induce apoptosis in infected cells | Unknown Block TAP transport of antigen, which in turn limits externalization of MHC class I Infection of these cell types leads to downregulation of response elements, especially in dendritic cells |
| Upregulation of interferon synthesis | Positive-feedback loop to limit infectability of nearby cells | vhs globally inhibits host gene expression; ICP0 blocks multiple transduction mechanisms of IFN signaling |

the nucleus to inhibit IRF3 and IRF7, as well as several other interferon-response nuclear proteins. In the cytoplasm, ICP0 appears to block the activity of ribonuclease L.

HSV also acts to reduce presentation of antigen to the adaptive immune system. ICP47 blocks entry of peptides to the ER. The activities that inhibit interferon actions also inhibit MHC class I expression. In addition, HSV exerts broad inhibitory activities when it invades dendritic cells, inducing downregulation of costimulatory surface proteins and adhesion molecules. This inhibition, mediated in part by the US3 protein kinase, most likely slows the response to HSV infection.

HSV proteins exert control even over apoptosis. Infection with HSV initially makes cells resistant to apoptosis by either the extrinsic or intrinsic pathways. However, later in infection in some cell types, apoptosis is induced by HSV. The pathogen appears to create a delicate balance between inhibition of apoptosis early in infection, prior to production of virions, and induction of apoptosis late in the infective cycle.

HSV is a prudent pathogen. It fails to completely inhibit the immune response, and local control of HSV infection is usually achieved with minimal lasting damage. By then the virus entered a latent infection in neurons, with only a small number of genes being transcribed. Neurons express low levels of MHC class I, which is further downregulated by HSV. Periodically, viral replication is turned on and viral particles are transported down the axon to its terminal near a

**TABLE 3.5** Antibody Classes and functions.

| Class | Location[a] | Structure | Function |
|---|---|---|---|
| IgD | Surface of B-cells only | 2 κ or λ light chains, 2 δ heavy chains | Unknown; expressed early in B cell differentiation along with IgM. |
| IgM | Plasma | 2 κ or λ light chains, 2 μ heavy chains, arranged in pentamers with 1 J chain | Activates complement; first functional immunoglobulin formed in immune response. |
| IgG | Widely distributed in extracellular fluid | 2 κ or λ light chains, 2 γ heavy chains | Complement activation, transfer to neonate via placenta, opsonization, neutralization of viruses and other pathogens. |
| IgA | Mucosal tissues, surfaces, and secretions | 2 κ or λ light chains, 2 α heavy chains arranged in dimers with 1 J chain | Important in mucosal immunity, has opsonizing activity. |
| IgE | Bound to mast cells and basophils | 2 κ or λ light chains, 2 ε heavy chains | Binds to and activates mast cells and basophils; important in defense versus multicellular parasites. |

[a]*All immunoglobulin classes are found on B-cells as antigen receptors.*

mucosal surface, where the virus can invade epithelial cells and initiate a lesion, with more opportunities for transmission to a new host.

## The African trypanosome and antibody diversity: dueling genomes

Blood is a tissue with many functions, many of them mediated by the cellular components of the blood—red blood cells, leukocytes, and platelets. However, the plasma contains a host of molecules involved in defense. These include the coagulation proteins, a variety of regulatory cytokines, as well as dedicated antimicrobial molecules such as antibodies and complement.

Antibodies are a major arm of the adaptive immune response. Antibodies consist of paired heavy and light polypeptide chains. They can be produced in large amounts, with extraordinary affinity and selectivity. Antibodies function to activate complement, direct effector cells to the pathogen, neutralize, and sequester microbes. The antibody has two structural domains that serve the functions of antigen recognition and effector engagement. The variable region binds antigen and varies extensively between antibody molecules. The antibody constant region recruits cells and molecules to eliminate the bound antigen and comes in five forms that determine the antibody class. Table 3.5 summarizes the structure and major functions of the five immunoglobulin classes: IgD, IgM, IgG, IgA, and IgE.

The bloodstream is an extremely hostile environment for microbes. Yet the agents of African sleeping sickness, *Trypanosoma brucei rhodesiense* and *Trypanosoma brucei gambiae*, establish and maintain extracellular, bloodstream infections that can last weeks, months, and occasionally years. Fig. 3.7 shows trypanosomes in the blood.

### Generation of antibody diversity: many ways of changing

Antibodies are the secreted form of the B-cell antigen receptor. Virtually any molecule can be the target of an antibody response, yet the genome is not large enough to contain a separate gene for an antibody to every antigen. Instead the vast repertoire is generated from a limited number of inherited gene segments that undergo alterations in individual B cells. Five different mechanisms generate antibody diversity (Fig. 3.8): (1) diversity of the variable-region sequences, (2) recombination of those regions into functional immunoglobulins, (3) combination of heavy and light chain variable regions to form an antigen-binding site, (4) junctional diversity introduced during joining, and (5) somatic hypermutation in activated B-cells.

The variable regions of one heavy and one light chain combine to form the antigen-binding site. During B-cell differentiation, V (variable) gene segments are recombined with J (joining) regions and C (constant) regions to form functional immunoglobulin genes. In heavy chain differentiation, a D region is also included. As B-cell differentiation proceeds, an antibody class-specific C region is attached to each V-J or V-D-J region to produce genes for IgD, IgM, IgG, IgA, or IgE.

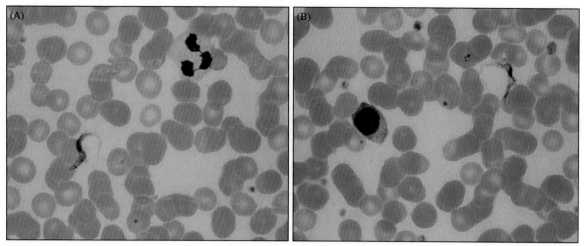

**FIGURE 3.7**  *Trypanosoma brucei* in the blood. The African trypanosome is typically seen in peripheral blood smears from infected patients. In the bloodstream, this extracellular parasite is exposed to complement, antibody, and white cells, but survives to produce a persistent infection (Wright-Giemsa stain, 1000×).

There is substantial diversity of variable region genes. Table 3.6 lists the numbers and combinations. In theory, the variable-region genes and their combinations can produce $1.9 \times 10^6$ different antigen-binding sites; in practice, many of these combinations are useless, unstable, or even recognize self-antigens. During the recombination of variable regions, additional diversity is generated. Two mechanisms add nucleotides at the V-J and V-D junctions. In the first, nucleotides are added by enzymatic reactions that create palindromic sequences. In the second mechanism, nucleotides are added randomly by terminal deoxynucleotide transferase. Another, less well-characterized mechanism may remove nucleotides. In combination, these add diversity in a semirandom way. Finally, somatic hypermutation of the variable regions occurs during proliferation of activated B-cells. Point mutations are targeted to the variable regions at a very high rate.

During B-cell development, antibody serves as the B-cell antigen receptor. Self-reactive cells are eliminated and there is selection for high-affinity receptors. Combined with mechanisms for generating diversity, an immune response of high specificity and increasing efficacy is produced. Antigenic stimulus is required to induce B-cells to produce antibody, but the mere presence of antigen is insufficient. Co-stimulation, either by CD4$^+$ helper T-cells or by particular antigens via interaction with a pattern-recognition receptor is required. Binding of complement to antigens activates coreceptors which enhance the response. Stimulation, selection, and differentiation result in maturation of the B-cell into a mature antibody-producing plasma cell.

### *Trypanosoma brucei* and evasion of the antibody response: diversity responds to diversity

African trypanosomes are unicellular, flagellate parasites carried by the tsetse fly (*Glossina* species). The fly injects the infectious metacyclic form into the host. The parasite invades the subcutaneous tissues, the regional lymph nodes, and finally the bloodstream. A salient characteristic of the trypanosome is a homogeneous glycophosphatidylinositol-linked surface protein called the variant surface glycoprotein (VSG). VSG is proinflammatory, is recognized by B-cells and T-cells, and generates an antibody response. Antibody, complement, and phagocytosis kill antibody-targeted trypanosomes.

A subpopulation changes to a new VSG structurally similar but antigenically distinct from the original one. Again, the host responds by producing antibody. Again, most of the flagellates are destroyed, but a few produce still another variation of the coat protein and the cycle continues. In patients, parasitemia can persist for months or even years. The ability to change surface proteins as the host immune system generates new antibodies is the fruit of genetic mechanisms hauntingly similar to the those that generate antibody diversity.

On a genomic level, the trypanosome contains hundreds of VSG genes and pseudogenes, only one expressed in a particular cell at a particular time. VSGs are transcribed from telomeric regions of the chromosome known as expression sites (ES). There are roughly 20 ES per cell. Most of the VSG genes not part of an ES are found at the telomeres of roughly 100 minichromosomes of 50–150 kb each of which most likely evolved to expand the VSG repertoire. Finally, a few VSG genes and large numbers of pseudogenes (which are truncated, contain frameshift or stop-codon mutations, or lack the biochemistry of expressed VSGs) are found in tandem-repeat clusters in sub-telomeric locations. The structure of the VSG repertoire is depicted in Fig. 3.9A.

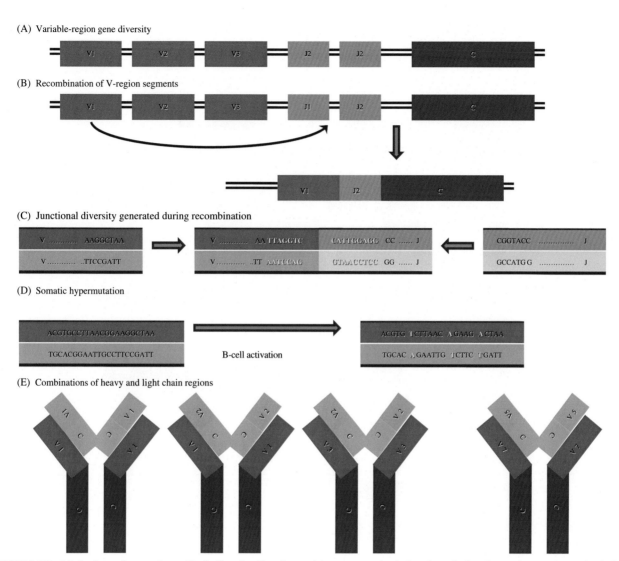

FIGURE 3.8   Mechanisms of generating antibody diversity. C regions and sequences are in shades of purple; J regions and sequences are in shades of green, and V regions and sequences are in shades of red. Altered or mutated sequences are in yellow. Designations of sequences (V1, etc.) are arbitrary and not meant to represent the actual arrangement of specific elements. (A) The inherent germline diversity of V and J regions provides some recognition diversity. (B) The combinations of V and J regions (V, D, and J in heavy-chains) provide additional diversity. (C) The V-J junctions undergo semirandom alterations during recombination, generating more variants. (D) In activated B-cells, the variable regions are hypermutated. (E) V regions of both light and heavy chains combine to form the antigen-recognition zone of the antibody. They can combine in different ways to provide still more variety of antigen recognition.

**TABLE 3.6  Variable-region gene diversity.**

| Immunoglobulin Class | Region | Number of Genes |
|---|---|---|
| **λ Light chains** | V | 30 |
| 120 total combinations | J | 4 |
| **κ Light chains** | V | 40 |
| 200 total combinations | J | 5 |
| **Heavy chains** | V | 40 |
| 6000 total combinations | D | 25 |
| | J | 6 |

Overall: $1.9 \times 10^6$ combinations

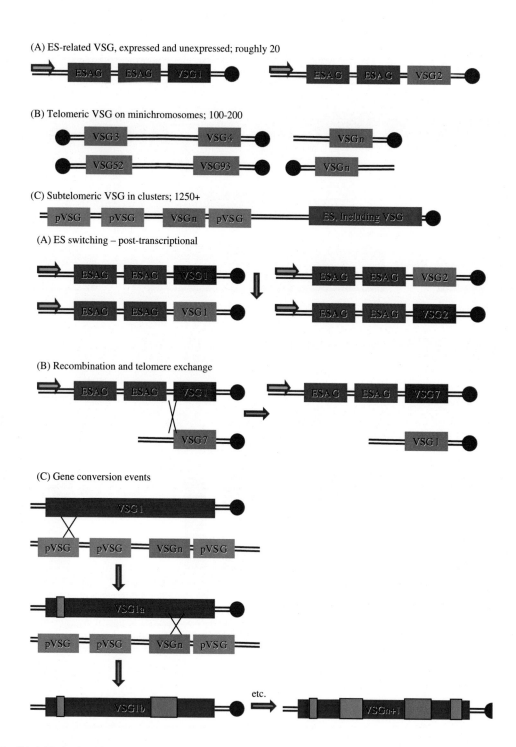

FIGURE 3.9   (A) A Mechanisms for generating variant surface glycoprotein diversity in trypanosomes: VSG genome structure. VSG sequences are in shades of red, others are purple. Silent VSG genes are dark red; expressed VSG genes are bright red, and VSG pseudogenes are pink. The *large dots* at the end of the chromosome represent telomeres. *Green arrows* are VSG promoters. ESAG are expression site–associated genes, non-VSG genes, which are part of the polycistronic transcript driven by the VSG promoter. Designations of sequences (VSG1, etc.) are arbitrary and not meant to represent the actual arrangement of specific elements. (a) ES-related VSG, expressed and unexpressed; roughly 20. (b) Telomeric VSG on minichromosomes; 100−200. (c) Subtelomeric VSG in Clusters; 1250+. (B) Mechanisms for generating variant surface glycoprotein diversity in trypanosomes: expressing new VSG. VSG sequences are in shades of red, others are purple. Silent VSG genes are dark red; expressed VSG genes are bright red, and VSG pseudogenes are pink. The *large dots* at the end of the chromosome represent telomeres. *Green arrows* are VSG promoters. The Xs represent recombination or gene conversion events. ESAG are expression site–associated genes, non-VSG genes, which are part of the polycistronic transcript driven by the VSG promoter. Designations of sequences (VSG1, etc.) are arbitrary and not meant to represent the actual arrangement of specific elements. (a) Post-transcriptional regulation causes different VSGs, located in alternative telomeric ESs, to be expressed. (b) Recombination can switch a VSG gene from a minichromosome or other telomere to an ES. (c) Gene conversion events can alter the sequence of VSGs located at ESs or elsewhere, drawing upon the sequence diversity not only of the silent VSGs but also of the VSG pseudogene pool.

Several mechanisms can lead to expression of a new VSG, as depicted in Fig. 3.9B. These include (1) activation of a new ES (there are roughly 20 per genome) with inactivation of the original ES, (2) homologous recombination of a VSG gene into the active ES with or without telomere exchange, and (3) segmental gene conversion of a portion a VSG gene. Complex chimeric VSG containing elements of one or more VSG genes or pseudogenes may be produced. The regulation of the ES in the living trypanosome remains mysterious. The ES promoter appears to be constitutively active and unregulated. Control of gene expression is mediated through post-transcriptional RNA processing and elongation. The mechanisms underlying activation and inactivation of a given ES are not fully understood, but appear to involve a competition for the transcription machinery, epigenetic silencing factors, telomere factors, and nuclear envelope associations.

If switching between VSGs were simply random, one would expect an initial wave of parasites and then a second, overwhelming wave containing all the possible VSG variants. However, this does not occur. There appears to be a hierarchy of switching mechanisms; the more probable switching events occur early, and VSGs generated by less probable mechanisms occur later in the infection. After switching between ESs, recombination of VSGs into an active ES is the next most commonly observed mechanism, followed by more-complex gene conversion and recombination events, producing a semi-programmed progression of surface coats in the population of organisms infecting a single host, which allows for prolonged parasitemia. These mechanisms also contribute to variation over historical and evolutionary time of the parasite's VSG repertoire.

## HIV: the immune guerilla

Our immune system has developed strategies to fight viral infections. In turn, viruses have evolved numerous ways to evade the host immune system. Since the early 1980s, a new infectious disease of epidemic proportion has successfully emerged and spread around the globe: acquired immune deficiency syndrome (AIDS).

Since the onset of the AIDS epidemic over 35 years ago, HIV has infected millions of people worldwide each year. In 2016, an estimated 36.7 million people lived with HIV globally, and 1 million patients died from AIDS. Since 1981, more than 35 million people have died from AIDS as a result of HIV infection (http://www.who.int/gho/hiv/en/ accessed on July 12, 2018).

HIV not only evades the immune response, but directly attacks the very effector cells that play a pivotal role in the fight against viruses, namely T-lymphocytes, macrophages, and dendritic cells. A paradox is that HIV elicits a broad immune response that is incompletely protective, while it causes progressive immune dysfunction on several levels. HIV infection is rarely terminated by the immune system, but continues for many years and slowly progresses to AIDS and death if left untreated. Since the discovery of HIV there has been an explosion of research aimed at deciphering the mechanism of infection, understanding why it cannot be controlled by our immune system, and at developing an effective vaccine.

### Structure and transmission of HIV—Small but deadly

HIV is a human retrovirus belonging to the lentivirus group. HIV is an RNA virus that utilizes reverse transcriptase and other enzymes to convert its genome from RNA into a chromosomally integrated proviral DNA. Its viral core contains the major capsid protein p24, nucleocapsid proteins, two copies of viral RNA, and three viral enzymes (protease, reverse transcriptase, and integrase). The viral particle is covered by a lipid bilayer derived from the host cell membrane. Two glycoproteins protrude from the surface: glycoproteins (gp)120 and gp41, which are critical for HIV infection of cells (Fig. 3.10). In contrast with the 150 kb HSV genome, HIV has to accomplish all its tasks with a genome of only 9.8 kb.

### Invasion of cells by HIV: into the lion's den

The high-affinity receptor used by HIV is the CD4 receptor, hence the major target for HIV is lymphoid tissue—more specifically, CD4$^+$ T-lymphocytes, macrophages, and dendritic cells (Fig. 3.11). The first encounter between HIV and the naïve host takes place in the mucosa and draining lymph node. Dendritic cells play an important role in the infectious process. They can present antigens via MHC class I and class II molecules, stimulating both T-helper and CTL responses. CD4 on host cells needs to be accompanied by one of two chemokine coreceptors: either CXCR4 or CCR5. R4 viruses utilize CXCR4, which is expressed on lymphocytes, but not on macrophages and are called lymphocyte-tropic or T-tropic. R5 viruses utilize CCR5, expressed on monocytes/macrophages, lymphocytes, and dendritic cells and

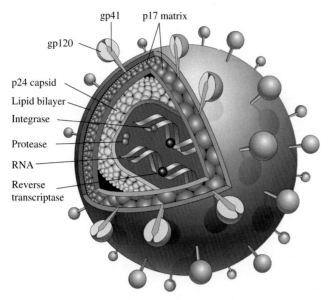

gp41   p17 matrix

gp120

p24 capsid
Lipid bilayer
Integrase
Protease
RNA
Reverse
transcriptase

**FIGURE 3.10**   The structure of the HIV virion. Schematic illustration of an HIV virion. The viral particle is covered by a lipid bilayer that is derived from the host cell. *Reprinted by permission from Elsevier Saunders. Robbins, and Cotran pathologic basis of disease 7th ed. p. 247, Copyright 2004.*

are called macrophage-tropic (M-tropic) viruses despite the fact that they can infect several cell types. Dual tropic viruses can use both CXCR4 and CCR5 as coreceptors. In the early phase of HIV infection, R5 (M-tropic) viruses dominate, but over the course of the infection the tropism often changes due to mutations in the viral genome, and R4 (lymphocyte-tropic) viruses increase in numbers.

The initial step in infection is the binding of gp120 to CD4, leading to a conformational change of the viral protein, which now recognizes the coreceptor. This then triggers conformational change of gp41, and fusion of the viral bilayer with the host cell membrane. The HIV genome enters the host cell and is reverse transcribed into cDNA (pro-viral DNA). The HIV reverse transcriptase is highly error prone, resulting in production of high genetic variability. HIV cDNA may remain in the cytoplasm in linear form, but in dividing host cells, the cDNA enters the nucleus and is integrated in the genome. In the infected T-cell, complete viral particles bud from the cell membrane. If there is extensive viral production, the host cell dies. Alternatively, HIV may remain silent, either in the cytoplasm or integrated as provirus, for months or even years (latent infection). Since macrophages and dendritic cells are relatively resistant to the cytopathic effect of HIV, they are likely important reservoirs of infection.

Clinically, the patient is asymptomatic or has flulike symptoms during this first phase of HIV infection characterized by widespread seeding of the lymphoid tissues, loss of activated CD4$^+$ T-cells, and the highest level of viremia at any time during infection,. The initial infection is controlled by the development of an HIV-specific CTL and humoral responses (Fig. 3.11), and the patient again becomes asymptomatic. The level of viremia and the viral load in the lymphoid tissue at the end of the acute HIV syndrome define the so-called 'set point', which differs between individual patients and has prognostic implications.

During clinical latency (also called the middle or chronic phase of HIV infection), the immune system is relatively intact. However, latently infected T-cells throughout the body, when activated by antigen contact or HIV itself, release intact virions and undergo apoptosis. Thus, it is misleading to talk about a latent infection in the context of HIV, since the definition of latency in the viral world implies a lack of viral replication. The clinical *latent phase* of HIV infection is actually a dynamic competition between the actively replicating virus and the immune system. Indeed, the potent adaptive immune response eventually selects for mutant viruses in key epitopes, a process termed immune escape.

In a fateful twist, the life cycle in latently infected T-cells comes to completion (and usually leads to cell death) at the very moment when the T-cell is needed most—upon activation. After T-cells are activated, signal transduction results in translocation of NF-κB into the nucleus and upregulation of the expression of several cytokines. Flanking regions of the HIV genome also contain similar NF-κB sites that are triggered by the same signal transduction molecule. Thus, the physiologic response of the T-cell stimulates virus production and leads to death of the infected cell. Infected CD4$^+$ T-cells are also killed by CTL cells that recognize HIV antigen; even uninfected CD4$^+$ T-cells, may be killed. Chronic activation of the immune system starts them down the pathway to apoptosis, in this case activation-induced.

The last phase of HIV infection is progression to full-blown AIDS. A vicious cycle of increasingly productive viremia, loss of CD4$^+$ cells, increased susceptibility to opportunistic infections, further immune activation, and progression

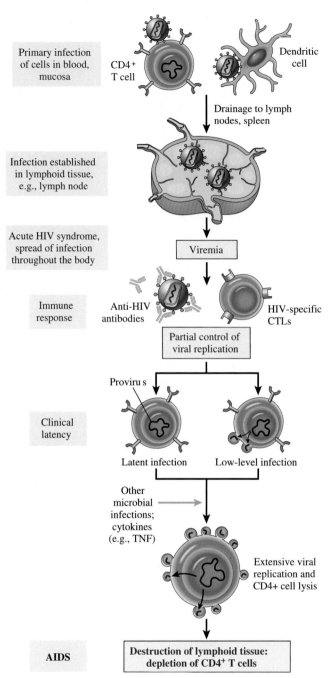

**FIGURE 3.11** Pathogenesis of HIV-1 infection. Pathogenesis of HIV-1 infection. Initially, HIV-1 infects T-cells and macrophages directly or is carried to these cells by Langerhans cells. Viral replication in the regional lymph nodes leads to viremia and widespread seeding of lymphoid tissue. The viremia is controlled by the host immune response, and the patient then enters a phase of clinical latency. During this phase, viral replication in both T-cells and macrophages continues unabated, but there is some immune containment of virus. Ultimately, CD4$^+$ cell numbers decline due to productive infection and other mechanisms, and the patient develops clinical symptoms of full-blown AIDS. *Reprinted by permission from Elsevier Saunders. Robbins, and Cotran pathologic basis of disease 7th ed. p. 248, Copyright 2004.*

of cell destruction develops. leading to a breakdown of host defense, a dramatic increase in circulating virus, and clinical disease. The onset of certain opportunistic infections such as invasive candidiasis, mycobacteriosis, or pneumocystosis (Fig. 3.12); secondary neoplasms; or HIV-associated encephalitis marks the beginning of AIDS.

Combination highly active antiretroviral therapy (HAART) regimens have transformed HIV from a progressive illness into a chronic manageable disease. The high mutation rate of HIV allows selection of drug-resistant viruses when treatments are used improperly. Mutation and variation both in viral antigens and in viral physiology play an important role in the pathogenesis of HIV disease. This is also one reason why the quest for an HIV vaccine has remained elusive. Finally, although HAART can arrest the progression of HIV infection, a cure in established HIV infections remains elusive, due to the persistence of truly latent virus in long-lived memory T-cells, as well as low-level replication in some treated patients and to persistence in protected anatomical sites.

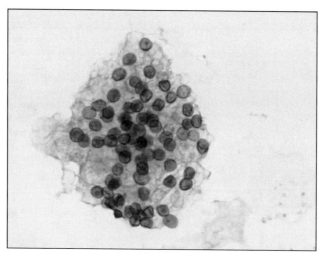

FIGURE 3.12   Opportunistic pathogen in AIDS. Cluster of *Pneumocystis jirovecii* cysts in bronchoalveolar lavage of an HIV-positive patient stained with toluidin blue (oil immersion, magnification 1000 ×).

**TABLE 3.7 Immune dysfunction in AIDS.**

Altered monocyte/macrophage functions
  Decreased chemotaxis and phagocytosis
  Decreased HLA class II antigen expression
  Decreased antigen presentation capacity
  Increased secretion of IL-1, IL-6, and TNFα
Altered T-cell functions in vivo
  Preferential loss of memory T-cells
  Susceptibility to opportunistic infections
  Susceptibility to neoplasms
Altered T-cell functions in vitro
  Decreased proliferative response to antigens
  Decreased specific cytotoxicity
  Decreased helper function for B-cell Ig synthesis
  Decreased IL-2 and IFN-γ production
Polyclonal B-cell activation
  Hypergammaglobulinemia and circulating immune complexes
  Inability to mount antibody response to new antigen
  Refractoriness to normal B-cell activation in vitro

Modified after Kumar V, Abbas AK, Aster JC, Robbins and Cotran pathologic basis of disease. 9th ed. Philadelphia (PA): Elsevier/Saunders; 2015.

The devastating clinical course of AIDS and the unique pathologic features of HIV infection, with significant viremia persisting for years, demonstrate the essential role of the CD4 T-cell in adaptive immunity (Table 3.7). While most CD4 T-cells have relatively modest effector function, they are the central regulators of the adaptive immune response. CD4 T-cells are essential for maturation and development of B-cells and CD8+ cytotoxic T-cells, as well as for activation of macrophages. Patients with late-stage HIV suffer from deteriorating antibody quality and production, loss of cell-mediated immunity, and decreased production and altered function of every kind of immune cell.

## Perspectives

The five pathogens discussed here fall far short of the breadth of microbial interactions with the host. In fact, these comparatively short discussions fail to cover the wondrous complexity of the interactions these few organisms have. Microbes, for the most part, do not enter the forbidding interior of mammalian hosts with a single, general-purpose toxin or strategy for causing disease. Instead, they come equipped with a variety of very specific molecular tools, each with a particular target, which allow them to survive, replicate, and be transmitted to a new host.

The interactions of pathogens with the host on the molecular level are extraordinarily subtle. Pathogens rarely eradicate an immune response. Rather, they attenuate, misdirect, and delay the host response sufficiently to accomplish their purpose of replication and transmission. Overall, the complexity and flexibility of the interaction benefits both the host and the pathogen. The pathogen benefits because, even if it is ultimately eliminated from the host, it survives and proliferates long enough and well enough to be transmitted. The host benefits because a temperate response is less likely to result in severe collateral tissue damage. In addition, if host responses were so rigid and forceful that the pathogen was forced to kill the host or die itself, the microbes, ultimately, would win, due to their rapid evolution. The mammalian immune system, for all its extraordinary complexity and power, is a compromise between metabolic cost and efficacy, between elimination and containment of pathogens, and often lives with what it cannot destroy.

## Key concepts

- Pathogens have evolved strategies to evade and manipulate host immune defenses.
- Innate immunity recognizes conserved markers of infection and induces rapid protective responses. B-cells and T-cells comprise the system of adaptive immunity and mount a slower pathogen-specific response. Effector mechanisms, directed by both innate and adaptive immunity, inactivate pathogens.
- *Staphylococcus aureus* utilizes multiple mechanisms to resist neutrophil-mediated killing.
- Mycobacterium tuberculosis invades the macrophage and manipulates the cellular response to allow it to escape destruction and proliferate intracellularly.
- Herpes simplex virus attenuates the immune response at multiple levels to facilitate viral survival and reproduction.
- The African trypanosome evades antibody responses by generating antigenic diversity at a genetic level.
- HIV invades critical elements of the immune system, the CD4+ T-cell, macrophages, and dendritic cells, and both evades their responses and degrades their function, ultimately resulting in immune failure.

## Suggested readings

[1] Brubaker SW, Bonham KS, Zanoni I, Kagan JC. Innate immune pattern recognition: a cell biological perspective. Annu Rev Immunol 2015;33:257−90.

[2] Iwasaki A, Medzhitov R. Control of adaptive immunity by the innate immune system. Nat Immunol 2015;16:343−53.

[3] Spaan AN, Surewaard BG, Nijland R, van Strijp JA. Neutrophils versus *Staphylococcus aureus*: a biological tug of war. Annu Rev Immunol 2013;67:629−50.

[4] Hmama Z, Pena-Diaz S, Joseph S, Av-Gay Y. Immunoevasion and immunosuppression of the macrophage by *Mycobacterium tuberculosis*. Immunol Rev 2015;264:220−32.

[5] Yu X, He S. The interplay between human herpes simplex virus infection and the apoptosis and necroptosis cell death pathways. Virol J 2016;13:77.

[6] Ma Y, He B. Recognition of herpes simplex viruses: toll-like receptors and beyond. J Mol Biol 2014;426:1133−47.

[7] Mugnier MR, Stebbins CE, Papavasiliou FN. Masters of disguise: antigenic variation and the VSG coat in Trypanosoma brucei. PLoS Pathog 2016;12:e1005784.

[8] Espíndola MS, Soares LS, Galvão-Lima LJ, Zambuzi FA, Cacemiro MC, Brauer VS, et al. HIV infection: focus on the innate immune cells. Immunol Res 2016;64:1118−32.

[9] Fraser C, Lythgoe K, Leventhal GE, Shirreff G, Hollingsworth TD, Alizon S, et al. Virulence and pathogenesis of HIV-1 infection: an evolutionary perspective. Science 2014;343:1243727.

[10] Murray AJ, Kwon KJ, Farber DL, Siliciano RF. The latent reservoir for HIV-1: how immunologic memory and clonal expansion contribute to HIV-1 persistence. J Immunol 2016;197:407−17.

Chapter 4

# Neoplasia

**William B. Coleman**

*American Society for Investigative Pathology, Rockville, MD, United States*

**Summary**

It is now recognized that cancer, in its simplest form, is a genetic disease or, more precisely, a disease of abnormal gene expression. Recent research efforts have revealed that different forms of cancer share common molecular mechanisms governing uncontrolled cellular proliferation, involving loss, mutation, or dysregulation of genes that positively and negatively regulate cell proliferation, migration, and differentiation (generally classified as proto-oncogenes and tumor suppressor genes). This chapter introduces basic and essential concepts related to neoplastic disease as a foundation for more detailed treatment of the molecular carcinogenesis of major cancer types. It provides an overview of cancer statistics and epidemiology, highlighting cancer types of importance to human health in the United States and worldwide, with a brief review of risk factors for the development of cancer. It discusses the classification of neoplasms, focusing on the general features of benign and malignant neoplasms, with an overview of nomenclature for human neoplasms, a description of pre-neoplastic conditions, and consideration of special subsets of neoplastic disease (cancers of childhood, hematopoietic neoplasms, and hereditary cancers). It also describes the distinguishing characteristics of benign and malignant neoplasms, with a focus on anaplasia and cellular differentiation, rate of growth, local invasiveness, and metastasis. The chapter further discusses the clinical aspects of neoplasia, giving an overview of cancer-associated pain, cancer cachexia, paraneoplastic syndromes, and methods for grading and staging of cancer.

## Introduction

Cancer does not represent a single disease. Rather, cancer is a myriad collection diseases with as many different manifestations as there are tissues and cell types in the human body, involving innumerable endogenous or exogenous carcinogenic agents, and various etiological mechanisms. What all of these disease states share in common are certain biological properties of the cells that compose the tumors, including unregulated (clonal) cell growth, impaired cellular differentiation, invasiveness, and metastatic potential. It is now recognized that cancer, in its simplest form, is a genetic disease or, more precisely, a disease of abnormal gene expression. Recent research efforts have revealed that different forms of cancer share common molecular mechanisms governing uncontrolled cellular proliferation, involving loss, mutation, or dysregulation of genes that positively and negatively regulate cell proliferation, migration, and differentiation (generally classified as proto-oncogenes and tumor suppressor genes). In the discussion that follows, basic and essential concepts related to neoplastic disease are presented as a foundation for more detailed treatment of the molecular carcinogenesis of major cancer types provided elsewhere in this book.

In this chapter, an overview of cancer statistics and epidemiology is provided, highlighting cancer types of importance to human health in the United States and worldwide, with a brief review of risk factors for the development of cancer. Subsequently, the classification of neoplasms is discussed, focusing on the general features of benign and malignant neoplasms, with an overview of nomenclature for human neoplasms, a description of pre-neoplastic conditions, and consideration of special subsets of neoplastic disease (cancers of childhood, hematopoietic neoplasms, and hereditary cancers). Next, the distinguishing characteristics of benign and malignant neoplasms are discussed in some detail, with a focus on anaplasia and cellular differentiation, rate of growth, local invasiveness, and metastasis. The chapter closes with a discussion of the clinical aspects of neoplasia, including an overview of cancer-associated pain, cancer cachexia, paraneoplastic syndromes, and methods for grading and staging of cancer.

Essential Concepts in Molecular Pathology. DOI: https://doi.org/10.1016/B978-0-12-813257-9.00004-8

# Cancer statistics and epidemiology

## Cancer incidence

Cancer is an important public health concern in the United States and worldwide. Because of the lack of nationwide cancer registries for all countries, the exact numbers of the various forms of cancer occurring in the world populations are unknown. Nevertheless, estimations of cancer incidence and mortality are generated on an annual basis by several domestic and world organizations, including the *American Cancer Society* (ACS; www.cancer.org), the National Cancer Institute's Surveillance, Epidemiology, and End Results program (SEER; www.seer.cancer.gov), and the *International Agency for Research on Cancer* (IARC; www.iarc.fr) and the *World Health Organization* (WHO; www.who.int). Monitoring of long-range trends in cancer incidence and mortality among different populations is important for investigations of cancer etiology. Given the long latency for formation of a clinically detectable neoplasm (up to 20–30 years) following initiation of the carcinogenic process (exposure to carcinogenic agent), current trends in cancer incidence probably reflect exposures that occurred many years (and possibly decades) before. Thus, correlative analysis of current trends in cancer incidence with recent trends in occupational, habitual, and environmental exposures to known or suspect carcinogens can provide clues to cancer etiology. Other factors that influence cancer incidence include the size and average age of the affected population. The average age at the time of cancer diagnosis for all tumor sites is approximately 65 years. As a higher percentage of the population reaches age 60, the general incidence of cancer will increase proportionally. Thus, as the life expectancy of the human population increases due to reductions in other causes of premature death (due to infectious and cardiovascular diseases), the average risk of developing cancer will increase.

### *General trends in cancer incidence*

The *ACS* estimates that 1,735,350 new cases of invasive cancer were diagnosed in the United States in 2018, reflecting 856,370 male cancer cases (49%) and 878,980 female cancer cases (51%). These estimates do not include carcinoma *in situ* occurring at any site other than in the urinary bladder and does not include common skin cancers. In fact, basal and squamous cell carcinomas of the skin represent the most frequently occurring neoplasms in the United States, with an estimated occurrence of >1 million total cases in 2018. Likewise, carcinoma *in situ* represents a significant number of new cancer cases with 63,960 newly diagnosed breast carcinomas *in situ* and 87,290 new cases of melanoma carcinoma *in situ*.

Estimated site-specific cancer incidence for both sexes combined is shown in Fig. 4.1. Cancers of the digestive system represent the largest group of newly diagnosed cancers in 2018 with 319,160 new cases., including colon (97,220

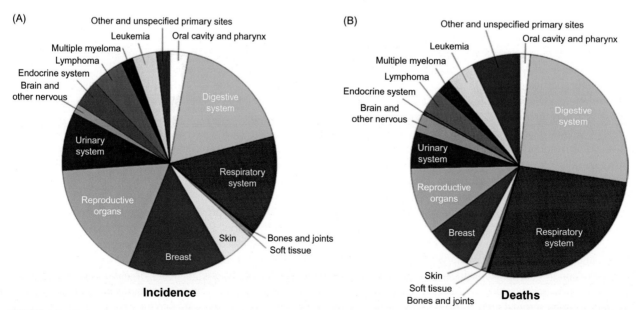

**FIGURE 4.1**   Cancer incidence and mortality by site for both sexes (United States, 2016). The relative contributions of the major forms of cancer to (A) overall cancer incidence and (B) cancer-related mortality (both sexes combined). Cancers of the reproductive organs include those affecting the prostate, uterine corpus, ovary, uterine cervix, vulva, vagina, testis, penis, and other organs of the male and female genital systems. Cancers of the digestive system include those affecting esophagus, stomach, small intestine, colon, rectum, anus, liver, gallbladder, pancreas, and other digestive organs. Cancers of the respiratory system include those affecting the lung, bronchus, larynx, and other respiratory organs.

new cases), pancreas (55,440 new cases), liver and bile duct (42,220 new cases), rectum (43,030 new cases), stomach (26,240 new cases), and esophagus (17,290 new cases), in addition to the other digestive system organs (small intestine, gallbladder, and others). The next most frequently occurring cancers originate in the reproductive organs (286,390 new cases), respiratory system (253,290 new cases), and breast (268,670 new cases). Cancers of the reproductive organs include prostate (164,690 new cases), uterine corpus (63,230 new cases), ovary (22,240 new cases), and uterine cervix (13,240 new cases), in addition to other organs of the genital system. Most new cases of cancer involving the respiratory system affect the lung and bronchus (234,030 new cases), with the remaining cases affecting the larynx or other components of the respiratory system. Other sites with significant cancer burden include the urinary system (150,350 new cases), lymphomas (83,180 new cases), skin (99,550 new cases, excluding basal and squamous carcinomas), leukemias (60,300 new cases), and the oral cavity and pharynx (51,540 new cases).

Among men, cancers of the prostate, respiratory system (lung and bronchus), and digestive system (colon and rectum) occur most frequently, accounting for 44% of all cancers diagnosed in men. Prostate is the leading site, accounting for 164,690 new cases and 21% of cancers diagnosed in men. Among women, cancers of the breast, respiratory system (lung and bronchus), and digestive system (colon and rectum) occur most frequently, accounting for 50% of all cancers diagnosed in women. Breast is the leading site for cancer affecting women, accounting for 268,670 new cases and 29% of all cancers diagnosed in women.

## General trends in cancer mortality in the United States

Mortality attributable to invasive cancers produced 609,640 cancer-associated deaths in 2018, reflecting 323,630 male cancer deaths (53% of total) and 286,010 female cancer deaths (47% of total). Estimated numbers of cancer deaths by site for both sexes are shown in Fig. 4.1. The leading cause of cancer death involves tumors of the respiratory system (158,770 deaths), the majority of which are neoplasms of the lung and bronchus (154,050 deaths). The second leading cause of cancer deaths involves tumors of the digestive system (160,820 deaths), most of which are tumors of the colorectum (50,630 deaths), pancreas (44,330 deaths), stomach (10,800 deaths), liver and intrahepatic bile duct (30,200 deaths), and esophagus (15,850 deaths). Together, cancers of the respiratory and digestive systems account for 53% of cancer deaths.

Trends in cancer mortality among men and women mirror in large part cancer incidence. Cancers of the prostate, lung and bronchus, and colorectum represent the three leading sites for cancer incidence and cancer mortality among men. In a similar fashion, cancers of the breast, lung and bronchus, and colorectum represent the leading sites for cancer incidence and mortality among women. While cancers of the prostate and breast represent the leading sites for new cancer diagnoses among men and women (respectively), the majority of cancer deaths in both sexes are related to cancers of the lung and bronchus — 27% of all cancer deaths among men and 26% among women. The age-adjusted death rate for lung cancer among men increased dramatically during the 20th century (from 1930 to 1990), whereas the death rates for other cancers (such as prostate and colorectal) remained relatively stable. The death rate for lung cancer among men peaked in the early 1990s and has been declining since. The lung cancer death rate for women has increased in an equally dramatic fashion since about 1960, becoming the leading cause of female cancer death in the mid-1980s after surpassing the death rate for breast cancer.

## Global cancer incidence and mortality

The IARC GLOBOCAN project estimates that 14,090,100 new cancer cases were diagnosed worldwide in 2012, corresponding to 7,427,100 male cancer cases (53%) and 6,663,000 female cancer cases (47%). Mortality attributed to cancer for the same year produced 8,201,600 deaths worldwide, reflecting 4,653,400 male cancer deaths (57%) and 3,548,200 female cancer deaths (43%). The leading sites for cancer incidence and mortality worldwide in 2012 included cancers of the lung, breast, colorectum, prostate, stomach, and liver. Lung cancer accounted for the most new cancer cases and the most cancer deaths during this period of time, with 1,824,700 new cases (13% of all cancers diagnosed) and 1,589,900 deaths (19% of all cancer-associated deaths) for both sexes combined. The leading sites for cancer incidence among males included lung (1,241,600 new cases), prostate (1,111,700 new cases), colorectum (746,300 new cases), stomach (631,300 new cases), and liver (554,400 new cases). Combined, cancers at these five sites account for 58% of all cancer cases among men. The leading causes of cancer death among men included tumors of the lung (1,098,700 deaths), liver (521,000 deaths), stomach (469,000 deaths), colorectum (373,600 deaths), and prostate (307,500 deaths). Deaths from these cancers account for 60% of all male cancer deaths. The leading sites for cancer incidence among females included breast (1,676,600 new cases), colorectum (614,300 new cases), lung (583,100 new cases), cervix uteri (527,600 new cases), and stomach (320,300 new cases). Combined, cancers at these five sites account for 56% of all cancer cases among women. The leading causes of cancer death among females directly mirror the leading causes of cancer incidence: breast (521,900 deaths), lung (491,200 deaths), colorectum (320,300 deaths), cervix uteri (265,700 deaths), and stomach (254,100 deaths). Combined, these five cancer sites account for 52% of female cancer deaths.

## Risk factors for the development of cancer

Risk factors for cancer can be considered anything that increases the chance that an individual will develop neoplastic disease. Individuals who have risk factors for cancer development are more likely to develop the disease at some point in their lives than the general population (lacking the same risk factors). However, having one or more risk factors does not necessarily mean that a person will develop cancer. It follows that some people with recognized risk factors for cancer will never develop the disease, whereas others lacking apparent risk factors for cancer will develop neoplastic disease. Although certain risk factors are clearly associated with the development of neoplastic disease, making a direct linkage from a risk factor to causation of the disease remains very difficult and often impossible. Some risk factors for cancer development can be modified, whereas others cannot. For instance, cessation of cigarette smoking reduces the chance that an individual will develop cancer of the lung, bronchus, or other tissues of the aerodigestive tract. In contrast, a woman with an inherited mutation in the *BRCA1* gene carries an elevated lifetime risk of developing breast cancer. Some of the major risk factors that contribute to cancer development include (1) age, (2) race, (3) gender, (4) family history, (5) infectious agents, (6) environmental exposures, (7) occupational exposures, and (8) lifestyle exposures.

### Age, race, and gender as risk factors for cancer development

Advancing age is the most important risk factor for cancer development for most people, and most malignant neoplasms are diagnosed in patients over the age of 65. According to the National Cancer Institute's SEER Statistics (www.seer.cancer.gov) the median age at diagnosis for lung/bronchus cancer is 70 years old (<2% of cases occur in people <45 years old), the median age at diagnosis for prostate cancer is 66 (only 10% of cases occur in men <55 years old), the median age at diagnosis for breast cancer is 62 years old (<2% of cases occur in women <35 years old), the median age at diagnosis for colorectal cancer is 68 years old (<6% of cases occur in people <45 years old), the median age at diagnosis for liver cancer is 63 years old (<4% of cases occur in people <45 years old), the median age at diagnosis for ovarian cancer is 63 years old (only 5% of cases occur in women <35 years old), and the median age at diagnosis of melanoma is 63 years old (<8% of cases occur in people <35 years old). The age-specific incidence and death rates for selected cancers for the period of 2009—2013 are shown in Fig. 4.2. The trends depicted in this figure clearly show that the majority of each of these cancer types occurs in individuals of advanced age. A notable exception to this relationship between advanced age and cancer incidence involves some forms of leukemia and other cancers of childhood. Acute lymphocytic leukemia (ALL) occurs with a bimodal distribution, with highest incidence among individuals <20 years of age, and a second peak of increased incidence among individuals of advanced age. The majority of ALL cases are diagnosed in children, with a median age of diagnosis of 15% and 57% of cases diagnosed in individuals under the age of 20. Despite the prevalence of this disease in childhood, a significant number of adults are affected. In fact, 19% of ALL cases are diagnosed in individuals over the age of 55 years of age. In contrast to ALL, the other major forms of leukemia demonstrate the usual pattern of age dependence observed with solid cancers, with large numbers of cases in older segments of the population.

Cancer incidence and mortality can vary tremendously with race (and/or factors that are associated with race) and ethnicity (www.seer.cancer.gov). In the United States, African-Americans and Caucasians are more likely to develop cancer than individuals of other races or ethnicity. African-American men demonstrate a cancer incidence for all sites combined of approximately 572 cases per 100,000 population, and Caucasian men exhibit a cancer incidence rate of 508 cases per 100,000 population. In contrast, Asian/Pacific Islander men show the lowest cancer incidence among populations of the United States with 317 cases per 100,000 population for all sites combined. Likewise, African-American women demonstrate a cancer incidence for all sites combined of approximately 401 cases per 100,000 population, and Caucasian women exhibit a cancer incidence rate of 423 cases per 100,000 population, whereas Asian/Pacific Islander women show the lowest cancer incidence with 297 cases per 100,000 population for all sites combined. These differences in race-related cancer incidence rates can be magnified when site-specific cancers are considered. The overall incidence of prostate cancer in the United States (for all men) is 129 cases per 100,000 men. African-American men have a significantly higher incidence rate (204 cases per 100,000 men) compared to Caucasian men (122 cases per 100,000). In contrast, Native American men have a significantly lower incidence of prostate cancer (64 cases per 100,000 men) compared to African-American and Caucasian men. The mechanisms that account for these

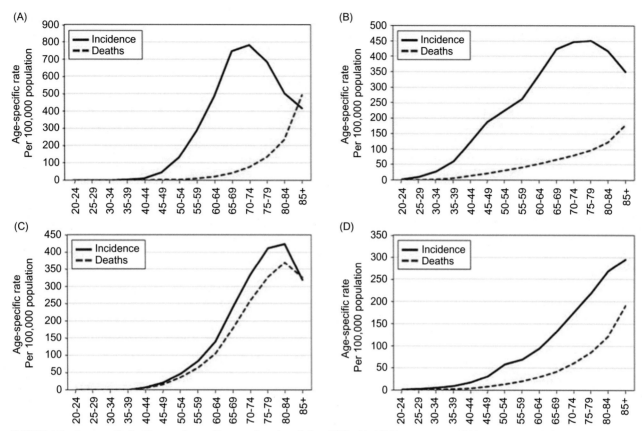

**FIGURE 4.2** Age-specific incidence and mortality rates for selected sites, 2009−13. (A) Prostate cancer, (B) breast cancer, (C) lung cancer, and (D) colorectal cancer. The age-specific rates for breast cancer incidence and mortality are for females only. The age-specific rates for lung cancer and colorectal cancer are combined for both sexes. These data were adapted from the NCI SEER Statistics Database (www.seer.cancer.gov). Rates are per 100,000 population and are age-adjusted to the 2000 standard population of the United States.

differences are not known but may be related to genetic factors or differences in various physiological factors (for instance, androgen hormone levels). Mortality due to cancer also differs among patients depending on their race or ethnicity. Similar to the cancer incidence rates, mortality due to cancer is higher among African-Americans (254 and 164 per 100,000 population for men and women, respectively) and Caucasians (203 and 144 deaths per 100,000 population for men and women, respectively) than other populations, including Asian/Pacific Islanders, American Indians, and Hispanics. Factors that are known to contribute to racial differences in cancer-related mortality include (1) differences in exposures (for instance, smoking prevalence), (2) access to regular cancer screening (for breast, cervical, and colon cancers), and (3) timely diagnosis and treatment. For both cancer incidence and mortality, racial and ethnic variations for all sites combined differ from those for individual cancer sites.

Gender is clearly a risk factor for cancers affecting certain tissues such as breast and prostate where there are major differences between men and women. However, there are numerous other examples of cancers that appear to develop preferentially in men or women, and/or where factors related to gender increase risk. For instance, liver cancer affects men more often than women. The ratio of male to female incidence in the United States is approximately 2.6:1, and worldwide is approximately 2.4:1. However, in high-incidence countries or world regions, the male to female incidence ratio can be as high as 8:1. This consistent observation suggests that sex hormones and/or their receptors may play a significant role in the development of primary liver cancer. Some investigators have suggested that hepatocellular carcinomas overexpress androgen receptors and that androgens are important in the promotion of abnormal liver cell proliferation. Others have suggested that the male predominance of liver cancer is related to the tendency for men to drink and smoke more heavily than women and are more likely to develop cirrhosis.

## Family history as a risk factor for cancer development

Familial cancers have been described for most major organ systems, including colon, breast, ovary, and skin. Hereditary cancers are typically characterized by (1) early age at onset (or diagnosis), (2) neoplasms arising in first-degree relatives of the index case, and (3) in many cases, multiple or bilateral tumors. Epidemiologic evidence has consistently pointed to family history as a strong and independent predictor of breast cancer risk. Thus, women with a first-degree relative (mother or sister) diagnosed with breast cancer are at elevated risk for development of the disease themselves. A substantial amount of research has led to the discovery of several breast cancer susceptibility genes, including *BRCA1*, *BRCA2*, and *p53*, which may account for the majority of inherited breast cancers. It has been estimated that 5%−10% of breast cancers occurring in the United States each year are related to genetic predisposition. Despite the recognition of multiple genetic and environmental risk factors for development of breast cancer, approximately 50% of affected women have no identifiable risk factors other than being female and aging. In many of these cases, the genetic predisposition to cancer development may be related to small but measurable risks associated with genetic variations (polymorphic variations) at multiple loci. Other risk factors for development of breast cancer include advancing age (over 50 years of age), early age at menarche, late age at menopause, first childbirth after age of 35, nulliparity, family history of breast cancer, obesity, dietary factors (such as high-fat diet), and exposure to high dose radiation to the chest before age 35. Many of these risk factors may interact with genetic polymorphisms and other genetic determinants to drive the development of breast cancer.

## Infectious agents as risk factors for cancer development

There are several excellent examples of infectious agents and cancer causation. These include (1) liver cancer and hepatitis virus infection, (2) cervical cancer and human papillomavirus (HPV) infection, and (3) Epstein−Barr virus (EBV) infection and Burkitt lymphoma. Human cancers are also associated with certain bacterial and parasitic infections.

Hepatitis B virus (HBV) and hepatitis C virus (HCV) infections are associated with development of hepatocellular carcinoma. Primary liver cancers are usually associated with chronic hepatitis, and 60%−80% of hepatocellular carcinomas occurring worldwide develop in cirrhotic livers, most commonly non-alcoholic post-hepatitic cirrhosis. However, hepatitis virus infection is not thought to be directly carcinogenic. Rather, HBV and/or HCV infection produces hepatocyte necrosis and regeneration (cell proliferation), which makes the liver susceptible to endogenous and exogenous carcinogens. Thus, pre-neoplastic nodules in the liver tend to occur in regenerative nodules of the injured liver (chronic hepatitis or cirrhosis). In certain geographic areas (such as China), large portions of the population are concurrently exposed to the hepatocarcinogen aflatoxin $B_1$ and HBV, which increases their relative risk for development of liver cancer. Some changes in the incidence of hepatocellular carcinoma have been observed in recent years, attributed to modification of risk factors by universal HBV immunization, successful treatment of HCV infection, and improved control over aflatoxin contamination of food products.

Cervical cancer is the fourth most common cancer of women worldwide (after breast, lung, and colorectum), with 527,600 new cases and 265,700 deaths in 2012. Infection with high-risk HPV types (in particular, HPV16 and HPV18) is specifically associated with development of cervical cancer (Fig. 4.3). Case-control and prospective epidemiological studies have shown that HPV infection precedes high-grade dysplasia and invasive cancer and represents the strongest independent risk factor for the development of cervical cancer. Molecular analyses of cervical cancers suggest that HPV plays a key role in cervical carcinogenesis because >90% of cervical cancer biopsies contain DNA sequences of high-risk HPV types. HPV vaccination is now available and widely implemented. Countries that implemented the HPV vaccine over 5 years ago have already experienced decreased prevalence of targeted HPV genotypes and associated diseases in women. Given the tight association between HPV infection and cervical cancer development, it is expected that the incidence of cervical cancer will decline over time as greater numbers of women are vaccinated for HPV.

EBV is a member of the human herpesvirus subfamily *Gammaherpesviridae*, a ubiquitous virus that infects >90% of the human population. Primary infection with EBV usually occurs in childhood, and in most cases there is no apparent clinical course. However, in a subset of infected individuals, primary EBV infection can result in infectious mononucleosis, a self-limited lymphoproliferative disease, particularly when primary infection is delayed into adolescence. EBV causes a latent infection in lymphoid cells, which persists for life in a small subpopulation of B-lymphocytes. Although EBV has a very high transforming potential in vitro, it rarely causes cancer in humans, suggesting that EBV-infected cells are subject to continuous surveillance by cytotoxic T-cells. However, EBV infection is also closely associated with the development of lymphoid and epithelial malignancies in apparently immunocompetent hosts. The human neoplasm that is classically associated with EBV infection is Burkitt lymphoma. On a worldwide scale, Burkitt lymphoma is a low incidence cancer. However, its prevalence shows striking geographic variations. For instance, Burkitt lymphoma occurs in an endemic form in equatorial Africa where it can represent the most common childhood tumor.

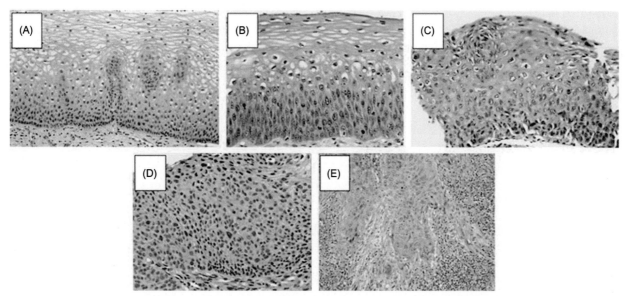

**FIGURE 4.3** Progression of dysplasia in the cervix. (A) Normal squamous epithelium of the cervix. The basal (bottom) layer of cells appears dark, and there is cellular maturation as the squamous cells move to the surface. (B) Low-grade squamous dysplasia. The basal layer appears thicker with mild nuclear pleomorphism and mitotic figures involving the lower third of the epithelium. In this image, several human papillomavirus–infected cells, known as koilocytes, are present in the middle third of the epithelium. A koilocyte has a wrinkled nucleus with cytoplasmic clearing. (C) Moderate squamous dysplasia. The dysplastic squamous cells and mitotic figures involve the lower two-thirds of the epithelium. There is still some maturation toward the surface of the epithelium. (D) High-grade squamous dysplasia. Dysplastic squamous cells and mitotic figures involve the full thickness of the epithelium. There is no invasion into the underlying stroma. (E) Invasive squamous cell carcinoma. Dysplastic squamous cells invade into the underlying stroma in a haphazard fashion. There is prominent inflammation in the stroma in response to the invasive carcinoma.

## Environmental and occupational exposures as risk factors for cancer development

The most well-studied hepatocarcinogen is a natural chemical carcinogen known as aflatoxin $B_1$ that is produced by the *Aspergillus flavus* mold. This mold grows on rice or other grains (including corn) that are stored without refrigeration in hot and humid parts of the world. Ingestion of food that is contaminated with *A. flavus* mold results in exposure to potentially high levels of aflatoxin $B_1$, which is a potent, direct-acting liver carcinogen in humans, and chronic exposure leads inevitably to development of hepatocellular carcinoma.

Another naturally occurring carcinogen is the radioactive gas radon, which has been suggested to increase the risk of lung cancer development. This gas is ubiquitous in the earth's atmosphere, creating the opportunity for exposure of vast numbers of people. However, passive exposure to the background levels of radon found in domestic dwellings and other enclosures is not sufficiently high to increase lung cancer risk appreciably. High-level radon exposure has been documented among miners working in uranium, iron, zinc, tin, and fluorspar mines. These workers show an excess of lung cancer (compared to non-miners) that varies depending on the radon concentration encountered in the ambient air of the specific mine.

Exposure to chemical carcinogens does not represent an important risk factor for most of the general population. Nevertheless, several chemicals, complex chemical mixtures, industrial processes, and/or therapeutic agents have been associated with development of malignant neoplasms in exposed human populations. These exposures may include therapeutic exposure to the radioactive compounds (such as thorium dioxide or thorotrast for the radiological imaging of blood vessels) and occupational exposures to certain industrial chemicals (such as vinyl chloride monomer, asbestos, bis[chloromethyl] ether, or chromium). Chemical carcinogens are classified as direct agents or indirect agents. Indirect agents are chemicals that require metabolic conversion to become an ultimate carcinogen. Examples of indirect carcinogenic agents include certain polycyclic hydrocarbons (such as those found in tobacco smoke and in smoked meats and fish) and azo dyes (such as β-naphthylamine). Because indirect carcinogens require metabolic activation for their conversion into ultimate carcinogens with genotoxic activity (DNA-damaging activity), a great deal of research has focused on the enzymatic pathways that are required for carcinogen activation. Many of these carcinogen activating pathways involve cytochrome P-450-dependent monooxygenase enzymes, which are encoded by highly polymorphic genes. Thus, the activity of these enzymes tends to vary among individuals depending on the specific form of the enzyme that is carried. Hence, susceptibility to a specific chemical carcinogen may depend in part on the specific form of the enzyme that is expressed, making it probable that molecular analyses to determine these genetic polymorphisms may become as important in the prediction of cancer risk in the future as consideration of exposures. Several other agents with carcinogenic potential in humans include vinyl chloride, arsenic, nickel, chromium, various insecticides, various fungicides, and polychlorinated biphenyls. These agents are encountered most frequently through occupational exposures. In addition, certain food preservatives (such as nitrites) are of significant concern as potential carcinogens. Nitrites can produce nitrosylation of amines in various foodstuffs, resulting in the formation of nitrosamine compounds, which are suspected to have carcinogenic potential in humans.

### Lifestyle exposures as risk factors for cancer development

Numerous risk factors for cancer development fall into the category of lifestyle exposures and generally reflect lifestyle choices. These exposures include (1) consumption of tobacco products, (2) consumption of excessive amounts of alcohol, and (3) excessive exposure to sunlight, among others.

Cancers of the lung, mouth, larynx, bladder, kidney, cervix, esophagus, and pancreas are related to consumption of tobacco products, including cigarettes, cigars, chewing tobacco, and snuff. Cigarette smoking alone is the suggested cause for one-third of all cancer deaths. Several lines of evidence strongly link cigarette smoking to lung cancer. Smokers have a significantly increased risk (11-fold to 22-fold) for development of lung cancer compared to non-smokers, and cessation of smoking decreases the risk for lung cancer compared with continued smoking. Furthermore, heavy smokers exhibit a greater risk than light smokers, suggesting a dose-response relationship between cigarette consumption and lung cancer risk. Numerous mutagenic and carcinogenic substances have been identified as constituents of the particulate and vapor phases of cigarette smoke, including benzo[a]pyrene, dibenza[a]anthracene, nickel, cadmium, polonium, urethane, formaldehyde, nitrogen oxides, and nitrosodiethylamine. There is also evidence that smoking combined with certain environmental (or occupational) exposures results in potentiation of lung cancer risk. Urban smokers exhibit a significantly higher incidence of lung cancer than smokers from rural areas, suggesting a possible role for air pollution in development of lung cancer.

Excessive alcohol consumption has been associated with increased risk of certain forms of cancer. For instance, a connection between alcohol consumption and increased risk of breast cancer has been established. Likewise, alcohol consumption is associated with development of cancers of the gastrointestinal tract. Given that the liver is a target for alcohol-induced damage, it is not surprising that chronic alcohol consumption is associated with an elevated risk for primary liver cancer. However, it is important to note that heavy sustained alcohol consumption is associated with risk of liver cancer, whereas moderate consumption of alcohol is not. Alcohol is not directly carcinogenic to the liver; rather it is thought that the chronic liver damage produced by sustained alcohol consumption (hepatitis and cirrhosis) may contribute secondarily to liver cancer formation. For some other major cancer sites (such as lung), the role of alcohol consumption as a cofactor in cancer development is not clear.

Most/all skin cancer is related to unprotected exposure to strong sunlight. All of the major forms of skin cancer (basal cell carcinoma, squamous cell carcinoma, and malignant melanoma) have been linked to sunlight exposure. The carcinogenic agent in sunlight that accounts for the neoplastic transformation of skin cells is ultraviolet (UV) radiation. Basal cell carcinoma is a malignant neoplasm of the basal cells of the epidermis that occurs predominantly in areas of sun-damaged skin. Thus, sun bathing and sun tanning using artificial UV light sources represent significant lifestyle risk factors for development of these cancers. Basal cell carcinoma is now diagnosed in some people at very young ages (second or third decade of life), reflecting increased exposures to UV irradiation early in life. Some researchers have suggested that the increasing frequencies of skin cancer can be partially attributed to depletion of the ozone layer of the earth's atmosphere, which filters out (thereby reducing) some of the UV light produced by the sun. Squamous cell carcinoma is a malignant neoplasm of the keratinizing cells of the epidermis. As with basal cell carcinoma, extensive exposure to UV irradiation is the most important risk factor for development of this cancer. Likewise, development of malignant melanoma occurs most frequently in fair-skinned individuals and is associated to some extent with exposure to UV irradiation. This accounts for the observation that Caucasians develop malignant melanoma at a much higher rate than individuals of other races and ethnicity.

## Classification of neoplastic diseases

The word "neoplasia" is derived from the Greek words meaning "*condition of new growth.*" The term "tumor" is commonly used to refer to a neoplasm. Tumor literally means "*a swelling.*" In the early 1950s, R.A. Willis provided a description of neoplasm that we still utilize today: "... *A neoplasm is an abnormal mass of tissue the growth of which exceeds and is uncoordinated with that of the normal tissues and persists in the same manner after the cessation of the stimuli which evoked the change...*" In similar fashion, Kinzler and Vogelstein describe tumors to be the result of a disease process in which a single cell acquires the ability to proliferate abnormally, resulting in an accumulation of progeny cells. Furthermore, Kinzler and Vogelstein define cancers to represent tumors that have acquired the ability to invade the surrounding normal tissues. This definition highlights one of the most important distinguishing factors in the overall classification of neoplasms – the distinction between benign and malignant tumors. The division of neoplastic diseases into benign and malignant categories is extremely important, both for understanding the biology of these neoplasms and for recognizing the potential clinical challenges for treatment. At the most basic level, neoplasms are classified as benign or malignant. Further sub-classification of malignant neoplasms draws distinctions to (1) cancers of childhood versus cancers that primarily affect adults, (2) solid cancers versus hematopoietic neoplasms, and (3) hereditary cancers versus sporadic neoplasms.

Development of neoplastic disease is a multistep process through which cells acquire increasingly abnormal proliferative and invasive behaviors. Neoplasia also represents a unique form of genetic disease, characterized by the accumulation of multiple somatic mutations in a population of cells undergoing neoplastic transformation. Genetic and epigenetic lesions represent integral parts of the processes of neoplastic transformation, tumorigenesis, and cancer progression. Several forms of molecular alteration have been described in human cancers, including gene amplifications, deletions, insertions, rearrangements, and point mutations. In many cases, specific genetic lesions have been identified that are associated with neoplastic transformation and/or cancer progression in a particular tissue or cell type. Aberrant epigenetic alterations (epimutations) in neoplastic disease include genome-wide hypomethylation of DNA (possibly resulting in induction of oncogene expression), gene-specific hypermethylation events (resulting in silencing of tumor suppressor genes), other changes in chromatin packaging, and aberrant posttranscriptional regulation of gene expression (related to abnormal microRNA expression). Statistical analyses of age-specific mortality rates for different forms of human cancer predict that multiple mutations or epimutations in specific target genes are required for the genesis and outgrowth of most clinically diagnosable cancers. In accordance with this prediction, it has been suggested that cancers grow through a process of clonal expansion driven by mutation or epimutations, where the first mutation/epimutation leads to limited expansion of progeny of a single cell, and each subsequent mutation/epimutation gives rise to a new clonal outgrowth with greater proliferative potential. The idea that carcinogenesis is a multistep process is supported by morphologic observations of the transitions between premalignant (benign) cell growth and malignant tumors. In colorectal cancer (and some other tumor systems), the transition from benign lesion to malignant neoplasm can be easily documented and occurs in discernible stages, including benign adenoma, carcinoma *in situ*, invasive carcinoma, and eventually local and distant metastasis (Fig. 4.4). Moreover, specific genetic alterations have been shown to correlate with each of these well-defined histopathologic stages of cancer development and progression. However, it is important to recognize that it is the accumulation of multiple genetic alterations in affected cells, and not necessarily the order in which these changes accumulate, which determines cancer formation and progression.

Both benign and malignant neoplasms are composed of (1) neoplastic cells that form the parenchyma, and (2) the host-derived non-neoplastic stroma that is composed of connective tissue, blood vessels, and other cells and that

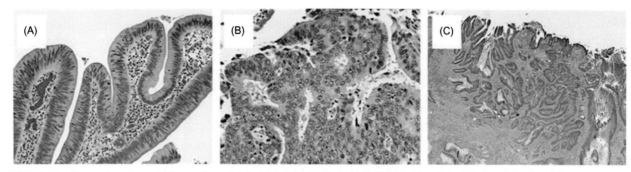

FIGURE 4.4 Progression of neoplastic transformation in the colon. (A) Low-grade glandular dysplasia. This image is of an adenomatous colon polyp, which is, by definition, polypoid low-grade dysplasia of the colonic glandular epithelium. The nuclei are hyperchromatic (dark) with pseudostratification (overlapping) and a higher nuclei to cytoplasmic ratio (the nucleus is occupying a larger percentage of cell volume than in a normal colon epithelial cell). (B) High-grade glandular dysplasia. Sometimes adenomatous polyps can harbor high-grade dysplasia. Features of high-grade dysplasia include both increased architectural complexity, including back-to-back glands without intervening stroma, and increased degree of nuclear pleomorphism. (C) Invasive adenocarcinoma. Glands, lined by dysplastic cells, are haphazardly invading through the muscularis mucosa and into the underlying submucosa.

supports the tumor parenchyma. The tumor stroma serves a critical function in support of the growth of the neoplasm by providing a blood supply for oxygen and nutrients. In nearly all cases, the parenchymal cells determine the biologic behavior (and clinical course) of the neoplasm. Furthermore, the parenchymal cell type of the neoplasm determines how the lesion is named.

## Benign neoplasms

The classification of neoplasms into benign and malignant categories is based on a judgment of the potential clinical behavior of the tumor, derived primarily from observations of the cellular features of the neoplasm, the growth pattern of the tumor, and various clinical findings. Benign neoplasms are characterized by features that suggest a lack of aggressiveness. The most important characteristic of benign neoplasms is the absence of local invasiveness (Fig. 4.5). Thus, while benign neoplasms grow and expand, they do not invade locally or spread to secondary tissue sites (remain localized), and are often amenable to surgical removal. However, it is important to note that benign neoplasms can cause adverse effects in the patient. Problems associated with benign neoplasms depend on (1) the size of the tumor, (2) the location of the tumor, and (3) secondary consequences related to presence of the neoplasm. Many benign neoplasms attain large size and impinge on important structures (such as nerves or blood vessels), resulting in various types of local effects. Consider a few examples. Many/most brain tumors are considered benign by virtue of the fact that they do not invade locally or produce distant metastases. However, as expanding space-filling lesions, these neoplasms can cause severe effects on the host due to the application of pressure to nearby aspects of the brain or brainstem. For instance, a benign meningioma can cause cardiac and respiratory arrest by compressing the medulla. Likewise, hemangiomas represent a benign neoplastic lesion of blood vessels, which creates a blood-filled cavity. Some hemangiomas (such as those affecting the liver) can become large and frequently impinge on the capsule of the organ. Lesions of this sort are subject to rupture, producing life-threatening bleeding in the patient. As these examples illustrate, despite a lack of invasive behavior, benign neoplasms can produce adverse effects in the patient and can have life-threatening consequences depending on their size and location.

Benign neoplasms are named by attaching the suffix *-oma* to the cell type from which the tumor originates. Thus, a benign neoplasm of fibrous tissue is termed a fibroma, a benign neoplasm of cartilaginous tissue is termed a chondroma, a benign neoplasm of osteoid tissue is termed an osteoma, a benign neoplasm arising from lipocytes is termed a lipoma, a benign neoplasm arising from blood vessels is termed a hemangioma, and a benign neoplasm arising from smooth muscle cells is termed a leiomyoma. The nomenclature for benign epithelial tumors is more complex. These neoplasms are classified either on the basis of their microscopic or macroscopic pattern, or according to their cells of origin. Thus, a benign epithelial neoplasm producing glandular patterns or a tumor arising from glandular cells is termed an adenoma, a benign epithelial neoplasm growing on any surface that produces microscopic or macroscopic fingerlike fronds is termed a papilloma, a benign epithelial neoplasm that projects above a mucosal surface to produce a macroscopically visible structure is termed a polyp, and cystadenoma refers to hollow cystic masses.

## Malignant neoplasms

Malignant neoplasms are collectively known as cancers. Malignant neoplasms display aggressive characteristics, can invade and destroy adjacent tissues, and metastasize to distant sites (Fig. 4.6). Adverse effects associated with malignant

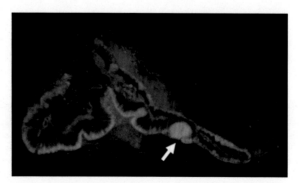

FIGURE 4.5   Adrenal adenoma. The normal cortex of the adrenal gland is yellow-gold in appearance. The *white arrow* points to a well-delineated, round lesion arising in the adrenal cortex. It does not appear to invade into the adjacent tissue.

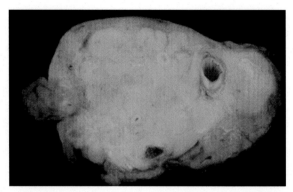

**FIGURE 4.6** Pancreatic adenocarcinoma. This is a cut section of a pancreatic tumor. It is poorly delineated (meaning it is difficult to identify the exact borders of the lesion) and appears to be infiltrating into adjacent adipose tissue (patchy bright yellow areas along the bottom of the specimen).

neoplasms are generally associated with tumor burden on the host once the cancer has spread throughout the body. Specific adverse effects can arise from (1) the size and location of the primary cancer, (2) consequences of local invasion and spread from the primary site, and (3) consequences associated with cancer colonization of tissue sites distant to the primary tumor. Most commonly, the cause of death associated with malignant neoplasms can be attributed to the metastatic spread of the cancer. Common sites of metastasis for malignant epithelial neoplasms include the lungs, liver, bone, and brain. The lung is the most common site for cancer metastasis which may be accomplished by hematogenous spread (through the blood), lymphatic spread, or by direct invasion. Likewise, cancers of the breast, lung, and colorectum are well known to spread to the liver, but cancers associated with any site in the body (including leukemias and lymphomas) can colonize the liver. Metastatic cancer found in the lungs or liver is characterized by the presence of multiple (often numerous) cancer nodules that can replace large percentages of the normal tissue and in liver can produce marked hepatomegaly. Adverse effects associated with lung metastasis include respiratory insufficiency and/or failure. Likewise, patient death related to liver metastasis results from various manifestations of liver insufficiency and/or failure.

The nomenclature for malignant neoplasms is very similar to that for benign neoplasms. Malignant neoplasms arising in mesenchymal tissues or its derivatives are called sarcomas. Sarcomas are designated by their histogenesis. Thus, a malignant neoplasm of fibrous tissue is termed a fibrosarcoma, a malignant neoplasm originating in cartilaginous tissue is termed a chondrosarcoma, a malignant neoplasm arising from lipocytes is termed a liposarcoma, a malignant neoplasm of osteoblasts is termed an osteosarcoma, a malignant neoplasm arising in blood vessels is termed an angiosarcoma, and a malignant neoplasm arising from smooth muscle cells is termed a leiomyosarcoma. Malignant epithelial neoplasms are called carcinomas. Carcinomas can be subclassified as adenocarcinoma and squamous cell carcinoma. Adenocarcinoma describes a malignant neoplasm in which the neoplastic cells grow in a glandular pattern. Squamous cell carcinoma describes a malignant neoplasm with a microscopic pattern that resembles stratified squamous epithelium. In each case, the nomenclature for a given tumor will specify the organ system of origin for the neoplasm (for instance, colonic adenocarcinoma or squamous cell carcinoma of the skin). In some cases, malignant neoplasms grow in an undifferentiated pattern that is inconsistent with a classification of adenocarcinoma or squamous cell carcinoma. In these cases, the neoplasm is termed a poorly differentiated carcinoma.

## Mixed cell neoplasms

Some neoplastic cells undergo divergent differentiation during tumor formation, giving rise to tumors of mixed cell type. Examples of mixed cell tumors include that of salivary gland origin and breast fibroadenoma. These neoplasms are composed of epithelial components dispersed throughout a fibromyxoid stroma that may contain cartilage or bone. In the case of mixed tumor of salivary gland origin, all of the cellular components of the neoplasm are believed to derive from epithelial and myoepithelial cells of the salivary glands. Therefore, the appropriate designation for these neoplasms is pleomorphic adenoma to reflect the differences in cellular differentiation among cells of common origin. Breast fibroadenoma is a benign neoplasm that contains a mixture of proliferating ductal elements (adenoma) contained in a loose fibrous tissue (fibroma). In contrast to most mixed cell neoplasms where all of the cell types observed in the lesion are neoplastic, in breast fibroadenoma only the fibrous component is thought to be neoplastic and the proliferating ductal elements are normal.

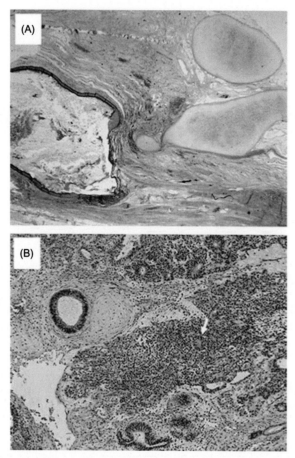

**FIGURE 4.7** Neoplasms of mixed cell type. (A) Mature teratoma. This example demonstrates mature cartilage (right) and skin (left). (B) Immature teratoma. This type of teratoma contains areas of primitive-appearing hyperchromatic cells, known as blastema. There is focal rosette formation (*white arrow*).

In contrast to mixed cell tumors where all of the cellular components of the neoplasm are believed to derive from the same germ layer, teratomas contain recognizable mature or immature cells or tissues representative of more than one germ cell layer, and sometimes all three. Teratomas typically originate from totipotential cells such as those found in the ovary and testis. In rare instances, teratomas originate from sequestered midline embryonic rests. The totipotent cells that give rise to teratomas have the capacity to differentiate into any cell type and can give rise to all of the tissues found in the adult body. It follows that teratomas are commonly composed of various tissue elements that can be recognized, including skin, hair, bone, cartilage, tooth structures, and others. Teratomas are further classified as benign (also referred to as mature) or malignant (also referred to as immature) (Fig. 4.7). As the designation implies, mature teratomas contain well-formed tissue elements that appear normal but arise in the abnormal context of the neoplasm. In contrast, immature teratomas are composed of abundant poorly differentiated (primitive) blast-like cells (blastema).

## Confusing terminology in cancer nomenclature

Some malignant neoplasms are conventionally referred to using terms that are suggestive of benign neoplasms based on the usual nomenclature for naming tumors. For example, lymphoma is a malignant neoplasm of lymphoid tissue, mesothelioma is a malignant neoplasm of the mesothelium, melanoma is a malignant neoplasm arising from melanocytes, and seminoma is a malignant neoplasm of the testicular epithelium. In each of these examples, the name given implies a benign neoplasm even though all of these are malignant neoplasms. Likewise, hepatocellular carcinomas are often called hepatomas. Unfortunately, these tumor designations are well established in medical terminology and are unlikely to be corrected. In addition to these tumor misnomers, several other examples of confusing terminology exist. For instance, hamartoma refers to a developmental malformation that presents as a mass lesion of disorganized tissue that is indigenous to that particular tissue site. Hamartomas are not neoplastic, even though the name implies a neoplasm.

These malformations can be found in any tissue, but a common site for these lesions is the lung. Hamartomas of the lung generally contain islands of cartilage, along with connective tissue and various types of cells.

## Pre-neoplastic lesions

Neoplastic disease develops in patients over long periods of time and typically is preceded by development of one or more pre-neoplastic lesions. Well-characterized pre-neoplastic lesions include (1) metaplasia, (2) hyperplasia, and (3) dysplasia.

Metaplasia represents a reversible change in tissues characterized by substitution of one adult cell type (epithelial or mesenchymal) by another adult cell type (Fig. 4.8). Metaplasia is a reactive condition, reflecting an adaptive replacement of cells that are sensitive to stress by cells that are resistant to the adverse conditions encountered by the tissue. Metaplasia is a well-recognized precursor for development of malignant neoplasms at various tissue sites. Examples of metaplasia include the adaptive changes that occur in the lungs of individuals who smoke cigarettes and the adaptive changes that occur in the esophagus of individuals with reflux disease. In cigarette smokers, columnar to squamous epithelial metaplasia occurs in the respiratory tract in response to chronic irritation caused by inhalation of cigarette smoke. The ciliated columnar epithelial cells of the normal trachea and bronchi become replaced by stratified squamous epithelial cells in a pattern that may be focal or more widely distributed. Squamous metaplasia in the respiratory tract is accompanied by loss of function secondary to loss of the ciliated epithelial cells. In addition, development of squamous cell carcinoma of the lung may originate in focal areas of squamous metaplasia. Barrett esophagus is an example of squamous to columnar metaplasia in response to refluxed gastric acid. The adaptive change is from stratified squamous epithelial cells to intestinal-like columnar epithelial cells, which are resistant to the effects of the gastric acid. Barrett esophagus is frequently the site of development of esophageal adenocarcinomas.

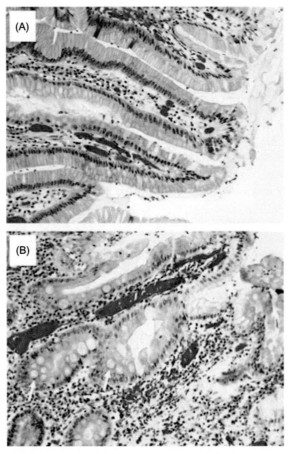

FIGURE 4.8 Metaplastic change in the stomach. (A) Normal glandular epithelium of the stomach. (B) Intestinal metaplasia. This image displays glandular epithelium of the stomach with goblet cells. Goblet cells contain a large, round intracytoplasmic vacuole of mucin and are normally found in the small and large intestine only.

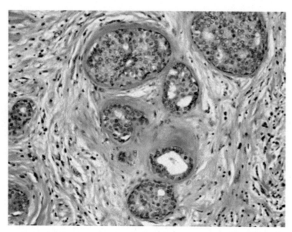

**FIGURE 4.9**   Epithelial hyperplasia of the breast. Normal mammary ducts are lined by a single layer of cuboidal epithelial cells (as in the duct second from the center bottom). The lumens of the remaining ducts in this image are filled with bland appearing-epithelium cells, representing prominent ductal epithelial hyperplasia.

Hyperplasia reflects an increase in the number of cells in an organ or tissue, typically resulting in an increased volume (or size) of the affected organ or tissue (Fig. 4.9). Hyperplasia can be physiologic or pathologic. Physiologic hyperplasia can be classified as (1) hormonal hyperplasia or (2) compensatory hyperplasia. Most forms of pathologic hyperplasia result from excessive (abnormal) hormonal or growth factor stimulation. Benign prostatic hyperplasia (BPH) is a commonly occurring condition among older men, which is typical of pathologic hyperplasia. In BPH, abnormal stimulation of the prostate tissue by androgen hormones results in a benign proliferation resulting in hypertrophy of the prostate gland. The abnormal cell proliferation that occurs in hyperplasia is controlled to the extent that on cessation of the stimulus (elimination of growth factor or hormone) cell proliferation will halt and the hyperplastic tissue will regress. Hyperplasia can precede neoplastic transformation, and many hyperplastic conditions are associated with elevated risk for development of cancer. For example, patients with endometrial hyperplasia (also referred to as endometrial intraepithelial neoplasia) are at increased risk for endometrial cancer.

Dysplasia is a proliferative lesion that is characterized by a loss in the uniformity of individual cells in a tissue and loss in the architectural orientation of the cells in a tissue. Thus, dysplasia can be simply described as a condition of disorderly but non-neoplastic cellular proliferation. Dysplastic cells show many alterations that are suggestive of their preneoplastic character, including cellular pleomorphism, hyperchromatic nuclei, high nuclear-to-cytoplasmic ratio, and increased numbers of mitotic figures (Fig. 4.10). In some cases, mitotic cells appear in abnormal locations within the tissue. For instance, mitotic figures may appear outside of the basal layers of dysplastic stratified squamous epithelium. Lesions characterized by extensive dysplastic changes involving the entire thickness of the epithelium but remaining confined within the normal tissue are classified as carcinoma *in situ* (Fig. 4.11). In many cases, dysplasia and/or carcinoma *in situ* are considered immediate precursors of invasive cancers. It follows that dysplastic changes are often found adjacent to invasive cancers. In the case of heavy cigarette smokers or individuals with Barrett esophagus, development of dysplasia often portends progression to invasive cancer. However, not all dysplastic lesions will develop into a malignant neoplasm.

## Cancers of childhood

Cancer is primarily a disease of adults and occurs much more frequently in older individuals. However, malignant neoplasms do occur in children. These neoplasms are rare but not uncommon. In 2016, an estimated 10,380 new cases of cancer (excluding benign/borderline brain tumors) and 1250 cancer-associated deaths occurred in children under 15 years old. The major childhood cancers include leukemia (30% of all childhood cancers), brain tumors and other neoplasms of the nervous system (26% of all childhood cancers), neuroblastoma (6% of all childhood cancer), Wilms' tumor (5% of all childhood cancer), non-Hodgkin lymphoma (6% of all childhood cancer), rhabdomyosarcoma (3.5% of all childhood cancer), retinoblastoma (3% of all childhood cancer), osteosarcoma (3% of all childhood cancer), and Ewing sarcoma (1.5% of all childhood cancer). ALL is the most common childhood leukemia, representing >80% of all childhood leukemias. The incidence of ALL peaks in children 2−5 years old. In contrast, the incidence of neuroblastoma and retinoblastoma peaks earlier, in children 1−2 years old. Neuroblastoma is a malignant neoplasm of the

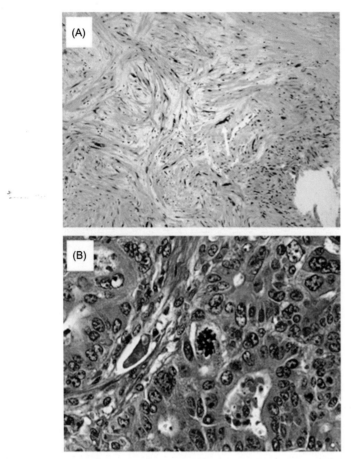

**FIGURE 4.10** Nuclear pleomorphism and abnormal mitotic figures in neoplastic cells. (A) Pleomorphism. This is an example of leiomyosarcoma. Several of the malignant stromal cells are very large and different in shape from neighboring cells. (B) Mitotic figure (center). Malignant neoplasms often have an increased number of mitotic figures.

sympathetic nervous system and is the most frequently occurring neoplasm among infants, with peak incidence among children <1 year old (Fig. 4.12). In fact, 40% of neuroblastomas are diagnosed in the first 3 months of life. Retinoblastoma is a tumor that originates in the retina of the eye. This neoplasm affects children as well as adults. Retinoblastoma occurring in children is typically associated with a genetic mechanism involving mutation of the *Rb1* gene. Wilms' tumor tends to occur in children <10 years old, with greatest incidence in children <5 years old. Wilms' tumor (also known as nephroblastoma) is the most commonly occurring pediatric kidney cancer. Astrocytomas represent the most frequently occurring brain tumor of children (52% of all childhood brain tumors), with ependymoma (9%), primitive neuroectodermal tumors (21%), and other gliomas (15%) representing most of the balance. Ependymomas and primitive neuroectodermal tumors occur most often in younger children (<5 years old), whereas astrocytomas are diagnosed with approximately the same frequency in children between birth and 15 years old. Rhabdomyosarcoma is the most common soft tissue sarcoma of children, typically occurring in children <10 years old. Osteosarcoma and Ewing sarcoma are the most commonly occurring bone tumors among children. Osteosarcomas derive from primitive bone-forming mesenchymal stem cells and most often occur near the metaphyseal portions of the long bones. There is a bimodal age distribution of osteosarcoma incidence, with peaks in early adolescence and in adults >65 years old. Ewing sarcoma is believed to be of neural crest origin and occur roughly evenly between the extremities and the central axis. Like osteosarcoma, Ewing sarcoma is a disease primarily of childhood and young adults, with highest incidence among children in their teenage years.

## Hematopoietic neoplasms

The majority of human cancers are classified as solid tumors (grouped as carcinomas or sarcomas). The exceptions to this classification include the malignant neoplasms of hematopoietic origin, including lymphoma, myeloma, and

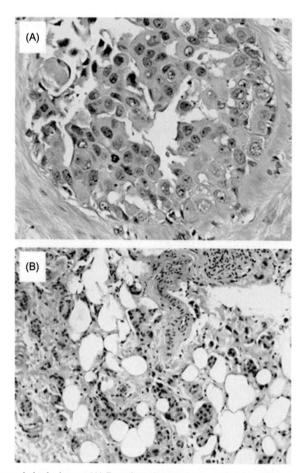

FIGURE 4.11  Progression of tumorigenesis in the breast. (A) Ductal carcinoma *in situ* of the breast. The lumen of this mammary duct is filled with pleomorphic cells with high nuclear-to-cytoplasmic ratios. Scattered mitotic figures are noted. This neoplastic process is *in situ* carcinoma because the dysplastic cells are confined to the lumen of the duct and have not invaded into the surrounding tissue. (B) Invasive adenocarcinoma of the breast. Glands lined by dysplastic cells haphazardly infiltrate into adipose tissue.

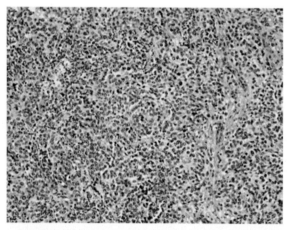

FIGURE 4.12  Childhood neuroblastoma. Neuroblastoma is considered one of the so-called small, round, blue cell tumors, as it is composed of small, hyperchromatic, monomorphic cells with scant cytoplasm.

leukemia (Fig. 4.13). In 2018, 174,250 new malignant neoplasms of hematopoietic origin were diagnosed, representing approximately 10.0% of all new cancers. These cancers include 83,180 cases of lymphoma, 30,770 cases of myeloma, and 60,300 cases of leukemia. Non-Hodgkin lymphoma occurs much more frequently than Hodgkin lymphoma and represents 89.5% of all lymphomas. The four major forms of leukemia (ALL, chronic lymphocytic leukemia, acute

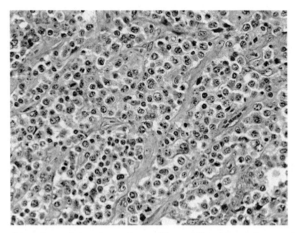

FIGURE 4.13   Lymphoma. This image of lymphoma demonstrates sheets of large cells with prominent nucleoli. Subclassification of the lymphoma depends on several additional laboratory tests.

myeloid leukemia, and chronic myeloid leukemia) account for 89% of all leukemias. The two most prevalent forms of leukemia are chronic lymphocytic leukemia and acute myeloid leukemia, which combine to account for 65% of all cases of leukemia.

## Hereditary cancers

A number of familial cancer syndromes and hereditary cancers have been recognized and characterized. Several rare genetic disorders involving dysfunctional DNA repair pathways are associated with elevated risk for cancer development. These disorders include xeroderma pigmentosum (XP), ataxia telangiectasia (AT), Bloom syndrome, and Fanconi anemia. Individuals affected by these conditions are prone to development of various malignancies when exposed to specific DNA-damaging agents. Patients with XP (MIM #278700, XPA; MIM #278730, XPD; MIM #278760, XPF; MIM #278780, XPG; and others) display hypersensitivity to UV light and increased incidence of several types of skin cancer, including basal cell carcinoma, squamous cell carcinoma, and malignant melanoma. XP represents a spectrum of diseases with differing genetic defects. XP patients are classified into one of seven complementation groups, which differ according to nucleotide excision repair pathway gene mutation. Patients with AT (MIM #208900) exhibit hypersensitivity to ionizing radiation and chemical agents and are predisposed to the development of B-cell lymphoma and chronic lymphocytic leukemias, and affected women demonstrate an increased risk of developing breast cancer. The genetic defect associated with AT is mutation of the *ATM* gene, which encodes a member of the phosphoinositide 3-kinase—related kinase (PIKK) family. In patients with mutated *ATM*, DNA repair is not properly activated after DNA-damaging events. Patients with Fanconi anemia (MIM #227650, FANCA; and others) demonstrate sensitivity to DNA cross-linking agents and are predisposed to malignancies of the hematopoietic system, particularly acute myelogenous leukemia. Sixteen different genes are known to contribute to Fanconi anemia, with *FANCA* being the most frequently mutated. FANCA and other Fanconi anemia genes encode proteins of a multiprotein core complex that is necessary for activation of DNA cross-link repair. Patients with Bloom syndrome (MIM #210900) demonstrate an increased incidence of several forms of cancer, including leukemia, skin cancer, and breast cancer. Bloom syndrome is associated with mutation of the *BLM* gene, which encodes DNA helicase RecQ protein-like-3 (RECQL3). These patients exhibit chromosomal instability manifested as abnormally high levels of sister chromatid exchange.

Colorectal cancer is a fairly common disease worldwide, and particularly in populations from Western nations. A substantial fraction of colorectal cancers exhibit a genetic component, and several familial colorectal cancer syndromes are recognized, including familial adenomatous polyposis (FAP; MIM #175100) and hereditary non-polyposis colon cancer (HNPCC; MIM #120436, Lynch syndrome I and Lynch syndrome II). Patients with Lynch syndrome type I exhibit colorectal cancers only, whereas patients with Lynch syndrome type II exhibit colorectal cancers as well as gynecologic cancers. Genes associated with each of these conditions have been identified and characterized. Of these familial colorectal cancer syndromes, HNPCC has been determined to be related to defective DNA repair. HNPCC is characterized by the occurrence of predominantly right-sided colorectal carcinoma with an early age of onset and an increased risk for the development of certain extra-colonic cancers, including cancers of the endometrium, stomach, urinary tract, and breast. Cancers associated with HNPCC exhibit a unique form of genomic instability, which represents a unique mechanism for a genome-wide tendency for instability in short repeat sequences (microsatellites), which was originally termed the replication error phenotype. The molecular defect responsible for microsatellite instability in HNPCC involves the genes that encode proteins required for normal mismatch repair.

Two addition familial cancer syndromes have been described that exhibit clinical features similar to that of HNPCC or FAP. The Muir-Torre syndrome (MIM #158320) is defined by the development of at least one sebaceous gland tumor and a minimum of one internal tumor, which is frequently colorectal carcinoma. This syndrome shares several features with HNPCC syndromes Lynch type I and Lynch type II, including the occurrence of microsatellite instability in a subset of cancers. This observation suggests the possible involvement of abnormal mismatch repair mechanisms in the genesis of a subset of Muir-Torre syndrome cancers. Turcot syndrome (MIM #276300) is defined by the occurrence of a primary brain tumor and multiple colorectal adenomas. The molecular basis for this syndrome has been suggested to involve mutation of the *APC* gene or mutation of a mismatch repair gene in tumors exhibiting microsatellite instability.

The Li-Fraumeni syndrome (MIM #151623) was initially characterized among several kindreds with excess cancer incidence. Patients with Li-Fraumeni syndrome develop various types of neoplasms, including breast cancer, soft tissue sarcomas and osteosarcomas, brain tumors, leukemias, and several others. Cancer susceptibility among individuals with Li-Fraumeni syndrome follows an autosomal dominant pattern of inheritance and is highly penetrant (90% by age 70), but many neoplasms develop early in life. It is now known that Li-Fraumeni syndrome is associated with germ-line mutations in the *p53* tumor suppressor gene.

It is well known that approximately 5%−10% of breast cancers are related to genetic predisposition. These breast cancers are typically associated with a strong family history of breast cancer development. A substantial amount of research has led to the discovery of several breast cancer susceptibility genes, including *BRCA1*, *BRCA2*, and *p53*, which may account for the majority of inherited breast cancers. In some cases, families at increased susceptibility for breast cancer also show elevated rates of ovarian cancer. Patients who are affected by familial breast and ovarian cancer syndrome tend to have germ-line mutation of *BRCA1*.

Familial melanoma is associated with (1) a family history of melanoma, (2) the presence of large numbers of common or atypical nevi, (3) a history of primary melanoma or other (non-melanoma) skin cancers, (4) immunosuppression, (5) susceptibility to sunburn, or (6) a history of blistering sunburn. Given the linkage between excess exposure to sunlight and development of skin cancers (including melanoma), it is not surprising that susceptibility to sunburn would confer an increased risk for melanoma. This susceptibility is particularly pronounced in individuals with fair complexion characterized by freckling, blue eyes, red hair, and skin that burns readily in response to sunlight and fails to tan. Two highly penetrant melanoma susceptibility genes have been identified: *CDKN2A* (which encodes cyclin-dependent kinase inhibitor 2A) and *CDK4* (which encodes cyclin-dependent kinase 4). *CDKN2A* is found on chromosome 9p21 and *CDK4* resides on 12q13. Germ-line inactivating mutations of the *CDKN2A* gene are the most common cause of inherited susceptibility to melanoma, whereas mutations of *CDK4* occur much more rarely. Nevertheless, germ-line mutations of *CDKN2A* are rare, and many account for a very small proportion of melanoma susceptibility among the general population. The *CDKN2A* gene encodes two important cell cycle regulatory proteins: $p16^{INK4A}$ and $p14^{ARF}$. Although other melanoma susceptibility loci have been mapped through genome-wide linkage analysis, the gene targets that contribute to familial melanoma predisposition remain undiscovered for a large proportion of recognized kindreds. Ongoing research is focused on the identification of low-penetrance melanoma susceptibility genes that confer a lower melanoma risk with more frequent variations. For instance, specific variants of the *MC1R* and the *OCA2* genes have been demonstrated to confer an increase in melanoma risk.

Familial cancers affecting a number of other tissues and organs have been described or suggested. Familial cancer of the pancreas has been reported and suggested to follow an autosomal dominant inheritance pattern. A familial form of gastric cancer has been suggested to represent approximately 10% of all gastric cancers and is associated with the *CDH1* gene on chromosome 16q22.1. A genome-wide scan of 66 high-risk prostate cancer families produced evidence of disease linkage to chromosome 1q24-25. Subsequently, the chromosome 1q24−25 susceptibility locus was connected to the *HPC1* gene.

## Characteristics of benign and malignant neoplasms

The distinction between benign and malignant neoplasms is based on observations related to cellular features of the cells composing the lesion, growth pattern of the neoplasm, and various clinical findings. Four fundamental features are particularly important: (1) cellular differentiation and anaplasia, (2) rate of growth, (3) presence of local invasion, and (4) metastasis.

## Cellular differentiation and anaplasia

In the evaluation of neoplasms, the extent of cellular differentiation and anaplasia of the parenchymal cells that constitute the neoplastic elements of the lesion are assessed. The extent of cellular differentiation describes the degree to which the neoplastic cells resemble their normal counterparts based on morphology and function. Benign neoplasms are typically composed of well-differentiated cells that closely resemble normal cells. Hence, a lipoma is composed of mature fat cells that contain cytoplasmic lipid vacuoles. Likewise, a chondroma is composed of mature cartilage cells that synthesize their usual cartilaginous matrix. In well-differentiated benign neoplasms, cellular proliferation rates are low and mitoses are infrequent. However, when observed, mitotic figures appear normal in these neoplasms. In contrast, malignant neoplasms exhibit an extremely wide range of cell differentiation. Many malignant neoplasms are very well differentiated, but it is not uncommon for these tumors to lack differentiated features and/or to appear completely undifferentiated. Malignant neoplasms of intermediate phenotype are designated as moderately differentiated.

In general, there is a direct relationship between the degree of cellular differentiation and the functional capabilities of the cells that compose the neoplasm. Thus, benign neoplasms (as well as some well-differentiated malignant neoplasms) of the endocrine glands can elaborate hormones characteristic of their origin. Likewise, well-differentiated squamous cell carcinomas (at various tissue sites) elaborate keratin (giving rise to histologically recognizable keratin pearls), and well-differentiated hepatocellular carcinomas synthesize bile salts. In contrast, many malignant neoplasms express genes and produce proteins or hormones that would not be expected from the cells of origin of the cancer. Some cancers synthesize fetal proteins that are not expressed by comparable cell types in adults. For instance, many hepatocellular carcinomas express $\alpha$-fetoprotein, which is not expressed in adult hepatocytes. Furthermore, malignant neoplasms of non-endocrine origin can excrete ectopic hormones producing various paraneoplastic syndromes. Certain lung cancers produce antidiuretic hormone (inducing hyponatremia in the patient), adrenocorticotropic hormone (resulting in Cushing syndrome), parathyroid-like hormone or calcitonin (both of which are implicated in hypercalcemia), gonadotropins (causing gynecomastia), serotonin and bradykinin (associated with carcinoid syndrome), or others.

Malignant neoplasms that are composed of undifferentiated cells are said to be anaplastic. Lack of cellular differentiation (or anaplasia) is considered a hallmark of cancer. The term anaplasia means "to form backward," which implies dedifferentiation (or loss of the structural and functional differentiation) of normal cells during tumorigenesis. However, it is now well recognized that many cancers originate from stem cells in tissues. Thus, the lack of cellular differentiation exhibited by these neoplasms results from a failure to differentiate rather than through a process of dedifferentiation of highly differentiated (specialized) cells. Other mechanisms that might account for loss of differentiation in malignant neoplasms include epithelial to mesenchymal transition, mesenchymal to epithelial transition, and transdifferentiation between cell differentiation states related to cellular plasticity. In general, malignant neoplasms that are composed of anaplastic cells (which are also typically rapidly growing) are unlikely to have specialized functional activities. Anaplastic cells tend to exhibit marked nuclear and cellular pleomorphism (extreme variation in nuclear or cell size and shape) (Fig. 4.14). The nuclei observed in anaplastic cells are typically hyperchromatic (darkly staining) and large, resulting in altered nuclear-to-cytoplasmic ratios (which may approach 1:1 instead of 1:4 or 1:6 as observed in normal cells). Very often malignant neoplasms contain giant cells (relative to the size of neighboring cells in the neoplasm), and these cells will contain an abnormally large nucleus or will be multinucleated. The nuclear pleomorphism observed in anaplastic cells of malignant neoplasms is characterized by nuclei that are highly variable in size and shape. These nuclei exhibit chromatin that appears coarse and clumped. Furthermore, the nucleoli may be very large relative to that observed in the nuclei of normal cells, possibly reflecting the extent of transcriptional activities taking place in these highly active cells. Given that malignant neoplasms often exhibit high rates of cell proliferation, numerous mitotic figures may be seen, and these mitotic figures are often abnormal. Typically, anaplastic cells will fail to organize into recognizable tissue patterns. This lack of cellular orientation reflects loss of normal cellular polarity, as well as a failure of normal structures to form.

## Rate of growth

In general, benign neoplasms grow more slowly than malignant neoplasms. However, there are exceptions to this rule, and in some cases benign neoplasms will display an elevated growth rate. For instance, some benign neoplasms exhibit changes in growth rate in response to hormonal stimulation or in response to alterations in blood supply. Leiomyomas are benign neoplasms of the uterus that originate in smooth muscle and are significantly influenced by circulating levels of estrogens. Thus, these neoplasms may display elevated cell proliferation and increase rapidly in size in response to the hormonal changes seen in pregnancy. Once hormonal levels normalize (after childbirth), the lack of sufficient

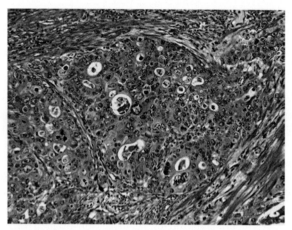

FIGURE 4.14 Cellular pleomorphism. The dysplastic cells in this adenocarcinoma demonstrate marked pleomorphism, meaning the cells vary widely in size and shape. The arrangement of the dysplastic cells is disordered, with only some residual gland formation.

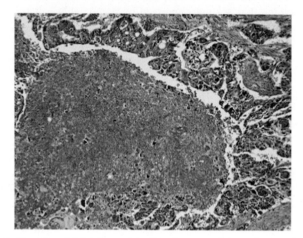

FIGURE 4.15 Tumor necrosis. Necrotic (dead tissue) debris is present within the center of the malignant glands in this adenocarcinoma.

hormone levels to sustain high rates of cellular proliferation results in greatly diminished neoplastic growth. Subsequently, with the elimination of hormones related to menopause, these tumors may become fibrocalcific. Despite variations in growth rate among neoplasms and some physiological exceptions (such as leiomyoma in pregnancy), most benign neoplasms proliferate slowly over time (months to years) and increase in size slowly.

The rate of cell proliferation (and lesion growth) of malignant neoplasms generally correlates with the extent of cellular differentiation of the cells that compose the cancer. Thus, poorly differentiated neoplasms exhibit high rates of cell proliferation and tumor growth, and well-differentiated neoplasms tend to grow more slowly. Nevertheless, there is considerable variability in the relative growth rate among malignant neoplasms. There is abundant evidence suggesting that most if not all malignant neoplasms take many years (and perhaps decades) to develop and emerge clinically. Many or most malignant neoplasms grow relatively slowly and at a constant rate. However, there may be several patterns of growth among malignant neoplasms. Some of these lesions grow slowly for long periods of time before entering into a phase characterized by more rapid expansion. In this case, the rapid expansion phase probably reflects the emergence of a more aggressive subclone of neoplastic cells. In certain rare instances, the proliferation of a malignant neoplasm will diminish to a level of undetectable cellular proliferation, or the neoplasm will regress spontaneously (due to widespread necrosis of the neoplastic cells). Most malignant neoplasms progressively enlarge over time in relation to their cellular growth rate. Rapidly growing malignant neoplasms tend to contain a central area of necrosis that develops secondary to ischemia related to inadequate blood supply (Fig. 4.15). Because the tumor blood supply is derived from normal tissues at the site of the neoplasm, the formation of new blood vessels to supply the expanding neoplasm may lag behind the proliferation of the neoplastic cells, resulting in inadequate supply of oxygen and other nutrients to the cancer mass.

## Presence of local invasion

Benign neoplasms, by definition, remain localized at their site of origin. These neoplasms do not have the capacity to infiltrate surrounding tissues, invade locally, or metastasize to distant sites. Benign neoplasms typically have smooth borders, are sharply demarcated from the normal tissue at the tumor site, and are frequently encapsulated by a fibrous capsule that forms a barrier between the neoplastic cells and the host tissue. However, not all benign neoplasms have a capsule, and the lack of a capsule around a neoplasm does not indicate that the neoplasm is malignant. The capsule of a benign neoplasm is primarily the product of the elaboration of tumor stroma. In addition, the tumor capsule may derive in part from the fibrous debris resulting from necrotic cell death of tissue cells adjacent to the neoplasm. Benign liver adenomas represent an example of a benign neoplasm that commonly exhibits a capsule. However, other benign neoplasms lack a well-developed capsule. Uterine leiomyomas do not infiltrate adjacent normal tissues and are typically discretely demarcated from the surrounding smooth muscle by a zone of compressed and attenuated normal myometrium, but these neoplasms do not elaborate a capsule. There are a few examples of benign tumors that are neither encapsulated nor discretely defined (for instance, some vascular benign neoplasms of the dermis).

The growth of malignant neoplasms is characterized by progressive infiltration, invasion, destruction, and penetration of surrounding normal (non-neoplastic) tissues (Fig. 4.6 and Fig. 4.11B). Malignant neoplasms do not form well-developed capsules. However, some slow-growing malignant neoplasms appear histologically to be encased in stroma resembling a capsule. The invasive (malignant) nature of these neoplasms is revealed by close microscopic examination, which reveals penetration of the margins of the stroma by neoplastic cells and invasion of adjacent tissue structures. The infiltrative or locally invasive nature of the growth of malignant neoplasms requires that broad margins of non-neoplastic tissue must be resected during surgical excision to ensure complete removal of all neoplastic cells. Local invasiveness and infiltration of adjacent tissue structures represent features that are strongly suggestive of a malignant neoplasm. These features represent the most reliable predictor of malignant behavior next to the development of distant metastasis.

## Metastasis

The term metastasis describes the development of secondary neoplastic lesions at distant tissue locations that are separated from the primary site of a malignant neoplasm. The clinical finding of metastasis provides a definitive classification of a primary neoplasm as malignant. A significant percentage of patients with newly diagnosed malignant neoplasms exhibit clinically evident metastases at the time of diagnosis ($\sim 30\%$), and others have occult metastases at the time of diagnosis ($\sim 20\%$). Some malignant neoplasms demonstrate a propensity for metastasis. Breast cancers tend to spread to bone, lung, liver, brain, and some other sites. The size of a primary breast cancer at the time of diagnosis is directly proportional to the probability for the development of distant metastasis over time. In general terms, malignant neoplasms that are large and anaplastic (less well differentiated) are more likely to metastasize. However, there are numerous exceptions; extremely small cancers may metastasize and have poor prognosis, whereas some large tumors may remain localized and not produce metastases. For some malignant neoplasms, initial diagnosis occurs at late stage after distant metastases have already developed. For example, osteogenic sarcomas typically metastasize to the lungs by the time the primary neoplasm is detected. However, not all malignant neoplasms exhibit a strong tendency for metastasis. For instance, basal cell carcinomas of the skin are highly invasive at the primary site of the tumor and only rarely form distant metastases. Likewise, many/most malignant neoplasms of the brain are highly invasive at their primary sites but rarely give rise to distant metastatic lesions.

Malignant neoplasms metastasize and spread to distant sites in the patient through (1) direct invasion or seeding within body cavities, (2) spread through the lymphatic system, or (3) hematogenous spread via the blood. Direct invasion or spreading by seeding occurs when a malignant neoplasm invades a natural body cavity. This pathway of tumor spread is characteristic of malignant neoplasms of the ovary. These cancers often widely disseminate across peritoneal surfaces, producing a significant tumor burden without invasion of the underlying parenchyma of the abdominal organs. In this example, the ovarian carcinoma cells demonstrate an ability to establish at new sites (previously uninvolved peritoneal surfaces) but lack the capacity to invade into new tissue sites. Pancreatic carcinoma tends to spread by direct invasion into the peritoneal cavity and can invade the stomach by direct extension. Likewise, lung carcinomas may invade into the plural cavity, directly invade the diaphragm, and eventually gain access to the peritoneal cavity after invasion through the diaphragm.

Malignant epithelial neoplasms (carcinomas) tend to metastasize through the lymphatic system (Fig. 4.16). Hence, lymph node involvement represents an important factor in the staging of many cancers. However, given the numerous

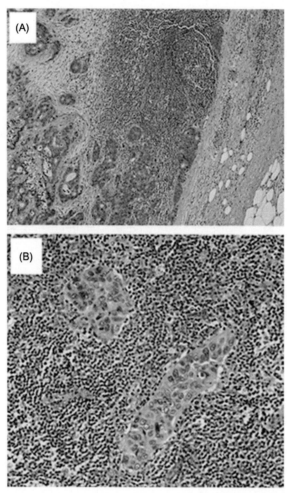

**FIGURE 4.16**   Metastatic spread of malignant cells through the lymphatic system. (A) Metastatic adenocarcinoma to a lymph node, low power. Residual lymph node tissue is present along the upper portion of the image. Infiltrating glands of adenocarcinoma are noted at the lower aspect of the image. (B) Metastatic carcinoma to a lymph node, high power. Two nests of plump malignant epithelial cells are present within the sea of smaller lymphocytes of the lymph node.

anatomic interconnections between the lymphatic and vascular systems, it is possible that some of these cancers may disseminate through either or both pathways. Although enlargement of lymph nodes near a primary neoplasm may arouse suspicion of lymph node involvement and metastatic spread of the primary malignant neoplasm, simple lymph node swelling does not accurately predict spread of neoplastic cells. Other processes can account for lymph node enlargement adjacent to malignant neoplasms. Reactive changes may occur when necrotic debris from a tumor encounters a lymph node. Patterns of lymph node seeding by metastatic cancer cells mainly depend on the site of the primary malignant neoplasm and the natural pathways of lymphatic drainage from that tissue site. Thus, lung cancers arising in the respiratory passages metastasize initially to regional bronchial lymph nodes and subsequently to the tracheobronchial and hilar nodes. In this case, the extent of spread can be inferred from which lymph nodes are positive for neoplastic cells. Breast cancer presents a slightly more complicated picture. Malignant neoplasms of the breast that arise in the upper outer quadrant of the breast will spread initially to axillary lymph nodes, whereas malignant tumors arising in the medial region of the breast may spread through the chest wall into the lymph nodes along the internal mammary artery. Seeding of the supraclavicular and infraclavicular lymph nodes by metastatic breast cancer cells can occur subsequent to the initial spread despite the initial route of tumor cell movement. In some cases, breast cancer cells appear to traverse lymphatic vessels without colonizing the lymph nodes that are immediately proximal to the site of the primary neoplasm but do implant in lymph nodes that are subsequently encountered (producing metastatic lesions that are referred to as skip metastases). Given the complexity and potentially complicated pattern of lymphatic spread by metastatic breast cancer, it is not surprising that sentinel node biopsy has become a useful procedure for surgical staging of this malignant neoplasm. A sentinel lymph node is defined as the first lymph node in a regional lymphatic basin that receives lymphatic drainage from the site of a primary neoplasm. It can be identified by injection of colored dyes (or radiolabeled tracers) into the site of the primary tumor with monitoring of the movement of the dye to the downstream lymph nodes. Histopathological evaluation of the sentinel lymph node provides a reliable indication of the extent of spread of a breast cancer and can be used to plan treatment.

Dissemination of a malignant neoplasm through the blood is referred to as hematogenous spread. Hematogenous spread of malignant mesenchymal neoplasms (sarcomas) occurs commonly, but this pathway is also readily utilized by malignant epithelial neoplasms (carcinomas). Invasion of the vascular system by neoplastic cells can involve either the arterial circulation (and associated vessels) or the venous circulation (and associated vessels). However, the arteries are penetrated less readily than are veins, making venous involvement more likely. In the process of venous invasion by a malignant neoplasm, the neoplastic cells in the blood follow the normal pattern of venous blood flow draining the primary site of the neoplasm. It is well recognized that the liver and lungs represent the most frequently involved secondary sites in hematogenous spread of malignant neoplasms. This is due to the fact that all portal venous blood drainage flows to the liver and all caval blood drainage flows to the lungs. In many cases, it is thought that metastatic cancer cells will arrest and colonize the first capillary bed that they encounter after gaining access to the venous circulation. However, numerous clinical and experimental observations combine to suggest that the anatomic location of the primary neoplasm and the natural pathways of venous blood flow from that site do not completely explain the patterns of metastatic disease observed with many forms of cancer. Thus, the so-called seed and soil hypothesis was developed to explain the tendency of certain cancers to preferentially spread to certain tissue sites. For example, prostatic carcinoma preferentially metastasizes to bone, bronchogenic carcinomas tend to spread to the adrenals and the brain, and neuroblastomas are prone to colonize the liver and bones. In contrast, whereas skeletal muscles represent a substantial organ system that is rich in capillary beds, it is rarely a site for metastatic tumor colonization. These observations suggest that factors intrinsic to the neoplastic cells and their secondary sites of involvement determine the ability of cancer cells to efficiently colonize a given tissue.

## Clinical aspects of neoplasia

As a malignant neoplasm develops, expands, and spreads, it can produce a number of effects on the host, including fever, anorexia, weight loss and cachexia, infection, anemia, and various hormonal and neurologic symptoms. Major manifestations of tumor effects on the host include (1) discomfort (pain), (2) cachexia, and (3) paraneoplastic syndromes.

### Cancer-associated pain

Pain is frequently associated with malignant neoplasms, particularly in advanced disease. Cancer-associated pain is often difficult to treat and unrelenting, producing major challenges for management and typically requiring the use of narcotic drugs. The cause of the pain can be related to destruction of tissue by the neoplasm, infection, stretching of internal organs (due to tumor involvement), pressure (from an expanding neoplasm), or obstruction (secondary to tumor impingement). Destruction of bone tissue by metastatic cancer results in characteristic bone pain. Many cancer patients have decreased immune function, resulting in infections that might not otherwise be encountered. For example, cancer patients with decreased immunity may develop infection by herpes zoster, which causes extreme nerve pain (causalgia). Pain associated with stretching of internal organs occurs with cancers of various tissues, including liver, pancreas, stomach, and other sites. The capsule of the liver is invested with blood vessels and nerves. Thus, with expansion of the tissue secondary to expansion of neoplastic growths (whether primary neoplasms or metastatic lesions), stretching of the capsule occurs, producing pain associated with perturbation of these nerves. Metastatic cancer of the liver is a major cause of hepatomegaly, and pain in the upper right quadrant is often an early symptom.

### Cancer cachexia

Cachexia refers to a progressive loss of body fat and lean body mass, accompanied by profound weakness, anorexia, and anemia. This condition affects many cancer patients with advanced disease. In fact, very often the size and extent of spread of the malignant neoplasm correlates with the severity of cachexia. However, the cachexic condition of the patient is not directly related to the nutritional needs of the neoplasm. Rather, cancer cachexia results from the action of soluble factors (such as cytokines) produced by the neoplasm. Cancer patients very often experience anorexia (loss of appetite), which can result in reduced caloric intake. Nevertheless, cancer patients often display increased basal metabolic rate and expend high numbers of calories in spite of reduced food caloric intake. It follows that as caloric needs increase (or remain high) in the patient, body fat reserves and lean tissue mass are consumed to meet the energy needs of the individual. The mechanistic basis for this metabolic disturbance has not been fully elucidated. However, some evidence suggests that certain cytokines (such as TNFα) may mediate cancer-associated cachexia. Other factors are probably involved in cancer cachexia as well, including proteolysis-inducing factor (which mediates breakdown of skeletal muscle proteins) and other molecules with lipolytic action. Cancer cachexia cannot be treated effectively; however, strategies for management of this aspect of cancer are being investigated. At the present time, the most effective treatment of cancer-associated cachexia is removal of the underlying cause (the malignant neoplasm).

## Paraneoplastic syndromes

Paraneoplastic syndromes refer to groupings of symptoms that occur in patients with malignant neoplasms that cannot be readily explained by local invasion or distant metastasis of the tumor, or the elaboration of hormones indigenous to the tissue of origin of the neoplasm. Paraneoplastic syndromes occur in 10%—15% of cancer patients. These paraneoplastic syndromes present challenges to management of the cancer patient, potentially leading to significant clinical problems that affect quality of life or contributing to potentially lethal complications. However, the symptoms associated with paraneoplastic syndromes may also represent an early manifestation of an occult neoplasm, presenting an opportunity for cancer detection and diagnosis. There are a number of different paraneoplastic syndromes that are associated with many different tumors. Well-characterized examples of paraneoplastic syndromes include (1) hypercalcemia, (2) Cushing syndrome, and (3) hypercoagulability (Trousseau syndrome). Hypercalcemia refers to a condition related to elevated plasma calcium concentrations. This condition occurs in 20%—30% of patients with metastatic cancer and is the most common paraneoplastic syndrome. Hypercalcemia affects multiple organ systems, can be life-threatening, and predicts a poor outcome for the patient. There are four types of hypercalcemia that have been recognized: (1) osteoclastic bone resorption due to metastasis to bone, (2) humoral hypercalcemia caused by systemic secretion of parathyroid hormone-related protein (PTHrP), (3) secretion of vitamin D, and (4) ectopic secretion of parathyroid hormone. Cushing syndrome is caused by ectopic secretion of corticotropin (hypercortisolism) and produces several major effects, including hyperglycemia, hyperkalemia, hypertension, and muscle weakness. Cushing syndrome is associated with several types of malignant neoplasm, including small cell lung carcinoma, pancreatic carcinoma, and various neural tumors. Trousseau syndrome is characterized by hypercoagulopathy and produces the major effect of venous thrombosis of the deep veins in the cancer patient. Small cell lung carcinoma and pancreatic carcinoma can cause Trousseau syndrome by elaborating platelet-aggregating factors and pro-coagulants from the cancer or its necrotic products.

## Grading and staging of cancer

In the clinical setting, it is important to be able to predict the relative aggressiveness of a given malignant neoplasm as a guide for designing treatment. To accurately predict the clinical course of a neoplasm and the probable outcome for the patient, clinicians make observations related to the histological aggressiveness of the cells that compose the neoplasm and the apparent extent and spread of the disease. This analysis yields a clinical description of the patient that reflects a score for the grade and stage of the malignant neoplasm.

The histological grade of a malignant neoplasm is based on (1) the degree of cellular differentiation of the neoplastic cells that compose the lesion and (2) an estimate of the growth rate of the neoplasm (based on mitotic index). In the past, it was thought that the degree of histological differentiation reflected the relative aggressiveness of a malignant neoplasm. However, it is now recognized that this notion oversimplifies the biology of neoplastic diseases. Nevertheless, the histological grade of certain cancers continues to have value for predicting clinical course. These cancers include those of the cervix, endometrium, colon, and thyroid. In general, histologic grade I refers to neoplasms that display 75%—100% differentiation, grade II reflects 50%—75% differentiation, grade III refers to 25%—50% differentiation, and grade IV describes tumors with <25% differentiation. Current methods also take into consideration the mitotic activity of the neoplastic cells, the degree of infiltration of adjacent tissues, and the amount of tumor stroma that is present.

Staging of malignant neoplasms is based on (1) the size of the primary lesion, (2) the extent of spread to regional lymph nodes, and (3) the presence or absence of distant metastases. Effective staging of neoplastic disease relies on clinical assessment of the patient, radiographic examination (using CT, MRI, or other technologies), and in some cases surgical exploration. The major convention for staging of malignant neoplasms is known as the TNM system. In this system, the T refers to the primary tumor (with T1, T2, T3, and T4 reflecting increasing size of the primary tumor), the N refers to regional lymph node involvement (with N0, N1, N2, and N3 reflecting progressively advancing node involvement), and M refers to the presence or absence of distant metastases (with M0 and M1 reflecting absence and presence of distant metastases, respectively). Staging guidelines vary for specific types of cancer. For instance, in primary hepatocellular carcinoma: T0 indicates no evidence of primary tumor; T1 indicates a solitary tumor, 2 cm or less in size, without vascular invasion; T2 indicates a solitary tumor, 2 cm or less in size, with vascular invasion or multiple tumors limited to one lobe (<2 cm) without vascular invasion; or a solitary tumor (>2 cm) without vascular invasion; T3 indicates a solitary tumor more than 2 cm in size with vascular invasion or multiple tumors limited to one lobe (<2 cm), with vascular invasion or multiple tumors limited to one lobe (>2 cm); and T4 indicates multiple tumors in more than one lobe of the liver, involving a major branch of the portal or hepatic vein(s) or invasion of adjacent organs; N0 indicates no regional lymph node metastasis; N1 indicates regional lymph node metastasis; M0 indicates no distant metastasis; and M1 indicates the presence of distant metastasis. In contrast to the staging of liver cancers, staging of colorectal cancer is fairly straightforward and primarily involves consideration of the depth of tumor invasion through the colon wall. Adenocarcinoma invasive into the submucosa of the colon is considered a T1 lesion, whereas invasion into but not through the muscularis propria is classified as T2. When the adenocarcinoma extends through the muscularis propria into the subserosa, it falls under the category of T3. T4 adenocarcinomas extend through the colon wall and directly invade adjacent organs or structures (such as bladder or another portion of the gastrointestinal tract). Like the staging classification for the liver, the N (nodal) and M (metastasis) status of the patient depends on the number of lymph nodes present and presence or absence of distant metastasis, respectively.

## Key concepts

- Cancer represents a major health problem in the United States and worldwide. In 2018, approximately 1.7 million new cases of invasive cancer were diagnosed in the United States, as well as over a million cases of common skin cancers, and there were over 600,000 cancer-associated deaths. In 2012, over 14 million new cases of invasive cancer were diagnosed worldwide and over 8.2 million cancer-associated deaths occurred.

- The development of cancer is a multistep, multifactorial process that occurs over a long period of time, through which cells acquire increasingly abnormal proliferative and invasive behaviors.

- Major risk factors that contribute to cancer development include: (1) age, (2) race, (3) gender, (4) family history, (5) infectious agents, (6) environmental exposures, (7) occupational exposures, and (8) lifestyle exposures. These risk factors interact with genetic polymorphisms and other genetic determinants to drive the development of cancer.

- The division of neoplastic diseases into benign and malignant categories is extremely important. Both benign and malignant neoplasms are composed of neoplastic cells and host-derived non-neoplastic stroma (composed of connective tissue and blood vessels). The classification of neoplasms into benign and malignant categories is based on a judgement of the potential clinical behavior of the tumor. Benign neoplasms are characterized by features that suggest lack of aggressiveness. The most important characteristic of benign neoplasms is the lack of invasiveness. Malignant neoplasms are collectively known as cancers. Malignant neoplasms display aggressive characteristics, can invade and destroy adjacent tissues, and spread to distant sites (metastasize).

- Neoplasms grow through a process of clonal expansion driven by mutation and/or epimutation, where the first mutation/epimutation leads to limited expansion of progeny of a single cell, and each subsequent mutation/epimutation gives rise to a new clonal outgrowth with greater proliferative potential.

- The idea that carcinogenesis is a multistep process is supported by morphologic observations of the transitions between premalignant (benign) cell growth and malignant lesions. For example, in colorectal cancer this transition from benign to malignant neoplasm can be easily documented and occursin discernable stages, including benign adenoma, carcinoma *in situ*, invasive carcinoma, and eventually local and distant metastasis. The stepwise development of neoplastic disease provides opportunity for detection of incipient disease, diagnosis at a definable stage, and appropriate therapeutic intervention.

- Staging of malignant neoplasms is based on (1) the size of the primary lesion, (2) the extent of spread to regional lymph nodes, and (3) the presence or absence of distant metastasis. The major convention for staging of malignant neoplasms is the TNM system, where T refers to the primary tumor (with T1, T2, T3, and T4 reflect increasing size of the primary tumor), N refers to regional lymph node involvement (with N0, N1, N2, and N3 reflect progressively advancing lymph node involvement), and M refers to the presence (M1) or absence (M0) of distant metastasis. Accurate staging of clinical neoplasms is important for management of the disease (application of appropriate treatment) and prediction of patient outcomes (prognosis).

# Suggested readings

[1] Howlader N, Noone AM, Krapcho M, Miller D, Bishop K, Altekruse SF, et al., editors. SEER cancer statistics review, 1975−2013. Bethesda, MD: National Cancer Institute; 2016. http://seer.cancer.gov/csr/1975_2013/. based on November 2015 SEER data submission, posted to the SEER wcb sitc.

[2] Siegel RL, Miller KD, Jemal A. Cancer statistics, 2018. CA Cancer J Clin 2018;68:7−30.

[3] Torre LA, Bray F, Seigel RL, et al. Global cancer statistics, 2012. CA Cancer J Clin 2015;65:87−108.

[4] Miller RA. Gerontology as oncology. Research on aging as the key to the understanding of cancer. Cancer 1991;68:2496−501.

[5] Simonetti RG, Camma C, Fiorello F. Hepatocellular carcinoma. A worldwide problem and the major risk factors. Dig Dis Sci 1991;36:962−72.

[6] Gayther SA, Pharoah PD, Ponder BA. The genetics of inherited breast cancer. J Mammary Gland Biol Neoplasia 1998;3:365−76.

[7] Ponder BA, Antoniou A, Dunning A. Polygenic inherited predisposition to breast cancer. Cold Spring Harb Symp Quant Biol 2005;70:35−41.

[8] Tiribelli C, Melato M, Croce LS, et al. Prevalence of hepatocellular carcinoma and relation to cirrhosis: comparison of two different cities of the world—Trieste, Italy, and Chiba, Japan. Hepatology 1989;10:998−1002.

[9] Kew MC. Hepatocellular carcinoma: epidemiology and risk factors. J Hepatocell Carcinoma 2014;1:115−25.

[10] Castle PE, Maza M. Prophylactic HPV vaccination: past, present, and future. Epidemiol Infect 2016;144:449−68.

[11] Irigaray P, Newby JA, Clapp R, et al. Lifestyle-related factors and environmental agents causing cancer: an overview. Biomed Pharmacother 2007;61:640−58.

[12] Willis RA. The spread of tumors in the human body. London: Butterworths; 1952.

[13] Kinzler KW, Vogelstein B. Introduction. In: Vogelstein B, Kinzler KW, editors. The Genetic Basis Of Human Cancer. 2nd ed. New York: McGraw-Hill; 2002. p. 3−6.

[14] Lengauer C, Kinzler KW, Vogelstein B. Genetic instabilities in human cancers. Nature 1998;396:643−9.

[15] Oey H, Whitelaw E. On the meaning of the word 'epimutation. Trends Genet 2014;30:519−20.

[16] Esteller M. Cancer epigenomics: DNA methylomes and histone-modification maps. Nat Rev Genet 2007;8:286−98.

[17] Calin GA, Croce CM. MicroRNA signatures in human cancers. Nat Rev Cancer 2006;6:857−66.

[18] Fearon ER, Vogelstein B. A genetic model for colorectal tumorigenesis. Cell 1990;61:759−67.

[19] Da Silva FC, Wernhoff P, Dominguez-Barrera C, Dominguez-Valentin M. Update on hereditary colorectal cancer. Anticancer Res 2016;36:4399−405.

[20] Gurzu S, Silveanu C, Fetyko A, et al. Systematic review of the old and new concepts in the epithelial-mesnechymal transition of colorectal cancer. World J Gastroenterol 2016;22:6764−75.

[21] Fokas E, Engenhart-Cabillic R, Daniilidis K, et al. Metastasis: the seed and soil theory gains identity. Cancer Metastasis Rev 2007;26:705−15.

# Chapter 5

# Basic concepts in human molecular genetics

Christine M. Koellner, Kara A. Mensink and W. Edward Highsmith, Jr.
*Department of Laboratory Medicine and Pathology, Mayo Clinic, Rochester, MN, United States*

**Summary**
Molecular diagnostics is the branch of laboratory medicine or clinical pathology that utilizes the techniques of molecular biology to diagnose disease, predict disease course, select treatments, and monitor the effectiveness of therapies. Molecular diagnostics is associated with virtually all clinical specialties and is a vital adjunct to several areas of clinical and laboratory medicine, but is most predominantly aligned with infectious disease, oncology, and genetics. The subject of this chapter is molecular genetics, which is concerned with the analysis of human nucleic acids as they relate to disease.

## Introduction

Molecular diagnostics is the branch of laboratory medicine or clinical pathology that utilizes the techniques of molecular biology to diagnose disease, predict disease course, select treatments, and monitor the effectiveness of therapies. Molecular diagnostics is associated with virtually all clinical specialties and is a vital adjunct to several areas of clinical and laboratory medicine, but is most predominantly aligned with infectious disease, oncology, and genetics. The subject of this chapter is molecular genetics, which is concerned with the analysis of human nucleic acids as they relate to disease.

Since the completion of the first working draft of the human genome sequence in 2000 and the completion of the refined sequence in 2003, progress in molecular genetics has been swift and shows no signs of abating. Relatively few gene tests were clinically available in the late 1990s, whereas over 5000 are available today. Further, molecular genetic testing has proven useful and robust enough to expand into population-based screening. Molecular testing serves as the final confirmatory test for several disorders included as part of expanded newborn screening programs, and in 2003 the *American College of Medical Genetics* and the *American Congress of Obstetricians and Gynecologists* recommended that population-based carrier screening for cystic fibrosis using molecular testing be implemented in the United States.

Molecular genetics as a discipline and as a clinical laboratory service does not exist in a vacuum. Rather, it is intimately tied to molecular and cell biology and the central paradigm of molecular biology—that genes code for proteins. Thus, it is through the analysis of genes that insight into the genesis of protein malfunction can be achieved. Such examination specifically entails an assessment of how the DNA sequence of a gene compares with its wild-type, or normal, sequence. Ultimately, protein malfunctions related to gene mutations lead to organ dysfunction and disease states. This chapter will review the fundamentals of molecular genetics and is divided into five sections that review concepts intrinsic to molecular genetics. The first section focuses on the molecular structure of DNA, DNA transcription, and protein translation. The second section focuses on molecular pathology, DNA replication, and DNA repair mechanisms. The third section provides a basic overview of transmission genetics. The fourth section highlights the relationship between genes, proteins, and phenotype and includes rationale for molecular genetic testing. The final section reviews allelic heterogeneity and corresponding choice of analytical methodology.

*Essential Concepts in Molecular Pathology.* DOI: https://doi.org/10.1016/B978-0-12-813257-9.00005-X

# Molecular structure of DNA, DNA transcription, and protein translation

The human genome is composed of 3 billion base pairs of DNA. This is not present as one continuous piece of double-stranded DNA, but is distributed among 22 pairs of autosomal chromosomes and 2 sex chromosomes. The DNA is associated with a large number of proteins (histones and others) that serve regulatory functions and package the genetic material into the large chromosomal units. Along the length of each chromosome, DNA is organized into linear domains consisting of genes (primarily non-repetitive DNA), repetitive elements, and apparently "functionless" regions, much like beads on a string (Fig. 5.1). Approximately half to two-thirds of the human genome consists of repetitive DNA, while the other half consists of non-repetitive sequence. Non-repetitive DNA includes regulatory sequences, protein coding (exon) sequence, and intronic sequences (the sequence between two exons). Protein coding regions account for a relatively small fraction of the human genome. In fact, it is estimated that only 2% of the human genome consists of protein coding, non-repetitive DNA.

Repetitive DNA tends to occur either in clusters of tandem repeats or as repetitive elements of various lengths dispersed throughout the genome. Repetitive DNA does not code for an apparently active RNA transcript or functional protein, but may have a role in protein folding and localization, DNA packaging and chromosome structure, and regulation of gene expression.

Genes are found among the non-repetitive DNA in the genome. Genes code for specific protein chains, each with a specific function in cell physiology. A gene is composed of regulatory elements, which determine where, when, and how a gene is transcribed, and coding regions, which are broken into segments, termed exons (expressed sequences). For example, the promoter, the site where gene transcription is initiated, is considered a regulatory element. The exons are separated by noncoding regions of DNA called introns (intervening sequences). The size of a gene may influence the molecular diagnostic laboratory's ability to design a clinical test for a particular disorder and certainly impacts the selection of the technology used to detect mutations.

Chemically, genes are composed of 2-deoxyribonucleic acid (DNA). DNA is a linear, non-branching polymer of nucleotides. Repeating ribose and phosphate subunits form a backbone, and attached to each of the ribose moieties is a purine (adenine, guanine) or pyrimidine (thymine or cytosine) base. Following standard nomenclature for the naming of ring containing compounds, the nitrogenous bases have their various carbon and heteroatom components numbered $10-6$ (for the pyrimidines) or $1-9$ (for the purines) and the ribose positions are indicated by numbers $1'-5'$. The bases are attached to the ribose subunits at the $1'$ position of the sugar molecule. The ribose subunits are joined by phosphodiester linkages between the $5'$ position of one ribose to the $3'$ position of the next (Fig. 5.2). Therefore, the molecule is not symmetrical and there is directionality implicit in a DNA strand. There is a $5'$ end of a DNA strand and a $3'$ end. Two DNA strands bind together to form the familiar double helical structure of double-stranded DNA (Fig. 5.3). In order for a double helix to be stable, there must be a complementary base on the opposite strand for every base on a strand of DNA. The complementary pairs of bases are adenosine and thymine (A:T) and guanine and cytosine (G:C). The two strands join in an antiparallel fashion (one strand is orientated $5'-3'$ and the other $3'-5'$). The ribose sugars form the scaffolding for the complementary nitrogenous bases connected by hydrogen bonds on the inside of the molecule. The DNA double helix is dynamic, and the weak hydrogen bonding between complementary bases allows for the DNA strands to easily separate and reassociate. In the laboratory, the process of separating (denaturing) double-stranded DNA and then allowing the complementary single strands of DNA to reassociate and return to a double-stranded configuration is called hybridization. The basis of many of the laboratory techniques central to molecular diagnostics hinge on hybridization and the remarkable specificity of a non-repetitive sequence of bases that make up a

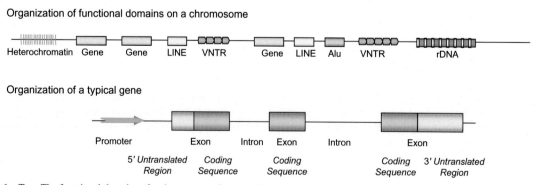

FIGURE 5.1 Top: The functional domains of a chromosome; Bottom: Organizational structure of a typical gene.

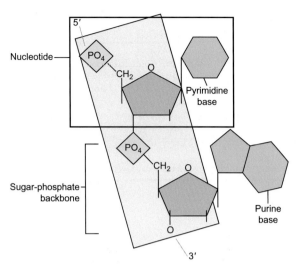

**FIGURE 5.2** Schematic view of nucleotide structure and how nucleotides join to form the DNA polymer.

**FIGURE 5.3** Schematic view of the double helical structure of double-stranded DNA. *Blue ribbons* represent the sugar-phosphate backbone. *Green/yellow* and *pink/lavender links* represent complementary purine/pyrimidine pairs.

single strand of DNA to bind to its complementary sequence and no other. In vivo, the denaturation and reassociation of double-stranded DNA is inherent to the process of gene transcription.

## DNA Transcription

Transcription is the first process in the cascade of events that lead from the genetic code contained in DNA to synthesis of a specific protein. The product of gene transcription is ribonucleic acid (RNA). The structure of RNA is similar to DNA, with three exceptions. First, the ribose sugar of RNA has two hydroxyl groups at the $2'$ and $3'$ carbons. Second, the base uracil (U) replaces thymine (T). And third, most RNA molecules are single rather than double stranded. There are four general types of RNA (Table 5.1). The specific type of RNA that results from the transcription of a structural gene is messenger RNA (mRNA). Transcription of DNA into RNA is catalyzed by RNA polymerase. RNA polymerase consists of multiple subunits that work together to recognize where the transcriptional complex should assemble,

**TABLE 5.1** Five types of RNA.

| Type of RNA | Summary |
| --- | --- |
| mRNA | The transcript product of a structural gene that encodes an amino acid sequence. |
| tRNA | Transfer RNA molecules recognize codons of mRNA and facilitate incorporation of each successive amino acid during protein synthesis. |
| rRNA | Integral component of the ribosomal machinery used for translating DNA transcript into protein. |
| Small RNA | Many small RNA molecules exist, and each has different functions in RNA modification. For example, snRNA assists with splicing intron transcripts out of precursor mRNA. An example of small RNA is snRNA. |
| microRNA | A species of small RNA molecules involved in gene regulation. |

synthesize the RNA single-stranded transcript, and dissociate from the DNA template once synthesis is complete. Under the influence of the gene promoter, various transcription factors are attracted to the upstream ($5'$) end of the gene. The transcription factors recruit the RNA polymerase and initiate transcription of the coding region of a gene into RNA. Simultaneous reading of the DNA template (anti-sense strand) and elongation of the RNA product by the RNA polymerase complex proceeds in the $5'-3'$ direction. Elongation ceases when the RNA polymerase complex recognizes the DNA terminator sequence and disassociates from the primary mRNA transcript and the double-stranded DNA. The mRNA transcript is complementary to the antisense strand and a replicate (with the exception of uracil replacing thymidine) of the sense strand (Fig. 5.4A).

Once the primary RNA sequence has been synthesized, the RNA transcript requires modification for stability and translational efficiency. The primary transcript (precursor mRNA or pre-mRNA) contains both the coding (exon) and noncoding (intron) sequences, and the intron material has to be removed prior to translation and protein synthesis. Sequences flanking the exons, the donor and acceptor splice sites, recruit a series of proteins that remove the introns from the transcript and splice the exons together to form the mature mRNA (Fig. 5.4B). Additional post-transcriptional modification includes the attachment of 7-methylguanosine CAP to the $5'$ end of the mRNA and the addition of a polyA tail that consists of a variable number (usually 80−250) of adenine nucleotides at the $3'$ end of the mRNA. Both the CAP and the polyA tail are thought to help stabilize the mRNA molecule, assist with its transport out of the nucleus into the cytoplasm, and may also help to regulate translation of mRNA into protein. Once splicing has occurred and the CAP and polyA tail have been added, RNA modification is complete and the mature mRNA transcript is exported to the cytoplasm for translation into protein (Fig. 5.4B).

## Protein translation

After the mature mRNA transcript is transported to the cytoplasm, it is translated into protein by the ribosomes in the endoplasmic reticulum. Protein synthesis is elegantly complex, with the many units of the ribosomal machinery acting in concert to achieve translation of the DNA transcript (mRNA) to an entirely different form of polymer—polymerized amino acids linked with peptide bonds. Ribosomes consist of two multiprotein subunits, each with an RNA component (rRNA) and several active centers.

Recall that mature mRNA essentially represents only the exonic, or coding, regions of a given gene. The base sequence within these coding regions is read by the ribosomal machinery in informational units of three bases, called codons. Each codon either codes for a specific amino acid or serves a regulatory function, such as stopping or starting protein chain synthesis. To initiate the process of translating mRNA into protein, the small ribosome subunit binds to mature mRNA at the CAP site and scans the mRNA sequence for the start codon, AUG. After the AUG codon is recognized, the large ribosome subunit binds a specific aminoacyl-tRNA, Met-tRNA, and the process of protein synthesis begins. An aminoacyl-tRNA (referred to as a *charged tRNA*) is an RNA molecule that carries a specific amino acid and has an anticodon sequence complementary to the mRNA codon. The specific amino acid that each charged tRNA carries is determined by the mRNA codon and is associated with the tRNA anticodon. As the ribosome translocates itself along mRNA in a $5'-3'$ direction, it catalyzes the successive binding of charged tRNAs to their associated mRNA codons. The ribosome catalyzes the chemical joining of amino acids together by creating peptide bonds between the amino and carboxyl groups of each successively added amino acid (Fig. 5.4C). It is this flow of genetic

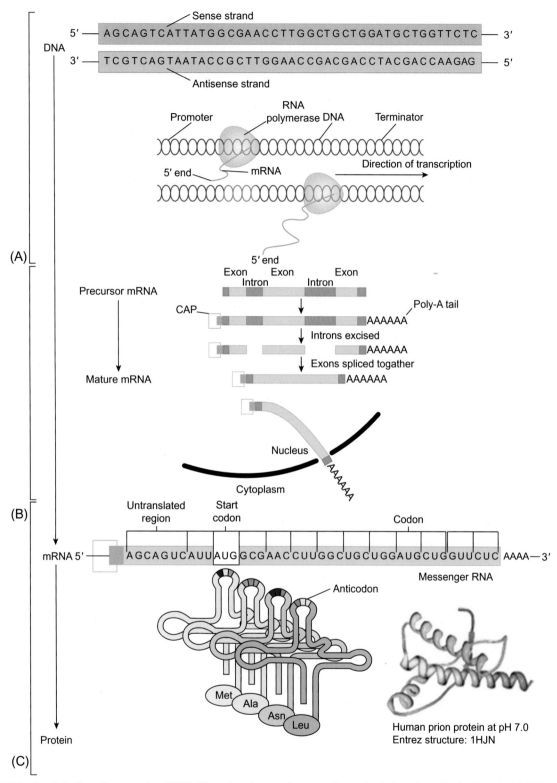

**FIGURE 5.4** Panel A: Top: Two strands of DNA illustrating the complementary bases that link to form the double-stranded DNA molecule. Bottom: Schematic of DNA transcription with initiation at the promoter, elongation of the mRNA product as the RNA polymerase translocates in the 5′−3′ direction, and termination at the termination codon. Panel B: Processing of mRNA illustrating the addition of the CAP and poly-A tail, splicing out of intron sequences, and transport of the mature mRNA molecule from the nucleus to the cytoplasm. Panel C: Using the human prion protein as an example, this schematic illustrates the translation of mature mRNA to protein. Translation begins at the start codon (AUG) with protein synthesis in the 5′−3′ direction. The structure of the charged tRNA molecule can be appreciated as each complementary tRNA anticodon recognizes its corresponding mRNA codon. The final result is a representation of the folded human prion protein as referenced in Entrez Structure http://www.ncbi.nim. nih.gov/sites/entrez?dp = structure (October, 2008).

information (DNA transcription to RNA, and RNA translation to protein) that is termed the central dogma (or paradigm) of molecular biology.

## Molecular pathology and DNA repair mechanisms

### Mutation and genetic variation

There is no single sequence of the human genome. Although the entire genome sequence from any given human is approximately 99.9% identical to the genome sequence of any other individual human, there are on the order of 3 million sequence variations between any two unrelated persons. It is the similarity of the genomes between individuals that defines them as human beings, and it is the differences that distinguish individuals. Although the majority of the sequence differences between individuals likely have no biological importance and do not contribute to physiological or observable differences, many clearly do have subtle effects and give rise to the remarkable diversity of the human race.

A large number of genetic variations occur at measurable frequencies in the population. Such variations are termed polymorphisms. Although often used to denote a nonpathogenic variation, the strict definition of the term polymorphism is a variation that is present at a frequency of 1% or greater in the population. The most common type of sequence variation is a difference between single nucleotides at a particular place in the genome or locus. For example, at a certain position, one individual may have a thymine residue, whereas another may have a cytosine. This type of variation is termed a single nucleotide polymorphism (SNP). To date, over 10 million different SNPs have been characterized and can be extremely useful tools for understanding genetic diversity and localizing disease genes. Another type of polymorphism involves not the substitution of one nucleotide for another, but variation in the number of copies of a string of nucleotides. One of the most common of this type of variation is the variation of the number of copies of a repetitive sequence at a given locus. When the length of the repetitive unit is small (one to tens of nucleotides), this type of polymorphism is termed a short tandem repeat (STR). When the length is longer, hundreds to thousands of nucleotides, they are termed variable number tandem repeats (VNTRs). STRs are a very common source of genetic differences between individuals and have been important in gene mapping studies. Currently, due to the high rate of heterozygosity, forensic laboratories utilize STR analysis extensively. Another type of polymorphism that has become appreciated through the use of comparative genome hybridization microarrays (CGH arrays) involves the deletions and duplications of regions of the genome. These regions can be quite large, up to several million bases in length, and may include genes. The role of these copy number variants in human variation and disease is a very active field of study in genetics, within both research and clinical laboratories.

A sequence variant that has a pathogenic effect is called colloquially a *genetic mutation*, although by the strict sense of the word, any alteration different from the *wild-type* sequence is considered a *genetic mutation*. In this chapter, a *genetic mutation* or *mutation* will denote a pathogenic sequence variant. Some mutations are relatively common in the population and meet the 1% population frequency criteria to be formally termed polymorphic—the common cystic fibrosis mutation deltaF508 in the Northern European population and the sickle cell anemia mutation HbS in the African populations are examples. However, most mutations are very rare in the population. Not infrequently, mutations are found that affect a single family and are termed private mutations. Pathogenic (disease-causing) mutations often involve changes in the base sequence that composes a codon (or coding unit). However, mutations can occur in regulatory elements such as splice sites and promoter regions as well. An alteration occurring in the portion of DNA that codes for a protein can result in (1) a change of one amino acid to another, (2) a change in an amino acid codon to one coding for a termination signal (stop), or (3) no change in the amino acid at that position. These types of changes are termed missense, nonsense, and silent mutations, respectively. A missense change can result in no change in the function of the protein, a total loss of function, a partial loss of function, or a change of function. Partial or total loss of function usually results in a pathological state, as does a change in function. The pathogenic effect of a loss of function of a gene product can be direct, such as the loss of chloride channel function that causes cystic fibrosis, or indirect, such as the loss of function of regulators of gene expression that can result in cancers. Examples of mutations that cause changes in function include those that cause constitutive activation of a function that is normally under regulation by the cell. It is important to note that not all losses (even complete losses) of protein function lead to an abnormal phenotype or disease.

Other types of mutations that can occur include deletions and insertions of nucleotides in and surrounding coding regions. These types of mutations, often abbreviated *indels*, can be small, from one to a few dozen bases, or large, covering large segments of chromosomes and including multiple genes. Since codons consist of a trio of bases, if a small indel occurs within the coding region of a protein that contains a number of bases that is divisible by three, the indel is said to be in-frame, as it will not shift the reading frame of the mRNA being translated into protein in the ribosome. If, on the other hand, the number of bases is not divisible by three, the indel will alter the reading frame of the mRNA and will typically result in the ribosome encountering a stop codon within a few dozen bases. Larger indels can involve whole exons, multiple exons, whole genes, or even multiple genes.

Abnormal expression of genes can also result from changes to the chemical structure of genes that are not a result of a change of the DNA sequence. Methylation of the nucleotide bases is a post-synthetic modification to DNA that affects the expression of genes. Abnormal patterns of DNA methylation can cause abnormal gene expression (i.e., affect transcription) and therefore create disease states. Repeat base sequences that have no apparent informational content regarding protein structure exist throughout the DNA. The expansion of the number of repeats in a gene has been associated with specific diseases, like the CGG repeats within the gene *FMR1*, the cause for Fragile X syndrome. Importantly, this change in the gene structure is heritable.

## DNA replication

DNA is synthesized as part of the DNA replication process that occurs during the S phase of the mitotic cell cycle and the first phase of meiosis. The DNA replication process involves multiple specialized enzymes (Table 5.2) that work together to synthesize two double-stranded daughter strands from one double-stranded parent strand. DNA polymerase synthesizes the daughter strands. The replication fork is the site at which double-stranded DNA is separated and DNA polymerase synthesizes the new daughter strands. Ahead of the replication fork is the parent double-stranded DNA; behind the replication fork are the newly synthesized daughter strands. DNA polymerases only synthesize in the $5'-3'$ direction; there are no DNA polymerases that synthesize in the $3'-5'$ direction. Thus, the DNA replication process is referred to as semi-discontinuous because, while one of each of the two new daughter strands is able to be replicated continuously in the $5'-3'$ direction (leading strand), the other strand (lagging strand) must be copied in short $5'-3'$ segments (Okazaki fragments) that are 100−1000 nucleotides in length. Okazaki fragments are joined together by the action of a ligase enzyme to complete the lagging strand. The fidelity of the replication is estimated to approach 99.98%. In the rare case that DNA replication incorporates an incorrect base, DNA proofreading and repair systems work to correct the error and prevent detrimental consequence.

## DNA repair

In the broadest sense, DNA repair mechanisms work to correct, or in some way mitigate, the effects of DNA replication inaccuracy and exogenous or endogenous genetic insult. Generally, when the integrity of wild-type DNA is compromised, the error is either corrected, overlooked, or programmed cell death occurs. A number of DNA repair pathways are known and can be roughly characterized into the following functional categories: (1) direct reversal, (2) excision repair, and (3) DNA double-strand break repair. A brief description of each follows. Although they are often studied separately, it is impossible to completely separate one from another because the various mechanisms are highly interconnected and act cooperatively as part of a large cellular arsenal with the common goal of genome integrity maintenance.

**TABLE 5.2** DNA replication enzymes.

| Replication enzymes | Function |
|---|---|
| Helicase | Breaks hydrogen bonds linking the two strands of the DNA double helix. |
| Topoisomerase | Mitigates the supercoiling effect that occurs in advance of the replication fork. |
| Single-strand binding protein | Acts as a retractor, preventing the single strands of the DNA double helix from rejoining. |
| RNA primase | Synthesizes the RNA primer that is required to initiate synthesis of the new daughter strands. |
| DNA polymerase | Synthesizes DNA daughter strands. Certain DNA polymerases also act as part of the DNA repair machinery. |
| DNA ligase | Links newly synthesized DNA fragments (Okazaki fragments). |

## *Direct reversal of DNA damage*

Correction of DNA damage by direct reversal is a type of DNA repair that predominantly involves action by a single enzyme repair system. Consider the enzymatic photo-reactivation reaction that works to repair damage induced by ultraviolet (UV) light. The formation of pyrimidine dimers (most commonly thymidine) is one type of pathologic cellular response to excess UV exposure. When present, the bulky pyrimidine dimers impede the DNA replication and transcription process. In a relatively simple light-dependent reaction, DNA photolyase acts to restore the pyrimidines to their correct monomer conformation. Direct reversal by DNA photolyase is not the only way the cell responds to UV damage. In fact, cellular response to UV damage also commonly involves one or more of the excision repair mechanisms.

## *Excision repair of DNA damage*

Correction of DNA damage by excision repair involves groups of proteins that act together to excise the incorrect base(s) or nucleotide(s), replace them with the correct sequence, and ligate the corrected strand back together. There are three DNA excision repair systems: mismatch repair (MMR), base excision repair (BER), and nucleotide excision repair (NER). Generally, these excision repair mechanisms can be distinguished by considering the context within which the error occurs, whether removal involves a base(s) or nucleotide(s), and the number of bases(s)/nucleotide(s) removed. Each type of excision repair system also invokes the use of unique proteins.

### Mismatch repair of DNA damage

DNA replication inaccuracy is the context within which the MMR pathway preserves genomic integrity. The primary purpose of MMR is to prevent mutations accrued during the DNA replication process from propagating and becoming the start of a mutant lineage by recognizing and excising the mismatched nucleotide, resynthesizing DNA, and ligating the broken strand back together.

### Base excision repair (BER) of DNA damage

BER involves the excision of a single base rather than the nucleotide and is most commonly used to repair damage caused by endogenous DNA insult and is especially important for cellular response to oxidative DNA damage. BER involves removing the base from the deoxyribose-phosphate chain by a specific glycosylase, endonuclease action, DNA polymerase beta, and either DNA ligase I or DNA ligase III/XRCC1 complex.

### Nucleotide excision repair (NER) of DNA damage

NER involves the excision of an oligonucleotide, rather than a single base (BER) or single nucleotide (MMR). NER is predominantly invoked in response to genomic damage caused by UV exposure and is also a substantially more complex process that includes at least 30 different proteins. Two subpathways of NER, termed global genome repair NER (GG-NER) and transcription-coupled NER (TC-NER), have been recognized. Typically, GG-NER is used when errors occur in non-transcribed areas of the genome and TC-NER, as the name implies, corrects errors that occur in areas of active gene expression.

## *DNA double-strand repair of DNA damage*

DNA double-strand repair is an important DNA repair mechanism that uses a number of proteins, many of which are similar to or the same as those used during meiotic recombination. DNA double-strand breaks (DSBs) can result from a number of exogenous and endogenous agents including ionizing radiation exposure, chemical exposure, and somatic DNA recombination or transposition events. Non-homologous end-joining (NHEJ) and homologous recombination (HR) are the two primary DNA double-strand repair mechanisms. Short homologous sequences (microhomologies) found on the single-stranded tails of the broken DNA are used to help rejoin the strands in NHEJ, whereas HR relies on homologous (or very close to homologous) sequence to repair the broken strands. HR is typically used when DNA replication is halted due to a single-strand break or another unrepaired lesion that causes collapse of the replication fork. Because it uses a homologous or near homologous template, HR is often thought to be more accurate than its NHEJ counterpart. However, both mechanisms show high accuracy, as well as imperfection.

# Modes of inheritance

A detailed family history provides the foundation for genetic diagnosis and risk assessment. Visually recorded using standardized symbols and nomenclature, the pedigree provides the tool by which inheritance patterns are elucidated and subsequent risk assessment is calculated. Observations made from controlled monohybrid and dihybrid crosses of pea-pod plants in the 1860s formed the foundation for Gregor Mendel's landmark laws of heredity that still govern basic pedigree interpretation today. The study of inheritance is now far more complex than Mendel himself may have imagined. This section of the chapter reviews modes of inheritance and factors that may influence pedigree interpretation.

## Mendelian inheritance

The central theory of Mendelian inheritance surrounds the idea that one gene/locus is associated with one trait. The concepts that two copies of a gene segregate from each other (law of segregation) and are transmitted unaltered (particulate theory of inheritance) from parents to their offspring help to explain the concepts of dominant and recessive traits. When the presence of one copy of a particular allele results in phenotypic expression of a particular trait, the trait is dominant. When two copies of a particular allele must be present for the phenotypic expression of a trait, the trait is recessive. Note that it is the phenotypic expression that is described as dominant or recessive, not the allele or gene itself. Thus, patterns of inheritance are distinguished by where the gene resides within the genome (autosome or sex chromosome) and whether or not phenotypic expression occurs in the heterozygous or homozygous state. Traditionally recognized Mendelian patterns of inheritance include autosomal dominant, autosomal recessive, X-linked recessive, X-linked dominant, and Y-linked (holandric). When each of these inheritance patterns is represented in a pedigree diagram, distinguishing features can be visually recognized (Table 5.3).

### Autosomal dominant inheritance

Autosomal dominant inheritance is designated when no difference in phenotypic expression is observed between heterozygous and homozygous genotypes. Visually, the autosomal dominant disease pedigree shows multiple affected generations in a vertical pattern, an equal distribution of males and females affected, and both males and females transmit the phenotype (including males transmitting the phenotype to other males). Typically, dominant disorders occur when a mutation confers an inappropriate activity (e.g. gain-of-function, loss-of-function, or dominant negative effect) on a gene product.

### Autosomal recessive inheritance

Autosomal recessive inheritance is designated when phenotypic expression is observed only when both copies of a gene are inactivated or mutated. Visually, the autosomal recessive pedigree typically shows a horizontal pattern where multiple affected individuals can be observed within the same sibship, and an equal number of males and females are affected. In instances of autosomal recessive inheritance, each parent of an affected individual has a heterozygous genotype composed of one copy of the mutated gene and one copy of the normal/functional gene. When a pedigree is analyzed, individuals who must be genetic carriers of the disorder in question, such as parents of an affected child, are termed obligate carriers. Other individuals in the pedigree may be at risk for being carriers. The risk to be a carrier is defined by each individual's position in the pedigree relative to affected individuals, or to the known carriers.

### X-linked recessive inheritance

X-linked recessive inheritance is designated when phenotypic expression is observed predominantly in males of unaffected, heterozygous mothers. All female offspring of affected males are obligate carriers. Visually, the pedigree typically shows a horizontal pattern of affected individuals with no instance of direct male-to-male transmission. However, males may transmit the disorder indirectly to a grandson, through a carrier female daughter.

It is not uncommon for X-linked recessive disorders to appear in a family such that before a certain generation the disease is not apparent, but is observed to be segregating in the family after that generation. This phenomenon is due to new mutations appearing de novo in an individual. This was explained by the American geneticist Haldane, and his theory is referred to as the Haldane hypothesis. If the reproductive fitness of a male affected with an X-linked recessive disorder is low or nil, then in a population one-third of all affected X chromosomes will be removed from the gene pool every generation. If the incidence of the disease is constant, then one-third of cases must be due to mutations arising de novo in a family. An example of decreased reproductive fitness among males is Duchenne muscular dystrophy.

**TABLE 5.3** Mendelian inheritance patterns.

| Inheritance pattern | Example pedigree | Clinical example |
|---|---|---|
| Autosomal dominant | | Huntington disease<br>Myotonic dystrophy<br>Retinoblastoma<br>Lynch syndrome<br>Neurofibromatosis I<br>TTR associated-amyloidosis<br>And many others |
| Autosomal recessive | | Cystic fibrosis<br>Galactosemia<br>Autosomal recessive (AR) deafness<br>AR epidermolysis bullosa<br>Tay-Sachs disease<br>Klippel-Fiel syndrome<br>And many others |
| X-Linked recessive | | Duchenne muscular dystrophy<br>Hemophilia A<br>X-linked ichthyosis<br>X-linked mental retardation<br>Optiz syndrome<br>Emery-Dreifuss muscular dystrophy<br>And many others |
| X-Linked dominant | | Vitamin D-resistant rickets<br>Coffin-Lowry syndrome and others |
| Y-Linked (Holandric) | | Hairy ears<br>Y-linked deafness<br>Very few others |

## X-linked dominant inheritance

X-linked dominant inheritance is designated when phenotypic expression is observed predominantly in females (ratio of about 2:1) and all daughters of affected males are affected and none of the sons of affected males are affected. Visually, the pedigree typically shows a vertical pattern of affected individuals, with no instance of direct male-to-male transmission. X-linked dominant conditions are substantially less common than X-linked recessive disorders.

## X-linked dominant male lethal inheritance

X-linked dominant male lethal inheritance is designated when phenotypic expression is observed only in females. Visually, the pedigree typically shows a vertical pattern with an increased rate of spontaneous abortion and where approximately 50% of the daughters from affected mothers are also affected.

## Y-linked or holandric inheritance

Y-linked or Holandric inheritance is designated when phenotypic expression is observed only in males with a Y chromosome. Visually, the pedigree shows only male-to-male transmission.

## Non-Mendelian inheritance

### Epigenetic inheritance—imprinting

When the phenotypic expression of a gene is essentially silenced dependent on the gender of the transmitting parent, the gene is referred to as imprinted. The phenomenon of imprinting renders the affected genes functionally haploid. In the case of imprinted genes, the functional haploid state disadvantages the imprinted gene because the gene is more susceptible to adverse effects of uniparental disomy, recessive mutations, and epigenetic (like DNA methylation-dependent gene silencing) defects. Visually, pedigrees that represent imprinting may appear similar to autosomal recessive or sporadic pedigrees and show a horizontal pattern. Imprinting disorders may also appear autosomal dominant and show a grandparental effect in the case of imprinting center mutations. Males and females are equally affected, and transmission is dependent on the gender of a parent. The two most well-known imprinting disorders are Prader-Willi and Angelman syndrome. Prader-Willi syndrome is caused by an absence of paternally contributed 15q11−13 (PWS/AS) region, whereas Angelman syndrome can be caused by an absence of maternal contribution at the same locus. In the case of PWS, lack of paternally contributed genes at 15q11−13 (regardless of mechanism) results in unmethylated, and therefore overexpressed, genes in this region. The same is true for Angelman syndrome. However, it is a lack of maternally contributed genes that causes the phenotype in for Angelman syndrome.

### Inheritance through mitochondrial DNA

The inheritance of mitochondrial disease is complicated by the fact that mitochondrial disease can be either the result of mutations in nuclear DNA and thereby subject to the Mendelian forms of inheritance described previously, or the result of mutations in organelle-specific mitochondrial DNA (mtDNA). Since the mitochondrial genome is maternally inherited, pedigrees demonstrating mitochondrial inheritance show an affected mother with all, or almost all, of her offspring (male and female) affected. The common phenomenon of heteroplasmy, where mtDNA mutations are present in only a portion of the mitochondria within a cell, can make laboratory analysis and clinical assessment difficult. It is estimated that only 10%−25% of all mitochondrial disease is the result of maternally inherited mutations in the mitochondrial genome. Therefore, mitochondrial disease should not always be equated with mitochondrial inheritance.

### Multifactorial inheritance

Sorting out whether a particular phenotype is predominantly the result of inherited genetic variation, environmental influence, or some combination therein can be difficult. When the combined effects of both inherited and environmental factors cause disease, the disorder is said to exhibit multifactorial inheritance. Multifactorial inheritance is associated with most, if not all cases of complex, common disease (cancer, heart disease, asthma, autism spectrum disorders, mental illness, and others). Typically, multiple loci or multiple genes are associated with the same complex disease phenotype. Such genetic heterogeneity is thought to work additively, such that the net effect of multiple mutations in multiple genes exacerbates and/or detracts from a particular clinical phenotype.

## Sporadic inheritance

Sporadic inheritance, where only one isolated case occurs within a family, is the most common pedigree pattern observed in clinical practice. Chromosomal abnormalities and new dominant mutations typically demonstrate sporadic inheritance. It is easy to imagine how autosomal recessive and X-linked recessive disorders can often appear sporadic, especially in situations where family size is small or clinical knowledge about extended family is limited. Thus, both Mendelian and non-Mendelian explanations for sporadic inheritance, each with its own recurrence risks, can apply. Non-inherited disorders are associated with virtually negligible recurrence risk as compared to those exhibiting Mendelian inheritance, chromosomal abnormalities, or new dominant mutations. It is important to make every effort to distinguish apparently sporadic cases from truly sporadic ones, but it is often not possible to make this determination and recurrence risk can be narrowed only to a broad range encompassing all possibilities.

## Differences in phenotypic expression can complicate pedigree analysis

The occurrence of reduced penetrance, variable expressivity, anticipation, and gender influence or limitation can confound pedigree analysis. Accurate recurrence risk is dependent on correct diagnosis and pedigree assessment.

## Genetic penetrance

The penetrance of a genetic disorder is measured by evaluating how often a particular phenotype occurs given a particular genotype or vice versa. Some disorders show 100% penetrance, where all individuals with a particular genotype express disease, while others show reduced penetrance, such that a proportion of individuals with a particular genotype never develop any features (even mild) of the associated clinical phenotype. Thus, penetrance is the probability that any phenotypic effects resulting from a particular genotype will occur. Certain factors are known to influence the gene penetrance for specific disorders. For example, phenotypic expression of a particular phenotype may be modified by age, termed *age-related penetrance*. Sometimes, as age increases, penetrance increases. Although less common, penetrance can also decrease with age, or be gender-related. Reduced penetrance can sometimes obscure an autosomal dominant inheritance pattern because, while some family members may have affected offspring, they themselves are not affected due to reduced penetrance of the disorder.

## Sex-influenced disorders

Sex-influenced disorders are disorders that demonstrate gender-related penetrance. When the probability of phenotypic expression is more likely given a specific gender, the disorder is said to be sex-influenced. Often the genetic mutation resulting in pathologic phenotype occurs in a gene that is predominately expressed or affects a gender specific organ, like the ovary or prostate.

## Sex-limited disorders

Sex-limited disorders refer to autosomal disorders that are non-penetrant for a particular gender. Very few sex-limited disorders have been documented.

## Variable expressivity

Variable expressivity refers to the difference in severity of disease among affected individuals, both between related and unrelated individuals. It is important to note that even between related individuals (with the same genotype), variable expressivity occurs. Variable expressivity is distinct from penetrance because it implies a degree of affectedness, not whether or not the individual is affected at all. The majority of inherited disease demonstrates some degree of variable expressivity. Variable expressivity can complicate pedigree analysis because individuals with subtle clinical manifestations can be mistaken for unaffected individuals It is hypothesized that genetic, epigenetic, and environmental factors play a role in the differing expressivity.

## Pleiotropy

Pleiotropy refers to disorders where multiple, seemingly unrelated organ systems are affected. For example, one individual in a pedigree may exhibit cardiac arrhythmia, whereas another individual with the same disorder in either the same or different pedigree shows muscle weakness and deafness. Since the manifestations of disease are so vastly and usually inexplicably different, disorders that show a high degree of pleiotropy are often difficult to diagnose. As a group, mitochondrial disorders typically show a high degree of pleiotropy, as any organ system can be affected, to almost any degree, with any age of onset.

## Anticipation

A disorder shows anticipation when an earlier age of onset or increased disease severity occurs in successive generations. Anticipation is predominantly associated with neurodegenerative trinucleotide repeat disorders (spinocerebellar ataxias, Huntington disease, myotonic dystrophy, and others). In such cases, the number of trinucleotide repeats expands through generations and is correlated with severity of disease and age of onset. However, not all disorders that exhibit anticipation are trinucleotide repeat disorders. Dyskeratosis congenita-Scoggins type shows anticipation via a mechanism of progressive telomere shortening in successive generations.

## Other factors that complicate pedigree analysis

### Genetic mosaicism

Mosaicism occurs when two or more genetically distinct cell lines are derived from a single zygote. The timing of the post-zygotic event(s) and tissues involved determine the clinical consequence and help to distinguish one type of mosaicism from another. Gonosomal mosaicism occurs early in embryonic development and is more likely to involve gonadal tissue and result in phenotypic expression. The clinical effects are often milder for mosaic individuals where only a proportion of cells carry a particular mutation, as compared to those who inherit germline mutations where all cells are affected. When mosaicism is confined to gonadal tissue, there are usually no clinical consequences to the individual with gonadal mosaicism. However, such individuals are at higher risk for having affected offspring. Thus, since individuals with gonosomal mosaicism have some proportion of mutant germ cells, they can (and do) have nonmosaic, affected offspring. There is no practical way to exclude the possibility of gonadal mosaicism, nor effectively test for it. This can cause a dilemma with respect to providing accurate recurrence risks to families. Gonadal mosaicism has been found to be more common for certain disorders, allowing for some empiric risk estimates to be determined.

### Consanguinity

Consanguinity is both a social and genetic concept. Generally, it refers to marriage or a reproductive relationship between two closely related individuals. The degree of relatedness between two individuals defines the proportion of genes shared between them. The offspring of consanguineous couples are at increased risk for autosomal recessive disorders due to their increased risk for homozygosity by descent. Consanguinity can complicate pedigree analysis when a provider is unaware of the presence of consanguinity at the time they are evaluating the pedigree, and what appears to be an autosomal dominant inheritance pattern is associated with an autosomal recessive disease phenotype.

### Preferential marriage between affected individuals

Increased reproductive risk can be the result of preferential marriage between affected individuals. Such selective mating (e.g. a couple that went to the same deaf high school or met at a support group for the particular condition) can increase the likelihood of pseudodominance within a pedigree because the mating environment is selected such that an autosomal recessive disorder appears more frequently than expected. An increased risk for autosomal dominant disorders is also present. In cases where both reproductive partners are affected with the same autosomal dominant condition recurrence risk ranges from 66% (when homozygous dominant inheritance is not compatible with life) to 75%.

## Other considerations for pedigree construction and interpretation

Clinical molecular genetics seeks to identify genetic variation and to determine whether or not the observed genetic variation has a phenotypic effect. Certainly, the latter cannot be accomplished without astute and thorough clinical evaluation and family history. Accurate and complete information is an imperative, as inaccurate and/or incomplete information can result in misinterpretation and, ultimately, misdiagnosis. Whenever possible, reported diagnoses must be confirmed with medical records.

# Central dogma and rationale for genetic testing

The clinical relevance of molecular genetics is fundamentally rooted in the central paradigm of molecular biology: genes encode proteins. Genes are the blueprint for the proteins that form the macromolecules of cellular structure and function. Cells, their respective functions, and the interactions between them translate to the observable characteristics, or clinical phenotype, of an organism. Endogenous and exogenous molecular, cellular, and organismal environments also play an important role in influencing clinical phenotype. So, the expression of DNA at the molecular level coupled with environmental effects leads to more tangible morphological and physiological traits at the level of the organism. However, organisms do not exist in isolation. Each organism functions as part of a population within a larger species and external environment. A species, and the organisms within it, is subject to evolutionary forces, including natural selection, genetic drift, and gene flow. Such forces ultimately impose, overlook, propagate, or extinguish genetic variation. The dynamic relationships between genetic variation, proteins, cells, organisms, populations, and environment(s) connect genetic laboratories to clinical practice, as evaluating for genetic variation (molecular genetics, cytogenetics), and/or its biochemical consequence (biochemical genetics) provides an explanation and/or causative evidence for clinical phenotype and diagnosis.

## Diagnostic and predictive molecular testing

The clinical applications of molecular genetic testing can be generalized into two groups based on whether the clinical information sought is intended for diagnostic or predictive purposes. Occasionally overlap between diagnostic and predictive testing occurs. Although commonly performed for the purpose of diagnosing a disorder in a symptomatic individual, diagnostic testing can also be informative for pre-symptomatic at-risk individuals. The degree of gene penetrance must be known in order for this diagnostic, yet predictive, testing to impart clinical value (Table 5.4).

The second broad group of molecular genetic tests includes those performed for the purpose of revising an already known risk. Predictive molecular testing typically employs molecular screening tests to more accurately determine the individual and familial/reproductive risks for an individual. It is important to recognize that while many predictive molecular screening tests are focused on evaluating at-risk individuals for autosomal recessive carrier status, other subgroups of predictive screens help to distinguish germline from somatic disease, or revise the prognosis or the risk related to complex disease based on presence or absence of disease-associated SNPs. In addition, the penetrance of the disorder may not be complete, thereby making it more difficult to recognize a specific inheritance pattern.

## Considerations for molecular testing

Benefits, risks and limitations of genetic testing should always be reviewed and openly discussed with patients as part of the informed consent process prior to testing. Disease and test-specific limitations of molecular testing are truly method, disease, and case specific, and it would be impossible to address each of them here.

## Benefits of molecular testing

The benefits of molecular genetic testing can be clinical and psychosocial. Clinical benefits of a molecular genetic diagnosis often include the ability for the care provider to recommend a preventive medicine and treatment plan based on the known natural history of a particular disorder. Genotype:phenotype correlations have been established for some particular mutations/disorders, such that a more individualized medical approach can be determined. Psychosocial benefits

**TABLE 5.4** Disease associated penetrance for common hereditary Cancer syndromes.

| Familial cancer syndrome | Gene | Lifetime penetrance |
|---|---|---|
| Hereditary breast/ovarian cancer | BRCA1 | 60%–80% |
| Hereditary breast/ovarian cancer | BRCA2 | 50%–70% |
| Retinoblastoma | RB1 | >99% |
| Familial adenomatous polyposis | APC | >99% by age 40 |
| Lynch syndrome | MLH1, MSH2, MSH6, PMS2 | 75% (may be slightly lower in females) |

of a confirmed molecular genetic diagnosis may include reduction of anxiety (1) if associated with a known versus unknown diagnosis; (2) if the diagnosis confirmed is considered by the patient to be less severe as compared to others being considered for the patient; and (3) if associated with a cease in the diagnostic odyssey that many patients with rare disorders experience (multiple medical consults, procedures, and laboratory tests associated with a continued search for a diagnosis over an often extended period of time). In addition, psychosocial benefits may accrue from implementation of a more individualized approach, as mentioned previously. For individuals undergoing pre-symptomatic testing, benefits may also include a sense of empowerment, regardless of their test result and a sense of relief if they test negative. Knowledge of one's risk for having children with a genetic disorder also assists with family planning, with individuals and couples being able to access genetic counseling and prenatal diagnosis.

## Risks associated with molecular testing

The risks of molecular genetic testing can be psychosocial and financial. Regardless of whether the result confirms the presence or absence of disease, the psychological impact of the result can be difficult. Coping with a positive result that reveals a life changing prognosis can be devastating. Sometimes, a definitive diagnosis produces more questions, resulting in anxiety. This is especially true in cases where gene penetrance, medical management recommendations, and/or natural history are not well defined. Coping with a true negative result can sometimes invoke survivor guilt, given the familial nature of genetic testing.

Financial risks associated with molecular genetic testing can include (1) minimal or lack of insurance to cover the cost of testing, and (2) the risk associated with the ability to secure life insurance at a reasonable premium, if at all. Given that many molecular genetic tests are relatively expensive due to the costs of processing and interpreting the sample, personal financial cost to the patient can be substantial.

## Limitations associated with molecular testing

Limitations of molecular genetic testing are usually related to confounding results, interpretive restrictions, or imperfections of the method used. While technical precision and accuracy of molecular genetic testing is high, the risk for a false-negative or false-positive result is always a possibility. Although rare, laboratory errors such as performing the wrong test or mislabeling samples can also occur.

Molecular genetic testing is often misunderstood as perfectly decisive. The identification of alterations for which the medical or functional significance is not clear, is not an uncommon limitation of molecular testing. These genetic alterations are termed *variants of uncertain significance* (VUS) and can be especially complicated to interpret.

## Considerations for selection of a molecular test

Selecting an appropriate molecular genetic test is dependent on the purpose for testing, the clinical information known, the sample(s) and testing methods available, and the clinical information sought. Molecular screening tests usually involve methods that investigate for common mutations (e.g., targeted analysis of the 23−100 most common cystic fibrosis mutations), whereas diagnostic testing methods are typically more comprehensive (e.g., DNA sequencing). The molecular methods used for the purpose of revising a known risk can sometimes be the same, but are often different than those used for diagnostic purposes. When evaluating the method to be used, the expected detection rate for individuals that are classically affected with the disorder in question and the clinical context of the patient being tested should be considered. To maximize the informative value of pre-symptomatic testing, in most cases, one must know the familial mutation(s). Practically, this translates into the necessity that an affected individual should be tested before pre-symptomatic testing is performed on at-risk family members. Preferred testing algorithms developed by expert clinicians and laboratorians are especially useful, though ultimately each clinical situation is different and should be considered within its own unique context.

# Allelic heterogeneity and choice of analytical methodology

The great majority of analyses performed in the clinical molecular genetics laboratory are based on the polymerase chain reaction (PCR). PCR is a technique for the rapid, in vitro amplification of specific DNA sequences. Knowledge of the sequence of the region of DNA flanking the area of interest is required for PCR. Two synthetic oligodeoxynucleotides (primers), typically 20−30 bases in length, are prepared (or purchased) such that one of the primers is complementary to an area on one strand of the target DNA 5' to the sequences to be amplified, and the other primer is complementary to the opposite strand of the target DNA, again 5' to the region to be amplified. To perform the amplification, one places the sample DNA in a tube along with a large molar excess of the two primers, all four deoxynucleotide triphosphates, buffer, magnesium ion, and a thermostable DNA polymerase. Successive rounds of heating to 93−95 °C to denature the DNA, cooling to 50−60 °C to allow annealing of the oligonucleotides, and heating to 72 °C (the temperature optimum for the DNA polymerase isolated from *Thermus aquaticus*) result in synthesis of the DNA that lies between the two primers. The amount of amplified DNA being synthesized doubles (approximately) with every temperature cycle, therefore the amount of DNA produced is exponential with respect to cycle number. After 30 cycles of denaturation, annealing, extension, $2^{30}$, or approximately $10^9$, copies of the DNA sequences lying between the two primers will have been generated. As each cycle takes 2−5 min, amplification of a specific sequence can easily be accomplished in several hours. After amplification, the DNA can be analyzed by one of several techniques, depending on the specific problem.

## Specific versus scanning methods

Analytical methods in molecular genetics for the detection of small mutations (point mutations, small indels) can be grouped into two broad categories: mutation detection techniques, which are used to investigate the actual base sequence at a particular locus, and quantitative methods, in which PCR-based techniques are used to quantify specific nucleic acid sequences. Mutation detection strategies can be further grouped into specific or scanning techniques.

### *Specific mutation detection*

Specific mutation detection entails straightforward, and largely routine, procedures that can be used to analyze DNA samples for previously identified mutations using an assay designed for maximum specificity. This approach targets known mutations in potentially large cohorts of patients or small panels of specific mutations in disorders characterized by one or a few common alleles. Results from these types of analyses may confirm or establish clinical diagnoses. Furthermore, in families at risk for a particular genetic disease, specific or targeted mutation detection allows for rapid screening of an entire family for the mutation identified in the proband (the first member of a family to be diagnosed with a genetic disorder), thereby permitting accurate carrier determinations that may aid reproductive decisions. Rapid testing of large numbers of patients permits an assessment of the frequency of a mutation among disease-causing alleles, thereby determining which mutations are most prevalent in different patient populations and guiding the creation of effective clinical mutation testing panels.

The specific mutation detection methods can themselves be divided into those that utilize electrophoretic- or hybridization-based methods (Table 5.5). Both types of platforms are robust, and in experienced hands yield reproducible results. Both types of systems are in widespread use in clinical and research laboratories. One criterion for choice between these general platforms is the cost incurred per sample analyzed. In the authors' experience, when the number of samples to be analyzed at one time (samples per batch) is low, electrophoretic methods are often the most cost effective to develop, validate, and implement. However, when the number of samples per batch is larger (greater than 8−12 samples), then the hybridization-based techniques, many of which can be adapted to 96 well microplate formats or real-time, are often more cost effective.

**TABLE 5.5 Examples of electrophoretic- and hybridization-based specific mutation detection methods.**

| Electrophoretic methods | |
| --- | --- |
| Restriction enzyme digestion<br>Amplification-refractory mutation system (ARMS, allele specific PCR)<br>Allele-specific primer extension<br>Triplet repeat | Typically lab developed<br>CF29v2 (Europe), Elucigene<br>Mass Array, Agena Biosciences<br>AmplideX, Asuragen |
| Hybridization methods | |
| Allele-specific hybridization<br>Allele-specific primer extension<br>Ligation-PCR | Resequencing arrays, Affymetrix<br>xTAG, Luminex<br>Golden Gate, Illumina |

## *Mutation scanning approaches*

Mutation scanning methods interrogate DNA fragments for all sequence variants present. By definition, these strategies are not predicated on specificity for specific alleles, but are designed for highly sensitive detection for all possible variants. In principle, all sequence variants present will be detected without regard to advance knowledge of their pathogenic consequences. Once evidence for a sequence variant is found, the sample must be sequenced to determine its molecular nature. The advantage of using a scanning method followed by sequencing of only positive PCR products is that the scanning methods are typically less costly to perform than DNA sequencing. Although a number of mutation scanning methods have been developed, they have been almost completely replaced by what is considered the gold standard mutation scanning method—DNA sequencing. There are a number of disease-associated genes that have high allelic heterogeneity or very few recurrent mutations in the population that are typically addressed for diagnostic purposes by whole gene sequencing. Only when they are combined with appropriate genetic data and in vitro functional studies can investigators distinguish disease-causing mutations from polymorphisms without clinical consequence. In the research laboratory, mutation screening is a critical and obligatory final step toward identifying genes that underlie genetic disease. In the clinical laboratory, these methods are applied toward the detection of mutations in diseases marked by significant allelic heterogeneity.

Given the number of laboratories offering whole-gene sequencing assays for an increasing number of genes, as well as so-called *panels* (testing that includes whole-gene sequencing for many different genes associated with a given phenotype), as well as whole-exome sequencing (WES) and even whole-genome sequencing (WGS), the amount of variation in coding regions is now appreciated to be significantly greater than previously thought. Unfortunately, the ability of the diagnostic laboratory to find these sequence alterations has vastly outstripped the ability of researchers to perform the necessary functional testing of the variants to help with the classification. This has the consequence that obtaining a previously unknown sequence variation in a patient sample is not uncommon, although there are concerted efforts, like ClinGen (the Clinical Genome Resource) to help laboratories share their data in the hopes that, by pooling their experiences, the clinical significance of any given variant can be agreed upon. The interpretation of such results is challenging and is not a solved problem.

## Interpretation of molecular testing results

Of the three types of coding region mutation caused by single nucleotide changes, two are often relatively straightforward to interpret. It is generally assumed that nonsense mutations (or indels giving rise to an in-frame stop codon) are deleterious and are likely to be associated with a disease phenotype, when the disease is caused by loss of protein function. Similarly, silent mutations are most often assumed to be benign. Exceptions exist, of course—silent mutations occurring at the first or last bases of an exon may influence RNA splicing. In addition, silent mutations may interrupt an exonic splice enhancer, again leading to altered splicing.

The interpretation of missense changes is challenging. Many examples (affecting many different genes) exist in which missense changes are either pathogenic or benign. The distinction typically requires the examination of multiple families carrying a given missense mutation, and/or functional studies of recombinant, mutant protein. When a novel missense change is encountered in a clinical laboratory setting, these studies are not available. Thus, novel missense changes are typically referred to as VUS.

There are two schools of thought with respect to how VUS should be reported. One school holds that unless the laboratory can give a clean interpretation and offer documentation as to whether a given variant is known to be pathogenic or benign, the report should simply indicate that a VUS was detected. Thus, the contribution of the genetic test to the management of the patient is nil—it is as if the test were not performed (and cannot be performed). Clearly, the advantage in this approach is that one is not tempted to over-interpret the results, potentially leading to an incorrect medical decision. The disadvantage is the frustration on the part of the patient (and healthcare provider) that a likely rather expensive test has been performed and no useful information was obtained. The other school of thought holds that the laboratory should use all the tools available and, when possible, make a probabilistic statement as to the potential effect of the variant. The advantage to this approach is that the final decision as to how the result will be used in guiding patient care remains with the patient and his/her healthcare provider. The clear disadvantage is the possibility that the result provided may lead to incorrect medical management. Because of this, interpretation of VUS should be done very carefully.

A number of tools to aid in the interpretation of missense changes have been developed. As an increasing number of species have had their complete genome sequence determined, it is possible to use a variety of sequence alignment tools to compare the amino acid found at a particular location in the human gene to that found in multiple other species. The rationale for this is the notion that if an amino acid is invariant across species, it is more likely to be important for protein function, and a missense change at a highly conserved residue is more likely to be pathogenic. On the other hand, if a given amino acid position is poorly conserved, a missense change may be more likely to be benign.

Several groups have developed algorithms quantifying the probability based on sequence conservation that a given missense change is pathogenic or benign. These tools are useful but are far from perfect. One mechanism by which sequence conservation strategies can be foiled occurs when a pathogenic change results in the substitution for an amino acid that is the normal sequence in another species. In addition to sequence conservation and global substitution probabilities, the chemical characteristics of the amino acids (like composition, polarity, and molecular volume), have been used to characterize missense changes. The sensitivity and specificity for all of the tools seem to be in the 70%−80% range. However, since each tool queries different properties of the gene and of amino acid substitution, it is possible that, when used together, the quality of the results may be improved.

The in-silico characterization of missense VUS changes is still in its infancy, and much more work needs to be done in this area. As we are in the era of WES, and have entered the time of WGS, the urgency of the need to characterize novel changes is increasing. Of critical note, however, is the fact that all of the prediction methods are just that—predictions. A careful laboratorian only uses the predictions obtained from any of the above in silico tools as one lone data point in their decision regarding the pathogenicity of any given variant, in addition to population frequency of the variant, disease-specific mutation databases, and the peer-reviewed literature available on the variant. Without any functional data, population frequency, case reports, etc., the in silico predictions are not sufficient to determine pathogenicity on their own. The interpretation of DNA sequence variants has increased in complexity, as the technology for detecting them has evolved. This challenge is ongoing, and has been the focus of a formal multidisciplinary workgroup comprised of clinical and laboratory experts since 2013. Joint consensus recommendations of the American College of Medical Genetics and Genomics (ACMG) and the Association for Molecular Pathology (AMP) provide guidance for the adoption of standardized terminology to describe variants identified in genes that cause Mendelian disorders, as well as offer recommendations for a process of classifying and reporting variants.

## Conclusion

Molecular genetics utilizes the laboratory tools of molecular biology to relate changes in the structure and sequence of human genes to functional changes in protein function, and ultimately to health and disease. Newer technology, such as next-generation sequencing, promises to greatly increase the reach and scope of molecular genetics. Indeed, some subspecialties, such as biochemical and cytogenetics may ultimately merge with molecular genetics and offer the medical community a more comprehensive and integrated approach to understanding the role of our genomic variation in health and disease. However, the interpretations of results from the clinical molecular genetics laboratory will always be rooted in the fundamentals of molecular and cell biology and in the central paradigm—that genes encode proteins. It will be from these roots that modern, personalized medicine will grow.

## Key concepts

- Deoxyribonucleic acid, DNA, is a double-stranded molecule consisting of two antiparallel polymers composed of ribonucleosides linked together by phosphodiester bonds. Weak hydrogen bonding between complementary bases allows for easy denaturing and reassociation of double-stranded DNA.
- Recognizing and correctly identifying the mode of inheritance of a disorder in a family provide powerful clues that can assist in establishing risk to family members.
- Genetic testing typically is sought for either diagnostic or predictive purposes.
- Risks and limitations of genetic testing should be discussed openly with the patient as part of the informed consent process *prior* to the initiation of testing.
- The selection of either a mutation scanning technology or a specific mutation detection technique generally depends upon the allelic heterogeneity for which the disorder is being tested.

## Suggested readings

[1] Bennett RL, Steinhaus KA, Uhrich SB, et al. Recommendations for standardized human pedigree nomenclature. Pedigree standardization task force of the national society of genetic counselors. Am J Hum Genet 1995;56:745−52.

[2] Tchernitchko D, Goossens M, Wajcman H, et al. In silico prediction of the deleterious effect of a mutation: proceed with caution in clinical genetics. Clin Chem 2004;50:1974–8.

[3] Chan PA, Duraisamy S, Miller PJ, et al. Interpreting missense variants: comparing computational methods in human disease genes CDKN2A, MLH1, MSH2, MECP2, and tyrosinase (TYR). Hum Mutat 2007;28:683–93.

[4] Goldgar DE, Easton DF, Deffenbaugh AM, et al. Breast Cancer information Core (BIC) steering committee. Integrated evaluation of DNA sequence variants of unknown clinical significance: application to BRCA1 and BRCA2. Am J Hum Genet 2004;75:535–44.

[5] Richards S, Aziz N, Bale S, et al. Standards and guidelines for the interpretation of sequence variants: a joint consensus recommendation of the American College of Medical Genetics and Genomics and the Association for Molecular Pathology. Genet Med 2015;17:405–24.

Chapter 6

# Understanding human disease in the post-genomic era

**William B. Coleman**

*American Society for Investigative Pathology, Rockville, MD, United States*

**Summary**

Molecular pathology represents the application of the principles of basic molecular biology to the investigation of human disease processes. Many genetic diseases have now been characterized that result from the mutation of a single gene or are associated with a specific chromosomal rearrangement. These genetic diseases include sickle-cell anemia, hemophilia, cystic fibrosis, Duchenne muscular dystrophy, Tay-Sachs disease, Down syndrome, Li-Fraumeni syndrome, Wilms tumor, Prader–Willi syndrome, Angelman syndrome, and many others. Our current understanding of the molecular basis of these diseases can be attributed, at least in part, to the complete sequencing of the human genome. Our ever broadening insights into the molecular basis of disease processes continues to provide an opportunity for the clinical laboratory to develop and implement new and novel approaches for diagnosis and prognostic assessment of human disease.

## Introduction

Early biochemists determined the essential building blocks of living cells and characterized their chemical nature, including nucleic acids (long-chain polymers composed of nucleotides). Nucleic acids were named based partly on their chemical properties and partly on the observation that they represent a major constituent of the cell nucleus. Remarkably, the recognition and acceptance that nucleic acids represent the chemical basis for transmission of genetic traits only occurred about 70 years ago. Prior to that time, there was considerable disagreement among scientists as to whether eukaryotic genetic information was contained in and transmitted by proteins or nucleic acids. Chromosomes had been shown to contain deoxyribonucleic acid (DNA) as a primary constituent, but it was not known if this DNA carried genetic information or merely served as a scaffold for some undiscovered class of proteins that carried genetic information. However, the demonstration that genetic traits could be transmitted through DNA formed the basis for numerous investigations focused on elucidation of the nature of the genetic code. During the last half-century, numerous investigators have participated in the scientific revolution leading to modern molecular biology. Of particular significance were the elucidation of the structure of DNA; determination of structure–function relationships between DNA and RNA; and acquisition of basic insights into the processes of DNA replication, RNA transcription, and protein synthesis.

Technical developments during the 1970s and 1980s set the stage for the launch of the Human Genome Project. These technical advances include the development of early DNA sequencing methods, amplification methods, cloning methods (particularly methods for cloning large segments of DNA), and accumulation of resources for chromosomal mapping studies. The Human Genome Project emerged from early studies (beginning in 1986–87) conducted by the United States Department of Energy (DOE). In 1988, the United States Congress provided funding for the National Institutes of Health (NIH) and the DOE to embark on a large-scale endeavor to "... coordinate research and technical activities related to the human genome..." and subsequently the National Center for Human Genome Research (NCHGR) branch of the NIH was born (in 1989). In 1990, the initial plans for the Human Genome Project were completed (https://www.genome.gov/10001477/human-genome-projects-fiveyear-plan-19911995/). The research plan set out specific goals for the first 5 years of what was anticipated to be a 15-year research effort. In 1993, a new 5-year plan for the Human Genome Project was released. In 1997, the NCHGR became the National Human Genome

*Essential Concepts in Molecular Pathology. DOI: https://doi.org/10.1016/B978-0-12-813257-9.00006-1*
© 2020 Elsevier Inc. All rights reserved.

Research Institute, and in 1998 a third 5-year plan for the Human Genome Project was released. In 2000, it was announced that the majority of the human genome had been sequenced, and the report containing 90% of the human genome sequence was published in early 2001. For the last decade and a half, we have lived in the so-called post-genomic era. During this time, the results of the Human Genome Project have been expanded and exploited to advance research related to specific diseases.

Molecular pathology represents the application of the principles of basic molecular biology to the investigation of human disease processes. Many genetic diseases have now been characterized that result from the mutation of a single gene or are associated with a specific chromosomal rearrangement. These genetic diseases include sickle-cell anemia, hemophilia, cystic fibrosis, Duchenne muscular dystrophy, Tay-Sachs disease, Down syndrome, Li-Fraumeni syndrome, Wilms tumor, Prader−Willi syndrome, Angelman syndrome, and many others. Our current understanding of the molecular basis of these diseases can be attributed, at least in part, to the complete sequencing of the human genome. Our ever broadening insights into the molecular basis of disease processes continues to provide an opportunity for the clinical laboratory to develop and implement new and novel approaches for diagnosis and prognostic assessment of human disease.

## Structure and organization of the human genome

The human genomic DNA is packaged into discreet structural units that vary in size and genetic composition. The structural unit of DNA is the chromosome, which is a large continuous segment of DNA. A chromosome represents a single genetically specific DNA molecule to which are attached a large number of protein molecules that are involved in the maintenance of chromosome structure and regulation of gene expression. Genomic DNA contains both coding and noncoding sequences. Noncoding sequences contain information that does not lead to the synthesis of an active RNA molecule or protein. This is not to suggest that noncoding DNA serves no function within the genome. On the contrary, noncoding DNA sequences have been suggested to function in DNA packaging, chromosome structure, chromatin organization within the nucleus, or in the regulation of gene expression. A portion of the noncoding sequences represent intervening sequences that split the coding regions of structural genes. However, the majority of noncoding DNA falls into several families of repetitive DNA whose exact functions have not been entirely elucidated. Coding DNA sequences give rise to all of the transcribed RNAs of the cell, including mRNA.

### DNA carries genetic information

DNA is a polymeric molecule that is composed of repeating nucleotide subunits. The order of nucleotide subunits contained in the linear sequence or primary structure of these polymers represents all of the genetic information carried by a cell. Each nucleotide is composed of (1) a phosphate group, (2) a pentose (5 carbon) sugar, and (3) a cyclic nitrogen-containing compound called a base. In DNA, the sugar moiety is 2-deoxyribose. Eukaryotic DNA is composed of four different bases: adenine, guanine, thymine, and cytosine. These bases are classified based on their chemical structure into two groups: adenine and guanine are double-ring structures termed purines, and thymine and cytosine are single-ring structures termed pyrimidines. Within the overall composition of DNA, the concentration of thymine is always equal to the concentration of adenine, and the concentration of cytosine is always equal to guanine. Thus, the total concentration of pyrimidines always equals the total concentration of purines. These monomeric units are linked together into the polymeric structure by 3′, 5′-phosphodiester bonds. Natural DNAs display widely varying sizes depending on the source. Relative molecular weights range from $1.6 \times 10^6$ Da for bacteriophage DNA to $1 \times 10^{11}$ Da for a human chromosome.

The structure of DNA is a double helix composed of two polynucleotide strands that are coiled about one another in a spiral. Each polynucleotide strand is held together by phosphodiester bonds linking adjacent deoxyribose moieties, and the two polynucleotide strands are held together by noncovalent interactions, including lipophilic interactions between adjacent bases and hydrogen-bonding between the bases on opposite strands. The sugar-phosphate backbones of the two complementary strands are antiparallel. That is, they possess opposite chemical polarity. As one moves along the DNA double helix in one direction, the phosphodiester bonds in one strand will be oriented 5′−3′, whereas in the complementary strand, the phosphodiester bonds will be oriented 3′−5′. This configuration results in base pairs being stacked between the two chains perpendicular to the axis of the molecule. The base pairing is always specific: adenine is always paired to thymidine, and guanine is always paired to cytosine. This specificity results from the hydrogen-bonding capacities of the bases themselves. Adenine and thymine form two hydrogen bonds, and guanine and cytosine form three hydrogen bonds. The specificity of molecular interactions within the DNA molecule allows one to predict the sequence of nucleotides in one polynucleotide strand if the sequence of nucleotides in the complementary strand is known. Although the hydrogen bonds themselves are relatively weak, the number of hydrogen bonds within a DNA molecule results in a very stable molecule that would never spontaneously separate under physiological conditions.

The genomes of any two people are more than 99% similar. Therefore, the small fraction of the genome that varies among humans is very important. These variations of DNA are what make humans unique. However, variations in DNA can occur in the form of genetic mutations in which a base is missing or changed. This results in an aberrant protein and can lead to disease. Through studies of the genetic variation of humans, it is hoped that we can gain insight into phenotypic variation and disease susceptibility. In addition, it is thought that DNA structure plays a role in certain human genetic diseases. Certain trinucleotide (CTG and CCG) repeat sequences have been shown to be found in genes whose aberrant expression leads to disease. The severity of the disease is associated with the number of repeats; diseased individuals have greater than 50 repeats, whereas normal individuals have very few repeats.

## General structure of the human genome

DNA packaging is mediated by histone proteins. The core nucleosome particle is composed of 147 bp of DNA wrapped around an octamer of four core histone proteins. These nucleosomes fold into 30 nm chromatin fibers, which are the components that make up a chromosome. The human genome is $3 \times 10^9$ bp or a length of about 1 m, which compacts into a nucleus that is only $10^{-5}$ m in diameter. Regulation of chromatin has profound consequences for the cell, as the ability to open and close the environment in which DNA is packaged is the primary mechanism by which the genes encoded within the DNA get expressed into proteins. The structure of chromatin is now well understood, but how chromatin is packaged into a chromosome is not. Chromosomes are clearly visible with dyes that react with DNA, which can then be visualized under a standard light microscope. The word chromosome is derived from Greek and describes a colored body, which reflects the ability to visualize dense regions. Dense, compact regions of chromosomes are referred to as heterochromatin consisting of mostly untranscribed and inactive DNA. Regions called euchromatin are less compact and consist of more highly transcribed genes. The genetic code is found in the DNA sequence, although the way in which the DNA is packaged into chromatin plays an important role in controlling and organizing the information that the DNA holds. When packaged into chromatin, some information is accessible and some is not, which depends on chemical modifications to the histone proteins. Chromatin is dynamic, and the accessible regions of DNA change during human development or different disease states. This process of altering gene expression in a stable, heritable manner without changing the DNA code is referred to as epigenetics.

## Chromosomal organization of the human genome

Our genome contains 46 chromosomes with 22 autosomal pairs and two sex chromosomes. These chromosomes differ about fourfold in size from chromosome 1 to chromosome 21, which is largest to smallest, respectively. Each of the 46 chromosomes in human cells contains a centromere (central region) and telomere (ends of the chromosome) composed of genes (2%); regulatory elements (1%−2%); noncoding DNA (50%), which includes chromosome structural elements, replication origins, and repetitive elements; and other sequences (45%). At the end of the 19th century it was accepted by numerous researchers that chromosomes formed the basis of inherited traits. There are approximately 60 trillion cells in the human body, which all originate from a single fertilized cell. The cells in the body undergo cell division, or mitosis, in which the chromosomes are condensed and genomic DNA is faithfully replicated. The nomenclature used to define the segments on a chromosome was determined by G-banding chromosomal staining, where the mitotic chromosomes were digested with trypsin and followed with Giemsa staining, which stains centromeric regions. The short arm region of the chromosome (usually displayed above the centromeres) is referred to as the p arm (for instance, 17p) and the long arm region of the chromosome (displayed below the centromeres) is called the q arm (for instance, 13q), with each band having a number associated with it. Chromosomal banding studies using Giemsa staining have shown that heterochromatin comprises 17%−20% of the human chromosome and consists of different families of alpha satellite DNAs and other higher order repeats.

## Subchromosomal organization of human DNA

### DNA features

The human genome contains a number of different repetitive DNA sequences, including Alu repeats, mammalian interspersed repeats (MIRs), medium reiteration repeats, long terminal repeats, and long interspersed nucleotide elements

(LINEs). Repetitive DNA such as short interspersed nuclear fragments, including MIRs and LINEs, are distributed throughout the genome, whereas satellite sequences are clustered in discrete areas (centromeres). Repetitive sequences have been recently proposed to be involved in genome compaction. Genomic regions of satellite DNA are condensed throughout the cell cycle, and there is evidence that LINEs are involved in X-chromosome condensation.

CpG islands appear in approximately 50% of human genes and are located preferentially at the promoter region of genes, flanking the transcription start site. It has been estimated that there are around 30,000—45,000 total CpG islands in the human genome. A CpG island is defined as a region with greater than 200 bp with a G+C percentage that is greater than 50% with an observed/expected CpG ratio that is greater than 0.6. CpG dinucleotides are sites for DNA methylation and in turn can downregulate gene expression. DNA methylation has been shown to be important during gene imprinting and tissue-specific gene expression. Gene inactivation by aberrant DNA methylation has been correlated with cancer in many different cell types. In addition, the identification of CpG islands throughout the genome can help predict promoter regions for human genes.

### Gene structure

Now that the human genome has been sequenced, one of the next steps is to utilize genomic tools to obtain a picture of how DNA is targeted by transcription factors and cofactors to regulate gene expression. These proteins (transcription factors and cofactors) control whether a gene is on or off. Transcription factor binding sites are thought to contain conserved sequence motifs of 6—20 bp. Transcription factor binding proteins bind to *cis*-acting elements including promoters, enhancers, silencers, splicing regulators, chromosome boundary elements, insulator elements, and locus control regions to control gene expression that regulates cell development and fate. The goal of the Encyclopedia of DNA Elements (ENCODE) Project is to identify and define all of these sequences in the human genome. Nuclease-hypersensitive sites are regions of DNA that interact with transcription factors in the chromatin environment in vivo. *Trans*-acting factors bind chromatin at DNase I hypersensitive sites (DHSs), which occur at accessible chromatin regions. Interestingly, CpG islands are associated with DHSs that are either constitutive or tissue-specific.

## Overview of the human genome project

The sequencing of the human genome was first proposed by Robert Sinsheimer (University of California at Santa Cruz) in 1985. This idea was met with some critiques from the scientific community, many thinking the idea was premature and unattainable. However, in 1988 project planning was launched with joint funding from the NIH and the DOE, and in 1990, the Human Genome Project was officially initiated, proposed to take 15 years, and had a budget of $3 billion. The first 5 years of the Human Genome Project sought to map the genetic and physical features of the human genome. Although the United States made the largest contribution to the Human Genome Project, it was an international effort with contributions from Britain, France, Germany, Canada, China, and Japan. Several species of bacteria and yeast had been completely sequenced in 1996, and this progress spurred the attempt at sequencing the human genome on a small scale. Eight years into the project in 1998, the plan included a sequencing facility to be built that would help sequence the human genome in only a 3-year period, ahead of schedule. The Human Genome Project agreed to release all sequences to the public and on June 26, 2000, a working draft of the human genome became available. Almost 3 years later on April 14, 2003, the Human Genome Project accomplished its ultimate goal and announced that the sequencing of the human genome was completed. Remarkably, this announcement occurred almost 50 years to the date of Watson and Crick's influential publication of the DNA double helix. Therefore, genomic science rapidly developed from the identification of the structure of DNA to the sequencing of the human genome (and many other organisms) in a span of 50 years.

## The human genome project—objectives and strategy

The primary goal of the Human Genome Project was to obtain the complete DNA sequence of the human genome by 2005. Through use of a whole-genome random shotgun method and a whole-genome assembly, along with a regional chromosome assembly, and through combination of sequence data from Celera (a private sector company that agreed to help sequence the human genome for profitable purposes) and the publicly funded genome center, a 2.91 billion bp consensus sequence was derived from the DNA of five individuals. At first the whole-genome shotgun approach proposed in 1997 by Weber and Mayers for the sequencing of the human genome was not well received. However, at that time (almost 8 years into the sequencing of the human genome), only 5% of the genome sequence had been completed, and it was clear that the goal of finishing by 2005 was unattainable. At that time PE Biosystems (now Applied Biosystems) developed a sequencer called the ABI PRISM 3700 DNA analyzer, which was going to be a part of Celera. Now with the ability to sequence with an automated, high-throughput capillary DNA sequencer, as well as new developments in tracking for whole-genome assembly, the chosen test case of the whole-genome assembly on a eukaryotic genome was *Drosophila melanogaster*. The *Drosophila* genome, comprising 120 Mb of euchromatic DNA, was sequenced over a 1-year period.

The Human Genome Project enrolled 21 donors and collected approximately 130 mL of blood from males and females from a variety of ethnic backgrounds. From the 21 donors, five were chosen, including two males and three females: two Caucasians, one Hispanic Mexican, one Asian Chinese, and one African American. In order for the shot-gun sequencing method to be fully utilized, the plasmid DNA libraries needed to be uniform in size, nonchimeric, and representative of the whole genome (rather than randomly representing the genome). Therefore, DNA from each donor was inserted into 2, 10, or 50 Kb plasmid libraries.

## Human genome project findings and current status

A complete and detailed analysis of the Human Genome Project was published by Venter et al. In the wake of the human genome sequence, there was considerable acceleration in the success of the identification of genes that were important for the development of disease. In 1990, fewer than 10 genes had been identified by positional cloning, but by 1997 that number grew to more than 100 genes.

The Human Genome Project defined 26,383 genes with confidence using a unique rule-based system called Otto. Regions of sequence that were likely gene boundaries were matched up with BLAST and partitioned by Otto, and grouped into bins of related sequence that may define a gene. Known genes were then matched to the corresponding cDNA and were annotated as a predicted transcript. However, the genome sequence has variations and frameshifts, and it was not always possible to predict a transcript perfectly agreed. Therefore, if a transcript matched the genome assembly for at least 50% of its length at greater than 92% identity, then the region was annotated by Otto. More recent analysis of the human genome sequence data suggests approximately 20,000 protein coding genes. It was predicted that an average gene in the human genome is approximately 27,874 bases. These variations in the human genome are being cataloged and will provide clues for the risk and diagnosis of common genetic diseases. Early results indicated that the human genome contained >2 million single nucleotide polymorphisms (SNPs). Now it is recognized that the human genome contains >10 million SNPs. Development of DNA arrays for parallel analysis of numerous genes at once has enabled determination of the gene expression patterns of as many as >10,000 genes at one time. Whole genome analyses enable all known and predicted genes to be analyzed concurrently. Next-generation sequencing methods are rapidly evolving to allow investigators to interrogate DNA sequences (including variations and mutations) in conjunction with examination of gene expression patterns.

The Human Genome Project examined the genome for regions that were gene-rich and regions that were devoid of genes. A gene-poor region was defined as a region greater than 500 Kb lacking an open reading frame. Using these criteria, about 20% of the genome represents gene-poor regions, and they were not evenly distributed throughout the genome. Gene-poor chromosomes were 4, 13, 18, X, reflecting 27.5% gene-free regions of the total 492 Mb, and gene-rich chromosomes were 17, 19, 22, having only 12% gene-free regions within their 171 Mb. The initial draft sequence of the human genome retained some gaps and the years following the sequencing of the human genome were spent closing these sequencing gaps for all chromosomes. Chromosomes 21 and 22 were completed first.

The Human Genome Project correlated CpG islands with gene start sites of computationally annotated genome transcripts and the entire human genome sequence. The Human Genome Project compared the variation of the CpG island computation with Larsen et al. and used two different thresholds of CG dinucleotide likelihood, including the original ratio of 0.6. The analysis showed a strong correlation between first coding exons and CpG islands. Genome-wide repeat elements were examined by the Human Genome Project. Approximately 35% of the human genome is composed of different repeat elements, with chromosome 19 having 57% repeat density, the highest repeat density as well as the highest gene density. Gene density and Alu repeat elements exhibit an association, whereas this was not observed with the other classes of repeat elements.

The human genome sequence and the variations contained within must be utilized to identify the genes responsible for hereditary diseases. Large-scale gene duplication was identified by the Human Genome Project. These included duplications that were known to be associated with proteins involved in bleeding disorders, developmental diseases, and cardiovascular conduction abnormalities. The duplications were located throughout the genome. However, there were gene families that were scattered in blocks within the genome, such as the olfactory receptor family. Chromosome

2 contains two very large duplications that are shared by two different chromosomes, 14 and 12. The first duplicated region is a block of 33 proteins spread in eight different regions spanning 20 Mb of 2p, and these genes are also found on chromosome 14 spanning 63 Mb. The second duplication is on 2q and chromosome 12. This duplication includes two of the four known Homeotic (*Hox*) gene clusters, and the other two *Hox* gene clusters are also seen as duplications on two different chromosomes. These *Hox* genes play a fundamental role in controlling embryonic development, X-inactivation, and renewal of stem cells. According to the Human Genome Project, SNPs occur frequently, but <1% affect protein assembly and function. These analyses were based on the potential of an SNP to impact protein function based on SNPs that are located within predicted gene coding regions. Interestingly, the frequency of SNPs is highest in intronic regions, followed by intergenic regions, and then exonic regions.

Sequences of known proteins were compared to predicted proteins by the Human Genome Project. Analysis demonstrated that out of the predicted 26,588 proteins, 12,809 (41%) of the gene products could not be classified and were termed proteins with unknown function. The remaining proteins were classified into broad groups based on at least two lines of evidence. Importantly, the molecular functions of the majority of the predicted proteins are transcription factors and proteins that regulate nucleic acid metabolism. Many other proteins were receptors, kinases, and hydrolases, as well as proto-oncogenes, and proteins involved in signal transduction, cell cycle regulators, and proteins that modulate kinase, G protein, and phosphatase activity. Large-scale analysis for characterizing proteins (proteomics) by their structure, function, modifications, localization, and interactions are being accomplished and utilized to gain understanding of their role in disease and cell differentiation.

A major challenge now that the human genome is sequenced is to understand how the DNA code is transcribed into biological processes that determine cell development and fate. The human genome sequence is only the first level of understanding and all functions of genes and the factors that regulate them must be defined. For example, in a disease such as diabetes mellitus, there may be up to 10 genes that result in an increased risk for the development of this very common disorder. Furthermore, with a known DNA sequence, small molecule drugs are being designed to block or stimulate specific gene products and pathways. For example, the main etiology for chronic myelogenous leukemia is a translocation between chromosomes 9 and 22, and a drug was designed to inhibit the kinase activity of the bcr-abl kinase, which is the protein that is produced as a result of this translocation. With advances in our understanding of the molecular underpinnings of human genetic diseases and cancer, the mechanisms for disease development may be elucidated, enabling targeted interventions as well as molecular diagnosis. These represent exciting possibilities, and as research escalates to help provide more preventative medicine and more early treatment options, the field moves ever closer to true personalized medicine.

## Impact of the human genome project on the identification of disease-related genes

The identification of disease-related genes is a laborious process. However, several experimental strategies have been employed in gene discovery. These strategies exploit what is known about candidate genes, including function of the protein/enzyme involved in the disease, the location of the gene within a chromosomal region, or animal models of the human disease in question. Linkage analysis, microsatellite markers, large DNA fragment-cloning techniques, and expressed sequence tags (ESTs) are important in the identification of genes responsible for human diseases. Sequence information and other molecular resources generated by the Human Genome Project have been utilized for the identification of disease-related genes. The human genomic DNA sequence provides the normal template against which disease-associated mutations are identified, as well as numerous (predicted) candidate genes for future studies. Presently, there are more genes than disease phenotypes, enabling the identification of many (if not all) genes associated with single gene disorders. The challenge now is to identify genes that are involved in polygenic disorders such as diabetes, hypertension, and most cancers.

### Positional gene cloning

Positional cloning is an approach that enables isolation/identification of a gene without understanding its function using knowledge of its physical location in the human genome. Positional cloning is slow and laborious and requires methods like chromosome walking and marker sequences (like ESTs). The first step in positional cloning is linkage analysis of the disorder in disease-prone families to determine chromosomal location, and subsequent isolation and testing of genes for mutations that segregate with the chromosomal location of this disorder. A typical chromosomal segment will contain 20–50 candidate genes, and the disease-causing gene is identified based upon presumptions of the disorder in question. Positional cloning has been employed to identify the genes associated with a number of inherited disorders and human cancers. In 1986, Orkin and colleagues reported success in positional cloning the X-linked gene for chronic granulomatous disease.

## Functional gene cloning

Functional cloning is an approach for gene identification that exploits what is known of the protein whose dysfunction is associated with disease without knowledge of the chromosomal location of the gene. The amino acid sequence of the protein (or peptides derived from the protein of interest) is used to predict the gene coding sequence, and cDNA libraries are screened with an oligonucleotide probe (degenerate oligonucleotides). In addition, polymerase chain reaction can be employed to amplify the cDNA using oligonucleotides designed from the amino acid sequence. Functional cloning has been used to identify genes causing human diseases such as phenylketonuria and sickle cell anemia.

## Candidate gene approach

In the candidate gene approach, the cloning of a specific gene depends on having some functional information about the disease and relies on the availability of information on genes that had been previously isolated. This approach depends upon some assumptions made about the kind of protein that may be responsible for the human disorder. Missense mutations in the *p53* genes were cloned using the candidate gene approach and were shown to be the cause of Li-Fraumeni syndrome, an inherited cancer disorder.

## Positional candidate gene approach

The positional candidate approach is employed when the disease-related gene has been mapped to a specific chromosomal location and the sequence of that chromosomal region contains candidate genes for cloning and testing. Candidate genes are analyzed by comparing their DNA sequence to the amino acid sequence of proteins with known functions, and affected individuals are examined to determine which gene (or genes) is responsible for the genetic disorder. The gene responsible for Marfan syndrome, an autosomal dominant disorder of connective tissue, was mapped using the positional candidate approach. Marfan syndrome was mapped to 15q by linkage analysis and this is the location of the fibrillin gene. When DNA from patients with Marfan syndrome was sequenced, mutations were found in the fibrillin gene. Another example of genes identified using the positional candidate approach are the four major genes found to encode proteins of the mismatch repair process that were implicated in hereditary nonpolyposis colon cancer.

# Sources of variation in the human genome

The DNA sequence between humans is >99.5% identical. Therefore, it is important to examine the sequence variation between individuals to gain insight into phenotypic variation, as well as disease susceptibility. SNPs; short tandem repeats; microsatellites and minisatellites; and small (<1 Kb) insertions, deletions, inversions, and duplications are responsible for most of the genetic variations in the human population. It is estimated that the human genome contains approximately 10 million SNPs. These genome variations can give rise to diseases through a gain or loss of dosage-sensitive genes. Through the sequencing of the human genome, new techniques such as genome-scanning arrays and comparative DNA sequence analysis have been developed to examine the composition of the human genome. These technologies have been important in finding copy number variants or segments of DNA that are 1 Kb or larger, including insertions, deletions, and duplications. Genomic disorders are influenced by the genome architecture around the recombination event and share a common mechanism for genomic rearrangement—non-allelic homologous recombination or ectopic homologous recombination between low-copy repeats that flank the rearranged DNA segment. Inversions are created by non-allelic homologous recombination events that occur between inverted low-copy repeats, whereas a non-allelic homologous recombination by direct low-copy repeats results in a duplication or deletion. In addition, nucleotide substitutions and point mutations cause alterations in protein sequence and can result in disease.

# Types of genetic diseases

## Genetic diseases associated with gene inversions

Chromosomal structural variants have been identified in the general population to be the cause of genetic disease in the offspring of parents who exhibit certain DNA inversions. Many of these are described in the Online Mendelian

Inheritance in Man (OMIM) database (https://www.omim.org). In patients with Williams-Beuren syndrome (OMIM #194050), there is a 1.5 Mb inversion at 7q11.23 that occurs in approximately one-third of the patients' parents with a 5% frequency of this inversion in the general population. This syndrome has an incidence of 1/20,000−50,000. An inversion that is 4 Mb at 15q12 is associated with Angelman syndrome (OMIM #105830), and about half of the parents of these patients have this variation as well as 9% of the general population. This syndrome has an incidence of 1/10,000−20,000. There are diseases in which inversions found in the patients affected have not been detected in the general population. Patients with hemophilia A (OMIM #300841) have a 400 Kb inversion in intron 22 in the factor VIII gene, and two copies are located 400 Kb telomeric in an inverted orientation. This nonallelic homologous recombination event results in inactivation of the factor VIII gene. A small inversion in the emerin gene in Emery-Dreifuss muscular dystrophy (OMIM #310300) has been identified. Hunter syndrome is an X-linked dominant disorder (OMIM #300823). Nonallelic homologous recombination between the iduronate 2-sulfatase gene (IDS) and an IDS pseudogene generates a genomic inversion resulting in a disruption of the functional IDS gene occurring in approximately 13% of Hunter syndrome patients. Within the Japanese population, Soto syndrome patients (OMIM #117550), carry a 1.9 Mb inversion variant at 5q35 that affects the *NSD1* gene. Constitutional translocations in the human genome mediated by a polymorphic inversion at olfactory-receptor gene clusters at loci 4p16 and 8p23 occur at frequencies of 12.5% and 26%, respectively. Heterozygous carriers of these translocations exhibit no phenotypic characteristics, whereas their offspring who inherit these translocations show phenotypes from mild dysmorphic features to Wolf-Hirschhorn syndrome (OMIM #194190) characterized by growth defects and severe mental retardation. These examples signify the importance in continuing to characterize inversions within the human genome in the general population in order to examine the risk these variations have on the carriers' offspring.

## Genetic diseases associated with gene deletions

Genomic disorders can be responsible for commonly occurring diseases. For instance, α-thalassemia (OMIM #141800) affects 5%−40% of the population in Africa and 40%−80% in South Asia and results from a homologous deletion of an approximately 4 Kb fragment that is flanked by two α-globin genes on 16q13.3. A nonallelic homologous recombination event between these two copies of the α-globin genes results in the deletion of one functional copy. Red and green pigment genes are located on Xq23, and individuals who have normal color vision have one copy of the red pigment gene and one or more copies of the green pigment gene. In red-green color blindness, which affects 4%−5% of males, deletions or fusions caused by nonallelic homologous recombination occur. In patients with incontinentia pigmenti (OMIM #308300), an 11 Kb deletion occurs by non-allelic homologous recombination between two low-copy repeats, with one in the diseased gene (*NEMO*) and one 4 Kb downstream of the gene. Hereditary neuropathy with liability to pressure palsy (OMIM #162500) is a common autosomal dominant neurological disorder that is caused by a 1.4 Mb deletion of a genomic fragment on 17p12. The gene NF1 that encodes for neurofibromatosis type 1 (OMIM #162200) is located on 17q11.2, and a 1.5 Mb deletion encompassing this gene accounts for 5%−22% of patients with this disease. Patients with DiGeorge syndrome/velocardiofacial syndrome (OMIM #188400) exhibit a 3 Mb deletion within a region-specific repeat unit, LCR22 that is flanked by LCR22A and LCR22D, or a 1.5 Mb deletion that is flanked by LCR22A and LCR22B located on chromosome 22q11.2. Patients with this congenital disease experience recurrent infection, heart defects, and known facial features. Smith-Magenis syndrome (OMIM #182290) effects 1/25,000 individuals and is caused by a 4 Mb deletion on several loci contained within chromosome 17 and depending on the loci involved the severity of mental retardation exhibited by the patient is determined.

## Genetic diseases associated with gene duplications

Charcot-Marie-Tooth disease (CMT; OMIM #601097) is an inherited autosomal dominant trait that occurs in about 1/25,000 individuals and is characterized by atrophy of the muscles in the legs, progressing over time to the hands, forearms, and feet. There are two clinical forms of CMT: type I and type II. In CMT type I (CMT1A), 75% of individuals have a duplication in one of the peripheral myelin protein 22 (*PMP22*) genes. Duplication on both chromosomes at 17p12 produces a severe form of CMT1A where essentially there are four copies of the *PMP22* gene. A central nervous system disorder affecting the myelin sheath covering the nerve fibers in the brain is called Pelizaeus-Merzbacher disease (OMIM #312080). The majority of patients with this disease have a duplication of the proteolipid protein gene (*PLP1*), which is found on Xq21-22.

# Genetic diseases and cancer

The Human Genome Project and the completion of the human genome sequence have significantly impacted the practice of medicine and molecular genetic research. The human genome sequence has led to advances in the development of designer drugs that act on specific molecular targets and pathways to disrupt diseases caused by single genes or combinations of gene products. It is beyond the scope of this chapter to provide a comprehensive review of genetic diseases and/or the genetic causes of cancer. However, a few examples of how the Human Genome Project advanced our understanding of human disease are included.

# Cystic fibrosis

The cystic fibrosis (CF) transmembrane regulator (*CFTR*) gene is located on chromosome 7q31.2 and was the first gene to be identified using resources from the Human Genome Project. Researchers defined the *CFTR* gene by the positional cloning approach. The function of the CFTR protein is to regulate chloride secretion and the inhibition of sodium absorption across the cell membrane. Approximately 1547 mutations of the *CFTR* gene have been described, with the most common mutation represented by a 3-bp DNA deletion that results in a loss of phenylalanine at position 508 occurring in 66% of CF patients. Out of the mutations described, only 23 have been shown directly to cause sufficient loss of *CFTR* to confer CF disease, and these mutations are seen in 85% of the diseased population. Interestingly, two or more *CFTR* mutations can be located in *trans* on two separate chromosomes and this will confer CF. However, mutations found in *cis* on the same chromosome are not associated with disease. Unfortunately, this distinction between *cis* and *trans* on chromosomes is not determined by most commercial testing laboratories. In addition, different mutations confer different CF phenotypes with some resulting in milder forms of the disease. Approximately 9.7% of genotyped individuals in the Cystic Fibrosis Foundation Patient Registry have at least one unidentified mutation, but the majority (90%) of *CFTR* mutations can be detected using regular screening methods. The discovery of the *CFTR* gene has given researchers a better understanding of the etiology, the genetic bases, and the pathobiology of CF.

CF is an autosomal recessive disease (OMIM #219700) that occurs in approximately 1 in 3500 newborns, is the most common lethal inherited genetic disease among the Caucasian population, and affects almost 30,000 Americans. Treatment advances for patients with CF have increased the survival age from mid-teens in the 1970s, to late-twenties to thirties in the 1990s, to more than 36 years old today. A CF diagnosis is based on several clinical characteristics, a familial history of CF or a positive CF newborn screening test, and a mutation in the *CFTR* gene and/or protein. Newborn screening has been implemented in all 50 states. CF is a disease that is caused by the improper regulation of the ion channels between the cell cytoplasm and the surrounding fluid, resulting in the inability of the exocrine epithelial cells to transport fluid and electrolytes in and out of the cells. CF patients have an abnormal accumulation of viscous, dehydrated mucus, and because of this cannot effectively clear inhaled bacteria resulting in an excessive inflammatory response to pathogens. Francis Collins summarized the important contribution of the Human Genome project in understanding genetic diseases such as CF: "… *Cystic fibrosis has become the paradigm for the study of genetic diseases and indeed, for the medicine of the future. The notion that it is possible to identify genes whose structure and function are unknown and to use that information to understand given disease and develop "designer" therapies is becoming the central paradigm of biomedical research, and cystic fibrosis is the disease that leads that charge…*"

# Phenylketonuria

Mutations in the phenylalanine hydroxylase gene (*PAH*) encoding the protein L-phenylalanine hydroxylase causes a mental retardation disease called phenylketonuria (PKU; OMIM #261600). The inability to hydrolyze phenylalanine to tyrosine leads to hyperphenylalaninemia, and elevated levels of phenylalanine produces toxic effects on the developing brain. In the 1980s, the *PAH* gene was cloned, sequenced, and mapped to chromosome 12q23.2. The PKU phenotype is not a simple disease, nor does it have a simple explanation. As with many other genetic diseases, each patient with PKU has to be treated differently. The locus for *PAH* covers 1.5 Mb of DNA and includes various features, including SNPs, repeat sequences, polymorphisms, and *cis* control elements embedded in the sequence, as well as harboring five other genes, providing for a wide range of disease-causing mutations. The *PAH* gene is expressed in the liver and kidney.

PKU caused a paradigm shift of attitudes about genetic disease by becoming one of the first disorders to show a treatment effect. PKU is an autosomal recessive inherited disease, causing mental retardation; a mousy odor; light pigmentation; peculiarities of sitting, standing, and walking; as well as eczema and epilepsy. The average incidence of PKU in the United States is 1 in 8000. PKU is one of the first genetic diseases to have an effective rational therapy. PKU can be identified with a biochemical test in newborns and is managed through a phenylalanine-free, tyrosine-supplemented diet, which permits normal or near-normal cognitive development. In the adolescent and adult patients, it was difficult to adopt the diet recommendations of PKU due to deficiencies in both organoleptic properties and nutrient content in the food. Fortunately, many diet deficiencies are being overcome, and diagnosis is occurring earlier so patients begin the recommended diets sooner, and are more aggressive throughout life.

## Breast cancer

The majority of hereditary breast and ovarian cancers are caused by mutations in the breast cancer-predisposing gene 1 or 2 (*BRCA1* or *BRCA2*). *BRCA1* was found by candidate gene approach in 1991, and *BRCA2* was located by linkage analysis and positional cloning in 1995 using familial breast cancer pedigrees with multiple cases of breast cancer in many generations. *BRCA1* is located on 17q21 encoding an 1863-amino acid polypeptide, and *BRCA2* is found on chromosome 13q12-13 encoding 3418 amino acids. *BRCA1* has been implicated in cell-cycle regulation, chromatin remodeling, protein ubiquitylation, and both proteins are involved in DNA repair. In the Ashkenazi Jewish population, there are founder mutations that occur at specific locations in *BRCA1* (185delAG and 5382insC) and *BRCA2* (617delT), but many mutations occur elsewhere in the gene, including frameshift or nonsense mutations as well as deletions or duplications. DNA-based methods have been recently employed to conduct analysis of both *BRCA1* and *BRCA2* for the presence of genomic rearrangements. The prevalence of genomic rearrangements in *BRCA1* is higher than that of *BRCA2*, accounting for 8%−19% of the total mutations in *BRCA1* and 0%−11% in *BRCA2* mutations.

Breast cancer affects one in eight women in the United States, and a woman born in the United States has an average lifetime risk of 13% for developing breast cancer. Familial breast cancer is associated with 10%−20% of all breast cancer cases. Mutations in the *BRCA1* and *BRCA2* genes in women have a 60%−80% increase of developing breast cancer. In addition, women who carry a mutation in the *BRCA1* gene have a 15%−60% lifetime risk for ovarian cancer, a greater risk than women with a mutation in the *BRCA2* gene (10%−27%). Women with *BRCA2* mutations tend to develop ovarian cancer after 50 years of age. Women who are under the age of 50 years and are *BRCA1* mutation carriers have a 57% chance of being diagnosed with breast cancer, and only a 28% chance of developing breast cancer if the *BRCA2* gene is mutated. Interestingly, men who harbor a *BRCA2* mutation are estimated to have a 6% chance of being diagnosed with breast cancer. *BRCA1*-mutated breast cancers are found to be more poorly differentiated, whereas *BRCA2*-mutated cancers tend to be higher-grade compared to sporadic breast cancer (lacking a hereditary component). *BRCA1*-mutated breast cancers are frequently triple-negative for estrogen receptor, progesterone receptor, and HER-2/Neu overexpression. It is recommended for women with *BRCA* mutations to begin monthly breast self-examinations at the age of 18 years and clinical breast examinations and annual mammograms beginning at 25 years of age.

## Non-polyposis colorectal cancer

Hereditary nonpolyposis colorectal cancer (HNPCC) is caused by mutations in the mismatch repair (MMR) genes *MLH1*, *MSH2*, *MSH6*, and *PMS2*. A hypermutation phenotype was discovered in 1993 in families with HNPCC similar to that observed in MMR-deficient bacteria and yeasts. Linkage analysis and positional cloning in HNPCC families subsequently identified *MSH2* and *MLH1* genes, and mutations in these genes account for 60%−80% of HNPCC diagnoses. Additionally, the MMR genes *PMS2* and *MSH6* are associated with HNPCC. A higher risk of colorectal cancer occurs in *MSH2* and *MLH1* mutation carriers as compared to *MSH6* or *PMS2* mutation carriers. The *MSH2* and *MSH6* genes are located on chromosome 2p22-p21 and 2p16, respectively. *MLH1* is found on 3p21.3, and *PMS2* is located on chromosome 7p22.

There are approximately 160,000 new cases of colorectal cancer diagnosed in the United States each year, with HNPCC accounting for 2%−7% of diagnosed colorectal cancer, affecting about 1 in 200 individuals. The average age of HNPCC diagnosis is 44 years old. HNPCC is also referred to as Lynch Syndrome. HNPCC is an autosomal dominant trait and exhibits phenotypic characteristics of less than 100 colonic polyps and early onset of multiple tumors in the colon. The Amsterdam criteria were established for the clinical designation of a family with HNPCC: (1) three or more relatives with colon cancer, one of them must be a first-degree relative (parent, child, sibling) of the other two; (2) at least two affected generations; (3) one or more members of a family must develop colon cancer before the age of 50 years; and (4) familial adenomatous polyposis should be excluded from the diagnosis.

# Perspectives

Millions of people around the world waited and watched for the completed human genome sequence to be released with the expectation that it would benefit humankind. Decades ago it was not anticipated that genomic disorders would represent such a common cause of human genetic disease. Now from this perspective humans are the best model organisms that we have to study the human genome, disease, and its associated phenotypes. Currently, large amounts of DNA sequence information have been generated from genome sequencing projects, including vertebrates and invertebrate species. One objective of the human genome sequence is to derive medical benefit from analyzing the DNA sequence of humans. It is undeniable that genomic science will begin to unlock more of the mysteries of complex hereditary factors in heart disease, cancer, diabetes, schizophrenia, and many more. Genetic tests have become available for individuals who have a strong family history or are more susceptible to a particular disorder, such as breast cancer or colorectal cancer. Healthcare professionals will become practitioners of genomic medicine as more genetic information about common illnesses is available and healthy individuals want to protect themselves from illness. Clinicians will have to grasp the understanding and advances of molecular genetics, and a group of physicians, nurses, and other clinicians called *The National Coalition for Health Professional Education in Genetics* has been organized to help prepare for the genomics era. Within the next decade, it is exciting to think that designer drugs will be available for diabetes mellitus, hypertension, mental illness, and many other genetic disorders. All cancers have a unique molecular fingerprint and the promise of individualizing medicine by tailoring therapeutic management to that person's unique molecular profile is becoming realized. Also within a decade or two, it may be possible to sequence the genome of an individual human with minimal laboratory cost (maybe less than $1000). If this becomes reality, we can imagine the possibilities for scientific research, clinical care, treatment options, and an overall dramatic change in the face of medicine.

# Key concepts

- In 1953, James D. Watson and Francis H.C. Crick discovered the double helical structure of DNA. This discovery single-handedly revolutionized molecular biology and the biological sciences, and formed the basis for currently accepted models for DNA replication, transcription of RNA, and DNA repair.
- The genomes of any two people are more than 99% similar. Therefore, the small fraction of the genome that varies among humans is very important. These variations of DNA sequence make humans unique. However, variations in DNA sequence can also occur in the form of mutations in which a base is missing or changed, resulting in the expression of aberrant protein products that can contribute to disease.
- The contribution of the Human Genome Project (HGP) to scientific research cannot be underestimated. Results from the HGP have significantly contributed to the current understanding of the genetic causation of disease, and the interactions between the environment and heritable traits defining human disease conditions.
- The post-genomic era is associated with an acceleration of success in identifying genes that are important in the development of disease.
- Single nucleotide polymorphisms (SNPs), short tandem repeats, microsatellite sequences, and small insertions, deletions, inversions, and duplications are responsible for most of the genetic variation in the human population.

# Suggested readings

[1]  Collins FS, Galas D. A new five-year plan for the U.S. Human Genome Project. Science 1993;262:43−6.
[2]  Collins FS, Patrinos A, Jordan E, et al. New goals for the U.S. Human Genome Project: 1998−2003. Science 1998;282:682−9.
[3]  Venter JC, Adams MD, Myers EW, et al. The sequence of the human genome. Science 2001;291:1304−51.
[4]  Lander ES, Linton LM, Birren B, et al. Initial sequencing and analysis of the human genome. Nature 2001;409:860−921.
[5]  The 1000 Genomes Project Consortium. A global reference for human genetic variation. Nature 2015;526:68−74.
[6]  Sudmant PH, Rausch T, Gardner EJ, et al. An integrated map of structural variation in 2,504 human genomes. Nature 2015;526:75−81.
[7]  The 1000 Genomes Project Consortium. An integrated map of genetic variation from 1,092 human genomes. Nature 2012;491:56−65.
[8]  The 1000 Genomes Project Consortium. A map of human genome variation from population-scale sequencing. Nature 2010;467:1061.
[9]  Collins FS, McKusick VA. Implications of the Human Genome Project for medical science. JAMA 2001;285:540−4.

Chapter 7

# The human transcriptome: implications for understanding, diagnosing, and treating human disease

Matthias E. Futschik[1], Markus Morkel[2], Reinhold Schäfer[3,4] and Christine Sers[2]

[1]School of Biomedical Sciences, Faculty of Medicine and Dentistry, Institute of Translational and Stratified Medicine (ITSMED), University of Plymouth, Plymouth, United Kingdom, [2]Laboratory of Molecular Tumor Pathology and Tumor Systems Biology, Charité − Universitätsmedizin Berlin, Berlin, Germany, [3]Charité Comprehensive Cancer Center, Charité − Universitätsmedizin Berlin, Berlin, Germany, [4]German Cancer Consortium (DKTK), German Cancer Research Center, Heidelberg, Germany

**Summary**

Transcriptomics is based on the fascinating capacity of analyzing simultaneously the entire set of RNA molecules or transcripts (messenger RNAs, microRNAs, long noncoding RNAs) produced in a population of cells or in tissues. While capturing transcript abundance on genome-wide level has become a routine task thanks to high-throughput technologies, the analysis of the transcriptome continues to be a challenging task due to the large heterogeneity and variability of the RNA content of a biological entity. The abundance of individual transcripts varies from a few copies to thousands of copies per cell. The kind and copy numbers of individual transcripts expressed at a given time depend on the developmental stage, external conditions, and environmental stimuli. Quantitative and qualitative alterations of RNAs can be directly linked to the molecular mechanism of disease, reflect the downstream consequences of the disease processes, or even possess prognostic and predictive value.

## Introduction

This chapter outlines the methodological prerequisites for transcriptome analysis and describes typical applications in molecular cell biology and pathology with a particular focus on cancer. During the last three decades, technology development and experimental approaches aiming at mRNA analysis were significantly fueled by molecular cancer research. One of the main reasons for progress in this area was the availability of relevant cell lines that could be propagated indefinitely and served as reproducible sources of RNA and of sufficient quantities of normal and diseased tissues. A strong motivation lay in the strong demand for distinguishing as many transcripts as possible in normal and tumorigenic cells to understand cancer-specific alterations in gene expression. While early work along these lines was mostly related to pathogenesis, more recent applications deal with diagnostic issues such as tumor outcome, prognosis, and therapy response prediction.

## Gene expression analysis from single genes to microarrays—early attempts to search for genes involved in normal physiological processes and pathogenesis

At present, transcriptome analysis through next-generation sequencing (NGS) is considered as the gold standard and has widely replaced the microarray-based expression profiling that dominated the field for more than a decade. Even prior to the advent of microarray technology, researchers used several classical approaches to study gene expression under normal physiological and disease conditions. In 1977 and 1980, researchers described the northern blot technique for transferring RNA, fractionated by electrophoresis, from an agarose gel to special paper strips, the coupling of the RNA to the paper surface, and the detection of specific RNA bands by hybridization with radioactive $^{32}$P-labeled DNA

*Essential Concepts in Molecular Pathology.* DOI: https://doi.org/10.1016/B978-0-12-813257-9.00007-3

probes followed by autoradiography. Several years later, other scientists used a more sophisticated approach for contrasting mRNA patterns of cellular material obtained from colon tumor biopsies by establishing a library of approximately 4000 complimentary DNA (cDNA) clones, obtained by reverse transcription of mRNA. The comparison of normal colonic mucosa with carcinomas showed expression alterations of approximately 7% of the cloned sequences and was extrapolated to the entire set of transcripts. The number of alterations was smaller between normal mucosa and benign adenomas indicating that transcriptional changes accumulate during cancer progression. Further advances in deciphering cancer-related transcripts were driven by increasingly efficient cDNA cloning and sequence techniques. Large collections of expressed sequence tags (ESTs) obtained from diverse cell types and tissues were deposited in expression databases. As a practical consequence of the global gene expression information provided by cDNA/EST databases, an approach called electronic Northern became feasible that allowed prediction of expression changes between normal and diseased tissues. These analyses were further boosted by advanced techniques including, suppressive subtraction hybridization, representational difference analysis, differential display, and serial analysis of gene expression.

Microarray technology provided an important technical breakthrough in transcriptome analysis in the 1990s and was pioneered by Pat Brown and colleagues at Stanford University, who not only published the first applications in addressing biological questions, but also laid open the necessary technical devices in detail. In parallel, a highly standardized microarray technology was developed by Affymetrix, Santa Clara, CA and other companies. The central element of the technique is that DNA-molecules, cDNA fragments or oligonucleotides (called probes), are arrayed and immobilized at defined positions on a solid support or matrix. The probes are hybridized with complementary and fluorescent dye-labeled RNA or DNA molecules (called targets) derived from the biological specimens under investigation. During scanning, the fluorescent intensity is obtained within the position of the probe and serves as measure for the abundance of the corresponding nucleotide sequence in the complex mixture of RNA/cDNA targets. Microarrays have established comprehensive gene expression profiling as a key method to understand a wide range of biological phenomena in vivo and under experimental conditions in vitro. A classic example of coordinated, yet complex, transcriptional regulation is observed in cultured fibroblasts entering the cell cycle after serum stimulation. Genes induced or repressed over time can be divided into several clusters with distinct patterns of regulation and different molecular functions, rendering a detailed picture of the cellular response to the external proliferative cue provided by growth factors in the serum (Fig. 7.1). Although microarray technology has ceased to be the gold standard for transcriptomics, it continues to be widely used and the produced data still make up the bulk of transcriptomics data in public repositories.

## Preparation of target RNA from biological and clinical specimens

While the preparation of RNA from experimental materials such as cell cultures is a straightforward task following standard techniques, extraction from surgically removed clinical materials requires several precautions to minimize RNA decay. In the latter case, one of the first steps is the interruption of the arterial blood supply. From this moment on, the tissue is exposed to hypoxia at body temperature. The duration between artery ligation and the final removal of a cancer specimen can vary considerably (warm ischemia) and is extremely difficult to standardize under clinical conditions. Following cancer resection, logistical constraints may lead to further considerable processing delays (cold ischemia) before the material is finally shock frozen at −80 °C. Overall, this lengthy process might lead to a considerable extent of target RNA degradation.

A widely preferred procedure for isolating biological materials for RNA processing and expression analysis starts with frozen tissues or fresh cell culture materials, although formalin-fixed, paraffin-embedded (FFPE) material can be used as well. Histological characterization can reveal the composition of the tissue, the percentage of the cell types of interest, necrotic areas, which contain degraded RNA, or fatty tissue from which RNA-extraction is difficult. Following RNA isolation, RNA-yield and quality are checked using Nanodrop, Qubit, or Bioanalyzer instruments. Fig. 7.2 shows a typical RNA check following the quality criteria of clear and well-defined 18S and 28S peaks of ribosomal RNA, low noise between the peaks and no or only minimal evidence for low-molecular-weight material. The RNA is then used for synthesis of the labeled sample nucleic acid, mostly cDNA or aRNA, which is again quality-controlled.

### Laser microdissection of tumor tissue

For tissue acquisition at very high specificity and resolution, laser capture microdissection to cut out the tissue areas of interest is the method of choice. If tumor material is contaminated with nontumor tissue such as stromal and immune

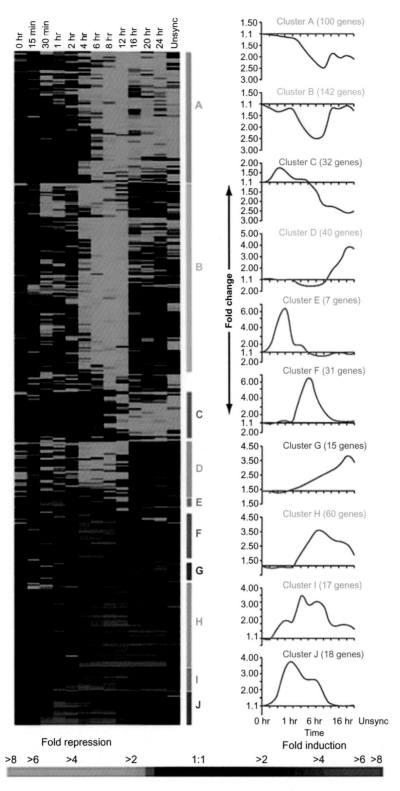

**FIGURE 7.1** Hierarchical clustering of genes induced or repressed during serum response in human fibroblasts. Ten gene clusters (A–J) harboring 517 genes, which show significant alterations in gene expression over time, are depicted. For each gene, the ratio of mRNA levels in fibroblasts at the indicated time intervals after serum stimulation compared to their level in the serum-deprived (time zero) fibroblasts is represented by a color code, according to the scale for fold-induction and fold-repression shown at the bottom. The diagram at the right of each cluster depicts the overall tendency of the gene expression pattern within this cluster. The term *unsync* denotes exponentially growing cells. *From Science 1999;283:83–87. Reproduced with permission from the AAAS.*

cells or if debris and necrotic areas occur, data quality may be severely hampered. Therefore, researchers take extra precautions to obtain homogeneous sample material. A convenient, although laborious, approach is laser-assisted microdissection (Fig. 7.3). Starting from a complex tissue architecture, areas with carcinoma cells only, stromal material, or any other area of interest, such as material located at the invasion front of the cancer, are obtained.

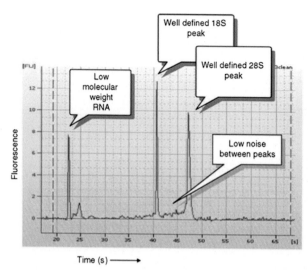

**FIGURE 7.2**    RNA quality assessment using bioanalyzer fluorescent spectroscopy.

# Transcriptome analysis based on RNA sequencing—technical aspects

The advent of RNA-sequencing (RNA-seq) technologies equipped researchers with powerful means to capture altera-tions in gene expression that occur in various pathologies, with an unprecedented wealth of details. RNA-seq outper-forms microarray platforms in sensitivity and specificity of transcript detection. However, similar to other technological advances in molecular biology, RNA-seq poses new challenges to its users. The most prominent one is the enormous volume of sequence data produced. Analysis and storage of such massive data sets require dedicated computer capacity even for experiments with a modest number of samples.

Typically, the main steps in transcriptome profiling by RNA-seq are (1) extraction of the RNA from samples and generation of cDNA, (2) amplification and sequencing of cDNA fragments of certain lengths, (3) mapping the obtained sequences (commonly called reads) to a known transcriptome or genome, and (4) assembly of mapped reads to tran-scripts in order to obtain a count for their abundance level in the sample. The last step is facilitated if the full genome annotation is available, as it is the case for human and mouse genomes. The underlying concept of this approach is that the number of mapped cDNA fragments of a gene should be proportional to the abundance of the corresponding tran-script. However, local sequence composition and secondary structure can affect the fragmentation and reverse transcrip-tion of transcripts. Thus, it is a common observation that reads are not equally distributed within gene sequences. Moreover, the assembly of reads into transcripts remains challenging for eukaryotes because of alternative gene splicing.

## Sequencing platforms

Various NGS technologies applicable for RNA-seq have been developed and introduced into the market. The term next-generation was introduced to distinguish these techniques from the first-generation automated Sanger sequencing method. Their underlying sequencing chemistry and data requisition differ considerably and result in distinct features, which make them more suitable for some particular applications than for others. Important criteria for choosing a cer-tain technology are the amount, quality, and price of sequence data generated. Depending on the technological platform, a typical instrument run can produce between several hundred thousand to up to billions of reads. Also, the read length plays an important role. Typical read lengths range between 35 and 1000 bp, although newer technologies are likely to extend the upper limit. Longer reads can enhance the fidelity of detected splice isoforms.

Currently, the most popular platform is constituted by sequencing machines from the company Illumina. The tech-nology is based on the spatially localized amplification of DNA molecules (derived from RNA in the case of RNA-seq) on a solid surface in a flow cell. Each of the millions of resulting spots contains only clones of one cDNA molecule. The sequencing itself follows the *sequencing by synthesis* concept. Fluorescently labeled nucleotides are incorporated in nascent DNA strains by DNA polymerases with the locally amplified DNA molecules as template. Through excitation with laser and optical recording of the emitted characteristic fluorescent signals, the incorporation of differently labeled

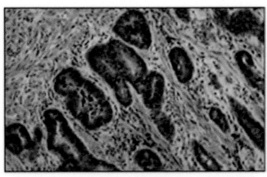
Before micro-dissection

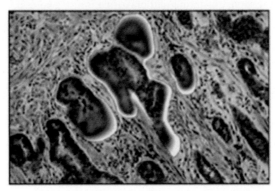
After micro-dissection

FIGURE 7.3 Laser microdissection of tissue samples. Colorectal carcinoma tissue prepared as 5 μm-thick sections were microdissected with the use of a laser beam. RNA was extracted from dissected material by a column-based procedure including DNAse digestion. Typically, microdissection of $5 \times 10^6 \ \mu m^2$ per specimen yields about 10−20 ng RNA in about 2 h working time.

nucleotides can be monitored and the synthesized sequence can be reconstructed. As this step occurs in a cyclic fashion simultaneously across the millions of spots, template libraries can be sequenced in a massively parallel process. An alternative technological concept is provided by Ion Torrent sequencers, which avoid the readout of signals using expensive optical devices. Instead, a semiconductor-based chip with millions of microwells is utilized. DNA fragments are amplified on beads and loaded onto the chip where each microwell can be occupied only by a single bead. Similar to Illumina technology, sequencing is performed by stepwise synthesis of a complementary strand. In contrast, the readout is the change of pH value in an individual microwell caused by liberation of a proton during nucleotide incorporation. The avoidance of optical detection systems allows the production of relatively cheap and small benchtop sequencers capable of processing selected sets of amplicons (gene panel sequencing). It needs to be emphasized that the NGS technology market remains highly dynamic, and some initially promising approaches (such as 454 technology) have already exited it, while new ones with great potential such as those based on nanopores are just entering.

## Sequencing workflow

RNA-seq requires as little as 100 ng of total RNA., Typical approaches involve enrichment of polyadenylated RNA and depletion of ribosomal RNA. Prior to sequencing, a library is prepared, which means that the extracted RNA is reversed transcribed to cDNA and fragmented. Usually, only fragments of a certain length are selected for further processing. Libraries can be barcoded, i.e., a specific marker sequences are attached to the cDNA fragments to be sequenced. This enables sequencing of multiple samples in a single, multiplexed run.

Some sequencing platforms such as those by Illumina allow the sequencing of both ends of cDNA fragments. This approach produces so-called paired-end reads in contrast to single-end reads resulting from sequencing only the 5' end of the fragments. Paired-end sequencing not only facilitates the reconstruction of alternatively spliced transcripts, but also the detection of transcripts resulting from chromosomal rearrangements. Since the cDNAs have been size-selected, the appearance of pairs of reads with very distant chromosomal locations, for instance, can indicate gene fusion events, where two or more genes are fused together. Gene fusions frequently occur in cancer.

## Bioinformatics I—basic processing of RNA-seq data

Finding meaningful structures and information in an ocean of numerical values obtained in transcriptome experiments is a formidable task and demands various approaches of data processing and analysis. Although the type of data analysis naturally depends on the research questions posed and the chosen technical platform, common first steps are data pre-processing and normalization to derive quantities and comparable measures for gene expression (Fig. 7.4). Subsequently, these measures are merged in a so-called gene expression matrix, which is basically a table with rows corresponding to specific transcripts and columns corresponding to samples. The constructed matrix holds two types of different expression profiles in a compact form. The set of expression values of the different genes measured in a sample constitutes the expression profile of the sample. Likewise, the expression of a gene across the different samples constitutes the expression profile of this gene. Thus, the columns of the gene expression matrix provide the profiles of the samples, while the rows provide the profiles of the genes. This matrix can then be scrutinized for the detection of genes with significant fold-changes in expression, clustering and classification of expression profiles of samples or genes, and functional profiling. In all these tasks, visualization of data plays an important role for quality control and knowledge

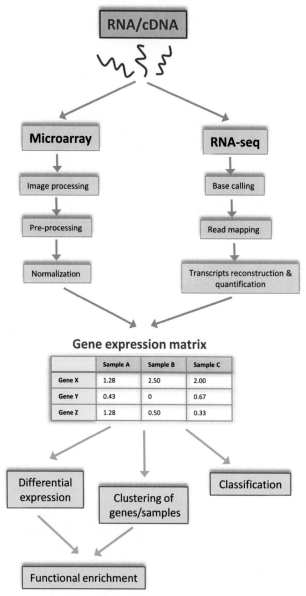

**FIGURE 7.4**   Bioinformatics workflow for transcriptomic analysis using microarrays or RNA-seq technology. Whereas data from both technologies require distinct preprocessing, higher level analyses can be carried by similar or even the same approaches.

discovery. It should be noted that the early analysis steps can influence subsequent examination. For example, the choice of preprocessing and normalization procedures can have considerable impact on the results of clustering and classification.

## Hardware requirements and software solutions for RNA-seq data analysis

For new users of RNA-seq technologies, the amount of data to be analyzed can be daunting. NGS data analysis typically requires the use of multiple CPUs, sufficient computer memory, and disk space up to terabytes even for a single experiment. Alternatives to in-house computational infrastructures are publicly accessible web-platforms such as Galaxy (https://usegalaxy.org/), or the use of commercial cloud computing. However, the cloud approach requires moving the data across the internet, which often presents a notorious bottleneck given the large file sizes. For researchers who carry out a few studies, it might be advisable to begin with web-tools, and then move to stand-alone tools if the required hardware resources are locally available. An excellent platform especially for follow-up analysis is provided by R/Bioconductor (http://www.bioconductor.org/), which offers numerous add-on packages for specific tasks such as detection of differential expression, functional enrichment analysis, clustering and classification, but also requires basic scripting knowledge.

## Base calling and sequencing quality

Base calling (converting measured intensity data into sequences and assessing the sequencing quality) is usually carried out by algorithms supplied by the vendor of the sequencing platform. The identified sequences and their corresponding quality scores are subsequently stored in files of *Fastq* format. The quality of base calling is presented by a so-called *Phred* score. Sequence or parts of sequences with low Phred scores indicate potential sequence errors and need to be removed. Also, reads need to be assessed for the presence of adapter sequences, which interfere with subsequent analysis.

## Read mapping and transcriptome reconstruction

To analyze and interpret the reads produced by RNA-seq, their position within a reference sequence must be determined, a process known as alignment or mapping. This is a challenging process not only due to the large number of reads to be aligned, but also due to sequencing errors or mutations in the sequence, which need to be coped within the alignment process. For mapping of short reads, numerous programs have been developed using different computational strategies. Several of them use the so-called Burrows-Wheeler transformation that was originally developed for file compression. It enables the indexing of the large genomes and its utilization for faster read mapping with reduced computer memory. Alternatively, parts of the reads termed seeds are first mapped to the reference after which the alignment is extended to the full read. Outputs of the aligners are files in Sequence Alignment/Map (SAM) or Binary Alignment/Map Bam (BAM) format, which present the chromosomal location along with the mapped sequences as text or binary encoding, respectively.

Ideally, one would like to use the transcriptome as reference (align the reads directly to the transcriptome). However, in practice reads are aligned to the genome, as complete transcriptomes are not (yet) available. This procedure adds a layer of complexity for the sequencing of eukaryotic RNA, as many genes undergo splicing. The removal of introns leads to transcript sequences that do not correspond to a continuous stretch on the genome, but are composed of sequences from distant exons. To reconstruct the exon structure of genes, alignment programs try to map reads, that could not be aligned in their full length to the genome, to known or predicted splice junctions (locations where two exons join), or splitting them and mapping the different read parts to different exons. Basically, reads overlapping the 5′ end sequence of one exon and the 3′ end sequence of another indicate that the two exons were spliced together. Based on the number of reads aligned to the exons and splice junctions, we can seek to quantify the different splice isoforms although this task has remained difficult and requires sufficient sequencing depth.

To enable comparison of gene expression within a sample or across different samples, a summarization and normalization step need to be carried out. Summarization provides the strength of gene expression, given all the reads mapped to its chromosomal region. For this quantification, the mapped reads are counted and divided by the gene length, as we expect longer genes to lead to more fragments and those to more reads, even if the transcript abundance stays the same. To enable comparison of RNA-seq runs with different number of total reads, a further normalization step is carried out. In the simplest version of normalization, this is achieved through an additional division by the total number of mapped reads producing RPKM (reads per kilo-base of exon model per million reads) values, as the number of reads mapped to a gene should be proportional to the total number of reads produced. Alternatively, other normalization procedures can be chosen, which for instance seek to keep the expression of house-keeping genes constant or minimize the overall fold-change between samples.

## Data visualization and inspection of read mapping

For visual display of the mapping of reads to the reference sequence, various software tools such the Integrative Genomics Viewer have been developed. As input, they use SAM or BAM files as well as available gene annotation. They help to inspect the coverage of specific genes or to discover genetic alterations. For instance, RNA-seq data can offer as a byproduct the accurate identification of single nucleotide polymorphisms (SNPs) in regions with high read coverage.

# Bioinformatics II—exploration and statistical evaluation of transcriptomics data

## Detection of differential expression

The most common task in RNA-seq data analysis is the detection of gene expression changes. In early microarray studies, a fixed threshold for fold-changes (e.g., twofold) was arbitrarily defined to identify differentially expressed genes. However, the setting of fixed thresholds can yield a large number of false positives. Since the measured intensity signals usually are noisy, genes may show differential expression purely due to random signal fluctuations. Particularly, signals related to weakly expressed genes are affected by background noise and therefore require selection based on a larger threshold than strongly expressed genes. To distinguish noise from meaningful changes in gene expression more stringently, statistical tests are commonly employed. Such tests assess the statistical significance of changes based on a set of assumptions about the distribution of the random errors (errors that are not correlated with any experimental variable) and cannot be corrected by normalization (unlike systematic errors). Random errors also set a limit of detectable changes of gene expression in transcriptome experiments. To estimate the random errors, biological and technical replicates are essential. After estimating the random error, statistical significance can be assigned to changes in gene expression in the framework of a statistical test. Including replicates is also helpful to assess the overall quality of the experiment. Ideally, the goal is a high degree of consistency between different replicates.

Statistical testing is based on the assessment of the validity of explicitly formulated hypotheses. In general, a *null hypothesis* $H_0$ (e.g., a gene is *not* differentially expressed) and an opposing alternative hypothesis $H_a$ (e.g., a gene is differentially expressed) are defined. The alternative hypothesis is supported, if there is evidence against the null hypothesis. The steps in hypothesis testing are as follows: (1) setting up $H_0$ and $H_a$, (2) use of a test statistic to compare the observed values with the values predicted by $H_0$, and (3) definition of a region for the test statistic for which $H_0$ is rejected in favor of $H_a$. The level of significance of a test is the probability that the test statistic falls in the rejection region, if $H_0$ is true. The incorrect rejection of $H_0$ is called a *type I error*, in contrast *to type II errors* where $H_0$ is not rejected although it is false. The probability $P$ that $H_0$ is true given the observed test statistic is called the $P$-value of the test.

A variety of statistical tests have been proposed for the identification of changes in gene expression. A classical test for comparing the mean gene expression values in two biological samples is the *Student's t-test*. However, since the number of samples is typically small, an accurate estimation of variability is generally not possible. To increase the power of the test under such conditions, the variability observed for other genes can be used to improve estimation of the variability for a given gene. Note that the *t*-test assumes a Gaussian distribution of the expression values. However, initial experiments indicated that RNA-seq data tend to follow a Poisson distribution. This distinction bears consequences for the testing of differential expression and more elaborate tests have been developed for RNA-seq data. Conversely, established methods for microarray data might be applied to RNA-seq data after appropriate transformation.

As alternative to parametric tests, *permutation tests* can be used as they do not assume any particular data distribution. Permutation tests rely solely on the observed data examples and can be applied with a variety of test statistics. The basic idea of a permutation test is simple: given labeled data, all permutations of the samples should be equally likely. Evaluating a chosen test statistic for all permutations, an empirical distribution of the test statistic can be derived. The percentage of random permutations that score higher than the actual observed case gives the significance level. However, a major restriction is that permutation tests can be computationally very intensive and are not as powerful in the detection of differential expression as parametric tests.

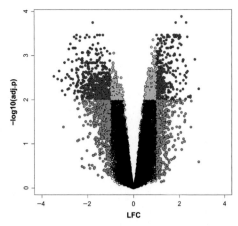

**FIGURE 7.5** The volcano plot is a graph that shows both fold-changes and statistical significance of recovered genes. The graph displays negative log10-transformed adjusted *P*-values against the log2 fold-changes (LFC). Volcano plots can be used for the selection of significant genes with a minimal required fold-change. Genes (displayed in blue and red) having statistically significant differential expression (adjusted $P < 0.01$) lie above a horizontal line. Genes (displayed in green and red) with absolute fold-changes larger than 2 lie outside a pair of vertical lines. Genes which fulfill both criteria are highlighted in red.

It is the nature of a RNA-seq experiment that generally thousands of transcripts are tested for differential expression. If multiple tests are performed in parallel, the level of significance for the whole set of tests does not equal the level of significance for the single tests. For example, the probability $P$ of rejecting a true null hypothesis in at least one of 1000 simultaneous tests with a significance level of 0.001 is 63%. Moreover, we would expect a large number of false positives even if there is no actual differential expression. Therefore, an adjustment of the overall significance level and the $P$-values is necessary. A popular approach to circumvent the problematic interpretation of $P$-values in multiple testing is the calculation of the false discovery rate (*FDR*), which is defined as the proportion of false positives among genes detected as significant. For instance, an *FDR* of 0.2 indicates that 20% of genes detected as significantly differentially expressed are likely to be false positives. For subsequent visualization of the results of statistical tests, *volcano* plots have become a popular mean. They offer the advantage of displaying both statistical significance and fold-changes observed (Fig. 7.5).

## Classification

In its widest definition, classification is the assignment of a set of objects to a set of classes. In transcriptomics, classification is commonly used to assign RNA specimens to different classes (distinguished types of cancer). Classification can be performed in an unsupervised and supervised manner. If class labels are not known in advance, the process is called unsupervised and is generally referred to as clustering. If class labels exist, the process of classification is called supervised. The aim of supervised classification methods is to correctly assign new examples based on a set of examples of known classes. Thus, a classifier should generalize from known class examples to new unclassified examples. In transcriptome analysis, the objects are provided as gene expression profiles and the classifiers have to identify decision boundaries between classes based on these given profiles (Fig. 7.6). To achieve this goal, classifiers are optimized in learning or training phase. After the optimization, the accuracy of the classifier can be tested using new examples of known class origin.

Supervised classification of tissue faces several major challenges. First, transcriptome data can contain a high level of noise. Experimental procedures such as tissue handling, RNA extraction, labeling, and amplification can introduce variability in the measured expression levels. For RNA-seq experiments, reads might be mapped to pseudogenes with high sequence similarity. Second, thousands of genes are monitored, while the number of RNA samples is usually restricted to hundreds or less. It is well known that classifiers generally perform poorly, when the number of examples is small compared to the number of genes used for classification. Third, tissue samples are frequently heterogeneous in their composition, i.e., different cell types are mixed in a single tissue sample used for RNA extraction. This heterogeneity can cloud the separation of the tissue classes of interest, e.g., the distinction of cancer and normal tissue.

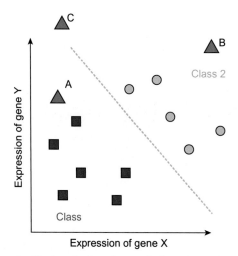

**FIGURE 7.6**   Extrapolation in classification: a classifier is trained on the sample from classes 1 and 2 based on the expression values of the two genes X and Y. The *dashed line* represents the *border line* derived by the classifier between the classes. Thus, new examples (represented by the *dashed line*) will be classified according to their gene expression values for X and Y. Example A will be assigned to class 1, whereas example B will be assigned to class 2. The classification of C remains problematic, since it is located close to the border line and different to previously seen examples. In this case, further tests are needed.

## Gene selection

Generally, large numbers of genes without changes in mRNA abundance present an intrinsic noise for classifiers and interfere with their performance. Gene selection aims at improving classification by excluding noninformative genes and thereby reducing the number of genes for the classifier. Genes are excluded if they only weakly contribute to the classification or not at all. Gene selection can be incorporated in a classification system in two different ways. First, gene selection and classification can be treated separately from the classification model. Genes are selected with respect to predefined criteria such as Pearson correlation or the significance in the Student's *t*-test. This approach often has the advantage of being computationally inexpensive and easy to process. However, the selected genes are frequently highly correlated to each other and are likely to be redundant. Alternatively, the selection of genes is determined by the classification methods themselves in an iterative manner. This constitutes an integrated approach, since an optimal set of features depends on the choice of the classifier.

## Classification methods

Numerous methods for classification have been applied in transcriptomics. One of the most basic methods is *the k-nearest neighbor method* (with *k* as a positive integer). The classification rule is simple: a new example is assigned to the class that is most common amongst its *k* nearest neighbors. The distance of the examples are calculated based on their similarity in the expression profiles. For instance, if *k* is 1, then the RNA sample is simply assigned to the class of its nearest neighbor. Other currently popular classifiers are *support vector machines* based on statistical learning theory and belonging to the class of kernel-based methods. The basic concept of support vector machines is the transformation of input vectors into a highly dimensional feature space, where a linear separation may be possible between the positive and negative class members. In this feature space the support vector learning algorithm maximizes the margin between positive and negative class members of the training set to achieve a good generalization.

## Cross-validation

Biological samples generally constitute only a small fraction of a larger sample cohort of interest. However, if a classifier is optimized based on a small number of examples, it shows frequently a decreased performance on new data, a phenomenon usually called *overfitting*. An approach to prevent overfitting is *k-fold cross-validation*. It splits the data in *k* segments of which *k* − *1* segments are used for the training and one segment for the testing of the classifier. This is repeated *k* times, so that every segment is used for testing. The classification error in the validation procedure is then the sum over the error in the $k$ tests. This approach has the advantage that a large part of the data can be retained for the training of the classifier, while the validation error is evaluated using all data examples equally. In the extreme case that $k$ equals the number of data objects, the cross-validation is also referred to as *leave-one-out* or jackknife method. If different models are compared by cross-validation, the model yielding the lowest validation error is generally selected.

## Visualization

Data visualization is also an important component in the assessment of class distributions. It provides a global picture of the separation of samples and help to identify potential outliers. However, a major challenge is the accurate representation of highly dimensional data, where samples are defined by the expression values of thousands of genes. In contrast, data plots are restricted to two or three dimensions. A standard method for representing high-dimensional transcriptome data is based on principal component analysis (PCA). The goal of this method is to find an optimal linear projection to a lower dimensional space. Practically, PCA leads a projection from the original gene expression values to an orthogonal basis of principal components. The principal components give the directions of the maximal variance in the data (Fig. 7.7).

## Clustering

Clustering, or unsupervised classification, has been studied for many decades in pattern recognition and related fields. Clustering methods generally aim at identifying subsets (clusters) in data sets based on the similarity between single objects. Similar objects are assigned to the same cluster, while dissimilar objects are assigned to different clusters. Cluster analysis, which can be understood as exploratory data analysis, is applied to search for patterns that may reveal relationships between individual examples. Frequently, the data structures detected by cluster analysis can give first insights into the underlying data producing mechanisms. It is especially useful, if prior knowledge is little or nonexistent, since it requires minimal prior assumptions. This feature has made clustering a widely applied tool in transcriptome data analysis. One of the main purposes of clustering is to infer the function of novel genes by grouping them with genes of well-known functionality. This method is based on the observation that genes with similar expression patterns (co-expressed genes) are often functionally related and are controlled by the same regulatory mechanisms (co-regulated genes). Therefore, expression clusters are frequently enriched by genes of certain related functions. If a novel gene of unknown function falls into such a cluster associated with a certain biological function, it seems likely that this gene also plays a role in the same process. This *guilt-by-association* principle enables assigning possible functions to a large number of genes by clustering of co-expressed genes.

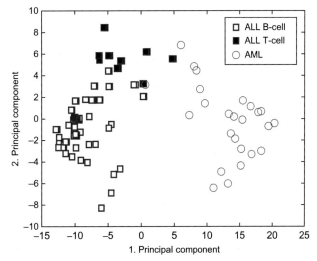

**FIGURE 7.7**   Principal component analysis of leukemia samples based on 100 genes that have the largest squared Pearson correlation with the two classes of leukemia, ALL and AML. The first two principal components include 63.3% of the total variance of the data. Most ALL and AML samples can be separated based on the first two principal components. However, note that the AML outlier makes a perfect separation difficult. *From Science 1999;286:531—37. Reproduced with permission from the AAAS.*

Clustering methods can be divided into hierarchical and partitional clustering. *Hierarchical clustering* creates a set of nested partitions, so that partitions on a higher hierarchical level comprise partitions on lower levels. The sequential partitioning is conventionally presented as a dendrogram, which displays the clusters in a tree structure. The length of the branches represents the similarity between clusters. The shorter the branches, the more similar the clusters are. Usually, hierarchical clustering is performed in a step-wise agglomerative manner, i.e., it starts with the single objects as singular clusters and gradually merges the clusters until all objects belong to a single cluster. To decide which clusters to merge, their similarity is calculated at every step of the clustering procedure. Clusters that show the largest similarity are subsequently joined. In transcriptomic data analysis, both genes and biological samples can be hierarchically clustered. If these tasks are performed simultaneously, the procedure is also referred to as two-way clustering. As an alternative approach, partitional clustering splits the data in several separate clusters without the definition of a cluster hierarchy. It is commonly used to detect temporal gene expression patterns in time series experiments. The most popular methods for partitional clustering is $k$-means clustering. It starts with $k$ randomly initiated cluster centers and splits the data in $k$ partitions (with an integer $k$ chosen by the user) based on the distance to the nearest cluster center. By repeated recalculation of the cluster centers and partitioning of the objects, it aims to iteratively minimize the within-cluster variation. A more robust version is $c$-means, which also can indicate the strength of association of genes to clusters.

Before clustering analysis, it is important to standardize the expression values, as co-expressed genes frequently show similar changes in expression but may differ in the overall expression rate. Therefore, the expression values of genes are usually adjusted to have a mean value of zero and a standard deviation of one. This ensures that genes with similar changes in expression have similar standardized expression values and will tend to cluster together.

A crucial question is how many clusters can be retrieved from the data. This is generally difficult to answer for gene expression data as the detected clusters frequently are nonhomogeneous and may show substructures, which can be interpreted as clusters themselves. While hierarchical clustering is able to indicate the different levels of clustering in the resulting dendrogram, partitional clustering algorithms lack the ability to indicate substructures in clusters. To critically assess their reliability, several measures for cluster validity have been introduced. Many of them assess the quality of clusters based on criteria such as compactness and isolation. Alternatively, detected clusters can be examined based on their robustness to noise. For this approach, one would artificially add noise to the data before clustering and compare the newly identified clusters with the original ones. Clusters that remain the same despite added noise are likely to be more reliable than clusters that vanish in the presence of noise.

## Functional profiling and other enrichment analyses

Frequently, large numbers of differentially expressed genes are detected in transcriptomics analysis making the overall interpretation of the results challenging. If further research is not focused on a few candidate genes, a helpful tool for understanding the complexity of the data set is functional profiling. This approach aims at identifying biologically informative classes of genes that were likely to be affected in the experiment. The underlying framework is given by Gene Ontology (GO), a popular database providing gene annotations in a systematic manner for various species (http://www.geneontology.org). In GO, genes have been assigned to a defined set of categories describing molecular functions, biological processes, and cellular compartments. The categories themselves are placed in a treelike structure with parent-child relationships. Categories at low levels are fairly general (e.g., "cell death") in contrast to more specific categories at higher levels (e.g., "regulation of caspases"). Since GO is computer-accessible, the assignment of annotations to list of genes has become much facilitated. After automatic gene annotation, functional profiling is performed by determining which GO category is represented more frequently than expected in the list of differentially expressed genes. Collecting involved GO categories frequently gives a more holistic picture than the inspection of individual genes. Nowadays, numerous software tools are available for functional profiling in transcriptomics analysis. They usually provide a list of significantly enriched GO categories associated with a particular experimental condition, disease state, etc. as a result. In a similar manner as for differentially expressed genes, groups of co-expressed genes found by clustering can be further scrutinized by functional enrichment analysis. Despite being well established in transcriptomics, there are important caveats with respect to functional profiling. Results can vary considerably between different software tools and often show a high grade of redundancy. In addition, while there is a considerable number of manually curated gene annotations in GO, many human or murine genes are still annotated solely by computational means.

The concept of functional profiling to examine enrichment of genes belonging to defined functional categories can be applied in a general way. For instance, pathway annotation as found in the KEGG database (http://www.genome.jp/kegg/pathway.html) can provide a basis for enrichment analysis and indications about the activity of specific pathways. Another example for enrichment analysis is the examination of the chromosomal location of differentially expressed

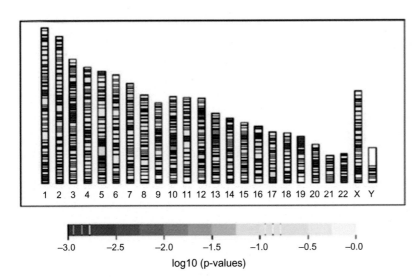

FIGURE 7.8 Chromosomal localization of genes exhibiting differential expression. The statistical significance for local enrichment of upregulated genes in a metastatic colorectal cancer cell line compared to a primary carcinoma line is shown. To detect possible changes in the chromosomal structure of the two related cell lines, differentially expressed genes were mapped to their corresponding chromosomal locus. Subsequent enrichment analysis using a sliding window technique indicted several potential chromosomal alterations.

genes. This strategy yields a first indication for potential underlying changes in the chromosomal structure, such as copy number alterations or deletions, and integrates transcriptomics and genomics (Fig. 7.8).

## Repositories for transcriptome data

Transcriptomics produces massive quantities of gene expression data. Therefore, it has become best practice (and typically requested by the journal editors prior to publication) to deposit generated data in publicly accessible databases. This allows independent researchers not only to scrutinize data obtained by others for their own interests, but also to validate original analyses. In fact, the practice of sharing transcriptomic data has allowed the community of bioinformaticians and statisticians to develop new methods and compare them with existing ones based on publicly accessible data sets. Such comparisons have been extremely valuable, since results from transcriptomic experiments rely not only on the raw data, but to a substantial part on the applied computational methods. For example, the interpretation of microarray experiments required a common forum providing various types of information on the examined samples and experimental conditions, arrayed genes, microarray platforms, and applied computational approaches. Therefore, standards for publishing microarray data were established. The most important one has been the Minimum Information about a Microarray Experiment (MIAME) standard. This standard requires deposition of raw and normalized expression data, sample annotation, experimental design, description of the array, and experimental conditions. Additionally, the development of large central transcriptome databases has facilitated data sharing. One of the first repositories was the Stanford Microarray Database (http://genome-www5.stanford.edu/) including a large collection of two-color microarray experiments. Currently, the two major public transcriptome databases are Gene Expression Omnibus (GEO) provided by the National Center for Biotechnology Information (http://www.ncbi.nlm.nih.gov/geo/) and Array Express provided by the European Bioinformatics Institute (http://www.ebi.ac.uk/arrayexpress/). The two databases follow the MIAME standard and provide several options to the user for depositing their own transcriptomic data and for accessing information from others. The Sequence Read Archive (https://www.ncbi.nlm.nih.gov/sra) and European Nucleotide Archive (http://www.ebi.ac.uk/ena) allow the deposit of raw sequence data from RNA-seq experiments, while the count data for transcripts can be placed in GEO or Array Express. Finally, there are also repositories and databases such as Oncomine (https://www.oncomine.org) and The Cancer Genome Atlas (TCGA) (https://cancergenome.nih.gov/), which are dedicated to cancer research and contained a large number of expression data sets for various cancer types.

## Trancriptome analysis—applications in basic research and translational medicine

A recent PubMed search revealed that the majority of gene expression profiling studies in medicine are devoted to some aspect of cancer. The terms "gene expression profiling" and "cancer" provide more than 58,000 entries. Cancer studies outnumber similar studies in cardiovascular diseases, neurodegenerative diseases, inflammation, and others

**TABLE 7.1** Number of published microarray and gene expression profiling applications in research.

| First keyword | Second keyword | | | |
|---|---|---|---|---|
| | None | Gene expression profiling | Microarray | RNA sequencing |
| None | – | 175,199 | 77,255 | 183,888 |
| Pharmacology | 5,665,893 | 38,166 | 16,602 | 29,069 |
| Diseases | 5,664,283 | 47,520 | 21,435 | 33,832 |
| Cancer | 3,423,145 | 58,339 | 27,064 | 26,175 |
| Pathology | 2,932,873 | 44,938 | 20,622 | 20,540 |
| Development | 2,311,384 | 43,654 | 17,933 | 32,126 |
| Cardiovascular diseases | 2,130,792 | 5489 | 2354 | 2728 |
| Immunology | 1,539,205 | 18,503 | 7464 | 15,420 |
| Infection | 1,434,751 | 11,180 | 4688 | 14,928 |
| Inflammation | 552,463 | 8504 | 4021 | 3746 |
| Drug development | 526,778 | 9285 | 3944 | 4597 |
| Nutrition | 370,558 | 2530 | 1158 | 1757 |
| Personalized medicine | 338,828 | 1372 | 520 | 378 |
| Neurodegenerative diseases | 260,858 | 2122 | 954 | 1482 |
| Toxicology | 159,060 | 2685 | 1287 | 1178 |

Results of a PubMed search dated January 20, 2017, using single keywords and combinations of two keywords without limits to publication years.

(Table 7.1). Therefore, we have chosen some exemplary and prominent applications of gene expression analysis in the field of cancer as paradigms to demonstrate the power of transcriptome analysis.

## Transcriptomes of healthy cells

While all cells of an individual contain largely the same genome, cellular transcriptomes are distinctive and a manifestation of cellular identity. Indeed, unique gene expression patterns discriminate every cell type arising during development and in the adult phases of life. Current RNA-seq technology has enabled us to document normal tissues transcriptomes in the human body regarding all major RNA species, such as protein-encoding mRNA and regulatory miRNAs and lncRNAs. Often, few genes dominate gene expression patterns in highly specialized cells. For instance, expression of the three hemoglobin genes contributes approximately 60% of all transcripts in blood. Cell types with high metabolic activity, such as kidney, muscle, and brain cells, express a large fraction of mitochondrial transcripts (more than 50% in kidney). Likewise, liver, testis, muscle, and brain have unique transcriptomes that are quite dissimilar from other solid tissues (Fig. 7.9). Protein-coding genes are most often expressed in multiple tissues; yet, patterns of alternative splicing are often tissue-specific. Furthermore, noncoding lncRNAs are often expressed in a tissue-specific manner.

While healthy tissue transcriptomes are similar between individuals, some RNAs and splice forms exist that display strong interindividual variation. These variable transcripts could be mediators of phenotypic variation, including disease susceptibility. Relatively few genes show sex-specific expression (less than 100 protein-coding genes), and these tend to be located on the X and Y sex chromosomes. As an exception, the autosomal gene *MMP3* encoding a matrix metalloprotease is expressed stronger in males and is linked to susceptibility to coronary heart disease. Other genes show expression that changes with age. This group is enriched in genes with functions in neurodegenerative disorders such as Parkinson and Alzheimer disease. Since large transcriptome data collections allowing statistical interpretations are becoming available, and costs of transcriptome analyses by RNA-seq are decreasing, it is expected that transcriptome analyses can aid in the identification of individual disease risks and in the interpretation of pathological states in the future. However, another recent publication suggests that transcription of protein-truncating gene variants is surprisingly frequent and well tolerated in humans, probably complicating the assessment of disease risk in transcriptome and gene variant data.

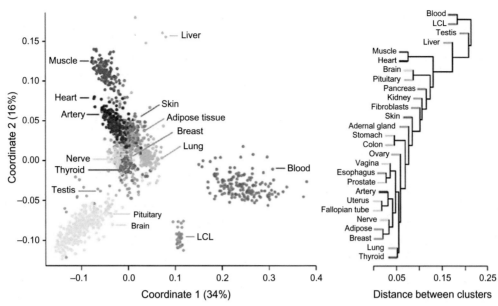

**FIGURE 7.9**    Sample and tissue similarity on the basis of gene expression profiles. Left: Multidimensional scaling. Right: Tissue hierarchical clustering. *From Science 2015;348:660—65. Reproduced with permission from the AAAS.*

**TABLE 7.2** Consortia providing large-scale Transcriptome data.

| Consortium | Resource | Key publications |
|---|---|---|
| FANTOM | RNAs in the mammalian cell | Carninci P, Kasukawa T, Katayama S, et al. The transcriptional landscape of the mammalian genome. Science 2005;309:1559—1563. |
| ENCODE | RNAs in relation to functional DNA elements | ENCODE Project Consortium. An integrated encyclopedia of DNA elements in the human genome. Nature 2012;489:57—74. |
| GTEx | Tissue-specific expression and quantitative trait loci | GETx Consortium. Human genomics. The Genotype-Tissue Expression (GTEx) pilot analysis: multitissue gene regulation in humans. Science 2015;348:648—660. |
| TCGA | Cancer genomes and transcriptomes | Weinstein JN, Collisson EA, Mills GB, et al. The cancer genome Atlas Pan-Cancer Analysis Project. Nat Genet 2013;45:1113—1120. |

Much has been learned from large-scale transcriptome analyses (Table 7.2). Systematic analyses of normal cellular transcriptomes and pathological counterparts have been conducted. Large-scale transcriptome data sets and bioinformatic tools for transcriptome interpretation are available for use in the scientific and medical communities. Important contributions came from the ENCODE and FANTOM consortia, focusing on functional annotation of the human and mammalian transcriptomes and genomes. In addition, the GTEx consortium builds a resource focusing on tissue-specific transcriptomes and differences between individuals, using clinical necropsies and specimens. Finally, The Cancer Genome Atlas (TCGA) network provides cancer genome and transcriptomes for use in the scientific community.

Pushing technical borders in transcriptome analysis, recent studies have concentrated on gene expression patterns in single cells to reveal cell-to-cell heterogeneity and rare differentiation states within cell populations, for instance in immune cells, in the intestinal epithelium, and in the pancreas, which are all tissues harboring a wide variety of cell types.

## Transcriptome analysis in cancer pathogenesis and diagnosis

Cancer has been termed a disease of the genome, due to the presence of point mutations, copy number changes, and structural variations in cancer cell genomes. Due to these alterations, cancer cells also display aberrant transcriptomes compared to the noncancerous cells of origin. Firstly, the presence of cancer-specific mutations can alter the activities of cancer-relevant signaling cascades and their target genes. Secondly, cancer cells exhibit aberrant differentiation, epigenetic and metabolic states compared to cells in the tissue of origin, with strong effects on the transcriptome. Thirdly, cancer cells are frequently aneuploid, resulting in higher or lower expression of genes residing in areas of chromosomal gain or loss, respectively. Finally, cancer tissue architecture often deviates from the tissue of origin, due to immune cell infiltration, the presence of stromal cells, or the formation of new blood vessels. Thus, cellular composition strongly shapes cancer tissue transcriptomes. Consequently, patterns of gene expression can reveal important molecular and functional traits of cancer.

The current themes of transcriptomics in cancer analysis are related to the mechanisms of pathogenesis, cancer classification, and outcome prediction. To elucidate the mechanisms of tumorigenesis and metastasis, particularly to study the complexity of the underlying processes, microarrays or RNA-seq are frequently used. Cancer classification based on transcriptome studies aims at identifying characteristics beyond anatomical site and histopathology. Outcome prediction attempts to overcome the limitations of current diagnostic procedures by establishing gene-based criteria to indicate and predict tumor prognosis and therapy response, even for individual cancer patients. Basically there are three types of large-scale transcriptome approaches. In *class comparison*, one tries to compare the expression profiles of two or more predefined classes; for example, two tissue samples, normal versus malignant cells or tissues, different developmental stages, or cells treated with drugs under different conditions. In *class discovery*, one tries to identify novel subtypes within an apparently homogenous population. In this case, the analysis is used to identify features that cannot be distinguished by other available tools. The starting point usually is a homogenous group of specimens, in which a concealed proportion behaves aberrantly or exhibits invisible or unknown features. The problem of cancer treatment falls into this category, since patients who are stratified into treatment groups according to standard clinical or histopathological criteria often respond differently to therapy. *Class prediction* seeks to identify a set of features that are predictive for a certain, predefined class. This is perhaps the most common application in transcriptomics. It is usually based on class discovery, but now the characterization of a novel class is intended. Rather, the idea is to establish a classifier as described above. A classifier is a set of features, like genes, proteins, mRNAs, or micro-RNAs that serve as surrogate markers for a certain class. This is the common approach to identify predictive gene sets or gene signatures that can predict clinical outcome or therapy response.

One important question is whether transcriptome analysis can distinguish cancers by tissue of origin or histological subtype. A meta-analysis of more than 3500 cancer specimens classified 12 major cancer types into 11 principal groups by gene expression (Fig. 7.10). The study found that most cancer gene expression groups were equivalent to known cancer types within an organ, such as the two transcriptome subtypes comprising of luminal or basal breast cancers. Some transcriptome subtypes encompass cancers of different origin but similar histology, such as a squamous cancer transcriptional subtype, composed of squamous lung, head and neck, and bladder cancers. Beyond their similar histology, these cancers also shared prevalence of mutations in the tumor suppressor gene *TP53* and amplification of *TP63*. The analysis also highlighted the limitations of cancer classification by transcriptome analysis: several cancers clustered as outliers to remote subgroups, such as a breast cancer sample with strong immune infiltration being classified with a cohort of acute myelogenous leukemia samples, probably due to a shared immune cell signature.

Characteristic differences in gene expression between cancer tissue types have been utilized to evaluate carcinomas of unknown primary (CUP). CUP patients comprise approximately 3% of advanced carcinoma cases. Their disease is highly aggressive with metastatic spread. However, the site of primary cancer is small and unrevealing. Identification of the primary cancer site is an unmet clinical requirement in these cases, since therapies can differ for carcinomas of different organs. In clinical tests, gene expression profiling of CUP tissue had a success rate of more than 95% in revealing the site of origin, in contrast to traditional immunohistochemistry, which revealed a tissue of origin in only 35%−55% of cases tested. However, transcriptome analysis is not established for CUP analysis in clinical routine today, mainly due to high cost and a lack of standardization.

## Lymphoma transcriptomes—from gene signatures to simple gene predictors

In 1999, a group of scientists from highly ranked medical schools in the United States assembled a specialized microarray representing genes preferentially expressed in lymphoid cells. The lymphochip harbored more than 17,000 cDNA probes specific for germinal center B cells, diffuse large B-cell lymphoma (DLBCL), follicular lymphoma (FL), mantle cell lymphoma, chronic lymphatic leukemia (CLL), genes induced or repressed upon T-cell or B-cell activation, lymphocyte and cancer genes. The consortium interrogated these arrays using targets prepared from normal cells and tumors to define signatures for the different immune cell types, under different conditions and developmental stages. Particularly, the researchers analyzed the most prevalent adult lymphomas using the lympho-chip. They identified

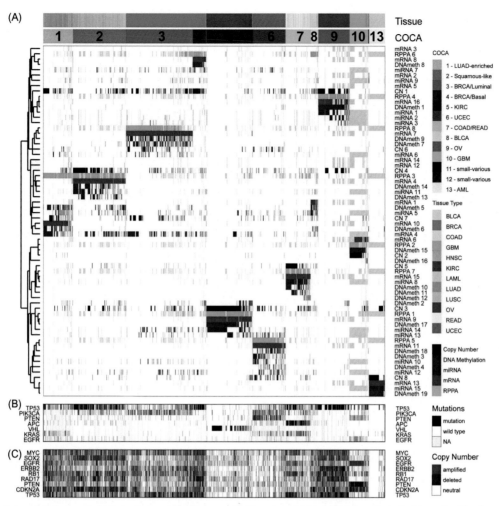

**FIGURE 7.10**  Integrated cluster-of-cluster assignments analysis reveals 11 major subtypes. (A) Integration of subtype classifications from five omic platforms resulted in the identification of 11 major groups/subtypes from 12 pathologically defined cancer types. The groups are identified by number and color in the second bar, with the tissue of origin specified in the top bar. The matrix of individual omic platform type classification/subtype schemes was clustered, and each data type is represented by a different color: copy number=black, DNA methylation=purple, miRNA=blue, mRNA=red and RPPA=green. (B) Mutation status for each of 10 significantly mutated genes coded as: wild-type=white, mutant=red, missing data=-gray. (C) Copy number status for each of nine important genes: amplified=red, deleted=blue, copy number neutral=white and missing data=gray. The color-coding schema is shown to the right. *From Cell 2014;158:929—44. Reproduced with permission from Cell Press.*

signatures for distinct types of DLBCL exhibiting a bad prognosis, FL exhibiting a low proliferation rate, and for CLL with slow progression ( > 20 years). In addition, profiles were obtained from normal lymphocytes (tonsil, lymph node) as well as from several lymphoma and leukemia cell lines. Clustering analysis placed the CLL and FL profiles close to those of resting B-cells, while genes of the so-called proliferation signature were weakly expressed in these tumors. DLBCL, the highly proliferative, more aggressive disease, had higher expression levels of proliferation-associated genes. A germinal center B-cell signature was identified that was clearly different from the resting blood B-cells and from activated B-cells in vitro. This indicated that germinal center B-cells represent a distinct stage of B-cells and do not simply resemble activated B-cells located in the lymph node.

When the scientists re-clustered all DLBCL cases, particularly considering the genes that define the germinal center B-cells, they could clearly separate two different subclasses of DLBCL. One of them strictly showed the signature of the germinal center B-cells, while the other one was clearly distinct and related to activated B-cells. These data suggested that a certain class of DLBCL was derived from germinal center B-cells and retained its differentiation signature even after malignant transformation. By investigating the genes exclusively expressed in either of the DLBCL types and re-clustering, the authors defined two signatures representative of either the germinal center-type (GC) and what they called the activated-B-cell DLBCL (ABC DLBCL) (Fig. 7.11). Analysis of the clinical follow-up showed that the

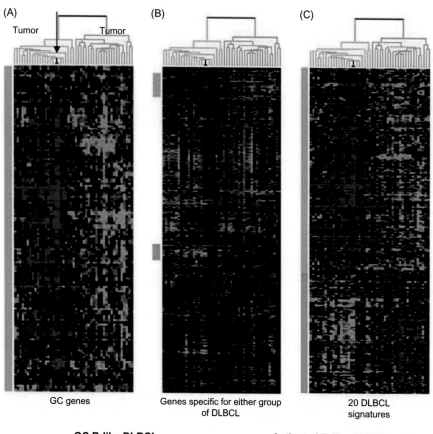

FIGURE 7.11 Gene signatures representing GC-like DLBCL and activated B-like DLBCL. (A) Genes characteristics for normal germinal center B-cells were used to cluster the tumor samples. This process defines two distinct classes of B-cell lymphomas, GC-like DLBCL and activated B-like DLBCL. (B) Genes that where selectively expressed either in GC-like DLBCL (*yellow bar*) or activated B-like DLBCL (*blue bar*) were identified in the tumor samples. (C) Result of hierarchical clustering that generated GC-like and an activated B-cell like DLBCL gene signatures. *From Nature 2000;403:503−11. Reproduced with permission from Macmillan Publishers Ltd.*

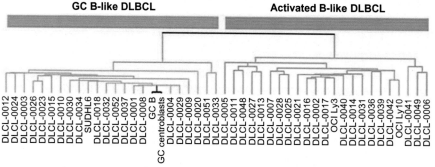

GC-like tumors have a much better prognosis than the ABC DLBCL (Fig. 7.12). Did the result of the microarray study provide novel information up to this stage of investigation? When the authors compared the microarray-based classification to the standard classifiers that define high clinical risk and low clinical risk, there was obviously no significant classification progress (Fig. 7.11B). However, when the low-risk patients initially classified conventionally are further stratified by subgrouping them into the GC DLBCL and ABC DLBCL types, the molecular classifier was superior. Subsequent functional classification of genes associated with ABC DLBCL revealed an NF-kB pathway signature that comprises several antiapoptotic genes. The functional studies culminated in the finding that inhibition of that pathway affected growth of ABC DLBCL, while CG DLBCL cells were insensitive. Several further microarray studies confirmed that gene signatures were associated with clinical outcome of diffuse B-cell lymphoma. However, there were disparities with the regard to the number and nature of informative genes.

The integration of focused transcriptome analysis to identify ABC DLBCL versus GC DLBCL in routine diagnostics turned out to be challenging due to high costs and low interlaboratory reproducibility. As a first step, data from transcriptome analysis were translated into immunohistochemical tests that were easier to be routinely performed on FFPE samples. However, algorithms using 3−5 antibodies such as the Hans, Choi, and Tally algorithms were reported to show unsatisfying reproducibility and low sensitivity in terms of correct classification when compared to gene

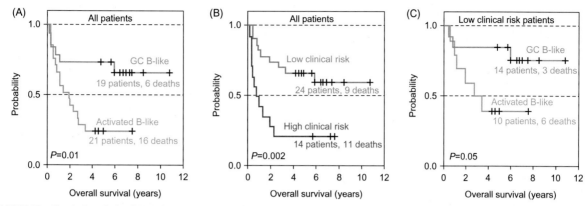

**FIGURE 7.12** Survival analysis of DLBCL patients distinguishable according to gene expression profiling, conventional clinical criteria, and a combination of both sets of criteria. (A) DLBCL patients grouped on the basis of gene expression profiling. The GC-like and the activated B-cell-like show clearly different survival probabilities. (B) DLBCL patients grouped according to the International Prognostic Index (IPI) form two groups with clearly different survival, independent of gene expression profiling. Low clinical risk patients (IPI score 0−2) and high clinical risk patients (IPI score 3−5) are plotted separately. (C) Low clinical risk DLBCL patients (IPI score 0−2) shown in B were grouped on the basis of their gene expression profiles and exhibited two distinct groups with different survival probabilities. *From Nature 2000;403:503−11. Reproduced with permission from Springer Nature.*

expression profiling. These problems prompted a search for alternative methods to integrate RNA profiling from FFPE samples into routine diagnostics. The Leukemia Lymphoma Molecular Profiling Project developed the Lymph2Cx test, a Nanostring-based approach comprised of 20 genes and being able to robustly classify ABC and GC-type DLBCL from FFPE samples with a high sensitivity and consistency between different laboratories. The test is currently used in a phase III clinical trial to select patients with ABC DLBCL for treatment with R-CHOP with or without lenalidomide (ROBUST trial; NCT02285062).

## Colorectal cancer transcriptomes: consensus subtypes predict clinical outcome

Colorectal cancer (CRC) is a heterogeneous disease regarding molecular features, therapeutic profiles, and outcome. However, available molecular markers can neither faithfully predict the course of the disease nor treatment outcome. Therefore, several studies have recently analyzed CRC transcriptomes to stratify CRC for molecular subtypes and therapy response.

In 2012, the TCGA consortium published an integrated genomic and transcriptomic analysis of 276 CRC specimens. Corroborating and extending previous studies, CRC transcriptomes revealed gene expression patterns associated with WNT and MAPK pathway activation, TGFβ pathway repression, and complex changes in MYC-directed transcription. CRCs could be divided into two main groups based on genomic and transcriptomic features, and these were termed hyper-mutated tumors enriched in microsatellite instable (MSI) specimens, activating *BRAF* gene mutations, strong CpG island genome methylation, and non-hyper-mutated tumors. The study did not attempt to subgroup or stratify CRC specimens based on transcriptome analyses alone.

Between 2012 and 2014, several independent studies provided in-depth analyses of CRC transcriptomes, and all studies identified CRC subgroups that were predictive for clinical course, prognostic for therapeutic response or enriched for biological features. However, it was unclear whether the three to six CRC subgroups presented by the various researchers were compatible. Therefore, the respective research groups teamed up for a meta-analysis of almost 4000 CRC samples. This effort resulted in the identification of four consensus molecular subtypes of CRC, termed CMS1-4, encompassing more than three quarters of the samples (the other samples remain unclassified) (Fig. 7.13). CMS1 encompasses mostly hyper-mutated MSI tumors and their transcriptome is dominated by infiltrating immune cell signatures. CMS2 and CMS3 are characterized by epithelial transcriptomes: CMS2 samples display high activities of WNT and MYC downstream targets, while CMS3 tumors are strongly influenced by multiple metabolic signatures such as those related to sugar metabolism, nucleotide, and fatty acid degradation. Finally, CMS4 transcriptomes are dominated by mesenchymal gene expression and high activity of TGFβ downstream targets.

When the transcriptome data were compared with the genomic analyses, the CMS1−4 classes displayed asymmetric enrichment of mutations: *BRAF* mutations were enriched in CMS1 samples, while *APC* mutations were underrepresented. *KRAS* mutations were most frequent in CMS3. *TP53* mutations were enriched in CMS2 and CMS4 compared to

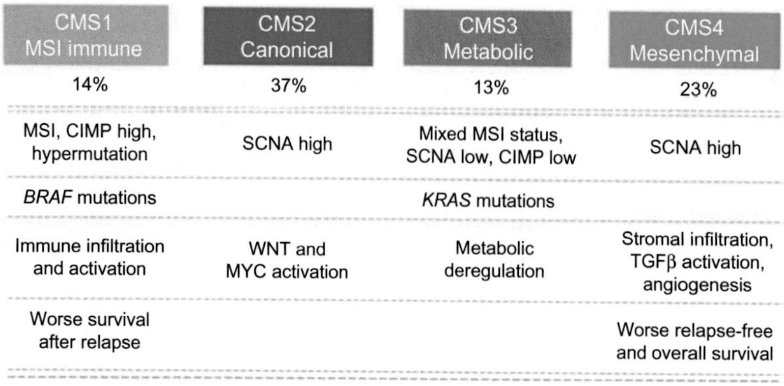

| CMS1<br>MSI immune | CMS2<br>Canonical | CMS3<br>Metabolic | CMS4<br>Mesenchymal |
|---|---|---|---|
| 14% | 37% | 13% | 23% |
| MSI, CIMP high, hypermutation | SCNA high | Mixed MSI status, SCNA low, CIMP low | SCNA high |
| *BRAF* mutations | | *KRAS* mutations | |
| Immune infiltration and activation | WNT and MYC activation | Metabolic deregulation | Stromal infiltration, TGFβ activation, angiogenesis |
| Worse survival after relapse | | | Worse relapse-free and overall survival |

**FIGURE 7.13** Proposed taxonomy of colorectal cancer reflecting significant biological differences in the gene expression—based molecular subtypes. *CIMP,* CpG island methylator phenotype; *MSI,* microsatellite instability; *SCNA,* somatic copy number alterations; *TGF,* transforming growth factor. *From Nature Medicine 2015;21:1350—56. Reproduced with permission from Springer Nature.*

CMS1 and CMS3. These findings suggest that distinct mutations can either directly influence gene expression patterns in CRC via activation or inactivation of transcription factor networks, or indirectly influence gene expression patterns in the cancer samples by altering the evolutionary course of tumor progression or tissue composition.

Importantly, the CMS1−4 classes also have predictive value, i.e., patients stratified for the transcriptional consensus groups differ in clinical outcome. Patients displaying the CMS4 transcriptome subtype had a worse overall and relapse-free survival compared to CMS1-3 patients. Furthermore, CMS1-type patients had the worst survival after relapse, while CMS2-type patients had a superior survival after relapse compared to the other groups. The outcome of patients differed significantly even when adjusting for clinicopathological features, for MSI status and frequency of *BRAF* and *KRAS* mutations. Intriguingly, it was discovered that the mesenchymal and TGFβ signaling component of the CMS4 transcriptome subtype does not stem from the epithelial cancer cells, but from the surrounding stromal fibroblasts. Functional studies following the initial transcriptome analyses suggested that the interaction between the cancer cells and the stromal microenvironment was responsible for cancer progression in these cases. The experiments also suggested that blockade of TGFβ signaling by inhibitors could halt interaction between tumor and stroma and consequently block disease progression in experimental cancer models. This example shows that transcriptome analyses are suited to guide functional experiments to understand disease mechanisms and can therefore contribute to the development of future therapies. Along these lines, a very recent study demonstrated substantial concordance between donor tumors, preclinical mouse xenograft models, and canceroid cultures and identified a novel gene signature for predicting the response of CRC toward inhibition of EGFR.

Despite these recent advances in transcriptome analysis, CRC RNA is not analyzed routinely in the clinic. One reason is that the predictive value of transcriptome signatures has not been rigorously compared to the current diagnostic standard that is testing of predictive *KRAS*, *NRAS*, and *BRAF* mutations. Moreover, RNA of tumor specimens is unstable compared to genomic DNA that is required for sequencing for predictive mutations. These characteristics of RNA raise concerns for the reproducibility and comparability of gene expression analyses in the clinic.

## Breast cancer transcriptomes—identification of hidden subtypes in breast cancer within apparently homogenous cancer cell populations

The group of Sorlie et al. identified 456 genes out of 8000 genes whose expression varied significantly between tumors of different patients, while this gene set showed only little variation upon comparison of successive tumor samples derived from an individual patient. These genes, called the intrinsic gene set, were then used on an array to discriminate between tumor subclasses in a cohort of 65 tumors from 42 breast cancer patients. Using hierarchical clustering, the researchers distinguished five distinct tumor groups characterized by their gene expression pattern: the basal epithelial cancer type, the luminal epithelial cancer types A−C, a group displaying expression of the breast cancer oncogene *ERBB2* (*HER2*), and a group without any known feature. There was yet another group showing features of normal breast epithelial cells (Fig. 7.14). In the next step of the analysis, the researchers addressed the question as to whether these different groups are characterized by distinct clinical parameters. Therefore, they compared the groups by certain statistical methods, among others by univariate statistical analysis, for either total survival or relapse-free survival monitored for up to 4 years (Fig. 7.15). The patient groups that were *ERBB2*-positive or were characterized by the basal-like breast cancer had the shortest survival times. While this information was not new for the *ERBB2*-positive tumors, the basal-like type breast cancers belong to a novel group with an obvious bad prognosis. One characteristic of this tumor

(A)

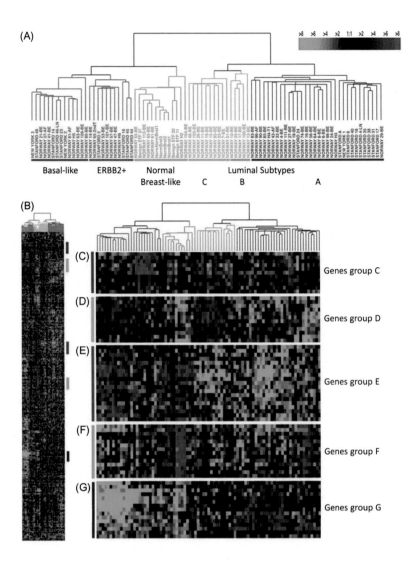

Basal-like    ERBB2+    Normal          Luminal Subtypes
                        Breast-like  C       B              A

(B)

(C)  Genes group C

(D)  Genes group D

(E)  Genes group E

(F)  Genes group F

(G)  Genes group G

**FIGURE 7.14**  Differential breast cancer gene expression. Gene expression patterns of 85 experimental samples (78 carcinomas, 3 benign tumors, 4 normal tissues) analyzed by hierarchical clustering using a set of 476 cDNA clones. (A) Cancer specimens were divided into six subtypes based on their differences in gene expression: luminal subtype A, dark blue; luminal subtype B, yellow; luminal subtype C, light blue; normal breast-like, green; basal-like, red; and ERBB2+, pink. (B) The full cluster diagram obtained after two-dimensional clustering of tumors and genes. The *colored bars* on the right represent the characteristic gene groups named C to G and are shown enlarged in the right part of the graph: (i) Group C: ERBB2 amplification cluster, (ii) Group D: novel unknown cluster, (iii) Group E: basal epithelial cell-enriched cluster, (iv) Group F: normal breast epithelial-like cluster, and Group G: luminal epithelial gene cluster containing ER. *From Proc Natl Acad Sci USA 2001;98:10869−874; Reprinted with permission from the National Academy of Sciences USA.*

type is the high frequency of *TP53* mutations. The tumor suppressor gene *TP53*, well known as the guardian of the genome, is lost or mutated in more than 50% of all advanced human cancers, and might be responsible for the bad prognosis. There was also a difference in clinical outcome between the luminal-type breast cancers. Most strikingly, luminal A cancers exhibited a very good clinical course at least within 4 years, while luminal B or C cancers were intermediate. Thus, this study opened the door to further screen many tumors for gene signatures indicative of the clinical performance of breast cancer patients.

In 2012, TCGA published the first integrated analysis of human breast cancers including more than 800 tumor samples and a median follow-up of 17 months. Three hundred forty-eight patient samples were analyzed on six technical platforms including DNA microarrays for gene expression and DNA methylation. Expression of miRNA was investigated via RNA-seq. Furthermore, exome sequencing and SNP array analysis were performed for identification of mutations and copy number alterations, respectively. Finally, to investigate signaling pathway patterns, 403 samples were tested on reverse phase protein arrays using 171 antibodies against selected proteins and phosphoproteins. Using this wealth of biological information, the researchers performed a Cluster-of-Cluster (C of C) approach to reveal higher order structures. Interestingly, all different platform-derived signatures for basal-like breast cancers clustered together, indicating the outstanding individual molecular characteristics of this breast cancer subtype. When mRNA expression signatures were omitted from the C of C analysis, 4−6 subtypes were identified. This analysis revealed that molecular information derived from copy number alterations, miRNAs, and DNA methylation is most likely reflected at the level of gene expression and protein function.

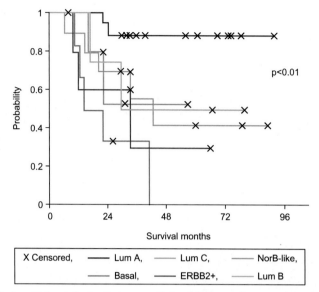

**FIGURE 7.15** Survival analysis (Kaplan—Meier plot) of patient groups distinguished according to gene expression profiling. The Y-axis shows the survival probability for each individual group, the X-axis represents the time scale according to patient follow-up data. All groups identified by gene expression profiling are shown. Luminal A, dark blue; luminal B, yellow; luminal C, light blue; normal-like, green; ErbB2-like type, pink; and basal-like, red. Patients with ErbB2-like or basal-like cancers had the shortest survival times, luminal A patients had the best prognosis. All others showed an intermediate probability and were not clearly distinguishable. *From Proc Natl Acad Sci USA 2001;98:10869—874; Reprinted with permission from the National Academy of Sciences USA.*

## Prediction of clinical outcome of breast cancer

With respect to cancer treatment, the most important issue is to find gene sets predictive for the susceptibility or resistance to therapy, particularly to widely used chemotherapies, and for the clinical outcome in the absence of other conventional indicators. Questions related to chemotherapy resistance and drug sensitivity have been extensively addressed by microarray studies, but have not been advanced to the clinical level.

Breast cancer patients with the same stage of the disease exhibit markedly different treatment responses and overall outcome. However, an adequate histopathological discrimination is not possible. The strongest predictors for metastases such as lymph node status and histological grade fail to classify breast tumors accurately according to their clinical behavior. A group from the Netherlands has pioneered gene array—based breast cancer diagnostics. This study was based on a well-characterized cohort of 117 breast cancer patients, including 78 sporadic primary invasive ductal and lobular breast carcinomas <5 cm in size. The tumor stages were T1 or T2, nodal status N0 (without axillary metastases), patient age was less than 55 years at diagnosis, and the patients did not have a history of previous malignancy. The patients received surgical treatment followed by radiotherapy but no adjuvant chemotherapy (except for five of them). The follow-up period of the patient cohort was 5 years. Tissue samples contained more than 50% tumor cells by pathological inspection; estrogen receptor (ER) and progesterone receptor (PR) status were known. The cohort was supplemented by 20 hereditary tumors carrying *BRCA1/BRCA2* mutations of similar histology as the sporadic carcinomas. Target RNA/cDNA was labeled and hybridized to an oligonucleotide array representing more than 24,000 human sequences and more than 1000 control sequences. The reference target used in this system was a pooled cRNA derived from an RNA mixture of all patients. This means that gene expression of each sample was determined relative to the pool of all samples. The hybridizations were performed in duplicate and approx. 5000 genes appeared significantly regulated more than two-fold with a *P*-value of less than 0.01.

In the first step of bioinformatic analysis, expression profiles of 98 cancer samples analyzed were clustered hierarchically according to similarities among the 5000 genes (Fig. 7.16A and B). This procedure revealed two distinct groups of cancers. In the upper group, 34 out of 62 patients had developed distant metastasis within 5 years, while in the lower group 25 out of 36 of patients (70%) exhibited invasion/metastasis. There was also a clear association with ER expression, which indicates a bad prognosis when lacking. Therefore, the researchers filtered out ER-negative tumors that did not express ER and also some of the known ER targets (Fig. 7.16C). In addition, the second group of tumors expressed a B-cell and T-cell gene signature. Hence, the tumors were characterized by lymphocyte infiltration and clearly separated from the ER-negative group (Fig. 7.16D).

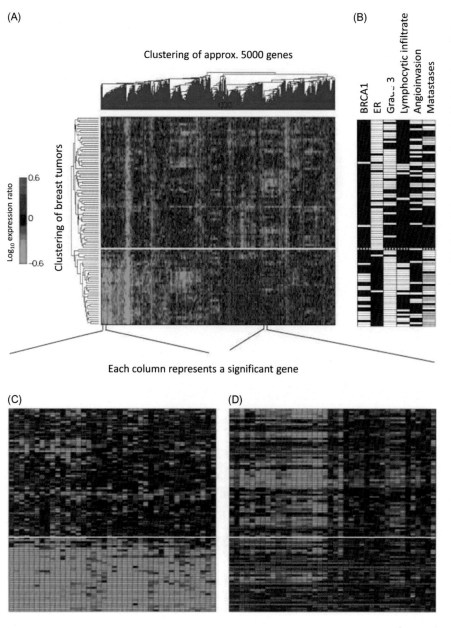

(A)

Clustering of approx. 5000 genes

Log₁₀ expression ratio

Clustering of breast tumors

0.6

0

−0.6

Each column represents a significant gene

(B)

BRCA1
ER
Grade 3
Lymphocytic infiltrate
Angioinvasion
Matastases

(C)

(D)

**FIGURE 7.16** Microarray-based prediction of breast cancer prognosis. Two-dimensional clustering of 98 cancer samples based on approximately 5000 significantly regulated genes. (A) Cluster analysis. (B) Molecular characteristics of the cancers (*BRCA1* mutation and estrogen receptor status (ER), grade, lymphocyte infiltration, blood vessel count, and distant metastases occurring within 5 years following diagnosis). The group above the *yellow line* is defined as the good prognosis group (34% of patients developed distant metastasis), the group below as the bad prognosis group (70% of patients developed distant metastasis). (C) Expression pattern of subgroup associated with estrogen receptor expression. (D) Subgroup exhibiting lymphocytic infiltration. *From Nature 2002;415:530−36. Reproduced with permission from Springer Nature.*

In a supervised classification procedure, the researchers from the Netherlands used the gene expression profiles obtained from the sporadic tumors only. In the first step of the classification procedure, the 5000 genes, which were significantly regulated in more than 3 of 78 tumors were selected from the 25,000 genes represented on the array. The correlation of each gene expression profile with the clinical outcome of patients was calculated, and 231 genes were found to be significantly associated with disease progression. In the second step, the 231 informative genes were rank-ordered according to their correlation coefficient. In the third step, researchers optimized the number of genes in this preliminary prognosis classifier by cross-validation, particularly by the leave-one-out procedure. The final result was a signature of 70 genes, which predicted the clinical outcome—distant metastasis within 5 years—with an accuracy of 83% (Fig. 7.17). This means that of 65 of 78 patients were assigned to the correct category—poor prognosis (cluster below the yellow line) or good prognosis (above) (Fig. 7.17). Five patients with poor prognosis and eight patients with good prognosis were misclassified. Van't Veer et al. used an independent set of 19 lymph-node negative breast tumors (Fig. 7.17C) to validate their classifier. This time, 2 of 19 patients were assigned to the wrong group. Thus, the classifier predictive of a short interval to distant metastases (poor prognosis signature) in patients without cancer cells in local

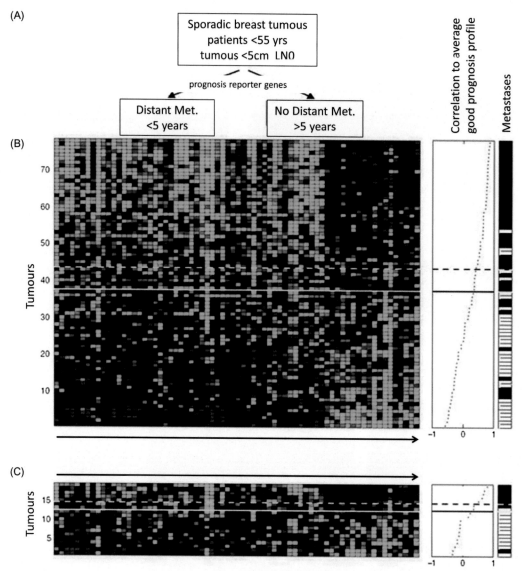

**FIGURE 7.17** Identification of the prognostic breast cancer gene set using a supervised approach. (A) The objective of expression profiling of node-negative breast cancers was to distinguish tumors developing distant metastases later than 5 years from those forming metastasis early on. Two independent patient cohorts (B) (n=78) and (C) (n=19) were analyzed. The 231 genes identified as being most significantly correlated to disease outcome were used to recluster the cancers. Each row represents a cancer and each column a gene. The genes are ordered according to their correlation coefficient with the two prognostic groups. The cancers are ordered according to their correlation to the average profile of the good prognosis group. The *solid line* marks the prognostic classifier showing optimal accuracy; the *dashed line* the classifier showing optimized sensitivity. Patients above the *dashed line* have a good prognosis signature, while patients below the *dashed line* have a poor prognosis signature. The metastasis status for each patient is shown at the right. *White bars* indicate patients who developed distant metastases within 5 years after the primary diagnosis; black indicates disease-free patients. *From Nature 2002;415:530–36. Reproduced with permission from Springer Nature.*

lymph nodes at diagnosis (lymph node-negative patients) showed a similar performance among this test set of cancers as compared to the training set.

Today, several gene expression–based prognostic breast cancer tests have been licensed for use. These include MammaPrint (Agendia BV, Amsterdam, the Netherlands), Oncotype DX (Genomic Health, Redwood City, California), EndoPredict (Sividon Diagnostics, Myriad Genetics), Genome Grade Index (MapQuant Dx, Ipsogen, France), and Prosigna (Nanostring Technologies, Seattle). All assays are based on gene expression profiling, using different and only partially overlapping gene sets. The tests also quantify ER expression and perform multivariate predictions based on their data. Mamma-Print measures 70 genes and stratifies patients into high-risk versus low-risk prognostic groups. The other tests interrogate 97 genes (Genome Grade Index), 50 genes (Prosigna), 16 genes (Oncotype DX), or 8 genes

(Endopredict), respectively. Endopredict additionally takes into account lymph node status and tumor size as common clinical parameters. Several assays have been approved by the FDA and European regulatory authorities, Endopredict and MammaPrint were also tested in clinical trials. The investigators stated that "…among women with early-stage breast cancer who were at high clinical risk and low genomic risk for recurrence, the avoidance of chemotherapy on the basis of the 70-gene signature led to a 5-year rate of survival without distant metastasis that was 1.5% points lower than the rate with chemotherapy…". This conclusion would indicate that a large proportion of these patients do not need chemotherapy, because it does not make a difference in the five-year survival, but rather improves the quality of life by avoiding side effects.

## Perspectives

Transcriptome analyses have become indispensable in basic research, translational, and clinical studies. Sequencing technologies have gradually replaced microarray techniques, due to the greater information content of the output, and also due to decreasing sequencing costs. Today, industrial platforms exhibit a high degree of standardization allowing reproducible analyses for academic core facilities and commercial service providers. Together with other omics technologies, transcriptomics are an essential component in world-wide efforts to understand normalcy and disease at many levels, from the single cell to the tissue and the patient.

In experimental and clinical pathology, transcriptomic approaches are now a standard strategy for class discovery and class prediction, in order to stratify cancer patients and select the best available therapies. Transcriptomic data collection and analysis in systems biology and systems medicine are instrumental for the development of future targeted therapies. The strong need for predictive markers in the clinic, the issue of personalized medicine, and the requirement to study the effects of old and novel drugs on the genetic program are expected to increase the application of transcriptomics even further. Areas such as alternative mRNA splicing, noncoding RNAs, and cell-to-cell variation in transcriptomes will increasingly be explored in development, normal physiology, and disease. Knowledge frontiers will be pushed towards deciphering the mechanisms of transcriptome deregulation by deciphering the master regulators of the transcriptome and the networks in which they function.

## Key concepts

- Comprehensive gene expression profiling, e.g., using microarrays or RNA sequencing, has become a key method to understand a wide range of biological phenomena *in vivo* and under experimental conditions *in vitro*
- Generation and processing of large scale gene expression data is a complex task: systematic errors at any experimental or bioinformatic stage can invalidate all subsequent analyses
- Public repositories for transcriptome data using standardized data formats are an important resource for scientists and health care professionals
- Transcriptional profiling has strongly contributed to our understanding of cancer formation, classification and outcome prediction.
- Characteristic differences in gene expression can distinguish cancers by tissue of origin or histological subtype. Transcriptomes can thus be used to characterize cancers of unknown origin.
- Transcriptional profiling of lymphoma, colorectal cancer and breast cancer resulted in the discovery of new clinical subtypes with prognostic or predictive value in the clinic.

## Suggested readings

### Classical studies

[1] Velculescu VE, Madden SL, Zhang L, et al. Analysis of human transcriptomes. Nat Genet 1999;23:387−8.
[2] Iyer VR, Eisen MB, Ross DT, et al. The transcriptional program in the response of human fibroblasts to serum. Science 1999;283:83−7.
[3] Zuber J, Tchernitsa OI, Hinzmann B, et al. A genome-wide survey of RAS transformation targets. Nat Genet 2000;24:144−52.

### Analysis of gene expression data and bioinformatics

[4] Loven J, Orlando DA, Sigova AA, et al. Revisiting global gene expression analysis. Cell 2012;151:476−82.
[5] Mortazavi A, Williams BA, McCue K, et al. Mapping and quantifying mammalian transcriptomes by RNA-Seq. Nat Methods 2008;5:621−8.
[6] Robinson MD, McCarthy DJ, Smyth GK. edgeR: a bioconductor package for differential expression analysis of digital gene expression data. Bioinformatics 2010;26:139−40.

[7] Eisen MB, Spellman PT, Brown PO, Botstein D. Cluster analysis and display of genome-wide expression patterns. Proc Natl Acad Sci USA 1998;95:14863−8.

[8] Rhee SY, Wood V, Dolinski K, Draghici S. Use and misuse of the gene ontology annotations. Nat Rev Genet 2008;9:509−15.

## Consortia providing large-scale Transcriptome data

[9] Carninci P, Kasukawa T, Katayama S, et al. The transcriptional landscape of the mammalian genome. Science 2005;309:1559−63.

[10] ENCODE Project Consortium. An integrated encyclopedia of DNA elements in the human genome. Nature 2012;489:57−74.

[11] GETx Consortium. Human genomics. the genotype-tissue expression (GTEx) pilot analysis: multitissue gene regulation in humans. Science 2015;348:648−60.

[12] Weinstein JN, Collisson EA, Mills GB, et al. The cancer genome atlas pan-cancer analysis project. Nat Genet 2013;45:1113−20.

## Cancer Transcriptomes

[13] Hoadley KA, Yau C, Wolf DM, et al. Multiplatform analysis of 12 cancer types reveals molecular classification within and across tissues of origin. Cell 2014;158:929−44.

[14] Alizadeh AA, Eisen MB, Davis RE, et al. Distinct types of diffuse large B-cell lymphoma identified by gene expression profiling. Nature 2000;403:503−11.

[15] The Cancer Genome Atlas Network. Comprehensive molecular characterization of human colon and rectal cancer. Nature 2012;487:330−7.

[16] Guinney J, Dienstmann R, Wang X, et al. The consensus molecular subtypes of colorectal cancer. Nat Med 2015;21:1350−6.

[17] The Cancer Genome Atlas Network. Comprehensive molecular portraits of human breast tumours. Nature 2012;490:61−70.

[18] Sorlie T, Perou CM, Tibshirani R, et al. Gene expression patterns of breast carcinomas distinguish tumor subclasses with clinical implications. Proc Natl Acad Sci USA 2001;98:10869−74.

[19] van't Veer LJ, Dai H, van de Vijver MJ, et al. Gene expression profiling predicts clinical outcome of breast cancer. Nature 2002;415:530−6.

[20] Cardoso F, Van't Veer LJ, Bogaerts J, et al. 70-gene signature as an aid to treatment decisions in early-stage breast cancer. N Engl J Med 2016;375:717−29.

Chapter 8

# The human epigenome—implications for the understanding of human disease

**María Berdasco[1] and Manel Esteller[1,2,3]**

[1]*Josep Carreras Leukaemia Research Institute (IJC), Barcelona, Spain,* [2]*Physiological Sciences Department, School of Medicine and Health Sciences, University of Barcelona (UB), Barcelona, Spain,* [3]*Institució Catalana de Recerca i Estudis Avançats (ICREA), Barcelona, Spain*

**Summary**

Epigenetic processes, defined as the heritable patterns of gene expression that do not involve changes in the sequence of the genome, and their effects on gene repression are increasingly understood to be such a way of modulating phenotype transmission and development. The patterns of DNA methylation, histone modifications, non-coding RNAs, and several chromatin-related proteins of sick cells usually differ from those of healthy cells, highlighting the importance of epigenetic regulation in most human pathologies. This chapter provides an overview of how epigenetic factors contribute to the development of human diseases, such as abnormal imprinting-causative pathologies, cancer malignancies, as well as autoimmune, cardiovascular, and neurological disorders. These studies have provided extensive information about the mechanisms that contribute to the phenotype of human diseases, but also provided opportunities for therapy.

## Introduction

Epigenetic processes, defined as the heritable patterns of gene expression that do not involve changes in the sequence of the genome, and their effects on gene repression are increasingly understood to represent mechanisms for modulating phenotype transmission and development. The patterns of DNA methylation, histone modifications, non-coding RNAs, and several chromatin-related proteins of sick cells usually differ from those of healthy cells, highlighting the importance of epigenetic regulation in most human pathologies. The aim of the present review is to provide an overview of how epigenetic factors contribute to the development of human diseases such as abnormal imprinting-causative pathologies; cancer; as well as autoimmune, cardiovascular, and neurological disorders. These studies have generated extensive information about the mechanisms that contribute to the phenotype of human diseases and have identified opportunities for therapy.

## Epigenetic regulation of the genome

### The human epigenome project

The study of epigenetics remains in infancy and a range of matters remains to be resolved, such as the relationships between the epigenetic players (the epigenetic code) and how the environment and/or aging modulate the epigenetic marks. A greater understanding of these phenomenon could be achieved by analyzing the epigenetic patterns on a genome-wide scale, an approach that has at last become possible thanks to recent technological advances. Several large-scale epigenomics projects have been developed in recent years. It is important to bear in mind that there is no single epigenome, but rather many different ones that are characteristic of normal and diverse human disorders, so it is essential to define the chosen starting material.

One of the first initiatives was the Human Epigenome Project, a multinational European project integrated by the Wellcome Trust Sanger Institute (United Kingdom), the Epigenomics AG (Germany), and the Centre National de Génotypage (France), which was designed to "... *identify, catalog, and interpret genome-wide DNA methylation*

*Essential Concepts in Molecular Pathology.* DOI: https://doi.org/10.1016/B978-0-12-813257-9.00008-5

*patterns of all human genes in all major tissues...".* Launched in 2008 by National Institutes of Health (NIH), the Roadmap Epigenomics Mapping Consortium is the largest epigenomics effort to date. The project has generated a data repository of genome-wide maps of CpG methylation, >30 histone modifications, chromatin accessibility, and mRNA expression across hundreds of human cell types and tissues. The NIH Roadmap is now integrated into the International Human Epigenome Consortium (IHEC). The IHEC grouped different projects and it coordinates the integration of reference maps of human epigenomes for key cellular states relevant to health and diseases with a final goal of at least 1000 epigenomes. The datasets generated by the IHEC are uploaded to a web portal providing access to >7000 reference epigenomic datasets, generated from >600 tissues, which have been contributed by seven international consortia: ENCODE, NIH Roadmap, CEEHRC, Blueprint, DEEP, AMED-CREST, and KNIH. Each project is focused on key cellular states, including mapping the epigenome of samples from distinct types of hematopoietic cells from healthy individuals and on their malignant leukemic counterparts (BLUEPRINT European Project) or the analysis of gastrointestinal epithelial cells, vascular endothelial cells, and cells of reproductive organs (The CREST/IHEC, Team Japan). Interestingly, IHEC also interacts with other international projects that are more genomics-associated, such as the International Cancer Genomic Consortium (ICGC) and ENCODE, to unravel the relationship between genetic and epigenetic variation worldwide. IHEC is not only a dataset platform, it also coordinates and improved the development of novel and robust bioinformatic tools, data models, and standardization of experimental protocols for epigenome research.

The information extracted from whole-genome assays will help us understand the role of the epigenetic marks and could have translational research benefits for diagnosis, prognostication, and therapeutic treatment. Defining human epigenomes associated with human disorders will help select patients who are likely to benefit from epigenomic therapies or prevention strategies, determine their efficacy and specificity, and lead to the identification of surrogate markers and end-points of its effects. Although the groundbreaking discoveries in the field of human disease were initially performed in cancer cells, the characterization of the error-bearing epigenomes underlying other disorders, such as neurological, cardiovascular, and immunological pathologies, are currently being broadly investigated.

# Genomic imprinting

## Epigenetic regulation of imprinted genes

Genomic imprinting is a genetic phenomenon by which epigenetic chromosomal modifications drive differential gene expression according to the parent-of-origin. Expression is exclusively due either to the allele inherited from the mother (such as the *H19* and *CDKN1C* genes) or to that inherited from the father (such as *IGF2*). It is an inheritance process that is independent of the classical Mendelian model. Most imprinted genes, which have been identified in insects, mammals, and flowering plants, are involved in the establishment and maintenance of particular phases of development. Nucleus transplantation experiments in mouse zygotes carrying reciprocal translocations carried out in the early 1980s suggested that imprinting may be fundamental to mammalian development. Assays confirmed that normal gene expression and development in mice require the contribution of both maternal and paternal alleles. However, it was not until 1991 that the first imprinted genes, insulin-like growth factor 2 (*IGF2*) and its receptor (*IGF2R*), were identified. Since then, hundreds of imprinted genes have been identified in mice and humans, about one-third of which are imprinted in both species. It has been predicted that 600 genes have a high probability of being imprinted in the mouse genome, and a similar genome-wide analysis predicts humans to have about half as many imprinted genes. As imprinting is a dynamic process and the profile of imprinted genes varies during development, regulation must be epigenetic. DNA methylation has been widely described as the major mechanism involved in the control of genes subjected to imprinting. One model for this regulation is based on the cluster organization of imprinted genes. This structure within clusters allows them to share common regulatory elements, such as non-coding RNAs and differentially methylated regions (DMRs). DMRs are up to several kilobases in size, rich in CpG dinucleotides (such as CpG islands), and may contain repetitive sequences. DNA methylation of DMRs is thought to interact with histone modifications and other chromatin proteins to regulate parental allele-specific expression of imprinted genes. Furthermore, the aforementioned regulatory elements usually control the imprinting of more than one gene, giving rise to imprinting control regions (ICRs). This cluster organization, observed in 80% of imprinted genes, and the specific DNA methylation patterns associated with DMRs are two of the main characteristics of imprinted genes. Deletions or aberrations in DNA methylation of ICRs lead to loss of imprinting (LOI) and inappropriate parental gene expression. Imprinted genes have diverse roles in growth and cellular proliferation, and specific patterns of genomic imprinting are established in somatic and germline cells. Recently, DMRs were identified in human placenta (not in mice) and are involved in the orchestrated imprinted expression following embryonic genome. Imprinting is erased in germline cells, and reprogramming involving a de novo methyltransferase is necessary to ensure sex-specific gene expression in the individual. Methyl groups are incorporated into most ICRs in oocytes, although only some ICRs are methylated during spermatogenesis. After fertilization, the specific methylation profiles of ICRs must be maintained during development in order to mediate the allelic expression of imprinted genes. In the primordial germline cells of the developed individual, these imprinting marks must be freshly erased by DNA demethylation to allow the subsequent establishment of new oocyte-specific and sperm-specific imprints. As a consequence, mammalian imprinting can be described as a development-dependent cycle based on germline establishment, somatic maintenance, and erasure.

## Imprinted genes and human genetic diseases

Since expression of imprinted genes is monoallelic, and thereby functionally haploid, there is no protection from recessive mutations that the normal diploid genetic complement would provide. For this reason, genetic and epigenetic aberrations in imprinted genes are linked to a wide range of diseases. Modulation of perinatal growth and human pregnancy has played a central role in the evolution of imprinting, and many of the diseases associated with imprinted genes involve some disorders of embryogenesis. This is the case of the hydatidiform mole disorder, where all nuclear genes are inherited from the father. In most cases, this androgenesis arises when an anuclear egg is fertilized by a single sperm, after which all the chromosomes and genes are duplicated. However, fertilization by two haploid sperms (diandric diploidy) may occasionally occur. Most cases are sporadic and androgenetic, but recurrent hydatidiform mole has biparental inheritance with disrupted DNA methylation of DMRs at imprinted loci. In contrast, the disorder of ovarian dermoid cysts arises from the spontaneous activation of an ovarian oocyte that leads to the duplication of the maternal genome. These abnormalities suggest that normal human development is possible only when the paternal and maternal genomes are correctly transmitted. Parent-of-origin effects involved in behavioral and brain disorders have been widely reported at the prenatal and postnatal stages of development. A postnatal growth retardation syndrome associated with the *MEST* gene expression is an illustrative example. The effect of introducing a targeted deletion into the coding sequence of the mouse *MEST* gene strongly depends on the paternal allele. When the deletion is paternally derived, *Mest+/−* mice are viable and fertile, but mutant mice show growth retardation and high mortality. *Mest−/+* animals, with a maternally derived deletion show none of these effects. This suggests that the phenotypic consequences of this mutation are detected only through paternal inheritance and are the result of imprinting. There is also evidence that some imprinting effects are associated with increased susceptibility to cancer. Absence of expression of a tumor suppressor gene could be the result of LOI or uniparental disomy (UPD) in imprinted genes. Conversely, LOI or UPD of an imprinted gene that promotes cell proliferation (an oncogene) may allow gene expression to be inappropriately increased. These aberrations could have a widespread effect if the aberrant imprinting occurs in an ICR, resulting in the epigenetic dysregulation of multiple imprinted oncogenes and/or tumor suppressor genes.

## Prader-Willi syndrome and Angelman syndrome

Prader-Willi syndrome (PWS; OMIM #176270) and Angelman syndrome (AS; OMIM #105830) are very rare genetic disorders with autosomal dominant inheritance in which gene expression depends on parental origin. The clinical features of both syndromes are quite similar—neurological disorders with mental retardation and developmental aberrations. PWS is characterized by diminished fetal activity, feeding difficulties, obesity, muscular hypotonia, mental retardation, poor physical coordination, short stature, hypogonadism, and small hands and feet, amongst other traits. AS is characterized by mental retardation, movement or balance disorder, characteristic abnormal behaviors, increased sensitivity to heat, absent or little speech, and epilepsy. The prevalence of the two syndromes is not accurately known, but is estimated to be between 1 in 12,000 and 1 in 15,000 live births, respectively. Most cases of PWS are caused by the deficiency of the paternal copies of the imprinted genes on chromosome 15 located in the 15q11-q13 region, while AS affects maternally imprinted genes in the same region. This deficiency could be due to the deletion of the 15q11-q13 region (3−4 Mb), parental uniparental disomy of chromosome 15, or imprinting defects. The imprinted domain on human chromosome 15q11-q13 is regulated by an ICR that is responsible for establishing the imprinting in the gametes and for maintaining the patterns during the embryonic phases. ICR regulates differential DNA methylation and chromatin structure, and in consequence, differential gene expression affecting the two parental alleles. The ICR in 15q11-q13 appears to have a bipartite structure; one part seems to be responsible for the control of paternal expression and the other for maternal gene expression. PWS is in effect a contiguous gene syndrome resulting from deficiency of the paternal copies of the imprinted *SNRF/SNRPN* gene, the necdin gene, and possibly other genes. It has been estimated that the region could contain more than 30 genes, so PWS probably results from a stochastic partial inactivation of important genes. While PWS appears to be more closely related to deficiencies caused by chromatin aberrations affecting several genes, AS is associated with mutations in single genes. For instance, the most common genetic defect leading to AS is a ~4 Mb maternal deletion in chromosomal region 15q11−13, which causes an absence of *UBE3A* expression in the maternally imprinted brain regions. Mutations in the gene encoding the ubiquitin-protein ligase E3A (*UBE3A*) have been identified in 25% of AS patients. The *UBE3A* gene is present on both the maternal and paternal chromosomes, but differs in its pattern of methylation. Paternal silencing of the *UBE3A* gene occurs in a brain-region−specific manner, with the maternal allele being active almost exclusively in the Purkinje cells, hippocampus, and cerebellum. Another maternally expressed gene, *ATP10C* (aminophospholipid-transporting adenosine triphosphatase [ATPase]), is also located in this region and has been implicated in AS. Like *UBE3A*, it exhibits imprinted, preferential maternal expression in human brain. Histone modifications are also altered in PWS: histone H3 Lys9 (H3K9) is methylated on the maternal allele and Lys4 (H3K4) methylated on the paternal allele. Indeed, the pharmacological inhibition of the histone methyltransferase G9a has been proposed as a novel therapeutic approach for PWS in a mouse model.

## Beckwith-Wiedemann syndrome

Beckwith-Wiedemann syndrome (BWS; OMIM #130650) is a well-characterized human disease involving imprinted genes that are epigenetically regulated, and studies of BWS patients have contributed much to the understanding of normal imprinting. BWS is a rare genetic or epigenetic overgrowth syndrome with an estimated prevalence of about 1 in 15,000 births and a high mortality rate in the newborn (about 20%). Neonatal patients are mainly characterized by exomphalos, macroglossia, and gigantism, but other symptoms may occur, such as organomegaly, adrenocortical cytomegaly, hemihypertrophy, and neonatal hypoglycemia. There is also an increased risk of developing specific tumors, such as Wilms' tumor and hepatoblastoma. The imprinted domain BWS-related genes are located on 11p15 and are regulated by a bipartite ICR. Two clusters of imprinted genes have been described: (1) *H19/IGF2* (imprinted, maternally expressed, untranslated mRNA/insulin-like growth factor 2) and (2) *p57KIP2* (a cyclin-dependent kinase inhibitor), *TSSC3* (a pleckstrin homology domain), *SLC22A1* (an organic cation transporter), *KvLQT1* (a voltage-gated potassium channel), and *LIT1* (*KCNQ1* overlapping transcript 1). Both clusters are regulated by DMRs: DMR1, which is responsible for *H19/IGF2* control and is methylated on the paternal but not the maternal allele, and DMR2, which is located upstream of *LIT1* and is normally methylated on the maternal but not the paternal allele. BWS can appear as a consequence of two separate mechanisms. First, some patients are characterized by UPD, which consists of the complete genetic replacement of the maternal allele region with a second paternal copy, and/or LOI affecting the *IGF2*-containing region and/or the *LIT1* gene, which causes a switch in the epigenotype of the *H19/IGF2* subdomain or the *p57/KvLQT1/LIT1* subdomain, respectively. When UPD affects *IGF2*, it yields a double dose of this autocrine factor, resulting in tissue overgrowth and increased cancer risk. The LOI mechanism involves aberrant methylation of the maternal *H19* DMR. Second, maternal replacements of the allele and localized abnormalities of allele-specific chromatin modification on *p57KIP2* (also known as *CDKN1C*) or *LIT1* could also contribute to the BWS phenotype. In conclusion, BWS is a model for the hierarchical organization of epigenetic regulation in progressively larger domains. Additionally, mutations in NSD1 (*nuclear receptor-binding SET domain-containing protein 1*), the major cause of the Sotos overgrowth syndrome, have been described in BWS patients, demonstrating the role this gene plays in imprinting the chromosome 11p15 region. Finally, it must be highlighted that the epigenetic defects affecting the expression of the imprinted genes within the 11p15.5 chromosomal region are also observed in the Silver-Russell syndrome (SRS) with opposite phenotype than BWS. It is necessary to better define the clinical diagnostic criteria and scoring systems as well as molecular causes in both SRS and BWS.

## Cancer epigenetics

Cancer encompasses a fundamentally heterogeneous group of disorders affecting different biological processes and is caused by abnormal gene/pathway function arising from specific alterations in the genome. Initially, cancer was thought to be solely a consequence of genetic changes in key tumor suppressor genes and oncogenes that regulate cell proliferation, DNA repair, cell differentiation, and other homeostatic functions. However, recent research suggests that these alterations could also be due to epigenetic disruption. The study of epigenetic mechanisms in cancer, such as DNA methylation, histone modification, nucleosome positioning, and non-coding RNA expression, has provided extensive information about the mechanisms that contribute to the neoplastic phenotype through the regulation of expression of genes critical to transformation pathways. These alterations and their involvement in tumor development are briefly reviewed in the following sections.

## DNA hypomethylation in cancer cells

The low level of DNA methylation in tumors compared with that in their normal-tissue counterparts was one of the first epigenetic alterations to be found in human cancer. It has been estimated that 3%−6% of all cytosines are methylated in normal human DNA, although cancer cell genomes are usually hypomethylated with malignant cells, featuring 20%−60% less genomic 5-methylcytosine than their normal counterparts. Global hypomethylation in cancer cells is generally due to decreased methylation in CpGs dispersed throughout repetitive sequences, which account for 20%− 30% of the human genome, as well as in the coding regions and introns of genes. A chromosome-wide and large-pro-moter−specific study of DNA methylation in a colorectal cancer cell line using the methyl-DIP approach has revealed extensive hypomethylated genomic regions located in gene-poor areas. Importantly, the degree of hypomethylation of genomic DNA increases as the lesion progresses from a benign cellular proliferation to an invasive cancer. From a functional point of view, hypomethylation in cancer cells is associated with a number of adverse outcomes, including chromosome instability, activation of transposable elements, and LOI (Fig. 8.1). Decreased methylation of repetitive sequences in the satellite DNA of the pericentric region of chromosomes is associated with increased chromosomal rearrangements, mitotic recombination, and aneuploidy. Intragenomic endoparasitic DNA, such as L1 (long interspersed nuclear elements) and Alu (recombinogenic sequence) repeats, are silenced in somatic cells and become reactivated in human cancer. Deregulated transposons could cause transcriptional deregulation, insertional mutations, DNA breaks, and an increased frequency of recombination, contributing to genome disorganization, expression changes, and chromosomal instability. DNA methylation underlies the control of several imprinted genes, so the effect on the LOI must also be considered. Wilms' tumor, a nephroblastoma that typically occurs in children, is the best-characterized imprinting effect associated with increased susceptibility to cancer. Other changes in the expression of imprinted genes caused by changes in methylation have been demonstrated in malignancies such as osteosarcoma, hepatocellular carcinoma, and bladder cancer. Finally, DNA methylation acts as a mechanism for controlling cellular differentiation, allowing the expression only of tissue-specific and housekeeping genes in somatic differentiated cells. It is possible that some tissue-specific genes became reactivated in cancer in a hypomethylation-dependent manner. Activation of *PAX2*, a gene that encodes a transcription factor involved in proliferation and other important cell activities, and *let-7a-3*, an miRNA gene, have been implicated in endometrial and colon cancer.

## Hypermethylation of tumor suppressor genes

Aberrations in DNA methylation patterns of the CpG islands in the promoter regions of tumor suppressor genes are accepted as being a common feature of human cancer (Fig. 8.1). The initial discovery of silencing was performed in the promoter of the retinoblastoma (*Rb*) tumor suppressor gene, but hypermethylation of genes like *VHL* (associated with von Hippel-Lindau disease), *p16INK4a*, *hMLH1* (a homologue of *Escherichia coli* MutL), and *BRCA1* (breast-cancer susceptibility gene 1) have also been described. The presence of CpG island promoter hypermethylation affects genes from a wide range of cellular pathways, such as cell cycle, DNA repair, toxic catabolism, cell adherence, apoptosis, and angiogenesis, among others, and may occur at various stages in the development of cancer. There is a CpG-island hypermethylation profile of human primary tumors, which shows that the CpG island hypermethylation profiles of tumor suppressor genes are specific to the cancer type. Each tumor type can be assigned a specific, defining DNA hypermethylome, rather like a physiological or cytogenetic marker. These marks of epigenetic inactivation occur not only in sporadic tumors but also in inherited cancer syndromes, in which hypermethylation may be the second lesion in Knudson's two-hit model of cancer development. However, to date 100−400 instances of gene-specific methylation have been noted in a given tumor. Sometimes the epigenetic alteration of a tumor suppressor gene has genetic consequences, for example, when a DNA repair gene (*hMLH1*, *BRCA1*, *MGMT*, or Werner syndrome gene) is silenced by promoter methylation and functionally blocked.

## Histone modifications of cancer cells

The histone modification network is very complex. Histone modifications can occur in various histone proteins (including H2B, H3, H4) and variants (such as H3.3) and affect different histone residues (including lysine, arginine, serine). Several chemical groups (methyl, acetyl, phosphate) may be added in different degrees (for example, monomethylation,

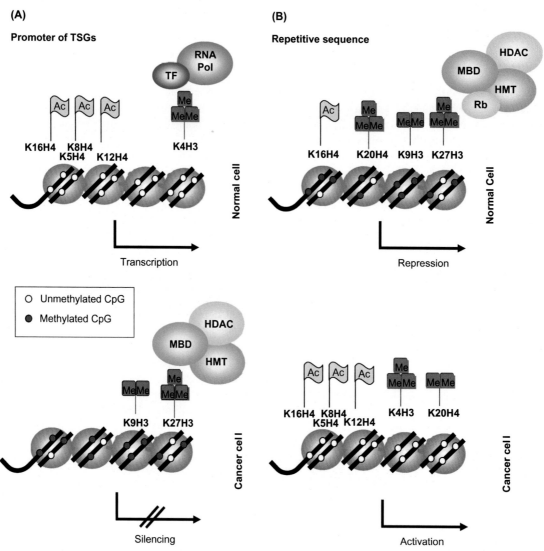

**FIGURE 8.1** A model for the disruption of histone modifications and DNA methylation patterns in cancer cells. Nucleosome arrays are located in the context of genomic regions that include (A) promoters of tumor suppressor genes (TSGs), (B) repetitive sequences in heterochromatin regions. Nucleosomes consisting of two copies of histones H2A, H2B, H3, and H4 are represented as gray cylinders. DNA (*black lines*) is wrapped around each nucleosome. In normal cells, TSG promoter regions are unmethylated and enriched in histone modification marks associated with active transcription, such as acetylation of histone H4 (at lysine K5, K8, K12, and K16) or trimethylation of histone H3 (at lysine K4). Transcription machinery recognizes these active marks and transcription of TSGs is allowed. In the same cells, the repetitive genomic DNA is silenced due to the high degree of DNA methylation and histone-repressive marks: histone H4 is densely trimethylated at lysine 20, and histone H3 is dimethylated at lysine 9 and trimethylated at lysine K27. This epigenetic profile is disrupted in transformed cells. TSG promoters are silenced by the loss of the histone-active marks and gain of promoter hypermethylation. Repetitive sequences are activated by replacement of the repressive marks, leading to activation of endoparasitic sequences, genomic instability, or loss of imprinting.

dimethylation, or trimethylation). Histone modifications also include non-classical marks, such as the conversion of an arginine to a citrulline (deimination), ADP-ribosylation, ubiquitination, and sumoylation or proline isomerization. Furthermore, the significance of each modification depends on the organism, the biological process, and the chromatin-genomic region. Due to this diversity of permutations and combinations, little is known about the patterns of histone modification disruption in human tumors. Results have shown that the CpG promoter-hypermethylation event in tumor suppressor genes in cancer cells is associated with a particular combination of histone markers: deacetylation of histones H3 and H4, loss of H3K4 trimethylation, and gain of H3K9 methylation and H3K27 trimethylation (Fig. 8.1). Increased acetylated histones H3 and H4, and H3/K4 at the *p21* and *p16* transcription start sites after genistein induction (an isoflavone found in the soybean with tumor suppressor properties), in the absence of *p21* promoter methylation, have been reported. The association between DNA methylation and histone modification aberrations in cancer also occurs at the global level. In human and mouse tumors, histone H4 undergoes a loss of monoacetylated and trimethylated lysines 16 and 20, respectively, especially in the repetitive DNA sequences. Subsequent studies showed that loss of trimethylation at H4K20 is involved in disrupting heterochromatic domains and may reduce the response to DNA damage of cancer cells. Immunohistochemical staining of primary prostatectomy tissue samples revealed that patterns of H3 and H4 were predictors of clinical outcome independently of tumor stage, preoperative prostate-specific antigen levels, and capsule invasion in prostate cancer. In the last years, several ChIP-Seq analyses covering the genome-wide sites for histone modifications have been described for several tumor types, providing information that will contribute to the deciphering of complex chromatin networks.

## Epigenetic regulation of non-coding RNAs in cancer

Non-coding RNAs (ncRNAs), defined as RNA sequences that do not specify any protein, play major biological roles in cellular biology, and development, and their deregulation is associated with human diseases, including cancer. In recent years, the concept of RNA as an intermediate between DNA sequence and proteins has been redefined and increasing evidence of the crucial functional roles of RNA in cell development and differentiation have accumulated. While most of the regulatory ncRNAs have yet to be functionally characterized, they have been classified based on their size into short ncRNA ($<200$ nt) and long ncRNA ($>200$ nt). The majority of investigations have focused on short ncRNAs (such as microRNAs), but the functional relevance of other ncRNAs such as long non-coding RNAs (lncRNAs) is also gaining prominence. ncRNAs show a fine-tuned expression in a temporal and spatially dependent manner. However, the mechanisms that govern the control of ncRNAs are largely unexplored. Multiple regulatory mechanisms, similar to those of coding regions, can be implicated including transcription factor binding, RNA splicing, RNA stabilization, and epigenetic regulation.

It has been widely described that miRNA expression profiles differ between normal and cancer tissues, and also among cancer types, whereby some microRNAs are downregulated in cancer (like tumor suppressor genes). This comparison suggests that miRNAs could be silenced by epigenetic mechanisms. DNA methylation has been shown to be the regulatory mechanism for at least two microRNAs, miR-127 and miR-124a. miR-127, which negatively regulates the protooncogene BCL6 (B-Cell Lymphoma 6), is usually expressed in normal cells but is silenced by DNA methylation in cancer cells. Similarly, the hypermethylation-dependent silencing of miR-124a results in an increase in expression of cyclin D-kinase 6 oncogene (CDK6) and is recognized as a common feature of a wide range of cancers. Similarly, although less well explored, epigenetic deregulation of lncRNAs has been detected in cancer. As one example, the lncRNAs Uc.160+, Uc283+A, and Uc.346+ are silenced by CpG island hypermethylation in cancer cells compared with expression in normal tissues. Furthermore, in recent years many studies have shown that lncRNAs can be regulated by microRNAs and that this process can be altered in cancer. In this way, miR-125 controls HOTTIP expression in hepatocellular carcinoma, miR-101 and miR-217 target MALAT1 in esophageal carcinoma, miR-211 controls the lncRNA loc285194 expression in colorectal cancer, and miR-1 regulates UCA1 in bladder cancer.

## Aberrations in histone-modifier enzymes

Aberrations in the epigenetic profiles, with respect to DNA methylation and histone modifications, could also be a consequence of genetic disruption of the epigenetic machinery. A preliminary set of genes involved in epigenetic modifications with mutations in cancer cells but not in the normal counterparts has been found. A list of genes involved in epigenetic modifications that are disrupted in human cancer is presented in Table 8.1. Genetic alterations affecting DNA methyltransferase enzymes (DNMTs), methyl-CpG-binding domain–containing proteins, and DNA hydroxylases are rarely observed in human cancers. The picture is different for histone modifier enzymes. In leukemia and sarcoma, chromosomal translocations that involve histone-modifier genes, such as histone acetyltransferases [such as cyclic AMP response element-binding protein (CREB)-binding protein-monocytic leukemia zinc finger (CBP-MOZ)] and histone methyltransferases (HMTs) (such as mixed-lineage leukemia 1 or MLL1), nuclear receptor–binding SET domain protein 1 (NSD1), and nuclear receptor–binding SET-domain protein 3 (NSD3) create aberrant fusion proteins. In solid cancers, both HMT genes such as EZH2, mixed-lineage leukemia 2 (MLL2), NSD3 or SETDB1, and a demethylase (designated Jumonji domain-containing protein 2C or JMJD2C/GASC1) are known to be amplified. Genetic aberrations also disrupt expression of histone deacetylases, such as histone deacetylase 2 (HDAC2), which could be affected by mutational frameshift inactivation in colon cancer; histone demethylases, such as UTX or JARID1C; and chromatin remodeling proteins, such as HLTF (helicase-like transcription factor), BRG1 (Brahma-related gene 1), and other components of the SWI/SNF family of proteins.

**TABLE 8.1** Disruption of genes involved in DNA methylation and histone modifications in Cancer.

| Gene | Alteration | Tumor type |
|---|---|---|
| Alterations affecting DNA methylation enzymes (DNMTs) | | |
| *DNMT1* | Overexpression | Various |
| *DNMT3b* | Overexpression, point mutations | Various |
| Alterations involving methyl-CpG-binding proteins (MBPs) | | |
| *MeCP2* | Overexpression, rare mutations | Various |
| *MBD1* | Overexpression, rare mutations | Various |
| *MBD2* | Overexpression, rare mutations | Various |
| *MBD3* | Overexpression, rare mutations | Various |
| *MBD4* | Mutations in microsatellite instable tumors | Colon, stomach, endometrium |
| Alterations affecting DNA demethylases | | |
| *TET1* | Translocations | Hematological malignancies |
| *TET2* | Point mutations | Hematological malignancies |
| Alterations disrupting histone acetyltransferases (HATs) | | |
| *P300* | Mutations in microsatellite instable tumors | Colon, stomach, endometrium |
| *CBP* | Mutations, translocations, deletions | Colon, stomach, endometrium, lung, leukemia |
| *pCAF* | Rare mutations | Colon |
| *MOZ* | Translocations | Hematological malignancies |
| *MORF* | Translocations | Hematological malignancies, leiomyomata |
| *TIP60* | Deletion | Hematological malignancies, head and neck, breast |
| *MOF* | Downregulation | Breast, medulloblastoma |
| Alterations disrupting histone deacetylases (HDACs) | | |
| *HDAC1* | Aberrant expression | Various |
| *HDAC2* | Aberrant expression, mutations in microsatellite instable tumors | Various |
| *SIRT2* | Downregulation | Gliomas |
| *SIRT3* | Downregulation | Hepatocellular carcinoma |
| *SIRT7* | Overexpression | Thyroid carcinoma, breast |
| Alterations affecting histone methyltransferases (HMTs) | | |
| *MLL1* | Translocation | Hematological malignancies |
| *MLL2* | Gene amplification | Glioma, pancreas |
| *MLL3* | Deletion | Leukemia |
| *NSD1* | Translocation | Leukemia |
| *NSD2* | Overexpression | Prostate, Hematological malignancies |
| *NSD3* | Amplification | Prostate, Breast |
| *EZH2* | Gene amplification, overexpression | Various |
| *RIZ1* | Promoter CpG-island hypermethylation | Various |
| *SETDB1* | Amplification | Melanoma, lung |
| Alterations affecting histone demethylases | | |
| *GASC1* | Gene amplification | Squamous cell carcinoma |
| *JARID1A* | Translocation | Prostate, Hematological malignancies |
| *JARID1C* | Point mutation | Renal |
| *UTX* | Deletion | Various |

Adapted from Simó-Riudalbas L, Esteller M. Cancer genomics identifies disrupted epigenetic genes. Hum Genet 2014;133:713—25.

# Human disorders associated with epigenetics

## Aberrant epigenetic profiles underlying immunological, cardiovascular, neurological, and metabolic disorders

The patterns of DNA methylation, histone modifications, microRNAs, and several chromatin-related proteins of sick cells usually differ from those of healthy cells, highlighting the importance of epigenetic regulation in most human pathologies. Most of our knowledge regarding human epigenetic diseases was first obtained from cancer cells, but currently there is increasing interest in understanding the role of epigenetic modifications in the etiology of human disease. DNA methylation is the best-characterized epigenetic modification in many pathways of immunology and is the source of much of our knowledge about the molecular network of the immune system. Classical autoimmune disorders, such as systemic lupus erythematosus (SLE), an autoimmune disease characterized by the production of a variety of antibodies against nuclear components and which causes inflammation and injury of multiple organs, and rheumatoid arthritis, a chronic systemic autoimmune disorder that primarily causes inflammation and destruction of the joints, are characterized by massive genomic hypomethylation. This decrease in DNA methylation levels is highly reminiscent of the global demethylation observed in the DNA of tumor cells compared with their normal tissue counterparts. How DNA hypomethylation in T-cells induces SLE is not well understood. It has been proposed that DNA demethylation induces overexpression of integrin adhesive receptors and leads to an autoreactive response. Identification of the full set of genes deregulated by DNA hypomethylation could help to explain these immunological disorders and will also enable the development of effective therapies to cure SLE. Additional pathologies with epigenetic regulation of immunology have been reported, including the epigenetic silencing of the ABO histo-blood group genes, the silencing of human leukocyte antigen class I antigens, and the melanoma antigen-encoding gene (MAGE) family. It is known that *MAGE* gene expression is epigenetically repressed by promoter CpG methylation in most cells, but *MAGE* genes may be expressed in various tumor types via CpG demethylation and can act as antigens that are recognized by cytolytic T-lymphocytes. Alterations of specific genomic DNA methylation levels have been described not only in the fields of oncology and immunology, but also in a wide range of biomedical and scientific fields. Epigenetic regulation controls neural differentiation as well as modulates crucial processes such as memory consolidation, learning, or cognition during healthy lifespan. As a result many neurodegenerative diseases are associated with epigenetic dysregulation. Changes of the brain profiles of 5-mC and 5-hmC in specific genes during the lifespan have been associated with the progression of Alzheimer disease. DUSP22, which dephosphorylates abnormal tau protein, is downregulated in Alzheimer disease brain samples as a result of hypermethylation of its promoter. Conversely, hypermethylation of the promoters of brain-derived neurotrophic factor (BDNF) and CREB protein, two genes involved in neural function, was found in frontal cortex of patients. miRNA deregulation is also frequently observed in Alzheimer disease: miR-106b and miR-153 are downregulated in Alzheimer disease (temporal cortex and frontal cortex, respectively) and one of its multiple targets is the mRNA of amyloid protein. Beyond this, DNA methylation changes are also known to be involved in cardiovascular disease, the biggest killer in Western countries. For example, aberrant CpG island hypermethylation has been described in atherosclerotic lesions.

The reversible nature of epigenetic factors and especially their role as mediators between the genome and the environment make them exciting candidates as therapeutic targets. The potential use of epigenetic enzymes as druggable targets to ameliorate the phenotype of diseases has been assayed in different experimental systems. Therapeutic assays have demonstrated that HDAC inhibitors can improve deficits in synaptic plasticity, cognition, and stress-related behaviors in a wide range of neurological and psychiatric disorders, including Huntington and Parkinson diseases, anxiety and mood disorders, and Rubinstein-Taybi and Rett syndromes. Abnormal histone modification patterns associated with specific gene expression have also been described in lupus CD4$^+$ T-cells, and HDAC inhibitors are able to reverse gene expression significantly.

## Genetic aberrations involving epigenetic genes

Genetic alterations of genes coding for enzymes that mediate chromatin structure could result in a loss of adequate regulation of chromatin compaction, and finally, the deregulation of gene transcription and inappropriate protein expression. Although the consequences in cancer have been widely described, in this section we extend the review to include other genetic diseases involving the function of several enzymes of the epigenetic machinery. The phenotype of these diseases also helps to clarify the role of various chromatin proteins in cell proliferation and differentiation. These include disorders arising from alterations in chromatin remodeling factors, alterations of the components of the DNA methylation machinery, and aberrations disturbing histone modifiers.

## Key concepts

- Epigenetics: The heritable patterns of gene expression that do not involve changes in the sequence of the genome.
- Epigenome: The overall epigenetic state of a cell.
- Genome Imprinting: The epigenetic marking of a locus on the basis of parental origin, which results in monoallelic gene expression.
- Chromatin: The complex of DNA and protein that composes chromosomes. Chromatin packages DNA into a volume that fits into the nucleus, allows mitosis and meiosis, and controls gene expression. Changes in chromatin structure are affected by DNA methylation and histone modifications.
- DNA methylation: The addition of a methyl group to DNA at the 5-carbon of the cytosine pyrimidine ring that precedes a guanine.
- Histone modifications: A set of reactions which introduce functional chemical groups into the histone tails. Posttranslational modifications of the histone tails include methylation, acetylation, phosphorylation, ubiquitination, sumoylation, citrullination and ADP-ribosylation.

## Suggested readings

[1] Bujold D, Morais DA, Gauthier C, et al. The international human epigenome consortium data portal. Cell Syst 2016;496−9.
[2] Berdasco M, Esteller M. Clinical epigenetics: seizing opportunities for translation. Nat Rev Genet 2019;20(2):109−27.
[3] Kim Y, Lee HM, Xiong Y, et al. Targeting the histone methyltransferase G9a activates imprinted genes and improves survival of a mouse model of Prader-Willi syndrome. Nat Med 2017;23:213−22.
[4] Feinberg AP. Phenotypic plasticity and the epigenetics of human disease. Nature 2007;447:433−40.
[5] Baujat G, Rio M, Rossignol S, et al. Paradoxical NSD1 mutations in Beckwith-Wiedemann syndrome and 11p15 anomalies in Sotos syndrome. Am J Hum Genet 2004;74:715−20.
[6] Abi Habib W, Brioude F, Azzi S, et al. 11p15 ICR1 partial deletions associated with IGF2/H19 DMR hypomethylation and Silver-Russell syndrome. Hum Mutat 2017;38:105−11.
[7] Simó-Riudalbas L, Esteller M. Cancer genomics identifies disrupted epigenetic genes. Hum Genet 2014;133:713−25.
[8] Eden A, Gaudet F, Waghmare A, et al. Chromosomal instability and tumors promoted by DNA hypomethylation. Science 2003;300:455.
[9] Herman JG, Baylin SB. Gene silencing in cancer in association with promoter hypermethylation. N Engl J Med 2003;349:2042−54.
[10] Bannister AJ, Kouzarides T. Regulation of chromatin by histone modifications. Cell Res 2011;21:381−95.
[11] Esteller M. Non-coding RNAs in human disease. Nat Rev Genet 2011;12:861−74.
[12] Guil S, Esteller M. RNA-RNA interactions in gene regulation: the coding and noncoding players. Trends Biochem Sci 2015;40:248−56.
[13] Calin GA, Croce CM. MicroRNA signatures in human cancers. Nat Rev Cancer 2006;6:857−66.
[14] Ley TJ, Ding L, Walter MJ, et al. DNMT3A mutations in acute myeloid leukemia. N Engl J Med 2010;363:2424−33.
[15] Langemeijer SM, Kuiper RP, Berends M, et al. Acquired mutations in TET2 are common in myelodysplastic syndromes. Nat Genet 2009;41:838−42.
[16] Pasqualucci L, Dominguez-Sola D, Chiarenza A, et al. Inactivating mutations of acetyltransferase genes in B-cell lymphoma. Nature 2011;471:189−95.
[17] De Jager PL, Srivastava G, Lunnon K, et al. Alzheimer's disease: early alterations in brain DNA methylation at ANK1, BIN1, RHBDF2 and other loci. Nat Neurosci 2014;17:1156−63.

# Chapter 9

# Clinical proteomics and molecular pathology

Lance A. Liotta[1], Justin B. Davis[1], Robin D. Couch[2], Claudia Fredolini[3], Weidong Zhou[1], Emanuel Petricoin[1] and Virginia Espina[1]

[1]*Center for Applied Proteomics and Molecular Medicine, George Mason University, Manassas, VA, United States,* [2]*Department of Chemistry and Biochemistry, George Mason University, Manassas, VA, United States,* [3]*SciLifeLab, School of Biotechnology, KTH — Royal Institute of Technology, Solna, Sweden*

**Summary**

Genomic and proteomic research is launching the next era of cancer molecular medicine. Molecular expression profiles can uncover clues to functionally important molecules in the development of human disease and generate information to subclassify human tumors and tailor a treatment to the individual patient. The next revolution is the synthesis of proteomic information into functional pathways and circuits in cells and tissues. Such synthesis must take into account the dynamic state of protein post-translational modifications; protein-protein or protein-DNA/RNA interactions; cross-talk between signal pathways; and feedback regulation within cells, between cells, and between tissues. This full set of information may be required before we can fully dissect the specific dysregulated pathways driving tumorigenesis. This higher level of functional understanding will be the basis for true rational therapeutic design that specifically targets the molecular lesions underlying human disease.

## Understanding cancer at the molecular level: an evolving frontier

Molecular expression profiles can uncover clues to functionally important molecules in the development of human disease and generate information to subclassify human tumors and tailor a treatment to the individual patient. The next revolution is the synthesis of proteomic information into functional pathways and circuits in cells and tissues. Such synthesis must take into account the dynamic state of protein post-translational modifications; protein-protein or protein-DNA/RNA interactions; cross-talk between signal pathways; feedback regulation within cells, between cells, and between tissues. Functional understanding will be the basis for true rational therapeutic design that specifically targets the molecular lesions underlying human disease (Table 9.1).

## Microdissection technology brings molecular analysis to the tissue level

Molecular analysis of enriched cell populations in their native tissue environment is necessary to understand the microecology of the disease process. Tissues are complicated three-dimensional structures composed of large numbers of different types of interacting cell populations. The cell subpopulation of interest may constitute a tiny fraction of the total tissue volume. For example, a biopsy of breast tissue harboring a malignant tumor usually contains the following types of cell populations: (1) fat cells in the abundant adipose tissue surrounding the ducts, (2) normal epithelium and myoepithelium in the branching ducts, (3) fibroblasts and endothelial cells in the stroma and blood vessels, (4) premalignant carcinoma cells in the in situ lesions, and (5) clusters of invasive carcinoma. If the goal is to analyze the genetic changes in the premalignant cells or the malignant cells, these subpopulations are frequently located in microscopic regions occupying less than 5% of the tissue volume. Culturing cell populations from fresh tissue is one approach to reducing contamination. However, cultured cells may not accurately represent the molecular events taking place in the actual tissue from which they were derived. Assuming methods are successful to isolate and grow the tissue cells of interest, the gene expression pattern of the cultured cells is influenced by the culture environment and can be quite

*Essential Concepts in Molecular Pathology.* DOI: https://doi.org/10.1016/B978-0-12-813257-9.00009-7

**TABLE 9.1** Opportunities, challenges, and potential solutions for use of proteomics for routine clinical practice and patient care.

| Opportunities | Challenges | Solutions |
|---|---|---|
| Proteomic multiplex<br>Cellular circuit analysis of clinical biopsy specimens<br>Patient stratification individualized therapy based on molecular profiling<br>Tailored combination therapy<br>Rational design of therapy following recurrence or outgrowth of metastasis | Platform sensitivity and accuracy<br>Heterogeneity of tissue populations<br>Perishability: requirement for immediate freezing or preservation of protein analytes<br>Formalin fixation unsuitable for protein or RNA extraction<br>Complex trial design and data analysis<br>Patient consent for serial biopsies<br>Low number of approved candidate-targeted agents<br>Lack of preclinical data<br>Unknown toxicity of combinations<br>Therapies from competing pharmaceuticals<br>Safety and justification for rebiopsy<br>Molecular profile of metastasis is different from the primary | Sensitive protein microarray<br>Microdissection<br>New protocols<br>Surrogate markers for tissue and blood molecular preservation<br>Extraction from formalin fixed state<br>New room temperature tissue and blood preservation technology<br>New classes of adaptive trial design<br>Dialogues with patient advocates and Institutional Review Board<br>Accelerate discovery of novel agents<br>New indicators for existing drugs<br>New classes of ex vivo treatment models<br>Restrict biopsy to accessible sites<br>Metastasis-specific tailored therapy |

different from the genes expressed in the native tissue state. Cultured cells are separated from the tissue elements that regulate gene expression and protein signaling, such as soluble factors, extracellular matrix molecules, and cell-cell communication. Thus, the problem of cellular heterogeneity has been a significant barrier to the molecular analysis of normal and diseased tissue. This problem can now be overcome by new developments in the field of laser tissue microdissection (Fig. 9.1).

The method of procurement of pure cell populations from heterogeneous tissue should fully preserve the state of the cell molecules if it is to allow quantitative analysis, particularly in sensitive amplification methods based on polymerase chain reaction (PCR), reverse transcription-PCR, or enzymatic function. Laser capture microdissection (LCM) has been developed to provide scientists with a fast and dependable method of capturing and preserving specific cells from tissue, under direct microscopic visualization. LCM allows investigators to determine specific molecular expression patterns from tissues of individual patients. Using multiplex analysis, investigators can correlate the pattern of expressed genes and post-translationally modified proteins with the histopathology and response to treatment. Microdissection can be used to study the interactions between cellular subtypes in the organ or tissue microenvironment.

## Beyond functional genomics to Cancer Proteomics

Whereas DNA is an information archive, proteins do all the work of the cell. The existence of a given DNA sequence does not guarantee the synthesis of a corresponding protein. The DNA sequence is also not sufficient to describe protein structure, protein-protein, protein-DNA interaction, and cellular location. Protein complexity and versatility stem from context-dependent post-translational processes, such as phosphorylation, sulfation, or glycosylation. Nucleic acid profiling (including microRNA) does not provide information about how proteins link together into networks and functional machines in the cell. In fact, the activation of a protein signal pathway, causing a cell to migrate, die, or initiate division, can immediately take place before any changes occur in DNA/RNA gene expression. Consequently, the technology to drive the molecular medicine revolution from the correlation to the causality phase is emerging from protein analytic methods.

The term *proteome* denotes all the proteins expressed by a genome. A major goal of investigators in this exciting field is to assemble a complete library of all proteins. To date, only a small percentage of the proteome has been cataloged. Because PCR for proteins does not exist, sequencing the order of 20 possible amino acids in a given protein remains relatively slow and labor intensive compared to nucleotide sequencing. Mass spectrometry has become the standard for protein characterization and discovery. It is possible to immediately highlight proteins that are differentially abundant in one state versus another (for instance tumor vs. normal, or before and after hormone treatment) using stable isotope labeling with amino acids in cell culture (SILAC), isobaric tags for relative and absolute quantitation

## History of microdissection

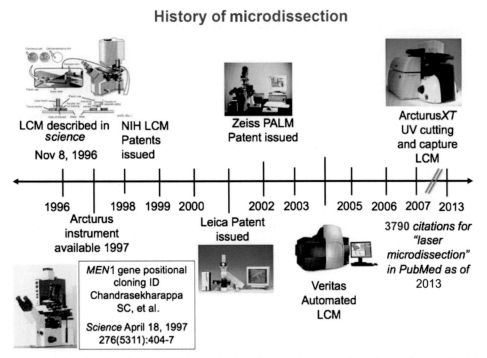

**FIGURE 9.1**  Laser capture microdissection (LCM). LCM is a technology for procuring pure cell populations from a stained tissue section under direct microscopic visualization. Tissues contain heterogeneous cellular populations (e.g., epithelium, cancer cells, fibroblasts, endothelium, and immune cells). The diseased cellular population of interest usually comprises only a small percentage of the tissue volume. LCM directly procures the subpopulation of cells selected for study, while leaving behind all of the contaminating cells. A stained section of the heterogeneous tissue is mounted on a glass microscope slide and viewed under high magnification. The experimenter selects the individual cell(s) to be studied using a joystick or via a computer screen. The chosen cells are lifted out of the tissue by the action of a laser pulse. The infrared laser, mounted in the optical axis of the microscope, locally expands a thermoplastic polymer to reach down and capture the cell beneath the laser pulse (insert). When the film is lifted from the tissue section, only the pure cells for study are excised from the heterogeneous cellular population. The DNA, RNA, and proteins of the captured cells remain intact and unperturbed. Using LCM, one to several thousand tissue cells can be captured in less than 5 min. Using appropriate buffers, the cellular constituents are solubilized and subjected to microanalysis methods. Proteins from all compartments of the cell can be readily procured. Protein conformation and enzymatic activity are retained if the tissues are frozen or fixed in ethanol before sectioning. The extracted proteins can be analyzed by any method that has sufficient sensitivity.

(iTRAQ), and multiple reaction monitoring. Not unexpectedly, cell culture does not provide an accurate picture of the proteins that are in use by cells in real tissue. Tissues are complicated structures composed of hundreds of interacting cell populations in specialized spatial configurations. The fluctuating proteins expressed by cells in tissues may bear little resemblance to the proteins made by cultured cells that are torn from their tissue context and reacting to a new culture environment.

Two classes of technologies were developed with the express purpose to address the tissue-context problem. The first technology is laser capture microdissection (LCM), used to procure specific tissue cell subpopulations under direct microscopic visualization of a stained frozen or fixed tissue section on a glass microscope slide. The second class of technology is protein microarrays. This multiplex analysis tool is sensitive enough to accurately measure the small concentration of proteins and post-translationally modified proteins in microdissected tissue samples.

While individualized treatments have been used in medicine for years, advances in cancer treatment have now generated a need to more precisely define and identify patients who will derive the most benefit from new-targeted agents. Molecular profiling using gene expression arrays has shown considerable potential for the classification of patient populations in all of these respects. Nevertheless, transcript profiling, by itself, provides an incomplete picture of the ongoing molecular network for a number of clinically important reasons. First, gene transcript levels have not been found to correlate significantly with protein expression or the functional (often phosphorylated) forms of the encoded proteins. RNA transcripts also provide little information about protein-protein interactions and the state of the cellular signaling pathways. Finally, most current therapeutics are directed at protein targets, and these targets are often protein kinases and/or their substrates. The activation state of proteins and their associated networks fluctuate constantly depending on the cellular microenvironment. Consequently, the source material for molecular profiling studies needs to shift from

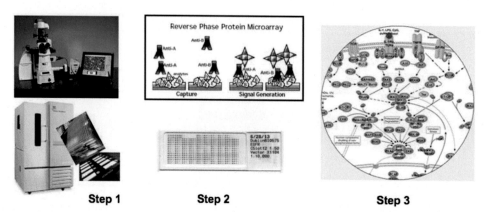

**FIGURE 9.2**    A roadmap for individualized cancer therapy. Following biopsy or needle aspiration, and laser capture microdissection, signal pathway analysis is performed using protein microarrays for phosphoproteomic analysis, and RNA transcript arrays. The specific signaling portrait becomes the basis of a patient-tailored therapeutic regime. Therapeutic assessment is obtained by follow-up biopsy, and the molecular portrait of signaling events is reassessed to determine if therapeutic selection should be modified further.

in vitro models to the use of actual diseased human tissue. Technologies that can broadly profile and assess the activity of the human kinome in a real biological context will provide a rich source of new molecular information critical for the realization of patient-tailored therapy (Fig. 9.2).

Although we have technology for quantifying post-translationally modified proteins, immunohistochemistry (IHC) enhances our ability to evaluate biochemical constituents in cellular context. IHC is a routine diagnostic assay that is easily adapted to laser capture microdissection. Immuno-LCM is laser capture microdissection on an antibody-stained tissue slide, allowing cells with a specific phenotype to be procured for analysis. Multiplexed IHC permits cell type identification, providing contextual specificity within a tissue section. Moreover, IHC reveals the spatial context of individual proteins. Spatial context of cells in relationship to each other is critical in assessing immune interaction, cell-to-cell communication, metastasis, and epithelial ductal integrity. The subcellular localization of proteins often determines protein function and activity. Many transcription factors (FOXO1, CREB, ChREBP) translocate between the cytosol and nucleus. This nuclear translocation is necessary to induce gene expression. Key regulatory proteins, have different regulatory functions depending on subcellular localization (cytosol vs. membrane). Thus, IHC provides the in situ protein/tissue relationship to supplement quantitative proteomic assays.

## Protein Microarray tools to guide patient-tailored therapy

Theoretically, the most efficient way to identify patients who will respond to a given therapy is to determine, prior to treatment initiation, which potential signaling pathways are actually activated in each patient. Ideally, this would come from analysis of tissue material taken from the patient through biopsy procurement. Protein microarrays represent powerful tools for drug discovery, biomarker identification, and signal transduction profiling of cellular material. The advantage of protein microarrays lies in their ability to provide a map of known cellular signaling proteins that can reflect, in general, the state of information flow through protein networks in individual specimens. Identification of critical nodes or interactions within the network is a potential starting point for drug development and/or the design of individual therapy regimens. Protein microarrays may be used to monitor changes in protein phosphorylation over time, before and after treatment, between disease and nondisease states, and responders versus nonresponders, allowing one to infer the activity levels of the proteins in a particular pathway in real-time to tailor treatment to each patient's cellular circuitry.

Protein microarrays with highly sensitive and specific antibodies are now able to achieve adequate levels of sensitivity for analysis of clinical specimens containing fewer than a few thousand cells.

At a basic level, protein microarrays are composed of a series of immobilized spots. Each spot contains a homogeneous or heterogeneous bait molecule. A spot on the array may display an antibody, a cell or phage lysate, a recombinant protein or peptide, or a nucleic acid. The array is queried with (1) a probe (labeled antibody or ligand) or (2) an unknown biologic sample (for instance, cell lysate or serum sample) containing analytes of interest. When the query molecules are tagged directly or indirectly with a signal-generating moiety, a pattern of positive and negative spots is generated. For each spot, the intensity of the signal is proportional to the quantity of applied query molecules bound to the bait molecules. An image of the spot pattern is captured, analyzed, and interpreted.

Protein microarray formats fall into two major classes, forward phase arrays (FPAs) and reverse phase arrays (RPPAs), depending on whether the analyte(s) of interest is captured from solution phase or bound to the solid phase. In FPAs, capture molecules are immobilized onto the substratum and act as the bait molecule. Each spot contains one type of known immobilized protein, fractionated lysate, or other type of bait molecule. In the FPA format, each array is incubated with one test sample (for instance, a cellular lysate from one treatment condition or serum sample from disease/control patients), and multiple analytes are measured at once. A number of excellent reviews summarize recent applications, obstacles, and new advances in FPA technology. Antibody array use is limited currently by the availability of well-characterized antibodies. A second obstacle to routine use of antibody arrays surrounds detection methods for bound analyte on the array. Current options include the use of specific antibodies recognizing analyte epitopes distinct from the capture antibodies (similar to a traditional sandwich-type enzyme-linked immunosorbent assay or ELISA), or the direct labeling of the analytes used for probing the array, both of which present distinct technical challenges.

In contrast to the FPA format, the RPPA format immobilizes an individual test sample in each array spot, such that an array is composed of hundreds of different patient samples or cellular lysates. Though not limited to clinical applications, the RPPA format provides the opportunity to screen clinical samples that are available in very limited quantities, such as biopsy specimens. Because human tissues are composed of hundreds of interacting cell populations, RPPAs coupled with LCM provide a unique opportunity for discovering changes in the cellular proteome that reflect the cellular microenvironment. The RPPA format is capable of extremely sensitive analyte detection with detection levels approaching attogram amounts of a given protein and variances of less than 10%. The sensitivity of detection for the RPPAs is such that low abundance phosphorylated protein isoforms can be measured from a spotted lysate representing fewer than 10 cell equivalents. This level of sensitivity combined with analytical robustness is critical if the starting input material is only a few hundred cells from a biopsy specimen. Since the RPPA technology requires only one antibody for each analyte, it provides a facile way for broad profiling of pathways where hundreds of phospho-specific analytes can be measured concomitantly. Most importantly, the RPPA has significantly higher sensitivity than bead arrays or ELISA, such that broad screening of molecular networks can be achieved from tissue specimens routinely procured in the physician's private office or hospital radiology center (such as a needle biopsy specimen).

## Molecular network analysis of human cancer tissues

Reverse-phase protein arrays for the analysis of human tissues demonstrate the potential for the technology to contribute valuable information for making therapeutic decisions. Reverse-phase array technology was first described when it was utilized to demonstrate that prosurvival proteins and pathways are activated during prostate cancer progression. Pathway mapping of a clinical study set of childhood rhabdomyosarcoma tumors using reverse phase protein arrays revealed that mammalian target of rapamycin pathway activation correlated with response to therapy. Moreover, the functional significance of suppressing this pathway was tested in xenograft models and shown to profoundly suppress tumor growth.

Reverse-phase protein arrays are also well suited to the analysis of clinical trial material in that they can provide signaling network information that complements standard histological analysis of patient specimens collected before, during, and after treatment. This technology is being applied to several ongoing clinical trials in a variety of cancers.

## Combination therapies

Increasing evidence demonstrates the promise and potential of combination therapies combining conventional treatments, such as chemotherapy or radiotherapy, and molecular-targeted therapeutics such as erlotinib (Tarceva, Roche) and trastuzumab (Herceptin, Roche) that interfere with kinase activity and protein-protein interactions in specific deregulated pathways. However, strategies that target multiple interconnected proteins within a signaling pathway have not been explored to the same extent. The view of individual therapeutic targets can be expanded to that of rational targeting of the entire deregulated molecular network, extending both inside and outside the cancer cell. Mathematical modeling of network-targeted therapeutic strategies has revealed that attenuation of downstream signals can be enhanced significantly when multiple upstream nodes or processes are inhibited with small molecule inhibitors compared with inhibition of a single upstream node. Also, inhibition of multiple nodes within a signaling cascade allows reduction of downstream signaling to desired levels with smaller doses of the necessary targeted drugs. While therapeutic strategies incorporating these lower dosages could lead to reduced toxicities and a broadened spectrum of available drugs, it must be recognized that testing these interacting drug modalities will necessitate clinical trials of complex design.

Molecular profiling of the proteins and signaling pathways produced by the tumor microenvironment, host, and peripheral circulation hold great promise in effective selection of therapeutic targets and patient stratification. For many of the more common sporadic cancers, there is significant heterogeneity in cell signaling, tissue behavior, and susceptibility to chemotherapy. Proteomic analysis is particularly useful in this area, given the ability to study multiple pathways simultaneously. Cataloguing of abnormal signaling pathways for large numbers of specimens will provide the data necessary for a rationally based formulation of combination therapy that, presumably, would be more effective than monotherapy and help to minimize the issues of tumor heterogeneity. The promise of proteomic-based profiling, different from gene transcript profiling alone, is that the resulting prognostic signatures are derived from drug targets (such as activated kinases), not genes, so the pathway analysis provides a direction for therapeutic mitigation. Thus, phosphoproteomic pathway analysis becomes both a diagnostic/prognostic signature as well as a guide to therapeutic intervention.

## Protein biomarker stability in tissue: a critical unmet need

The promise of tissue protein biomarkers to provide revolutionary diagnostic and therapeutic information will never be realized unless the problem of tissue protein biomarker instability is recognized, studied, and solved (Fig. 9.3). There is a critical need to develop standardized protocols and novel technologies that can be used in the routine clinical setting for seamless collection and immediate preservation of tissue biomarker proteins, particularly those that have been posttranslationally modified, such as phosphoproteins. This critical need transcends the large research hospital environment and extends most acutely to the private practice, where most patients receive therapy. While molecular profiling offers tremendous promise to change the practice of oncology, the fidelity of the data obtained from a diagnostic assay applied to tissue must be monitored and ensured; otherwise, a clinical decision may be based on incorrect molecular data. To date, clinical preservation practices routinely rely on protocols that are decades old, such as formalin fixation, and are designed to preserve specimens for histologic examination. Under the current standard of care, tissue is procured for pathologic examination in three main settings: (1) surgery in a hospital-based operating room, (2) biopsy conducted in an outpatient clinic, and (3) image-directed needle biopsies or needle aspirates conducted in a radiologic suite.

Fig. 9.3 depicts the two categories of variable time periods that define the stability intervals for human tissue procurement. Time point A is defined as the moment that tissue is excised from the patient and becomes available ex vivo for analysis and processing. The post excision delay time (EDT) is the time from time point A to the time that the specimen is placed in a stabilized state (immersed in fixative or snap-frozen in liquid nitrogen), herein called time point B. Given the complexity of patient-care settings, during the EDT, the tissue may reside at room temperature in the operating room or on the cutting board of the pathologist, or it may be refrigerated in a specimen container. The second variable time period is the processing delay time. At the beginning of this interval, the tissue is immersed in a preservative solution or stored in a freezer. At the end of this interval, time point C, the tissue is subject to processing for molecular analysis. In addition to the uncertainty about the length of these two time intervals, a host of known and unknown

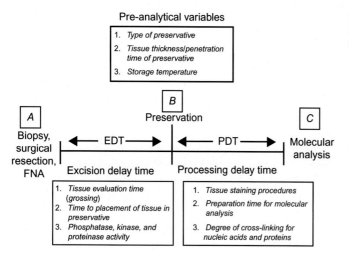

**FIGURE 9.3** Preanalytical variables during tissue acquisition. Excision delay time, processing delay time, and the type of preservation chemistry selected will affect the quality of molecular analyses.

variables can influence the stability of tissue molecules during these time periods. These include (1) temperature fluctuations prior to fixation or freezing; (2) preservative chemistry and rate of tissue penetration; (3) size of the tissue specimen; (4) extent of handling, cutting, and crushing of the tissue; (5) fixation and staining prior to microdissection; (6) tissue hydration and dehydration; and (7) the introduction of phosphatases or proteases from the environment at any time. In the face of these uncertainties, it is essential to develop a standardized specimen collection procedure for routine clinical profiling. Even if a strict protocol is followed, there is no ultimate assurance that processing variables are free from compromise up to the time that the molecular profile data are collected.

### Recognition that the tissue is alive and reactive following procurement

While investigators have worried about the effects of vascular clamping and anesthesia, prior to excision, a much more significant and underappreciated issue is the fact that excised tissue is alive and reacting to ex vivo stresses (Fig. 9.4). The instant a tissue biopsy is removed from a patient, the cells within the tissue react and adapt to the absence of vascular perfusion, ischemia, hypoxia, acidosis, accumulation of cellular waste, absence of electrolytes, and temperature changes. In as little as 30 min post-excision, drastic changes can occur in the protein signaling pathways of the biopsy tissue as the tissue remains in the operating room suite or on the pathologist's cutting board. In response to wounding cytokines, vascular hypotensive stress, hypoxia, and metabolic acidosis, it would be expected that a large surge of stress-related, hypoxia-related, and wound repair-related protein signal pathway proteins and transcription factors will be induced in the tissue immediately following procurement. Over time the levels of candidate proteomic markers (or RNA species) would be expected to widely fluctuate upward and downward. This fluctuation will significantly distort the molecular signature of the tissue compared to the state of the markers in vivo. Moreover, the degree of ex vivo fluctuation could be quite different between tissue types and influenced by the pathologic microenvironment. This physiologic fact must be taken into consideration as we plan to implement tissue protein biomarkers in the real world of the clinic, where the living, reacting tissue may remain in the collection basin or on the cutting board for hours.

### Phosphoprotein stability: the balance between kinases and phosphatases

Phosphoproteins offer a unique minute-by-minute record of ongoing signal pathway events of high functional relevance to therapeutic target selection and the prediction of toxicity. Phosphorylation and dephosphorylation of structural and regulatory proteins are major intracellular control mechanisms. Protein kinases transfer a phosphate from a nucleotide triphosphate, such as adenosine triphosphate, to a specific serine, threonine, or tyrosine residue of a substrate protein. Phosphatases remove the phosphoryl group and restore the protein to its original dephosphorylated state. Hence, the

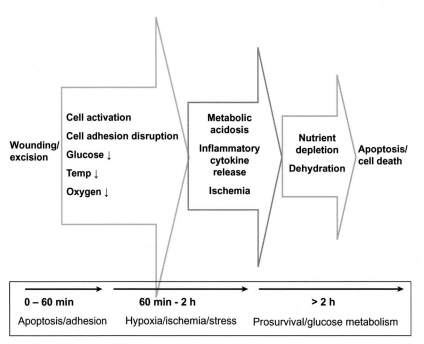

FIGURE 9.4 Molecular stages and timeline of tissue cell death. Post tissue excision, cascades of cellular kinases are activated and deactivated as tissue reacts to wounding, ischemia, inflammation, environmental stresses, hypoxia, and nutrient depletion. *Adapted from Espina VA, Edmiston KH, Heiby M, et al. A portrait of tissue phosphoprotein stability in the clinical tissue procurement process. Mol Cell Proteom 2008;7:1998—2018.*

phosphorylation-dephosphorylation cycle can be regarded as a molecular on-off switch. At any point in time within the cellular microenvironment, the phosphorylated state of a protein is a function of the local stoichiometry of associated kinases and phosphatases specific for the phosphorylated residue. During the ex vivo time period, if the cell remains alive, it is conceivable that phosphorylation of certain kinase substrates may transiently increase due to the persistence of functional signaling, activation by hypoxia, or some other stress-response signal. On the other hand, the availability of ubiquitous cellular phosphatases would be expected to ultimately destroy phosphorylation sites, given enough time.

## Phosphatases determine phosphoprotein stability

Protein phosphatases (PPs) have been classified into three distinct categories: (1) serine/threonine (Ser/Thr)-specific, (2) tyrosine-specific, and (3) dual-specificity phosphatases (DSP). Based on biochemical parameters, substrate specificity, and sensitivity to various inhibitors, Ser/Thr PPs are divided into two major classes. Type I phosphatases, which include PP1, can be inhibited by two heat-stable proteins known as Inhibitor-1 (I-1) and Inhibitor-2 (I-2). They preferentially dephosphorylate the β-subunit of phosphorylase kinase. Type II phosphatases are subdivided into spontaneously active (PP2A), $Ca^{2+}$-dependent (PP2B), and $Mg^{2+}$-dependent (PP2C) classes of phosphatases. They are insensitive to heat-stable inhibitors and preferentially dephosphorylate the α-subunit of phosphorylase kinase. Protein tyrosine phosphatases (PTPs) remove phosphate groups from phosphorylated tyrosine residues of proteins. PTPs display diverse structural features and play important roles in the regulation of cell proliferation, differentiation, cell adhesion and motility, and cytoskeletal function. With chemistry knowledge in mind, fixatives have been rationally designed to preserve phosphoproteins and histomorphology without freezing.

# Serum proteomics: an emerging landscape for early-stage cancer detection

Recognizing cancer as a product of the proteomic tissue microenvironment with diverse, interconnected communication networks has important implications. First, it shifts the emphasis away from therapeutic targets being directed solely against individual molecules within pathways and focuses the effort on targeting nodes in multiple pathways, inside and outside the cancer cell, that cooperate to orchestrate the malignant phenotype. Second, the tumor-host communication system may involve unique enzymatic events and sharing of growth factors. Consequently, the microenvironment of the tumor-host interaction could be a source for biomarkers that could ultimately be shed into the serum proteome.

## Application of serum proteomics to early diagnosis

Cancer is too often diagnosed and treated too late, when the tumor cells have already invaded and metastasized. At this stage, therapeutic modalities are limited in their success. Detecting cancers at their earliest stages, even in the premalignant state, means that current or future treatment modalities might have a higher likelihood of a true cure. Ovarian cancer is a prime example of this clinical dilemma. More than two-thirds of cases of ovarian cancer are detected at an advanced stage, when the ovarian cancer cells have spread away from the ovary surface and have disseminated throughout the peritoneal cavity. Although the disease at this stage is advanced, it rarely produces specific or diagnostic symptoms. Consequently, ovarian cancer is usually treated when it is at an advanced stage. The resulting 5-year survival rate is 35%−40% for patients with late-stage disease who receive the best possible surgical and chemotherapeutic intervention. By contrast, if ovarian cancer is detected when it is still confined to the ovary (stage I), conventional therapy produces a high rate (95%) of 5-year survival. Thus, early detection of ovarian cancer, by itself, could have a profound effect on the successful treatment of this disease. Unfortunately, early stage ovarian cancer lacks a specific symptom or a specific biomarker and accurate and reliable, noninvasive diagnostic procedures. A clinically useful biomarker should be measurable in a readily accessible body fluid, such as serum, urine, or saliva. Clinical proteomic methods are especially well suited to discovering such biomarkers. Serum or plasma has been the preferred medium for discovery because this fluid is a protein-rich information reservoir that contains the traces of what has been encountered by the blood during its constant perfusion and percolation throughout the body. Until recently, the search for cancer-related biomarkers for early disease detection has been a one-at-a-time approach. However, biomarker discovery is moving away from the idealized single cancer-specific biomarker. Despite decades of effort, single biomarkers have not been found that can reach an acceptable level of specificity and sensitivity required for routine clinical use for the detection or monitoring of the most common cancers. Most investigators believe that this is due to the patient-to-patient molecular heterogeneity of tumors. A second level of population heterogeneity exists for tumor location, size, histology, grade, and stage. Moreover, an individual patient's organ may harbor co-existence of multiple stages in the same tissue (such as in situ and invasive cancer). Epidemiologic heterogeneity, including differences in age, sex, and genetic background, is a third level of patient-to-patient variability that can reduce cancer biomarker specificity. Taking a cue from gene arrays, the hope is that panels of tens to hundreds of protein and peptide markers may transcend the heterogeneity to generate a higher level of diagnostic specificity. While an individual biomarker candidate may be specific and sensitive only for a certain stage or molecular etiology, combinations of many markers, screened for sensitivity and specificity concomitantly, may be able to bracket across the heterogeneity to reach a higher level of specificity and sensitivity in the aggregate. Thus, while marker A may work for 50% of the population, marker B for 30%, and marker C for 20%, combining A, B, and C has the potential to cover the entire population. The overall specificity and sensitivity required for clinical use depends entirely on the intended use of the marker(s). Markers for general population screening for rare diseases may have to approach 100% specificity to be accepted. On the other hand, markers that are used for high-risk screening or for relapse monitoring can have much lower specificity but require high sensitivity.

The low-molecular-weight (LMW) range of the serum proteome (generally defined here as peptides less than 50,000 Da) is called the peptidome due to the abundance of protein peptides and fragments. While some dismissed the peptidome as noise, biological trash, or too small and unstable to be biologically relevant, others have proposed that just the opposite is the case—it may contain a rich, untapped source of disease-specific diagnostic information. Tissue proteins that are normally too large to passively diffuse through the endothelium into the circulation can still be represented as fragments of the parent molecule. The information in the peptidome resides in multiple dimensions: (1) the identity of the parent protein, (2) the peptide isoform identity, (3) post-translational modifications (glycosylation/phosphorylation sites, and others), (4) the specific size and cleavage ends of the peptide, and (5) peptide quantity.

Cancer is a product of the tissue microenvironment. While normal cellular processes (and the peptide content generated by these processes) are also a manifestation of the tissue microenvironment, the tumor microenvironment, through the process of aberrant cell growth, cellular invasion, and altered immune system function, represents a unique constellation of enzymatic (for example kinases and phosphatases) and protease activity (for example matrix metalloproteases), resulting in changed stoichiometry of molecules within the peptidome itself compared to the normal milieu. Interactions between the precancerous cells, the surrounding epithelial and stromal cells, vascular channels, the extracellular matrix, and the immune system are mediated by various enzymes, cytokines, extracellular matrix molecules, and growth factors. An added benefit for cancer biomarkers is the leaky nature of newly formed blood vessels and the increased hydrostatic pressure within tumors. This pathologic physiology would tend to push molecules from the tumor interstitium into the circulation (Fig. 9.5). As cells die within the microenvironment, they will shed the degraded products. The mode of death, apoptosis versus necrosis, would be expected to generate different classes of degraded cellular constituents. As a consequence, the blood peptidome may reflect ongoing recordings of the molecular cascade of communication taking place in the tissue microenvironment (Fig. 9.5). Combinations of peptidome markers representing the specific interactions of the tumor tissue microenvironment at the enzymatic level can achieve a higher specificity and a higher sensitivity for early-stage cancers. This optimism is in part based on the concept that the biomarkers are derived from a population of cells that comprise a volume that is greater than just the small precancerous lesion itself. In this way, the peptidome can potentially supersede individual single biomarkers and transcend the issues of tumor and population heterogeneity.

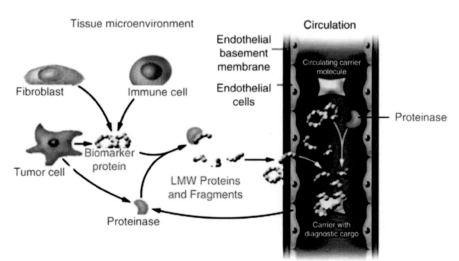

**FIGURE 9.5** The peptidome hypothesis: circulating peptides and protein fragments are shed from all cell types in the tissue microenvironment. Proteolytic cascades within the tissue generate fragments that diffuse into the circulation. The identity and cleavage pattern of the peptides provide two dimensions of diagnostic information.

## Methods for discovering and validating candidate protein biomarkers

Researchers can choose from a series of separation, chromatography, electrophoresis, and Mass Spectrometry (MS)-based methodologies useful for discovering the low-molecular-weight peptidome. Methods are available for profiling, harvesting, purifying, enriching, and sequencing the peptidome (Table 9.1). Each methodology has advantages and disadvantages. MS profiling technology is useful for rapidly obtaining an ion fingerprint of a fluid sample. MS profiling can also be conducted in an electrospray mode using inline liquid chromatography followed by ESI-MS. This can be done before or after trypsinization of the sample. If the sample is trypsinized, then the identity of the trypsin fragments can be scored, but the original peptide fragment size information may be lost. An intermediate system is represented by a solid particle or bead capture system. These methods are used to harvest peptides by fairly nonspecific hydrophobic binding. The harvested peptides eluted from the beads can be profiled by MALDI or electrospray MS, followed by MS sequencing and identification of abundant peaks. Native carrier proteins, such as albumin, constitute an endogenous in vivo affinity chromatography system for harvesting peptides. Carrier protein harvesting is a facile method for obtaining high-resolution ion profiles. Furthermore, the captured peptides can be eluted and sequenced with excellent yield. The disadvantages of the bead and carrier protein capture systems are low throughput and minimal prefractionation.

The complex low molecular weight (LMW) peptidome appears to exist bound to albumin and other high-abundant proteins. Many investigations for biomarker discovery begin with depleting the blood of these high-abundance proteins, a great deal of caution is warranted in using this approach for LMW candidate discovery. One-dimensional gel preparative electrophoresis exhibits good yield in the low mass range <20 kDa, but is also hampered by slow throughput. Slow throughput and large volume requirements are also a drawback of tagging systems such as ICAT, and these tagging systems inherently ignore those molecules that are truly off-on in disease states and have the highest predictive value. Thus, a researcher planning an LMW peptidome biomarker discovery project must carefully select the proper sample preparation method tailored to the volume of the available samples, the resolution desired, the required throughput, and the need for identification of the candidate biomarker peptides.

A recommended staging process for biomarker discovery and validation is depicted in Fig. 9.6. The starting point for rigorous analysis is the development of a discovery study set consisting of a population of serum or plasma samples from patients who have (1) histologically verified cancer, (2) benign or inflammatory nonneoplastic disease, and (3) unaffected, apparently healthy controls or hospital controls, depending on the intended use. Including specimens for initial discovery from patients with no evidence of cancer but with inflammatory conditions, reactive disease, and benign disorders is of critical importance to ensure that specific markers are enriched for from the outset. This issue is critical for cancer research, especially since the disease almost always occurs in the background of inflammatory processes that are part of the disease pathogenesis itself. The peptidome, as a mirror of the ongoing physiology of the entire

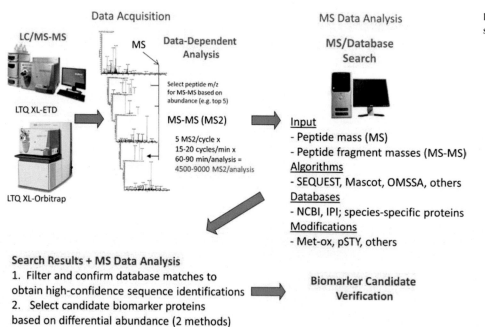

FIGURE 9.6 Proteomics mass spectrometry analytical workflow.

individual, may be especially sensitive to these processes, which is why care must be taken to at least minimize the chance that nonspecific markers are selected. As an additional level of rigor to reduce or minimize bias, the blood collection standard operating protocol for all patients should be exactly identical in methodology, handling, and storage. Prior to the initial peptidome analysis, the biomarker researcher should work with an epidemiologist to develop a discovery subset that is matched in every possible epidemiologic and physiologic parameter. This includes age, sex, hormonal status, treatment and hospitalization status, clinic location, and any other variable that can be known. Reduction of the potential upfront bias is critical prior to undertaking the discovery phase of the research. Because of the overall lability of the proteome and the resultant peptidome, all specimens should be handled with identical standard operating procedures, processed and stored in identical fashion, and frozen/stored as rapidly as possible after removal from the body.

The next stage is LMW peptidome fractionation, separation, isolation and enrichment, concentration, and MS−based identification (Fig. 9.6). At this stage it is essential that iterative and repetitive MS-based analysis and MS/MS sequencing be conducted on each sample. Candidate peptides identified repetitively over many iterations within a sample and within a study set have a higher likelihood of being correct. The researcher ends up with a list of candidate diagnostic markers that are judged to be differentially abundant in the cancer versus the control populations. The next step is to find or make specific antibodies or other ligands for each candidate peptide marker. After each antibody is validated for specificity using a reference analyte, the antibody can then be used to validate the existence of the predicted peptide marker in the disease and non-diseased discovery set samples. Multiple reaction monitoring offers the potential to quantify and identify peptides with such high confidence that antibody validation may not be required (Fig. 9.7).

Clinical validation of the candidate biomarkers starts with ensuring the sensitivity and precision of the measurement platform. The antibodies developed in discovery phase can now be applied as capture or detection reagents in the analytical platform. While some immunoassays can utilize antibodies that recognize variant analytes whereby a neo-epitope is formed by post-translational medications, or even recognize fragment-specific termini (such as cleaved caspases), most size-specific discriminations require alternate technologies to be employed, such as immuno-MS. Once the measurement platform is proven to be reliable and reproducible, then the clinical validation can proceed.

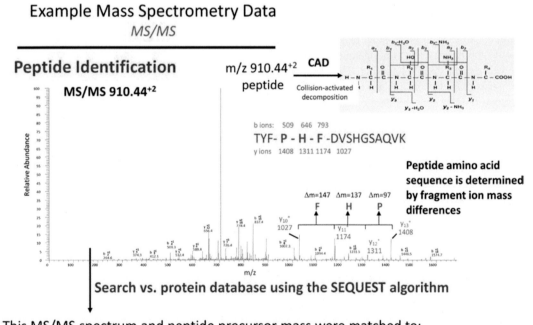

FIGURE 9.7  Mass spectrometry multiple reaction monitoring (MRM) is a technology for quantifying and specifically identifying protein analytes based on highly accurate mass profiling of a protease peptide fragment. This technology is used to validate biomarker candidates without the requirement for a specific antibody.

The final and most critical stage of research clinical validation is blinded testing of the biomarker panel using independent (not used in discovery), large clinical study sets that are ideally drawn from at least three geographically separate locations. The required size of these test sets for adequate statistical powering depends on both the performance of the peptide analyte panel in the platform validation phase and the intended use of the analyte in the clinic. For example, markers for general population screening require more patients than those for high-risk screening, which in turn require more patients than markers for recurrence/therapeutic monitoring. Employing previously verified controls and calibrators, under standard clinical chemistry guidelines (Clinical Laboratory Standards Institute) for immunoassays, the sensitivity and specificity can be determined for a test population. It is important to emphasize that sensitivity and specificity in an experimental test population does not translate to the positive predictive value that would be seen if the putative test is used routinely in the clinic. The true positive predictive value is a function of the indicated use and the prevalence of the cancer (or other disease condition within the target population). The percentage of expected cancer cases in a population of patients at high genetic risk for cancer is higher than the general population. Consequently, the probability of false positives in the latter population would be much higher. For this reason, the ultimate adoption of a peptidome-based or proteome-based test will be strongly dependent on the clinical context of its use.

Finally, the information content of the peptidome will never be fully realized unless blood collection protocols and reference sets are standardized, new instrumentation for measuring panels of specific fragments are proven to be reproducible and sensitive, and extensive clinical trial validation is conducted under rigorous and reproducible regulatory guidelines (College of American Pathologists [CAP]/Clinical Laboratory Improvement Amendments [CLIA]).

## Frontiers of nanotechnology and medicine

Nanotechnology will have a significant impact on early diagnosis and targeted drug delivery. The development of inorganic nanoparticles that bind specific tumor markers that exist at very low concentrations in serum may be able to be used as serum harvesting agents. In the future, patients may be injected with such nanoparticles that seek out and bind tumor or disease markers of interest. Once the nanoparticles have bound their targets, they can be harvested from the serum to enable diagnosis or to monitor disease progression.

Luchini et al. have created core shell hydrogel nanoparticles to address these fundamental roadblocks to biomarker purification and preservation. The nanoparticles simultaneously conduct molecular sieve chromatography and affinity chromatography, in one step, in solution (Fig. 9.8). The molecules captured and bound within the affinity matrix of the particles are protected from degradation by exogenous or endogenous proteases. These smart nanoparticles conduct enrichment and encapsulation of selected classes of proteins and peptides from complex mixtures of biomolecules such as plasma, purify them away from endogenous high-abundance proteins such as albumin, and protect them from degradation during subsequent sample handling. The particles have a molecular sieving shell surrounding a specific bait core. Each class of smart particle contains a core bait molecule specific for a general category of analyte. Thus, each class of particle will sequester and concentrate a class of all analytes below a molecular weight cutoff that selectively recognizes the bait.

Moreover, it is possible to concentrate the biomarkers captured by hydrogel particles into a small volume that is a small fraction of the starting volume. Particles incubated with serum trap the target analytes and are isolated by centrifugation. Candidate biomarkers are then released from particles by means of elution buffers. The ratio of the volume of elution buffer to the original starting solution establishes the concentration amplification factor. This concentration step is a fundamental point for biomarker discovery and measurement because it provides a means to effectively raise the concentration of rare biomarkers that become the input for a measurement system such as an immunoassay platform or MS.

## Future of cancer clinical proteomics

Proteomics offers a detailed examination of the identity, abundance, and activity of proteins in tissue and serum that is unavailable from genomic analyses alone. However, the future of proteomics can greatly be enhanced by the adoption of metabolomics. Metabolomics, the study of the metabolites or small molecules within an organism, provides key insights into the motivations behind changes in protein activity. The identity and quantity of metabolites present within a cell or biofluid reflects not only the fitness of the related specific organ or tissue, but systemic status as well. Carbohydrates, lipids, protein cofactors, small cations, free phosphorylated nucleotides, and other small molecules are integral to cell homeostasis and cell signaling, but these molecules are readily ignored by genomic and proteomic

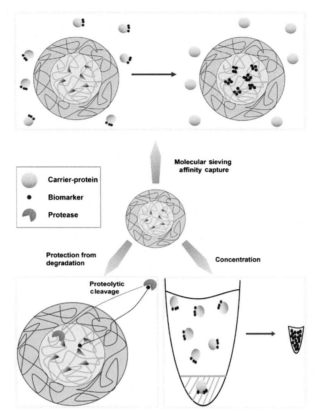

**FIGURE 9.8**  Schematic representation of particle structure and function. Particles are constituted by a bait containing core, surrounded by a sieving shell. When introduced into a complex solution, such as serum, core-shell particles remove low-molecular-weight proteins from carrier albumin with affinity capture and perform molecular weight sieving with total exclusion of high-molecular-weight proteins. Depending on the starting volume of the fluid sample, the particles can effectively amplify low abundance biomarkers 100-fold to 1000-fold.

investigations. However, these metabolites regularly act as protein agonist and antagonists, not only regulating kinase and phosphatase activity, but affecting gene expression and protein localization.

Metabolomics has become more widely adopted due to the advancements in analytical chemistry techniques and increased research interest. Volatile organic compounds can be readily analyzed through headspace solid-phase microextraction coupled to gas chromatography MS, with solid-phase chemistries available to match desired analytes. Nonvolatiles can be extracted from serum or tissue lysates and analyzed by liquid chromatography coupled to MS. The greatest limitation lies in the limited size of spectral databases required to identify detected metabolites. MS matches based only on mass to charge ratio, in the absence of tandem MS/MS fragmentation patterns, require known standards to confirm metabolite identity. This challenge not only limits sample throughput, but places limitations on true discovery. As metabolomics investigations become more prevalent, opportunities will arise to directly correlate changes in protein identity and activity to changes in the metabolome. These insights may present new understandings for the causations behind proteomic changes and offer a whole new paradigm for treatment.

The pathologist of the future will detect early manifestations of disease using proteomic patterns of body-fluid samples combined with genomics and metabolomics, providing the primary physician a diagnosis based on multi-omic network signatures as a complement to histopathology. He or she will be able to scrutinize a patient's individual tumor molecularly, identifying the specific regulatory pathways that are disturbed in the cell cycle, differentiation, apoptosis, and invasion and metastasis. Based on this knowledge, recommendations will be made for an individualized selection of therapies that best strike the entire disease-specific protein network of the tumor. The pathologist and the radiologist are essential members of the clinical team to perform concurrent assessments of therapeutic efficacy and toxicity. Proteomic, genomic, and metabolomic analyses of recurrent tumor lesions could be the basis for rational redirection of therapy, as they could reveal changes in the diseased protein network that are associated with drug resistance. The paradigm shift to integrated multi-omic analysis will directly affect clinical practice, as it has an impact on all of the crucial elements of patient-centered care and management.

## Key concepts

- The proteome, all proteins expressed by a genome, provides functional and phenotypic information including identity, abundance, and activity of proteins.
- Cancer is a product of the tissue microenvironment. Molecular analysis of cell populations in the context of their native tissue environment is necessary to understand the influence of the tissue microenvironment in the disease process.
- Tissue heterogeneity requires the use of laser capture microdissection to procure enriched cell populations for molecular analyses.
- Protein biomarker instability may be due to pre-analytical variables (warm and cold ischemia times), tissue fixation methods, or an imbalance of protein kinases and phosphatases post collection.
- Protein post-translational modifications can be quantified using reverse phase protein arrays.
- Mass spectrometry enables biomarker discovery, identification of proteins, quantitative comparisons across proteomes and samples, and antibody-free protein/peptide quantification.
- Methods to harvest and concentrate biomarkers, such as hydrogel nanoparticles, enable detection of proteins that are below the current detection limits of laboratory instruments and assays.
- Validation and verification of biomarkers is essential for advancing new discoveries to the clinic. Rigor and reproducibility in clinical research may be enhanced by adhering to clinical laboratory professional guidelines (College of American Pathologists and Clinical Laboratory Standards Institute) and federal laws (Clinical Laboratory Improvement Act).

## Suggested readings

[1] Emmert-Buck MR, Bonner RF, Smith PD, et al. Laser capture microdissection. Science 1996;274:998–1001.

[2] Ma XJ, Salunga R, Tuggle JT, et al. Gene expression profiles of human breast cancer progression. Proc Natl Acad Sci USA 2003;100:5974–9.

[3] Celis JE, Gromov P. Proteomics in translational cancer research: toward an integrated approach. Cancer Cell 2003;3:9–15.

[4] Gygi SP, Rist B, Gerber SA, et al. Quantitative analysis of complex protein mixtures using isotope-coded affinity tags. Nat Biotechnol 1999;17:994–9.

[5] Paweletz CP, Charboneau L, Bichsel VE, et al. Reverse phase protein microarrays which capture disease progression show activation of pro-survival pathways at the cancer invasion front. Oncogene 2001;20:1981–9.

[6] Haab BB. Antibody arrays in cancer research. Mol Cell Proteom 2005;4:377–83.

[7] Espina VA, Edmiston KH, Heiby M, et al. A portrait of tissue phosphoprotein stability in the clinical tissue procurement process. Mol Cell Proteom 2008;7:1998–2018.

[8] Petricoin EF, Bichsel VE, Calvert VS, et al. Mapping molecular networks using proteomics: a vision for patient-tailored combination therapy. J Clin Oncol 2005;23:3614–21.

[9] Fend F, Emmert-Buck MR, Chuaqui R, et al. Immuno-LCM: laser capture microdissection of immunostained frozen sections for mRNA analysis. Am J Pathol 1999;154:61–6.

[10] Sato S, Jung H, Nakagawa T, et al. Metabolite regulation of nuclear localization of carbohydrate-response element-binding protein (ChREBP): role of AMP as an allosteric inhibitor. J Biol Chem 2016;291:10515–27.

[11] Petricoin 3rd EF, Espina V, Araujo RP, et al. Phosphoprotein pathway mapping: Akt/mammalian target of rapamycin activation is negatively associated with childhood rhabdomyosarcoma survival. Cancer Res 2007;67:3431–40.

[12] Araujo RP, Petricoin EF, Liotta LA. A mathematical model of combination therapy using the EGFR signaling network. Biosystems 2005;80:57–69.

[13] Arteaga CL, Baselga J. Clinical trial design and end points for epidermal growth factor receptor-targeted therapies: implications for drug development and practice. Clin Cancer Res 2003;9:1579–89.

[14] Espina V, Wulfkuhle J, Calvert VS, et al. Reverse phase protein microarrays for monitoring biological responses. In: Fisher P, editor. Cancer genomics and proteomics: methods and protocols. Totowa (NJ): Humana Press; 2008.

[15] Neel BG, Tonks NK. Protein tyrosine phosphatases in signal transduction. Curr Opin Cell Biol 1997;9:193–204.

[16] Mueller C, Edmiston KH, Carpenter C, et al. One-step preservation of phosphoproteins and tissue morphology at room temperature for diagnostic and research specimens. PLoS One 2011;6:e23780.

[17] Lowenthal MS, Mehta AI, Frogale K, et al. Analysis of albumin-associated peptides and proteins from ovarian cancer patients. Clin Chem 2005;51:1933–45.

[18] Luchini A, Geho DH, Bishop B, et al. Smart hydrogel particles: biomarker harvesting: one-step affinity purification, size exclusion, and protection against degradation. Nano Lett 2008;8:350–61.

[19] Ross PL, Huang YN, Marchese JN, et al. Multiplexed protein quantitation in Saccharomyces cerevisiae using amine-reactive isobaric tagging reagents. Mol. Cell. Proteomics 2004;3:1154–69.

[20] Picotti P, Aebersold R. Selected reaction monitoring-based proteomics: workflows, potential, pitfalls and future directions. Nat. Methods 2012;9:555–66.

[21] Wulfkuhle JD, Berg D, Wolff C, et al. Molecular analysis of HER2 signaling in human breast cancer by functional protein pathway activation mapping. Clin. Cancer Res. 2012;18:6426–35.

[22] Mann M, Hendrickson RC, Pandey A. Analysis of proteins and proteomes by mass spectrometry. Annu Rev Biochem 2001;70:437–73.

Chapter 10

# Integrative systems biology

Gregory J. Tsongalis[1,2]

[1]The Audrey and Theodor Geisel School of Medicine at Dartmouth, Hanover, NH, United States, [2]Laboratory for Clinical Genomics and Advanced Technology (CGAT), Department of Pathology and Laboratory Medicine, Dartmouth Hitchcock Medical Center, Lebanon, NH, United States

**Summary**

Systems biology focuses on the investigation of complex biological interactions. The promises of systems biology as it relates to the practice of medicine is the hope of defining disease processes as a perturbation of a normal system. This chapter highlights why a systems-based approach to understanding human disease is necessary for continued improvement in health and wellness. Systems biology comprises three challenges: (1) how to generate sufficient quantities of data to assess variability in networks, (2) how to properly integrate data from multiple sources into a usable corpus of knowledge, and (3) how to use that corpus of knowledge to model and optimize a component system. These models will allow us to better understand systems behavior and how they can be changed to better the practice of clinical medicine.

## Introduction

Systems biology refers to the study of complex biological systems thru computational and mathematical modeling. To date, most of what we know about biological systems has been determined through a reductionist method of analysis. This approach has allowed for much discovery and innovation, but the complexity of all biological systems now requires a new approach to better understand the interactions of all of the pieces of the puzzle. This is especially true now in light of the translational sciences and the more immediate impact of this knowledgebase on patient care and wellness. With respect to human disease, one could argue that the Human Genome Project represented the first significant systems approach by intertwining all aspects of biology and genomic sciences with engineering, computational biology and informatics. This is better viewed as the beginning of an approach to further understand the basic fundamentals leading to a systems biology focus and setting the stage for an integrative systems approach to the study of human disease. The goal of these systems approaches is to better understand and model how individual biological properties interact to create networks and signaling pathways that maintain homeostasis of living things and to understand the results of those properties when they go astray.

With respect to systems biology, living things can be described as collections of various forms of cells consisting of numerous networks and pathways, each composed of interacting collections of molecules which can be further characterized into individual elements. In the space of molecular pathology, these networks can be reduced to pathways, proteins, genes, nucleotides, etc. However, it is our understanding of the macro network that is lacking in the study of human disease. Networks and systems biology have generated significant interest in part due to the advances in technologies that allow for better modeling and prediction of their actions. This ability has given us new perspectives on biological systems through integration of existing knowledge. Systems biology focuses on the investigation of complex interactions and networks in biological systems. In studying human disease, the promise of systems biology would be a better understanding of the etiology and development of disease as a function of an abnormal system. Human diseases impact multiple tissues and cell types, different organ systems and biological processes and many involve somatic and/ or constitutional changes in nucleic acids. A mathematical and computational approach to study the interactions of these biological processes is critical to the success of truly understanding human wellness and disease. Classic examples of the need for a systems approach to studying human disease can be seen in such diseases as cystic fibrosis where mutations in a single gene impact many organ systems and multifactorial events impact patient outcomes. In human cancers, multiple pathways impact a specific population of cells through genomic and epigenetic aberrations that lead to

Essential Concepts in Molecular Pathology. DOI: https://doi.org/10.1016/B978-0-12-813257-9.00010-3

abnormal signaling and interactions with the microenvironment. These findings often leave us asking whether we truly understand the disease and more importantly whether we truly understand the patient. The latter leading us to a critical need for understanding the "normal" system.

Systems, in the sense of biological networks and pathways, interact with each other and are impacted differently by genetic and environmental factors. Thus, much of the variability in severity of disease phenotypes seen in patients is most likely due to the variations in these systems. Systems can impact single cells and single organs or be more broad in scope whereby multiple organs are affected by the dysfunction of a system. Careful modeling of these different relationships within the phenome (the set of all phenotypes of an organism) or within the interactome (the set of all interactions of an organism) can greatly increase the power to identify functional genomic variation important in disease phenotypes and lead to a better understanding of how to readjust the system to return to a normal state.

Systems biology comprises three challenges: (i) how to generate a sufficient quantity of data to analyze variability in networks, (ii) how to properly integrate data from multiple disparate sources into a usable knowledgebase, and (iii) how to use that knowledgebase to model a component system. Modeling allows predictions of the behavior of a system and, ultimately, an understanding of how to change that behavior in predictable ways. Because the exact definition of a system is not fixed, anything from a particular biochemical pathway up to the level of an entire organism can be considered a system, applications of systems modeling in medicine are limitless. Population-level biosciences, such as epidemiology and population genetics, have long practiced a form of systems biology, but their focus has always been on a collection of individuals as the system of interest. Current systems biology approaches model systems of interest to clinical outcomes involving metabolic pathways, individual cells, and organs. Modeling at this level has the appeal of allowing translation of results from decades of population-based laboratory research to be applied to precision medicine. One such approach utilizes the Biological Universal Modeling Language (BioUML, http://www.biouml.org).

The field of systems biology borrows from control theory, which deals with the behavior of dynamic systems (systems that fluctuate in a dependent fashion). Control theory proposes the idea that systems can best be modeled as a cycle of controllers which modify inputs to produce the desired outputs, and sensors which provide feedback to the system, producing a steady-state system in which equilibrium is achieved by balancing the input and output concentrations and the feedback signals. In a similar fashion, systems biology includes the definition and measurement of the components of a system, formulation of a model, and the systematic perturbation (either genetically or environmentally) of and remeasurement of the system. The experimentally observed responses are then compared with those predicted by the model, and new perturbation experiments are designed and performed to distinguish between multiple or competing models. This cycle of test-model-retest is repeated until the final model predicts the reaction of the system under a broad range of perturbations.

## Systems biology as a paradigm shift

The study of systems in biology is not new. Early studies of enzyme kinetics employed a similar cycle of testing and modeling, followed by retesting of new hypotheses. In a similar fashion, modeling of neurophysiological processes like the propagation of action potentials along a neuron are illustrative of early forays into mathematical modeling of cellular processes. However, like all next-generation methodological shifts, systems biology involves a rethinking of basic principles—a return to a more basic understanding or a more simplistic framework in which to generate hypotheses. Traditional science follows a bottom-up paradigm—that is, break a problem down into its individual components (reductionism), learn everything there is to know about those individual components, and then integrate the information together to get information about a system. Systems biology can work within this paradigm, creating models for individual components, and then integrating the models (creating, in essence, models of models, to explain the behavior of the original system). However, this approach requires that knowledge of the individual components is sufficient to explain the behavior of the system. Apart from a few exceptions, current reductionist research models have not been successful at explaining the large-scale behavior of biologic systems. To this end, systems biology endorses a more top-down approach, allowing modeling to occur at a higher level of organization (not at the level of the components, but at the level of the organ or the individual). In this point of view, a better understanding of the system as a whole can create a better understanding of the components. Noble wrote that "... *systems biology ... is about putting together rather than taking apart, integration rather than reduction. It requires that we develop ways of thinking about integration that are as rigorous as our reductionist programs, but different...*" More simply, if you were to describe all of the details of an airplane and its functions, that still would not tell you how the airplane flies. This has led to the recognition that a new paradigm (integration rather than reduction) may be necessary to understand and model systems with multiple interacting quantitative components.

# Data generation

Central to the success of any modeling endeavor is the generation of large quantities of data. Because of the increased interest in systems biology, the theme of the late 1990s and early 21st century reflected exactly this necessity, exemplified by the fact that several papers were published on methods of dealing with the so-called "data deluge." Genomic technologies (i.e. gene expression profiling and DNA sequencing) were the first to enter the high-throughput realm and subsequently the first to produce large quantities of high-quality data. Coupled with a concurrent explosion in computing power, high-throughput data generation made realistic modeling feasible. New technologies produced during this time vastly increased the ability to query an entire system at once and led to the advent of the major international efforts of the early 21st century, such as the Human Genome Project (the effort to sequence the entire complement of human chromosomes). The HapMap Project (the effort to catalog common genetic variation across multiple human ethnic populations), the ENCODE Project (basically a merger of the Human Genome and HapMap projects—an effort to re-sequence particular portions of the human genome in multiple individuals from different ethnic populations to explore common and rare genetic variation at a higher resolution), the 1000 Genomes Project (an effort to completely sequence 1000 human genomes), and The Cancer Genome Atlas (TCA) (an effort to identify all DNA alterations in the more than 200 types of human cancer and they interact to drive the disease). Genomic data coupled with advances in mass spectroscopy, combinatorial chemistry, and robotics created an explosion of data and unique visualizations of cellular processes.

## Microarrays

Microarray technology grew out of a confluence of trends and technologies. On the experimental level, microarrays are most similar to Southern blotting, where fragmented DNA is attached to a substrate and then probed (hybridized) with a known gene or fragment to identify complementary sequences. The trend through the 1990s was to increase capacity in individual experiments, allowing the analysis of more and more samples (or more and more markers) in a single experiment. With the advent of paper-blotting techniques (allowing DNA molecules to be immobilized or spotted on special paper, then hybridized), and subsequently glass-blotting techniques, coupled with advances in robotic pipetting (allowing smaller and smaller volumes of liquid to be spotted, in greater and greater densities), microarrays became feasible. Initial microarray experiments began with small numbers of immobilized probes, owing mostly to limited knowledge about the genes that make up a genome. This was quickly followed by arrays with hundreds, thousands, tens of thousands, and now (currently) millions of probes as our understanding of the components of genomes grew (owing, in large part, to early efforts to identify and catalog all expressed genes in organisms and to large-scale efforts like the Human Genome and ENCODE projects).

Microarrays have now found use in multiple experimental and clinical venues. Most commonly, the molecules being immobilized on the array are DNA molecules, allowing measurement of expression levels or detection of polymorphisms (such as single-nucleotide polymorphisms or SNPs). A DNA microarray consists of thousands of microscopic spots (or features), each containing a specific DNA sequence. Each feature is typically labeled with a fluorescent tag and then used as probes in hybridization experiments with a DNA or RNA sample (the target), which is labeled with a complementary fluorescent label. Hybridization is quantified by fluorescence-based scanning of the array, allowing determination of the relative abundance of nucleic acid sequences in the target by characterization of the relative abundance of each fluorophore (Fig. 10.1). This technology has provided large datasets to address numerous experimental questions and has also generated equally large datasets by routine testing of patients with various disease conditions. For example, the American College of Medical Genetics has the set the standard of practice for all children with developmental delay to be screened by chromosomal microarray analysis as a first tier of testing. Using the FDA approved

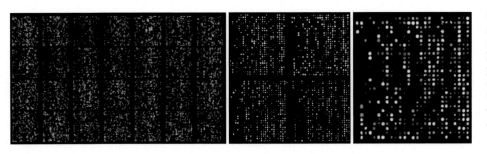

FIGURE 10.1 Gene expression microarray. A spotted array is shown following a two-color hybridization experiment. The left panel provides example of an approximately 40,000 probe spotted oligonucleotide microarray with enlarged inset to show detail. *Image from Wikimedia Commons.*

Thermofisher Cytoscan HD™ microarray, laboratories can create a virtual karyogram by digitizing the human genome based on microarray probe hybridization data for more than 2.6 million markers.

## Transcriptomics

Transcriptomics is the study of relative RNA transcript abundances, using microarray technologies. Chips that are specialized for this purpose are known as RNA microarrays and are typically prepared with a library of transcripts of known origin (representative tags for a known complement of genes, for example—the current human RNA arrays contain approximately 60,000 probes, representative of a majority of known RNA species from the 20,000 or so human genes). These are then interrogated with RNA (typically reverse transcribed into cDNA) from two different samples, labeled with different dyes (commonly green and red dyes). This allows the relative abundance (in one sample versus the other) of each RNA transcript to be assessed. Experiments of this type are routinely used to determine which RNAs are upregulated or downregulated in a disease sample versus a normal sample. However, note that the utility of the information generated from transcriptomics studies is highly variable, as transcription levels are potentially influenced by a wide range of factors, including disease phenotypes. To be of general use, transcriptomic data must be generated under a wide variety of conditions and compared, to eliminate the trivial sources of variation.

Implementation of next generation (NGS) or massively parallel sequencing technologies has allowed for the transcriptome to be more precisely defined using assays such as RNA-Seq or whole transcriptome sequencing. NGS can reveal the presence and quantity of RNA in a biological sample at any given moment in time and thus can be used to analyze the continually changing cellular transcriptome. RNA-Seq can detect alternative gene spliced transcripts, post-transcriptional modifications, gene fusions, and mutations/SNPs. In addition to mRNA transcripts, RNA-Seq can also look at different populations of RNA including total RNA, miRNA, tRNA, and ribosomal profiling.

## Genotyping

One of the earliest explosions of data on a whole-genome scale came from early genetic linkage studies. The first whole-genome genetic linkage scans were performed in the early 1990s using panels of ~350 genetic markers scattered throughout the autosomal chromosomes, as well as the X-chromosome. Subsequent advances in mapping techniques and marker discovery increased the number of markers in the genetic maps, and included both sex chromosomes, as well as the mitochondrial genome. The earliest genetic markers to be used (easily observable phenotypes such as eye color, sex, handedness, and others) were generally treated as simple binary traits. The first protein biomarkers (blood group protein polymorphisms and HLA polymorphisms) were complex, having many alleles with complicated ethnic and regional variations in frequency. Restriction fragment length polymorphisms (or RFLPs) having only two alleles indicating the presence or absence of a recognition site for restriction enzymes were the next marker type to be in vogue, and were the first to be suggested as usable for whole-genome analysis because of their frequency throughout the genome. This was followed by the discovery of DNA-length polymorphisms, in the form of tandem repeated regions (called Variable Numbers of Tandem Repeats or VNTRs), which are complex polymorphisms. Recognition of the existence of tandem repeats led to the discovery of microsatellite markers—basically short tandem repeats (STRs), on the order of 2−5 base pairs repeated several times. The frequency of STRs in the genome enabled generation of the first whole-genome maps. This was followed in quick succession by the discovery of single nucleotide polymorphisms (or SNPs), which initially were described as having only two alleles, and so continue in use as simple markers (though in fact many SNPs have >2 alleles). Due to their abundance (estimated at 1 SNP per 100 base pairs of DNA) and the relative ease of genotyping these markers, the development of DNA microarray-based genotyping technologies for SNPs occurred quickly. SNPs are now the current standard for genome-wide association studies. When SNP content on genotyping arrays became sufficiently dense (and also due to findings from other genomic technologies like array-based comparative genomic hybridization (arrayCGH) techniques), the widespread occurrence of copy number variations (deletions and duplications) was detected, heralding a return to more complex (multiallelic) markers again. Current genotyping arrays now contain a mixture of simple polymorphic markers (SNPs) and so-called copy-number probes, allowing the assessment of copy-number changes across the genome, at a density which provides excellent coverage for most ethnic groups (~1,000,000 SNPs and nearly that many copy number probes, meaning each array carries almost 2 million features). It is expected that chip densities will continue to grow, allowing for higher-throughput genotyping. The Oncoscan™ microarray provides data on copy number variation, loss of heterozygosity and common somatic mutations in over 900 cancer related genes. However, at the same time rapid low-cost sequencing technologies are becoming available that may well allow for affordable whole-genome sequence-based assessment of genomic variation, replacing any need for increased array densities.

Next generation sequencing (NGS) allows for the analysis of numerous target sequences from multiple individuals simultaneously. Its ability to assess multiple genetic variants concurrently (including point mutations, small INDELs, copy number alterations, and rearrangements or gene fusions), generates a mutation profile of the individual tumor that is used in selecting the most efficacious management strategy. Accurate assessment of human cancer, which correlates with best patient outcomes, needs to be performed in such a way as to provide the most clinically actionable information to the provider. Our understanding of the complexity of human cancer, including driver, passenger, sensitizing, and resistance mutations, has become the target of many novel therapies that now require profiling as a companion diagnostic. These datasets have shed light onto the complexity of human cancers both biologically with tumor heterogeneity and biochemically with the numerous pathways involved. A systems biology approach to understanding the integration of mutation profiles with pathways and therapeutics is critical as no tumors have a single mutations in a gene and combinations of events are the most likely reason for less than optimal drug efficacy.

## Other omic disciplines

Recognizing that many properties of biological systems are emergent, or a result of complex interactions between components that are not understandable at the level of individual system components, scientists in the 21st century turned to analysis of different levels of organization in an organism. Since the genome is the entire set of genes of an organism, genomics represents the whole set of genes of a biological system. Various methods, such as genotyping, sequencing and transcriptomics can be used to interrogate the genome. Proteomic methods include traditional techniques such as mass spectroscopy—identifying compounds based on the mass-charge ratio of ionized particles. But just as microarrays are a modification of traditional genomic techniques for high-throughput experiments, proteomics modifies the techniques of mass spectroscopy in a similar fashion—creating experimental and informatics pipelines that allow a sample to be partitioned into various fractions, have those fractions examined via mass spectroscopy, and have the results of the mass spectroscopy experiments compared to other mass spectroscopy profiles to allow for identification of individual components.

Other omic disciplines (Fig. 10.2) are similarly named— (i) metabolomics—the study of the entire range of metabolites taking part in a biological process; (ii) interactomics—the study of the complete set of interactions between

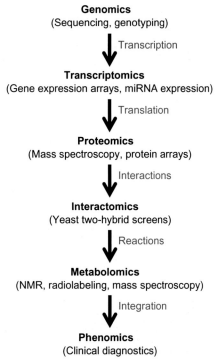

FIGURE 10.2   Omics technologies generate data on numerous levels. Various omics fields are displayed along with the laboratory techniques used to generate the data, as well as the relationship of data from one level to another. *Adapted from Fischer HP. Towards quantitative biology: integration of biological information to elucidate disease pathways and drug discovery. Biotechnol Annu Rev 2005;11:1−68.*

proteins or between these and other molecules; (iii) localizomics—the study of the localization of transcripts, proteins, and other molecules; and (iv) phenomics—the study of the complete set of phenotypes of a given organism.

## Data integration

While data generation is certainly of paramount importance for systems biology, because the technologies that generate the data have their own unique (and often proprietary) formats for storing the data, systems biologists must also concern themselves with integrating data that span multiple resources. Since 1998, the number of recognized online databases related to biological information has increased 10-fold. Resources like the Bioinformatics Links Directory (http://bioinformatics.ca/links_directory/) extend this further by collecting links to molecular resources and tools as well as databases. However, the problem is more than just quantity. Issues of data quality, lack of standards, lack of interfaces allowing for integration, and longevity (or lack thereof) continue to plague online biological data resources, and make cross-resource querying or integration difficult. Nonetheless, tools continue to appear to assist in the integration of data from disparate sources.

## Semantic web technologies

Many of the problems inherent to integration of biological data resources are similar to those being faced by the larger community of World Wide Web users. The Semantic Web is a vision for how to have computers infer information relating one web page (or element) to another. It is an extension of the current web protocols (primarily the HyperText Transport Protocol or HTTP), which allows for meaning to be imbedded together with content, such that automated agents can make associations between data without needing user input. In essence, creating the Semantic Web involves recasting the information in the World Wide Web (which currently stores relationships in the form of hyperlinks, which can link anything to anything, and so do not represent information content or meaning) into a format which allows relationships to be represented. The eXtensible Markup Language (or XML) is an early example of this type of recasting—allowing content to be stored with representative tags that describe the content. Resource Description Framework (or RDF) builds on XML by using triples (subject-predicate-object) to represent the information in XML tags (or in hyperlinks), and defining a standard set (or schema) of RDF triples to describe a particular object. Each part of a triple names a resource using either a Uniform Resource Identifier (URI) or Uniform Resource Locator (URL), or a literal. The advantage of this format is that RDF schemas are predefined so that meaning can be imbedded in the definition, or by using a hierarchy or ontology, describing the relationship within and between schemas. Also, since the RDF triple can contain a location as well as attribute-value pair, the components of a schema do not need to be located in the same place. Thus, if we have a schema representing a web page in which the sections (header, footer, left panel, right panel, title, and others) are defined by RDF triples, these web pages can easily integrate data from multiple sources.

The representation of information by RDF allows the location of data and data resources to be independent of the location of user interfaces or analytic resources. In essence, this allows the integration of data from multiple repositories and the querying of the integrated data. Coupled with the concept of RDF schemas to accurately describe knowledge about objects and ontologies to organize the relationship of objects to one another, the use of RDF can solve several of the problems mentioned for data integration (namely, the lack of standards and the lack of interfaces for integration). Another current trend—the use of wikis (a website that allows collaborative editing of its content by its users)—addresses the problem of data quality by allowing users to mark data as reliable or not, or by allowing users to update out-of-date or incorrect data.

## Modeling systems

Once the appropriate types of data have been generated, and the various sources of data collected and integrated, the resulting information is turned into knowledge by interpreting what the data actually mean, and how they address questions that need to be answered. Data modeling is used to understand the relationships important in defining the system. As noted previously, systems biology draws heavily from control theory, which itself is derived from mathematical modeling of physical systems. Although the mathematical derivation of control problems is complex, the fundamental concepts governing the formulation of control models are more intuitive and relatively limited. Three basic concepts can be thought of as central to forming control models: (i) the need for control (regulation or feedback), (ii) the need for fluctuation, and (iii) the need for optimization. A simple control model is shown in Fig. 10.3.

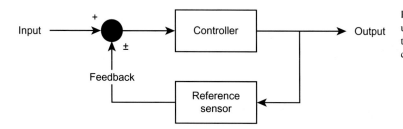

**FIGURE 10.3** Example of a control module. This module represents a simple feedback loop, with output from the sensor either upregulating or downregulating the process that converts the input into output.

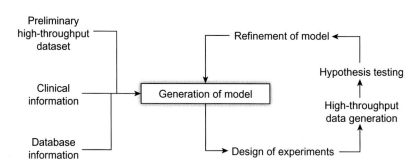

**FIGURE 10.4** The iterative nature of systems biology research. Note that every refinement of the model needs additional data generation for retesting of specific hypotheses. *Adapted from Studer SM, Kainski N. Towards systems biology of human pulmonary fibrosis. Proc Am Thorac Soc 2007;4:85—91.*

Key characteristics of this model include a controller (responsible for the conversion between input and output) and sensor (responsible for determining the degree and direction of the feedback)—each of which can be the result of a single or multiple elements. An initial model of this form can often be derived from an initial data generation step, which can measure a baseline set of conditions for the system and also provide an idea of the components of the system. For example, an expression network (groups of genes that are co-regulated in some fashion) that regulates a biochemical pathway can be thought of as a control model. A single transcriptomics experiment can give you information on genes that are potentially co-regulated (sets of genes that are overexpressed compared to control, and so which might be providing the function of a controller), as well as information about potential sensors (genes that differ in response—one being overexpressed, the other underexpressed). However, the problem with a single experiment (or a single snapshot of the transcriptome) is that the information derived is not sufficient to disentangle the true positives (elements that truly should be components of the model) from the false positives (random variation which transiently mimics the behavior of the model). This reflects the need for fluctuation. By perturbing (changing an aspect of the system to produce a predictable outcome) and re-measuring the system, additional confidence regarding the true positives can be achieved assuming the new measurements reflect the predictions of the model. If not, another model which explains the previous data measurements is selected, and another test is devised. This cycle continues until the final model accurately represents the data measured in all tests. In this manner, the most optimized model is selected (Fig. 10.4).

Data modeling also requires large amounts of computational power. Due to the size of the high-throughput data sets currently in use (expression data on 30,000 genes for multiple time points; SNP genotypes for millions of markers; occurrence of all known protein-protein interactions; computational prediction and annotation of all known protein-DNA binding sites; and others), the advanced mathematical and visualization frameworks currently in use, and the iterative nature of analysis, large amounts of computing power are necessary, often pushing the limits of even parallelized or grid-enabled computational clusters. Continued growth in computing power will be necessary to support the ongoing use of systems biology as the models in use become more and more elaborate and incorporate information from greater numbers of high-throughput methodologies.

## Implications for understanding disease

The sequencing of the human genome resulted in new data and technologies that could be used to enhance patient care as part of a precision medicine initiative. However, the complexity of genotype-phenotype correlations became apparent when it was found that millions of variations exist in the DNA sequence of individual humans, with an infinite number of possible variations. Genetic studies revealed that many chronic inflammatory diseases, such as Crohn's disease, ulcerative colitis, chronic pancreatitis, rheumatologic disease, heart diseases, and others do not have a single genetic factor causing the disease, but rather several dozen common genetic variations (polygenic) that, in the context of many possible interacting environmental factors (multifactorial), cause a disease that is defined by the location of inflammation and associated signs and symptoms. Unfortunately, the simplicity and early success of the germ theory of disease, in which a single pathological agent was responsible for a specific disease with characteristic signs and symptoms, lulled physicians into thinking that all disease would follow a similar pattern. This reductionist-like approach in medicine allowed physicians and scientists from different disciplines to triangulate on the same target using different methods and perspectives. It has identified numerous components of individual systems—many of which are reused over and over, leading to a modular view of system components—and has provided numerous insights into human disease. Despite these advances, the reductionist approach in medicine has been less successful in identifying the many complex interactions between disease components, and in explaining how system properties (like disease phenotypes) emerge—a concept which becomes important when considering manipulation of systems to produce predictable and desirable outcomes, as required for precision medicine. As such, paradigms requiring new approaches and new methods are needed to better understand disease and wellness.

Personalized or precision medicine can be thought of as an application of systems biology-type approach to medicine. Personalized medicine represents a transition from population-based thinking (describing risks on a population level) to an individual-based approach (describing risks for an individual based on his or her personal genomic/proteomic/metabolomic/phenomic profiles). Reductionist approaches have successfully identified many of the biomarkers required to define disease states in complex disorders on a population level (that is attributing population risks to various changes), but have not been able to describe how individual exposures or biomarkers (or combinations of these components) interact to create the disease state in individuals. To achieve this transition, systemic models which explain the function of systems (cells, organs, etc.) are needed, from which the implications of changes in particular exposures or biomarkers (genetic changes, proteomic changes, and others) can be understood on the system-level.

## Redefining human diseases

The first major hurdle in transitioning from allopathic medicine to personalized medicine is to redefine common human diseases. In allopathic medicine, diseases are typically defined by characteristic signs and symptoms that occur together and meet accepted criteria. Chronic inflammatory diseases are defined by the location of the inflammation, the persistence of inflammation, and the presence of tissue destruction or scarring. These definitions indicate that the specific mechanism causing a specific organ to develop and to sustain an inflammatory reaction is not known. Furthermore, it has been impossible to identify the cause when using the traditional scientific methods used to identifying a single, causative infectious agent. Instead, the possibility that disease affecting one or more organs can occur through any one of multiple pathways, and that each pathway has multiple steps and regulatory components, and that multiple effects on multiple systems and multiple environmental exposures may be required before disease is manifest must be embraced to transition to personalized medicine. Indeed, physicians recognized that diseases located in specific organs share some common systemic features such as the type of inflammatory response (autoimmune or fibrosing), or chronic pain since they use a limited number of anti-inflammatory or pain medications for a variety of diseases. Additionally, studies which have looked at the clustering of disease phenotypes based on their representative genomics (associated gene/SNP polymorphisms or expression profiles) have demonstrated that even disease phenotypes we once thought were roughly homogeneous (for instance, disease subtypes like Crohn's disease with ileal involvement in Caucasian-only populations) are in fact associated with different genetic loci (heterogeneous). In human cancers, the dynamics of cell biology can be integrated with cell behavior to model the disease using a systems biology approach. Personalized medicine must deconvolute systemic and tissue-specific pathways and reconstruct them in a way that leads to precise, patient-specific treatments.

## The transition to personalized medicine

The effect of approaching complex disorders as a single disease rather than as a complex process is illustrated through the comparison of multiple small studies by meta-analysis. While some of the variance in estimated effect sizes from small genetic studies can be attributed to random chance, the possibility also exists that populations from which the samples were taken were not equivalent, and that different etiologies will lead to the same end-stage signs and symptoms through different, parallel pathways. In cases where a candidate gene is critical to some pathologic pathway, but not others, the wide variance between small genetic studies may reflect the fraction of subjects that progress to disease through that specific gene-associated pathway.

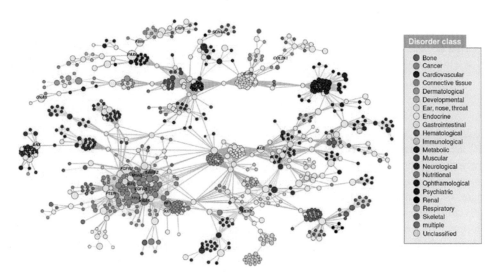

**FIGURE 10.5** Disease gene network. Each node is a single gene, and any two genes are shown linked if implicated in the same disorder. In this network map, the size of each node is proportional to the number of specific disorders in which the gene is implicated. *Reproduced with permission from Loscalzo J, Kohane I, Barabási AL. Human disease classification in the postgenomic era: a complex systems approach to human pathobiology. Mol Syst Biol 2007;3:124.*

In order to favorably impact precision medicine efforts, very large association studies require the recruitment of subjects from many large medical centers in different regions and different countries. This approach will tend to obscure mechanism-based heterogeneity and converge on only the most common features of the disease (population-level effects), even though other factors may have a stronger biological effect in a limited number of patients, while being irrelevant in others. This argues for more detailed phenotyping of study populations (phenomics) and careful analysis of context-dependent effects (interactomics). The application of systems biology to this data would allow for a better understanding of the various pathways leading to disease states, and so a better understanding of the possible interactions (or context-dependent effects) that accurately predict disease in a single individual.

## Applications of systems biology to medicine

The realization of personalized medicine requires not only a more systems-like perspective regarding risk factors, it also requires a rethinking of existing classifications, which are largely derived from observation and reductionist approach (common observable phenotypes should be the result of common underlying factors). For example, reclassification of the disease phenome, the space of all associations between disease phenotypes, can be addressed using the explosion of current information on genetic disease associations. Disease-associated genes can be clustered based on information about gene-gene or protein-protein interactions (from transcriptomics or interactomics studies), and superimposed with a representation of the organ system of the associated disease phenotype. The resulting network graph (Fig. 10.5) demonstrates the association of different genes together in modules, many of which represent associations of proteins in macromolecular complexes (protein-protein interactions) or associations of proteins in metabolic or regulatory pathways.

This view of the network of disease genes can at the same time inform us about novel associations we might not have been aware of (such as PAX6 in the ophthalmological cluster, or PTEN and KIT in the cancer cluster), and also direct us toward pathways (clusters) that might be of most benefit in understanding the network more completely (that is, those that are involved in multiple disease states or that have the most connections to them). This network-centric view of disease can also inform clinical medicine in terms of the choice of medications (indicating that medications used in one disease might also function appropriately in another, based on clustering of associated genetic factors). The therapeutic implications of a systems biology approach to common yet complex diseases such as cancer are significant as these diseases are not defined by a single aberrant gene or pathway. Cancer is a disease of multiple genes, multiple pathways, and numerous external factors. In addition, the biology of tumor heterogeneity and development of resistance mechanisms adds to the complexity and must be integrated into these approaches. Next generation sequencing data which identifies numerous variants in tumor DNA is beginning to impact patient management and outcomes.

## Convergence

As seen in the example of cancer genomic testing, simply identifying gene variants or abnormal protein expression is not enough to adequately impact patient outcomes. Not only must we approach these complexities through the study of systems biology, we must also practice convergence science. Convergence science goes beyond systems biology and traditional interdisciplinary collaborations. Convergence science integrates traditional research fields with new and emerging technologies. To understand the complexity of biological processes requires the integration of new tools, skill sets, knowledge bases, and problem solving approaches.

## Discussion

We have outlined many of the reasons that a systems biology-based approach is needed for understanding complex disorders. Once the key components are organized into a logical series, then formal modeling of the disease process can be applied, tested, additional experimental data added, the system calibrated and retested. The optimal model or models are yet to be determined experimentally for any system, but with the continued rapid increase in high-throughput data generation and the continued increase in computational power, large-scale integrations and models can be achieved. The challenge will be to accurately anticipate the information necessary to be included in the model, as accurate assessment of the context is essential to the appropriate understanding of the effect of individual variation and the integration of that understanding into clinical practice. It is not enough to know that certain changes affect a phenotype in a population, as these overall effects are typically very small. However, with a proper understanding of the context in which each variant is important, we will better understand disease and wellness so that the most appropriate interventions with the best outcomes can be achieved.

## Key concepts

- Systems biology focuses on the investigation of complex interactions (or networks) in biological systems.
- Three challenges face researchers interested in systems biology: (a) how to generate sufficient amounts of data for modeling of whole systems, (b) how to integrate data from multiple data sources, and (c) how to use integrated data to model a system and optimize the model.
- Various "omics" disciplines (genomics, proteomics, transcriptomics) have grown out of technological advances associated with the Human Genome Project and other large-scale biological projects of the early 21st century. Microarray technologies, in particular, have greatly expanded the quantity and quality of data generation for modeling purposes. NGS promises to do the same.
- Semantic web technologies, including XML and RDF, coupled with a greater understanding of the importance of ontologies, have provided much needed data integration capabilities.
- Control theory can be used to derive general frameworks for model construction, in which effects are organized into either regulatory (controlling) or monitoring (feedback) components, and in which derived models are tested via a cycle of prediction-perturbation-measurement, allowing hypothesis tests to be conducted and leading to model refinements and the generation of more appropriate models.
- Implications of systems biology for medicine include the development of a better understanding of diseases and their associations with one another. As disease models are built that explain conglomerations of diseases grouped by systems-level thinking (rather than by clinical measures), individual risk factors (i.e. components of the controller or feedback components of the model) become more easily identifiable, leading to better understanding of pathological mechanisms, and better drug and intervention targets.

## Suggested readings

[1] Auffray C, Balling R, Barroso I, et al. Making sense of big data in health research: towards an EU action plan. Genome Med 2016;8:71–5.
[2] Cejovic J, Radenkovic J, Mladenovic V, et al. Using semantic web technologies to enable cancer genomics discovery at petabyte scale. Can Inform 2018;17:1–7.
[3] El-Deiry WS, Goldberg RM, Lenz HJ, et al. The current state of molecular testing in the treatment of patients with solid tumors, 2019. CA: Cancer J Clin 2019;. Available from: https://doi.org/10.3322/caac.21560.
[4] ENCODE Project Consortium, Birney E, Stamatoyannopoulos JA, et al. Identification and analysis of functional elements in 1% of the human genome by the ENCODE pilot project. Nature 2007;447:799–816.
[5] Fallmann J, Videm P, Bagnacani A, et al. The RNA workbench 2.0: next generation RNA data analysis. Nucleic Acids Res 2019;47:W511–15.
[6] Faratian D, Clyde RG, Crawford JW, Harrison DJ. Systems pathology: taking molecular pathology into a new dimension. Nat Rev Clin Oncol 2009;6:455–64.
[7] Grabowski P, Rappsilber J. A primer on data analytics in functional genomics: how to move from data to insight? Trends Biotech Sci 2019;44:21–32.

[8]   Human Genome Project: http://www.ornl.gov/sci/techresources/Human_Genome/home.shtml

[9]   Jung HS, Lefferts JA, Tsongalis GJ. Utilization of the OncoScan microarray assay in cancer diagnostics. Appl Cancr Res 2017;37:1−8.

[10]  Karczewski KJ, Snyder MP. Integrative omics for health and disease. Nat Rev Genet 2018;19:299−310.

[11]  Kim J, Campbell AS, Esteban-Fernandez de Avila B, Wang J. Wearable biosensors for healthcare monitoring. Nat Biotech 2019;37:389−406.

[12]  Kolpakov F, Akberdin I, Kashapov T, et al. BuioUML: an integrated environment for systems biology and collaborative analysis of biomedical data. Nucl Acids Res 2019;1−9.

[13]  Lappalainen T, Scott AJ, Brandt M, Hall IM. Genomic analysis in the age of human genome sequencing. Cell 2019;177:70−84.

[14]  Manning M, Hudgins L, for the Professional Practice and Guidelines Committee. Array-based technology and recommendations for utilization in medical genetics practice for detection of chromosomal abnormalities. Genet Med 2010;12:742−5.

[15]  Nussinov R, Tsai CJ, Shehu A, Jang H. Computational structural biology: successes, future directions and challenges. Molecules 2019;24:1−12.

[16]  Noble D. The music of life: biology beyond the genome. Oxford: Oxford University Press; 2006. p. 21.

[17]  Remondini D. Systems biology approaches to cancer: towards new therapeutical strategies and personalized approaches. Mol Cell Oncol 2019;6:3.

[18]  The International HapMap Consortium. A haplotype map of the human genome. Nature 2005;437:1299−320.

[19]  1000 Genomes Project: http://www.1000genomes.org/page.php.

[20]  Tomczak K, Czerwinska P, Wiznerowicz M. The Cancer Genome Atlas (TCGA): an immeasurable source of knowledge. Contemp Oncol 2015;19:A68−77.

[21]  Van Dijk EL, Jaszczyszyn Y, Naquin D, Thermes C. The third revolution in sequencing technology. Trends Genet 2018;34:666−81.

[22]  Wainberg M, Merico D, Delong A, Frey BJ. Deep learning in biomedicine. Nat Biotech 2018;36:829−38.

[23]  Yin H, Xue W, Anderson AG. CRISPR_Cas: a tool for cancer research and therapeutics. Nat Rev Clin Oncol 2019;16:281−95.

[24]  Yohe S, Thyagarajan B. Review of clinical next generation sequencing. Arch Pathol Lab Med 2017;141:1544−57.

Chapter 11

# Pathology: the clinical description of human disease

**William K. Funkhouser**
*Department of Pathology and Lab Medicine, UNC School of Medicine, Chapel Hill, NC, United States*

**Summary**
Pathology is that field of science and medicine concerned with the study of diseases, specifically their initial causes (etiologies), their step-wise progressions (pathogenesis), and their effects on normal structure and function. This chapter will consider the history of relevant discoveries and technologies that have led to our current understanding of diseases, as well as the Pathologist's current role in the diagnosis, prognosis, and prediction of response of human diseases.

"Future discoveries will not *likely be made by morphologists ignorant of molecular biologic findings, or by biologists unaware or scornful of morphologic data, but by those willing and capable of integrating them through a team approach . . .*"
Rosai J. Rosai and Ackerman's surgical pathology. 9th ed. St. Louis, MO: Mosby; 2004.

## Introduction

This chapter will discuss the fundamental concepts, terminology, and practice of pathology as the discipline dedicated to the understanding of causes, mechanisms, and effects of diseases. A section on key terms, definitions, and concepts is followed by sections on historical human approaches to diseases, an overview of current diagnostic practice, and a vision for new interface with applied molecular biology.

## Terms, definitions, and concepts

*Pathology* (from the Greek word *pathología*, meaning the study of suffering) refers to the specialty of medical science concerned with the cause, development, structural/functional changes, and natural history associated with diseases. *Disease* refers to a definable deviation from a normal phenotype (observable characteristics due to genome and environment), evident via patient complaints (symptoms), and/or the measurements of a careful observer (signs). The cause of the disease is referred to as its *etiology* (from the Greek word meaning the study of cause). One disease entity can have more than one etiology, and one etiology can lead to more than one disease. Each disease entity develops through a series of mechanistic chemical and cellular steps. This stepwise process of disease development is referred to as its *pathogenesis* (from the Greek word meaning generation of suffering). Pathogenesis can refer to the changes in the structure or function of an organism at the gross/clinical level, and it can refer to the stepwise molecular abnormalities leading to changes in cellular and tissue function.

The presentation of a disease to a clinician is in the form of a human patient with variably specific complaints (*symptoms*), to which the examining physicians can add diagnostic sensitivity and specificity by making observations (screening for *signs* of diseases). These phenotypic (measurable characteristic) abnormalities reflect the interaction of the genotype (cytogenetic and nucleic acid sequence/expression) of the patient and his/her environment. Patient *workup* uses present illness history with reference to past medical history, review of other organ systems for other abnormalities, review of family history, physical examination, radiographic studies, clinical laboratory studies (for example, peripheral blood or CSF specimens), and anatomic pathology laboratory studies (for example, tissue biopsy or pleural fluid cytology specimens). As you will see from other chapters in this book, the ability to rapidly and inexpensively

Essential Concepts in Molecular Pathology. DOI: https://doi.org/10.1016/B978-0-12-813257-9.00011-5

screen for chromosomal translocations, copy number variation, genetic variation, and abundance of mRNA and miRNA is adding substantial molecular correlative information to the workup of diseases.

The *differential diagnosis* represents the set of possible diagnoses that could account for symptoms and signs associated with the condition of the patient. The conclusion of the workup generally results in a specific diagnosis which meets a set of diagnostic criteria, and which explains the patient's symptoms and phenotypic abnormalities. Obviously, arrival at the correct diagnosis is a function of the examining physician and pathologist (fund of knowledge, experience, alertness), the prevalence of the disease in question in the particular patient (age, race, sex, site), and the sensitivity/specificity of the screening tests used (physical exam, vital signs, blood solutes, tissue stains, genetic assays). The *pathologic diagnosis* represents the best estimate currently possible of the disease entity affecting the patient, and is the basis for downstream follow-up and treatment decisions. The diagnosis implies a natural history (course of disease, including chronicity, functional impairment, and survival) that most patients with this disease are expected to follow. Be aware that not all patients with a given disease will naturally follow the same disease course, so differences in patient outcome do not necessarily correspond to incorrect diagnosis. Variables that independently correlate with clinical outcome differences are called *independent prognostic variables* , and are routinely assessed in an effort to predict the natural history of the disease in the patient. It is also important to note that medical therapies for specific diseases do not always work. Variables that independently correlate with (predict) responses to therapy are called *independent predictive variables*.

Diagnosis of a disease and development of an effective therapy for that disease do not require knowledge of the underlying etiology or pathogenesis. For example, granulomatous polyangiitis (née Wegener's granulomatosis) was understood by morphology and outcomes to be a lethal disease without treatment, yet responsive to cyclophosphamide and corticosteroids, before it was found to be an autoimmune disease targeting neutrophil cytoplasmic protein PR3 (Fig. 11.1). However, understanding the molecular and cellular pathogenesis of a disease allows development of screening methods to determine risk for clinically unaffected individuals, as well as mechanistic approaches to specific therapy.

The Pathologist is that physician or clinical scientist who specializes in the art and science of medical risk estimation and disease diagnosis, using observations at the clinical, gross, body fluid, light microscopic, immunophenotypic, ultrastructural, cytogenetic, and molecular levels. Clearly, the pathologist has a duty to master any new concepts, factual knowledge, and technology that can aid in the estimation of risk for unaffected individuals, the statement of accurate and timely diagnosis, accurate prognosis, and accurate prediction of response to therapy for affected individuals.

# A brief history of approaches to disease

The ability of *H. sapiens* to adapt and thrive has been due in part to the ability of humans to remember the past, respect tradition, recognize the value of new observations, develop tools/symbols, manipulate the environment, anticipate the future, and role-specialize in a social structure. The history of human understanding of diseases has progressed at variable rates, depending on the good and bad aspects of these human characteristics.

## Concepts and practices before the scientific revolution

Our understanding of ancient attitudes toward diseases is limited by the historical written record. Thus, the start point for written medical history corresponds to around 1700 BCE for Mesopotamian rules in the code of Hammurabi, and around 1550 BCE for the analogous Egyptian rules in the Ebers papyrus. By definition, these philosophers, theologians,

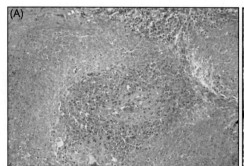
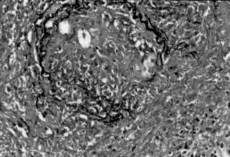

FIGURE 11.1 Granulomatous polyangiitis (née Wegener's granulomatosis) of the lung. (A) Hematoxylin and eosin staining of granulomatous polyangiitis of lung. Necrosis, granulomatous inflammation, and vasculitis are identified. (B) Elastin stain of granulomatous polyangiitis of lung. Elastica disruption of the arterial wall supports a diagnosis of vasculitis.

and physicians had access and assets to allow a written record, and materials and storage sufficient for the written records to survive. The Mesopotamian records indicate a deity-driven, demon-driven theory and empirical practice by recognized professional physicians. In this context, the prevailing thought was that *"Disease was caused by sprit invasion, sorcery, malice, or the breaking of taboos; sickness was both judgment and punishment . . ."*

The Greek medical community evolved a theory of disease related to natural causes and effects, with less emphasis on deity-driven theory. The Hippocratic Corpus includes "On the Sacred Disease" (circa 400 BCE), which rejected a divine origin for diseases, and postulated a natural rather than supernatural basis for disease etiology (*". . . nowise more divine nor more sacred than other diseases, but has a natural cause . . . like other affections. . ."*). Aristotle (384−322 BCE) wrote broadly on topics including logic, biology, physics, metaphysics, and psychology. To Aristotle, observations led to a description of causes, or first principles, which in turn could be used logically in syllogisms to predict future observations. We would agree with these basic notions of induction and deduction. However, there was a different background philosophical construct regarding the nature of matter and causality (four elements, four humors, and four causes, including a final or teleological purpose). We would recognize Aristotle's "efficient" cause of a disease as its etiology. Alexander the Great's conquest of Egypt in the 4th century BC led to Greek (Ptolemaic) leadership of Egypt from 305 BCE to 30 BCE, with development of the Alexandrian library and University. Faculty such as Euclid developed geometric models of vision (*Optica*), and Herophilus described human anatomy by direct dissection and observation (Greek medicine apparently allowed dissection in Alexandria, including vivisection of the condemned). During the Roman imperial era, Galen (129−207 CE) used dissection and observation of other animals such as the macaque (human dissection was illegal) to extrapolate to human anatomy and physiology. Like Aristotle's approach, Galen's approach to patients, diseases, and treatments was guided by philosophical constructs of four humors (blood, phlegm, yellow bile, and black bile) and the resulting temperaments (buoyant, sluggish, quick-tempered, and melancholic) due to humoral imbalances. It is thought that many of Galen's texts were destroyed with the Alexandrian library before the 7th century AD, but a subset was preserved and translated by Middle Eastern scholars. These ancient classic texts were then retranslated into Latin and Greek when printing houses developed in the 15th century (for instance, Hippocrates'*De Natura Hominis*, circa 1480 CE, and Galen's *Therapeutica*, circa 1500 CE).

The historical picture of the Greco-Roman understanding of disease is one of empirical approaches to diseases based on inaccurate understanding of anatomy, physiology, and organ/cellular pathology. Greek medicine became less superstitious and more natural cause-and-effect oriented, yet philosophy still trumped direct observation, such that evidence was constrained to fit the classic philosophical constructs. Some of the concepts sound familiar; for example, normal represents equilibrium and disease represents disequilibrium. However, we would differ on what variables are in disequilibrium (the historical humors, numbers, and opposites versus contemporary chemical and kinetic equilibria).

Following the collapse of the Western Roman Empire in 476 CE, the classic texts of Aristotle, Hippocrates, and Galen were protected, translated, and built upon in the Byzantine and Arab societies of the near East, and in Spain during the Muslim/Moorish period through the 11th century. During these middle ages for Western Europe outside Spain, there was apparently a retreat to pre-Hellenistic beliefs in supernatural forces that intervened in human affairs, with protecting saints and relics for disease prevention and therapy. Centers of medical learning following the Spanish Muslim model developed in Montpellier, France, and in Salerno, Italy, beginning in the 11th century.

## The scientific revolution

Aristotle's concept of induction from particulars to general first principles, then use of syllogistic logic to predict particulars, evolved into the scientific method during the Renaissance. Ibn Alhazen's (al-Haitham's) work on the physics of optics in the 11th century challenged Euclid's concepts of vision from the Alexandrian era. (Euclid thought that the eye generated the image, rather than light reflected from the object being received by the eye). In the 13th century, Roger Bacon reinforced this use of observation, hypothesis, and experimentation. The printing press (Gutenberg, 1440 CE) allowed document standardization and reproduction, such that multiple parallel university and city libraries could afford to have similar collections of critical texts, facilitating scholarly publications in journals. Access to publications in libraries and universities led to a system of review, demonstration, discussion, and consensus regarding new scientific findings.

The concept of the body as an elegant machine was captured by not only 15th and 16th century Renaissance artists like da Vinci and Michelangelo, but also by anatomists and pathologists interested in the structure/function of health and disease. The ancient models of Aristotle and Galen had become sacrosanct, and newer, evidence-based models were considered heretical to some degree. So it was somewhat revolutionary when Vesalius dissected corpses, compared them with Galenic descriptions, and published *De humani corporis fabrica libri septem* (1543 CE; "Seven Books on the Structure of the Human Body"; the *Fabrica*), challenging and correcting 16th century understanding of normal human anatomy. Vesalius's successors (Colombo, Fallopius, and Eustachius) further improved the accuracy of human anatomic detail. Thus, correction of Galen's anatomical inaccuracies (including the rete mirabile at the base of the brain, blood vasculature, the five lobed liver, and curved humerus) required at least 13 centuries for challenge, scientific disproof, and eventual medical community acceptance.

The Scientific Revolution describes the progressive change in attitude of scientists and physicians toward understanding of the natural world, health, and disease. This revolution began circa 1543 CE, when Copernicus published arguments for a heliocentric universe and Vesalius published the *Fabrica* series on human anatomy. By the 17th century, Galileo, Kepler, Newton, Harvey, and others had used this observation-based, matter-based, and mathematical law-based perspective to develop a scientific approach similar to our own modern approach of testing hypotheses with experimental data and statistics. In human biology, the investigation of structure led to studies of function, initially of human cardiovascular physiology, for example, Harvey's *Anatomical Exercise Concerning the Motion of the Heart and Blood in Animals* (1628). Whereas Galen conceived of parallel but unconnected arteries and veins, with continuous blood production in the liver and continuous blood consumption in the periphery, Harvey demonstrated that blood was pumped by the heart through arteries, through tissue capillaries, to veins, and then back to the heart in a circle (circulation). Correction of these and other Galenic physiological inaccuracies (such as nasal secretions representing the filtrate of cerebral ventricle fluid) thus required at least 14 centuries before challenge, scientific disproof, and eventual medical community acceptance.

The scientific method facilitates empirical, rational, and skeptical approaches to observational data, and minimizes human dependence on non-evidence-based traditional models. In spite of the scientific method, physicians are still human, and the medical community still shows an inertial reluctance to adapt to new information when it disrupts traditional paradigms. Recent examples would include reluctance to accept an etiologic role for the *H. pylori* bacterium in peptic ulcer disease, and reluctance to offer less than radical mastectomy for primary breast carcinoma.

## Discovery of the microscopic world

Before the use of lenses to magnify objects, it was physically impossible to make observations on objects smaller than the resolving limit of the human eye (about 0.1−0.25 mm). Thus, prokaryotic and eukaryotic cells, tissue architecture, and comparisons of normal and disease microanatomy were philosophical speculation until description of the mathematics of optics and lens design. In a real sense, optical technology was rate-limiting for the development of the fields of tissue anatomy, cellular biology, and microbiology. Concepts of optics were written as early as 300 BCE in Alexandria (*Optica*, Euclid). Clear glass (crystallo) was developed in Venice in the 15th century. A compound microscope was invented by Janssen in 1590 CE. Microanatomy and structural terminology was begun by Malpighi (1661 CE), who examined capillaries in frog lung, trachea tubes for airflow in silkworms, and stomata in plant leaves. Robert Hooke used a compound microscope to describe common objects in *Micrographia* (1665 CE). Antony Van Leeuwenhoek used self-made simple magnifying lenses to count the threads in cloth in a Dutch dry-goods store, then later published descriptions of bacteria (termed *animalcules*), yeast, and algae, beginning in 1673 CE. Yet the relevance of these observations in microanatomy and microbiology to human diseases required changes in conceptual understanding of the etiology and pathogenesis of diseases. For example, it was 200 years after Van Leeuwenhoek that a *Streptococcus* sp. was recognized by Koch and Pasteur as the likely etiologic agent of puerperal fever in post-partum women.

## Critical developments during the 19th century

### Cellular pathology, germ theory, and infectious etiologic agents

The relevance of microanatomy and microbiology to human disease required expansion of conceptual understanding to include morphologic changes in diseased cells and tissues, as well as recognition of an etiologic role for microorganisms. Rokitansky's gross correlates with clinical disease (*A Manual of Pathological Anatomy*, 1846), Paget's surgical perspective on gross pathology, and Virchow's morphologic correlates with clinical disease were critical to the development of clinico-pathologic correlation, and served to create a role for pathologists to specialize in autopsy and tissue diagnosis. Virchow's description of necrotizing granulomatous inflammation, the morphologic correlate of infections caused by mycobacteria such as TB and leprosy, preceded the discovery of the etiologic agents years later by Hansen (*M. leprae*, 1873) and Koch (*M. tuberculosis*, 1882).

The causal relationship between microorganisms and clinical disease required scientific demonstration and logical proof before medical community acceptance. For example, identification of the cause-and-effect relationship between

Streptococcus sp. and puerperal fever required an initial recognition of unusual clinical outcomes (clusters of post-partum deaths), then correlation of puerperal fever clusters with obstetrician habits, then Semmelweiss's experimental demonstration in 1847 that hand-washing reduced the incidence of puerperal fever, then the demonstration that particular bacteria (*Streptococci*) are regularly associated with the clinical disease (Koch, circa 1870), and finally by culture of the organism from the blood of patients with the disease (Pasteur, 1879).

'Germ theory' articulates this causal relationship between microorganisms and clinical diseases in animals and humans. Technical improvements in microscopes (Abbe condenser, apochromatic lenses, oil immersion lenses), the development of culture media, and the development of histochemical stains no doubt made it possible for Koch to identify *M. tuberculosis* in 1882. To be emphasized is the process of recognition, first of all of the variables associated with the clinical disease, then the scientific demonstration of a causal relationship between one or more of these variables with the clinical disease. This latter step was enunciated as Koch's postulates (1890): (i) the bacteria must be present in every case of the disease, (ii) the bacteria must be isolated from a diseased individual and grown in pure culture, (iii) the specific disease must be reproduced when a pure culture is inoculated into a healthy susceptible host, and (iv) the same bacteria must be recoverable from the experimentally infected host.

Viruses, like bacteria, were understood and manipulated clinically prior to their isolation. Manipulation of viruses as vaccines is traced to description of smallpox variolation in Turkey by at least 1718 (described by Lady Montagu). Smallpox variolation was used in the US continental army in the 1770s. Local availability of cowpox, a non-lethal pox-virus, allowed Jenner to demonstrate (cross-) vaccination against smallpox (1796). Vaccination against smallpox was used in Napolean's army in 1812, and was mandated for Massachusetts schoolchildren by 1855. The success of these vaccination programs prompted development of vaccines against other human viruses through the 19th and 20th centuries, including rabies (Pasteur, 1885), yellow fever virus (Theiler, 1936), influenza A/B (1942), polio virus (Salk 1955, Sabin 1960), rubeola (1962), rubella (1969), hepatitis B virus (1981), varicella/zoster (1981). Transmission electron microscopy (Ruska, Knoll, 1931) allowed visualization of viruses, and large-scale sequencing starting in the 1990s allowed publication of viral genome sequences. Vaccination efforts are one of the great public health success stories of human history, and have reduced (rubella, rubeola, mumps, influenza, polio) or eliminated (smallpox, 1980) several of these viruses worldwide. Rapid identification of viruses in infected individuals is now possible, e.g. influenza virus ID by reverse-transcriptase PCR detection of type-specific RNA, or by immunosassay detection of type-specific proteins.

## Organic chemistry

Prior to 1828, organic (carbon-containing) compounds were thought to derive from living organisms, and it was thought that they could never be synthesized from nonliving (inorganic) material. This concept ('vitalism') was disproven by the *in vitro* synthesis of urea by F. Wohler in 1828. Work from this era initiated the field of organic chemistry. Predictable rules for *in vitro* and *in vivo* organic reactions, structural theory, modeling, separation technologies, and accurate measurement subsequently allowed the chemical description of natural products, and the chemical synthesis of both natural products and synthetic compounds. In addition to setting the stage for a systematic understanding of cellular biochemistry and physiology, organic chemistry set the stage for laboratory synthesis of natural products, such as dyes, vitamins, hormones, proteins, and nucleic acids. In that era, the textile industry was the main consumer of dyes. Access to imported natural product dyes from plants was predictable for seafaring nations, but not for landlocked nations. In parallel with natural product extraction and purification was the recognition that aniline from coal (Perkin, 1856) could be modified to generate a spectrum of colors, catalyzing the development of the German dye industry in the last half of the 19th century. Some of these synthetic dyes were found useful for histochemical staining.

## Histotechnology

Morphologic diagnosis requires thin (3−5 μm), contrast-rich (requiring dyes) sections of chemically fixed (cross-linked or precipitated) tissue. Thin sections allow the passage of light through the tissue, but reduce overlapping of cells in the light path. Thus, technologies had to develop for cutting and staining of thin fixed tissue sections. Work leading to our current technique for tissue solidification in paraffin wax was first described by Klebs in 1869. Prototypes of our current mechanical microtome for making thin (~5 μm thick) tissue sections was developed by Minot in 1885. Precursor work leading to our current technique for tissue fixation with diluted formalin was first described by Blum in 1893.

Histochemical stains developed in parallel with dye technology for the textile industry. Botanists used carmine as a stain in 1849, and subsequently Gerlack applied carmine to stain brain tissue in 1858. The current hematoxylin dye used in tissue histochemistry was originally extracted from logwood trees from Central America for the dye industry (to compete with indigo). Metallic ion mordants made oxidized hematoxylin (hematein) colorfast in textiles, and a protocol for tissue staining was published by Boehmer in 1865. Similar to the hematoxylin story, semisynthetic analine dye technology was adapted by histochemists from 1850 to 1900. Many of these dyes are still routinely used for recognition of tissue structure, peripheral blood cells, and microorganisms, including hematoxylin, eosin, methylene blue, Ziehl-Neelsen, Gram, van Gieson, Mallory trichrome, and Congo red.

The most commonly used stains for general tissue diagnosis are the hematoxylin and eosin (H&E) stains, which provide a wealth of nuclear and cytoplasmic detail not visible in an unstained section. Supplemental histochemical stains demonstrate specific structures and organisms: collagen and muscle (trichome), elastin (Verhoff-von Giessen), glycogen/mucin (periodic acid-Schiff, PAS), mucin (PAS diastase, mucicarmine, alcian blue), fungi (Gemori methenamine silver, GMS), mycobacteria (Ziehl-Neelsen, Fite), and bacteria (Gram, Warthin-Starry). Each of these stains is inexpensive ($10−$50), fast (minutes to hours), and automatable, making them extremely valuable for service diagnostic pathology use.

Light microscopy lens technology matured during the last half of the 19th century. Critical were Abbe's introductions of the apochromatic lens system to eliminate chromatic aberration (different focal lengths for different wavelengths of visible light) in 1868, a novel condenser for compound microscopes (to provide better illumination at high magnification) in 1870, and an oil immersion lens in 1878.

By 1900, maturation of tissue fixation chemistry, histochemical stain protocols, and light microscope technology had evolved into the workhorse technique for evaluation of morphologic abnormalities in tissue examination in anatomic pathology labs, and for the evaluation of morphologic features of microorganisms in microbiology labs. The scope of this chapter is limited to pathology, but it should be clear to the reader that the momentous discoveries of deep general anesthetics (Long, 1841; Morton, 1846), commercial electricity (Edison, 1882), and radiography (Roentgen, 1895) also contributed to the development of the modern medical specialties of diagnostic pathology and laboratory medicine.

## Developments during the 20th century

### Humoral and cellular immunology

The development of antisera in the 20th century for therapeutic purposes (for instance, treatment of diptheria) led to progressive understanding of the antibody, the efferent arm of the humoral immune response. Similarly, tissue transplantation experiments led to the recognition of cellular rejection due to thymus-derived T-cells. Antibodies and T-cells cooperate to react to foreign (non-self) molecules, common examples being allergic responses, viral infections, and organ transplants. Antibodies were found to be made by B-cells and plasma cells, and were found to be exquisitely specific for binding to their particular antigens (ligands) either in solid phase or in solution. The analogous T-cell receptor (TCR) recognizes a ligand made up of a 15−20mer peptide presented by self MHC (HLA in human) molecules on the surface of antigen-presenting cells (macrophages, dendritic cells, activated B-cells). B-cell and T-cells activate and proliferate when exposed to non-self proteins, but not to self proteins, attesting to tolerance to self. When B-cell and T-cell self-tolerance breaks down, autoimmune diseases can result (including myasthenia gravis, Grave's disease, and lupus erythematosus). Antibodies (immunoglobulins) were found to be heterodimers of 50 kD heavy chains and 25 kD light chains, folded together so that a highly variable portion defines the antigen-binding site, and a constant portion defines the isotype (IgM, IgD, IgG, IgE, or IgA). Similarly, T-cell receptors were found to be heterodimers of immunoglobulin-like molecules with a highly variable portion for ligand binding, and a constant portion, but without different isotypes. The range of antigen-binding specificities in a normal mammal is extensive, perhaps infinite, subtracting out only those self proteins to which the animal is tolerant. The genes encoding immunoglobulins and T-cell receptors were sequenced and, surprisingly, the extensive variation of specificities was due to a unique system of gene rearrangements of polymorphic V, D, and J gene segments with random nucleotide addition at the junctions. This system allows generation of literally billions of different Ig and TCR binding specificities. Although the Ig and TCR molecules define the specificity of antigen (ligand) binding, we now understand that the probability of T lymphocyte activation is adjustable, based on activating (e.g. CD28:B7) or inhibitory (e.g. PD-1:PD-L1) receptor:ligand interactions.

Polyclonal antibodies raised in other species (goat, mouse, rabbit) against an antigen can be used to detect the antigen in diffusion gels (Ochterlony, western blot), in solution (ELISA), and in tissue sections. Use of fluorescent-tagged antibodies for frozen section immunohistochemistry was developed first, and immunofluorescence (IF) is still routinely used in renal pathology and dermatopathology. Peroxidase-tagged secondary antibodies and DAB chemistry were developed to generate a stable chromogen in the tissue, and this is now the primary method for detecting antigens in formalin-fixed tissue sections. Improved antibody binding specificity and industrial production required monoclonal antibodies, which in turn required the development of mouse plasmacytomas/myelomas and cell fusion protocols. The net result is that commercial antibodies are now available for antigens of both clinical and research interest, including antibodies specific enough to distinguish minimally modified variant antigens (for instance, phosphorylated proteins or proteins with single amino acid substitutions).

## Natural product chemistry and the rise of clinical laboratories

Diseases due to dietary deficiencies (like scurvy) and hormonal imbalances (like diabetes mellitus) were described clinically long before they were understood pathologically. Dietary deficiency diseases prompted searches for the critical metabolic cofactors, so-called 'vital amines' (vitamins). Xerophthalmia was linked to retinol (vitamin A) deficiency in 1917 (McCollum). Rickets was linked to calcitriol (vitamin D) deficiency in 1926. Beri-beri was linked to a deficiency of thiamine (vitamin B1) in 1926. Scurvy was linked to ascorbic acid (vitamin C) deficiency in 1927. Pellagra in the United States was linked to niacin deficiency in 1937. Pernicious anemia was linked to cobalamin (vitamin B12) deficiency in 1948. It is currently unusual to see morphologic features of these diseases in this country.

Diseases due to non-dietary physiologic imbalances prompted isolation of circulating molecules with systemic effects, i.e. the hormones. Parathyroid hormone was isolated (Berman, Collip) in the 1920s. Thyroxine and cortisone were isolated (Kendall) in 1915, and table salt was iodized starting in 1917. Insulin was isolated (Banting, Best) in 1921, non-human insulin was industrially purified and marketed soon thereafter, and recombinant human insulin was marketed starting in 1982.

These examples highlight the ability of 20th century chemists to fractionate, purify, synthesize, measure bioactivity, and manufacture these compounds for safe use by humans. Study of diseases due to deficiencies and excesses of single molecule function led to a mechanistic understanding of biochemistry and physiology, with resultant interconnected reaction pathways of byzantine complexity (now referred to as systems biology). Clinical demand for body fluid levels of ions (such as sodium, potassium chloride, and bicarbonate), glucose, creatinine, hormones (such as thyroxine and parathyroid hormone), albumin, enzymes (related to liver and cardiac function), and antibodies (reactive to ASO, Rh, ABO, and HLA antigens) led to the development of clinical laboratories in chemistry, endocrinology, immunopathology, and blood banking. Functional assays for coagulation cascade status were developed, as were methods for estimating blood cell concentration and differential, leading to coagulation and hematology laboratories. Serologic and cell activation assays to define HLA haplotype led to HLA laboratories screening donors and recipients in anticipation of bone marrow and solid organ transplants. Culture medium-based screening for infectious agents led to dedicated clinical microbiology laboratories, which are beginning to incorporate nucleic acid screening technologies for speciation and prediction of treatment response. The clinical laboratories now play a critical, specialized role in inpatient and outpatient management, and their high test volumes (a 700-bed hospital may perform 5 million tests per year) have catalyzed computer databases for central record-keeping of results.

## Natural product chemistry: nucleic acids

The previous vignettes indicate a scientific approach to natural products of the steroid and protein types, but do not indicate how proteins are encoded, what accounts for variation in the same protein in the population, or how inherited diseases are inherited. It turns out that the instruction set for protein sequence is defined by DNA sequence. The role of nucleoproteins as a genetic substance was alluded to by Miescher in 1871, and was shown by Avery to be the pneumococcal transforming principle in 1944. The discovery of X-ray crystallography in 1912 made it possible for Franklin, Wilkins, and Gosling to study DNA crystal structure, and led to the description of the antiparallel double helix of DNA by Watson and Crick in 1953. This seminal event in history facilitated dissection of the instruction set for an organism, with recognition that 3-base codons specified amino acids in 1961, and description of the particular codons encoding each amino acid in 1966. Demonstration of *in situ* hybridization in 1969 made it possible to localize specific DNA or RNA sequences within the cells of interest. Recognition and purification of restriction endonucleases and DNA ligases, and the development of cloning vectors, made it possible to clone individual sequences, leading to methods for synthesis of natural products (such as recombinant human insulin in 1978). Chemical methods were developed for sequencing DNA, initially with radioisotope-tagged nucleotide detection in plate gels, then with fluorescent nucleotide detection in 1986. Subsequent conversion to capillary electrophoresis and computer scoring of sequence output allowed high throughput protocols which generated the human genome sequence by 2001. The development of polymerase chain reaction (PCR) chemistry in 1986 has made it possible to quickly screen for length polymorphisms of microsatellite (identity testing, donor:recipient ratio after bone marrow transplant, and microsatellite instability), coding sequence abnormalities in genes (translocations, rearrangements, insertions, deletions, substitutions), and epigenetic control of transcription (promoter methylation) in a targeted fashion. Quantitative PCR methods using fluorescent detection of amplicons has made it possible to study DNA and RNA copy number, and to mimic Northern blots/oligo microarrays in estimating RNA transcript abundance for cluster analysis. Massively parallel solid-state ("next-generation") sequencing chemistries now allow rapid alignment of hundreds of independent sequencing reactions, facilitating rapid comparison of germline and somatic sequence, as well as identification of low-abundance mutations.

## Current practice of pathology

Diseases can be distinguished from each other based on differences at the molecular, cellular, tissue, fluid chemistry, and/or individual organism level. One hundred and sixty years of attention to the morphologic and clinical correlates of diseases has led to sets of diagnostic criteria for the recognized diseases, as well as a reproducible nomenclature for rapid description of the changes associated with newly discovered diseases. Sets of genotypic and phenotypic abnormalities in the patient are used to determine a diagnosis, which then implies a predictable natural history and can be used to optimize therapy by comparison of outcomes among similarly afflicted individuals. The disease diagnosis becomes the management variable in clinical medicine, and management of the clinical manifestations of diseases is the basis for day-to-day activities in clinics and hospitals nationwide. The pathologist is responsible for integration of the data obtained at the clinical, gross, morphologic, and molecular levels, and for issuing a clear and logical statement of diagnosis.

Clinically, diseases present to front-line physicians as patients with sets of signs and symptoms. Symptoms are the patient's complaints of perceived abnormalities. Signs are detected by examination of the patient. The clinical team, including the pathologist, will work up the patient based on the possible causes of the signs and symptoms (the differential diagnosis). Depending on the differential diagnosis, the workup typically involves history-taking, physical examination, radiographic examination, fluid tests (blood, urine, sputum, stool), and possibly tissue biopsy.

Radiographically, abnormalities in abundance, density or chemical microenvironment of tissues allow distinction from surrounding normal tissues. Traditionally, the absorption of electromagnetic waves by tissues led to summation differences in exposure of silver salt photographic film. Tomographic approaches such as computerized tomography (CT, 1972) and nuclear magnetic resonance (NMR, MRI, 1973) complemented summation radiology, allowing finely detailed visualization of internal anatomy in any plane of section. In the same era, ultrasound technology allowed visualization of tissue with density differences, such as a developing fetus or gallbladder stones. More recently, physiology of neoplasms can be screened with positron emission tomography (PET, 1977) for decay of short half-life isotopes such as fluorodeoxyglucose. Neoplasms with high metabolism can be distinguished physiologically from adjacent low-metabolism tissues, and can be localized with respect to normal tissues by pairing PET with standard CT. The result is an astonishingly useful means of identifying and localizing new space-occupying masses, assigning a risk for malignant behavior and, if malignant, screening for metastases in distant sites. This technique is revolutionizing the preoperative decision-making of clinical teams, and improves the likelihood that patients undergo resections of new mass lesions only when at risk for morbidity from malignant behavior or interference with normal function.

Pathologically, disease is diagnosed by determining whether the morphologic features match the set of diagnostic criteria previously described for each disease. Multivolume texts are devoted to the gross and microscopic diagnostic criteria used for diagnosis, prognosis, and prediction of response to therapy. Pathologists diagnose disease by generating a differential diagnosis, then finding the best fit for the clinical presentation, the radiographic appearance, and the pathologic (both clinical lab and morphologic) findings. Logically, the Venn diagram of the clinical, radiologic, and pathologic differential diagnoses should overlap. Unexpected features expand the differential diagnosis and may raise the possibility of previously undescribed diseases. For example, Legionnaire's disease, human immunodeficiency virus (HIV), Hantavirus pneumonia, and severe acute respiratory syndrome (SARS) are examples of newly described diagnoses during the last 40 years. The mental construct of etiology (cause), pathogenesis (progression), natural history (clinical outcome), and response to therapy is the standard approach for pathologists thinking about a disease. A disease may have one or more etiologies (initial causes, including agents, toxins, mutagens, drugs, allergens, trauma, or genetic mutations). A disease is expected to follow a particular series of events in its development (pathogenesis), and to follow a particular clinical course (natural history). Disease can result in a temporary or lasting change in normal function, including patient death. Multiple diseases of different etiologies can affect a single organ, for example, infectious and neoplastic diseases involving the lung. Different diseases can derive from a single etiology, for example, emphysema, chronic bronchitis, and small cell lung carcinoma in long-term smokers. The same disease (for instance, emphysema of the lung) can derive from different etiologies (emphysema from $\alpha$-1-antitrypsin deficiency or cigarette smoke).

Modern surgical pathology practice hinges on morphologic diagnosis, supplemented by special stains, immunohisto-chemical stains, cytogenetic/molecular data, and other clinical laboratory findings, as well as on the clinical and radiographic findings. Findings that meet all of these criteria are diagnostic for the disease. If some, but not all, of the criteria are present to make a definitive diagnosis, the pathologist must either equivocate or make an alternate diagnosis. Thus, a firm grasp of the diagnostic criteria, and the instincts to rapidly create and sort through the differential diagnosis, must be possessed by the service pathologist.

The tissue diagnosis has to make sense, not only from the morphologic perspective, but from the clinical and radiographic vantage points as well. It is both legally risky and professionally erosive to make a clinically and pathologically impossible diagnosis. In the recent past, limited computer networking meant numerous phone calls to gather the relevant clinical and radiographic information to make an informed morphologic diagnosis. For example, certain diseases such as squamous and small cell carcinomas of the lung are extremely rare in non-smokers. Thus, a small cell carcinoma in the lung of a non-smoker merits screening for a non-pulmonary primary site. Fortunately for pathologists, computing and networking technologies now allow access to preoperative clinical workups, radiographs/reports, clinical laboratory data, and prior pathology reports. All of these data protect pathologists by providing them with the relevant clinical and radiographic information, and protect patients by improving diagnostic accuracy. Just as research scientists "*... ignore the literature at their peril ...*", diagnostic pathologists "*... ignore the presentation, past history, workup, prior biopsies, and radiographs at their peril...*"

There are limitations to morphologic diagnosis by H&E stains. First, lineage of certain classes of neoplasms (including small round blue cell tumors, clear cell neoplasms, spindle cell neoplasms, and undifferentiated malignant neoplasms) is usually clarified by immunohistochemistry, frequently by cytogenetics (when performed), and sometimes by electron microscopy. Second, there are limitations inherent in a snapshot biopsy or resection. Thus, the etiology and pathogenesis can be obscure or indeterminate, and rates of growth, invasion, or timing of metastasis cannot be inferred. Third, the morphologic changes may not be specific for the underlying molecular abnormalities, particularly the rate-limiting (therapeutic target) step in the pathogenesis of a neoplasm. For example, *RET* gain of function mutations in a medullary thyroid carcinoma will require DNA level screening to determine germline involvement, familial risk, and presence or absence of a therapeutic target. Fourth, the same morphologic appearance may be identical for two different diseases, each of which would be treated differently. For example, there is no morphologic evidence by H&E stain alone to distinguish host lymphoid response to Hepatitis C viral (HCV) antigens from host lymphoid response to allo-HLA antigens in a liver allograft. This is obviously a major diagnostic challenge when the transplant was done for HCV-related cirrhosis, and when the probability of recurrent HCV infection in the liver allograft is high.

Paraffin section immunohistochemistry has proven invaluable in neoplasm diagnosis for clarifying lineage, improving diagnostic accuracy, and guiding customized therapy. If neoplasms are poorly differentiated or undifferentiated, the lineage of the neoplasm may not be clear. For example, sheets of undifferentiated malignant neoplasm with prominent nucleoli could represent carcinoma, lymphoma, or melanoma. To clarify lineage, a panel of immunostains is performed for proteins that are expressed in some of the neoplasms, but not in others. Relative probabilities are then used to lend support (rule in) or exclude (rule out) particular diagnoses in the differential diagnosis of these several morphologically similar undifferentiated neoplasms. The second role is to make critical distinctions in diagnosis that cannot be accurately made by H&E alone. Examples of this would include demonstration of myoepithelial cell loss in invasive breast carcinoma but not in its mimic, sclerosing adenosis (Fig. 11.2), or demonstration of loss of basal cells in invasive

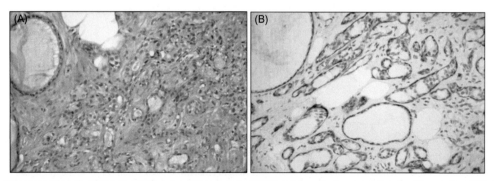

FIGURE 11.2  Sclerosing adenosis of breast. (A) Hematoxylin and eosin (H&E) staining of sclerosing adenosis of breast. By H&E alone, the differential diagnosis includes infiltrating ductal carcinoma and sclerosing adenosis. (B) Actin immunostain of sclerosing adenosis of breast. Actin immunoreactivity around the tubules of interest supports a diagnosis of sclerosing adenosis and serves to exclude infiltrating carcinoma.

prostatic adenocarcinoma (Fig. 11.3). The third role of immunohistochemistry is to identify particular proteins, such as nuclear estrogen receptor (ER) (Fig. 11.4) or the plasma membrane HER2 proteins (Fig. 11.5), both of which can be targeted with inhibitors rather than generalized systemic chemotherapy. Morphology remains the gold standard in this diagnostic process, such that immunohistochemical data support or fail to support the H&E findings, not vice versa.

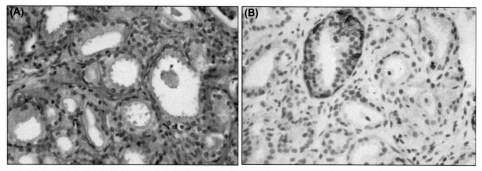

FIGURE 11.3   Invasive adenocarcinoma of prostate. (A) Hematoxylin and eosin (H&E) staining of invasive adenocarcinoma of prostate. By H&E alone, the differential diagnosis includes invasive adenocarcinoma and adenosis. (B) High molecular weight cytokeratin immunostain of invasive adenocarcinoma of prostate. Loss of high molecular weight cytokeratin (34βE12) immunoreactivity around the glands of interest supports a diagnosis of invasive adenocarcinoma.

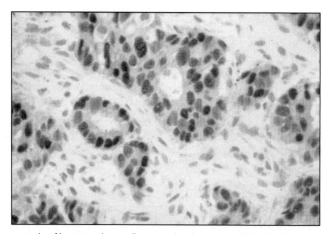

FIGURE 11.4   Estrogen receptor immunostain of breast carcinoma. Strong nuclear immunoreactivity for ER is noted, guiding use of ER inhibitor therapy.

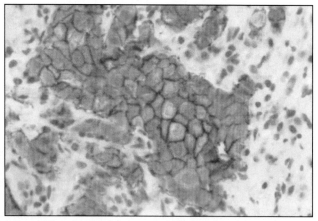

FIGURE 11.5   HER2/c-erbB2 immunostain of breast carcinoma. Strong plasma membrane immunoreactivity for c-erbB2/HER2 is noted, guiding use of either anti-HER2 antibody or HER2 kinase inhibitor therapy.

Probability and statistics are regular considerations in immunohistochemical interpretation, since very few antigens are tissue-specific or lineage-specific. Cytokeratin is positive in carcinomas, but also in synovial and epithelioid sarcomas. This example may imply aspects of the lineage of these two sarcomas that may be helpful in our categorization of these neoplasms. Another example would be the diagnosis of small cell carcinoma in the lung of a nonsmoker. Because lung primary small cell carcinoma is extremely uncommon, in non-smokers, this diagnosis would prompt the pathologist to inquire about screening results for other, non-pulmonary, sites. Likewise, immunohistochemistry results are always put into the context of the morphologic, clinical, and radiographic findings. For example, an undifferentiated CD30+ neoplasm of the testis supports embryonal carcinoma primary in the testis, whereas a lymph node effaced by sclerotic bands with admixed CD30+ Reed-Sternberg cells supports nodular sclerosing Hodgkin's disease.

Demand for both diagnostic accuracy and report promptness has increased as hospitals come under increasing financial pressure to minimize length of patient stay. Hospitals now manage all but the sickest patients as outpatients. Minimally invasive approaches for the acquisition of tissue samples for diagnosis use flexible endoscopic biotomes or hollow needles that sample 1−2 mm diameter tissue specimens. Multidisciplinary conferences function almost real-time with respect to the initial biopsies. Together, these changes have forced modern pathologists to make critical diagnoses on progressively smaller biopsy specimens, sometimes bordering on the amounts seen in cytopathology aspirates, and to do this in a timely fashion. This requires a clear understanding of the limitations to development of an accurate diagnosis, and a willingness on the part of the pathologist to request repeat biopsy for additional tissue when it is necessary for accurate diagnosis.

Diagnostic criteria involving electron microscopic ultrastructure found relevance for the evaluation of neoplasms described as small round blue cell tumors, spindle cell tumors, melanocytic tumors, and neuroendocrine/neuroblastic tumors, as well as delineation of ciliary ultrastructural abnormalities in primary ciliary dyskinesia. Current approaches to these neoplasms are now generally approached using paraffin section immunohistochemistry. Electron microscopy is now currently used mainly for nephropathology, platelet morphology, ciliary axoneme morphology, and for rare cases where immunohistochemistry is not diagnostic and where demonstration of premelanosomes, neuroendocrine granules, or amyloid is diagnostic.

Adequate sampling of a lesion is critical to making an accurate diagnosis. Undercall diagnostic discrepancies are frequently due to sampling of a small portion of a large lesion that is not representative of the most abnormal portion of the lesion. Insufficient sampling can result in an equivocal diagnosis or, worse, an inaccurate diagnosis. Empirical rules have been adopted over the decades to ensure statistically adequate sampling of masses and organs, e.g. transurethral resections of prostate, soft tissue sarcomas, and heart allograft biopsies.

In spite of the limitations and statistical uncertainties relating to morphologic diagnosis, a wealth of information is conveyed to a service pathologist in a tried-and-true H&E section. Analogous to the fact that a plain chest X-ray is the sum total of all densities in the beam path, the morphologic changes in diseased cells and tissues are the morphologic sum total of all of the disequilibria in the abnormal cells. For most neoplastic diseases, morphologic criteria are sufficient to predict the risk of invasion and metastasis (the malignant potential), the pattern of metastases, and the likely clinical outcomes. For example, the etiology and pathogenesis in small cell lung carcinoma can be inferred (cigarette smoking, with carcinogen-induced genetic mutations) and the outcome predicted (early metastasis to regional nodes and distant organs, with high probability of death within 5 years of diagnosis). New molecular data for both neoplastic and non-neoplastic diseases will most likely benefit unaffected individuals by estimating disease risk, and will most likely benefit patients by defining the molecular subset for morphologically defined diagnostic entities, thus guiding individualized therapy.

## The future of diagnostic pathology

Diagnostic pathology will continue to use morphology and complementary data from protein (immunohistochemical) and nucleic acid (cytogenetic, *in situ* hybridization, DNA sequence, epigenetic, and RNA abundance) screening assays. Improvements in current technologies should improve current test performance, reduce test cost, and shorten turnaround time. Development of new technologies will lead to better diagnostic algorithms, improved diagnostic accuracy and prognosis, improved patient assignment to prospective randomized clinical trials, improved prediction of response to customized therapy, and improved sensitivity for measurement of residual subclinical disease.

### Individual identity at the molecular level

For transplant candidates, major histocompatability complex (MHC, HLA in human) screening is evolving from cellular assays and serology toward sequencing of the alleles of the class I and II HLA loci. Rapid sequencing of these alleles in newborn cord blood would allow creation of a database of the population's haplotypes, facilitating perfect matches for required bone marrow or solid organ transplants.

## Rapid cytogenetics

Current uses of *in situ* hybridization to screen for viruses (such as EBV), light chain restriction (in B lymphomas), and copy number variation (for instance, HER2 gene amplification) demonstrate the benefit of *in situ* nucleic acid hybridization assays. Interphase *in situ* hybridization with fluorescent probes (FISH) is now being used in the initial diagnostic workup for certain diseases, e.g. sarcoma-specific translocations, ploidy analysis in hydatidiform moles, and copy number variation analysis for detection of locus amplification and deletion.

## Rapid DNA sequence, RNA sequence, and RNA abundance screening

Current uses of DNA screening for *BCR-ABL* translocation, donor:recipient ratios after allogeneic bone marrow transplant, microsatellite instability, quantitative viral load (e.g. EBV, BK, CMV), single gene mutations (e.g. *CFTR*, *Factor 2*, *SERPINA1* (α-1-antitrypsin), *HFE*), and promoter methylation (e.g. *MLH1*, *MGMT*) demonstrate the benefit of nucleic acid screening in diagnosis and management. It is likely that each new malignant neoplasm will be promptly inventoried for chromosomal ploidy, gene translocations, gene copy number variation, DNA base substitutions/insertions/deletions, annotation of non-synonymous coding sequence and splice-site changes as either benign polymorphisms or pathogenic mutations, local and global promoter methylation status, and RNA expression cluster subset. This inventory at the time of initial diagnosis, prior to therapy, should allow molecular subgrouping, individualized therapy, and targeted residual disease screening. As the cost of genome sequencing approaches the $1000 mark, sequencing of newborns for germline mutations that predispose to subsequent developmental or neoplastic diseases now seems practical for the purposes of early diagnosis, prompt management, and routine follow-up.

## Computer-based prognosis and prediction

Current uses of morphology, immunohistochemistry, and molecular pathology demonstrate their benefit through improved diagnostic accuracy. However, diagnosis, extent of disease, and molecular subsets are currently imperfect estimators of prognosis and response to therapy. Relational databases which correlate an individual's demographic data, family history, and concurrent diseases with the neoplasm's morphologic features, immunophenotype, and molecular subset, and which integrate disease prevalence by age, sex, and ethnicity using Bayesian probabilities, should improve accuracy of prognosis and prediction of response to therapy. As risk correlates are developed, it is possible that healthy individuals can be screened and given risk estimates for development of different diseases, as desired. It is worth noting here that unaffected patients in inherited-disease kindreds such as Huntington's disease may prefer not to know what the future holds with respect to the disease that runs in their extended family, so germline screening should hinge on pre test patient consent.

## Normal ranges and disease risks by ethnic group

Current uses of normal ranges for serum chemistry assume similar bell-curve distributions across ages, sexes, and ethnicity. This may not be true for all analytes, so normal ranges should be measured and compared, with publication of analytes whose normal ranges differ by age, sex, or ethnic group. Computer reference databases facilitate this sort of normal range database, stratified by age/sex/ethnicity of individual patients. Similarly, familial risk for an inherited disease may vary by ethnic group, and this variation may be used in Bayesian calculations to define risk for unaffected at-risk family members.

## Individual metabolic differences relevant to drug metabolism

Current uses of liver and renal impairment to guide drug dosage demonstrate the benefit of using patient physiology to customize therapy. We are now starting to use gene haplotype data for specific genes that encode specific drug-metabolizing enzymes, so as to guide starting doses for particular drugs (e.g. warfarin, tamoxifen, or clopidogrel/Plavix) prior to treatment initiation.

## Serum biomarkers and residual disease testing

Current uses of prostate specific antigen (PSA) to screen for prostate carcinoma and its recurrence demonstrate the benefit of serum biomarkers in common neoplasms. It is likely that high-sensitivity screening for single and clustered serum analytes using proteomic and metabolomic technologies will lead to improved methods for early detection of neoplasms, autoimmune diseases, and infections. Residual disease testing for malignancies with clonal mutations or translocations can now be performed via assays based on circulating tumor cells or circulating cell-free tumor DNA (so-called liquid biopsies). Similar high-sensitivity testing for disease-specific analytes may also improve diagnosis, treatment, and residual disease testing for autoimmune diseases, and infections.

## Treatment by pathway, rather than by tissue diagnosis

Traditional medical oncology chemotherapy strategies for malignant neoplasms hinged on the tissue diagnosis and the organ of origin. This is changing, as current knowledge of signaling pathway disruption by gain-of-function mutations, e.g. EGFR or BRAF, points to specific treatment targets. The pharmaceutical industry is able to create small-molecule inhibitors and humanized antibodies that allow specific treatment of specific mutant enzyme isoforms. Current data suggests that many, if not most, patients with driver gain-of-function mutations treated with specific inhibitors will show an initial brisk response, although later recurrence is possible due to outgrowth of subclones with different driver mutations. The challenge for the next generation of scientists and oncologists is how to rescue homozygous loss-of-function mutations that are etiologic in human diseases, e.g. *CFTR* or *P53*.

## Conclusions

Pathologists consider each disease to have a natural, mechanical, physicochemical basis. Each disease has an etiology (initial cause), a pathogenesis (stepwise progression), and a natural history with effects on normal function (clinical outcome). Pathologists collect the data needed to answer patients' and clinicians' questions, simply phrased as "what is it?" (diagnosis), "how it going to behave?" (prognosis), and "how do I treat it?" (prediction of response to therapy). Instincts and diagnostic criteria, as well as the optical, mechanical, chemical, and computing technologies described previously, are the basis for modern service pathology. As the human genome is deciphered, and as the complex interactions of cellular biochemistry are refined, risk of disease in unaffected individuals will be calculable, disease diagnosis will be increasingly accurate and prognostic, and molecular subsets of morphologically defined disease entities will be used to guide customized therapy for individual patients. It is a great time in history to be a pathologist.

## Key concepts

- Clinically, diseases present to front-line physicians as patients with sets of signs and symptoms. Symptoms are the patient's complaints of perceived abnormalities. Signs are detected by examination of the patient. The clinical team (including the pathologist) evaluate the patient based on the possible causes of the signs and symptoms (the differential diagnosis).
- Pathologically, disease is diagnosed by determining whether the morphologic features match the
- set of diagnostic criteria previously described for each disease. Pathologists diagnose disease by generating a differential diagnosis, then finding the best fit for the clinical presentation, the radiographic appearance, and the pathologic (both clinical lab and morphologic) findings.
- Etiology describes the causes of a disease. One disease entity can have more than one etiology, and a single etiology can lead to more than one disease. For example, emphysema, chronic bronchitis, and small cell lung carcinoma can all occur in long-term smokers (different diseases derived from a single etiology). Likewise, the same disease (for instance, emphysema of the lung) can derive from different etiologies (emphysema from a-1-antitrypsin deficiency or cigarette smoke).
- The pathogenesis of a disease describes its stepwise progression after initiation in response to a specific etiologic factor (or factors). Pathogenesis can refer to the changes in the structure or function of an organism at the gross/clinical level, and it can refer to the stepwise molecular abnormalities leading to changes in cellular and tissue function.
- The natural history of a disease describes the expected course of disease, including chronicity, functional impairment, and survival. However, not all patients with a given disease will naturally follow the same disease course, so differences in patient outcome do not necessarily correspond to incorrect diagnosis. Variables that

independently correlate with clinical outcome differences are called independent prognostic variables, and are assessed routinely in an effort to predict the natural history of the disease in the patient.

- Variables that independently correlate with response to therapy are called independent predictive variables, and are assessed routinely in an effort to optimize therapeutic response for each patient.

## Suggested readings

[1] Porter R. The greatest benefit to mankind: a medical history of humanity. 1st ed. New York, NY: W.W. Norton & Co; 1998.

[2] Marshall BJ, Warren JR. Unidentified curved bacilli in the stomach of patients with gastritis and peptic ulceration. Lancet 1984;1:1311–15.

[3] Jay V. The legacy of Karl Rokitansky. Arch Pathol Lab Med 2000;124:345–6.

[4] Paget J. Lectures on surgical pathology. London: Brown, Green, and Longmans; 1853.

[5] Virchow R. Cellular pathology. Berlin: August Hirschwald; 1858.

[6] Hansen G. Investigations concerning the etiology of leprosy. Norsk Mag Laegervidenskaben 1874;4:1–88.

[7] Koch R. Die Äetiologie der Tuberkulose (The etiology of tuberculosis). Berl Klin Wochenschr 1882;15:221–30.

[8] Turk JL. Rudolf Virchow—father of cellular pathology. J R Soc Med 1993;86(12):688–9.

[9] Holmes O. Contagiousness of puerperal fever. N Engl Q J Med 1843;1:503–30.

[10] Dunn PM. Oliver Wendell Holmes (1809–1894) and his essay on puerperal fever. Arch Dis Child Fetal Neonatal Ed 2007;92:F325–7.

[11] Raju TN. Ignac Semmelweis and the etiology of fetal and neonatal sepsis. J Perinatol 1999;19:307–3103.

[12] Pasteur L. On the germ theory. Science 1881;2:420–2.

[13] Wohler F. Ueber kunstliche bildung des harnstoffs. Ann Phys Chem 1828;88:253–6.

[14] Gal AA. In search of the origins of modern surgical pathology. Adv Anat Pathol 2001;8:1–13.

[15] Medawar PB. The immunology of transplantation. Harvey Lect 1956;144–76.

[16] Early P, Huang H, Davis M, et al. An immunoglobulin heavy chain variable region gene is generated from three segments of DNA: VH, D and JH. Cell 1980;19:981–92.

[17] Leder P, Max EE, Seidman JG, et al. Recombination events that activate, diversify, and delete immunoglobulin genes. Cold Spring Harb Symp Quant Biol 1981;45(Pt 2):859–65.

[18] Kroczek R. Emerging paradigms of T-cell co-stimulation. Curr Opin Immunol 2004;16:321–7.

[19] Coons AH, Kaplan MH. Localization of antigen in tissue cells; improvements in a method for the detection of antigen by means of fluorescent antibody. J Exp Med 1950;91:1–13.

[20] Kohler G, Milstein C. Continuous cultures of fused cells secreting antibody of predefined specificity. Nature 1975;256:495–7.

[21] Wilkins MH, Stokes AR, Wilson HR. Molecular structure of deoxypentose nucleic acids. Nature 1953;171:738–40.

[22] Franklin RE, Gosling RG. Molecular configuration in sodium thymonucleate. Nature 1953;171:740–1.

[23] Watson JD, Crick FH. Molecular structure of nucleic acids; a structure for deoxyribose nucleic acid. Nature 1953;171:737–8.

[24] Crick FH, Barnett L, Brenner S, et al. General nature of the genetic code for proteins. Nature 1961;192:1227–32.

[25] Nirenberg M, Caskey T, Marshall R, et al. The RNA code and protein synthesis. Cold Spring Harb Symp Quant Biol 1966;31:11–24.

[26] Gall JG, Pardue ML. Formation and detection of RNA-DNA hybrid molecules in cytological preparations. Proc Natl Acad Sci USA 1969;63:378–83.

[27] Pardue ML, Gall JG. Molecular hybridization of radioactive DNA to the DNA of cytological preparations. Proc Natl Acad Sci USA 1969;64:600–4.

[28] Goeddel DV, Kleid DG, Bolivar F, et al. Expression in *Escherichia coli* of chemically synthesized genes for human insulin. Proc Natl Acad Sci USA 1979;76(1):106–10.

[29] Sanger F, Nicklen S, Coulson AR. DNA sequencing with chain-terminating inhibitors. Proc Natl Acad Sci USA 1977;74:5463–7.

[30] Maxam AM, Gilbert W. A new method for sequencing DNA. Proc Natl Acad Sci USA 1977;74:560–4.

[31] Smith LM, Sanders JZ, Kaiser RJ, et al. Fluorescence detection in automated DNA sequence analysis. Nature 1986;321:674–9.

[32] Lander ES, Linton LM, Birren B, et al. Initial sequencing and analysis of the human genome. Nature 2001;409:860–921.

[33] Venter JC, Adams MD, Myers EW, et al. The sequence of the human genome. Science 2001;291:1304–51.

[34] Mullis K, Faloona F, Scharf S, et al. Specific enzymatic amplification of DNA in vitro: the polymerase chain reaction. Cold Spring Harb Symp Quant Biol 1986;51(Pt 1):263–73.

[35] Mills S. Sternberg's diagnostic surgical pathology. Philadelphia: Lippincott; 2004.

[36] Fletcher C. Diagnostic histopathology of tumors. Churchill Livingstone; 2007.

[37] Rosai J. The continuing role of morphology in the molecular age. Mod Pathol 2001;14:258–60.

Chapter 12

# Understanding molecular pathogenesis: the biological basis of human disease and implications for improved treatment of human disease

William B. Coleman[1] and Gregory J. Tsongalis[2]

[1]*American Society for Investigative Pathology, Rockville, MD, United States*, [2]*Laboratory for Clinical Genomics and Advanced Technology (CGAT), Department of Pathology and Laboratory Medicine, Dartmouth Hitchcock Medical Center, Lebanon, NH, United States*

**Summary**

Disease has been a feature of the human existence since the beginning of time. Descriptions of diseases and therapeutic interventions were recorded in the earliest written histories of medicine. Over time, our knowledge of science and medicine has expanded and with it our understanding of the biological basis of disease. In this regard, the biological basis of disease implies that more is understood about the disease than merely its clinical description or presentation. In the last several decades, we have moved from causative factors in disease to studies of molecular pathogenesis. Molecular pathogenesis takes into account the molecular alterations that occur in response to environmental insults and other contributing factors, to produce pathology. By developing a deep understanding of molecular pathogenesis, we will uncover the pathways that contribute to disease, either through loss-of-function or through gain-of-function. By understanding the involvement of specific genes, proteins, and pathways, we will be better equipped to develop targeted therapies for specific diseases. Continued expansion of our knowledge base with respect to underlying mechanisms of disease has resulted in unprecedented patient management strategies. Identification of genetic variants in genes once associated with the diagnosis of a disease process is now being reevaluated as it may impact new therapeutic options. In this chapter we describe three disease entities: Hepatitis C virus (HCV) infection, acute myeloid leukemia (AML), and cystic fibrosis (CF). These examples are representative of our increased understanding of molecular pathology and how novel therapeutics are being introduced into clinical practice.

## Introduction

Disease has been a feature of the human existence since the beginning of time. Descriptions of diseases and therapeutic interventions were recorded in the earliest written histories of medicine. Over time, our knowledge of science and medicine has expanded and with it our understanding of the biological basis of disease. In this regard, the biological basis of disease implies that more is understood about the disease than merely its clinical description or presentation. In the last several decades, we have moved from causative factors in disease to studies of molecular pathogenesis. Molecular pathogenesis takes into account the molecular alterations that occur in response to environmental insults and other contributing factors, to produce pathology. By developing a deep understanding of molecular pathogenesis, we will uncover the pathways that contribute to disease, either through loss-of-function or through gain-of-function. By understanding the involvement of specific genes, proteins, and pathways, we will be better equipped to develop targeted therapies for specific diseases. Continued expansion of our knowledge base with respect to underlying mechanisms of disease has resulted in unprecedented patient management strategies. Identification of genetic variants in genes once associated with the diagnosis of a disease process is now being reevaluated as it may impact new therapeutic options.

In this chapter we describe three disease entities: Hepatitis C virus (HCV) infection, acute myeloid leukemia (AML), and cystic fibrosis (CF). These examples are representative of our increased understanding of molecular pathology and how novel therapeutics are being introduced into clinical practice.

Essential Concepts in Molecular Pathology. DOI: https://doi.org/10.1016/B978-0-12-813257-9.00012-7

# Hepatitis C virus infection

HCV infection represents the most common chronic viral infection in North America and Europe, and a common viral infection worldwide. In the United States, the Centers for Disease Control and Prevention has periodically conducted a National Health and Nutrition Examination Survey (https://www.cdc.gov/nchs/nhanes/index.htm). This survey initially estimated that 3.9 million people had detectable antibodies to HCV (for the period 1988−94), indicating a prior exposure to the virus, and 75% of these individuals were positive for HCV RNA, suggesting an active infection. During this same period, 2.7 million individuals were estimated to have chronic HCV infection. For the period of 1999−2002, 3.2 million people in the United States were estimated to have chronic HCV infection (representing 1.3% of the population). Most recently (2003−10), 2.7 million people in the United States were found to have chronic HCV infection (representing 1.0% of the population). Globally, 177.5 million people are infected with HCV (representing 2.5% of the population). Prevalence varies significantly by geographical location, with the highest rates of HCV infection in Africa and Asia, and lower rates of infection in Europe and North America. In Egypt, the prevalence of HCV infection is estimated to be 18%. Likewise, the prevalence of HCV infection in Mongolia is estimated at 10%. In contrast, Canada has a very low prevalence of HCV infection, estimated at <1%. HCV infection has been found to be more common in certain populations, including prison inmates and homeless people, where the prevalence of infection may be as high as 40%.

## Identification of the Hepatitis C virus

HCV was first recognized in 1989 using recombinant technology to create peptides from an infectious serum that were then tested against serum from individuals with non-A, non-B hepatitis. This approach resulted in the isolation of a section of the HCV genome. Subsequently, the entire HCV genome was sequenced. HCV is a member of the family of flaviviridae. Flaviviruses are positive, single-stranded RNA viruses. The HCV genome encodes a gene for production of a single polypeptide chain of approximately 3000 amino acids. This polypeptide gives rise to a number of specific proteins. The Env proteins are among the most variable parts of the peptide chain and are associated with multiple molecular forms in a single infected person. The mutations affecting this portion of the HCV genome (and the encoded Env protein) seem to be critical for escape of the virus from the host immune response. The HCV protein NS5a contains an interferon-response element. Evidence from several studies suggest that mutational variation in the HCV genome encoding this protein are associated with resistance to interferon, the main antiviral agent used in treatment of HCV. Other proteins encoded by the HCV genome include the NS3 region that codes for a protease and the NS5b region that codes for an RNA polymerase. Drugs that target the HCV protease or polymerase are now undergoing trials as therapeutic agents to treat HCV infection.

There are several HCV strains that differ significantly from each other. The nomenclature adopted to describe these HCV strains is based on division of the HCV RNA into three major levels: (1) genotypes, (2) subtypes, and (3) quasispecies. There are six recognized genotypes of HCV that are numbered from 1 to 6. Among these HCV genotypes there is <70% homology in the nucleotide sequence. HCV subtypes typically display 77%−80% homology in nucleotide sequence, while quasispecies have >90% nucleotide sequence homology. Infection of an individual with HCV involves a single genotype and subtype (except in rare instances). However, infected individuals will carry many quasispecies of HCV because these RNA viruses do not contain a proofreading mechanism and acquired mutations in the HCV genome over time are common. HCV genotype 1 is the most prevalent world-wide (49%), followed by genotype 3 (18%), genotype 4 (17%), genotype 2 (11%), and genotypes 5/6 (<5%).

## Risk factors for Hepatitis C virus infection

There are a number of recognized risk factors for HCV. Among the most common risk factors for HCV infection are (1) the use of injectable drugs, and (2) blood transfusion or organ transplant recipient before 1992. A significant percentage of people who used recreational injectable drugs in the 1960s and 1970s became infected with HCV. Less commonly, HCV infection can be transmitted by dialysis, by needle stick injury, through sexual contact, and through vertical transmission from an infected mother to her child. The likelihood of infection from needle stick injury or vertical transmission is estimated to be 3%−5%.

## Hepatitis C infection

The primary target cell type for HCV infection is the mature hepatocyte, although there is some evidence that infection can also occur in other cell types, particularly circulating mononuclear cells. Following the initial HCV infection, there is a latency period of 2−4 weeks before viral replication is detectable. In most cases, there is no clinical evidence of the infection even after viremia develops. In fact, only 10%−30% of individuals with HCV infection will develop the clinical symptomology of acute hepatitis. When acute HCV infection develops, patients display symptoms of fever, loss of appetite, nausea, diarrhea, and specific liver symptoms, including discomfort and tenderness in the right upper abdomen, jaundice, dark urine, and pale-colored stools. Typically, these symptoms occur 2−3 months after the initial HCV infection and then gradually resolve over a period of several weeks. During this time, liver enzymes such as alanine aminotransferase (ALT) and aspartate aminotransferases (AST) are found at elevated levels in the blood, reflecting hepatocyte injury and death. In acute HCV infection, these enzymes are typically increased from 10-fold to 40-fold the upper reference limit of normal. In the majority of individuals infected with HCV, there are no signs or symptoms that accompany the initial infection. In most of these cases, a chronic HCV infection develops, resulting in chronic hepatitis (ongoing inflammation in the liver). In general, chronic HCV infections can be clinically silent for many years without obvious symptomology associated with the infection or liver injury, or produce only mild, nonspecific symptoms such as fatigue, loss of energy, and difficulty performing tasks that require concentration. The major end-stage diseases that result from chronic hepatitis include cirrhosis and hepatocellular carcinoma. It has been estimated that 20%−30% of individuals with chronic HCV infection will progress to cirrhosis after 20 years of infection, although the fibrotic changes in the liver progress at different rates in different individuals. Cirrhosis due to HCV infection has now become the most common indication for liver transplantation in the United States.

## Testing for Hepatitis C virus infection

Most clinical testing for HCV infection begins with detection of antibodies against HCV proteins. The sensitivity of the anti-HCV assay is reported to be in the range of 97%−99% for detecting HCV infection. Most false-negative results of the anti-HCV assay occur in the setting of immunosuppression, such as with human immunodeficiency virus (HIV) infection, or in renal failure. Anti-HCV antibodies are detectable after 10−11 weeks of infection (on average) using the second-generation anti-HCV assays, but the third-generation anti-HCV assays show improved sensitivity with positive detection of anti-HCV antibodies by 7−8 weeks after the initial infection. At the time of clinical presentation with acute HCV infection, >40% of patients lack detectable anti-HCV. In the current clinical laboratory setting, the major method employed to determine the presence of active HCV infection is HCV RNA measurement. With acute HCV infection, HCV RNA becomes detectable 2−4 weeks after infection, and viral loads climb rapidly. Average HCV viral loads are approximately 2−3 million copies per mL. Qualitative assays are designed to determine the presence or absence of HCV RNA, without consideration of actual viral load. Two primary methodologies are used in this type of assay: (1) reverse-transcriptase polymerase chain reaction (RT-PCR) and (2) transcription-mediated amplification (TMA). The detection limit for assays of this type is <50 IU/mL. The approaches employed for qualitative determination of HCV RNA utilize a known amount of a synthetic standard to enable quantitative measurement of HCV RNA through comparison of the amounts of HCV amplified and the amount of standard amplified using a calibration curve. Determination of HCV viral load has become standard of care in evaluating patients before and during treatment for chronic HCV infection. Real-time PCR allows for reduced carryover amplification, more rapid detection of amplification, increased low-end sensitivity, and a wider dynamic range for detection and quantification.

Several techniques have been developed to determine the particular genotype and subtype of HCV causing infection in an individual infected patient. These techniques typically target the 5′-untranslated and/or core regions of the HCV genome, which represent the most highly conserved regions. Because most amplification methods for HCV RNA also target the 5′-untranslated region, qualitative PCR methods can provide amplified RNA for use in determination of HCV genotype. The most widely used technique is a commercial line probe assay. In this assay, a large number of oligonucleotide sequences are immobilized on a membrane, incubated with amplified RNA, and then detected using a colorimetric reagent that detects areas of hybridization. The line probe assay enables recognition and identification of most HCV types and subtypes accurately, although there are several subtypes that cannot be distinguished from one another.

## Clinical course of Hepatitis C virus infection

Although anyone who is infected with HCV will experience an initial infection incident, in most cases this phase of the infection will be clinically silent without obvious symptoms. Acute infection with HCV is most likely to be detected when it occurs following a needle stick exposure from a person with known HCV, or when the infection arises under other circumstances but produces symptomatic infection and jaundice (estimated to occur in less than one-third of all cases). There is some evidence to suggest that patients who develop clinical jaundice are actually more likely to clear the infection and not progress to chronic HCV infection. During the initial incubation period approximately 2 weeks following infection, HCV RNA is either undetectable or can be detected only intermittently. Subsequently, there is a period of rapid increase in the amount of circulating HCV, with an estimated doubling time of <24 h. HCV viral loads reach very high levels during this period of time, typically reaching values of $10^7$ IU/mL and occasionally higher. Evidence of liver injury appears after an additional 1−2 months of HCV infection. This liver injury can be detected secondary to increased serum levels of ALT and AST. Approximately 40%−50% of individuals with acute HCV infection that are clinically diagnosed are detected during this stage of infection, prior to the development of anti-HCV. By 7−8 weeks following infection, anti-HCV becomes detectable using the third-generation immunoassays. However, detection of anti-HCV using the second-generation immunoassay cannot be accomplished until 10−12 weeks following infection. At the time of this seroconversion, the HCV viral loads decrease, sometimes to undetectable levels. In most individuals who will progress to chronic hepatitis (and in some that eventually clear the infection) the HCV viral load remains detectable, but at reduced levels. In a person suspected of having acute HCV infection, the most reliable test for proving exposure is HCV RNA. Because of the high viral loads seen, either qualitative or quantitative assays would be acceptable for this purpose. Detectable HCV RNA in the absence of anti-HCV is a strong evidence of recent HCV infection.

## Treatment of Hepatitis C infection

In contrast to other chronic viral infections such as those associated with hepatitis B virus or HIV, treatment has been successful in eradicating replicating HCV and halting progression of liver damage. Interferon alpha-2 is the agent of choice for treatment of chronic HCV infection. There are currently two potential approaches to treatment of chronic HCV: (1) interferon alone or (2) a combination of interferon plus ribavirin. While ribavirin is ineffective as a single agent for treating HCV, it increases the effectiveness of interferon. Application of ribavirin in combination with interferon increases the number of patients who respond to therapy by two-fold to three-fold. For many years, the only form of interferon available was standard dose interferon. Using standard dose interferon, large doses (typically 3 million units) were delivered to patients infected with HCV several times each week. The short half-life of this interferon produced widely fluctuating interferon levels in these patients, diminishing its therapeutic effectiveness. In 2001, a longer-acting form of interferon was approved for use in treating HCV infection. The longer-acting interferon was modified by attachment of polyethylene glycol (pegylated interferon), which resulted in increased half-life for the administered drug. Use of pegylated interferon results in sustained high levels of interferon in the patient, reducing the number of required administrations to a single injection each week. There was also an improvement in response rates among patients treated with the pegylated interferon. Currently, the preferred treatment for chronic HCV infection is the combination of pegylated interferon plus ribavirin.

Recent developments in the treatment of HCV infection employ direct acting agents against targets that emerge from the HCV life cycle. For example, the NS3/4 A protease represents a major target for antiviral intervention because loss of NS3/4 A impairs the HCV life cycle by inhibiting maturation of the viral polyprotein. Likewise, replication of HCV genetic material is a target for antiviral drugs that function to inhibit NS5B (including both nucleotide analogues and nonnucleoside inhibitors of NS5B). Through the use of direct-acting agents, 90% of HCV infections can be cured, although concerns related to development of resistance remain.

## Guided treatment of Hepatitis C virus

The appropriate duration of treatment for HCV infection varies depending on the HCV strain (genotype) that infects the patient. HCV genotypes 2 and 3 respond much better to standard treatment regimens. Thus, only 24 weeks of therapy are needed to achieve maximum benefit, compared to 48 weeks in persons infected with other HCV genotypes. In current clinical practice, treatment is offered to all patients with HCV infection except those with decompensated cirrhosis, where treatment may lead to worsening of the patient's condition. Once treatment is initiated, the most reliable means to determine efficacy is to evaluate the response by measuring HCV RNA. Successful treatment is associated with at least two different phases of viral clearance. The first phase, which occurs rapidly over the course of days, is thought to reflect HCV RNA clearance from a circulating pool through the antiviral effect of interferon. In the second phase of clearance, infected liver cells (the major site of viral replication) undergo cell turnover and are replaced by uninfected cells. The second phase of clearance is more variable in duration. First-phase clearance is less specific for detecting success of antiviral treatment; therefore, it is necessary to evaluate whether second-phase clearance has occurred.

## Summary

HCV infection represents a relatively recently identified infectious agent that has a varied natural history from patient to patient. Intensive research efforts have characterized the phases of HCV infection and the clinical symptomology of acute and chronic HCV infection. Through improved understanding of the biology of the HCV virus and its life cycle in the infected host, effective and sensitive diagnostic tests have been developed. Unlike some other chronic viral infections, HCV infection can be effectively treated using interferon in combination with ribavirin. However, it is now recognized that effective therapy of the patient depends on knowing the genotype of the HCV causing the infection. With continued advances in the understanding of the pathogenesis of HCV infection, new treatments and/or new modes of administration of known anti-HCV drugs will emerge that provide effective control of the viral infection with minimal adverse effects for the patient.

## Acute myeloid leukemia

The human leukemias have been classified as a distinct group of clinically and biologically heterogeneous disorders that are a result of genetic abnormalities that affect specific chromosomes and genes. In the United States, 62,130 new cases of leukemia were diagnosed in 2017, including 21,400 cases of AML (34% of all leukemias). In 2017, AML was associated with 10,600 cancer-related deaths. AML is characterized by accumulation of neoplastic immature myeloid cells, consisting of ~30% myeloblasts in the blood or bone marrow and classified on the basis of their morphological and immunocytochemical features. AML can arise (1) *de novo*, (2) in a setting of a preexisting myelodysplasia, or (3) secondary to chemotherapy for another disorder.

### Chromosomal abnormalities in acute myelogenous leukemia

Various cytogenetic and/or molecular abnormalities have been associated with various types of AML. The World Health Organization recognizes several AML subtypes based upon their cytogenetic abnormalities. Chromosomal translocations are the most common form of genetic abnormality identified in acute leukemias. Typically, these translocations involve genes that encode proteins that function in transcription and differentiation pathways. As a result of chromosomal translocation, the genes proximal to the chromosome breakpoints are disrupted, and the 5′-segment of one gene is joined to the 3′-end of a second gene to form a novel fusion (chimeric) gene. When the chimeric gene is expressed, a novel protein product is produced from the chimeric mRNA. Other genetic alterations such as point mutations, gene amplifications, and numerical gains or losses of chromosomes can also be identified in the acute leukemias. The clinical heterogeneity seen in AML may be due in part to differences in the number and nature of genetic abnormalities that occur in these cancers. However, these same molecular differences define various prognostic and therapeutic characteristics associated with the specific disorder in a given patient.

A major chromosomal translocation in AML involves chromosomes 15 and 17. This genetic abnormality, t(15;17) (q21;q21), occurs exclusively in acute promyelocytic leukemia (APL). APL represents 5%−13% of all *de novo* AMLs. The presence of the t(15;17) translocation consistently predicts responsiveness to a specific treatment utilizing all-*trans*-retinoic acid (ATRA). Retinoic acid is a ligand for the retinoic acid receptor (RAR), which is involved in the t(15;17). ATRA is thought to overcome the block in myeloid cell maturation, allowing the neoplastic cells to mature (differentiate) and be eliminated. Approximately 75% of patients with APL present with a bleeding diathesis, usually the result of one or more processes including disseminated intravascular coagulation, increased fibrinolysis, and thrombocytopenia, and secondary to the release of pro-coagulants or tissue plasminogen activator from the granules of neoplastic promyelocytes. This bleeding diathesis may be exacerbated by standard cytoreductive chemotherapy. Two morphologic variants of APL have been described, typical (hypergranular) and microgranular, both of which carry the t(15;17) translocation. In the typical or hypergranular variant, the promyelocytes have numerous azurophilic cytoplasmic granules that often obscure the border between the cell nucleus and the cytoplasm. Cells with numerous Auer rods in bundles are common. In the microgranular type the promyelocytes contain numerous small cytoplasmic granules that are difficult to discern with the light microscope but are easily seen by electron microscopy.

### Consequence of the t(15;17) translocation in acute myelogenous leukemia

The t(15;17) is a balanced and reciprocal translocation in which the *PML* (for promyelocytic leukemia) gene on chromosome 15 and the *RARα* gene on chromosome 17 are disrupted and fused to form a hybrid gene. The *PML-RARα* fusion gene, located on chromosome 15, encodes a chimeric mRNA and a novel protein. On the derivative chromosome 15, both the *PML* and *RARα* genes are oriented in a head-to-tail orientation. The function of the normal *PML* gene is poorly understood. However, the gene is ubiquitously expressed and encodes a protein that contains a dimerization domain and is characterized by an N-terminal region with two zinc finger—like motifs (known as a ring and a B-box). Given its structural features, the PML protein is thought to be involved in DNA binding. Furthermore, the normal PML protein appears to have an essential role in cell proliferation. The *RARα* gene encodes a transcription factor that binds to DNA sequences in *cis*-acting retinoic acid-responsive elements. High-affinity DNA binding also requires heterodimerization with another family of proteins, the retinoic acid X receptors. The RARα protein contains transactivation, DNA binding, heterodimerization, and ligand-binding domains. The normal RARα protein plays an important role in myeloid differentiation.

There are three major forms of the *PML-RARα* fusion gene, corresponding to different breakpoints in the *PML* gene. The breakpoint in the *RARα* gene occurs in the same general location in all cases, involving the sequences within intron 2. Approximately 40%—50% of cases have a *PML* breakpoint in exon 6 (the so-called long form, termed *bcr1*), 40%—50% of cases have the *PML* breakpoint in exon 3 (the so-called short form, termed *bcr3*), and 5%—10% of cases have a breakpoint in *PML* exon 6 that is variable (the so-called variable form, termed *bcr2*). In each form of the translocation, the PML-RARα fusion protein retains the 5'-DNA binding and dimerization domains of PML and the 3'-DNA binding, heterodimerization, and ligand (retinoic acid)-binding domains of RARα. Recent studies indicate that the different forms of *PML-RARα* fusion mRNA correlate with clinical presentation or prognosis. In particular, the *bcr3* type of *PML-RARα* correlates with higher leukocyte counts at time of presentation. Both higher leukocyte counts and variant morphology are adverse prognostic findings, and the *bcr3* type of *PML-RARα* does not independently predict poorer disease-free survival.

## Detection of the t(15;17) translocation in acute myelogenous leukemia

A number of methods may be used to detect the t(15;17) translocation. Conventional cytogenetic methods detect the t(15;17) in 80%—90% of APL cases at time of initial diagnosis. Suboptimal clinical specimens and poor-quality metaphases explain a large subset of the negative results. Fluorescence *in situ* hybridization (FISH) is another useful method for detecting the t(15;17) in APL. Different methods employ probes specific for either chromosome 15 or chromosome 17 (or both), and commercial kits are available. Southern blot hybridization is another method to detect gene rearrangements that result from the t(15;17). The chromosomal breakpoints consistently involve the second intron of the *RARα* gene, and therefore, probes derived from this region are the most often utilized. Virtually all cases of APL can be detected by Southern blot analysis using two or three genomic *RARα* probes. RT-PCR is a very convenient method for detecting the *PML-RARα* fusion transcripts. Primers have been designed to amplify the potential transcripts, and each type of transcript can be recognized. Results using this method are equivalent to or better than other methods at time of initial diagnosis.

Polyclonal and monoclonal antibodies reactive with the PML and RARα proteins have been generated, and immunohistochemical studies to assess the pattern of staining appear to be useful for diagnosis. PML or RARα immunostaining correlates with the presence of the t(15;17). APL cells immunostaining for either PML or RARα reveals a microgranular pattern. The fusion protein may prevent PML from forming normal oncogenic domains, since treatment with ATRA allows PML reorganization into these domains. For the diagnosis of residual disease or early relapse after therapy, conventional cytogenetic studies, Southern blot analysis, and immunohistochemical methods are limited by low sensitivity. Quantitative RT-PCR and FISH methods are very useful. The sensitivity and rapid turnaround time of RT-PCR makes this method very useful for monitoring residual disease after therapy.

## Summary

Acute promyelocytic leukemia is a distinct subtype of acute myeloid leukemia that is cytogenetically characterized by a balanced reciprocal translocation between chromosomes 15 and 17 [t(15;17)(q21;q21)], which results in a gene fusion involving *PML* and *RARα*. This disease is the most malignant form of acute leukemia with a severe bleeding tendency and a fatal course of only weeks in affected individuals. In the past, cytotoxic chemotherapy was the primary modality for treatment of APL, producing complete remission rates of 75%—80% in newly diagnosed patients, a median duration of remission from 11 to 25 months, and only 35%—45% of the patients were cured. However, with the introduction of ATRA in the treatment and optimization of the ATRA-based regimens, the complete remission rate increased to 90%—95% and 5-year disease-free survival improved to 74%.

# Cystic fibrosis

CF is a clinically heterogeneous disease that exemplifies the many challenges of complex genetic diseases and the causative underlying mechanisms. CF is the most common lethal autosomal recessive disease in individuals of European decent with a prevalence of 1:2500 to 1:3300 live births. While CF occurs most commonly in the Caucasian population, members of other racial and ethnic backgrounds are also at risk for this disease. The highest prevalence of CF occurs in Irish populations where this disease occurs in 1:1400 live births. In contrast, CF occurs rarely in Hispanics (1:4000 to 1:10,000 live births) and African-Americans (1:15,000 to 1:20,000), with even lower incidence rates in individuals of Asian descent. In the United States, 850 individuals are newly diagnosed on an annual basis, and 30,000 children and adults are affected by the disease. The majority of CF diagnoses are made in individuals who are less than 1 year of age (http://www.genetests.org/).

## Cystic fibrosis transmembrane conductance regulator gene

The Cystic Fibrosis Transmembrane Conductance Regulator (*CFTR*) gene is responsible for CF. This gene is large, spanning 230 kb on chromosome 7q, and consists of 27 coding exons. The *CFTR* mRNA is 6.5 kb and encodes a CFTR membrane glycoprotein of 1480 amino acids with a mass of $\sim 170,000$ Da. CFTR functions as a cAMP-regulated chloride channel in the apical membrane of epithelial cells. To date over 1000 unique mutations in the CFTR gene have been described (Cystic Fibrosis Mutation Data Base, http://www.genet.sickkids.on.ca/cftr/). The most common *CFTR* mutation is the deletion of phenylalanine at position 508 ($\Delta$F508). This mutation affects 70% of patients worldwide. The allelic frequency of *CFTR* mutations varies by ethnic group. For example, the $\Delta$F508 *CFTR* mutation is only present in 30% of the affected Ashkenazi Jewish population.

## Diagnosis of cystic fibrosis

A diagnosis of CF in a symptomatic or at-risk patient is suggested by clinical presentation and confirmed by a sweat test. In the presence of clinical symptoms (such as recurrent respiratory infections), a sweat chloride above 60 mmol/L is diagnostic for CF. Although the results of this test are valid in a newborn as young as 24 h, collecting a sufficient sweat sample from a baby younger than 3 or 4 weeks old is difficult. The sweat test can also confirm a diagnosis of CF in older children and adults, but is not useful for carrier detection. Mutations in the *CFTR* gene are grouped into six classes, including (i) Class I, characterized by defective protein synthesis where there is no CFTR protein at the apical membrane; (ii) Class II, characterized by abnormal/defective processing and trafficking where there is no CFTR protein at the apical membrane; (iii) Class III, characterized by defective regulation where there is a normal amount of nonfunctional CFTR at the apical membrane; (iv) Class IV, characterized by decreased conductance where there is a normal amount of CFTR with some residual function at the apical membrane; (v) Class V, characterized by reduced or defective synthesis/trafficking where there is a decreased amount of functional CFTR at the apical membrane; and (vi) Class VI, characterized by decreased stability where there is a functional but unstable CFTR at the apical membrane. Of the CFTR mutations, classes I–III are the most common and are associated with pancreatic insufficiency. The $\Delta$F508 CFTR mutation (which is most common worldwide) represents a class II mutation, with varying frequency between ethnic groups.

## Abnormal function of CFTR in cystic fibrosis

CFTR is a member of an ATP-binding cassette family with diverse functions such as ATP-dependent transmembrane pumping of large molecules, regulation of other membrane transporters, and ion conductance. Mutations in the *CFTR* gene can lead to an abnormal protein with loss or compromised function resulting in defective electrolyte transport and faulty chloride ion transport in apical membrane epithelial cells affecting the respiratory tract, pancreas, intestine, male genital tract, hepatobiliary system, and the exocrine system, resulting in complex multisystem disease. The loss of CFTR-mediated anion conductance explains a variety of CF symptoms including elevated sweat chloride, due to a defect in salt absorption by the sweat ducts, and meconium ileus, a defect in fluid secretion by intestinal crypt cells. The malfunction of CFTR as a regulator of amiloride-sensitive epithelia $Na^+$ channel leads to increased $Na^+$ conductance in CF airways, which drives increased absorption of $Cl^-$ and water. Most of the symptoms associated with CF, such as meconium ileus, loss of pancreatic function, degeneration of the vas deferens, thickened cervical mucus, and failure of adrenergically mediated sweating are due to the role CFTR plays in Cl-driven fluid secretion.

CFTR is an anion channel that functions in the regulation of ion transport. It plays multiple roles in fluid and electrolyte transport, including salt absorption, fluid absorption, and anion-mediated fluid secretion. Defects in this protein lead to CF, the morbidity of which is initiated by a breach in host defenses and propagated by an inability to clear the resultant infections. Since inflammatory exacerbations precipitate irreversible lung damage, the innate immune system plays an important role in the pathogenesis of CF. Respiratory epithelial cells containing the CFTR also provide a crucial environmental interface for a variety of inhaled insults. The local mucosal mechanism of defense involves mucociliary clearance that relies on the presence and constituents of airway surface liquid (ASL). The high salt in the ASL found in CF patients interferes with the natural antibiotics present in ASL such as defensins and lysozyme. The role of CFTR in the pathogenesis of CF-related lung disease can be categorized by dividing patients into two groups. The first describes defects in CFTR that result in altered salt and water concentrations of airway secretions. This then affects host defenses and creates a milieu for infection. The second is associated with CFTR deficiency that results in biologically and intrinsically abnormal respiratory epithelia. These abnormal epithelial cells fail as a mechanical barrier and enhance the presence of pathogenic bacteria by providing receptors and binding sites or failing to produce functional antimicrobials.

Much debate exists regarding the relative biologic activity of antibacterial peptides such as beta-defensins and cathelicidins in human ASL and their role in the pathogenesis in CF-related lung disease. It is possible that the innate immune system provides a first line of host defense against microbial colonization by secreting defensins, small cationic antimicrobial peptides produced by epithelia. The innate antibiotics are thought to possess salt-sensitive bacteriocidal capabilities. Hence, these innate antibiotics demonstrate altered (impaired) function in the lungs of CF patients. Mannose-binding lectin represents another antimicrobial molecule that is present in ASL and is thought to be inactivated by high salt concentrations in the lungs of CF patients. Mannose-binding lectin, an acute phase serum protein produced in the liver, opsonizes bacteria and activates complement. Common variations in the mannose-binding lectin gene (MBL2) are associated with increased disease severity, increased risk of infection with *B. cepacia*, poor prognosis, and early death. The understanding that such naturally occurring peptide antibiotics exist has resulted in the pharmacologic development of these peptides for therapeutics.

## Pathophysiology of cystic fibrosis

The occurrence of CF leads to clinical, gross, and histologic changes in various organ systems expressing abnormal CFTR, including the pancreas, respiratory, hepatobiliary, intestinal, and reproductive systems. In addition, pathologic changes have been observed in organ systems that do not express the *CFTR* gene (such as the rheumatologic and vascular systems). The current age of individuals affected with CF ranges from 0 to 74 years, and the predicted survival age for a newly diagnosed child is 33.4 years. The increasing age of survival of CF has led to increased manifestation of pulmonary and extrapulmonary disorders (gastrointestinal, hepatobiliary, vascular, and musculoskeletal) associated with the disease. The extent and severity of disease tends to correlate with the degree of CFTR function. Although all these organ systems are affected, the pulmonary changes are the most pronounced and the major cause of mortality in most cases.

Lung infection remains the leading cause of morbidity and mortality in CF patients. It is currently recognized that CF-related lung disease is the consequence of chronic pulmonary consolidation by the well-known opportunistic pathogens *Pseudomonas aeruginosa* (mucoid and nonmucoid), *Burkholderia cepacia*, *Staphylococcus aureus*, and *Haemophilus influenzae*. Morbidity and mortality due to persistent lung infection despite therapeutic advances focus attention toward the expanding microbiology of pulmonary colonizers. These increasingly prevalent flora include *Burkholderia cepacia* complex (genomovar I-IX), methicillin-resistant *Staphylococcus aureus*, *Stenotrophomonas maltophilia*, *Achromobacter xylosoxidans*, *Mycobacterium abscessus*, *Mycobacterium-avium* complex, *Ralstonia* species, and *Pandoraea* species. Inflammatory exacerbation precipitates progressive irreversible lung damage, of which bronchiectases are the landmark changes. Bronchial mucous plugging facilitates colonization by microorganisms. Repetitive infections lead to bronchiolitis and bronchiectasis. Other pulmonary changes include interstitial fibrosis and bronchial squamous metaplasia. Often, subpleural bronchiectatic cavities develop and communicate with the subpleural space with resultant spontaneous secondary pneumothorax, the incidence of which increases later in life.

Exocrine pancreas insufficiency is present in the majority of patients with CF. This clinically manifests by failure to thrive and fatty bulky stools owing to deficiency of pancreatic enzymes. However, pancreatic lesions vary greatly in severity, and the pancreas may be histologically normal in some patients who die in infancy. Early in the postnatal development of the pancreas, patients with CF have a deficiency of normal acinar development. Increased secretory material within the ducts and increased duct volume also contribute to progressive degradation and atrophy of pancreatic acini. These factors result in duct obstruction and progressive pancreatic pathology. Exocrine pancreatic disease appears to develop as a result of deficient ductal fluid secretion due to decreased anion secretion. Coupled to normal protein load derived from acinar cell secretion, this then leads to pancreatic protein hyperconcentration within the pancreatic ducts. The protein hyperconcentration increases susceptibility to precipitation and finally obstruction of the duct lumina. Hence, the characteristic lesion is cystic ductal dilation, atrophy of pancreatic acini, and severe parenchymal fibrosis.

The manifestation of CF in the hepatobiliary system is directly related to CFTR expression. The liver disease in CF is considered inherited liver disease due to impaired secretory function of the biliary epithelium. While defective CFTR may be expressed, males are more likely to be affected than females and the risk for developing liver disease is between 4% and 17% as assessed by yearly exams and biochemical testing. CFTR is expressed in epithelial cells of the biliary tract. Therefore, any or all cells of the biliary tree may be affected. While a variety of liver manifestations exist, including fatty infiltration (steatosis), common bile duct stenosis, sclerosing cholangitis, and gallbladder disease, the rare but characteristic liver lesion in CF is focal biliary cirrhosis, which develops in a minority of patients and is usually seen in older children and adults. With the increasing life expectancy in patients with CF, liver-related deaths have increased and may become one of the major causes of death in CF. The associated liver disease usually develops before or at puberty, is slowly progressive, and is frequently asymptomatic. There is negligible effect on nutritional status or severity of pulmonary involvement. Only a minority of patients go on to develop a clinically problematic liver disease with rapid progression. Abnormal bile composition and reduced bile flow ultimately lead to intrahepatic bile duct obstruction and focal biliary cirrhosis. Diagnosis of CF-associated liver disease is based on clinical exam findings, biochemical tests, and imaging techniques. Although liver biopsy is the gold standard for the diagnosis of most chronic liver diseases, only rarely is it employed in the diagnostic workup, mainly due to sampling error.

The gastrointestinal manifestations of cystic fibrosis are seen mainly in the neonatal period and include meconium ileus, distal intestinal obstruction syndrome (DIOS), fibrosing colonopathy, strictures, gastroesophageal reflux, rectal prolapse, and constipation in later childhood. Throughout the intestines CFTR is the determinant of chloride concentration and secondary water loss into the intestinal lumen. Decreased water content results in viscous intestinal contents, with a 10%−15% risk of developing meconium ileus in babies born with cystic fibrosis. This also accounts for DIOS and constipation in older children. DIOS (formerly meconium ileus equivalent) is a recurrent partial or complete obstruction of the intestine in patients with CF and pancreatic insufficiency.

Arthritis is a rare but recognized complication of cystic fibrosis that generally occurs in the second decade. Three types of joint disease are described in patients with cystic fibrosis: (1) cystic fibrosis arthritis or episodic arthritis, (2) hypertrophic pulmonary osteoarthropathy, and (3) co-existent or treatment-related arthritis. The most common form, episodic arthropathy, is characterized by episodic, self-limited polyarticular arthritis with no evidence of progression to joint damage. Histologic features are minimal with prominent blood vessels and interstitial edema occurring most commonly, or rarely lymphocytic inflammation.

Infertility is an inevitable consequence of cystic fibrosis in males occurring in >95% of patients and is due to congenital bilateral absence or atrophy of the vasa deferentia (CBAVD) and/or dilated or absent seminal vesicles. Spermatogenesis and potency remain normal. Mutations in the *CFTR* gene are present in up to 70% of the patients with CBAVD. Diagnosis of obstructive azoospermia may be diagnosed by semen analysis; however, it must be confirmed by testicular biopsy and no other reason for azoospermia. Fertility in females may be impaired due to dehydrated cervical mucus, but their reproductive function is normal. Advances in techniques such as microscopic epididymal sperm aspiration (MESA) and intracytoplasmic sperm injection have allowed males with cystic fibrosis the ability to reproduce.

## Summary

CF is a complex multi-organ system disease that results from mutation in the CFTR gene. Advances in the understanding of the pathogenesis of this disease and related complications (such as recurrent lung infection) have led to improvement in diagnosis and treatment of affected individuals, resulting in improved life expectancy. With continued expansion of our understanding of the molecular pathogenesis of this disease and the variant manifestations of CF-related disorders, it is expected that new treatments will emerge that attempt to counteract or correct the pathologic consequences of *CFTR* mutation.

# Key concepts

- Molecular pathogenesis describes the molecular alterations that occur in response to environmental insults, exogenous exposures, genetic predispositions, and other contributing factors, to produce pathology. By developing a complete understanding of molecular pathogenesis, the pathways that contribute to disease through loss of function or gain of function will be elucidated. A greater understanding of the involvement of specific genes, proteins, and pathways in the causation of specific diseases will facilitate the development of targeted therapies for particular diseases.

- The natural history of HCV infection varies from patient to patient. Intensive research efforts have characterized the phases of HCV infection and the clinical symptomatology of acute and chronic HCV infection. Through improved understanding of the biology of the HCV virus and its life cycle in the infected host, effective and sensitive diagnostic tests have been developed. Unlike some other chronic viral infections, HCV infection can be effectively treated using interferon in combination with ribavirin. However, effective therapeutic treatment of the individual patient requires knowledge of the genotype of the HCV associated with the infection. With continued advances in the understanding of the pathogenesis of HCV infection, new treatments and/or new modes of administration of known anti-HCV drugs will emerge that provide effective control of the viral infection with minimal adverse effects for the patient.

- Acute promyelocytic leukemia (APL) is a distinct subtype of acute myeloid leukemia that is cytogenetically characterized by a balanced reciprocal translocation between chromosomes 15 and 17 [t(15;17)(q21;q21)], which results in a gene fusion involving PML and RARa. This disease is associated with a severe bleeding tendency and a fatal course of only weeks in affected individuals. Cytotoxic chemotherapy was once the primary modality for APL treatment, producing complete remission rates of 75–80% in newly diagnosed patients, a median duration of remission from 11 to 25 months, but only 35–45% of the patients were cured. However, with the introduction of all-trans retinoic acid (ATRA), the complete remission rate increased to 90–95% and five-year disease free survival improved to 74%.

- Cystic fibrosis is a complex multiorgan system disease that results from mutation in the CFTR gene. Advances in the understanding of the pathogenesis of this disease and related complications (such as recurrent lung infection) have led to improvement in diagnosis and treatment of affected individuals, resulting in improved life expectancy. With continued expansion of our understanding of the molecular pathogenesis of this disease and the variant manifestations of CF-related disorders, it is expected that new treatments will emerge that attempt to counteract or correct the pathologic consequences of CFTR mutation.

- Understanding of the molecular pathogenesis of disease creates opportunities for the development of new molecular diagnostics and targeted treatments that combine to improve the available modalities for treating affected individuals.

## Suggested readings

[1] Alter MJ, Kruszon-Moran D, Nainan OV, et al. The prevalence of hepatitis C virus infection in the United States, 1988 through 1994. N Engl J Med 1999;341:556–62.

[2] Petruzziello A, Marigliano S, Loquercio G, et al. Global epidemiology of hepatitis C virus infection: an up-date of the distribution and circulation of hepatitis C virus genotypes. World J Gastroenterol 2016;22:7824–40.

[3] Bostan N, Mahmood T. An overview about hepatitis C: a devastating virus. Crit Rev Microbiol 2010;36:91–133.

[4] Kim WR. The burden of hepatitis C in the United States. Hepatology 2002;36:S30–4.

[5] Mengshol JA, Golden-Mason L, Rosen HR. Mechanisms of disease: HCV-induced liver injury. Nat Clin Pract Gastroenterol Hepatol 2007;4:622–34.

[6] Pawlotsky JM. Hepatitis C virus resistance to direct-acting antiviral drugs in interferon-free regimens. Gastroenterology 2016;151:70–86.

[7] Scheel TK, Rice CM. Understanding the hepatitis C virus life cycle paves the way for highly effective therapies. Nat Med 2013;19:837–49.

[8] Manns MP, Cornberg M. Sofosbuvir: the final nail in the coffin for hepatitis C? Lancet Infect Dis 2013;13:378–9.

[9] Manns MP, Buti M, Gane E, et al. Hepatitis C virus infection. Nat Rev Dis Prim 2017;3:17006.

[10] Siegel RL, Miller KD, Jemal A. Cancer statistics, 2017. CA Cancer J Clin 2017;67:7–30.

[11] Swerdlow SHN, Jaffe E, Pileri S, et al. WHO classification of tumours of the haematopoietic and lymphoid tissues. 4th ed. Lyon, France: IARC; 2008.

[12] Emandi A, Karp JE. The clinically relevant pharmacogenomics changes in acute myelogenous leukemia. Pharmacogenomics 2012;13:1257–69.

[13] Warrell Jr. RP, de The H, Wang ZY, et al. Acute promyelocytic leukemia. N Engl J Med 1993;329:177–89.

[14] Wang ZY, Chen Z. Acute promyelocytic leukemia: from highly fatal to highly curable. Blood 2008;111:2505–15.

[15] Brennan AL, Geddes DM. Cystic fibrosis. Curr Opin Infect Dis 2002;15:175–82.

[16] O'Sullivan BP, Freedman SD. Cystic fibrosis. Lancet 2009;373:1891–904.

[17] Farrell PM. The prevalence of cystic fibrosis in the European Union. J Cyst Fibros 2008;7:450–3.

[18] Ratjen F, Bell SC, Rowe SM, et al. Cystic fibrosis. Nat Rev Dis Prim 2015;1:15010.

[19] Ratjen F, Doring G. Cystic fibrosis. Lancet 2003;361:681–9.

[20] Curry MP, Hegarty JE. The gallbladder and biliary tract in cystic fibrosis. Curr Gastroenterol Rep 2005;7:147–53.

Chapter 13

# Integration of molecular and cellular pathogenesis - a bioinformatics approach

**Jason H. Moore[1] and Michael D. Feldman[2]**

[1]*Division of Informatics, Department of Biostatistics and Epidemiology, Institute for Biomedical Informatics, The Perelman School of Medicine, University of Pennsylvania, Philadelphia, PA, United States,* [2]*Professor of Pathology, The Perelman School of Medicine, University of Pennsylvania, Philadelphia, PA, United States*

**Summary**

As the field of molecular pathology moves forward in the genomics age with big data and challenging analytical questions about disease processes, bioinformatics is playing an increasingly important role. The interdisciplinary field of bioinformatics blends computer science and biostatistics with biomedical sciences such as epidemiology, genetics, genomics, and proteomics. In combination, these approaches facilitate the management, analysis, and interpretation of data from biological experiments and observational studies. The goal of this chapter is to introduce some of the important concepts in bioinformatics that must be considered when planning and executing a modern molecular pathology study. This article reviews database resources as well as data mining software tools.

## Introduction

Historically, the touchstone of diagnostic pathology has been the histologic identification of diseased and normal tissues using conventional stains such as hematoxylin and eosin (H&E). Of course, the microscopic appearances of tissues, both diseased and normal, are the result of differential expression of approximately 20,000 protein-coding genes in the human genome. However, until the development of immunohistochemical techniques in the 1960s no information on protein expression was available. The advent of immunohistochemistry revolutionized both diagnostic pathology as well as the scientific study of disease. Today, we are on the verge of a similar revolution, which is bringing with it understanding of the roles of DNA level changes in molecular pathology.

Differential expression of specific proteins is, of course, due in large part to changes at the DNA level, including changes in transcription and mRNA splicing, and alterations in other aspects of the complex regulation of RNA metabolism and translation. With the sequencing of the human genome and recent advances in polymerase chain reaction (PCR)-based and array-based technologies, quantitation of specific RNA levels, simultaneous detection of tens or hundreds of thousands of different genetic transcripts, and DNA-based technologies for looking at genetic and chromosomal changes are becoming available. With these changes, bioinformatics is playing an increasingly important role in the modern molecular pathology research strategy.

The tremendous wealth of information now available on the molecular changes associated with normal physiology and disease processes is posing a considerable challenge to the field of molecular pathology. In the practice of conventional H&E pathology, coordinate expression of thousands of genes result in the creation of distinctive microscopic appearances, which are recognized by trained pathologists. Thus, physical appearances serve as the basis for diagnostic pathology and the scientific study of disease. With the advent of immunohistochemistry these subtle and complex images were supplemented with information about the expression of a few specific proteins, one for each stain. Now, as DNA-based and RNA-based technologies are being introduced into the general practice of pathology, two things are happening. First, information about specific genetic alterations or changes in specific RNA levels is being added to the information available about diseased tissue. The wealth of new insights provided by these techniques rivals that which became available when immunohistochemistry was introduced. Conceptually, the integration of this information into the mainstream of pathology is straightforward and similar to what has been done in the past. However, unlike the

*Essential Concepts in Molecular Pathology.* DOI: https://doi.org/10.1016/B978-0-12-813257-9.00013-9

previous technologies, nucleic acid—based technologies have been incorporated into arrays, which provide thousands or hundreds of thousands of individual pieces of information—genotype information, information about DNA methylation, data about gene copy number, or RNA expression levels. We are now squarely in the era of next-generation sequencing that provides a more complete picture of the entire genome, its regulation, and its expression. The analysis of this avalanche of big data, which can be generated from a single biopsy or autopsy specimen, requires bioinformatics techniques to glean meaningful information.

For example, consider the information currently available from a biopsy of a high-grade glioma such as a glioblastoma (GBM). The diagnosis is made today, as it has been for decades, based on the H&E appearance of the lesion—on the presence of endothelial proliferation and tumor necrosis in what is histologically a malignant glial lesion with the nuclear morphology typical of an astrocytic tumor. For many generations that comprised all the pathologist could tell the clinician about the tumor, and it was all the clinician needed to know to select a therapy and treat the patient. It had been recognized since the 1940s that these tumors fell into two distinct but overlapping clinical categories: (1) primary GBMs that tended to be found in older patients with a shorter clinical course, which appeared to arise as GBMs and (2) secondary GBMs, which were found in slightly younger patients and arose by progression from lower grade astrocytic tumors. This distinction was of no therapeutic importance, and there were no pathologic correlates that could be used to distinguish a primary from a secondary GBM. The use of immunohistochemical techniques did not change the situation for this particular tumor. Stains for glial fibrillary acidic protein could be used to demonstrate the astrocytic character of the lesion and markers for cycling cells. For example, the Ki-67 antigen could be used to demonstrate the relatively high proliferative index of the tumor, but it rarely provided new information that was clinically important. In the mid-1990s, studies using nucleic acid—based technologies demonstrated that primary and secondary GBMs have distinctive clinical histories, and, despite being histologically identical, tumors were completely different with distinctive molecular signatures, including frequent epidermal growth factor receptor (EGFR) amplification and mutation in the primary tumors, and p53 gene and isocitrate dehydrogenase 1 mutations in the secondary GBMs. At the time, the distinction provided limited clinical significance, but it helped establish the idea that despite a pathologist's inability to separate the GBMs into subtypes based on H&E histology, they were a molecularly heterogeneous group of tumors. Today, specific molecularly targeted therapies directed, for example, against tumors overexpressing EGFR are entering clinical trials. A new WHO classification of brain tumors is being presented in 2016 and integrates the histopathology with genetic markers to create a more comprehensive and genomically-aligned nomenclature.

Although these single marker candidate gene studies of GBM did a great deal to elucidate the molecular pathogenesis of these tumors and define molecularly distinct subsets of GBM, it was clear to most pathologists that the separation of GBM into primary and secondary tumors did not adequately capture the complexity of the situation. Within the last several years, studies using RNA expression arrays and classification based on unsupervised clustering of thousands of mRNA levels have suggested that GBMs may be best thought of as being of three types: those with (1) proneural, (2) proliferative, or (3) mesenchymal molecular signatures. Other studies based on high-resolution copy number analysis using oligonucleotide-based array comparative genomic hybridization have also identified three subsets of GBM, one that seems to correspond to the classically defined primary GBM and two others that represent secondary GBMs. The integration of these RNA and DNA studies into a single unified classification system remains to be done, but recent studies are contributing useful information. For example, epigenetic silencing of the gene for the DNA-repair enzyme MGMT has been shown to influence the response of these tumors to conventional therapies. Stratification of GBM patients based on MGMT promoter methylation status is now part of NCCN guidelines for GBM adjuvant therapy. It is also becoming standard clinical practice to recognize a subset of tumors, which are histologically indistinguishable from other GBMs, but which, like oligodendrogliomas, have loss of the short arm of chromosome 1 and/or the long arm of chromosome 19 and have a better prognosis than the usual GBM. These studies are now being refined using next-generation sequencing that looks at DNA sequence variation on a genome-wide scale.

These examples demonstrate the potential for molecular biotechnology to significantly impact our ability to use molecular pathology to understand disease processes. However, our ability to exploit these new technological resources will depend critically on our ability to make sense out of mountains of data collected for a set of pathology samples. The remainder of this chapter will introduce bioinformatics and the resources that are available to pathologists for making full use of genetics, genomics, and proteomics.

## Overview of bioinformatics

Bioinformatics is an interdisciplinary field that blends computer science and biostatistics with biomedical sciences such as epidemiology, genetics, genomics, and proteomics. Emerging as an important discipline shortly after the development of high-throughput DNA sequencing technologies in the 1970s, it was the momentum of the Human Genome Project that spurred the rapid rise of bioinformatics as a formal discipline. The word *bioinformatics* did not start appearing in the biomedical literature until around 1990, but it quickly caught on as the descriptor of this important new field. An important goal of bioinformatics is to facilitate the management, analysis, and interpretation of data from biological experiments and observational studies. Thus, much of bioinformatics can be categorized as database development and implementation, data analysis and mining, and biological interpretation and inference. The goal of this chapter is to review each of these three areas and provide some guidance on getting started with a bioinformatics approach to molecular pathology investigations of disease susceptibility.

The need to interpret information from whole-genome sequencing projects in the context of biological information acquired in decades of research studies prompted the establishment of the National Center for Biotechnology Information (NCBI) as a division of the National Library of Medicine (NLM) at the National Institutes of Health (NIH) in the United States in November 1988. When NCBI was established, it was charged with (1) creating automated systems for storing and analyzing knowledge about molecular biology, biochemistry, and genetics, (2) performing research into advanced methods of computer-based information processing for analyzing the structure and function of biologically important molecules and compounds, (3) facilitating the use of databases and software by biotechnology researchers and medical care personnel, and (4) coordinating efforts to gather biotechnology information worldwide. Since 1988, the NCBI has fulfilled many of these goals and has delivered a set of databases and computational tools that are essential for modern biomedical research in a wide range of different disciplines including molecular epidemiology. The NCBI and other international efforts, such as the European Bioinformatics Institute (EBI) that was established in 1992, have played very important roles in inspiring and motivating the establishment of research groups and centers around the world that are dedicated to providing bioinformatics tools and expertise. Some of these tools and resources will be reviewed here.

## Database resources

One of the most important pre-study activities is the design and development of one or more databases that can accept, store, and manage molecular pathology data. There are eight steps for establishing an information management system for genetic studies. These are broadly applicable to many different kinds of studies. The first step is to develop the experimental plan for the clinical, demographic, sample, and molecular/laboratory information that will be collected. What are the specific needs for the database? The second step is to establish the information flow. That is, how does the information find its way from the clinic or laboratory to the database? The third step is to create a model for information storage. How are the data related? The fourth step is to determine the hardware and software requirements. How much data needs to be stored? How quickly will investigators need to access the data? What operating system will be used? Will a freely available database such as mySQL (http://www.mysql.com) serve the needs of the project, or will a commercial dataset solution such as Oracle (http://www.oracle.com) be needed? The fifth step is to implement the database. The important consideration here is to define the database structure so that data integrity is maintained. The sixth step is to choose the user interface to the database. Is a web page portal to the data sufficient? The seventh step is to determine the security requirements. Do HIPAA regulations (http://www.hhs.gov/ocr/hipaa) need to be followed? Most databases need to be password protected at a minimum. The eighth and final step outlined is to select the software tools that will interface with the data for summary and analysis. Some of these tools will be reviewed below.

Although most investigators choose to develop and manage their own database for security and confidentiality reasons, there are an increasing number of public databases for depositing data and making it widely available to other investigators. The tradition of making data publicly available soon after it has been analyzed and published can largely be attributed to the community of investigators using gene expression microarrays. Microarrays represent one of the most revolutionary applications derived from knowledge of whole genome sequences. Extensive use of this technology has led to the need to store and search expression data for all the genes in the genome acquired in different genetic backgrounds or in different environmental conditions. This resulted in a number of public databases such as the now retired Stanford Microarray Database, the Gene Expression Omnibus (GEO) (http://www.ncbi.nlm.nih.gov/geo), ArrayExpress (http://www.ebi.ac.uk/arrayexpress), and others. ArrayExpress, for example, contains gene expression data for more than 65,000 experiments totaling more than 42 terabytes of data. Both GEO and ArrayExpress accept gene expression data derived from high-throughput sequencing studies. The nearly universal acceptance of the data sharing culture in this area has yielded a number of useful tools that might not have been developed otherwise. The need to define standards for the ontology and annotation of microarray experiments led to proposals such as the Minimum Information about a Microarray Experiment (MIAME), which provided a standard that greatly facilitates the storage, retrieval, and sharing of data from microarray experiments. MIAME standards provided an example for other types of data such as single nucleotide polymorphisms (SNPs) and protein mass spectrometry spectra. See Brazma et al. for a comprehensive review of standards for data sharing in genetics, genomics, proteomics, and systems biology. The success of these different databases depends on the availability of methods for easily depositing data and tools for searching the databases often after data normalization.

Despite that acceptance of data sharing in the genomics community, the same culture does not yet exist in molecular pathology. One of the few such examples is the Pharmacogenomics Knowledge Base of PharmGKB (http://www.pharmgkb.org). PharmGKB was established with funding from the NIH to store, manage, and make available molecular data in addition to phenotype data from pharmacogenetic and pharmacogenomic experiments and clinical studies. It is anticipated that similar databases for molecular epidemiology will appear and gain acceptance over the next few years as the NIH and various journals start to require data from public research be made available to the public.

In addition to the need for a database to store and manage molecular pathology data collected from experimental or observational studies, there a number of database resources that can be very helpful for planning a study. A good starting point for database resources are those maintained at the NCBI (http://www.ncbi.nlm.nih.gov). Perhaps the most useful resource when planning a molecular pathology study is the Online Mendelian Inheritance in Man (OMIM) database (http://www.ncbi.nlm.nih.gov/omim). OMIM is a catalog of human genes and genetic disorders with detailed summaries of the literature. The NCBI also maintains the PubMed literature database with more than 25 million indexed abstracts from published papers in more than 4700 life science journals. The PubMed Central database (http://www.ncbi.nlm.nih.gov/pmc) is quickly becoming an indispensable tool with more than 3 million full text papers from numerous journal indexed in PubMed.

Rapid and free access to the complete text of published papers significantly enhances the planning, execution, and interpretation phases of any scientific study. The new Books database (http://www.ncbi.nlm.nih.gov/books) provides free access for the first time to electronic versions of many textbooks and other resources such as the NCBI Handbook that serves as a guide to the resources that NCBI has to offer. This is a particularly important resource for students and investigators that need to learn a new discipline such as genomics. One of the oldest databases provided by the NCBI is the GenBank DNA sequence resource (http://www.ncbi.nlm.nih.gov/Genbank). DNA sequence data for many different organisms has been deposited in GenBank for more than two decades now totaling more than 100 million sequences. GenBank is a common starting point for the design of PCR primers and other molecular assays that require specific knowledge of gene sequences. For example, curated information about genes, their chromosomal location, their function, their pathways, etc., can be accessed through the Gene database (http://www.ncbi.nlm.nih.gov/gene).

Important emerging databases include those that store and summarize DNA sequence variations and functional genomic elements. NCBI maintains the dbSNP (http://www.ncbi.nlm.nih.gov/projects/SNP/) database for single-nucleotide polymorphisms or SNPs. dbSNP provides a wide range of different information about SNPs including the flanking sequence primers, the position, the validation methods, and the frequency of the alleles in different populations. As with all NCBI databases it is possible to link to a number of other datasets such as PubMed and OMIM as well as ClinVar that aggregates genomic variation information with known relationships with clinical endpoints. Another useful database is the Encyclopedia of DNA Elements (ENCODE) (http://www.encodeproject.org) that catalogs genomic sequences that have a biological function other than coding for proteins (e.g., regulatory sequences). These studies are carried out in different cell lines allowing the results to be used for studies of the human tissues the cells are derived from.

In addition to databases for storing raw data, there are a number of databases that retrieve and store knowledge in an accessible form. For example, the Kyoto Encyclopedia of Genes and Genomes (KEGG) database stores knowledge on genes and their pathways (http://www.genome.jp/kegg). The Pathway component of KEGG currently stores knowledge on more than 480 pathway maps. While the Pathway component documents molecular interaction in pathways, the Brite database stores knowledge on higher-order biological functions. One of the most useful knowledge sources is the Gene Ontology (GO) project that has created a controlled vocabulary to describe genes and gene products in any organism in terms of their biological processes, cellular components, and molecular functions (http://www.geneontology.org). GO descriptions and KEGG pathways are both captured and summarized in the NCBI databases. For example, the description of p53 in Gene includes KEGG pathways such as cell cycle and apoptosis. It also includes GO descriptions such as protein binding and cell proliferation.

In general, a good place to start for information about available databases is the annual Database issue and the annual Web Server issue of the journal *Nucleic Acids Research*. These special issues include annual reports from many of the commonly used databases.

# Data analysis

Once data are collected and stored in a database, an important goal of molecular pathology is to identify biomarkers or molecular/environmental predictors of disease endpoints. Statistical methods in bioinformatics provide a good starting point for the analysis of molecular pathology data. For example, this may include commonly used methods such as *t*-tests, analysis of variance, linear regression, and logistic regression, or it may include more advanced data mining and machine learning methods such as cluster analysis or neural networks. Although many of these methods require special training in mathematics, statistics, or computer science, the good news is that most simple and advanced analysis methods are easily implemented in one or more freely available software packages. We briefly review several of these below.

## Data mining using R

R is perhaps the one software package that all investigators should have in their bioinformatics arsenal. R is an open-source freely available programming language and data analysis and visualization environment that can be downloaded from http://www.r-project.org. According to the webpage, R includes (1) an effective data handling and storage facility, (2) a suite of operators for calculations on arrays, in particular matrices, (3) a large, coherent, integrated collection of inter-mediate tools for data analysis, (4) graphical facilities for data analysis and display either on-screen or on hardcopy, and (5) a well-developed, simple and effective programming language that includes conditionals, loops, user-defined recursive functions, and input and output facilities. A major strength of R is the enormous community of developers and users who ensure just about any analysis method you need is available. Perhaps the most useful contribution to R is the Bioconductor project (http://www.bioconductor.org). According to the Bioconductor web page, the goals of the project are to (1) provide access to a wide range of powerful statistical and graphical methods for the analysis of genomic data; (2) facilitate the integration of biological metadata (e.g., PubMed, GO) in the analysis of experimental data; (3) allow the rapid development of extensible, scalable, and interoperable software; (4) promote high-quality and reproducible research; and (5) provide training in computational and statistical methods for the analysis of genomic data.

There are numerous packages for machine learning and data mining that are either part of the base R software or can be easily added. For example, the neural package includes routines for neural network analysis (http://cran.r-project.org/web/packages/nnet). Others include *arules* for association rule mining (http://cran.r-project.org/web/packages/arules), *cluster* for cluster analysis (http://cran.r-project.org/web/packages/cluster), *genalg* for genetic algorithms (http://cran.r-project.org/web/packages/genalg), *som* for self-organizing maps (http://cran.r-project.org/web/packages/som), and *tree* for classification and regression trees (http://cran.r-project.org/web/packages/tree). Many others are available. A full list of contributed packages for R can be found at http://cran.r-project.org/web/packages/.

The primary advantage of using R as your data mining software package is its power. However, the learning curve can be challenging at first. Fortunately, there is plenty of documentation available on the web and in published books.

## Data mining using Weka

One of the most mature open-source and freely available data mining software packages is Weka (http://www.cs.waika-to.ac.nz/ml/weka). Weka is written in Java and thus will run in any operating system (e.g., Linux, Mac, Sun, Windows). Weka contains a comprehensive list of tools and methods for data processing, unsupervised and supervised classification, regression, clustering, association rule mining, and data visualization. Machine learning methods include classification trees, k-means cluster analysis, k-nearest neighbors, logistic regression, naïve Bayes, neural networks, self-organizing maps, and support vector machines, for example. Weka includes a number of additional tools such as search algorithms and analysis tools including cross-validation and bootstrapping. A nice feature of Weka is that it can be run from the command line, making it possible to run the software from Perl or even R (see http://cran.r-project.org/web/packages/RWeka). Weka includes an experimenter module that facilitates comparison of algorithms. It also includes a knowledge flow environment for visual layout of an analysis pipeline. This is a very powerful analysis package that is relatively easy to use.

## Data mining using Orange

Orange is another open-source and freely available data mining software package (http://orange.biolab.si) that provides a number of data processing, data mining, and data visualization tools. What makes Orange different and in some way preferable to other packages, such as R, is its intuitive visual programming interface. With Orange, methods and tools are represented as icons that are selected and dropped into a window called the canvas. For example, an icon for loading a dataset can be selected along with an icon for visualizing the data table. The file load icon is then *wired* to the data table icon by drawing a line between them. Double-clicking on the file load icon allows the user to select a data file. Once loaded, the *wire* automatically transfers the data file to the data table icon. Double-clicking on the data table icon brings up a visual display of the data. Similarly, a classifier such as a classification tree can be selected and wired to the file icon. Double-clicking on the classification tree icon allows the user to select the settings for the analysis. Wiring the tree viewer icon then allows the user to view a graphical image of the classification tree inferred from the data. Orange facilitates high-level data mining with minimal knowledge of computer programming. A wide range of data analysis tools is available. A strength of Orange is its visualization tools for multivariate data. Recent additions to Orange include tools for microarray analysis and genomics such as heat maps and GO analysis.

### Interpreting data mining results

Perhaps the greatest challenge of any statistical analysis or data mining exercise is interpreting the results. How does a high-dimensional statistical pattern derived from population-level data relate to biological processes that occur at the cellular level? This is an important question that is difficult to answer without a close working relationship between pathologists, for example, and statisticians and computer scientists. Fortunately, there are a number of emerging software packages that are designed with this in mind. GenePattern (http://www.broad.mit.edu/cancer/software/genepattern/ ), for example, provides an integrated set of analysis tools and knowledge sources that facilitate this process.

## The future of bioinformatics

We have only scratched the surface of the numerous bioinformatics methods, databases, and software tools that are available to the pathology community. We have tried to highlight some of the important software resources such as Weka and Orange, which might not be covered in other reviews that focus on more traditional methods from biostatistics. While there are an enormous number of bioinformatics resources today, the software landscape is changing rapidly as new technologies for high-throughput biology emerge. Over the next few years we will witness an explosion of novel bioinformatics tools for that analysis of whole-genome sequence data and, more importantly, the joint analysis of sequence variations data with other types of data such as gene expression and proteomics data. Each of these new data types and their associated research questions will require special bioinformatics tools and perhaps special hardware such as faster computers with bigger storage capacity and more memory. Some of these datasets will easily require more than a terabyte or more of memory or more for analysis and could require thousands of processors or more to complete a data mining analysis in a reasonable amount of time. The challenge will be to scale our bioinformatics tools and hardware such that a genome-wide dataset can be processed as efficiently as we can process a candidate gene dataset today. Only then can molecular pathology truly arrive in the genomics age.

## Key concepts

- An important challenge for the practice of pathology is to integrate the tremendous wealth of information now available on the molecular changes associated with normal physiology and disease processes.
- Bioinformatics is an interdisciplinary field that blends computer science and biostatistics with biomedical sciences such as epidemiology, genetics, genomics, and proteomics.
- One of the most important pre-study activities is the design and development of one or more databases that can accept, store, and manage molecular pathology data.
- Once the data are collected and stored in the database, an important goal of molecular pathology is to use biostatistics and data mining to identify biomarkers or molecular/environmental predictors of disease end points.
- The challenge for the practice of pathology in the future will be to scale bioinformatics software and hardware such that information from the entire genome or proteome, for example, can be harnessed in an efficient manner.

## Suggested readings

[1]   Ashburner M, Ball CA, Blake JA, et al. Gene ontology: tool for the unification of biology. Gene Ontol Consortium Nat Genet 2000;25:25−9.

[2]   Barrett T, Suzek TO, Troup DB, et al. NCBI GEO: mining millions of expression profiles--database and tools. Nucleic Acids Res 2005;33: D562−6.

[3]    Brazma A, Hingamp P, Quackenbush J, et al. Minimum information about a microarray experiment (MIAME)-toward standards for microarray data. Nat Genet 2001;29:365−71.

[4]    Benson D, Boguski M, Lipman DJ, Ostell J. The National Center for Biotechnology Information. Genomics 1990;6:389−91.

[5]    Benson D, Lipman DJ, Ostell J. GenBank. Nucleic Acids Res 1993;21:2963−5.

[6]    Boguski MS. Bioinformatics. Curr Opin Genet Dev 1994;4:383−8.

[7]    Ceccarelli M, Barthel FP, Malta TM, et al. Molecular profiling reveals biologically discrete subsets and pathways of progression in diffuse glioma. Cell 2016;164:550−63.

[8]    ENCODE Project Consortium. The ENCODE (encyclopedia of DNA elements) project. Science 2004;306:636−40.

[9]    Ewens WJ, Grant GR. Statistical methods in bioinformatics. Springer; 2001.

[10]   Gene Ontology Consortium. The gene ontology (GO) project in 2006. Nucleic Acids Res 2006;34:D322−6.

[11]   Gentleman R, Carey VJ, Huber W, Iri RA, Dudoit S. Bioinformatics and computational biology solutions using R and bioconductor. New York, NY: Springer; 2005.

[12]   Hamosh A, Scott AF, Amberger J, Valle D, McKusick VA. Online Mendelian inheritance in man (OMIM). Hum Mutat 2000;15:57−61.

[13]   Hastie T, Tibshirani R, Friedman J. The elements of statistical learning. New York, NY: Springer; 2001.

[14]   Kanehisa M, Sato Y, Kawashima M, Furumichi M, Tanabe M. KEGG as a reference resource for gene and protein annotation. Nucleic Acids Res 2016;44:D457−62.

[15]   Landrum MJ, Lee JM, Riley GR, et al. ClinVar: public archive of relationships among sequence variation and human phenotype. Nucleic Acids Res 2014;42:D980−5.

[16]   Maher EA, Brennan C, Wen PY, et al. Marked genomic differences characterize primary and secondary glioblastoma subtypes and identify two distinct molecular and clinical secondary glioblastoma entities. Cancer Res 2006;66:11502−13.

[17]   Moore JH, Williams SW. Traversing the conceptual divide between biological and statistical epistasis: systems biology and a more modern synthesis. BioEssays 2005;27:637−46.

[18]   Ogata H, Goto S, Sato K, et al. KEGG: kyoto encyclopedia of genes and genomes. Nucleic Acids Res 1999;27:29−34.

[19]   Ohgaki H, Kleihues P. Genetic pathways to primary and secondary glioblastoma. Am J Pathol 2007;170:1445−53.

[20]   Phillips HS, Kharbanda S, Chen R, et al. Molecular subclasses of high-grade glioma predict prognosis, delineate a pattern of disease progression, and resemble stages in neurogenesis. Cancer Cell 2006;9:157−73.

[21]   Reimers M, Carey VJ. Bioconductor: an open source framework for bioinformatics and computational biology. Methods Enzymol 2006;411:119−34.

[22]   Rosner B. Fundamentals of biostatistics. Duxbury; 2000.

[23]   Schena M, Shalon D, Davis RW, Brown PO. Quantitative monitoring of gene expression patterns with a complementary DNA microarray. Science 1995;270:467−70.

[24]   Wheeler DL, Barrett T, Benson DA, et al. Database resources of the National Center for Biotechnology Information. Nucleic Acids Res 2006;34:D173−80.

[25]   Whitten IH, Frank E. Data mining. Boston, MA: Elsevier; 2005.

Chapter 14

# Molecular basis of cardiovascular disease

Avrum I. Gotlieb

*Department of Laboratory Medicine and Pathobiology, Faculty of Medicine, University of Toronto, Laboratory Medicine Program, University Health Network, Toronto, ON, Canada*

**Summary**

A new era of innovative research is transforming the study of cardiovascular pathobiology from static histopathology research to dynamic mechanistic cell and molecular biology investigation. The development of reliable cell culture methods to harvest, maintain, and study the cells of the cardiovascular system now allows for the investigation of dynamic structure and function relationships at the cellular and molecular levels. Genetic studies to identify monogenic and polygenic effects on disease and to understand the regulation of gene function through DNA and epigenetic analysis offer insights into diagnosis, prevention and treatment of cardiovascular disease. Stem/precursor/pluripotential populations of cells have been identified and isolated from both humans and experimental animal models that are both renewable and differentiate along cardiovascular lineages to regulate regeneration and repair. Bioactive molecules are being identified as diagnostic biomarkers and therapeutic targets. Studies of microRNAs have shown them to be important regulators of many processes in the pathogenesis of cardiovascular disease where they also may facilitate communication between cells and function as biomarkers and therapeutic targets. New knowledge on pathogenesis is being gained from both in vivo and in vitro molecular studies especially where genes are modified and gene expression is manipulated. These advances bring new hope to be able to prevent and/or diagnose cardiovascular diseases early and to repair injured cardiovascular tissue and restore normal function by using precision medicine to target pathogenic molecules and pathways and to utilize cell therapy and tissue engineering to regenerate cardiovascular tissue.

## The cells of cardiovascular organs

### Vascular endothelial cells

Vascular ECs, embryologically derived from splanchnopleuric mesoderm, form a thromboresistant selective permeability barrier at the blood-vessel wall interface and are metabolically active altering their function as their microenvironment both in the blood stream and/or the vessel wall change. Normal ECs show paracrine communication with underlying SMCs and secrete extracellular vesicles to communicate with other vascular cells such as myeloid cells (Table 14.1). ECs participate actively in thrombosis, hemostasis and inflammation by providing surfaces for molecular interactions and by expressing coagulation and platelet factors, cytokines, chemokines, and leukocyte adhesion molecules. Regulation of normal physiologic functions maintains homeostasis of the endothelium while the presence of endothelial dysfunction is associated with pathobiology and disease (Table 14.2).

ECs are quiescent but have the ability to migrate and proliferate once appropriate genes are activated in response to endothelial injury and/or vascular disease in order to reestablish a normal endothelium. ECs do undergo senescence and lose their ability to proliferate and regenerate in association with systemic aging.

A major role of ECs is to transduce hemodynamic shear stress from a physical force to a biochemical signal that regulates signaling pathways, gene expression, and/or protein secretion of bioactive agents. Normal laminar flow creates a set point for gene expression to regulate normal EC function however oscillating shear stress, alteration in flow or development of turbulence induce changes in the expression of these genes. These shear stress—activated molecules include vasoactive compounds, extracellular matrix proteins and degradation enzymes, growth factors, coagulation and anti-coagulation factors and pro and anti-inflammatory factors (Table 14.3). The process of mechanotransduction involves the cell coat, the cell membrane, ion channels, the cell cytoskeleton and associated integrins, cell junctions and focal adhesion plaques. ECs also show epigenetic plasticity in response to biomechanical stress.

*Essential Concepts in Molecular Pathology.* DOI: https://doi.org/10.1016/B978-0-12-813257-9.00014-0

**TABLE 14.1 The cells of the cardiovascular system.**

Cardiac myocytes
Cardiac interstitial fibroblasts
Valve interstitial cells (VICs)
Valve endothelial cells (VECs)
Vascular endothelial cells (ECs)
Vascular smooth muscle cells (SMCs)
Pericytes
Adventitial fibroblasts
Endothelial cells and smooth muscle cells of vasa vasorum
Endothelial progenitor cells (EPCs)
Mesenchymal stem cells
Bone marrow–derived stem cells
Tissue–resident stem cells
Dendritic cells
Macrophages/foam cells (atherosclerosis)
Lymphocytes
Mast cells
Giant cells

**TABLE 14.2 Endothelial function.**

| Physiologic function | Endothelial dysfunction |
| --- | --- |
| Platelet resistant | Platelet adhesion |
| Anti-coagulation | Pro-coagulation |
| Fibrinolysis | Anti-fibrinolysis |
| Quiescent | Migration/proliferation |
| Leukocyte resistant | Leukocyte adhesion |
| Anti-inflammatory | Pro-inflammatory |
| Selective impermeability | Enhanced permeability |
| Quiescent SMC | SMC activation |
| Vasomotion | Disrupted vasomotion |
| Matrix stability | Matrix remodeling |
| Vascular stability | Angiogenesis |

Endothelial-to-mesenchymal transition (EndMT) is the process by which ECs lose their cell-specific markers and morphology and acquire a mesenchymal cell-like phenotype. EndMT regulates endocardial cushion formation and embryonic heart valve development. This process reappears in adult disease to be important in the pathogenesis of adult cardiovascular disease as well such as in vein graft stenosis following surgical revascularization.

## Vascular smooth muscle cells

Vascular SMCs are the major cell type of the media with a few present in the normal human intima. Adult vascular SMCs are differentiated and highly specialized, but they are not terminally differentiated and do display considerable plasticity. SMCs maintain vascular tone through contractile proteins that regulates blood vessel diameter, blood pressure, and the distribution of blood flow. The master molecular regulator of the contractile phenotype is myocardin. Nitric oxide and prostacyclin are important regulators of dilation while endothelin and thromboxane promote vasoconstriction. Some of these vasoactive compounds are derived from EC and thus this regulation of wall contractility is a prime example of the functional relationship between SMC and EC. EC also regulate SMC growth in part and are involved in intercellular signaling via EC secreted exosomes.

**TABLE 14.3** Selected shear stress–regulated factors in endothelium.

**Vasoactive compounds**
Angiotensin-converting enzyme (ACE)
NO—endothelial nitric oxide synthase (eNOS)
NO—induced nitric oxide synthase (iNOS)
Prostacyclin
Endothelin-1
**ECM/ECM degradation enzymes**
Matrix metalloproteinases
Collagen
Thrombospondin
**Growth factors**
Epidermal growth factor (EGF)
Platelet derived growth factor (PDGF)
Basic fibroblast growth factor (bFGF)
Granulocyte monocyte-colony stimulating factor (GM-CSF)
Insulin-like growth factor–binding protein (IGFBP)
**Coagulation/fibrinolysis**
Thrombomodulin
Tissue factor
Tissue plasminogen activator (tPA)
Protease-activated receptor 1—thrombin receptor (PAR-1)
**Inflammation factors**
Monocyte chemoattractant protein (MCP-1)
Vascular cell adhesion molecule (VCAM-1)
Intercellular adhesion molecule (ICAM-1)
E-Selectin, P-Selectin
NF-κB
**Others**
Klf2, Klf4 (transcription factors)
Extracellular superoxide dismutase (ecSOD)
Sterol regulatory element–binding protein (SREBP)
Bone morphogenic protein-4 (BMP-4)
Platelet/endothelial cell adhesion molecule (PECAM-1)
microRNAs (miRNAs)

Molecular markers of SMCs include smooth muscle alpha actin, smooth muscle heavy chain, h1-calponin, smoothelin, and smooth muscle 22α. SMCs maintain the normal matrix of the vascular wall and are quiescent in the media. Upon injury, the cells undergo phenotypic transformation as they become proliferating, secreting, and migrating SMCs with a capacity to become myofibroblast-like cells and participate in repair. These cells lose most of their muscle markers and transformation is regulated by several factors including platelet-derived growth factor (PDGF), transforming growth factor-beta (TGF-β), and oxidized low-density lipoprotein (ox-LDL). Downregulation of microRNAs (miRNAs) such as miR 23b also promotes switching. Krüppel-like factor 4 (KLF4), a gene critical in the maintenance of pluripotency in embryonic stem cells, is reexpressed in SMCs undergoing phenotypic switching. SMCs may become foam cells through ingestion of lipids. SMCs are important regulators of vascular remodeling with secretion of matrix metalloproteinases (MMPs), membrane targeted-MMPs (MT1-MMPs), and tissue inhibitors of metalloproteinases (TIMPs). In response to vascular injury or disease SMCs may undergo transdifferentiation by losing classic SMC marker expression while expressing a macrophage phenotype or osteogenic and chondrogenic phenotypes.

## Valve endocardial cells

Valve endocardial (endothelial) cells (VECs) form a single cell layer of adherent cells that cover the surface of the valve and function as a thromboresistant surface and a selective permeability barrier. Disruption of these functions lead to valvular disease. The cells are normally quiescent, but upon injury, they migrate and proliferate to reconstitute the confluent thromboresistant surface. VECs are heterogeneous and show important differences when compared to vascular ECs including phenotypic differences in response to shear. VECs have been shown to differentially express 584 genes on the aortic side versus the ventricular side of normal adult pig aortic valves. These differences could help explain the vulnerability of the aortic side of the valve cusp to calcification seen in calcific aortic stenosis (CAS). Since calcification occurs within the valve tissue, it is likely that VECs may be playing more of a transducing role, regulating valve interstitial cell (VIC) function.

## Valve interstitial cells

The term valve fibroblasts, although still used in the literature, should be abandoned and replaced by the term valve interstitial cells (VICs) since these cells have specific features that are context-dependent to the heart valve and show differences when compared to fibroblasts in other tissues and organs.

VICs are embedded in extracellular matrix in all three layers of the leaflet: the fibrosa, spongiosa, and ventricularis. Compartmentalization occurs at late gestation. However, between 20 and 39 weeks the valves only have two layers. Although it is not known how remodeling of the valve into individual layers occurs, it is likely that physical forces play a role since the three layers present in the adult leaflet are only established by early adulthood.

Cell cultures of VICs have been characterized (Fig. 14.1) and have provided information on the cell and molecular biology of these cells. Currently, five phenotypes best represent the VIC family of cells (Fig. 14.2). Each phenotype exhibits specific sets of cellular functions present in normal valves and in pathobiological conditions (Table 14.4). The phenotypes are referred to as embryonic progenitor endothelial/mesenchymal cells, quiescent VICs (qVICs), activated VICs (aVICs), progenitor VICs (pVICs), and osteoblastic VICs (obVICs). The embryonic progenitor endothelial/mesenchymal cells undergo endothelial-to-mesenchymal transformation that initiates the process of valve formation in the endocardial cushion of the embryonic heart tube. The qVICs are at rest in the adult valve and maintain normal valve physiology. The aVICs regulate the pathobiological response of the valve in disease and injury. These cells are a type of myofibroblast cell similar to those found at sites of wound repair in a variety of tissues. In vitro studies on the regulation of the transformation to the myofibroblastic phenotype, which expresses α-smooth muscle actin, suggest that TGFβ, fibroblast growth factor-2 (FGF2), and wingless integrated 3α/β-catenin (Wnt3a/β-catenin) are involved. The pVICs are poorly defined and likely consist of a heterogeneous population of stem, progenitor and pluripotent cells that may be important in repair. These cells have the potential of self-renewal and multi-lineage differentiation and may be present in the inactive state in the valve tissue; they may enter the valve tissue through the endocardial surface from the blood stream derived from bone marrow derived endothelial progenitor cells or mesenchymal stem cells. They may also differentiate from resident aVICs and qVIC. The obVICs regulate chondrogenesis and osteogenesis. VICs have been shown to have osteogenic, chondrogenic, and adipogenic potential in tissue culture models. obVICs may be derived from aVICs or even qVICs since loss of β-catenin may promote chondrogenic differentiation and subpopulations of osteoprogenitor cells have been identified in valve tissue. Although these five phenotypes may exhibit plasticity and convert from one phenotype to another (Fig. 14.2), characterizing VIC function by distinct phenotypes brings clarity to our understanding of the complex VIC biology and pathobiology. New phenotypes will likely be discovered to further explain VIC function.

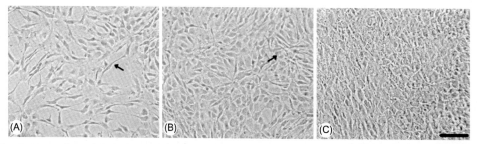

**FIGURE 14.1** Phase contrast photomicrographs of VICs in monolayer culture at moderate (A), confluent (B), and superconfluent (C) densities. Note the elongated morphology in (A) and the overlapping growth pattern in (B) as indicated by *arrows*. Scale bar represents 20 μm. Magnification 200×. *Reprinted from Liu AC, Joag VR, Gotlieb AI. The emerging role of valve interstitial cell phenotypes in regulating heart valve pathobiology. Am J Pathol 2007;171:1407–18 with permission from the American Society for Investigative Pathology.*

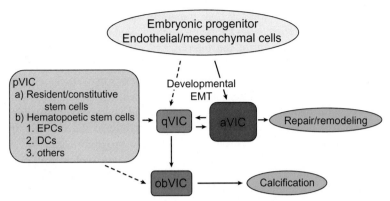

**FIGURE 14.2** Numerous VIC functions can be conveniently organized into five phenotypes: embryonic progenitor endothelial/mesenchymal cells, quiescent VICs (qVICs), activated VICs (aVICs), stem cell−derived progenitor VICs (pVICs), and osteoblastic VICs (obVICs). These represent specific sets of VIC functions in normal valve physiology and pathophysiology. Embryonic progenitor endothelial/mesenchymal cells undergo endothelial-mesenchymal transformation in fetal development to give rise to aVIC and/or qVICs resident in the normal heart valve. The VICs undergoing the transformation do have features of aVICs, including migration, proliferation, and matrix synthesis. When the heart valve is subjected to an insult, be it abnormal hemodynamic/mechanical stress or pathological injury, qVICs become activated, giving rise to aVICs, which participate in repair and remodeling of the valve. pVICs including bone marrow-derived stem/ pluripotential cells, circulating stem/ pluripotential cells, and resident valvular stem/progenitor cells are additional sources of aVICs in the adult. The relationship between bone marrow, circulating, and resident pVICs is not well understood. Under conditions promoting value calcification, such as in the presence of osteogenic and chondrogenic factors, qVICs can undergo osteoblastic differentiation into obVICs. It is possible that obVICs are derived from pVICs. obVICs actively participate in the valve calcification process. Compartmentalizing VIC function into distinct phenotypes recognizes the transient behavior of VIC phenotypes. The hatched arrows depict possible transitions for which there is no solid evidence currently. *Reprinted from Liu AC, Joag VR, Gotlieb AI. The emerging role of valve interstitial cell phenotypes in regulating heart valve pathobiology. Am J Pathol 2007;171:1407−18 with permission from the American Society for Investigative Pathology.*

**TABLE 14.4** Heart valve interstitial cell phenotypes.

| Cell Type | Location | Function |
| --- | --- | --- |
| Embryonic progenitor Endothelial/mesenchymal cells | Embryonic cardiac cushions | Give rise to resident qVICs, possibly through an activated stage and promoted by FGF2 and TGFβ. EMT can be detected by the loss of endothelial and the gain of mesenchymal markers |
| qVICs | Valve cusp leaflet | Maintain physiologic valve structure and function and inhibit angiogenesis in the leaflets |
| aVICs | Valve cusp leaflet | VICs with α-SMA signifying activation of cellular repair processes including proliferation, migration, apoptosis, matrix remodeling, and upregulation of TFGβ |
| pVICs | Bone marrow, circulation, within valve leaflet/cusp | Enter injured valve or are resident in valve to provide aVICs to repair valve, may be CD34−, CD133−, and/or S100-positive |
| obVICs | Within valve cusp/leaflet, bone marrow, circulation | Calcification, chondrogenesis, and osteogenesis of valve. Secrete alkaline phosphatase, osteocalcin, osteopontin, bone sialoprotein |

Modified from Liu AC, Joag VR, Gotlieb AI. The emerging role of valve interstitial cell phenotypes in regulating heart valve pathobiology. Am J Pathol 2007;171:1407−18.

## Myeloid cells

Myeloid cells are not prominent in normal cardiovascular tissue but play important roles in cardiovascular disease. There are occasional resident macrophages, dendritic cells and lymphocytes in the normal vessel wall, especially in the intima. The adventitia contains some macrophages, dendritic cells and lymphocytes.

Endothelial dysfunction due to injury promotes monocytes to enter the wall, become activated tissue macrophages that promote vessel dysfunction and further injury. Inflammation is a key process in the pathogenesis of the atherosclerosis plaque. Macrophages are an important cell type in the atherosclerotic plaque along with dendritic cells and less common lymphocytes, and mast cells. Macrophages may transform into plaque foam cells. Neutrophil extracellular traps are shown to be associated with macrophage cytokine production in plaque development.

Polymorphonuclear leukocytes are prominent at the interface of necrotic and intact myocardium in a myocardial infarction. In the early stages of some vasculitis polymorphs are present often in association with fibrinoid necrosis. Activated T-lymphocytes and B-lymphocytes are present in diseased arteries as well. T-cells predominate in atherosclerotic plaques. B-cells are involved in antibody response such as to oxyLDL.

Macrophages are associated with repair of vessels, myocardium and heart valves. In several experimental models macrophages appear as at least two populations which is likely a response to the microenvironment they find themselves in. Activated M1 macrophages exhibit a strong inflammatory program while M2 macrophages have anti-inflammatory activity and promote tissue remodeling and repair. Macrophages switch from one functional phenotype to another in response to changes in the microenvironment. It is likely that there are several additional macrophage subtypes that may represent a spectrum of phenotypes along a continuum.

## Vascular progenitor/stem cells

Mesenchymal stem cells derived from the bone marrow stroma can develop both vascular and cardiomyocyte phenotypes. The stem or progenitor cells resident in tissue are usually rare within a population of cells and specific techniques are required to isolate the cells and then expand the population. The cells are identified by specific cell-surface markers.

Endothelial progenitor cells (EPCs) are a specialized subset of hematopoietic cells found in the adult bone marrow and peripheral circulation arising from hemangioblasts prenatally. The cell number and migratory activity of circulating EPCs is decreased in patients with stable coronary artery disease and is inversely correlated with the number of coronary risk factors in patients. They are phenotypically characterized by surface antigens including CD133, CD34, c-kit, VEGFR2, CD144, and Sca-1. EPCs are immature cells, which have the capacity to proliferate, migrate, and differentiate into endothelial lineage cells, but have not yet acquired characteristics of mature ECs, including surface expression of vascular endothelial cadherin (VE-cadherin) and von Willebrand Factor, and loss of CD133. The discovery of circulating EPCs in the adult changed the view that new blood vessel growth occurs exclusively by angiogenesis postnatally. Vasculogenesis that occurs in the developing embryo is thought to be activated in the adult when EPCs are mobilized and recruited to areas of neovascularization to form new blood vessels.

EPC transplantation is designed to promote collateral circulation in order to treat myocardial and other tissue ischemia. On the basis of the promising outcomes in animal models, clinical applications of EPCs for ischemic diseases are in progress. Intra-myocardial injection of autologous bone marrow EPCs into patients with myocardial infarction improved left ventricular global function and ejection fraction, as well as cardiac perfusion after 3−9 months. However, the life of transplanted EPCs is often short, so many investigators interpret the improved cardiac function to be due to paracrine effects of EPCs on resident cells due to the secretion of bioactive molecules such as cytokines.

## Cardiac progenitor/stem cells

The adult heart is considered to be a postmitotic organ comprised of fully differentiated cardiomyocytes that survive a lifetime without replenishment. Immature cardiomyocytes, capable of mitotic division, have been identified at the infarct border zone of myocardial infarction by some investigators, whereas most others failed to observe this. Much research is ongoing to develop cell-based therapy to replace dead myocardium with functional myocytes to improve diminished cardiac function that occurs post myocardial infarction. These cells must be able to integrate, survive, and electronically couple to the damaged myocardium. Researchers are utilizing populations of pluripotent stem cells derived from embryonic stem cells or from induced pluripotential stem cells (iPSCs) exposed to defined culture conditions, which give rise to new cardiomyocytes. Differentiation protocols aimed at generating cardiomyocytes have evolved to use defined conditions with timed applications of specific growth factors known to be important in cardiac development, e.g., Activin A, BMP4, bFGF, Wnts. Some protocols include manipulations of extracellular matrix.

Improvement of cardiac function post myocardial infarction has been reported, often attributed to the release of paracrine factors from the transplanted cells that promote cell survival. Some cell types and protocols result in problems with cardiac differentiation and with integration into the surrounding electromechanical cardiac environment The arrhythmogenicity of transplanted cells poses a serious risk to the patient. Although the progress with differentiation protocols has been striking, a closer examination of the resulting cell populations reveals that significant cellular heterogeneity remains which could adversely impact clinical outcomes. The techniques and protocols need to be further refined including using iPSCs to derive specific myocardial subtypes that mimic atrial cells, ventricular cells, and conduction cells.

Another process under investigation is transdifferentiation, a term to describe change in the cell type, in this case, to produce cardiomyocytes directly from somatic cell sources independent of the iPSCs process. The source cells for transdifferentiation can be adult stem cells or terminal cells, while the target cells after transdifferentiation are cells of a different type from the original cells. In vitro transdifferentiation of myocardial fibroblasts into cardiomyocyte-like cells could be achieved by the overexpression of transcription factors GATA4, MEF2C, TBX5, and HAND2. One week after these transcription factors were introduced into myocardial fibroblasts, the expression of myocardial-specific markers was detected. The transdifferentiated myocardial cells had ionic calcium current and exhibited spontaneous contraction.

# Atherosclerosis

Atherosclerosis is a chronic vascular disease initially developing in the intima of elastic and larger muscular arteries and characterized by the presence of fibro-inflammatory lipid plaques (atheromas), which grow in size to develop complicated plaques, lumenal stenosis, and focal weakening of vessel walls, especially the aorta. Atherosclerosis leads to local aneurysm and/or rupture, and to ischemic heart disease, cerebral vascular disease, and peripheral vascular disease. Epidemiologic studies have identified environmental and genetic conditions that increase the risk of developing atherosclerosis (Table 14.5) Risk modification is an important medical approach to prevention and treatment of atherosclerosis. To provide a greater predictive capability beyond the traditional biomarkers such as low density lipoprotein-C (LDL-C), additional biomarkers of endothelial dysfunction are being studied including vascular cell adhesion molecule-1 (VCAM-1), intercellular adhesion molecule-1 (ICAM-1), high sensitivity C-reactive protein, interleukin-1 (IL-1), IL-6, and tumor necrosis factor $\alpha$ (TNF$\alpha$).

The interplay between an individual's genetic disposition and the environment adopted by the individual may result in an imbalance between pro-atherogenic and anti-atherogenic factors and processes that then leads to initiation and growth of the atherosclerotic plaque. There is no single atherogenic gene that explains pathogenesis. Instead, multiple genes (polygenic), including clusters of genes forming networks regulating specific cell functions, are likely to interact with the environment and with each other to promote atherogenesis. Genome-wide association studies (GWAS) show a strong association of coronary artery disease with about 60 SNPs at present including a locus on chromosome 9p21.3. The pathobiological role of the 9p21.3 locus is not understood at present. Much effort has gone into identifying genetic variants of specific genes that are described as candidate genes because they have been shown to be important in health and disease of the artery wall. Examples include ACE insertion/deletion, *APOE*, *APOE2*, *APOE3*, and *APOE4*. Most of these genetic association studies are not robust and further studies are needed. Extensive epigenetic DNA hypomethylation has been associated with atherosclerotic plaques and epigenetic DNA methylation itself is flow-dependent. Many miRNAs, short non-coding RNAs and long noncoding RNAs are associated with cardiovascular disease and have specific effects on the regulation of pro-atherogenic and anti-atherogenic genes.

The development of atherosclerosis may begin as early as the fetal stage, with the formation of intimal cell masses, or perhaps shortly after birth, when fatty streaks begin to evolve. The characteristic lesion, which is initially not clinically significant, requires as long as 20−30 years to form. Then serious acute events may occur, and/or complicated lesions may emerge after several more years of plaque growth.

The evolution of the clinical plaque in humans can be divided into three stages; (1) a plaque initiation and formation stage, (2) a plaque adaptation stage, and (3) a clinical stage.

**TABLE 14.5 Risk factors for atherosclerosis.**

| | |
|---|---|
| LDL receptor mutations | Obesity |
| Hyperlipidemias | Family history |
| Diabetes mellitus | Gender |
| Hypertension | Advancing age |
| Cigarette smoke | |

## Stage I: Plaque initiation and formation

Endothelial injury is an early event and may be due to several conditions including hypercholesterolemia, hemodynamic shear stress, microorganisms, toxins, hyperlipidemia, hypertension, diabetes mellitus and immunologic events. Normal laminar flow establishes a physiological shear stress set point which is anti-atherogenic. The intimal lesions initially occur at sites where the time-averaged wall shear stress is low and the direction of flow changes during the cardiac cycle (oscillating) disrupting physiologic flow. This occurs at branch points and curves which exhibit endothelial dysfunction. The accumulation of subendothelial SMCs, as occurs in an intimal cell mass (eccentric intimal thickening, intimal cushion) distal to branch points and at other sites in certain vessels, particularly the coronary arteries, is also considered prone to plaque formation since it provides a readily available source of SMCs. It is thought that these intimal thickenings are physiological adaptations to mechanical forces.

Low and oscillating shear induce the appearance of activated cell adhesion molecules on the surface of ECs which promote monocyte attachment, an early step in inflammation. The leukocytes first roll along the endothelium mediated by P-selectin and E-selectin and then adhere due to chemokine-induced EC activation and integrin interactions with cell adhesion molecules such as VCAM-1 and ICAM-1. The leukocytes penetrate the endothelial barrier at interendothelial sites, regulated by platelet EC adhesion molecule (PECAM) (CD31). Low shear also disrupts normal repair following endothelial injury, thus exposing the denuded endothelial surface to blood flow for longer periods of time. Hemodynamic forces induce gene expression of several biologically active molecules in ECs that promote atherosclerosis, including FGF-2, tissue factor (TF), plasminogen activator, and endothelin (Table 14.4). However, shear stress also induces gene expression of agents that are considered anti-atherogenic, including nitric oxide synthase (NOS) and plasminogen activator inhibitor 1 (PAI-1) (Table 14.3).

LDL plays a prominent causal role in atherosclerosis. Lipid accumulation is enhanced by transcytosis of LDL particles and disruption of the integrity of the endothelial barrier due to disruption of cell-cell adhesion junctions, cell loss, and/or cell dysfunction. Lipid induced oxidative stress in ECs and macrophages leads to cellular dysfunction and damage. Some monocyte/macrophages that have entered the vessel wall become foam cells, due in part to the uptake of ox-LDL via scavenger receptors. They undergo necrosis and release lipids. A change in the types of connective tissue and proteoglycans synthesized by the SMCs in the intima renders these sites prone to lipid accumulation due to high binding affinity of lipoproteins for chondroitin sulfate–rich proteoglycans, versican and biglycan.

Innate immunity plays a prominent role especially in the early stage of atherogenesis. Macrophages promote inflammation by secreting cytokines and chemokines and also releasing growth factors, thereby promoting further accumulation of SMCs. Oxidized lipoproteins induce tissue damage and further macrophage accumulation. Macrophages secrete monocyte chemoattractant protein-1 (MCP-1), which promotes macrophage accumulation. Macrophages secrete reactive oxygen species. Macrophages synthesize platelet derived growth factor (PDGF), FGF, TNF, IL-1, IL-6, interferon-γ (IFN-γ), and TGFβ, each of which can modulate the growth of SMCs and ECs. SMCs undergo phenotypic switching from contractile quiescent cells to secretory SMCs that proliferate and migrate. There are also processes that inhibit growth. MicroRNA-29 b inhibits SMC growth as does IFN-γ and TGFβ. ADAMTS-7 inhibits reendothelialization of the injured artery wall. Cytokines IL-1 and TNF stimulate ECs to produce platelet-activating factor, TF, and PAI. TF expression is also upregulated by oxidized lipids. Thus, the vascular surface changes from anticoagulant to procoagulant.

As the lesion progresses, small mural thrombi may develop on the intimal surface as platelets and the coagulation cascade are both activated. Since thrombosis also initiates fibrinolysis and inhibition of coagulation factors, the thrombus may alternatively lyse if anti-thrombotic factors supersede. TGFβ, which regulates secretion of collagen, matrix proteins, and differentiation of SMC into myofibroblasts, helps promote recanalization and organization of the thrombus and incorporation into the plaque.

The deeper parts of the thickened intima are poorly nourished and this sets the stage for the development of the necrotic core. Cell death is promoted by proteolytic enzymes released by macrophages and by tissue damage caused by ox-LDL and other reactive oxygen species. The macrophages and SMCs undergo ischemic necrosis and apoptosis. Impaired clearance of apoptotic cells by phagocytosis, called efferocytosis, enhances necrotic core growth. Hypoxia promotes hypoxia-inducible 1 alpha translocation to the nucleus of SMCs and macrophages, which bind to the promoter-specific hypoxia response element, leading to the transcriptional activation of VEGF, and other target genes. VEGF and tissue hypoxia initiate angiogenesis with new vessels forming in the plaque derived from the vasa vasorum. Some investigators regard the presence of neovascularization as a condition that establishes permanency to the plaque and prevents significant regression.

**TABLE 14.6** Composition of atherosclerotic plaque.

**Cells**
    Endothelial
    Smooth muscle
    Macrophages
    Foam
    Dendritic cells
    Lymphocytes
    Mast cells
    Giant cells
**Matrix**
    Collagens
    Proteoglycans—biglycan, versican, perlecan
    Elastin
    Glycoproteins
**Lipids and lipoproteins, cholesterol crystals**
**Serum proteins**
**Platelet and leukocyte products**
**Necrotic debris**
**Microvessels**
**Hydroxyapatite crystals**

The fibroinflammatory lipid plaque is formed, with a central necrotic core and a fibrous cap that separates the necrotic core from the blood in the lumen (Table 14.6, Fig. 14.3). The plaque is heterogeneous with respect to inflammatory cell infiltration, lipid deposition, and matrix organization. TGFβ is an important regulator of plaque remodeling and extracellular matrix deposition including collagen, fibronectin, and proteoglycans. It inhibits proteolytic enzymes that promote matrix degradation by enhancing the expression of protease inhibitors.

Adaptive immunity plays a role in atherogenesis where activated T cells promote inflammation in mature lesions. The expression of HLA-DR antigens on both ECs and SMCs in plaques implies that these cells have undergone some type of immunological activation, perhaps in response to IFN-γ released by activated T-cells in the plaque. The presence of T-cells reflects an immune response that is important for the progression of atherosclerotic lesions through T-cell secretion of cytokines that modulate inflammation in the plaque. Possible plaque antigens include ox-LDL to which antibodies have been identified in the plaque.

## Stage II: Adaptation stage

As the plaque encroaches upon the lumen (in coronary arteries), the wall of the artery undergoes remodeling. Changes in hemodynamic shear stress and presence of turbulent flow around the plaque as the plaque grows in size modulates remodeling regulated through mechanotransduction signaling by the ECs. These changes activate the expression of a variety of genes that encode for proteins that promote remodeling such as MMPs, collagens, and TGFβ. SMC turnover (proliferation) and apoptosis and matrix synthesis and degradation modulate remodeling to maintain patency and blood flow in the lumen. Once a plaque encroaches upon half the lumen, compensatory remodeling can no longer maintain normal patency, and the lumen of the artery becomes narrowed (stenosis). The initial adaptation may delay diagnosis of atherosclerosis since the plaque is clinically silent. Yet, even though the plaque is small, plaque rupture through the fibrous cap may occur with catastrophic results.

## Stage III: Clinical stage

Plaque growth continues as the plaque encroaches further into the lumen. Hemorrhage into a plaque due to leakage from the small fragile vessels of neovascularization may not necessarily result in actual rupture of the plaque but may still increase plaque size. Complications develop in the plaque, including surface erosion, ulceration, fissure formation, mural thrombosis, calcification, and aneurysm formation. Calcification is driven by chondrogenesis and osteogenesis, regulated in part by TGFβ, osteogenic progenitor cells, and bone-forming proteins. The osteogenic cells may be derived

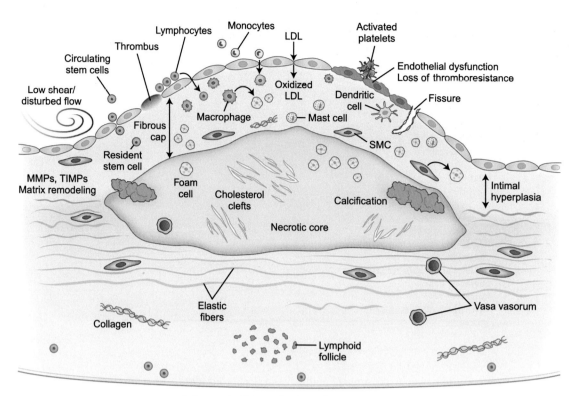

**Advanced atherosclerotic plaque in aorta**

**FIGURE 14.3**   Advanced complicated atherosclerotic plaque in aorta with fibrous cap and central necrotic core. *Reprinted from Cardiovascular Pathology, 2016, ed. Buja LM and Butany J, Elsevier, p. 95.*

---

**TABLE 14.7  Plaque rupture.**

Endothelial erosion, ulceration, fistula
Thin fibrous cap
Decreased smooth muscle cells in cap
Inflammation—macrophages
Foam cells
Hemodynamic shear stress (low, oscillating, turbulence)
Imbalance in matrix synthesis/degradation (metalloproteinases, tissue inhibitors of metalloproteinases)
Nodular calcification

---

from several places including from rare resident osteogenic cells in the vessel wall, transdifferentiation of SMCs, and from mesenchymal stem cells derived from bone marrow which enter the vessel. Activated mast cells are found at sites of erosion and may release pro-inflammatory mediators and cytokines. Continued plaque growth leads to severe stenosis or occlusion of the lumen. An acute myocardial infarction may occur due to acute plaque rupture through the fibrous cap, and ensuing lumen thrombosis since the necrotic core is highly thrombogenic. (Table 14.7).

# Ischemic heart disease

Ischemic heart disease (coronary heart disease) may occur clinically as stable angina and acute coronary syndromes, which include unstable angina, non-ST elevation myocardial infarction, and ST-elevation myocardial infarction. Reliable biomarkers are available to identify cardiac damage, especially sensitive cardiac troponin assays.

Ischemia of the myocardial cell leads to a series of intracellular structural and biochemical changes, which begin almost immediately after onset and evolve over time. Initiation of ATP depletion begins within seconds. Initially the

**TABLE 14.8** Myocardial infarction: stages of healing.

**Ischemic injury phase**
  Necrosis
  Hemorrhage
**Inflammation and vascular phase**
  Vasodialation
  Edema
  Polymorphonuclear leukocytes
  Mononuclear leukocytes
**Granulation tissue phase**
  Neovasculature (Angiogenesis)
  Fibroblast activation, myofibroblast differentiation, migration, proliferation, and extracellular matrix secretion
**Matrix remodeling phase**
  Prominent fibrosis
  Scar
  Blood vessel remodeling and regression

reaction of the cells results in reversible injury. However, by 20—30 min the myocardial cells become irreversibly injured and undergo necrosis. Cardiomyocytes in the subendocardium are most at-risk for ischemia and irreversible injury occurs first in the subendocardium and progresses as a wave front toward the epicardium, resulting in a transmural myocardial infarct. Progression of necrosis will involve the full cardiac bed supplied by the occluded coronary artery and usually is complete by 6 h. The repair of the infarcted myocardium follows a well-characterized sequence of necrosis, inflammation, granulation tissue, remodeling, and scar formation (Table 14.8). Cardiac muscle does not regenerate.

## Aneurysms

Aneurysms may occur in any vessel, however, the thoracic and abdominal aorta, and cerebral arteries are common sites. Aneurysms are most often a complication of atherosclerosis. However some abdominal aortic aneurysms (AAA) are considered to be the result of complex multifactorial factors with an unknown genetic component which through GWAS and candidate gene studies have been identified as susceptibility loci located on several chromosomes. These aneurysms may result in rupture as wall tension consistent with the Law of Laplace is the force expanding the aneurysm. The pathogenesis of non-atherosclerotic AAA is not well understood. However, connective tissue degradation, inflammation, and loss of SMCs are characteristic features. Rupture is considered to be due to collagen degradation secondary to increased MMP2, MMP9, and cysteine collagenase. Cytokines are involved in AAAs, including IL1β, TNFα, MCP-1, IL-8, granulocyte colony stimulating factor (GCSF), macrophage colony stimulating factor (MCSF) and IL-13. Altered expression of miRNAs including miR-133b, miR-133a, miR-30c-2, and miR-204 may contribute to pathogenesis.

Dissecting aneurysms, both syndromic (genetic) and non-syndromic (idiopathic), usually occur in the ascending and descending thoracic aorta and blood dissects along the long axis of the media resulting in a channel filled with blood. There is an intimal tear, the entry point, and often a distal exit tear back into the lumen. Common nonspecific histologic findings are found including medial degeneration characterized by fragmentation and loss of elastic fibers, accumulation of proteoglycans, and depletion of SMCs. Thoracic dissections are classified based on involvement of the ascending aorta: Type A (ascending aorta involved) and Type B (distal, sparing the ascending aorta). Associated conditions include hypertension, bicuspid aortic valve, and idiopathic aortic root dilation.

Aneurysms and dissections are linked to genetic syndromes, including Marfan, Ehlers-Danlos (Type IV), Loeys-Dietz syndromes and to filamin A mutations. In addition, these conditions may also occur as an inherited autosomal dominant condition with decreased penetrance and variable expression without the syndromes. Mutations have been identified in genes for fibrillin-1 (*FBN1*), TGFβ receptor 2 (*TGFβR2*), TGFβ receptor1 (*TGFβR1*), SMC specific β myosin (*MYH11*), and α-actin (*ACTA2*). Epigenetic factors appear to play a role in Marfan and non-Marfan thoracic aortic aneurysms. An increase in TGFβ signaling is important in the pathogenesis of Marfan and Loeys-Dietz syndromes. Single gene mutations in fibrillin-I disrupt the contractile functions of vascular SMCs, which leads to activation of the stress and stretch pathways of the SMC. It has been postulated that the stretch pathways promote increased levels of MMPs (especially MMP2 and MMP9) and proteoglycans, and promote proliferative agents such as insulin growth factor-1 (IGF-1), TGFβ, and macrophage inflammatory protein1α (MIP1α) and MIP1β.

## Vasculitis

Vasculitis is inflammation of the blood vessel wall. There are several ways of classifying these conditions, based on the size of the vessel affected, anatomic site, microscopic morphology of the lesions, and/or clinical course. There are overlaps in the characteristics between different types of vasculitis. Large vessel vasculitis includes giant cell and Takayasu arteritis; medium includes polyarteritis nodosa and Kawasaki disease; and small vessel includes Wegner granulomatosis, Churg-Strauss syndrome, and microscopic polyarteritis. Although the pathogenesis is still poorly understood for most vasculitis, putative theories on pathogenesis are based on immunologic and infectious etiology. Infections may be direct or indirect and involve all types of microorganisms. Immunologic pathogenesis has been characterized as (1) immune complex conditions, which include those induced by infection, serum sickness, drugs, systemic lupus erythematosis, rheumatoid arthritis, and Henoch-Schönlein purpura; (2) anti-neutrophil cytoplasmic antibody conditions (ANCA) including Wegner granulomatosis (autoantigen is proteinase 3), microscopic polyarteritis (autoantigen is myeloperoxidase), and Churg-Strauss syndrome (myeloperoxidase); (3) anti-glomerular basement membrane antibody, as in Goodpasture Syndrome; (4) anti-endothelial cell antibodies, as in Kawasaki disease and systemic lupus erythematosis; (5) cell-mediated response, as in allograft rejection. ANCA titers in patients are associated with a recurrence of active disease; ANCA has been shown to activate neutrophils, monocytes, and ECs and cause arteritis and glomerulonephritis in experimental animal models.

Polymorphisms of candidate genes have been explored. These studies are not robust. The general conceptual approach is that vasculitides are the result of complex interactions between environmental factors and genetically determined host responses. How this works is not yet known. Epigenetics including microRNA biology is also being studied.

## Valvular heart disease

### Mitral valve prolapse

Mitral valve prolapse (MVP) is characterized by progressive thinning of the mitral leaflet, causing leaflets to billow backward during ventricular contraction, prolapsing into the left atrium beyond their normal position of closure. MVP valves show myxomatous degeneration and greatly increased type III collagen, with less of an increase of type I and V collagens, and an accumulation of dermatan sulfate. The accompanying loss in elastin and reduction in VICs is similar to the histological changes described previously in dissecting aneurysms.

The natural history of primary (non-syndromic) MVP disease is extremely heterogeneous. It can be asymptomatic. It can vary from benign, with a normal life expectancy, to adverse, with significant morbidity and mortality attributed to the development of valvular insufficiency. Complications, including heart failure, mitral regurgitation, bacterial endocarditis, thromboembolism, and atrial fibrillation, are extremely uncommon, affecting less than 3% of patients with MVP.

MVP is also a feature of many syndromes. Ehler-Danlos syndrome (Type IV) is rare and part of a heterogeneous group of connective tissue heritable disorders and is associated with mutations in the *COL3A1* gene, which encodes type III procollagen, thus promoting MVP.

Marfan syndrome is an autosomal dominant genetic disorder of the connective tissue associated with mutations in *fibrillin-1*, a major component of the microfibrils that form a sheath surrounding amorphous elastin. *Fibrillin-1* is required for the proper formation of the extracellular matrix including the biogenesis and maintenance of elastic fibers. The extracellular matrix is critical for both the structural integrity of connective tissue but also serves as a reservoir for growth factors that are essential for normal maintenance and regulation of repair.

TGFβ is a well-known regulator of extracellular matrix deposition and remodeling and plays a key role in MVP associated with syndromes. It is secreted by numerous cell types including heart VICs. It promotes differentiation of mesenchymal cells into myofibroblasts and to regulate multiple aspects of the myofibroblast phenotype through transcriptional activation of alpha-smooth muscle actin, collagens, matrix metalloproteinases, and other cytokines, such as connective tissue growth factor and basic fibroblast growth factor. TGFβ is secreted in a latent complex containing active TGFβ and latency-associated protein (LAP) (Figs. 14.4 and 14.5). This latent complex is tethered through latent TGFβ binding proteins (LTBPs) to matrix proteins to allow cells to tightly regulate TGFβ bioavailability and create

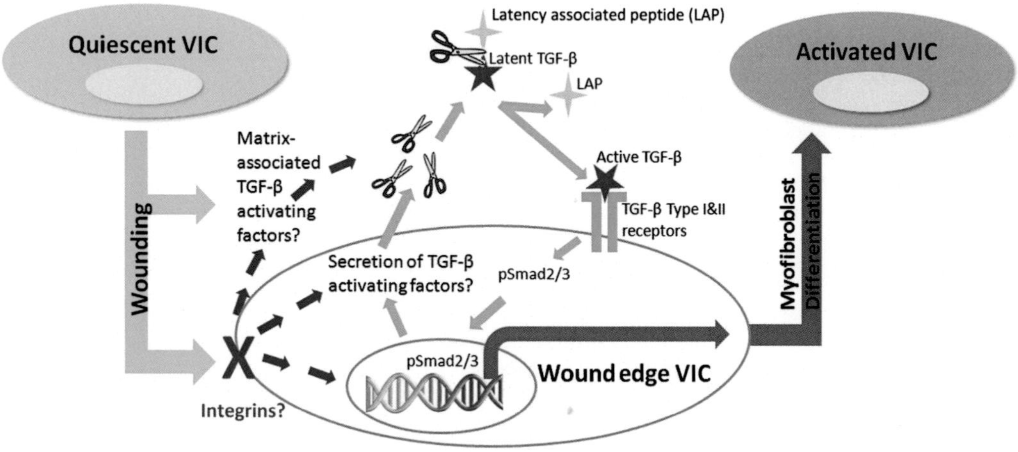

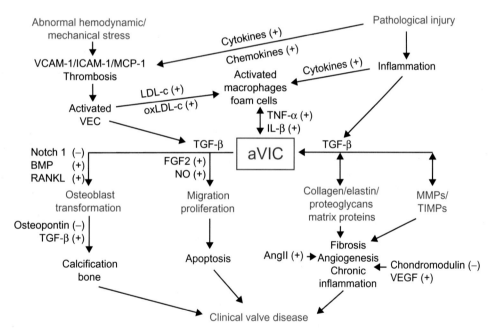

**FIGURE 14.4** Model of latent TGFβ activation and TGFβ signaling in VICs upon experimental wounding in a cell culture model. VICs in non-wounded confluent monolayer cultures proliferate at a very low rate. Wounding leads to activation of latent TGFβ present in the extracellular environment of VICs at the wound edge, possibly through changes in integrins or secretion of TGFβ activation proteases. Active TGFβ modulates VICs to myofibroblast differentiation giving rise to aVICs at the wound edge to repair the wound.

**FIGURE 14.5** The normal adult heart valve is well adapted to its physiological environment, able to withstand the unique hemodynamic/mechanical stresses under normal conditions. Under conditions of pathological injury or abnormal hemodynamic/mechanical stresses, VICs become activated through activation of VECs and by inflammation characterized by cytokine and chemokine secretion by myeloid cells. Macrophages are activated and aVICs increase matrix synthesis, upregulate expression of matrix remodeling enzymes, migrate, proliferate, and undergo apoptosis; and may undergo osteoblast transformation. These processes are regulated by a variety of factors, several secreted by the aVIC. If the aVICs continue to promote these cellular processes, angiogenesis, chronic inflammation, fibrosis, and calcification result, leading to progressive clinical valve disease. *Reprinted from Liu AC, Joag VR, Gotlieb AI. The emerging role of valve interstitial cell phenotypes in regulating heart valve pathobiology. Am J Pathol 2007;171:1407−18 with permission from the American Society for Investigative Pathology.*

special gradients in the cell microenvironment. Fibrillin-1 regulates TGFβ activation. Fibrillin-1 interacts with LTBPs to sequester latent TGFβ at specific locations in the matrix and stabilizes the inactive large latent complex (TGFβ, LAP, and LTBP), rendering it less prone to activation. Reduced or mutated fibrillin-1 leads to increased TGFβ activation and subsequent elevated levels of TGFβ signaling, resulting in cellular responses such as extensive degradation and remodeling of the extracellular matrix.

**TABLE 14.9 Similarities between calcific aortic stenosis and atherosclerosis.**

Histopathology features
Inflammation
Reactive oxygen species
Renin-angiotensin system; angiotensin II
Calcification, chondrogenesis, osteogenesis, mineralization proteins
Remodeling of matrix
Risk factors
ApoE4 polymorphism risk
Hypercholesterolemic rabbit develops calcific aortic stenosis
Retrospective clinical studies, statins delay progression of calcific aortic stenosis
Prospective studies, statins beneficial in atherosclerosis, yet to be shown in calcific aortic stenosis
$LDLr^{-/-}ApoB100/100$ mouse shows mineralization of valve

The importance of the TGFβ pathway in MVP pathogenesis is further highlighted by the discovery of Loeys-Dietz syndrome, which has many similar clinical features to Marfan syndrome and is caused by mutations in the genes encoding TGFβR1 or TGFβR2.

Another condition, X-linked valvulopathy, expresses MVP-like phenotypes and may exert its effect through the interaction of *filamin A* with molecules in the TGFβ pathway by augmenting TGFβ signaling through its interaction with Smad proteins such as Smad-2 and Smad-5. Thus, defective Smad-mediated TGFβ signaling due to *filamin A* mutations appears to underlie the cause of X-linked valvulopathy.

## Calcific aortic stenosis

Calcific aortic stenosis (CAS) is a complication of progressive leaflet fibrosis, which in 2%−3% of cases leads to calcification and symptomatic disease. The cause of most cases of CAS is unknown. However, a congenital bicuspid aortic valve may develop CAS later in life attributed to abnormal hemodynamic and biomechanical forces impinging on the leaflets. Since CAS shares common morphological changes with atherosclerosis (Table 14.9), there may be a common pathogenic mechanism. This may mean that both conditions have similar therapeutic targets and may respond to similar drugs such as 3-hydroxy-3-methylglutaryl coenzyme A reductase inhibitors (statins). However, recent studies using statins did not show a reduction in CAS progression. Thus, either lipid lowering is not a significant therapeutic strategy or the statins were given too late in the evolution of the disease condition. However intensive lipid lowering with rosuvastatin showed no effect on the progression of aortic stenosis in asymptomatic patients with mild to moderate stenosis.

## Cardiomyopathies

Cardiomyopathies are a heterogeneous group of human cardiac diseases with primary dysfunction of the cardiomyocytes, the etiology of which is either known or unknown. The former are usually referred to by their etiology and includes ischemia, hypertension, heart valve abnormalities, alcoholism, as well as systemic diseases such as diabetes, hemochromatosis, and amyloidosis. Those of unknown etiology are classified into dilated, hypertrophic, restrictive, arrhythmogenic right ventricular cardiomyopathy and channelopathies based on clinical and gross and microscopic features. Non-compaction cardiomyopathy (NCC) is considered a distinct cardiomyopathy but may also appear as a trait common to several types of cardiomyopathies.

As the cell and molecular biology is better understood, many turn out to be primary disorders of the myocardium due to one of hundreds of inherited gene mutations encoding cytoskeletal or sarcomere proteins. Hypertrophic cardiomyopathy (HCM) is considered a disease of the sarcomere; dilated cardiomyopathy (DCM), a disease of sarcomere-sarcolemma; and arrhythmogenic right ventricular cardiomyopathy (ARVC), a disease of desmosomes. Although there may be overlap in gene mutation among different cardiomyopathies, the molecular pathogenesis of each cardiomyopathy is unique and in most cases it is not known how the specific mutation results in leaflet dysfunction. Although genetic testing is available for several monogenic forms of cardiomyopathy, physicians and medical geneticists need to discuss the benefits and limitations of such testing with their patients.

## Cardiomyocyte structure and function

Cardiomyocytes are joined in series through intercalated discs containing gap junctions, adherens junctions, and desmosomes. Cardiomyocytes are surrounded by specialized plasma membranes, the sarcolemma, and contain bundles of longitudinally arranged myofibrils. The myofibrils are formed by repeating sarcomeres, the basic contractile units of cardiac muscle, composed of interdigitating thin actin filaments and thick myosin filaments. The thin filaments contain alpha-tropomyosin and troponins, while the thick filaments contain myosin-binding proteins. These myofibrils compose the excitation-contraction function of the cardiomyocyte in which rhythmical electrical stimulation drives cardiac mechanical force. Myofibers also contain a third filament type formed by the large filamentous protein, titin, which acts as a molecular template for the layout of the sarcomere. The extrasarcomeric cytoskeleton provides structural support for the sarcomere and other subcellular structures and transmits mechanical and chemical signals within and between myocytes. For example, desmin intermediate filaments form a three-dimensional scaffold throughout the extrasarcomeric cytoskeleton, allowing longitudinal connections to adjacent sarcolemma and lateral connections to subsarcolemmal costameres. Costameres are interconnections between the various cytoskeletal networks linking the sarcomere and sarcolemma and functioning as an anchor site for stabilization of the sarcolemma and integration of pathways involved in mechanical force transduction. Costameres contain focal adhesion-type complexes, spectrin-based complexes, and the dystrophin/dystrophin-associated protein complexes (DAPCs). Voltage-gated sodium channels and potassium channels co-localize with dystrophin proteins in DAPCs.

## Molecular genetics and pathogenesis of hypertrophic cardiomyopathy

HCM is primarily inherited as an autosomal dominant trait and is characterized by marked thickening of the left ventricular wall in the absence of increased external load. The myocardium shows cardiomyocyte disarray and cardiac fibrosis. Clinical complications include heart failure, arrhythmias, and sudden death. Sixty percent of HCM is due to more than 450 different mutations within eight sarcomeric genes. Most HCM mutations occur in two genes, *MYH7* and *MYBPC3*, encoding the beta-myosin heavy chain and myosin-binding protein C, respectively. Mutations in genes that encode cardiac troponin T and cardiac troponin I, essential myosin light chain, regulatory myosin light chain, alpha-tropomyosin, and cardiac actin occur but are less common in HCM patients. Rare mutations in cardiac troponin C and alpha-myosin heavy chain also cause HCM. Mutations in titin, muscle LIM protein, telethonin, and myozenin, which are Z-disc proteins forming a framework to connect sarcomere units to each other, can also cause HCM.

It is unclear whether the molecular pathogenesis of HCM is due to impaired sarcomere function or a gain of new sarcomere function. Findings of mutant myosin fragments and skeletal muscles carrying HCM mutations show decreased actin sliding velocity, supporting the hypothesis that mutations in the myosin components in HCM cause a decrease in motor function of sarcomeres and a compensatory hypertrophic response of the myocardium. Mouse and rabbit models of HCM carrying mutations in myosin heavy chain, myosin-binding protein C, and troponin T show enhanced actin-activated myosin ATPase activity, increased force generation, and accelerated actin sliding velocity that result in increased cardiac performance often evident in patients with HCM. It appears that the heterogeneity of mutant structural proteins within the sarcomere uncouples the normal mechanical coordination between myosin heads and the associated enhanced ATPase activity, resulting in higher levels of energy consumption which may trigger hypertrophy. This combined with decreased energy supply due to impaired blood flow to the hypertrophied heart results in myofibrillar disarray, leading to cardiomyocyte death and fibrosis.

Intracellular $Ca^{2+}$ is a key regulator of signaling pathways that link sarcomere function and dysfunction to cardiomyocyte biology and pathobiology. $Ca^{2+}$ homeostasis is disrupted early in HCM. A proposed mechanism is that sarcomeric release of $Ca^{2+}$ at the end of systole is impaired, leading to $Ca^{2+}$ accumulation within the sarcomere and $Ca^{2+}$ release from the sarcoplasmic reticulum into the cytoplasm. Elevated cytoplasmic diastolic $Ca^{2+}$ levels induce hypertrophic responses of cardiomyocytes through the calcineurin-nuclear factor of activated T-cells (NFAT) signaling pathway. Increased intracellular $Ca^{2+}$ results in calmodulin saturation and calcineurin activation, leading to dephosphorylation of the transcription factor in the cytoplasm. Upon dephosphorylation, NFAT translocates to the nucleus, where it induces expression of proliferative and growth-related genes for cardiomyocytes in coordination with other transcription factors.

## Molecular genetics and pathogenesis of dilated cardiomyopathy

DCM is the most common form of cardiomyopathy, involving about 90% of all clinical cardiomyopathies. It is characterized by increased ventricular size, reduced ventricular contractility, left or biventricular dilation and heart failure.

Twenty to forty percent of DCM patients have familial forms of the disease, with autosomal dominant inheritance being most common. Autosomal recessive, X-linked, and mitochondrial inheritance of the disease is also found. The majority of mutations are in genes encoding either cytoskeletal or sarcomeric proteins. Mutations in the cytoskeletal proteins lead to defects of force transmission, while mutations in the sarcomeric proteins lead to defects of force generation in the myocardium. In the majority of instances, the mechanism through which a loss or a defect in these proteins alters function is not well understood. However, identification of these genetic mutations can be used as biomarkers to carry out risk stratification to improve management of cardiomyopathies.

## Cytoskeletal defects in DCM

Mutations in numerous genes have been identified in autosomal dominant DCM, including desmins, delta-sarcoglycan, metavinculin, alpha-actinin-2, ZASP, actin, troponin T, beta-myosin heavy chain, titin, and myosin-binding protein C. The majority of mutations are in genes encoding desmin, delta-sarcoglycan, and metavinculin.

Desmin forms intermediate filaments and is found at the intercalated discs and functions to attach and stabilize adjacent sarcomeres. Desmin networks provide a scaffold allowing connections to adjacent sarcolemma and to sarcolemmal costameres.

Delta-sarcoglycan is involved in stabilization of the sarcolemma and in signal transduction. In the absence of delta-sarcoglycan, the remaining sarcoglycans (beta, gamma, sigma) cannot assemble properly in the endoplasmic reticulum.

Mutations in the gene metavinculin which encodes vinculin and its splice variant metavinculin also disrupt cytoskeletal function. Vinculin is localized to subsarcolemmal costameres in the heart, and interacts with alpha-actinin, talin, and gamma-actin to form a microfilamentous network linking cytoskeleton and sarcolemma. Vinculin and metavinculin are also present in adherens junctions, in intercalated discs, and participate in cell-cell adhesion.

## Sarcomeric defects in DCM

Mutations in sarcomeric genes may produce DCM, HCM, or restrictive cardiomyopathy, as well as left ventricular non-compaction. In DCM, sarcomeric mutations giving rise to autosomal dominant inheritance are mainly in genes encoding actin, alpha-tropomyosin, and troponins. Cardiac actin is a sarcomeric protein in sarcomeric thin filaments interacting with tropomyosin and troponin complexes. Mutations in sarcomeric thin filament proteins cardiac actin, alpha-tropomyosin, cardiac troponin T, and troponin I all give rise to DCM as do mutations in the thick filament protein beta-myosin heavy chain. These mutations perturb the actin-myosin interaction and force generation and alter cross-bridge movement during myocardial contraction.

## Molecular genetics and pathogenesis of arrhythmogenic right ventricular cardiomyopathy

ARVC is characterized by right ventricular fibro-fatty replacement of myocardial tissue. ARVC presents with palpitations or syncope as a result of ventricular tachyarrhythmias and is an important cause of sudden death at young ages.

ARVC is familial in about 50% of cases and is mainly autosomal dominant with some autosomal recessive inheritance. Mutations in ARVC have been identified in desmosomal proteins such as plakophilin-2, desmoplakin, plakoglobin, desmoglein-2, and desmocolin-2. Patients carrying mutations in the *PKP2* gene encoding plakophilin-2, an important desmosomal protein linking cadherins to intermediate filaments, present with ARVC at an earlier age than those without *PKP2* mutations. Over 50 independent *PKP2* mutations, the majority of which are truncation mutations, are found in 70% of familial cases of ARVC.

The molecular mechanisms by which desmosomal mutations lead to ARVC are believed to be related to defects in desmosome composition and function, abnormalities of intercalated discs, and defective Wnt/β-catenin signaling. Mutations in genes encoding components of the desmosomal complex lead to disturbed formation or reduced numbers of functional desmosomes. Desmosomes then cannot protect other junctions in the intercalated discs from mechanical stress leading to loss of cell-cell contacts and cardiomyocyte death. Gap junctions are perturbed with smaller and fewer gap junctions present and reduced protein connexin 43 at the intercalated discs. This results in heterogeneous conduction, a contributor to the characteristic arrhythmogenesis in ARVC.

Desmosomal components participate in the Wnt/β-catenin pathway involved in developmental processes. Desmosomal dysfunction results in nuclear translocation of the desmosomal protein plakoglobin, resulting in competition between plakoglobin and β-catenin. This leads to inhibition of Wnt/β-catenin signaling and a shift in cell fate from cardiomyocyte to adipocyte. Plakophilins are also localized to the nucleus and may be involved in transcriptional regulation or in regulating β-catenin activity in Wnt-signaling. Defective desmosomes are believed to be unable to maintain tissue integrity under excessive mechanical stress, such as in the thinnest areas of the right ventricle. This predisposes to damage in these areas, leading to disruption and subsequent degeneration of cardiomyocytes followed by replacement by fibro-fatty tissue.

## Molecular genetics and pathogenesis of non-compaction cardiomyopathy

Non-compaction cardiomyopathy (NCC) is a rare congenital cardiomyopathy. In NCC, myocardial development is hindered during embryogenesis beginning around 8 weeks post-conception. At this stage, the myocardium is normally spongelike. As the embryo grows, the myocardium compacts and matures. In NCC, the myocardium fails to fully compact, resulting in a spongiform appearance with poor heart pumping, tachyarrhythmias, thromboembolisms, and sudden death.

NCC is inherited in an autosomal dominant manner or through X-linked inheritance. The most common gene responsible for NCC is *TAZ*, an X-linked gene encoding taffazin, which is involved in the biosynthesis of cardiolipin, an essential component of the mitochondrial inner membrane. Mutations in *TAZ* also cause Barth syndrome, a metabolic condition with DCM, with or without non-compaction, neutropenia, skeletal myopathy, and 3-methylglutaconicaciduria. Mutations in cytoskeleton and sarcomere-related genes *DTNA* and *LDB3* can also give rise to NCC. *DTNA* encodes alpha-dystrobrevin, a dystrophin-associated protein involved in maintaining the structural integrity of the sarcolemma. *LDB3* encodes the sarcomeric Z-band protein, LIM domain−binding 3 protein.

## Channelopathies

Channelopathies are a group of cardiac conditions that display defects in ion channel and transporter function. Most are due to inherited mutations that disrupt ion channel biophysical properties. Other defects arise from disruptions in ion channel membrane trafficking and post-translational modifications. The most serious clinical complications resulting from arrhythmogenic channelopathies are sudden cardiac death and ventricular tachyarrhythmias.

Congenital long QT syndromes, which affect about 1 in 3000 persons, are a potentially lethal group of cardiac conditions characterized by delayed repolarization of the myocardium and QT prolongation. This arrhythmogenic disorder is characterized by a significant increased risk of syncope, seizures, and sudden cardiac death. Since several mutations in genes that encode ion channels or their associated proteins account for about 80% of cases, postmortem genetic testing for sudden unexplained death in the young is useful. In addition, new cardiac channel gene mutations and defects in associated proteins continue to be identified.

## Lymphatic circulation

The lymphatic vascular network is a complex system that regulates and maintains tissue fluid balance, participates in immune surveillance, especially for trafficking of antigen-presenting cells from tissues to lymph nodes, and provides a pathway for the absorption of fatty acids in the gut and for reverse cholesterol transport from peripheral tissue to the liver. These functions occur as the thin-walled lymphatic vessels regulate the forward flow of lymph propelled by tonic and sporadic contractions with back flow prevented by valves. The lymph flows through lymph nodes, which is important in immunosurveillance and leukocyte recirculation, and drains into the internal jugular veins and the venous circulation at the junctions of the left and right subclavian veins.

The lymphatic EC phenotype differs from that of blood vessel ECs in that they do not form adherences or tight junctions, are devoid of a basement membrane, and are not surrounded by pericytes. Transcription factor prospero homeobox-1 (PROX-1) is essential for the initial steps in lymphatic formation characterized by lymphatic progenitor cells budding from the anterior cardinal vein and forming lymph sacs. Studies in zebrafish suggest that lymphatic ECs acquire their lymphatic fate when lymphatic specifications are induced in a restricted population of angioblasts in the posterior cardinal vein under the control of Wnt5b. Prox1 is a downstream target of Wnt5b. The VEGF-C/VEGFR-3 pathway is required in development to mediate lymphatic endothelial cell migration, proliferation, and survival. Transcriptional factor mafba is a downstream transcriptional effector of VEGF-C signaling promoting migration of lymphatic precursors once initial sprouting occurs from the posterior cardinal vein as seen in zebrafish. Then a primary lymphatic plexus is formed under the control of several factors including podoplanin and neuropilin-2, BMP2, and RAFI/MEK/ERK signaling cascades. In addition to PROX-1 and VEGFR-3, Fox2, the forkhead transcription factor, is required in the latter stages of embryonic development to regulate valve morphogenesis and maintain the lymphatic capillary phenotype. The potent bioactive lipid sphingosine-1-phosphate, which is associated with inflammation and angiogenesis, regulates lymphangiogenesis by S1P1/Gi/PLC/Ca$^{2+}$ signaling pathways by which inflammation and lymphangiogenesis may be linked.

Missense mutations in *VEGFR-3* result in lymphedema and lymphatic hypoplasia. In some patients, primary lymphedema has been shown to be associated with *VEGFR-3* mutations, *FOXC2* mutations, and SRY-related HMG-box 18 (SOX18).

## Key concepts

- Inflammation, immune function, thrombosis and mesenchymal tissue repair are key pathobiological processes in cardiovascular disease.
- Although cells of the cardiovascular system have unique properties and can be identified by unique biomarkers, dedifferentiation and transdifferentiation of adult cells does occur in disease conditions.
- Cell function is regulated by the combined actions of several molecules, some that promote and others that inhibit the same cellular process. It is the balance between the bioactivity of all these molecules that dictates the specific function of the cell at any given moment in time.
- Depending on conditions, the same molecule may both promote or inhibit a given cellular function, generally by directly or indirectly acting on signaling molecules that regulate the given function.
- Different signaling pathways may interact by sharing downstream molecules.
- Microenvironments are important in autocrine and paracrine regulation of cardiovascular cell function as many cells regulate each other's function.
- Cell-extracellular matrix interactions are critical in normal physiology and in pathogenesis of disease.
- Physical forces, both hemodynamic and biomechanical, regulate cardiovascular cell functions, in blood vessels, heart valves and myocardium.
- The genetic contribution to the pathogenesis of cardiovascular diseases may be monogenic or polygenic. In the latter numerous independent risk alleles present in the genome contribute to the total heritability of the disease.
- Mechanisms of gene regulation are complex and involve epigenetic mechanisms including DNA base modifications, histone modifications and RNA modifications involving non-coding microRNA and long non-coding RNA. This complex regulation may provide unique opportunities for the design of therapeutic interventions.
- The reexpression of genes and the reactivation of molecules and pathways in disease conditions that were once active during development in the embryo and then repressed in the adult is being investigated to understand pathogenesis of adult cardiovascular diseases.
- The response to tissue injury paradigm is a useful model to study vascular, cardiac, and valvular disease.
- In vitro and in vivo experimental models of cardiovascular disease have yielded valuable molecular knowledge that leads to strategies to diagnose, prevent and treat human disease.
- Stem and progenitor cells are important in cardiovascular biology and pathobiology. Bone marrow derived stem and pluripotent cells, embryonic stem cells (ESCs) and induced pluripotent stem cells (iPSCs) are studied to develop cell therapy for treatment of cardiovascular disease and for regeneration and repair of injured and diseased myocardium and blood vessels.

## Suggested readings

[1] Bondue A, Arbustini E, Bianco A, et al. Complex roads from genotype to phenotype in dilated cardiomyopathy: scientific update from the working group of myocardial function of the European Society of Cardiology. Cardiovasc Res 2018;114:1287—303.
[2] Cunningham KS, Gotlieb AI. The role of shear stress in the pathogenesis of atherosclerosis. Lab Invest 2005;85:9—23.
[3] Davis FM, Rateri DL, Daugherty A. Abdominal aortic aneurysm: novel mechanisms and therapies. Curr Opin Cardiol 2015;30:566—73.
[4] Elliott P, Andersson B, Arbustini E, et al. Classification of the cardiomyopathies: a position statement from the European Society of Cardiology working group on myocardial and pericardial disease. Eur Heart J 2008;29:270—6.
[5] Klotz L, Norman S, Vieira JM, et al. Cardiac lymphatics are heterogeneous in origin and respond to injury. Nature 2015;522:62—7.
[6] Kumar S, Woo Kim C, Simmons RD, Jo H. Role of flow-sensitive microRNAs in endothelial dysfunction and atherosclerosis—"mechanosensitive athero-miRs". Arterioscler Thromb Vasc Biol 2014;34:2206—16.
[7] Laffont B, Rayner KJ. MicroRNAs in the pathobiology and therapy of atherosclerosis. Can J Cardiol 2017;33:313—24.
[8] Loeys BL, Schwarze U, Holm T, et al. Aneurysm syndromes caused by mutations in the TGF-beta receptor. N Engl J Med 2006;355:788—98.
[9] McPherson R, Tybjaerg-Hansen A. Genetics of coronary artery disease. Circ Res 2016;118:564—78.

[10]  Olivotto I, d'Amati G, Basso C, et al. Defining phenotypes and disease progression in sarcomeric cardiomyopathies: contemporary role of clinical investigations. Cardiovasc Res 2015;105:409—23.

[11]  Poe AJ, Knowlton AA. Exosomes and cardiovascular cell-cell communication. Essays Biochem. 2018;62:193—204.

[12]  Prondzynski M, Mearini G, Carrier L. Gene therapy strategies in the treatment of hypertrophic cardiomyopathy. Pflugers Arch 2018;471:807—15.

[13]  Schoen FJ, Gotlieb AI. Heart valve health, disease, replacement, and repair: a 25-year cardiovascular pathology perspective. Cardiovasc Pathol 2016;25:314—52.

[14]  Smith AST, Macadangdang J, Leung W, et al. Human iPSC-derived cardiomyocytes and tissue engineering strategies for disease modeling and drug screening. Biotechnol Adv 2017;35:77—94.

[15]  Tabas I, Bornfeldt KE. Macrophage phenotype and function in different stages of atherosclerosis. Circ Res 2016;118:653—67.

[16]  Vimalanathan AK, Ehler E, Gehmlich K. Genetics of and pathogenic mechanisms in arrhythmogenic right ventricular cardiomyopathy. Biophys Rev 2018;10.1007/s12551-018—0437-0.

[17]  Xu S, Bendeck M, Gotlieb AI. Vascular pathobiology: atherosclerosis and large vessel disease. In: Buja LM, Butany J, editors. Cardiovascular Pathology. 4th ed. New York, NY: Elsevier; 2016.

Chapter 15

# Molecular basis of hemostatic and thrombotic diseases

Karlyn Martin, Alice D. Ma and Nigel S. Key

*Department of Medicine, Division of Hematology/Oncology, University of North Carolina, Chapel Hill, NC, United States*

**Summary**

The coagulation system, composed of cells and soluble protein elements, leads to hemostasis (physiologic blood clotting) at the site of blood vessel injury. Thrombosis refers to the process of excessive clotting that results in adverse cardiovascular events, including myocardial infarction, thrombotic stroke, and venous thromboembolism. Hemostasis is normally regulated by antithrombotic mechanisms that serve to prevent excess clot formation. Disorders of hemostasis or thrombosis can occur when components of the coagulation system are missing or dysfunctional. Hemostatic defects may lead to bleeding via the following mechanisms: (1) *Deficient thrombin generation* on the proper cellular surface due to deficiencies of factor VIII or IX (hemophilia A and B) required for the propagation phase of thrombin generation, deficiencies of factors in the final common pathway of thrombin generation (factors II, V, and X), deficiency of factor XI (required for 'over-drive' of coagulation), defect in fibrin polymerization via either deficiencies or abnormalities of fibrinogen, or deficiency of Factor XIII (required for cross linking fibrin); (2) *Defect in primary hemostasis* due to Von Willebrand disease or platelet dysfunction or (3) *Abnormal fibrinolysis*. Defects in the anticoagulant system lead to thrombosis via the following mechanisms: (1) *Unopposed or excess generation of thrombin* via antithrombin deficiency or prothrombin G20210A mutation and (2) *Insufficient inactivation of procoagulant proteins* via activated protein C resistance (including that due to the factor V Leiden mutation) and deficiencies of Protein C or Protein S.

## Introduction and overview of coagulation

Blood coagulation is the process whereby cells and soluble protein elements interact to form an intravascular blood clot. When this occurs in response to vessel injury, it is an important protective mechanism that functions to seal vascular bleeds, and thereby prevent excessive hemorrhage. This physiological process is generally referred to as hemostasis. In pathological situations, blood coagulation may be triggered by a variety of stimuli to form a maladaptive intravascular clot or thrombus that may obstruct blood flow to or from a critical organ, and/ or embolize to a distal site through the circulatory system. This process is known as thrombosis or thromboembolism, and it may affect either the arterial or venous circulations. A comprehensive review of the molecular basis of all of the defects leading to disorders of hemostasis and thrombosis is beyond the scope of this chapter. We will instead focus on broad themes, particularly defects in the soluble coagulation factors, defects in platelet number or function, other defects leading to hemorrhage, and inherited defects predisposing to thrombosis.

Coagulation can be conceptualized as a series of steps occurring in overlapping sequence. Primary hemostasis refers to the interactions between the platelet and the injured vessel wall, culminating in formation of a platelet plug. The humoral phase of clotting (secondary hemostasis) encompasses a series of enzymatic reactions, resulting in a hemostatic fibrin plug. Finally, fibrinolysis and wound repair mechanisms are recruited to restore normal blood flow and vessel integrity. Each of these steps is carefully regulated, and defects in any of the main components or regulatory mechanisms can predispose to either hemorrhage or thrombosis. Depending on the nature of the defect, the hemorrhagic or thrombotic tendency can be either profound or subtle.

Primary hemostasis begins at the site of vascular injury, with platelets adhering to the subendothelium, utilizing interactions between molecules in the vessel wall, such as collagen and von Willebrand factor (VWF), with glycoprotein and integrin receptors on the platelet surface. Specifically, the primary platelet receptor for subendothelial VWF

*Essential Concepts in Molecular Pathology.* DOI: https://doi.org/10.1016/B978-0-12-813257-9.00015-2

under high shear conditions is the glycoprotein (GP) Ib/V/IX complex, while the primary platelet receptors for direct collagen binding are GP VI and the integrin $\alpha_2\beta_1$. Platelet spreading is followed by activation and release of granular components, and exposure to the cocktail of agonists exposed at a wounded vessel amplifies the process of platelet activation. Through a process known as inside-out signaling, the integrin $\alpha_{2b}\beta_3$ (also known as GP IIbIIIa) undergoes a conformational change in order to be able to bind fibrinogen, which cross-links adjacent platelets and leads to platelet aggregation. Activated GP IIbIIIa can also bind VWF under certain circumstances (such as high shear). Secretion of granular contents is also triggered by external activating signals, further potentiating platelet activation. Lastly, the membrane surface of the platelet changes to serve as a scaffold for the series of enzymatic reactions that result in thrombin generation. This process is primarily dependent on the exposure of negatively charged phospholipids that are normally confined to the inner leaflet of the plasma membrane bilayer.

Our understanding of the process by which fibrin is ultimately generated by thrombin cleavage of soluble fibrinogen has undergone several iterations over the past half century. The waterfall (or cascade) model of coagulation was developed by two groups nearly simultaneously and included an extrinsic, intrinsic, and a common pathway leading to fibrin formation. In this model, the intrinsic pathway can be initiated by interaction with negatively charged surfaces; *in vivo*, several candidate negatively charged molecules may activate the intrinsic pathway, including platelet-derived polyphosphates (polyP), cell-free DNA and/or RNA, or misfolded proteins. *In vitro*, the negatively charged inorganic compounds such as kaolin or ellagic acid are used to activate the pathway in the activated partial thromboplastin test (aPTT). The negatively charged surface activator interacts with the contact factors (pre-kallikrein or PK, high molecular weight kininogen or HMWK, and factor XII) to generate the activated form of factor XII, which is denoted XIIa. Factor XIIa then cleaves and activates factor XI. Factor XIa activates factor IX to generate IXa, which cooperates with its cofactor, factor VIIIa, to form the so-called tenase complex. This complex then proteolytically cleaves factor X to generate factor Xa. Factor Xa, along with its cofactor Va, forms the prothrombinase complex, which generates thrombin from its zymogen (factor II or prothrombin). The thrombin formed by this complex cleaves fibrinogen to allow it to polymerize into insoluble fibrin strands. Factor XIII is also activated by thrombin, and the role of factor XIIIa, a transglutaminase, is to cross-link fibrin strands to provide additional clot strength and stability. The extrinsic pathway requires tissue factor (TF) in complex with factor VIIa to form the tenase complex. TF, unlike other procoagulant clotting factors, is not active in solution, but rather is present as a transmembrane protein on certain cell types. Additionally, TF/VIIa can activate factor IX to IXa, serving as an alternate method of activating the intrinsic tenase complex (Fig. 15.1).

While the cascade hypothesis explains the prothrombin time (PT) and the activated partial thromboplastin time (aPTT) tests as they are performed *in vitro*, it fails to explain the relative intensity of the bleeding diathesis seen in individuals deficient in factors XI, IX, and VIII, as well as the lack of bleeding in those deficient in factor XII, HMWK, or PK. The first reconciliation of this paradox came with the discovery that the tissue factor/VIIa complex activates both

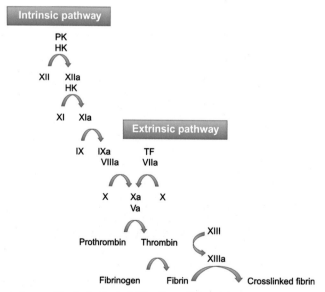

**FIGURE 15.1**   Classic view of hemostasis according to the "waterfall" or "cascade" model.

Molecular basis of hemostatic and thrombotic diseases **Chapter | 15** **231**

factor IX, as well as factor X. More recently, a cell-based model of hemostasis was proposed to address these deficiencies, and to integrate the role of cell surface-bound coagulation reactions in hemostasis. In this model, a TF-bearing cell such as a fibroblast or activated monocyte serves as the site for generation of a small amount of thrombin and factor IXa. The initial thrombin burst is quickly limited by the interaction of the resultant factor Xa with tissue factor pathway inhibitor (TFPI), a Kunitz-like inhibitor present in plasma and at cell surfaces. The binary complex of Xa-TFPI then forms an inhibitory quaternary complex with TF-VIIa. The small amount of thrombin generated by this initiation step is insufficient to cleave fibrinogen, but is adequate to activate platelets and proteolytically release circulating factor VIII from its non-covalently associated VWF, allowing for the formation of VIIIa. The factor IXa already formed on the TF-bearing cell cooperates with VIIIa to form the tenase complex on the surface of the activated platelet. The Xa thus formed interacts with the Va generated on the platelet surface (via thrombin activation of factor V) to form the pro-thrombinase complex. This complex generates a large burst of thrombin that is now sufficient to cleave fibrinogen, activate factor XIII, and activate the thrombin activatable fibrinolysis inhibitor (TAFI), thus allowing for formation of a stable fibrin clot (Fig. 15.2). In this model, XI also binds to the surface of activated platelets where it is activated by higher concentrations of thrombin. Factor XIa so formed further boosts the activity of the intrinsic tenase complex, but is not necessary for adequate thrombin generation under most circumstances in which the hemostatic process is triggered.

Fibrinolysis leads to clot dissolution, thereby restoring normal blood flow and vascular remodeling. Plasminogen is activated to plasmin by the action of either tissue plasminogen activator (t-PA) or urokinase plasminogen activator (u-PA). Both plasminogen and the plasminogen activators bind to free lysine residues exposed on fibrin, and in this way fibrin acts as the cofactor for its own destruction. Plasmin is capable of degrading both fibrin and fibrinogen, and can thus dissolve both formed clot as well as its soluble precursor. However, non-fibrin-bound plasmin is inhibited by a number of circulating inhibitors, of which $\alpha_2$-plasmin inhibitor ($\alpha_2$-PI) is the most significant. Plasminogen activation is also inhibited by a number of molecules; chief among them is plasminogen activator inhibitor-1 (PAI-1), a serine protease inhibitor (SERPIN) that irreversibly binds to—and inactivates—the plasminogen activators. Lastly, cellular receptors act to localize and potentiate or clear plasmin and plasminogen activators.

## Disorders of soluble clotting factors

While classic hemophilia/hemophilia A (factor VIII deficiency) and Christmas disease/hemophilia B (factor IX deficiency) are the best known examples of the clotting factor disorders, the following overview will discuss defects, including deficiencies (Table 15.1), of each clotting factor in the numerical order ascribed by the Roman numeral classification system. Note that fibrinogen is rarely referred to as factor I even though this is how it is designated using strict nomenclature; similarly, tissue factor is rarely referred to by its designated number (factor III). No tissue factor

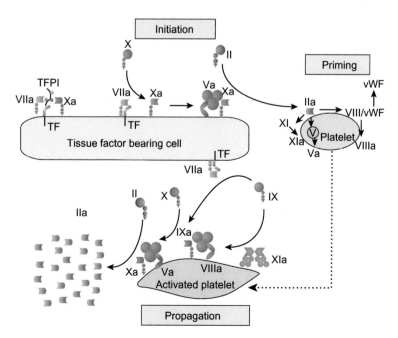

**FIGURE 15.2** The cell-based model of hemostasis. *Courtesy of Dr. Dougald Monroe.*

**TABLE 15.1 Clinical features of inherited coagulation factor deficiency state.**

| Defect | | Inheritance pattern | Bleeding manifestations | PT | PTT | TCT | BT | Treatment |
|---|---|---|---|---|---|---|---|---|
| **Fibrinogen abnormalities** | Afibrinogenemia | Autosomal | Severe, but less so than severe Hemophilia A and B | Infinite | Infinite | Infinite | Prolonged | Fibrinogen concentrate, Cryoprecipitate |
| | Dysfibrinogenemia | Autosomal | Variable bleeding and/or clotting | Prolonged | Prolonged | Prolonged or shortened | Normal | Crycprecipitate |
| **Prothrombin deficiency** | | Autosomal | Varies with prothrombin levels | Prolonged | Prolonged | Normal | Normal | PCCs |
| **Factor V deficiency** | | Autosomal | Mild-moderate | Prolonged | Prolonged | Normal | Prolonged | FFP, potential need for exchange transfusion |
| **Factor VII deficiency** | | Autosomal | Moderate-severe | Prolonged | Normal | Normal | Normal | Recombinant-activated factor VII |
| **Hemophilia A** | | X-linked recessive | Variable, depending on Factor VIII level | Normal | Prolonged | Normal | Normal | Factor VIII concentrates, DDAVP in mild cases |
| **Hemophilia B** | | X-linked recessive | Variable, depending on Factor IX level | Normal | Prolonged | Normal | Normal | Factor IX concentrates |
| **Factor X deficiency** | | Autosomal | Variable, depending on Factor X level | Prolonged | Prolonged | Normal | Normal | Plasma, PCCs, or Factor X concentrate |
| **Factor XI deficiency** | | Autosomal | Variable, but NOT dependent on Factor XI levels | Normal | Prolonged | Normal | Normal | Plasma or Recombinant-activated factor VII |
| **Deficiency of Factor XII, prekallikrein, or high molecular weight kininogen** | | Autosomal | None | Normal | Prolonged | Normal | Normal | None needed |
| **Factor XIII deficiency** | | Autosomal | Severe | Normal | Normal | Normal | Normal | Factor XIII concentrate, Cryoprecipitate |
| **Deficiency of alpha-2 plasmin inhibitor or plasminogen activator inhibitor-1** | | Autosomal | Severe | Normal | Normal | Normal | Normal | Antifibrinolytic agents (epsilon aminocaproic acid or tranexamic acid) |

**TABLE 15.2** Useful websites for documented mutations in individual coagulation factor deficiency States.

| | |
|---|---|
| Fibrinogen | www.geht.org/databaseang/fibrinogen/ |
| Prothrombin | https://www.isth.org/?MutationsRareBleedin |
| | www.ncbi.nlm.nih.gov/gene |
| Factor V | https://www.isth.org/?MutationsRareBleedin |
| | www.hgmd.org |
| Factor VII | http://www.umd.be/F7/W_F7/index.html |
| Factor VIII | http://www.factorviii-db.org/index.php |
| Factor IX | http://www.factorix.org/ |
| Factor X | https://www.isth.org/?MutationsRareBleedin |
| Factor XI | www.factorXI.org |
| Factor XIII | www.f13-database.de/(xhgmobrswxgori45zk5jre45)/index.aspx |
| VWF | http://www.vwf.group.shef.ac.uk/ |
| Protein S | https://www.isth.org/?ProteinSDeficiency |
| Protein C | http://www.itb.cnr.it/procmd/ |
| Antithrombin | https://www1.imperial.ac.uk/departmentofmedicine/divisions/experimentalmedicine/haematology/coag/antithrombin/ |
| Vitamin K dependent factors | https://c.ymcdn.com/sites/www.isth.org/resource/resmgr/publications/fvkdf_mutations-2011.pdf |

deficiency states have been described in humans. In addition, factor IV is reagent calcium, and there is no factor VI in the current scheme. While hemophilia A occurs in approximately 1:10,000 of the population, and hemophilia B in about 1:40,000, the remaining factor deficiencies (with the possible exception of factor XI in Ashkenazi Jewish populations) typically occur at a frequency between 1:500,000 and 1:2,000,000 of the general population, and are thus usually known collectively as the rare inherited bleeding disorders.

Rather than attempt to provide an exhaustive list of mutations affecting all coagulation proteins in this text, we have provided a number of useful websites containing described mutations in these proteins in Table 15.2. In addition, a summary of all described mutations in factors II, V, VII, X, XI, XIII, and combined V/VIII deficiency can be found at https://www.isth.org/?SummaryofMutations [1]. The Human Gene Mutation [2] database, available at www.hgmd.org, also contains detailed information on many genetic mutations in a variety of inherited disorders, including those affecting the coagulation pathways. Table 15.3 summarizes the essential biochemical data on each of the important inherited deficiency states.

## Fibrinogen abnormalities

Fibrinogen abnormalities are inherited in an autosomal pattern and occur in two main patterns: hypo/afibrinogenemia and dysfibrinogenemia. Afibrinogenemia is a very rare disorder that occurs when any one of the three genes coding for the alpha, beta, or gamma chains that make up the fibrinogen molecule is mutated. If the mutation is sufficient to disrupt formation or secretion of any of the three chains, afibrinogenemia results. Afibrinogenemic patients have a severe bleeding disorder manifest by bleeding after trauma into subcutaneous and deeper tissues that may result in dissection. Bleeding from the umbilical stump at birth and intra-cranial hemorrhage occur frequently, as does early miscarriage in affected women. Though hemarthroses do occur in these patients, they are less frequent than in the severe forms of hemophilia A and B. Less severe mutations in one of the fibrinogen chains lead to reduced but measurable circulating levels of fibrinogen, or hypofibrinogenemia. Patients with hypofibringenemia manifest a variable but concordant reduction in fibrinogen antigen and activity levels in plasma.

Dysfibrinogenemia is also rare but is more common than afibrinogenemia. The dysfibrinogens are the result of missense, nonsense, or splice junction mutations, with the majority of patients being heterozygous for the disorder. Several hundred mutations have been recorded, many of which result in neither a hemorrhagic nor thrombotic state. However, other dysfibrinogens are associated with bleeding episodes, while a few may be associated with venous or arterial thrombosis.

## Prothrombin (factor II) deficiency

Inherited prothrombin deficiency is rare, with fewer than 100 distinct mutations being reported. It is an autosomal recessive disorder, and heterozygotes have no bleeding symptoms. Symptomatic patients may be homozygous or doubly

TABLE 15.3 Summary of biochemical features and hemostatic levels of coagulation factors.

| | Chromosome | MW (Da) | Number of Chains (Active) | T ½ (Half-Life) | Plasma Concentrations (mg/l) | Number of Gla Residues | Number of Described Mutations | Prevalence of Deficiency State | Target Hemostatic Levels (% ref. Range) |
|---|---|---|---|---|---|---|---|---|---|
| FII | 11 | 72,000 | 2 | 3–4 d | 100 | 10 | ~60 | $1/2 \times 10^6$ | 20%–40% |
| FV | 1 | 330,000 | 2 | 36 h | 10 | – | ~140 | $1/10^6$ | 15%–25% |
| FVII | 13 | 50,000 | 2 | 4–6 h | 0.5 | 10 | ~250 | 1/500,000 | 15%–20% |
| FVIII | X | 260,000 | | 10–14 h | 0.1 | – | >2000 | 1/10,000 | 60–100% depending on site of bleed[a] |
| FIX | X | 57,000 | | 25 h | 5 | 12 | ~1100 | 1/40,000 | 60–100% depending on site of bleed[a] |
| FX | 13 | 59,000 | – | 40–60 h | 10 | 11 | >100 | $1/10^6$ | 15%–20% |
| FXI | 4 | 160,000 | 2 | 40–72 h (52 h) | 5 | – | >200 | $1/10^6$ | 15%–20% |
| FXIII (2 genes) | 6 (A chain) 1 (B chain) | 320,000 | 4 | 11–14 d | 30 | – | >100 | $1/2 \times 10^6$ | 2%–5% |
| FV/FVIII ERGIC 53 MCFD2 | 18 2 | 330,000 (FV) (FVIII) | 2 (FV) 3 (FVIII) | 36 h (FV) 10–14 h (FVIII) | 10 (FV) 0.1 (FVIII) | – | >30 (ERGIC) ~20 (MCFD2) | $1/2 \times 10^6$ | 15%–20% |
| Fibrinogen (3 genes) | 4 (α, β, and γ chains) | 340,000 | 6 | 2–4 d | 1500–3500 | – | >900 | $1/10^6$ | 50 mg/dL |

[a]For more detailed guidelines on goal factor levels in the management of hemophilia, see World Hemophilia Federation Treatment Guidelines at http://www.wfh.org/image/7-publications-en/Treatment_Guidelines_Table_7.1.jpg.

heterozygous for causative mutations. By convention, patients with hypoprothrombinemia manifest a concordant reduction in circulating prothrombin antigen and functional activity, whereas those with hypodysprothrombinemia classically have a reduced concentration of prothrombin antigen with a discordantly low functional activity. Bleeding in affected patients varies from mild to severe, depending on the functional prothrombin activity level. The complete absence of prothrombin probably leads to embryonic lethality.

## Factor V deficiency

Factor V is a large molecule (330 kDa) that shares significant structural homology with FVIII and ceruloplasmin. Factor V deficiency is an autosomal recessive disorder that results from mutations in the factor V gene. Heterozygotes are generally asymptomatic, while homozygotes or combined heterozygotes may have mild to moderately severe bleeding symptoms. About 100 mutations in the factor V gene have been reported. Bleeding manifestations are similar to those seen in classic hemophilia, except that they tend to be milder, and hemarthroses are less common.

## Factor VII deficiency

Approximately 1% of the total factor VII in plasma circulates as the active serine protease (FVIIa, 10–100 pmol/L). Controversy still exists as to which enzyme is responsible for basal activation of factor VII *in vivo*, although there is evidence implicating FIXa. In the absence of its cofactor TF, FVIIa is a very weak enzyme.

Factor VII deficiency is an autosomal recessive bleeding disorder that occurs in mild, moderate, and severe forms. It is generally considered to be the most common of the rare bleeding disorders, with a prevalence of 1:300,000 to 1:500,000 in most populations. More than 200 mutations in the gene for factor VII have been reported. Bleeding manifestations vary, but in severely affected patients, bleeding can be as severe as that seen in severe classic hemophilia and may include crippling hemarthroses. Factor VII levels of 10% of normal are probably sufficient to control most bleeding episodes, but occasionally higher levels may be required for hemostasis. Some patients with almost no measurable factor VII activity may express very few hemorrhagic manifestations. Furthermore, there are reports of thrombotic events occurring in patients with FVII deficiency, such that it has been unclear whether distinct mutations may actually be prothrombotic. However, at present, registry analysis suggests that while factor VII deficiency does not seem to predispose to thrombosis, neither is it protective [34].

## Hemophilia A and hemophilia B (classic hemophilia and Christmas disease)

Hemophilia A and B result from deficiencies of factor VIII and IX, respectively. Factor VIII and IX are necessary for the sustained generation of factor Xa (and ultimately thrombin) to form a normal hemostatic plug in response to vascular injury. Hemophilia A and B are the only two soluble clotting factor deficiencies that are inherited as X-linked recessive disorders. Over a thousand distinct mutations in each gene have been reported. These mutations result in mild, moderate, and severe forms of hemophilia, and the clinical manifestations of hemophilia A and B are, for all practical purposes, indistinguishable. In the severe form (<1% basal activity), both disorders are characterized by recurrent hemarthroses that result in chronic crippling arthropathy unless treated by replacing the deficient factor on a scheduled prophylactic basis. Central nervous system hemorrhage is especially hazardous and remains one of the leading causes of death. Retro-peritoneal hemorrhage and bleeding into the pharynx may also be life-threatening.

## Factor X deficiency

Like factor VII deficiency, factor X deficiency is inherited in an autosomal recessive fashion and can be mild, moderate, or severe. Numerous mutations have been recorded. Severely affected patients have symptoms similar to severe classic hemophilia, including hemarthroses and chronic crippling hemarthropathy. About one-third of affected individuals exhibit a type 1 deficiency, with reduction in both antigen and activity levels, while the rest have a type 2 pattern, with factor X antigen levels that are preserved, or at least significantly above the level of factor X activity.

## Factor XI disorders

Factor XI deficiency is an autosomal recessive disorder that commonly occurs in patients of Ashkenazi Jewish descent. In these communities, homozygotes may be as prevalent as 1:500, compared to a frequency of about 1:1,000,000 in most other populations. Three mutations account for most cases of factor XI deficiency in Jewish patients. Factor XI-deficient patients have normal levels of factor VIII and IX to form the tenase complex and normal levels of factors V and X to form the prothrombinase complex. While FXI deficient patients do not bleed spontaneously, they can be susceptible to injury-induced bleeding because of the role of factor XI as both a procoagulant (factor IX activating) and an antifibrinolytic agent. Without factor XIa activation of factor IX, there is a reduction in potentiation of thrombin generation after the initial thrombin burst. Factor XIa indirectly activates thrombin activatable fibrinolysis inhibitor (TAFIa), which cleaves free lysine residues on fibrin to prevent fibrin-dependent activation of plasminogen by plasminogen activators. With FXI deficiency, TAFI activation is reduced and down-regulation of fibrinolysis is diminished, and indeed, factor XI deficient patients tend to bleed at sites of enhanced fibrinolytic activity such as the nasopharyngeal or genitourinary tracts.

More severely affected factor XI deficient patients (FXI:c <15%) have been shown to be relatively protected from ischemic stroke but not myocardial infarction, while elevated levels of factor XI have been found to be a risk factor for venous thromboembolism and ischemic stroke [28–33]. Recently, lowering factor XI levels with targeted antisense oligonucleotides has been shown to lower venous thrombosis risk after knee replacement surgery without increasing bleeding risk [26,27].

## Deficiencies of factor XII, prekallikrein (PK), and high molecular weight kininogen (HK)

Deficiencies of factors XII, PK, and HK (the so-called contact factors) are inherited in an autosomal recessive fashion. These defects cause a marked prolongation of the aPTT, but other screening tests of coagulation are normal. They are not associated with bleeding even after trauma or surgery, although the prolonged aPTT may cause a great deal of consternation among those not familiar with these defects. A good history revealing the absence of bleeding in these patients and their family members despite a long aPTT is the best indication that one is dealing with one of these defects. The relationship between FXII and venous and arterial thrombosis remains unclear with conflicting evidence, but an elevated factor XII level does not appear to be a risk factor for thrombosis, nor does a factor XII deficiency appear to protect against clot formation. The kallikrein-kinin system may be more physiologically important in inflammation, blood pressure regulation, and fibrinolysis than in hemostasis.

## Factor XIII deficiency

FXIII is activated by thrombin in the presence of calcium. Factor XIIIa is a plasma transglutaminase that covalently cross-links fibrin alpha and gamma chains through γ-glutamyl-ε-lysine bonds to form an impermeable fibrin clot. Although a clot may form in the absence of factor XIII and be held together by hydrogen bonds, this clot is excessively permeable and is easily dissolved by the fibrinolytic system. The clot formed in the absence of factor XIII does not form a normal framework for wound healing, and abnormal scar formation may occur.

Factor XIII consists of two A chains and two B chains. Factor XIII-A is synthesized in megakaryocytes, monocytes, and macrophages, whereas factor XIII-B is synthesized in hepatocytes. The complete molecule is an A2B2 tetramer with the A chains containing the active site and the B chains acting as a carrier for the A subunits. Platelet alpha granules contain A chains but not B chains. Factor XIII deficiency may result from mutations in the genes encoding either the A or B chains, with A chain mutations being more common. Autosomal genes govern hepatic synthesis of the factor, and the disease is expressed as a recessive disorder. Typically, only patients with severe deficiency of factor XIII (<3%–5%) are symptomatic.

Bleeding manifestations are generally severe, and hemorrhage can occur into any tissue. Umbilical stump bleeding in the neonatal period is common in factor XIII deficiency. Intracranial hemorrhage is also a relatively common manifestation that may mandate long-term prophylactic factor replacement.

## Multiple clotting factor deficiencies

The two most common multiple clotting factor deficiencies are a combined deficiency of factors V and VIII and a combined deficiency of the vitamin K-dependent factors (factors II, VII, IX, X, and Protein C and S).

A combined deficiency of factors V and VIII is inherited in an autosomal recessive fashion and can be distinguished from a combined inheritance of mild classic hemophilia and mild factor V deficiency by family studies or by genetic analysis. The disorder is due to defects in one of two genes: the *LMAN1* gene and a gene called the multiple clotting factor deficiency 2 (*MCFD2*) gene. The products of both genes play a critical role in the transport of factors V and VIII from the endoplasmic reticulum to the Golgi apparatus and are necessary for normal secretion of these factors. The disorder results in a mild to moderate bleeding tendency with factor V and VIII levels ranging from 5% to 30% of normal.

Inherited combined deficiencies of the vitamin K dependent factors can be due to defects in either the gene for vitamin K-dependent carboxylase or the gene for vitamin K epoxide reductase. These are autosomal recessive disorders that may be associated with severe deficiency of prothrombin, factor VII, IX, and X, as well as Protein C and S. The diagnosis must be distinguished from surreptitious ingestion of coumarin drugs (including rodenticides), which is an acquired disorder with bleeding manifestations of recent onset.

## Von Willebrand disease (VWD)

The most common hereditary bleeding disorder arises from abnormalities in von Willebrand factor (VWF). VWF occurs in plasma as multimers of a 240,000 Dalton subunit, with molecular weights ranging from 1 million to 20 million Daltons. The principal functions of VWF are to act as a carrier for clotting factor VIII and to mediate platelet adhesion to the injured vessel wall. The larger molecular weight multimers are the most effective at mediating platelet adhesion. VWF binds to glycoprotein Ib on the platelet surface and also to collagen in the vessel wall. VWF is particularly important for platelet adhesion under high shear stress vascular beds. It has been elegantly demonstrated that shear stress leads to unfolding of the globular collagen-bound VWF multimers, thereby revealing the molecular domain responsible for binding of platelets to VWF. VWF also cross-links platelets via binding to glycoprotein IIb-IIIa.

There are three major types of von Willebrand disease (VWD): type 1, 2, and 3 (Table 15.4). Type 1 is autosomal dominant and represents a partial quantitative deficiency of VWF. Generally, it is explained by reduced synthesis of VWF, and analysis of multimers in plasma reveals a global decrease in multimers of all sizes. More recently, it has been appreciated that the pathophysiology of some cases of type 1 VWF is accelerated clearance of VWF from plasma. The prototype of this phenotype is the so-called Vicenza R1205H mutation, which causes a moderate to severe variant of von Willebrand disease. Type 3 VWD is a severe quantitative deficiency in which there is a near absence of VWF, which results from an autosomal recessive inheritance of homozygous or compound heterozygous genetic mutations. Type 2 VWD is characterized by qualitative abnormalities in VWF function, and occurs in four major forms: 2A, 2B, 2N, and 2M. All forms occur as an autosomal dominant disorder except 2N, which is an autosomal recessive disorder. Types 2A and 2B are characterized by absence of the higher molecular weight multimers of VWF in plasma. Type 2B

TABLE 15.4 Von Willebrand disease subtypes.

| | Inheritance pattern | Bleeding manifestations | Diagnostic testing | | | | Treatment |
| --- | --- | --- | --- | --- | --- | --- | --- |
| | | | VWF Ag | VWF Activity | Factor VIII activity | VWF multimers | |
| **Type 1** | Autosomal dominant | Generally mild | Low | Low | Low | Normal | DDAVP, factor VIII concentrates rich in VWF |
| **Type 2** | | | | | | | |
| 2 A | Autosomal dominant | Mild-moderate | Low | Lower than antigen | Variable | Absent high molecular weight forms | Factor VIII concentrates rich in VWF |
| 2B | Autosomal dominant | Mild-moderate | Low | Lower than antigen | Variable | Absent high molecular weight forms | Factor VIII concentrates rich in VWF |
| 2 N | Autosomal recessive | Mild | Normal | Normal | Low | Normal | Factor VIII concentrates rich in VWF |
| 2 M | Autosomal dominant | Mild-moderate | Normal | Lower than antigen | Normal | Normal | DDAVP, factor VIII concentrates rich in VWF |
| **Type 3** | Autosomal recessive | Severe | Near absent | Near absent | Near absent | Absent | Factor VIII concentrates rich in VWF |

is also associated with thrombocytopenia as a result of a gain of function mutation resulting in a VWF molecule with higher affinity for the GPIb receptor, thus enhancing platelet agglutination and accelerated clearance. Type 2 M patients show reduced binding of VWF to GPIb, although they have normal VWF multimeric composition in plasma.

Type 2N VWD is a rare disorder arising from a mutation in the factor VIII binding site on the VWF molecule. Without the protection provided by VWF binding, factor VIII has a markedly decreased half-life resulting in reduced plasma levels. VWF multimers and antigen and activity levels may be normal, while the factor VIII activity levels are low enough to be confused with mild classic hemophilia. These two disorders can be distinguished by an ELISA-based factor VIII binding assay or by direct genotyping. Clinically affected patients are either homozygous for one of several gene mutations in the D0 or D3 domain of mature VWF or are combined heterozygous for a 2N mutation and a type 1 mutation. Specific diagnosis requires either demonstration of the lack of binding of VWF to factor VIII or genetic analysis.

Although VWD is a defect in a soluble clotting factor, bleeding in patients with this disorder is more similar to that produced by a defect in platelet number or function. The bleeding manifestations tend to be more of the "oozing and bruising" variety, with hematoma formation being rare. Bleeding in types 1 and 2 VWD is usually mild to moderate, although severe bleeding may occur with trauma and surgery. Some patients with type 1 VWD may be relatively asymptomatic. Table 15.4 lists the diagnostic features of the various types of VWD.

# Disorders of platelet number or function
## Disorders of platelet production

Inherited disorders causing thrombocytopenia are a heterogeneous group of conditions. Some are associated with a profound thrombocytopathy, some are associated with other somatic changes, while others manifest thrombocytopenia only.

### The MYH9-associated disorders

The May-Hegglin anomaly is the prototype of a family of disorders due to a defect in the MYH9 gene. Other MYH9 gene disorders include Sebastian syndrome, Fechtner syndrome, and Epstein syndrome. These are autosomal dominant macrothrombocytopenias that are distinguished by different combinations of clinical and laboratory findings, such as sensorineural hearing loss, cataracts, nephritis, and polymorphonuclear inclusions known as Döhle-like bodies. Mutations in the MYH9 gene encoding for the non-muscle myosin heavy chain IIA (NMMHC-IIA) have been identified in all of these syndromes.

### Defects in transcription factors

Alterations in megakaryocyte development due to defective transcription factors underlie a large number of the familial thrombocytopenias. Derangements in development of other cell types as well as other somatic mutations can also occur. Mutations in *HOXA11* have been described in two unrelated families with bone marrow failure and skeletal defects.

The Paris-Trousseau syndrome is an autosomal dominant condition characterized by macrothrombocytopenia with giant alpha granules. It is caused by hemizygous loss of the *FLI1* gene due to deletion at 11q23. Lack of FLI1 protein leads to lack of platelet production due to arrested megakaryocyte development. The 11q23 deletion is also seen in patients with Jacobsen's syndrome who also have congenital heart disease, trigonocephaly, dysmorphic facies, mental retardation and multiple organ dysfunction as well as macrothrombocytes with abnormal alpha granules.

Mutations in *GATA-1* lead to the X-linked congenital dyserythropoietic anemia and thrombocytopenia syndrome. The platelets are large and exhibit defective collagen-induced aggregation. X-linked thrombocytopenia without anemia is due to mutations within *GATA-1* that disrupt FOG-1 (Friend of GATA) interactions while leaving DNA-binding intact. The GATA-1 transcription factor has two zinc fingers. The C-terminal finger binds DNA in a site-specific fashion, while the N-terminal finger stabilizes the DNA binding as well as interacts with FOG-1 (Friend of GATA). Mutations within *GATA-1* may alter DNA-binding, FOG-1 interactions, or both, and phenotypes may differ depending on the site of mutation. X-linked thrombocytopenia without anemia is due to mutations within *GATA-1* that disrupt FOG-1 interactions while leaving DNA-binding intact. By contrast, *GATA-1* mutations, which affect binding to DNA while not interrupting FOG interactions, lead to a thalassemic phenotype. The acute megakaryoblastic anemia seen in conjunction with Down's syndrome can be associated with mutations in *GATA-1*.

Mutations in *RUNX1* lead to familial thrombocytopenic syndromes with a predisposition to development of myeloid malignancies, including acute myelogenous leukemia and myelodysplastic syndrome. RUNX1 mutations cause an arrest in megakaryocyte development with an expanded population of progenitor cells, and the platelets that are produced show defects in aggregation. The development of acute leukemia likely requires a second mutation within RUNX1 or another gene.

## Defects in platelet production

Congenital amegakaryocytic thrombocytopenia (CAMT) is due to defects in the *c-MPL* gene encoding the thrombopoietin receptor. Children born with this disorder have severe thrombocytopenia and may go on to develop deficiencies in other cell types.

Thrombocytopenia with absent radii (TAR) is a syndrome characterized by severe congenital thrombocytopenia along with absent or shortened radii. The platelets produced show abnormal aggregation. Although thrombopoietin levels are elevated, no defect in *c-MPL* has been identified. While the majority of cases of TAR syndrome has been associated with a 1q21.1 deletion, other associated mutations have been found in the *RBM8A* gene. Abnormal intracellular signaling pathways are postulated as the cause of this rare disorder.

Perhaps the most common hereditary thrombocytopenia with small platelets is the Wiskott-Aldrich syndrome (WAS), a disorder associated with the triad of immune deficiency, eczema, and thrombocytopenia. This syndrome is X-linked and results from mutations in the gene for Wiskott-Aldrich syndrome protein (WASP). Platelets as well as T-lymphocytes show defective function, and clinical manifestations vary widely. As opposed to the macrothrombocytopenic defects, WAS platelets are small and defective in function.

## Disorders of platelet function

### Defects in platelet adhesion

Bernard Soulier Syndrome (BSS) is a severe bleeding disorder characterized by macrothrombocytopenia, decreased platelet adhesion to VWF, abnormal prothrombin consumption, and reduced platelet survival. Deficient platelet binding to subendothelial von Willebrand factor is due to abnormalities (either qualitative or quantitative) in the GP-Ib-IX-V complex. Mutations in GPIbα binding sites for P-selectin, thrombospondin-1, factor XI, factor XII, aMb2, and high molecular weight kininogen may mediate variations in the phenotype seen. The product of four separate genes (*GPIBA*, *GPIBB*, *GP9*, and *GP5*) assemble within the megakaryocyte to form the GP-Ib-IX-V on the platelet surface. Defects in any of the genes may lead to BSS. Classically, affected platelets from affected individuals aggregate normally in response to all agonists except ristocetin.

Platelet-type von Willebrand disease is due to a gain of function mutation such that plasma VWF binds spontaneously to platelets, and the platelets exhibit agglutination in response to low dose ristocetin. Mutations generally lie within the *GPIBA* gene. High molecular weight multimers of VWF bound to platelets are cleared from the circulation, which may result in bleeding. The phenotype is identical to that seen in type 2B VWD, in which the mutation lies within the VWF rather than its receptor. It can therefore be quite difficult to distinguish between type 2B VWD and platelet-type VWD. Gene sequencing of the *VWF* gene, the *GPIBA* gene, or both may be required.

### Defects in platelet aggregation

Glanzmann thrombasthenia is a rare, autosomal recessive disorder characterized by absent platelet aggregation. It is due to absent or defective GPIIbIIIa on the platelet surface. Patients have severe mucocutaneous bleeding, which becomes refractory to platelet transfusions as alloantibodies form to transfused platelets. Though demonstration of absent platelet aggregation in response to all agonists (with the exception of ristocetin) will suggest the diagnosis, definitive diagnosis relies on showing absence of functional GPIIbIIIa on the platelet surface, either by flow cytometry or by electron microscopy using immuno-gold labeled fibrinogen imaging. A database of described mutations in the GPIIb and GPIIIa genes that may result in Glanzmann thrombasthenia may be found at http://sinaicentral.mssm.edu/intranet/research/glanzmann/menu [3]. Acquired Glanzmann thrombasthenia has been described in patients who develop autoantibodies against GPIIbIIIa. These patients may have underlying immune thrombocytopenic purpura, but the severity of their bleeding is out of proportion to platelet number.

Some patients exhibit defects in aggregation in response to specific agonists. Platelets from these patients may show defects in either platelet receptors or in the downstream intracellular signaling pathways leading to activation. These disorders must be distinguished from the effects of drugs such as aspirin, whose ingestion can produce similar effects on platelet function.

## Disorders of platelet secretion: the storage pool diseases

Platelets contain two types of intracellular granules: alpha and delta (or dense) granules. Alpha granules contain proteins either synthesized within the megakaryocytes or endocytosed from the plasma, including fibrinogen, factor V, thrombospondin, platelet-derived growth factor, multimerin, fibronectin, factor XIII-A chains, high molecular weight kininogen, and VWF among others. Their membrane contains molecules such as P-selectin and CD63 that are translocated to the outer plasma membrane after secretion and membrane fusion. Dense granules contain ATP and ADP as well as calcium and serotonin, and any deficiency of dense granules thus leads to a defective secondary wave of platelet aggregation.

## Defects in alpha granules

The gray platelet syndrome (GPS) is an autosomal recessive condition that leads to a mild bleeding diathesis. It may be recognized by examination of a Wright- Giemsa stained peripheral blood smear showing platelets that appear gray without the usual red-staining granules. Electron microscopy showing a depletion of alpha granules is a more sensitive method to diagnose the syndrome. GPS is thus classified with the other platelet secretion defects, but may also be classified with the macrothrombocytopenias, since platelets may be slightly larger than usual, albeit not as large as those seen in the giant platelet disorders described previously. Furthermore, the platelet count is only moderately depressed, and bleeding symptoms are mild. Patients with GPS may also develop early onset myelofibrosis, a probable consequence of the impaired storage of growth factors such as PDGF.

The Quebec platelet disorder (QPD) is associated with a normal to slightly low platelet count with a mild bleeding disorder. Pathophysiologically, this is due to abnormal proteolysis of alpha granule proteins that appears to be mediated by ectopic intragranular production of urokinase plasminogen activator (u-PA). QPD was first recognized as a deficiency of platelet factor V associated with normal concentrations of plasma factor V; however, subsequently it has been characterized as a deficiency of multiple alpha granule proteins, including platelet factor V. The platelets appear normal on peripheral blood smears under the light microscope, and diagnosis depends on showing decreased alpha granule proteins. It is inherited in an autosomal dominant fashion. Tandem duplications of the urokinase plasminogen activator gene, *PLAU*, on chromosome 10 have been implicated in QPD, though the mechanisms by which the dramatic increase in intra-granular uPA occurs and the platelet count is lowered remain unknown.

## Defects in dense granules

The Hermansky-Pudlak syndrome (HPS) is the association of delta storage pool deficiency with oculocutaneous albinism and increased ceroid in the reticuloendothelial system. Granulomatous colitis and pulmonary fibrosis are also part of the syndrome. The syndrome is inherited in an autosomal recessive pattern, and there are several subtypes of the HPS resulting from distinct mutations. Mutations in at least eight genes (*HPS-1* through *HPS-8*) lead to defects in HPS proteins responsible for organelle biosynthesis and protein trafficking. Described mutations accounting for the Hermansky-Pudlak syndrome are collated in available databases [4] at http://liweilab.genetics.ac.cn/hpsd/.

The Chediak-Higashi syndrome is also associated with storage pool deficiency and is characterized by oculocutaneous albinism, neurologic abnormalities, immune deficiency with a tendency to infections, and giant inclusions in the cytoplasm of platelets and leukocytes. The disorder is rare, and bleeding manifestations are relatively mild. The syndrome is due to mutations in the *LYST* (lysosomal trafficking regulator) gene; these are listed in the Chediak-Higashi database at http://structure.bmc.lu.se/idbase/LYSTbase/index.php?content=index/IDbases [5]. Affected patients are homozygous, while heterozygotes are phenotypically normal.

## The Scott syndrome

In this disorder, activated platelets cannot translocate phosphatidylserine (PS) from the inner to the outer platelet membrane when the flip-flop of the membrane leaflet occurs. Mutations in the gene *ANO6*, which encodes transmembrane protein 16F (TMEM16F), have recently been identified. TMEM16F is a membrane protein essential for calcium-dependent scrambling of PS. Because of this defect, factors Xa and Va are unable to efficiently bind to the membrane to assemble the prothrombinase complex, and thrombin generation on the platelet surface is impaired. Scott syndrome is characterized by a mild bruising and bleeding tendency. It can be detected using flow cytometry with antibodies against annexin V (as a marker of PS exposure).

## Disorders of platelet destruction

Disorders of platelet destruction are too numerous to discuss here. Therefore, the discussion in the section will be limited to those disorders where the molecular pathogenesis is known at a greater level of detail. A more comprehensive description of disorders characterized by platelet destruction can be found in recent reviews.

### Antibody-mediated platelet destruction

Neonatal alloimmune thrombocytopenia (NAIT) is a bleeding disorder caused by transplacental transfer of maternal antibodies directed against human fetal platelet antigens (HPA) inherited from the father. In Caucasians, the antigens most frequently implicated include HPA- 1a (PLA1, 80%) and HPA-5b (Bra,15%). In Asians, HPA-4a and HPA-3a account for the majority of NAIT cases. NAIT occurs with a lower frequency in Caucasians than is expected by the incidence of HPA-1a negativity in the population, suggesting that other factors influence antibody development. One such factor may be HLA class II allele; there is a strong association between HLA Class II antigen DRB3*01:01 and women who produce anti-HPA-1 antibodies, and it may be that immune response to HPA 1a antigen presentation is optimal with HLA DRB3*01:01. Additionally, NAIT mediated by antibodies against HPA-1a is clinically more severe, perhaps because these antibodies may also block platelet aggregation, since HPA-1a is an antigen expressed on platelet GPIIIa. Mothers who do not express the fetal platelet antigen can develop antiplatelet antibodies that can cross the placenta and result in clearance of fetal platelets leading to severe fetal thrombocytopenia. Intracranial hemorrhage is a feared and devastating complication. Even first pregnancies can be affected by NAIT, and subsequent pregnancies have a near 100% rate of NA IT.

Post-transfusion purpura (PTP) is associated with thrombocytopenia resulting from a mismatch between platelet antigens in transfused blood and recipient platelets. In this condition, patients previously sensitized against certain platelet antigens (similar antigens that lead to NAIT, most commonly HPA-1a) develop acute, severe thrombocytopenia 5−14 days after transfusion. Though packed red cells are most commonly associated with PTP, transfusion of any blood component may precipitate this disorder. These blood components contain platelet microparticles that express the offensive platelet antigen, leading to an anamnestic production of antibodies. However, paradoxically, patients develop antibodies directed against their own platelets in addition to the offending antigen. The means by which alloantibodies destroy autologous platelets is not clear, but postulated mechanisms include fusion of the exogenous microparticles with their own platelets, or by a process in which exposure to foreign platelets leads to formation of autoantibodies.

### Thrombotic microangiopathies

Thrombotic thrombocytopenic purpura (TTP) is an acute disorder that usually presents in previously healthy subjects. TTP is a thrombotic microangiopathy characterized by microvascular thrombi and hemolytic anemia leading to thrombocytopenia and often organ damage, classically with renal failure and neurologic consequences. It is highly lethal unless treated promptly, which generally entails plasma exchange with or without additional immunosuppression. In 1982, Moake made the seminal observation that the plasma of patients with TTP contained ultralarge (UL) multimers of VWF, which were absent in normal plasma. He hypothesized that TTP could be due to the absence of a protease or depolymerase responsible for cleaving the UL VWF multimers. The protease was identified in 1996 by the groups of Tsai and Furlan, its gene was cloned, and the enzyme named ADAMTS-13, when it was found to be a member of the "*a d*isintegrin-like *a*nd *m*etalloprotease with *t*hrombospondin repeats" family of metalloproteases. ADAMTS-13 levels are low in patients with both familial and acquired TTP. However, hereditary TTP (also called Upshaw-Schulman syndrome) is caused by an inherited mutation in the ADAMTS-13 gene, leading to a severe deficiency of ADAMTS-13, while acquired TTP is caused by an IgG auto-antibody inhibitor directed against ADAMTS-13. In both cases, ULVWF multimers are not cleaved due the deficiency of ADAMTS-13, and these ULVWF multimers induce platelet aggregation that leads to microvascular thrombotic angiopathy, which ultimately results in the clinical consequences of the disease.

The hemolytic uremic syndrome (HUS) shares many clinical features with TTP, including microangiopathic hemolytic anemia, thrombocytopenia, and renal insufficiency. Renal findings are more prominent and neurologic findings less so than in TTP. HUS is divided into diarrhea-associated HUS (D+HUS) and atypical (diarrhea-negative) HUS. Diarrhea-positive HUS is triggered by infection with a Shiga-toxin-producing bacteria and is much more commonly encountered in children. *Escherichia coli* O157:H7 is implicated in 80% of cases, but other bacteria including other *E. coli* subtypes and *Shigella dysenteriae* serotype 1 can cause D+HUS. Shiga toxin is internalized via binding to the glycosphingo- lipid receptor globotriaosylceramide (Gb3) on the surface of renal mesangial, glomerular, and tubular epithelial cells. The toxin impairs protein synthesis through inhibition of 60S ribosomes, and cell death ensues. Plasma from patients with HUS demonstrates markers of abnormal thrombin generation. As compared with TTP, ADAMTS-13 levels are typically normal in patients with HUS, and the fibrin microthrombi do not contain VWF strands.

Atypical HUS (aHUS) occurs in patients without a diarrheal prodrome and instead is caused by increased activation of the alternative complement pathway. Mutations that result in the loss or impairment of proteins that regulate complement occur most commonly. Factor H (CFH) and membrane cofactor protein (MCP or CD46) have been implicated in aHUS, and both regulate complement factor I (CFI), a serine protease that cleaves and inactivates surface-bound C3b and C4b. Mutations in CFH and/or MCP cause unregulated activity of protease CFI, which ultimately leads to increased complement activation. Acquired autoantibodies against these proteins have been reported, as have activating mutations in complement proteins B, C3, and factor I [6].

### Heparin-induced thrombocytopenia

Heparin-induced thrombocytopenia (HIT) is a common iatrogenic immune-mediated thrombocytopenic disorder that can paradoxically lead to thrombosis. It occurs in 1%−5% of patients treated with standard unfractionated heparin for at least 5 days and in <1% of those treated with low molecular weight heparin. Approximately 50% of patients develop venous and/or arterial thromboses, and reported mortality rates are between 10% and 20%. HIT is caused by autoantibodies directed against neoepitopes of platelet factor 4 (PF4) that are induced by heparin and other anionic glycosaminoglycans (GAGs). PF4 is an abundant protein stored in the alpha granules of platelets in complex with chondroitin sulfate (CS). Upon platelet activation, PF4/CS complexes are released and bind to the platelet surface. Heparin can displace CS, forming PF4/heparin complexes. Binding of IgG antibodies to the PF4/heparin complexes leads to Fcγ receptor-mediated clearance of platelets but also leads to platelet activation and generation of procoagulant microparticles via FcγRIIA. PF4/heparin complexes can also form on the surface of monocytes and endothelial cells, and antibody binding on those surfaces leads to tissue factor-driven thrombin generation and clot formation. The PF4/heparin complexes are most antigenic when PF4 and heparin are present at equimolar concentrations, where they form ultralarge molecular complexes. Low molecular weight heparin forms these ultra-large complexes less efficiently and at concentrations that are supratherapeutic, perhaps explaining the lower frequency of HIT in patients treated with LMWH as opposed to standard unfractionated heparin.

In addition to heparin, PF4 can form complexes with other anions such as bacterial LPS and nucleic acids. Recently, it has been proposed that HIT may be a misdirected anti-bacterial immune mechanism in which the rapid generation of IgG antibodies that serve to rapidly clear bacteria coated with PF4-bacterial LPS complexes goes awry and instead targets platelets coated with PF4−heparin complexes [7].

## The thrombophilias

An understanding of how the coagulation system is physiologically regulated is necessary when seeking to determine how it can become deranged. Thrombosis can result from excessive activation of coagulation and/or impaired endogenous regulation. This section will focus on the two major natural anticoagulant pathways that serve to inhibit thrombin generation − the protein C/S pathway and the antithrombin pathway − and that when disturbed can lead to thrombosis.

### The protein C/S pathway and thrombosis

Protein C (PC) is a vitamin K-dependent protein that is activated by thrombin. When bound to the endothelial cell surface protein thrombomodulin, thrombin changes its substrate specificity, losing its ability to cleave fibrinogen and activate platelets. Instead, the thrombin-thrombomodulin complex proteolytically activates zymogen PC to form activated protein C (APC). Activated protein C and its cofactor protein S (another vitamin K-dependent protein) inactivate factors Va and VIIIa, thereby inhibiting thrombin generation. APC also exerts cytoprotective effects by downregulating inflammatory pathways, stabilizing endothelial barriers, and inhibiting p53-mediated apoptosis of ischemic brain endothelium. The endothelial protein C receptor (EPCR), localized on the surface of endothelial cells, serves to bind PC and thereby enhance its activation by thrombin-thrombomodulin 5-fold. EPCR is also found in a soluble form in plasma, and its levels are enhanced in such systemic conditions as disseminated intravascular coagulation and systemic lupus erythematosus. APC binding to EPCR also, shifts its substrate specificity to favor activation of the protease activated receptor-1 (PAR-1). This pathway thereby facilitates crosstalk between the coagulation system and inflammatory cell, endothelial, and platelet functions.

Heterozygous protein C deficiency is a recognized risk factor for venous thromboembolism, with an odds ratio of 6.5–8 [8–10]. Most of the causative mutations are of the type I variety, with a concordant decrease in activity and antigen. These mutations affect protein folding and lead to unstable molecules that are either poorly secreted or are degraded more rapidly. Type II defects lead to activity levels that are reduced disproportionately to the antigen levels and result in dysfunctional molecules with ineffective protein-protein interactions. Heterozygous protein C deficiency has a prevalence of 0.2%–0.4% in the general population and is found in approximately 4%–5% of patients with confirmed venous thromboembolism. Protein C deficient individuals with personal and family histories of thrombosis may have a second thrombophilic defect, such as factor V Leiden to account for the thrombotic tendency. Venous thromboembolic disease (VTE) occurs in 50% of heterozygous individuals in affected families by the age 45, and half of the VTE events are spontaneous. Venous thrombosis at unusual sites (e.g. cerebral sinus and intra-abdominal) is a clinical hallmark. Arterial thrombosis is rare, though reported. Homozygous protein C deficiency with levels <1% generally presents with neonatal purpura fulminans and massive thrombosis in affected infants. Individuals with protein C deficiency are predisposed to develop warfarin skin necrosis when anticoagulated with vitamin K antagonists such as warfarin. Since protein C has a much shorter half-life (8 hours) than the procoagulant vitamin K-dependent factors such as prothrombin and factor X (24–48 hours), a transient hypercoagulable state can occur in patients treated with vitamin K antagonists in the absence of an alternate bridging anticoagulant such as heparin. This risk is magnified in patients with underlying deficiency of either protein C or vitamin K.

Protein S is a vitamin K-dependent protein that is not a serine protease; rather, it acts as a cofactor for APC and functions to accelerate the inactivation of factors Va and VIIa. In normal plasma, 60% of protein S is bound to C4b-binding protein (C4BP), and the remainder is present in the free form. Only the free form of protein S can function as the cofactor for APC. Protein S also exhibits anticoagulant activities independent of APC by binding to and inactivating factor Xa. Most recently, it has been shown that protein S is a cofactor for TFPI-mediated inactivation of tissue factor. Protein S deficiency exists in three forms: (i) type I has equal decrements of antigen and activity; (ii) type II has low activity but normal antigen levels, and (iii) type III shows low free protein S levels, with total protein S levels in the low to normal range, and low activity levels. The odds ratio for VTE with protein S deficiency has been variably reported as 1.6, 2.4, 8.5, and 11.5 [9–12]. More than 50% of VTE events are unprovoked. Arterial thromboses occur at higher frequency, especially among smokers or those with other thrombotic risk factors. Laboratory testing needs to be interpreted with caution. Normal levels vary with age and gender; premenopausal women have lower baseline levels than men, with further reductions occurring as a result of estrogen therapy or pregnancy. Measured protein S activity can be falsely low in patients with inherited resistance to activated protein C. Acquired protein S deficiency occurs in a variety of conditions, including acute thrombosis, inflammation, liver disease, nephrotic syndrome, vitamin K deficiency, disseminated intravascular coagulation, and in association with the lupus anticoagulant. Antibodies to protein S can be seen in children with varicella or other viral illnesses.

Addition of APC to plasma normally causes a prolongation of the aPTT. In 1993, Dahlback reported a novel familial thrombophilia in which the plasma of the proband and affected relatives exhibited resistance to APC, with much less prolongation of the aPTT than control (pooled normal) plasma [15]. Mixing studies showed this defect to be due to a defect in factor V, and the genetic defect responsible for APC resistance was shown to be a mutation at the major cleavage site of APC on factor Va, which was mutated from arginine to glutamine (R506Q). This mutation, now known as factor V Leiden, after the Dutch city in which it was first discovered, is the most prevalent inherited mutation leading to thrombophilia. It is found in approximately 5% of Caucasian populations and is thought to be the result of a founder mutation in a single ancestor 21,000 to 34,000 years ago. The mutant factor Va is inactivated by APC 10-fold more slowly, thereby leading to excessive thrombin generation.

Factor V Leiden is estimated to account for 20%–25% of inherited thrombophilia. Heterozygosity for this mutation confers a relatively low risk for VTE in younger patients (OR=1.2 in those aged 40–50), but the risk increases steeply with age (OR=6 for those older than 70) [13]. Approximately 90% of affected individuals do not suffer any venous thromboembolic events during their lifetime. On the other hand, homozygotes have an OR for VTE of 50–100, and half of such individuals will experience thrombosis during their lives [14]. Coronary artery thrombosis may also occur with greater frequency in young men and women with other risk factors, such as smoking. In general, however, factor V Leiden is not considered to be a major risk factor for arterial thrombosis. The risk for venous thrombosis in individuals with factor V Leiden is greatly magnified when other risk factors for thrombosis are present. These risks may be either genetic or acquired, including PC deficiency, PS deficiency, the prothrombin G20210A mutation, elevated levels of factor VIII, antiphospholipid antibodies, hyperhomocysteinemia, prolonged immobility, surgery, malignancy, pregnancy, or use of oral contraceptives. An acquired form of APC resistance may be caused by conditions other than factor V Leiden, including pregnancy, lupus anticoagulants, inflammation, and use of anticoagulants. Testing for APC resistance is best performed using factor V deficient plasma, which will eliminate interference from the preceding conditions. Genetic testing for factor V Leiden, generally using a PCR-based assay, is also available and is sensitive and specific for the disorder.

A mutation found in 1% of Caucasians is the second most frequent cause of inherited thrombophilia. This mutation in the 3′-untranslated region of the pro- thrombin gene (*G20210A*) results in elevated prothrombin synthesis. Thrombotic risk is probably a result of increased thrombin generation and/or decreased fibrinolysis mediated by enhanced activation of TAFI. The relative risk for first episode of VTE in heterozygotes is between 2 and 5.5, and 4%−8% of patients presenting with their first VTE will be found to have this mutation [16−21]. Homozygosity for the mutation appears to confer a higher risk of VTE. Venous clots in odd locations, as well as arterial clots, are found with increased frequency, especially in patients younger than 55, and especially in those with other thrombotic risk factors [22,23]. PCR amplification of the pertinent region, followed by DNA sequencing, is required for the diagnosis. Measurement of factor II levels is neither sensitive nor specific for the disorder.

## Antithrombin deficiency

Antithrombin (AT) is a SERPIN that inactivates thrombin and clotting factors Xa, IXa, and XIa by forming irreversible 1:1 complexes in reactions accelerated by glycosaminoglycans such as pharmacologically administered heparin or heparan sulfate on the surface of endothelial cells. Deficiency of antithrombin therefore results in potentiation of thrombosis. In type I deficiency, the antigen and activity levels are decreased in parallel, whereas in type II deficiency, a dysfunctional molecule is present. Type IIa mutations affect the active center of the inhibitor, which is responsible for complexing with the active site of the protease. Type IIb mutations target the heparin-binding site, and type IIc mutations are heterogeneous. Severe antithrombin deficiency with levels <5% is rare, resulting from one of several IIb mutations, and leads to severe recurrent arterial and venous thrombosis. The odds ratio for venous thrombosis in heterozygotes is approximately 10−20 [9,24]. Lower extremity deep vein thrombosis is common, and clots in unusual sites have been reported. Clots tend to occur at a younger age, with 70% presenting before age 35, and 85% before age 50 [25]. Some patients with AT deficiency exhibit resistance to the anticoagulant effects of heparin. Other conditions associated with reduced levels of AT include treatment with heparin, acute thrombosis, disseminated intravascular coagulation, nephrotic syndrome, liver disease, treatment with the chemotherapeutic agent L-asparaginase, and pre-eclampsia.

# Key concepts

- Defects in hemostasis may lead to bleeding via the following mechanisms:
  a. *Deficient thrombin generation on the proper cellular surface due to*
     i. Deficiencies of factor VIII or IX (hemophilia A and B) required for propagation phase of thrombin generation
     ii. Deficiencies of factors in the final common pathway of thrombin generation (factors II, V, and X)
     iii. Deficiency of factor XI (required for "overdrive" of coagulation)
  b. *Defect in fibrin polymerization*
     i. Deficiencies of abnormalities of fibrinogen
     ii. Deficiency of factor XIII (required for cross linking fibrin)
  c. *Defect in primary hemostasis*
     i. Von Willebrand disease
     ii. Platelet dysfunction (adhesion or aggregation defect)
     iii. Thrombocytopenia
  d. *Abnormal fibrinolysis*
- Defects in the anticoagulant system lead to thrombosis via the following mechanisms:
  a. *Unopposed or excess generation of thrombin*
     i. Antithrombin deficiency
     ii. Prothrombin G20210A mutation
- *Insufficient inactivation of procoagulant proteins*
  i. Activated protein C resistance including that due to the factor V Leiden mutation
  ii. Deficiencies of protein C or protein S

# References

[1] "Summary of Mutations Causing Rare Bleeding Disorders." International Society on Thrombosis and Haemostasis 2018. Available at https://www.isth.org/general/custom.asp?page = SummaryofMutations. Accessed 11/2/2016.

[2] *The Human Gene Mutation Database* at the Institute of Medial Genetics in Cardiff. Available at http://www.hgmd.cf.ac.uk/ac/index.php. Accessed 11/2/2016.

[3] "Glanzmann Thromboasthenia Database." Medical College of Wisconsin 2016—2018. Available at https://glanzmann.mcw.edu. Accessed 10/16/2019.

[4] "Mutations of the Hermansky-Pudlak Syndrome-1 gene (HPS1) associated with Hermansky-Pudlak Syndrome." International Albinism Center, University of Minnesota. Available at http://www.ifpcs.org/albinism/hps1mut.html. Accessed 10/16/2019.

[5] Schaafsma, Gerard. "LYSTbase: Variation registry for Chediak-Higashi syndrome." ID bases, http://structure.bmc.lu.se/idbase/LYSTbase/index.php?content = index/IDbases. Accessed 10/16/2019.

[6] Maga TK, et al. Mutations in alternative pathway complement proteins in American patients with atypical hemolytic uremic syndrome. Hum Mutat 2010;31(6):E1445—60.

[7] Krauel K, et al. Platelet factor 4 binds to bacteria, [corrected] inducing antibodies cross-reacting with the major antigen in heparin-induced thrombocytopenia. Blood 2011;117(4):1370—8.

[8] Folsom AR, Aleksic N, Wang L, et al. Protein C, antithrombin, and venous thromboembolism incidence: A prospective population-based study. Arterioscler Thromb Vasc Biol 2002;22:1018—22.

[9] Martinelli I, Mannucci PM, De Stefano V, et al. Different risks of thrombosis in four coagulation defects associated with inherited thrombophilia: A study of 150 families. Blood. 1998;92:2353—8.

[10] Koster T, Rosendaal FR, Briet E, et al. Protein C deficiency in a controlled series of unselected outpatients: An infrequent but clear risk factor for venous thrombosis (Leiden Thrombophilia Study). Blood. 1995;85:2756—61.

[11] Faioni EM, Valsecchi C, Palla A, et al. Free protein S deficiency is a risk factor for venous thrombosis. Thromb Haemost. 1997;78:1343—6.

[12] Simmonds RE, Ireland H, Lane DA, et al. Clarification of the risk for venous thrombosis associated with hereditary protein S deficiency by investigation of a large kindred with a characterized gene defect. Ann Intern Med 1998;128:8—14.

[13] Ridker PM, Glynn RJ, Miletich JP, et al. Age-specific incidence rates of venous thromboembolism among heterozygous carriers of factor V Leiden mutation. Ann Intern Med 1997;126:528—31.

[14] Rosendaal FR, Koster T, Vandenbroucke JP, et al. High risk of thrombosis in patients homozygous for factor V Leiden (activated protein C resistance). Blood. 1995;85:1504—8.

[15] Dahlback B, Carlsson M, Svensson PJ. Familial thrombophilia due to a previously unrecognized mechanism characterized by poor anticoagulant response to activated protein C: Prediction of a cofactor to activated protein C. Proc Natl Acad Sci USA 1993;90:1004—8.

[16] Leroyer C, Mercier B, Oger E, et al. Prevalence of 20210 A allele of the prothrombin gene in venous thromboembolism patients. Thromb Haemost. 1998;80:49—51.

[17] Salomon O, Steinberg DM, Zivelin A, et al. Single and combined prothrombotic factors in patients with idiopathic venous thromboembolism: Prevalence and risk assessment. Arterioscler Thromb Vasc Biol 1999;19:511—18.

[18] Margaglione M, Brancaccio V, Giuliani N, et al. Increased risk for venous thrombosis in carriers of the prothrombin G—>A20210 gene variant. Ann Intern Med 1998;129:89—93.

[19] Hillarp A, Zoller B, Svensson PJ, et al. The 20210 A allele of the prothrombin gene is a common risk factor among Swedish outpatients with verified deep venous thrombosis. Thromb Haemost. 1997;78:990—2.

[20] Cumming AM, Keeney S, Salden A, et al. The prothrombin gene G20210A variant: Prevalence in a U.K. anticoagulant clinic population. Br J Haematol 1997;98:353—5.

[21] Brown K, Luddington R, Williamson D, et al. Risk of venous thromboembolism associated with a G to A transition at position 20210 in the 30-untranslated region of the prothrombin gene. Br J Haematol 1997;98:907—9.

[22] Zawadzki C, Gaveriaux V, Trillot N, et al. Homozygous G20210A transition in the prothrombin gene associated with severe venous thrombotic disease: Two cases in a French family. Thromb Haemost. 1998;80:1027—8.

[23] Howard TE, Marusa M, Channell C, et al. A patient homozy- gous for a mutation in the prothrombin gene 30-untranslated region associated with massive thrombosis. Blood Coagul Fibrinolysis. 1997;8:316—19.

[24] van Boven HH, Vandenbroucke JP, Briet E, et al. Gene-gene and gene-environment interactions determine risk of thrombosis in families with inherited antithrombin deficiency. Blood. 1999;94:2590—4.

[25] Hirsh J, Piovella F, Pini M. Congenital antithrombin III deficiency. Incidence and clinical features. Am J Med 1989;87:34S—8S.

[26] Buller HR, Gailani D, Weitz JI. Factor XI antisense oligonucleotide for venous thrombosis. N Engl J Med 2015;372(17):1672.

[27] Zhang H, et al. Inhibition of the intrinsic coagulation pathway factor XI by antisense oligonucleotides: a novel antithrombotic strategy with lowered bleeding risk. Blood 2010;116(22):4684—92.

[28] Salomon O, et al. Reduced incidence of ischemic stroke in patients with severe factor XI deficiency. Blood 2008;111(8):4113—17.

[29] Cushman M, et al. Coagulation factors IX through XIII and the risk of future venous thrombosis: the Longitudinal Investigation of Thromboembolism Etiology. Blood 2009;114(14):2878—83.

[30] Meijers JC, et al. High levels of coagulation factor XI as a risk factor for venous thrombosis. N Engl J Med 2000;342(10):696—701.

[31] Suri MF, et al. Novel hemostatic factor levels and risk of ischemic stroke: the Atherosclerosis Risk in Communities (ARIC) Study. Cerebrovasc Dis 2010;29(5):497—502.

[32]   Siegerink B, et al. Intrinsic coagulation activation and the risk of arterial thrombosis in young women: results from the Risk of Arterial Thrombosis in relation to Oral contraceptives (RATIO) case-control study. Circulation 2010;122(18):1854–61.

[33]   Yang DT, et al. Elevated factor XI activity levels are associated with an increased odds ratio for cerebrovascular events. Am J Clin Pathol 2006;126(3):411–15.

[34]   Mariani G, Herrmann FH, Dolce A, et al. Clinical phenotypes and factor VII genotype in congenital factor VII deficiency. Thromb Haemost. 2005;93:481–7.

## Suggested readings

Amirlak I, Amirlak B. Haemolytic uraemic syndrome: an overview. Nephrology (Carlton) 2006;11:213–18.

Buller HR, Gailani D, Weitz JI. Factor XI antisense oligonucleotide for venous thrombosis. N Engl J Med 2015;372:1672.

Choi SH, Smith SA, Morrissey JH. Polyphosphate is a cofactor for the activation of factor XI by thrombin. Blood 2011;118:6963–70.

Curtis BR. Recent progress in understanding the pathogenesis of fetal and neonatal alloimmune thrombocytopenia. Brit J Haematol 2015;171:671–82.

Folsom AR, Aleksic N, Wang L, Cushman M, Wu KK, White RH. Protein C, antithrombin, and venous thromboembolism incidence: a prospective population-based study. Arterioscler Thromb Vasc Biol 2002;22:1018–22.

Greinacher A. Clinical practice. Heparin-induced thrombocytopenia. N Engl J Med 2015;373:252–61.

Kitchens CKC, Konkle B. Consultative hemostasis and thrombosis. Philadelphia, PA: Elsevier; 2013.

Key NS. Epidemiologic and clinical data linking factors XI and XII to thrombosis. Hematology Am Soc Hematol Educ Program 2014;2014:66–70.

Krauel K, Potschke C, Weber C, et al. Platelet factor 4 binds to bacteria, [corrected] inducing antibodies cross-reacting with the major antigen in heparin-induced thrombocytopenia. Blood 2011;117:1370–8.

Levy GG, Nichols WC, Lian EC, et al. Mutations in a member of the ADAMTS gene family cause thrombotic thrombocytopenic purpura. Nature 2001;413:488–94.

Levy JH, Hursting MJ. Heparin-induced thrombocytopenia, a prothrombotic disease. Hematol Oncol Clin North Am 2007;21:65–88.

McCrae KR e. Thrombocytopenia. New York: Taylor and Francis; 2006.

Moake JL, Rudy CK, Troll JH, et al. Unusually large plasma factor VIII: von Willebrand factor multimers in chronic relapsing thrombotic thrombocytopenic purpura. N Engl J Med 1982;307:1432–5.

Mosnier LO, Zlokovic BV, Griffin JH. The cytoprotective protein C pathway. Blood 2007;109:3161–72.

Nichols WL, Hultin MB, James AH, et al. von Willebrand disease (VWD): evidence-based diagnosis and management guidelines, the National Heart, Lung, and Blood Institute (NHLBI) Expert Panel report (USA). Haemophilia 2008;14:171–232.

Peyvandi F, Cattaneo M, Inbal A, De Moerloose P, Spreafico M. Rare bleeding disorders. Haemophilia 2008;14:202–10.

Roberts HR, Hoffman M, Monroe DM. A cell-based model of thrombin generation. Semin. Thromb. Hemostasis 2006;32:32–8.

Sadler JE. What's new in the diagnosis and pathophysiology of thrombotic thrombocytopenic purpura. Hematol Am Soc Hematol Educ Program 2015;2015:631–6.

Schafer AI, Levine MN, Konkle BA, Kearon C. Thrombotic disorders: diagnosis and treatment. Hematol Am Soc Hematol Educ Program 2003;520–39.

Seri M, Pecci A, Di Bari F, et al. MYH9-related disease: May-Hegglin anomaly, Sebastian syndrome, Fechtner syndrome, and Epstein syndrome are not distinct entities but represent a variable expression of a single illness. Medicine (Baltimore) 2003;82:203–15.

Zhu S, Travers RJ, Morrissey JH, Diamond SL. FXIa and platelet polyphosphate as therapeutic targets during human blood clotting on collagen/tissue factor surfaces under flow. Blood 2015;126:1494–502.

# Chapter 16

# Molecular basis of lymphoid and myeloid diseases

Joseph R. Biggs and Dong-Er Zhang

*Department of Pathology and Division of Biological Sciences, University of California San Diego, La Jolla, CA, United States*

**Summary**

All types of mature blood cells are produced by differentiation of hematopoietic stem cells found in the bone marrow. This process is controlled by transcription factors, which in turn are regulated by other transcription factors, microRNAs, epigenetic modifications, and extracellular signals transmitted by cell surface receptors. Mutation of genes encoding these hematopoietic transcription factors, microRNAs, epigenetic modifiers, or receptors (or deregulated expression) can initiate transformation of normal cells into leukemic cells, or cause cell death and anemia. Diseases such as leukemia, lymphoma, and anemia are usually genetically complex, a large number of different mutations contribute to disease development, and most patients acquire more than one mutation. This genetic complexity has resulted in proposed new treatments based on individual patient leukemia genomes and/or immunotherapy.

## Introduction

All types of mature blood cells are produced by differentiation of hematopoietic stem cells found in the bone marrow. This process is controlled by transcription factors, which in turn are regulated by other transcription factors, microRNAs, epigenetic modifications, and extracellular signals transmitted by cell surface receptors. Mutation of genes encoding these hematopoietic transcription factors, microRNAs, epigenetic modifiers, or receptors (or deregulated expression) can initiate transformation of normal cells into leukemic cells, or cause cell death. Diseases such as leukemia, lymphoma, and anemia are usually genetically complex. A large number of different mutations contribute to disease development, and most patients acquire more than one mutation. This genetic complexity has resulted in proposed new treatments based on individual patient leukemia genomes and/or immunotherapy.

This chapter will begin with an overview of normal hematopoietic development—the process by which hematopoietic stem cells differentiate into the various types of mature hematopoietic cell. This process depends on a series of transcription factors acting sequentially during the successive stages of proliferation and differentiation. The level and timing of hematopoietic transcription factor activity is controlled by signal transduction pathways that regulate transcription and translation of the genes encoding the transcription factors. After providing a description of normal hematopoiesis and hematopoietic organs, the different types of leukemia will be described individually, along with known genetic mutations that contribute to the development of the different forms of leukemia.

## Development of the blood and lymphoid organs

### Hematopoietic stem cells

All hematopoietic cells are derived from hematopoietic stem cells (HSCs) that are capable of both self-renewal and differentiation to all blood cell lineages. Hematopoiesis occurs first in the yolk sac and generates an initial wave of primitive erythroid progenitors expressing fetal hemoglobin, macrophages, and megakaryocytes. A later wave produces erythroid-myeloid progenitors. Subsequently, the first HSCs are produced in the aorta-gonad-mesonephros (AGM) region of the embryo. The HSCs then leave the AGM for the placenta and fetal liver, where they undergo transient proliferation, and finally the HSCs colonize the bone marrow, the site of adult hematopoiesis.

In the standard model of hematopoiesis (Fig. 16.1), HSCs differentiate into intermediates with less self-renewal potential, thcn to multipotent progenitors (MPPs). MPPs segregate into common myeloid progenitors (CMPs) and common lymphoid progenitors (CLPs). CMPs differentiate into granulocyte-monocyte progenitors (GMPs) and megakaryocyte-erythroid progenitors (MEPs). GMPs and MEPs differentiate into the various types of mature myeloid and erythroid cells (Fig. 16.1). CLPs differentiate into B-cells, T-cells, and natural killer (NK) cells. The appearance of various types of hematopoietic cell is shown in Fig. 16.2. A recent study has suggested an alternative to the standard

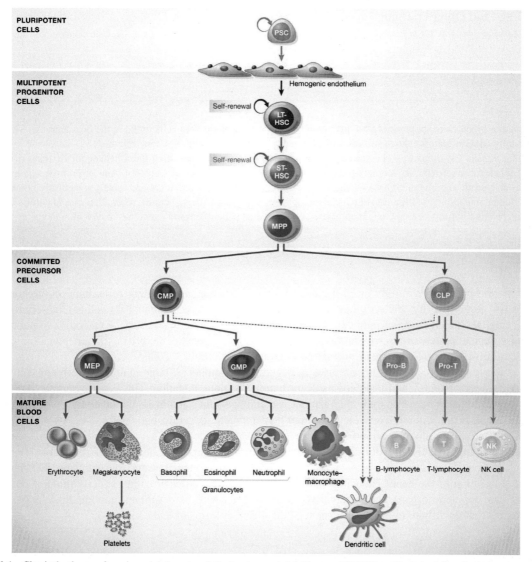

FIGURE 16.1 Classical scheme of murine adult hematopoietic development. Multipotent LT-HSCs, with their ability for long-term reconstitution potential, can further differentiate toward ST-HSCs and also MPPs in the bone marrow. Upon subsequent differentiation, MPPs give rise to either CMPs, which have the ability to differentiate into the myeloid lineage, or CLPs, able to generate the lymphoid lineage. Following these committed progenitors, both MEPs and GMPs are able to form all differentiated cells of the myeloid lineage in the bone marrow, whereas CLPS further differentiate into pro-T-cells and T-cells by positive-negative selection in the thymus. Generation of B-cells is ensured also by CLPs in the bone marrow following B-cell transition. *CLP*, common lymphoid progenitor; *CMP*, common myeloid progenitor; *GMP*, granulocyte macrophage progenitor; *HSC*, hematopoietic stem cell; *LT*, long term; *MEP*, megakaryocyte elytroid progenitor; *MPP*, multipotent progenitor; *ST*, short term. *Reproduced from Ackermann M, Liebhaber S, Klusmann JH, Lachmann N. Lost in translation: pluripotent stem cell-derived hematopoiesis. EMBO Mol Med 2015;7:1388−402, with permission.*

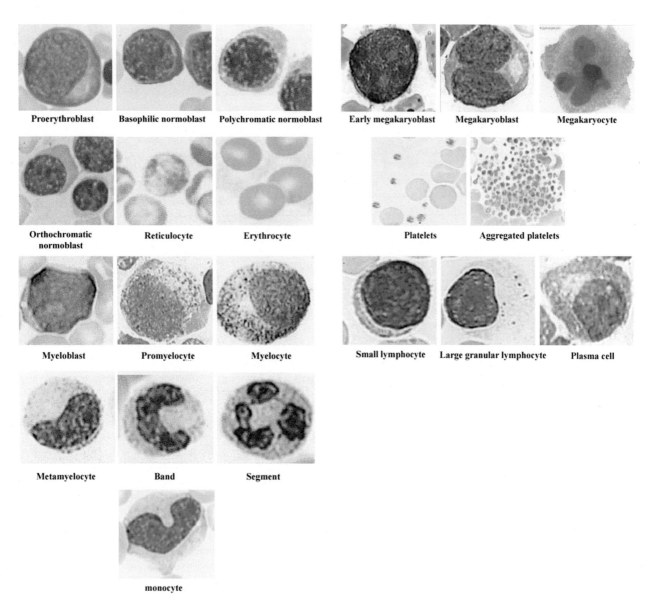

**FIGURE 16.2** Normal hematopoietic cells. Upper left panels show the stages of erythrocyte (red blood cell) development. Lower left panels show the stages of granulocyte/monocyte development. Upper right panels show differentiation of megakaryocytes and platelets. Lower right panels show lymphocytes. *Courtesy of Amos Cohen, M.D., Rabin Medical Center, Tel Avive, Israel.*

model, in which unipotent progenitor cells (which differentiate only to erythroid, myeloid, or lymphoid cells) are produced directly from multipotent progenitors. Further studies will determine whether the standard model of hematopoiesis needs to be revised.

## Bone marrow niche

After HSCs migrate to the bone marrow, they enter a special microenvironment known as the bone marrow niche. Interactions between the HSCs and other types of cells that compose the niche maintain the HSCs in a quiescent state that protects them from genotoxic insults. Responding to external signals, HSCs may leave the quiescent state, proliferate, and differentiate into linage-committed progenitor cells. Quiescence and exit from quiescence are regulated by an exchange of signals between HSCs and other cells of the niche, as illustrated in Fig. 16.3.

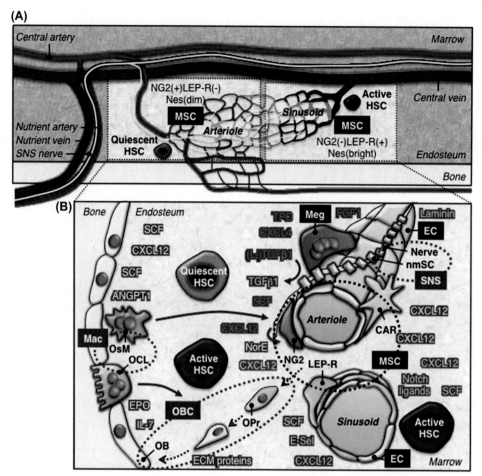

**FIGURE 16.3** Organization of the HSC Niche. (A) Overall anatomy of the bone marrow cavity depicting the sympathetic innervation and the vasculature and highlighting the interconnection between arteriole and sinusoid blood vessels. Each of these regions (*dotted box*) is enriched for a particular subset of perivascular MSCs, which controls a different HSC functional state. Quiescent HSCs are G0 dormant cells. Active HSCs are cells that have just exited quiescence or are already actively cycling or migrating. (B) Enlargement of the essential (black) and accessory (gray) HSC niche cells with their respective secreted and/or cell-bound factors (color-coded) that regulate HSC functional states. Dotted circles group cells with either similar origin (i.e., perivascular MSC subsets and differentiating OBCs) or similar function (i.e., specialized macrophages, designated as Mac, and SNS components). *Black arrows* highlight MSC progeny that are differentiating into bone-lining OBs and forming the OBC compartment. *Gray arrows* indicate the long-range, indirect effects of several accessory HSC niche cells. *CAR*, CXCL12$^{bright}$ MSCs; *E-Sel*, E-selectin; *LEP-R*, NG2 − LEP-R$^+$Nes$^{bright}$ MSCs; *NG2*, NG2$^+$LEP-R − Nes$^{dim}$ MSCs; *nmSC*, nonmyelinating Schwann cells; *NorE*, norepinephrine; *OCL*, osteoclasts; *OPr*, osteoprogenitors; *OsM*, osteomacs. *Reproduced from Ackermann M, Liebhaber S, Klusmann JH, Lachmann N. Lost in translation: pluripotent stem cell-derived hematopoiesis. EMBO Mol Med 2015;7:1388−402, with permission.*

## Spleen

The spleen is a lymphoid organ that also serves as a blood filter. The arteries of the spleen are ensheathed by lymphocytes, which form the white pulp; the white pulp is further subdivided into a T-cell domain and a B-cell domain. The spleen, along with the lymph nodes, is a major repository for lymphocytes and a major site of adaptive immune response to foreign antigens. The remaining internal portion of the spleen is composed of red pulp, which is designed to filter foreign matter from the bloodstream, including damaged blood cells.

## Thymus

Mature mammalian T-cells originate in the bone marrow or fetal liver as pluripotent precursors, which then migrate to the thymus, where they proliferate extensively and differentiate into the various mature T-cell lineages. The sole function of the thymus is to serve as the site of T-cell differentiation. Beginning at puberty, the thymus involutes and shrinks, until it eventually consists of groups of epithelial cells depleted of lymphocytes.

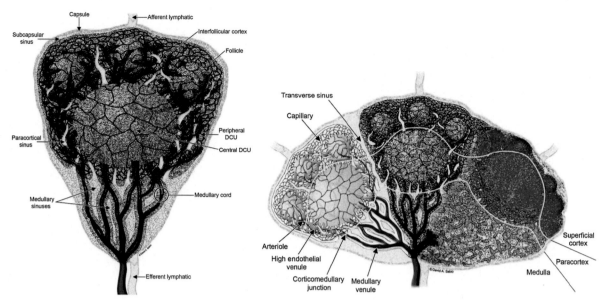

**FIGURE 16.4** Lymph node structure. The left panel shows the simplest possible lymph node containing a single lobule. Lymph from the afferent lymphatic vessel spreads over the apical surface in the subcapsular sinus and then flows through medullary sinuses and exits via the efferent lymphatic vessel. The sinuses are spanned by a reticular meshwork, and the lobule contains a denser meshwork indicated by darker, more condensed background. The meshwork provides a scaffold for lymphocytes, antigen-presenting cells, and macrophages to interact. B-cells home to follicles in the superficial cortex, where they interact with dendritic cells. Three follicles are shown as small spheres. Follicles are surrounded and separated by nterfollicular cortex. In the deep cortex (paracortex) T-cells home to the deep cortical unit (DCU) where they interact with dendritic cells. The right panel shows an idealized section of a small lymph node containing three lymphoid lobules. Taken together, the follicles and interfollicular cortex of these lobules constitute the superficial cortex of the mode; their deep cortical units, the paracortex; and their medullary cords and sinuses, the medulla. Left lobe shows arterioles (red), venules (blue), and capillary beds (purple). Center lobe as in left panel. Right lobe shows a micrograph from a rat mesenteric lobule as it appears in histological section. *Reproduced from Willard-Mack CL. Normal structure, function, and histology of lymph nodes. Toxicol Pathol 2006;34:409—24, with permission.*

## Lymph nodes

Lymph nodes are small glands located in many parts of the body, mainly in the neck, under the arms, and in the groin. Lymph vessels drain fluid from tissue, which then enters the lymph nodes via afferent lymphatic vessels. Lymph nodes are composed of multiple lymphoid lobules surrounded by lymph-filled sinuses and enclosed by a capsule (Fig. 16.4). The smallest nodes may contain only one lobule, while the largest contain a great number of lobules. The lobules are divided into regions containing spherical follicles separated by interfollicular cortex and regions containing deep cortical units (DCUs). Immature B-cells originating in the bone marrow home to follicles where they interact with follicular dendritic cells (FDCs). FDCs trap antigen-antibody complexes that may be collected from lymph carried into the follicle. If a B-cell encounters its antigen displayed on an FDC, it is stimulated to proliferate, and the proliferating B-cells form distinctive germinal centers that are referred to as secondary follicles. A large number of B-cells undergo apoptosis during this process of proliferation and differentiation. In contrast to B-cells, T-cells migrate to the paracortex and interfollicular cortex and survey dendritic cells. The dendritic cells that interact with T-cells form a separate class from those that interact with B-cells. These cells collect and process antigens in tissue, then migrate to the lymph nodes. The interfollicular cortex and DCU serve as corridors for the movement of B-cells and T-cells. Several afferent lymphatic vessels enter the lymph node, but each delivers a stream of lymph to a specific lobule. After passing over the lobules, all the lymph streams exit the lymph node through a single efferent vessel. Thus, individual lobules are exposed to different sets of antigens and cells collected from a specific drainage area by an individual afferent vessel. The constant flow of lymph-containing cells and antigens collected from tissue allows the lymph nodes, like the spleen, to serve as a major site of interaction between foreign antigens and lymphocytes.

## Hematopoietic differentiation and the role of transcription factors

All types of mature blood cells are produced by linage-restricted differentiation of HSCs. This process is believed to be regulated by a relatively small group of transcription factors, some required for HSC formation and others for differentiation. Each stage of this process is controlled by specific transcription factors. As with HSC transcription factors, the identity of these transcription factors has been determined largely by the study of conventional or conditional gene knock-outs in mice and other model organisms. The fact that most of these factors show lineage and stage-restricted expression also provides information about their function. Human HSCs and hematopoietic cells have been studied using in vitro assays and engraftment in immune-deficient mice.

Transcription factors essential for the formation of HCSs include SCL/TAL-1 and its partner LMO2, as well as RUNX1 and its partner core binding factor-beta (CBFβ). The histone methyltransferase myeloid/lymphoid leukemia (MLL, also known as KMT2A—lysine-specific methyltransferase 2A), which is necessary to maintain HOX gene expression, also has a vital role in hematopoiesis. In the absence of SCL/TAL-1 and LMO2, failure of both primitive and definitive hematopoiesis is observed. In the absence of RUNX1 or MLL, HSCs do not appear in the AGM region of the mouse embryo. A striking observation is that this set of transcription factors controlling HSC development account for the majority of known leukemia-associated translocations in patients. These translocations either deregulate the expression of the locus or generate chimeric fusion proteins. A second set of transcription factors is required for differentiation of HSCs into specific types of mature blood cell, and the transcription factors involved in the development of HSCs also have roles in later hematopoietic development. Like the factors that control HSC development, these lineage-specific factors have been identified largely through the study of gene knockout models. As examples, loss of the factor GATA-1 or its cofactor FOG results in failure of erythroid and megakaryocytic differentiation, while mice deficient in the transcription factor C/EBPα lack GMPs and granulocytes.

Hematopoietic transcription factor levels can be controlled by both transcriptional and post-transcriptional mechanisms. Studies on microRNAs (miRNAs) suggest that they provide an additional mechanism for controlling hematopoietic transcription factor levels. MiRNAs bind to the 3′-untranslated region of mRNAs and suppress translation. It can be difficult to perform gene knockout studies on miRNAs because many miRNAs exist as duplicates and can occur in mammals as gene clusters containing multiple similar miRNAs. However, several individual miRNAs have been identified as important regulators of hematopoiesis through gene knockout experiments. A survey of miRNA expression in hematopoietic cells (27 phenotypic cell populations from stem/progenitor cells to mature cells) showed an average of 96 miRNAs expressed in each population. Numerous studies using cultured cells and animal models provided additional evidence that many of these miRNAs are involved in the regulation of hematopoietic differentiation.

## Hematopoietic differentiation and the role of signal transduction

During the processes of proliferation and differentiation, cells respond to external signals such as growth factors or cell-cell contacts. Growth factors act by binding to a specific cell surface receptor and activating intracellular cascades that stimulate or suppress downstream transcription factors. Many of the receptors that regulate normal hematopoiesis are receptor tyrosine kinases (RTKs), such as colony-stimulating factor 1 receptor, also known as macrophage colony-stimulating factor or cellular McDonough Feline Sarcoma tyrosine kinase (C-FMS), FMS-related tyrosine kinase (FLT3) (receptor for FLT3-ligand), receptor for stem cell factor (C-KIT), and platelet-derived growth factor receptor (PDGFR). Ligand binding activates the tyrosine kinase activity of these RTKs, which then phosphorylate tyrosine residues on associated proteins, thereby triggering cascades in which intracellular kinases are sequentially phosphorylated and activated, until finally the signal is transmitted to nuclear transcription factors. Hematopoietic RTKs usually activate several such cascades, including the RAS/RAF/ERK pathway, the PIK3/AKT pathway, and the JAK/STAT pathway. The activation of these pathways usually favors cell proliferation and survival. Fig. 16.5 illustrates several of the pathways commonly activated in response to receptor tyrosine kinases.

Many leukemias are associated with mutations that cause constitutive activation of RTKs (the receptor tyrosine kinase is continuously active, even in the absence of ligand). This results in a continuous signal to the cell favoring growth and survival. Two observations about leukemic mutations have led to the proposal of the two-hit model of leukemogenesis. First, many leukemia patients contain two types of mutation, one affecting a hematopoietic transcription factor, such as RUNX1 (formerly acute myeloid leukemia [AML]1), and the other affecting a receptor tyrosine kinase or signal-transduction molecule, such as FLT3. Secondly, studies in model systems have shown that many of the leukemia-associated mutations found in patients are unable to induce leukemia by themselves, but can induce leukemia in combination with other mutations. Therefore the two-hit model proposes that induction of leukemia requires the presence of two types of mutation: a class I mutation in a receptor or signal-transduction molecule that confers a proliferative or survival advantage, and a class II mutation in a hematopoietic transcription factor that impairs differentiation.

Advances in DNA sequencing technology have given researchers the opportunity to test the two-hit model by sequencing the cancer genomes from large numbers of patients. In one study, bone marrow samples from 197 AML

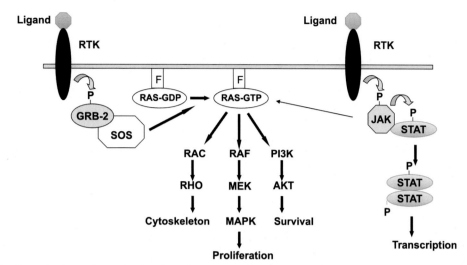

**FIGURE 16.5**  Signal transduction pathways involved in leukemia. The drawing on the left shows signaling of a receptor tyrosine kinase (RTK) through RAS. Ligand binding causes phosphorylation of GRB-2 by the RTK and formation of a GRB-2/SOS complex. Interaction of GRB-2/SOS with farnesylated (F) RAS-GDP causes conversion to active RAS-GTP, which in turn phosphorylates RAC, RAF, and PI3K, leading to stimulation of their respective pathways. The drawing on the right shows the JAK/STAT pathway. Ligand binding causes RTK phosphorylation of JAK, which may then activate the RAS pathway and phosphorylate STATs. The STATs form homodimers or heterodimers with other STATs and translocate to the nucleus where they activate transcription of specific target genes.

patients were used to sequence 51 genes previously identified as mutated in myeloid neoplasms. The samples were also tested for the presence of transcripts from fusion genes generated by leukemic translocations, such as *RUNX1-ETO* (or *RUNX1-RUNX1T1*) and *CBFB-MYH11*. Mutations were found in 44 of the 51 genes. However, only five genes (*FLT3*, *NPM1*, *CEBPA*, *DNMT3A*, and *C-KIT*) were mutated in more than 10% of the AML patients. One hundred eighty three patients (92.9%) carried mutations in at least one of the genes tested, and the average number of mutations per patient was $2.56 \pm 0.11$. Forty one patients carried the *RUNX1-ETO* fusion gene and 14 the *CBFB-MYH11* fusion. The largest number of mutations (138/197 patients) occurred in Class II genes—altered transcription factors that block differentiation (including *CEBPA*, *RUNX1*, *GATA2*, and the *RUNX1-ETO* and *CBFB-MYH11* fusion genes). One hundred sixteen of one hundred ninety-seven patients had mutations in Class I genes, which encode receptors or signal transduction proteins, and when mutated confer a growth advantage (including *FLT3*, *C-KIT*, *N/KRAS*, *PTPN1*, *JAK1/3*, and *TP53*). However, many patients (91/197) also carried mutations in genes that regulate epigenetic modifications, such as DNA methylation (*ASXL1*, *ATRX1*, *EZH2*, *TET2*, *PBRM1*, *DNMT3A*, *IDH1/2*, *KDM6A*, *MLL*, and *DOT1L*). This suggests that loss of normal epigenetic modifications can contribute to leukemogenesis. Many patients also carried mutations in genes regulating cell adhesion and RNA splicing, which suggests that disruption of these processes may also contribute to leukemogenesis. Most patients carried multiple mutations from different classes, supporting the idea that in most AML patients, the disease is caused by disruption of multiple pathways. It is likely that epigenetic mutations can affect the expression of both Class I and Class II genes. Many AML patients were observed to have epigenetic mutations paired with mutations in Class I or Class II genes (Fig. 16.6). It now seems clear that while the two-hit model of leukemia is not completely incorrect, it does not describe the true genetic complexity of diseases like AML.

## Myeloid disorders

### Anemia

Anemia is a condition in which the blood contains a lower than normal number of red blood cells (RBCs) or RBCs that do not contain enough hemoglobin (the normal range is 4.2–6.1 million RBCs per microliter and 12.1–17.2 g of hemoglobin per deciliter). Anemia may be caused either by lower than normal production of RBCs, or higher than normal rates of RBC destruction. Nongenetic causes of anemia include blood loss, iron deficiency, lack of folic acid (vitamin B-12), or chronic disease, all of which can impede the production of RBCs. Higher than normal rates of RBC destruction can be caused by inherited disorders such as sickle cell anemia and thalassemia and certain enzyme deficiencies. Hemolytic anemia occurs when the immune system mistakenly attacks RBCs. Anemia may also be caused by myelodysplastic syndromes (MDS), defined as one or more secondary blood cytopenias (cell loss) caused by bone marrow dysfunction.

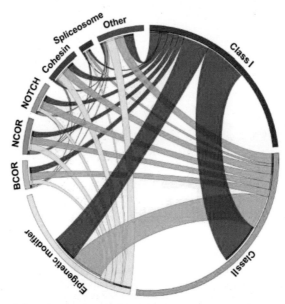

**FIGURE 16.6** Circos plot showing association of mutated genes in AML according to conceptual classification and function. Overlap mutations between Class I, Class II, and epigenetic modifying genes were frequently observed. These major mutations often coexisted with mutations in other gene families, such as the cohesin complex, the BCOR family, or spliceosome genes. *Reproduced from Kihara R, Nagata Y, Kiyoi, H et al. Comprehensive analysis of genetic alterations and their prognostic impacts in adult acute myeloid leukemia patients. Leukemia 2014;28:1586—95, with permission.*

## Neutropenia

Neutropenia may occur as chronic idiopathic neutropenia or severe congenital neutropenia. Chronic idiopathic neutropenia is defined as any unexplained reduction in neutrophil count to below average. The criteria for diagnosis are absolute neutrophil counts below $1.5-1.8 \times 10^9$/L blood lasting more than 3 months, absence of evidence of underlying disease associated with the neutropenia, no history of exposure to radiation or chemicals that might cause neutropenia, normal bone marrow karyotype, and serum negative for anti-neutrophil antibodies. Neutropenia caused by anti-neutrophil antibodies is known as primary autoimmune neutropenia and is most common in newborns where it shows a tendency to resolve spontaneously. Chronic idiopathic neutropenia (CIN) is believed to result from impaired bone marrow granulopoiesis, but the precise molecular mechanism remains unknown.

Severe congenital neutropenia (SCN) is characterized by life-long neutropenia with an absolute neutrophil count under $0.5 \times 10^9$/L, recurrent bacterial infections, and arrest of neutrophil maturation at the promyelocyte stage. Approximately 60% of patients with SCN carry mutations in the neutrophil elastase (ELANE) gene. These patients fall into the categories of dominant inheritance of the disease or spontaneous acquisition of the disease. Mutations in the ELANE gene are also present in patients with cyclic hematopoiesis, in which the number of neutrophils and other blood cells oscillates in weekly phases. Neutrophil elastase is a protease found in the granules of mature neutrophils. The ELANE mutations found in patients with SCN or cyclic hematopoiesis induce the unfolded protein response (UPR) and apoptosis. Protein folding occurs in the lumen of the endoplasmic reticulum. Misfolded proteins trigger the UPR, which leads to attenuation of translation, expression of ER-resident chaperones, and ER-associated degradation pathways. If this adaptive response is overwhelmed, apoptosis is induced. Specific ELANE mutations are associated with either SCN or cyclic hematopoiesis, and it has been hypothesized that cyclic hematopoiesis is caused by ELANE mutations that cause a less drastic activation of the UPR. However, patients with SCN, unlike those with cyclic hematopoiesis, also display a deficiency of the transcription factor LEF-1, leading in turn to reduced levels of the LEF-1 targets C/EBPα, cyclin D1, C-MYC, and survivin. The LEF-1 deficiency (not coupled to a mutation in the gene) is present in SCN patients with either the *ELA2* mutation or the *HAX-1* mutation. This suggests that LEF-1 deficiency may synergize with the *ELA2* or *HAX-1* mutations to promote neutropenia. In contrast to other SCN patients, those who acquired the disease through recessive inheritance lack mutations in the *ELANE* gene but carry mutations in the HS-1-associated protein X (*HAX-1*) gene. This form of neutropenia was first described by Rolf Kostmann and is also known as Kostmann disease. HAX-1 is a mitochondria-targeted protein, containing BCL-2 homology domains, and is critical for maintaining the inner mitochondrial membrane potential. Loss of HAX-1 function causes increased apoptosis in myeloid cells. It therefore appears that mutations in either ELANE or HAX-1 contribute to neutropenia by causing enhanced levels of apoptosis in myeloid precursor cells.

Granulocyte colony stimulating factor (G-CSF) therapy is widely used for the treatment of SCN. Ninety-five percent of SCN patients respond to treatment with G-CSF (also called CSF3) with increased neutrophil counts and reduced mortality. However, the increased life expectancy conferred by G-CSF treatment brings increased risk (21% of patients) of developing MDS or leukemia. This risk is associated, in part, with mutations in the *CSF3R* gene, which encodes the receptor for G-CSF/CSF3. These mutations result in truncated CSF3R receptors that hamper neutrophil differentiation and confer increased proliferative response to G-CSF. Seventeen percent of SCN patients who have not received G-CSF treatment carry *CSF3R* mutations, while they are present in 34% of patients who have undergone treatment. *CSF3R* mutations are present in 78% of SCN patients who have developed MDS/leukemia. These studies suggest that while *CSF3R* mutations are not strongly associated with initial development of SCN, they are involved in the progression from SCN to MDS/leukemia.

## Myelodysplastic syndromes

MDS are diagnosed at a rate of 3.6/100,000 people in the United States. MDS occurs primarily in older patients ( > 60 years), but occasionally in younger patients. Anemia, bleeding, easy bruising, and fatigue are common, and splenomegaly or hepatosplenomegaly may occasionally be present. MDS is characterized by abnormal bone marrow and blood cell morphology. The bone marrow is usually hypercellular, but approximately 15% or patients have hypoplastic bone marrow. Circulating granulocytes are often severely reduced and hypogranular or hypergranular. Early, abnormal myeloid progenitors are identified in the marrow in varying percentages, depending on the type of MDS. Abnormally small megakaryocytes may be seen in the marrow and hypogranular or giant platelets in the blood.

MDS is classified according to cellular morphology, etiology, and clinical features. The morphological classification is largely based on percent myeloblasts in the bone marrow and blood, the type of myeloid dysplasia, and the presence of ringed sideroblasts. Beginning in the late 1970s, MDS was classified according to the French-American-British (FAB) classification scheme under the direction of the FAB Cooperative Group. The FAB classification scheme divides MDS into (1) refractory anemia, (2) refractory anemia with ringed sideroblasts, (3) refractory anemia with excess blasts, (4) refractory anemia with excess blasts in transformation, and (5) chronic myelomonocytic leukemia. In 1997, a working group of pathologists under the direction of the World Health Organization (WHO) agreed to a new classification scheme for hematopoietic and lymphoid malignancies. The WHO classification system was further modified in 2008. The WHO scheme attempts to correct weaknesses in the FAB scheme such as failure to take cytogenetic findings into account. The WHO scheme divides MDS into (1) refractory cytopenia with unilineage dysplasia, (2) refractory anemia with ring sideroblasts (RARS), (3) refractory cytopenia with multi-lineage dysplasia, (4) refractory anemias with excess blasts, (5) MDS associated with isolated del(5q), and (6) MDS, unclassifiable.

Whole genome sequencing studies have recently revealed a great deal of new information on the molecular pathogenesis of MDS. The most common mutations in MDS occur in genes involved in RNA splicing, including *SF3B1*, *SRSF2*, *U2AF1*, and *ZRSR2*. *SF3B1* is mutated in 28% of MDS patients and >70% of patients with ring sideroblasts. Fifty-seven percent of MDS patients have a spliceosome mutation. These mutations are usually mutually exclusive and highly specific to MDS. The spliceosome mutations are believed to be gain-of-function, and recent studies have shown that the *SF3B1* mutation affects expression levels and exon choice in a large number of genes. Other genes frequently mutated in MDS include epigenetic regulators (*TET2*, *DNMT3A*, and *ASXL1*) and genes involved in signal transduction and transcriptional regulation (*RUNX1*, *TP53*, and *JAK2*). *JAK2* mutations are rare except in patients with refractory anemia with ringed sideroblasts associated with thrombosis (RARS-T). Fifty percent of these patients carry the JAK2V617F mutation. In addition to point mutations, recurring chromosomal abnormalities have been observed in approximately 50% of MDS patients. Among the most common are del(5q) (deletion within the long arm of chromosome 5), monosomy 7/del(7q), del(20q), del(17p), and del(11q). For the most part, it has not been determined if these chromosomal abnormalities are driver events in MDS. Del(5q) occurs in 10%−20% of MDS, and MDS with isolated del(5q) is a distinct class of refractory anemia. The deleted region has been narrowed to 5q23, and candidate genes in this region include *RPS14*, *SPARC*, and *CSNK1A1*. A small number of patients have mutations in *CSNK1A1*, and other studies have shown that haploinsufficiency of the *RPS14* ribosomal protein causes p53 activation and is crucial to anemia development.

## Myelodysplastic/myeloproliferative overlap diseases and myeloproliferative neoplasms

Myelodysplastic/myeloproliferative overlap diseases have features of both MDS and myeloproliferative neoplasms. A greater than normal number of stem cells develop into one or more types of more mature cells, and the blood cell number increases, but there is also some degree of failure to mature properly. The three main types of myelodysplastic/myeloproliferative diseases are chronic myelomonocytic leukemia (CMML), juvenile myelomonocytic leukemia (JMML), and atypical chronic myeloid leukemia (aCML). Little is currently known about the molecular pathology of aCML. A myelodysplastic/myeloproliferative disease that does not match any of the previous types is referred to as myelodysplastic/mycloproliferative disease, unclassifiable (MDS/MPD-UC). Diseases classified simply as myeloproliferative neoplasms include chronic myeloid leukemia (CML), chronic neutrophilic leukemia (CNL), polycythemia vera (PV), essential thrombocythemia (ET), primary myelofibrosis (PMF), and mastocytosis. These diseases may evolve into MDS or AML, but overall have a better prognosis than the other diseases.

## Chronic myelomonocytic leukemia

CMML is characterized by the overproduction of myelocytes and monocytes, as well as immature blasts. Gradually these cells replace other cell types, such as red cells and platelets in the bone marrow, leading to anemia or easy bleeding. The specific pathologic features of CMML include persistent monocytosis of greater than $1 \times 10^9$/L in the peripheral blood, no Philadelphia chromosome (BCR/ABL fusion gene), fewer than 20% blasts in the blood or bone marrow, and dysplasia involving one or more myeloid lineages. The CMML bone marrow may exhibit hypercellularity (75% of cases), a blast count of less than 20%, granulocytic and monocytic proliferation, micromegakaryocytes or megakaryocytes with lobulated nuclei (80% of cases), and fibrosis (30% of cases).

Point mutations in at least 17 genes have been identified in CMML patients. The most common (percentage of CMML patients in parentheses) are *NRAS/KRAS* (30%−40%), *TET2* (30%−60%), *ASXL1* (∼40%), *SRSF2* (45%), and *RUNX1* (15%−37%). Analysis of mutational data and studies using mouse models suggest that *RAS*, *TET2*, and *ASXL1* may be driver mutations for CMML. TET2 and ASXL1 regulate DNA methylation. Disruption of these genes alters the DNA methylation pattern in CMML cells and alters expression of genes that control self-renewal and apoptosis.

A study of 81 CMML patients revealed that 37% carried mutations in the *RUNX1* gene that would be predicted to produce truncated Runx1 proteins. There was no difference in overall survival between patients with and without *RUNX1* mutations, but *RUNX1* mutation positive patients had a higher risk of progressing to AML, especially if the mutations were located in the C-terminal protein-coding region.

## Juvenile myelomonocytic leukemia

JMML accounts for 2% of all childhood leukemias. The three required criteria for a diagnosis of JMML are no Philadelphia chromosome (*BCR/ABL* fusion gene), peripheral blood monocytosis greater than $1 \times 10^9$/L, and fewer than 20% blasts in the blood and bone marrow. The presence of two or more of the following minor criteria is also required: Fetal hemoglobin increased for age, immature granulocytes in the peripheral blood, a white blood cell count greater than $1 \times 10^9$/L, a clonal chromosomal abnormality, and granulocyte-macrophage colony-stimulating factor (GM-CSF) hypersensitivity of myeloid progenitors. The bone marrow of JMML patients may show hypercellularity with granulocytic proliferation, hypercellularity with erythroid precursors (some patients), monocytes comprising 5%−10% of marrow cells, minimal dysplasia, and reduced numbers of megakaryocytes.

A distinctive characteristic of JMML leukemic cells is their spontaneous proliferation in vitro due to their hypersensitivity to GM-CSF. This hypersensitivity has been attributed to altered Ras pathway signaling as a result of mutually exclusive mutations affecting one of the pathway regulatory molecules. More than 90% of JMML patients carry a mutation in one of the following genes: *NRAS*, *KRAS*, *PTPN11*, *CBL*, or *NF1*. Approximately 35% of JMML patients display *PTPN11* mutations, 20%−25% display *N/KRAS* mutations, 11% have *NF1* mutations, and approximately 15% have *CBL* mutations. RAS is a GTP-dependent protein (G-protein) localized at the inner side of the cell membrane and transduces signal from growth factor receptors to downstream effectors. *PTPN11* (Protein-Tyrosine Phosphatase, Nonreceptor-type, 11) encodes the SHP-2 protein, which transmits signals from growth factor receptors to RAS. *NF1* (Neurofibromatosis, type 1) is a tumor suppressor gene that inactivates RAS through acceleration of RAS-associated GTP hydrolysis. Activating mutations in *RAS* or *PTPN11* or inactivating mutations in *NF1* in JMML cells all result in enhancement of signaling through the RAS pathway and increased stimulus to proliferate (Fig. 16.5). This finding has stimulated interest in molecules that inhibit the RAS pathway as possible therapeutic agents. However, JMML patients are currently treated by allogenic HSC transplantation.

## Chronic myeloid leukemia

Chronic phase CML is characterized by less than 10% blasts and promyelocytes in peripheral blood and bone marrow. The transition from chronic phase to the accelerated phase and later blast phase may occur gradually over a period of 1 year or more, or it may appear suddenly (blast crisis). Signs of impending progression of CML include progressive leukocytosis, thrombocytosis or thrombocytopenia, anemia, increasing splenomegaly or hepatomegaly. Accelerated phase CML is characterized by 10%−19% blasts in either the peripheral blood or bone marrow, and blast phase CML by 20% or more blast cells in the peripheral blood or bone marrow. Blast crisis is defined as 20% or more blasts plus fever, malaise, and progressive splenomegaly

The leukemic cells of almost all CML patients contain a distinctive cytogenetic abnormality, the Philadelphia chromosome. The Philadelphia chromosome is formed by a reciprocal translocation between the long arms of chromosomes 9 and 22 and results in the fusion of the *ABL* gene on chromosome 9 to the *BCR* gene on chromosome 22. The resulting fusion gene, *BCR-ABL*, produces a fusion protein containing the oligomerization and serine/threonine kinase domains of BCR at the amino terminus and most of the ABL protein at the carboxyl-terminus. ABL is a nonreceptor tyrosine kinase, and its activity is normally tightly regulated in cells. The fusion of *BCR* sequences constitutively activates the ABL tyrosine kinase, transforming ABL into an oncogene. BCR-ABL, with the aid of mediator proteins, associates with RAS and stimulates its activation. Through stimulation of the RAS-RAF pathway, BCR-ABL increases growth factor-independent cell growth (Fig. 16.7). BCR-ABL also activates the phosphatidylinositol-3-kinase (PI3K) pathway, suppressing programmed cell death or apoptosis.

Knowledge of the role of BCR-ABL in the development of CML led to the discovery of imatinib, a small molecule ABL kinase inhibitor, a highly effective therapy for early phase CML. However, some patients develop resistance to imatinib, usually caused by point mutations in the kinase domain of *BCR-ABL* that reduce sensitivity to imatinib. To overcome this problem, second- and third-generation BCR-ABL inhibitors have been developed. Inhibitors that target oncogenic signaling pathway downstream of BRC-ABL have also been considered.

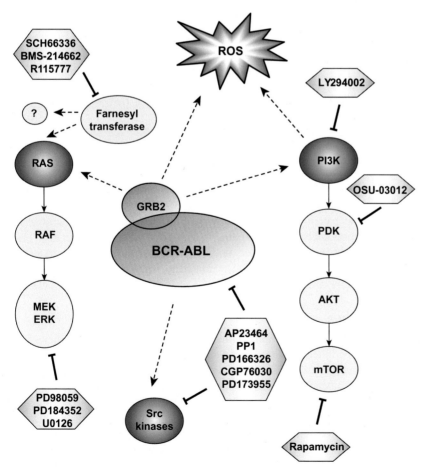

**FIGURE 16.7** Targeting signaling pathways of BCR-ABL. The BCR-ABL oncoprotein chronically activates many different downstream signaling pathways to confer malignant transformation in hematopoietic cells. For example, efficient activation of PI3K, Ras, and reactive oxygen species (ROS) requires autophosphorylation on Tyr177, a GRB-2 binding site in BCR-ABL. Also, activation of SRC family tyrosine kinases have been implicated in the BCR-ABL−related disease process. A selection of some inhibitors and pathways discussed in the text are illustrated. *Reproduced from Walz C, Sattler M. Novel targeted therapies to overcome imatinib mesylate resistance in chronic myeloid leukemia (CML). Crit Rev Oncol Hematol 2006;57:145−64, with permission.*

## Chronic neutrophilic leukemia

CNL is a rare disorder characterized by peripheral blood neutrophilia (greater than $25 \times 109/L$) and hepatosplenomegaly. The bone marrow is hypercellular, and there is no significant dysplasia in any cell lineage. Cytogenetic studies are normal in 90% of patients. Approximately 20% of CNL patients are positive for the *JAK2V617F* mutation. JAK2 is a member of the Janus family of tyrosine kinases (JAK1, JAK2, JAK3, and TYK2), which are cytoplasmic kinases that mediate signaling downstream of cytokine receptors (Fig. 16.5). Activation of a JAK-cytokine receptor complex results in recruitment and phosphorylation of STAT proteins, which then translocate to the nucleus and induce target gene transcription. The *JAK2V617F* mutation is also found in other myeloproliferative disorders.

## Polycythemia vera

In PV too many RBCs are made in the bone marrow, and the blood becomes thickened with RBCs (erythrocytosis). The extra red cells may collect in the spleen causing it to swell, or may cause bleeding problems and clots. Approximately 96% of polycythemia vera patients have the *JAK2V617F* mutation (3% have another activating JAK2 mutation). JAK2 is one of a family of cytoplasmic tyrosine kinases that mediate signaling by growth factor receptors. The V617F mutation changes JAK2 amino acid 617 from valine to phenylalanine and creates a constitutively active form of JAK2. JAK2V617F renders cells hypersensitive to the growth-stimulating effects of the erythroid growth factor erythropoietin and other growth factors. Constitutive activation of the JAK/STAT, PI3K, ERK, and AKT signal-transduction pathways is also observed in the presence of *JAK2V617F*, all of which may promote cell proliferation.

The presence of *JAK2V617F* mutations in both CNL and PV (as well as ET and PMF) raises the question of how the presence of the *JAK2V617F* mutation can result in several different diseases. Among the theories suggested are transformation of different types of hematopoietic stem or progenitor cell, allelic dose of *JAK2V617F*, or the effects of additional somatic mutations.

## Essential thrombocythemia

ET causes an abnormal increase in the number of platelets in the blood and bone marrow (thrombocytosis). This may inhibit blood flow and lead to problems such as stroke or heart attack A high percentage (approximately 50%) of ET patients carry the *JAK2V617F* mutation. An additional 25%−35% of ET patients have mutations in the calreticulin (*CALR*) gene. *CALR* encodes a component of the machinery that ensures proper protein folding, and the mutant CALR protein found in ET cells appears to activate STAT signaling (recall JAK and STAT are part of the same signal transduction pathway; Fig. 16.5).

## Primary myelofibrosis

PMF is characterized by the production of too few RBCs and too many white cells and platelets. An important constant is the production of too many megakaryocytes, which results in overproduction of platelets and cytokine release in the bone marrow. The cytokines stimulate the development of fibrous tissue in the marrow. Megakaryocytes can become so abnormal that platelet production is decreased in some patients. Approximately 50% of patients with myelofibrosis carry the *JAK2V617F* mutation. The *JAK2V617F* mutation is also found in patients with polycythemia vera or ET. The disease that develops as a result of the mutation is thought to depend on genetic background of the patient or the presence of secondary mutations. As already suggested, the three diseases are also related to some degree. About 10%−15% of cases of myelofibrosis begin as either polycythemia vera or ET.

## Chronic eosinophilic leukemia/hypereosinophilic syndrome

In chronic eosinophilic leukemia (CEL), a clonal proliferation of eosinophilic precursors results in persistently increased numbers of eosinophils in the blood, bone marrow, and peripheral tissues. In 2008, the WHO established a semimolecular classification scheme of eosinophilia subtypes including: (1) myeloid and lymphoid neoplasms with eosinophilia and abnormalities of platelet-derived growth factor receptor A (PDGFRA), platelet-derived growth factor receptor B (PDGFRB), or fibroblast growth factor receptor 1 (FGFR1), (2) CEL, not otherwise classified, and (3) idiopathic hypereosinophilia syndrome. The mutations found in the first category all take the form of fusion genes that cause constitutive activation of the PDGFRA, PDGFRB, or FGFR1 tyrosine kinase activity. *FIP1F1-PDGFRA* is the most common fusion gene, found in 10%−20% of patients with CEL, while other fusion genes are relatively rare.

The *FIP1F1-PDGFRA* fusion causes constitutive activation of the PDGFRA tyrosine kinase activity by disruption of the PDGFRA autoinhibitory juxtamembrane motif. The active FIP1F1-PDGFRA kinase stimulates the JAK/STAT, PI3K, ERK, and AKT signal-transduction pathways leading to increased survival and proliferation.

## Systemic mastocytosis

SM is a rare disease in which too many mast cells are found in skin, bones, joints, lymph nodes, liver, spleen, and the gastrointestinal tract. The 2008 WHO classification of mastocytosis includes indolent systemic mastocytosis (ISM), aggressive SM (ASM), SM associated with a clonal non-MC linage disease (SM-AHNMD), and mast cell leukemia (MCL). Most patients with ISM, ASM, or SM-AHNMD carry mutations in the *C-KIT* gene. The oncogene *C-KIT* encodes a receptor tyrosine kinase. C-KIT and its ligand stem cell factor (SCF) are required for the growth and survival of normal mast cells. SCF ligation to C-KIT activates the RAS/RAF/ERK cascade, the PI3K/AKT pathway, the SHP/RAC/JNK/C-JUN pathway, and the NF-κB pathway. Activation of the PI3K/AKT and NF-κB pathways have been shown to be necessary for mast cell proliferation. Systemic mastocytosis patients carry mutations that cause ligand-independent activation of C-KIT, by far the most common being *KITD816 V*, found in 78% of ISM patients, 82% of ASM patients, and 60% of SM-AHNMD patients. Other *C-KIT* mutations are found in small numbers (<5%) of patients. It is not currently clear if activating *C-KIT* mutations are sufficient for onset of mastocytosis, or why a small number of patients do not carry *C-KIT* mutations. Other oncogenic mutations recently identified in mastocytosis patients include those in *TET2* and *NRAS*, but the pathogenic role of these mutations is unclear.

## Acute myeloid leukemia

Normal myeloid stem cells eventually develop into granulocytes, macrophage/monocytes, and megakaryocytes. In AML, myeloid stem cells usually develop into a type of immature white blood cell called myeloblasts. Myeloblasts are abnormal and do not differentiate. For many years, the different categories of AML were described by the FAB classification scheme. The eight FAB subtypes are: M0 (undifferentiated AML), M1 (myeloblastic, without maturation); M2 (myeloblastic, with maturation); M3 (promyelocytic), or acute promyelocytic leukemia (APL); M4 (myelomonocytic), M4eo (myelomonocytic together with bone marrow eosinophilia); M5 monoblastic leukemia (M5a) or monocytic leukemia (M5b); M6 (erythrocytic) or erythroleukemia; and M7 (megakaryoblastic). Beginning in 1997, WHO developed a new classification scheme for acute myeloid leukemias that attempts to incorporate morphology, cytogenetics, and molecular genetics. The WHO scheme also reduced the required blast percentage in the blood or bone marrow for a diagnosis of AML from 30% to 20%. The category of AML with characteristic genetic abnormalities is associated with high rates of remission and favorable prognosis.

## AML with t(8;21) ((q22;q22))

AML with translocation (8;21) (q22;q22) is one of the most common genetic abnormalities in AML and accounts for 5%−20% of all cases of AML (the incidence of this form of AML varies with age and ethnic background). AML with translocation (8;21) (q22;q22) previously fell into FAB classification M2 and is characterized by large blasts often containing azurophilic granules, auer rods (found in mature neutrophils), smaller blasts in the peripheral blood, dysplasia in the bone marrow (with promyelocytes, myelocytes, and mature neutrophils), abnormal nuclear segmentation, increased eosinophil precursors, reduced monocytes, and normal erythroblasts and megakaryocytes.

t(8;21) generates the fusion gene *RUNX1-ETO*, which fuses sequences coding the amino-terminal portion of the transcription factor *RUNX1* (formerly *AML1*) to almost the entire coding region of *RUNX1T1* (formerly *ETO* or *MTG8*). The resulting fusion protein contains the DNA-binding domain (runt domain) of RUNX1 fused to the RUNX1T1 corepressor protein. Numerous studies in model systems have demonstrated that expression of the RUNX1-ETO protein alone is insufficient to induce leukemia, but can induce leukemia in cooperation with other mutations. Expression of RUNX1-ETO does lead to some inhibition of myeloid, lymphoid, and erythroid differentiation, as well as promotion of stem cell self-renewal. This is thought to predispose HSCs to leukemia development. It was formerly believed that RUNX1-ETO changed gene expression patterns by dominant-negative suppression of RUNX1 target genes. However, subsequent gene expression studies found that RUNX1-ETO activated as many genes as it repressed, suggesting that RUNX1-ETO promotes leukemogenesis by complex effects on gene expression. Among the oncogenic proteins known to promote leukemia development in cooperation with RUNX1-ETO expression are the TEL-PDGFRβ fusion protein and the FLT3 internal tandem duplication (FLT/ITD), both of which stimulate growth-promoting signal-transduction pathways. RUNX1-ETO will also promote leukemia when expressed in cells lacking the cell cycle inhibitor p21/WAF1/CDKN1A.

## AML with inv(16)(p13q22) or t(16;16)(p13q22)

AML with inv(16)(p13q22) or t(16;16)(p13q22) comprises 10%−12% of all cases of AML and is predominant in younger patients. This type of AML was formerly classified as FAB type M4 and is characterized by monocytic and granulocytic differentiation, abnormal eosinophils with immature granules often with eosinophilia, auer rods in myeloblasts, and decreased neutrophils in the bone marrow.

Both inv(16)(p13q22) or t(16;16)(p13q22) result in the fusion of the *CBFβ* gene located at 16q22 to the smooth muscle myosin heavy chain (*MYH11*) gene at 16p13. CBFβ has no DNA-binding domain, but forms a heterodimer with the RUNX1 transcription factor and stabilizes RUNX1 binding to DNA. Since the RUNX1 and CBFβ proteins function as a heterodimeric transcription factor, the leukemic fusion protein RUNX1-ETO and the CBFβ-MYH11 fusion protein are predicted to disrupt expression of a similar set of target genes. CBFβ-MYH11 binds to RUNX1 with a much higher affinity than CBFβ, and two mechanisms have been proposed by which CBFβ-MYH11 may disrupt normal RUNX1/CBFβ activity. CBFβ-MYH11 may sequester RUNX1 in the cytoplasm through the interaction of the MYH11 region with the actin cytoskeleton, or the MYH11 sequences may recruit corepressors when bound with RUNX1 to promoters in the nucleus. It is not yet clear if CBFβ-MYH11 utilizes one or both mechanisms.

As observed with *RUNX1-ETO*, expression of *CBFβ-MYH11* in model systems was not sufficient for leukemogenesis unless secondary mutations were introduced. Secondary mutations that can produce AML in cooperation with *CBFβ-MYH11* include loss of the cell cycle inhibitors *p14ARF*, *p16INK4a*, or *p19ARF*, or coexpression of *FLT3-ITD*. 60%−70% of patients with inv(16)(p13q22) or t(16;16)(p13q22) contain activating mutations in one of the following: *FLT3*, *C-KIT* (receptor tyrosine kinases), *NRAS*, or *KRAS* (signal-transduction proteins). This suggests that each of these mutations can in fact cooperate with *CBFβ-MYH11* to induce leukemia.

## Acute promyelocytic leukemia with t(15;17)(q22q12)

APL comprises 5%−8% of all cases of AML and is found as typical APL or microgranular APL. Common features of typical APL include promyeloctyes with kidney-shaped or bilobed nuclei, cytoplasm densely packed with large granules, bundles of Auer rods in the cytoplasm, larger Auer rods than other types of AML, strongly positive myeloperoxidase reaction in all leukocytes, and only occasional promyelocytes in the blood. Features of microgranular APL include bilobed nuclei, scarce or absent granules, a small number of abnormal promyelocytes with visible granules and bundles of Auer rods, high leukocyte count in the blood, and a strongly positive myeloperoxidase reaction in all promyelocytes. APL was formerly classified as FAB type M3.

In over 98% of cases, the retinoic acid receptor alpha (*RARα*) gene at 17q12 is fused to the *PML* gene at 15q22, t(15;17)(q22q12). In rare cases, *RARα* is fused to another gene, including *PLZF*, *NUMA1*, *NPM*, or *STAT5B*. Retinoid signaling is transmitted by two families of nuclear receptors, retinoic acid receptor (RAR) and retinoid X receptor (RXR), which form RAR/RXR heterodimers. In the absence of ligand, the RAR/RXR heterodimer binds to target gene promoters and represses transcription. When a ligand (such as retinoic acid) binds to the complex, it induces a conformational change, which transforms the heterodimer into a transcriptional activator. The PML-RARα fusion protein binds to RAR/RXR target genes and acts as a potent transcriptional repressor that is not activated by physiological concentrations of ligand. This is due to the fact that all the oncogenic fusion partners of RARα provide a dimerization domain, which results in a dimerized fusion protein with two corepressor-binding sites instead of the one found in the RAR/RXR complex. However, recent studies suggest that the PML-RARα fusion protein must have other oncogenic properties, since enforced corepressor binding onto RARα does not initiate APL in model systems. Recent models suggest the PML-RARα leukemogenesis combined enhanced corepressor recruitment and relaxed target specificity to both enhance repression of some genes and target genes not normally bound by RAR/RXR. This disruption of normal gene expression is thought to affect two pathways—myeloid progenitor cell self-renewal and promyelocyte differentiation.

APL is highly sensitive to treatment with all-trans retinoic acid (ATRA), which overcomes the enhanced repression by PML-RARα and induces differentiation of leukemic cells.

## AML with 11q23 (MLL) abnormalities

AML with 11q23 (MLL) abnormalities comprises 5%−6% of all cases of AML. Two groups of patients show a high frequency of this type of AML—infants and adults with therapy-related AML (usually occurring after treatment with topoisomerase inhibitors). Common morphologic features include monoblasts and promyelocytes predominant in the bone marrow and showing strong positive nonspecific esterase reactions. AML due to 11q23 abnormalities can be associated with acute myelomonocytic, monoblastic, and monocytic leukemias (FAB M4, M5a, and M5b classifications) and more rarely with leukemias with or without maturation (FAB M2 and M1).

The *MLL* (myeloid/lymphoid leukemia) gene encodes a DNA-binding protein that methylates histone H3 lysine 4 (H3K4) and is also known as the lysine-specific methyltransferase 2A (*KMT2A*) gene. *MLL* knockout studies indicate that *MLL* is necessary for proper regulation of *HOX* gene expression. *HOX* genes are a family of transcription factors that regulate many aspects of tissue development. The precise mechanism by which *MLL* regulates gene expression has not yet been determined. All *MLL* translocations contain the first 8–13 exons of *MLL* and a variable number of exons from a fusion partner gene. At least 52 *MLL* fusion partner genes have been described, and these fusion partners have diverse functions. All MLL fusion proteins have lost the domain necessary for H3K4 methylation. It is believed that leukemogenesis mediated by MLL fusion proteins involves disruption of normal gene expression patterns regulating stem cell differentiation and self-renewal. In some cases the *MLL* fusion is believed to reactivate the self-renewal program in committed myeloid progenitors. The protein domains contributed by the MLL fusion partners are believed to contribute to leukemogenesis through their effects on transcription, chromatin remodeling, and protein-protein interactions.

## Cytogenetically normal AML

Almost half of AML patients display no cytogenetic abnormalities. Numerous studies have attempted to establish prognostic subgroups for cytogenetically normal AML (CN-AML) based on gene mutation/expression patterns. One study of 197 AML patients identified at least one patient mutation in 44 out of 51 oncogenes tested. However, only five genes (*FLT3*, *NPM1*, *CEBPA*, *DNMT3A*, and *C-KIT*) were mutated in more than 10% of patients. Seven percent of patients had no mutations in any of the genes tested, suggesting the existence of additional gene mutations that contribute to the development of AML. This is confirmed by data released by The Cancer Genome Atlas Network after sequencing 200 AML patient samples (whole genome sequencing for 50 samples, sequencing of the protein coding region for the remainder). This study identified 260 genes mutated in two or more patient samples (of which 23 were termed significantly mutated) and a further 1623 genes mutated in only one sample. The average number of mutations per patient was 13, with an average of five genes recurrently mutated in AML. In this study, *FLT3*, *NPM1*, and *DNMT3A* were mutated in more than 10% of patients, and significantly mutated genes included *IDH1/2*, *TET2*, *RUNX1*, *CEBPA*, and *C-KIT*. Both studies of AML patient mutations were able to group the mutated genes into functional categories: transcription-factor fusions, *NPM1*, tumor-suppressor genes, DNA-methylation-related genes, activated signaling genes, myeloid transcription-factor genes, cohesin-complex genes, and spliceosome-complex genes.

Mutations in the *FLT3* gene are among the most common in AML. Activating mutations in the FMS-like tyrosine kinase (*FLT3*) are present in 20%–30% of all cases of de novo AML. Although *FLT3* mutations can be associated with all the major leukemic translocations (*RUNX1-ETO*, *CBFβ-MYH11*, *PML-RARα*, *MLL* fusions), the majority of cases of AML with *FLT3* mutations are cytogenetically normal. Other common clinical features are leukocytosis and monocytic differentiation. Two major types of *FLT3* mutation are found in AML patients. An internal tandem duplication of the region of the gene encoding the juxtamembrane domain is found in 25%–35% of adult and 12% of childhood AML. The second type of mutation is a missense mutation in the activation loop of the tyrosine kinase domain, with codons D835 and D1836 commonly affected. Both types of mutation result in constitutive phosphorylation of the FLT3 receptor in the absence of ligand and activation of downstream signaling pathways including the PI3K/AKT pathway and the RAS/RAF/ERK pathway.

Nucleophosmin 1 (*NPM1*) mutations are found in 30% of all AML patients and in 50%–60% of patients with cytogenetically normal AML. *NMP1* encodes a phosphoprotein involved in cell proliferation and apoptosis. *CEBPA* encodes a transcription factor (CEBPα), which is necessary for neutrophil differentiation. *CEBPA* mutations are found in approximately 10% of all AML patients and are more common in cytogenetically normal AML. Approximately two-thirds of patients with *CEBPA* mutations have biallelic mutations, and thus loss of CEBPA protein expression.

Forty-four percent of AML patients exhibit mutations in genes that regulate DNA methylation, such as *DNMT3A* (24.6%), *IDH1* (9.6%), and *IDH2* (10.2%). DNMT3A (DNA methyltransferase 3A) catalyzes the addition of methyl groups to cytosine, including cytosine found in CpG islands, regulatory elements found in gene promoters. Hypermethylation of these CpG islands suppresses gene transcription. Loss of DNMT3A in mouse HSCs causes changes in methylation (up or down) at distinct loci, changes in gene expression, and expansion of the HSC population and loss of peripheral blood differentiation after transplant. However, the role of *DNMT3A* mutations in leukemogenesis is currently unclear; studies have reached no clear consensus.

IDH1 and IDH2 (mitochondrial homolog of IDH1) are enzymes that normally catalyze the oxidative decarboxylation of isocitrate, producing α-ketoglutarate (α-KG). α-KG is used by the enzyme TET2, which demethylates cytosine residues in DNA—DNMT3A catalyzes methylation and TET2 demethylation of the same site on cytosine. The mutated

*IDH1/2* genes found in AML patients produce altered proteins with new and distinct functions. These mutant proteins catalyze conversion of α-KG to 2-hydroxyglutarate (2HG). 2HG acts as a competitive inhibitor of α-KG dependent reactions, including demethylation of DNA by TET2. *TET2* mutations are also found in AML patients, and *TET2* mutations and *IDH1/2* mutations appear to be mutually exclusive, suggesting overlapping epigenetic effects. *IDH1/2* mutations produce a distinct hypermethylation pattern in AML leukemic cells, causing altered expression of a number of genes involved in myeloid differentiation of hematopoietic stem/progenitor cells (suppression of *GATA1* and upregulation of *C-KIT*, for example). These changes are believed to contribute to leukemogenesis.

### Therapy-related AML and MDS

This class includes both AML and MDS that arise after chemotherapy or radiation therapy. These diseases are classified according to the mutagenic agents used for treatment, but it can be difficult to attribute a secondary AML to a specific agent because treatment often involves multiple mutagenic agents.

**Alkylating agent-related AML**: Alkylating agent—related AML usually occurs 5—6 years after exposure to the agent. Typically it is first observed as an MDS with bone marrow failure. Some cases evolve into AML, which may correspond to acute myeloid leukemia with maturation (FAB class M2), acute monocytic leukemia (M5b), AMML (M4), erythroleukemia (M6a), or acute megakaryoblastic leukemia (M7). Cytogenetic abnormalities are observed in more than 90% of cases of therapy-related AML/MDS. Complex abnormalities are the most common finding, often including chromosomes 5 and 7. Many patients with therapy-related MDS or AML carry point mutations in *p53* (24% of cases), *RUNX1* (16% of cases), or various other oncogenes. An association between *p53* point mutations and chromosome 5 aberrations and between *RUNX1* mutations and chromosome 7 aberrations was observed, suggesting that these sets of mutations may cooperate in the development of therapy-related AML/MDS.

**Topoisomerase II inhibitor-related AML**: Topoisomerase II inhibitor—related AML may develop in patients treated with the topoisomerase II inhibitors etoposide, teniposide, doxorubicin, or 4-epi-doxorubicin. Development of AML is observed approximately 2 years after treatment and is most commonly diagnosed as acute monoblastic or myelomonocytic leukemia. AML resulting from treatment with topoisomerase poisons such as etoposide is predominantly associated with translocations of the *MLL* gene at 11q23. 5%—10% of *MLL*-associated leukemias are therapy-related. Translocations involving other genes associated with leukemogenesis, such as *RUNX1*, *CBFβ*, and *PML-RARα*, have also been observed.

# Lymphocyte disorders

Disorders of lymphocytes include deficiency of lymphocytes (lymphopenia) and overproliferation of lymphocytes. Overproliferation of lymphocytes is due to either reactive proliferation of lymphocytes (lymphocytosis) or to neoplastic problems.

## Lymphopenia

Lymphopenia is defined by less than 1500 lymphocytes/microliter of blood in adults and less than 3000 lymphocytes/microliter of blood in children. Lymphopenia is relatively rare compared to other leukopenias involving granulocytic cells. Some lymphopenias are due to genetic abnormalities, which are categorized as congenital immunodeficiencies. Most lymphopenias are due to viral infection, chemotherapy, radiation, under nutrition, immunosuppressant drug reaction, and autoimmune diseases.

## Lymphocytosis

Lymphocytosis can be divided into relative lymphocytosis and absolute lymphocytosis. Normally 20%—40% of human white blood cells are lymphocytes. When the percentage exceeds 40%, it is recognized as relative lymphocytosis. When the total lymphocyte count in blood is more than 4000/μL in adults, 7000/μL in older children, and 9000/μL in infants, the patient is diagnosed with absolute lymphocytosis.

The best known type of lymphocytosis is infectious mononucleosis. This disease is due to an infection of Epstein-Barr virus (EBV). EBV infection at an early age will not show any specific symptoms. However, infection in adolescents and young adults can cause more severe problems (Kissing Disease), such as fever, sore throat, lymphadenopathy, splenomegaly, hepatomegaly, and increased atypical lymphocytes in blood. EBV is a member of the herpesvirus family. EBV infects B-lymphocytes. In a minority of infected B-cells, EBV infection occurs in the lytic form, which induces cell lysis and virus release. In a majority of cells, EBV infection is nonproductive and the virus is maintained in latent form. The cells with latent viruses are activated and undergo proliferation, and also produce specific antibodies against the virus. The massive expansion of monoclonal or oligoclonal cytotoxic CD8+ T-cells presented as atypical lymphocytes in peripheral blood is the major feature of infectious mononucleosis. Such strong humoral and cellular responses to EBV eventually highly restrict EBV infection.

## Neoplastic problems of lymphocytes

Lymphocytic leukemia and lymphoma are the two major groups of lymphoid neoplasms. Leukemia and lymphoma do not have a very clear distinction, and the use of the two terms can be confusing. In general, the term lymphocytic leukemia is used for neoplasms involving the general area of the bone marrow and the presence of a large number of tumor cells in the peripheral blood. In contrast, lymphomas show uncontrolled growth of a tissue mass of lymphoid cells. However, quite often, especially at the late stage of lymphoma, tumor cells originating from the lymphoma mass may spread to peripheral blood and produce a phenotype similar to leukemia. According to the World Health Organization classification, lymphocytic neoplasms are divided into five major categories: (1) precursor B-cell neoplasms, (2) peripheral B-cell neoplasms, (3) precursor T-cell neoplasms, (4) peripheral T-cell neoplasms, and (5) Hodgkin lymphoma. It is important to mention that all lymphoid neoplasms develop from a single transformed lymphoid cell. Furthermore, neoplastic transformation happens after the rearrangement of antigen receptor genes, including T-cell receptors and immunoglobulin heavy and light chains. Therefore, antigen receptor patterns are generally used to distinguish monoclonal neoplasms from polyclonal reactive lymphadenopathy. Although there are many different lymphocytic malignancies, the majority of adult lymphoid neoplasms are one of four diseases: follicular lymphoma, large B-cell lymphoma, chronic lymphocytic leukemia/small lymphocytic lymphoma, and multiple myeloma. Likewise, the majority of childhood lymphoid neoplasms are one of the two diseases: acute lymphoblastic leukemia/lymphoma and Burkitt lymphoma.

### Acute lymphoblast leukemia/lymphoma

Acute lymphoblast leukemia/lymphoma (ALL) is most common between the ages of 2 and 5 years although it affects both adults and children. The majority of ALL is pre-B-cell leukemia. Pre-T-cell leukemia is often reported in adolescent males. Morphologically, it is difficult to separate T and B lineage ALL. Furthermore, patients with T-ALL and B-ALL also present similar symptoms. Therefore, flow cytometry studies to identify the expression of specific cell surface markers are generally used to distinguish the lineage and differentiation of ALL.

Various chromosomal locus translocations are associated with the development of ALL. The most common are those involving the *MLL* gene (chromosome 11q23) and a diverse range of fusion partners, t(1:19) *TCF3-PBX1* (*E2A-PBX1*) and t(12:21) *ETV6-RUNX1* (*TEL-AML1*). *MLL* gene translocations are observed in 6% of ALL patients, the *TCF3-PBX1* translocation in 4%, and the *ETV6-RUNX1* translocation in 22%.

The critical fusion protein generated from the *ETV6-RUNX1* translocation contains 336 amino acids from the N-terminal region of ETV6 and almost the entire RUNX1 protein. Quite frequently, another allele of *ETV6* is also lost in t(12;21) ALL patient samples. This finding suggests that *ETV6* is a potential tumor suppressor gene. ETV6-RUNX1 can form dimers via the ETV6 helix-loop-helix domain and contains the RUNX1 DNA—binding domain. Therefore, it is believed that ETV6-RUNX1 affects the expression of RUNX1 target genes to promote leukemia development. Interestingly, the *ETV6-RUNX1* fusion gene has been identified in neonatal blood spots of children who developed leukemia between 2 and 5 years of age, suggesting that t(12;21) is not sufficient for leukemogenesis without additional malignant promoting factors.

The Philadelphia chromosome caused by t(9;22) is the most frequently identified chromosomal translocation in adult ALL. The Philadelphia chromosome encodes the fusion protein BCR-ABL. The constitutive activation of the ABL tyrosine kinase and the interaction of this fusion protein with various signaling regulators and protooncogene products promote B-ALL development (Fig. 16.6). More than 70% of patients with *BCR-ABL* lymphoid leukemia also carry mutations in the *IKZF1* gene.

More than 50 recurring deletions or amplifications have been identified in ALL patients, many of which involve a single gene or few genes. Using high-resolution single-nucleotide polymorphism arrays and genomic DNA sequencing to study 242 B-ALL patient samples, the *PAX5* gene has been identified as the most frequent target of somatic mutation. Deletion or point mutation of the *PAX5* gene occurs in 31.7% of patients and results in decreased expression or partial loss of its function. PAX5 is also known as B-cell specific activating protein, which plays a crucial role during B lineage commitment and differentiation. The lymphoid transcription factor *IKZF1* (IKAROS) is altered by focal deletions or sequence mutation in 15% of pediatric B-ALL patients. *IKZF1* encodes a transcription factor required for development of all lymphoid lineages. *IKZF1* alterations are associated with a high risk of treatment failure.

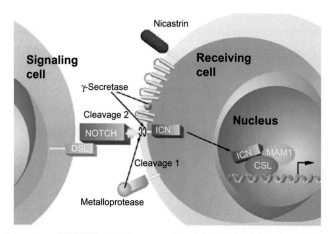

**FIGURE 16.8** NOTCH signaling. Interaction of NOTCH and delta serrate ligand (DSL) stimulates proteolytic cleavage of NOTCH by metalloproteinases and g-secretase. This leads to the release of the intracellular ICN domain, which translocates to the nucleus where it interacts with the DNA binding protein CSL, displaces corepressors, and recruits coactivators (MAM1), thereby converting CSL from a repressor to an activator of gene expression. *Reproduced from Armstrong SA, Look AT. Molecular genetics of acute lymphoblastic leukemia. J Clin Oncol 2005;23:6306—15, with permission.*

T-ALL patients have their own distinct array of chromosome rearrangements and mutations. The NOTCH signaling pathway plays important roles during hematopoiesis, especially in T-cell lineage development. The interaction of cell surface NOTCH receptors and their ligands of the Delta-Serrate-LAG2 family induces two-step proteolytic cleavage of the NOTCH protein and generates the intercellular domain of NOTCH (ICN) fragment. ICN translocates to the nucleus and activates target gene expression via interaction with the DNA-binding transcription factor CSL, displacement of transcription repressors, and recruitment of transcription activators to the DNA-binding complexes (Fig. 16.8). A NOTCH activating mutation involving somatic alteration of the *NOTCH1* gene has been identified in over 50% of T-ALL patients. Furthermore, the FBW7 ubiquitin E3 ligase responsible to the degradation of ICN is also mutated in T-ALL patient samples and cell lines, which increases the cellular concentration of ICN and further enhances NOTCH signaling. The best known NOTCH target gene related to cancer development is the *MYC* oncogene.

Half of T-ALL patients carry translocations involving the fusion of T-cell receptor genes to various oncogenes (including oncogenes *TLX1/HOX11/TCL3, TLX3/HOX11L2, TAL1/SCL/TCL5, LYL1, LMO1,* and *LMO2*). These T-cell receptor/oncogene fusions are often associated with inactivating/deletion mutations in the plant homeodomain finger 6 (*PHF6*) gene. Sixteen percent of childhood and 38% of adult T-ALL patients have mutations in the *PHF6* gene. The *PHF6* gene product is an RNA-interacting protein and component of the nucleosome remodeling and deacetylation complex, but its role in leukemogenesis is poorly understood.

## Chronic lymphocytic leukemia/small lymphocytic lymphoma

Chronic lymphocytic leukemia (CLL) is characterized by the presence of over 5000/µL of mature-appearing lymphocytes in peripheral blood and a specific range of immunophenotypes. Small lymphocytic lymphoma refers to a small percentage of cases in which the tumor cells have a similar immunophenotype to CLL, but are restricted to lymph nodes without blood and bone marrow involvement. Due to currently unclear genetic factors, CLL is rare in Asian populations, but is the most common form of leukemia in North America and Europe.

The cytogenetic abnormalities detected by fluorescence in situ hybridization in CLL are mainly chromosome trisomies and deletions. Analysis of 325 CLL patient samples identified trisomy 12 (18%), deletions on chromosome 13q (55%) and chromosome 11q (16%), and a deletion on chromosome 17p (7%). Importantly, chromosome 11q and chromosome 17p deletions are related to poor prognosis and an advanced stage of this disease. The well-known tumor suppressor gene *p53* is located in the deleted region of chromosome 17. The ATM kinase that regulates p53 activity is located in the deleted region of chromosome 11. Since p53 is a critical inhibitor of cell cycle progression and most chemotherapy drugs target p53-dependent pathways, it is valuable to evaluate cytogenetic conditions before treating CLL patients.

Detailed analysis of CLL deletions in the 13q14 region identified the deleted tumor suppressor gene as a cluster of two microRNAs, miR-15a and miR-16-1. Loss of miR-15/16 results in overexpression of the *BCL2* gene, which promotes cell survival. This finding has encouraged the development of ABT-199, a potent oral drug that inhibits BCL2, and which shows promise as a CLL treatment. Currently, when treatment of CLL becomes necessary, first-line therapy usually involves a combination of chemotherapy and anti-CD20 monoclonal antibodies.

## Follicular lymphoma

The neoplastic cells of follicular lymphoma are derived from germinal center B-cells or cells differentiated toward germinal center B-cells. Furthermore, they present in either a pure follicular pattern or mixed with follicular and diffused areas. The occurrence of this disease is also affected by genetic background. It is one of the most common lymphocytic neoplasms in North America and Europe, but less common in Asia. Follicular lymphoma is a disease of late life with a peak of detection between 60 and 70 years of age.

*BCL2* is highly expressed in neoplastic cells in over 90% of follicular lymphoma patients. Therefore, BCL2 immunostaining is used to distinguish normal follicles from follicular lymphoma. This high expression of BCL2 is due to a specific chromosomal translocation [t(14; 18)(q32;q21)] that generates a fusion between the immunoglobulin heavy chain enhancer on chromosome 14 and the *BCL2* gene on chromosome 18. This was one of the earliest discovered chromosomal translocations related to cancer development. The translocation breakpoint on chromosome 14 is at the functional diversity region-joining region joint, indicating that mistaken recombination involving the recombination enzymes is the molecular mechanism of generating this translocation.

BCL2 is a strong antiapoptotic factor. Normally, most B-cells should be terminated via apoptosis if they are not challenged by specific antigens. With the overexpression of BCL2, follicular lymphoma cells are able to overcome normal apoptotic signals and avoid termination. Therefore, the prolonged life span of follicular cells due to this defect in apoptotic elimination contributes to the development of follicular lymphoma. However, additional cytogenetic lesions besides t(14;18) (q32;q21) are generally observed in most follicular lymphoma cells, which include trisomies, monosomies, deletions, amplifications, and chromosome translocations. These observations suggest that additional mutations beyond the overexpression of BCL2 are required for the development of follicular lymphoma.

## Diffuse large B-cell lymphoma

The name diffuse large B-cell lymphoma (DLBCL) is based on the morphology and behavior of this group of malignant cells. They typically express B-cell markers but lack terminal deoxytransferase. DLBCL cells are large and diffusely invade the lymph nodes and extranodal areas. However, this is a highly heterogeneous disease. DLBCL is generally identified in older patients with a median age >60 years and with almost equal distribution between male and female patients. This is the most common lymphocytic malignancy in adults. Cases of DLBCL are currently divided into three main subtypes based on gene expression profiling: the germinal center B-cell (GCB)-like subtype, the activated B-cell (ABC)-like subtype, and the primary mediastinal B-cell lymphoma subtype. These subtypes arise from B-cells at different stages of differentiation and have different clinical outcomes. Exome (the exon part of the genome) sequencing has been conducted with over 200 DLBCL samples. The results revealed that each DLBCL subtype has a specific pattern of chromosome translocations, deletions, and gene mutations. Although DLBCL subtypes have similarities, the pattern of mutations is also highly heterogeneous from patient to patient.

The GCB-DLBCL subtype is characterized by a more favorable clinical outcome. The t(14:18)(q32:q21) translocation, which causes overexpression of *BCL2* (a strong antiapoptotic factor) and is also observed in follicular lymphoma, is found only in this subtype (34% of GCB-DLBCL patients). This subtype also has relatively high rates of somatic mutation of the *MLL2* and *EZH2* genes (>20% of GCB-DLBCL patients). Both genes regulate histone methylation and are also mutated at high rates in follicular lymphoma patients.

The ABC-DLBCL subtype has a less favorable clinical outcome and is characterized by a higher incidence of 3q27 translocations, which cause overexpression of the antiapoptotic factor *BCL6* (24% in ABC-DLBCL vs. 10% in GCB-DLBCL). However, a high rate of *BCL6* somatic mutation is also observed (more than 70% in GCB-DLBCL, 44% in ABC-DLBCL), and levels of *BCL6* mRNA and protein do not correlate with the *BCL6* translocations. The ABC-DLBCL subtype also shows a high rate of somatic mutation in the *MYD88* (30%−37% of ABC-DLBCL patients) and *MLL2* genes. MYD88 normally activates NF-κB signaling. Mutations of *MYD88* found in DLBCL patients are believed to cause increased activation of NF-κB signaling.

The PMLB-DLBCL subtype is characterized by amplification of the *JAK2* gene in nearly half of cases, and recurrent deletion of *SOCS1*, which suppresses JAK-STAT signaling. Both mutations lead to activation of JAK-STAT signaling. PMLB-DLBCL is a rare subtype that tends to occur in the anterior mediastinum of young female patients. PMLB-DLBCL shows enriched expression of IL13 pathway genes, which is also seen in Hodgkin lymphoma.

## Burkitt lymphoma

In 1958, Denis Burkitt reported a special type of jaw tumors in African children and these tumors were later named Burkitt lymphoma. Burkitt lymphoma cells are generally monomorphic medium-sized cells (bigger than ALL cells and smaller than DLBCL cells) and with round nuclei and multiple nucleoli. These cells are extremely hyperproliferative and are also highly apoptotic. Close to 100% of Burkitt lymphoma cells are positive for the proliferation marker Ki-67. According to the WHO, Burkitt lymphoma can be divided into three categories: endemic, sporadic (nonendemic), and immunodeficiency-associated. The common feature of Burkitt lymphoma is the chromosomal translocation-induced overexpression of the *C-MYC* protooncogene. The most common form of translocation is t(8;14) (q24;q32), which leads to immunoglobulin heavy chain regulatory element—directed expression of the *C-MYC* gene. Interestingly, *C-MYC* was the first gene known to be involved in a chromosome translocation—associated neoplasm via the study of t(8;14)(q24; q32) in Burkitt lymphoma. Since *C-MYC* is also overexpressed in other forms of leukemia and lymphoma, it is believed that other genetic lesions also play critical roles in the development of Burkitt lymphoma.

Genomic DNA sequencing of Burkitt lymphoma samples has uncovered some of the oncogenic mutations that cooperate with *C-MYC* overexpression in the pathogenesis of Burkitt lymphoma. The *C-MYC* gene itself is the most recurrently mutated gene (70% of Burkitt lymphoma cases). Mutated C-MYC proteins may have enhanced transforming ability. Inactivating mutations in the *TP53* gene (encoding the proapoptotic protein p53) are also common, occurring in 35% of cases. Mutations in the *CCND3* gene occur in 38% of sporadic Burkitt lymphoma patients. *CCND3* encodes cyclin D3. These mutations stabilize the cyclin D3 protein, and the resulting higher cellular levels of cyclin D3 are believed to promote cell cycle progression. Seventeen percent of patients have inactivating mutations in the *CDKN2A* gene, which encodes the cell-cycle inhibitors p16(INK4a) and p14(ARF). Most significantly, all three subtypes of Burkitt lymphoma carry mutations in the *TCF-3* (*E2A*) gene (10%—25%) and/or its negative regulator, *ID3* (35%—58%). Mutations that disrupt normal TCF-3/ID3 activity are found in 70% of patients with sporadic Burkitt lymphoma. TCF-3 is a transcription factor; TCF-3 DNA-binding activity is inhibited by heterodimerization with ID3. The mutations found in Burkitt lymphoma disrupt ID3, but produce an active TCF-3 protein with reduced or absent affinity for ID3. All Burkitt lymphoma cell lines depend on TCF-3 for survival and proliferation, and TCF-3 also has a key role in normal B-cell development.

Sporadic Burkitt lymphoma shows no geographic preference. Furthermore, it also has less age restrictions and is detected in adults. The lymph nodes and terminal ileum are the common sites of this type of lymphoma.

HIV infection has been related to the development of various forms of lymphoma, including Burkitt lymphoma, DLBCL, low-grade B-cell lymphoma, peripheral T-cell lymphoma, primary effusion lymphoma, and classical Hodgkin lymphoma. Depending on the difference in pathological classification, Burkitt lymphoma is either the most (35%—50%) or the second most (after DLBCL) common lymphoma in HIV patients. Soluble Tat protein encoded by HIV-1 can be released by the infected cells and then taken up by uninfected cells. This protein has been reported to interact with the RB family member RB2 and inactivate the normal function of RB2. Experimental evidence also suggests that Tat preferentially targets B-cells. RB2 is one of the three RB family members, which plays important roles during cell cycle progression by controlling E2F activity during G1-S phase transition (Fig. 16.9). Therefore, Tat may behave synergistically with *C-MYC* overexpression to promote Burkitt lymphoma development.

## Multiple myeloma

Multiple myeloma is the most important and common plasma cell neoplasm. Plasma cells are mature immunoglobulin producing cells. Plasma cell neoplasms are a group of neoplastic diseases of terminally differentiated monoclonal immunoglobulin producing B-cells. They are generally referred to as myeloma. The monoclonal immunoglobulin produced by these cells is considered the M factor of myeloma. In normal plasma cells, the production of immunoglobulin heavy chain and light chain is well balanced. Under neoplastic conditions, normal balance may not be maintained, resulting in the over production of either heavy chain or light chain. The free light chains are known as *Bence Jones proteins*. Multiple myeloma is generally preceded by a premalignant condition called monoclonal gammopathy of undetermined significance (MGUS). MGUS is quite common in older people. About 20% of patients with MGUS will develop myeloma, generally multiple myeloma. Multiple myeloma is a disease with multiple masses of neoplastic plasma cells in the skeletal system, which is generally associated with pain, bone fracture, and renal failure. It occurs mainly in people of older age (> 45 years of age) and affects more men than women. Furthermore, the rate of multiple myeloma in African Americans is about twice that of the rest of the population in the United States.

Like other lymphoid malignancies, multiple myeloma is related to the overexpression of various regulators of cell proliferation and survival due to chromosomal translocations, which place these genes under the control of

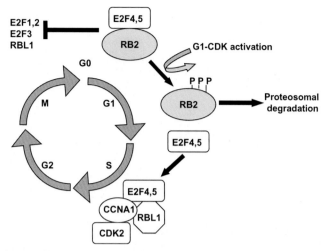

**FIGURE 16.9** RB2 function. In quiescent (G0) cells, the nuclear E2F-RB complex represses several cellular promoters. After activation of the cell cycle, RB2 is phosphorylated by G1 cyclin-dependent kinases (CDKs) and then degraded by the proteosome. This results in the derepression of various genes, including RBL1. RBL1 is then able to interact with E2F4 and E2F5, which have been released from RB2, and associate with cyclin A (CCNA1) and CDK2. *Adapted from Bellan C, Lazzi S, De Falco G et al. Burkitt's lymphoma: new insights into molecular pathogenesis. J Clin Pathol 2003;56:188—92.*

immunoglobulin regulator elements, mainly the immunoglobulin heavy chain locus on chromosome 14. The five most common chromosomal translocations involving the immunoglobulin heavy chain locus are those involving the cyclin D1 gene (*CCND1*) on chromosome 11q13 (16%), the cyclin D3 gene (*CCND3*) on chromosome 6p21 (3%), the *MAF* gene on chromosome 16q23, the *MAFB* gene on chromosome 20q12, and the *MMSET* and *FGFR3* genes on chromosome 4p16 (15%). The well-established MAF targets, integrin 7 and cyclin D2, are important in communication with the cellular microenvironment and in regulating cell cycle progression, respectively. MMSET is a histone methyl transferase and is likely involved in the regulation of chromatin structure and protein-protein interactions to regulate gene expression. FGFR3 is a receptor tyrosine kinase and its activation directly promotes cell proliferation and survival.

## Hodgkin lymphoma

In contrast to non-Hodgkins lymphoma, Hodgkin lymphoma starts from a single lymph node or a chain of nodes and spreads in an orderly way from one node to another. Further microscopic analysis revealed that Hodgkin lymphoma presents a very unique type of malignant cells, called Reed-Sternberg cells (Fig. 16.10). The classical morphology of these cells is large size (20—50 μm), relative abundance, amphophilic and homogeneous cytoplasm, and two mirror-image nuclei (owl eyes) with one eosinophilic nucleolus in each nucleus. Reed-Sternberg cells only occupy a small portion of the tumor mass. The majority of the cells in the tumors are reactive lymphocytes, macrophages, plasma cells, and eosinophils, which are attracted to the surrounding malignant Reed-Sternberg cells by their secreted cytokines. Hodgkin lymphoma affects people of relatively young age. It is one of the most common forms of cancer in young adults, with an average age at diagnosis of 32 years. Currently, highly developed radiation therapy and chemotherapy treatments have made Hodgkin lymphoma a curable cancer. However, about 20% of patients still die from this disease. Furthermore, successfully treated patients have a higher risk of dying from late toxicities, such as secondary malignancies and cardiovascular diseases.

Genomic sequencing of Hodgkin and Reed Sternberg (HRS) cells purified by flow sorting revealed mutations in genes regulating antigen presentation, chromosome integrity, transcriptional regulation, and ubiquitination. Beta-2-microglobulin (*B2M*) is the most commonly mutated gene in HRS cells with inactivating mutations being found in about 70% of patient samples. Loss of B2M leads to loss of major histocompatibility complex class 1 (MHC-1) cell surface expression. Lower B2M expression is associated with lower stage disease and better prognosis.

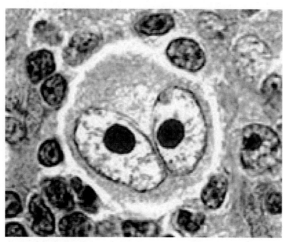

**FIGURE 16.10** Reed-Sternberg cell. The image shows the typical characteristics of the Reed-Sternberg cell: large size (20–50 μm), amphophilic and homogeneous cytoplasm, and two-mirror image nuclei (owl eyes) with one eosinophilic nucleolus in each nucleus. Reed-Sternberg cells only occupy a small portion of the tumor mass. *Image courtesy of the Department of Pathology, Stanford University (http://hematopathology. stanford.edu/).*

## Key concepts

- Leukemias and lymphomas occur in response to genetic mutations (point mutations, deletions, or translocations) that disrupt the normal function of genes that regulate the development and maintenance of normal hematopoietic cells. Many different types of genes are responsible for the regulation of normal hematopoietic cells, including genes that encode transcription factors, kinases, DNA methylases, splicing factors, and microRNAs. For example, leukemia or lymphoma may be induced by mutations that produce a constitutively active kinase, which signals the cell to proliferate in the absence of normal growth signals.

- Mutations that alter transcription factors can result in abnormal patterns of gene expression that stimulate growth and/or block the normal apoptotic response to DNA damage, chromosomal abnormalities, or uncontrolled growth. Normal gene expression patterns are the result of a combination of transcription factor activity, methylation (and other modifications) of DNA and chromatin proteins, and posttranscriptional regulation by microRNAs, splicing factors, and other proteins. Therefore, mutations in any of these gene expression regulators may disrupt the normal gene expression pattern, producing effects on cell growth and apoptosis.

- Recent studies have shown that leukemia and lymphoma, like other cancers, are genetically complex, containing multiple mutations in different types of genes. For example, AML patients often contain mutations in genes encoding transcription factors, kinases, and DNA methylases. It is believed that the combined effects of these different classes of mutation are necessary to produce the leukemia. Since many specific mutations are found in more than one type of leukemia or lymphoma, the precise combination of different mutations is also believed to determine the type of leukemia or lymphoma that is produced.

- The genetic complexity of leukemia, lymphoma, and other cancers presents a great challenge when developing new cancer treatments. Researchers are attempting to meet this challenge by targeted therapies based on individual cancer genomes and by therapies that harness the power of the immune system.

## Suggested readings

[1]  Doulatov S, Notta F, Laurenti E, Dick JE. Hematopoiesis: a human perspective. Cell Stem Cell 2012;10:120–35.

[2]  Ackerman M, Liebherr S, Klasmann J-H, Lachmann N. Lost in translation: pluripotent stem cell-derived hematopoiesis. EMBO Mol Med 2015;7:1388–402.

[3]  Notta F, Zandi S, Takayama N, et al. Distinct routes of lineage development reshape the human blood hierarchy across ontogeny. Science 2016;351:aaab2116.

[4]  Schepers K, Campbell TB, Passegue E. Normal and leukemic stem cell niches: insights and therapeutic opportunities. Cell Stem Cell 2015;16:254–67.

[5]  Sive JI, Gottgens B. Transcription network control of normal and leukaemic haematopoiesis. Exp Cell Res 2014;329:255–64.

[6]  Khalaj M, Tavakkoli M, Stranahan AW, Park CY. Pathogenic microRNA's in myeloid malignancies. Front Genet 2014;2014:1–18.

[7]  Kihara R, Nagata Y, Kiyoi H, et al. Comprehensive analysis of genetic alterations and their prognostic impacts in adult acute myeloid leukemia patients. Leukemia 2014;28:1586−95.

[8]  Pellagatti A, Boultwood J. The molecular pathogenesis of the myelodysplastic syndromes. Eur J Haematol 2015;95:3−15.

[9]  Mullighan CG, Goorha S, Radtke I, et al. Genome-wide analysis of genetic alterations in acute lymphoblastic leukaemia. Nature 2007;446:758−64.

[10]  Okosun J, Bodor C, Wang J, et al. Integrated genomic analysis identifies recurrent mutations and evolution patterns driving the initiation and progression of follicular lymphoma. Nat Genet 2014;46:176−81.

[11]  Afghahi A, Sledge GW. Targeted therapy for cancer in the genomic era. Cancer J 2015;21:294−8.

[12]  Zheng H, Vasekar M, Liu X, Vrana KE. Novel immunotherapies for hematological malignancies. Curr Mol Pharmacol 2016;9:1−8.

Chapter 17

# Molecular basis of diseases of immunity

David O. Beenhouwer[1,2]

[1]Department of Medicine, David Geffen School of Medicine at University of California, Los Angeles, CA, United States, [2]Division of Infectious Diseases, Veterans Affairs Greater Los Angeles Healthcare System, Los Angeles, CA, United States

**Summary**

The human immune system has evolved over millions of years to protect against infection by microorganisms. It is remarkable in both its complexity and its effectiveness. Without a functioning immune system, humans cannot survive past the first few months of life. Despite this highly evolved system, infectious diseases were by far the most common cause of death until the 20th century. The increase in life expectancy seen in the past 100 years or so reflects progress in control of infectious diseases, including improved hygiene, vaccines, and antimicrobial drugs. However, the improved control of infections in modern society may have come at the expense of increasing the risk for autoimmune and allergic diseases. Following a brief review of some of the more important cells and molecules of the immune system. This chapter covers the pathophysiology of several major syndromes of immune dysfunction including diseases of deficient immunity (primary and acquired immunodeficiencies), hyperactive immunity (hypersensitivity), and dysregulated immunity (autoimmune diseases).

## Introduction

The human immune system has evolved over millions of years to protect against infection by microorganisms. It is remarkable in both its complexity and its effectiveness. Without a functioning immune system, humans cannot survive past the first few months of life. Despite this highly evolved system, infectious diseases were by far the most common cause of death until the 20th century. The increase in life expectancy seen in the past 100 years or so reflects progress in control of infectious diseases, including improved hygiene, vaccines, and antimicrobial drugs. Casanova and Abel have proposed that recent medical progress has masked widespread inherited defects in immunity. However, the improved control of infections in modern society may have come at the expense of increasing the risk for autoimmune and allergic diseases.

This chapter covers major syndromes of immune dysfunction. These include diseases of deficient immunity, hyperactive immunity (hypersensitivity), and dysregulated immunity (autoimmune diseases). To appreciate the pathophysiology of these syndromes, one needs to have an understanding of the normal immune system. However, a detailed description of the immune system and how it functions is beyond the scope of this chapter.

## Immune responses

### Innate immune responses

The immune system must be capable of responding immediately to a foreign invader. This response, mediated by soluble and cell-attached pathogen recognition receptors (PRRs), is called the innate immune response. Important cells of the innate immune response include macrophages, natural killer (NK) cells, and granulocytes. Soluble molecules such as antimicrobial peptides and complement also play important roles. Recognition of a foreign invader provokes release of cytokines, in particular tumor necrosis factor (TNF), as well as chemokines. This leads to infiltration of the infection site with neutrophils, macrophages, and dendritic cells as well as lymphocytes. Most breaches in host defenses are controlled by the innate immune response.

**Essential Concepts in Molecular Pathology.** DOI: https://doi.org/10.1016/B978-0-12-813257-9.00017-6

## Adaptive immune responses

The adaptive immune response primarily involves lymphocytes with their wide diversity of antigen receptors. The adaptive immune response continues to be adjusted during infection and is set in motion following activation of the innate response. Antigen presenting cells (APCs), particularly dendritic cells, ingest pathogens and associated antigens and travel from the site of infection to lymph nodes where they encounter dense collections of T-cells and B-cells. Dendritic cells process antigens and present them complexed with major histocompatibility complex (MHC) class II. T-cells that bind these peptide-MHC complexes are activated to proliferate and produce cytokines that further direct immune responses. This process leads to clonal amplification of pathogen-specific T-cells. Specialized APCs, called follicular dendritic cells, do not process antigens into fragments, but carry larger fragments and even entire pathogens on their cell surface, where they can be presented to B-cells, which are then clonally amplified as well. B-cells can also process antigen and present it to T-cells in the context of MHC II. T-cells recognizing antigen in this context can then provide signals to the B-cell leading to maturation of the antibody response, including isotype switching and somatic hypermutation.

An important caveat of lymphocyte activation is that it requires two signals: (1) antigen binding, which signals through the antigen receptor and (2) a second signal provided by another cell. For T-cells, this second signal is provided by APCs. For B-cells, the second signal is usually provided by activated T-cells. The second signal is delivered through the interaction of certain cell surface molecules, such as CD40 and CD40 ligand or CD28 and B7, known as costimulatory molecules. If lymphocytes receive signaling only through the antigen receptor without co-stimulation, the lymphocyte either becomes anergic (immunologically inactive) or dies via apoptosis. This process ensures that the powerful adaptive immune system is only activated by foreign pathogens initiating an inflammatory response and not to inert or host antigens.

Another important feature of the adaptive immune response is the development of antigenic memory or recall. The initial adaptive response to a pathogen, known as the primary response, usually takes 7–10 days to reach full effectiveness. In a process that is still rather unclear, during the primary (and subsequent) response, some clonally amplified lymphocytes, known as memory cells, are produced that remain circulating in the host for many years. Upon re-exposure to the same pathogen, these cells are then stimulated, leading to quick and efficient responses and eradication of infection. This response is also known as a secondary response and requires only a few days to reach maximum intensity and efficacy. For most antigens, memory does fade over time. This is why many vaccines require booster immunizations to remain effective.

In the past, there has been a tendency to think about the adaptive immune response as two separate entities—humoral immunity (antibody) and cell-mediated immunity (T-cell and APC). It is now clear that effective responses to most infections require both types of responses and that this dichotomy is largely a didactic one.

## Hypersensitivity reactions

In certain cases, harmless environmental antigens may induce an adaptive immune response, and upon re-exposure to the same antigen, an inflammatory state ensues. These antigens are referred to as allergens, and the hypersensitivity responses to them are also known as allergic reactions. While the development of an allergic reaction is divided into two phases for instructive purposes (the sensitization phase and the effector phase), these often occur practically simultaneously. Allergic reactions may range from mildly itchy skin to significant illnesses such as asthma and to life-threatening situations such as anaphylactic shock. Gell and Coombs divided hypersensitivity reactions into four main types for didactic purposes (Table 17.1), although distinguishing these practically may be difficult, as antibody-mediated and cell-mediated immune responses may overlap or occur simultaneously. Also, this classification system tends to maintain the humoral versus cellular dichotomy, which has little place in modern immunology. More recently, a clinical classification of hypersensitivity has been proposed by the *European Academy of Allergology and Clinical Immunology* (EAACI). Rajan has further proposed that looking at these phenomena as hypersensitivity reactions tends to overlook the reason they exist in the first place, namely to protect the host against infection. Keeping these limitations in mind, the following discussion of the hypersensitivity reactions will utilize the classical scheme proposed by Gell and Coombs.

### Type I or immediate hypersensitivity

Immediate hypersensitivity is classically mediated by IgE and may occur as a local (such as asthma) or systemic (anaphylaxis) reaction. The initial phase, occurring within minutes following exposure to allergen, is characterized by vasodilation, vascular leakage, smooth muscle spasm, and glandular secretions. A late phase reaction, lasting several days may then occur and is characterized by infiltration of tissues with eosinophils and CD4+ T-cells, as well as tissue destruction.

**TABLE 17.1 Hypersensitivity reactions.**

| Type | Antigen | Mediator | Effector mechanism | Examples |
|------|---------|----------|---------------------|----------|
| I Immediate | Soluble antigen, allergen | IgE | FcεRI activation of mast cells | Asthma, anaphylaxis |
| II Antibody-mediated | Cell-associated antigen | IgM, IgG | Fcγ-R-bearing cells, complement | Drug reaction |
| III Immune-complex | Soluble antigen | IgG | Fcγ-R-bearing cells, complement | Serum sickness, Arthus reaction |
| IV Delayed-type hypersensitivity | Soluble antigen, cell-associated antigen | T-cells | Macrophage activation, direct cytotoxicity | Tuberculin reaction, contact dermatitis |

During the sensitization phase of a type I response, allergen is presented to T-cells by APCs, and signals are generated that cause differentiation of naïve CD4+ T-cells into Th2 cells. Interleukin (IL)-4 and IL-13 secreted by Th2 cells stimulate B-cells to produce IgE. Allergen-specific IgE then binds to high-affinity FcεRI on mast cells and basophils. During the effector phase, the allergen binds to cytophilic IgE cross-linking FcεRs and leading to activation of mast cells and basophils. This is followed by immediate mast cell degranulation with release of preformed mediators such as histamine and proteolytic enzymes, which induce smooth muscle contraction, increase vascular permeability, and break down tissue matrix proteins. At the same time, there is induction of synthesis of chemokines, cytokines, leukotrienes, and prostaglandins. IL-5 promotes eosinophil production, activation, and chemotaxis. IL-4 and IL-13 promote and amplify Th2 responses. TNF, platelet activating factor (PAF), and macrophage inflammatory proteins (MIP-1α and MIP-1β) induce efflux of effector leukocytes from the bloodstream into tissues. Leukotriene B4 is chemotactic for eosinophils, neutrophils, and monocytes. Leukotrienes C4 and D4 are potent ( > 1000-fold more active than histamine) mediators of vascular permeability and smooth muscle contraction. Prostaglandin D2 causes increased mucous secretion.

While mast cells initiate the Type I hypersensitivity response, eosinophils play a major role in the Type I hypersensitivity response, particularly in the late phase. Eosinophils produce major basic protein and eosinophil cationic protein, which are enzymes that can cause tissue destruction. Epithelial cells also produce cytokines (such as TNF and IL-6) and chemokines (like IL-8 and eotaxins) that amplify the inflammatory response.

Susceptibility to Type I hypersensitivity reactions, termed atopy, is familial. However, the genetic basis for this predisposition has not been clearly established and is probably polygenic. Most allergens that trigger a type I reaction are low-molecular weight, highly soluble proteins that enter the body via the mucosa of the respiratory and digestive tracts.

## Type II or antibody-mediated hypersensitivity

Type II hypersensitivity is mediated by IgM or IgG targeting membrane-associated antigens. A sensitization phase leads to production of antibodies that recognize substances or metabolites that accumulate in cellular membrane structures. In the effector phase, target cells become coated with antibodies, a process termed opsonization, which leads to cellular destruction by three mechanisms: (1) phagocytosis, (2) complement-dependent cytotoxicity (CDC), and (3) ADCC. First, IgG or IgM antibodies coating target cells can bind to Fc receptors present on cells such as macrophages and neutrophils and mediate phagocytosis. IgG or IgM antibodies can also activate complement via the classical pathway. This leads to deposition of C3b, which can mediate phagocytosis. Complement activation also leads to production of the MAC, which forms pores in the cellular membrane resulting in cytolysis (complement-dependent cytotoxicity). Finally, IgG antibodies can bind FcγRIII on NK cells and macrophages, mediating release of granzymes and perforin and resulting in cell death by apoptosis (antibody-dependent cell-mediated cytotoxicity).

The most common cause of type II reactions are medications including penicillins, cephalosporins, hydrochlorothiazide, and methyldopa, which become associated with red blood cells or platelets leading to anemia and thrombocytopenia. The mechanisms involved in type II hypersensitivity also play a role in cellular destruction by autoantibodies.

## Type III or immune complex reaction

In this reaction, antibodies bind antigen to form immune complexes, which become deposited in tissues where they elicit inflammation. It is important to note that the formation of immune complexes occurs during many infections and is an effective method of antigen clearance. However, under certain circumstances these complexes may escape clearance by the reticuloendothelial system and are deposited in tissues including the kidney, joints, and blood vessels. The immune complexes fix complement and bind to leukocytes via Fc receptors. Activation of the complement cascade leads to production of vasoactive mediators C5a and C3a. Other vasoactive mediators, such as platelet activating factor (PAF), are also released following Fc receptor engagement. The result is tissue edema and deposition of immune complexes in the vessel walls and surrounding tissue. At the same time, chemotactic factors are produced leading to neutrophil and monocyte recruitment, which upon activation cause tissue destruction.

Examples of Type III reactions include serum sickness and the Arthus reaction. Serum sickness was originally described in patients suffering from diphtheria who were treated with immune horse serum (antiserum). The condition is characterized by rash, joint pain, lymph node swelling, and fever, with typical onset about 10 days after initial serum exposure. Currently, the most common causes of serum sickness—like illness are antibiotics (e.g., cephalosporins and sulfonamides) and blood products. The Arthus reaction is a local immune complex—mediated vasculitis, usually observed as edema and necrosis in the skin occurring several hours following antigen exposure. As with Type II reactions, the mechanisms involved in Type III hypersensitivity also play a role in certain autoimmune diseases.

## Type IV or delayed-type hypersensitivity

While the first three types of hypersensitivity are mediated primarily by antibodies, delayed-type hypersensitivity (DTH) is mediated by T-cells. During the sensitization phase, naïve CD4+ T-cells are exposed to antigens with an induction of an adaptive Th response. In the effector phase, antigen is carried by APCs to lymph nodes and presented to memory T-cells, which become activated and then travel back to the site of antigen deposition where they may be stimulated to secrete interferon (IFN) $\gamma$, thereby initiating a Th1 response and tissue inflammation mediated primarily by macrophages.

The classic example of DTH is the tuberculin skin reaction, which occurs 24—48 h following intradermal injection of a purified protein derivative prepared from *Mycobacterium tuberculosis*. Contact dermatitis from poison ivy represents another common example of DTH.

# Immunologic deficiencies

The first inherited immunodeficiency was described by Bruton over 50 years ago. Many of these diseases have devastating pathological consequences for afflicted individuals. An understanding of the underlying mechanisms responsible for these disease processes has been invaluable in our current understanding of immune system function. There are currently over 200 described primary immunodeficiencies (PIDs). A systematic review of these deficiencies is well beyond the scope of this chapter. However, a few of the more common and important PIDs will be discussed. In addition, there are acquired immunodeficiencies, which historically were iatrogenic in nature. In the late 1970s, infection with human immunodeficiency virus (HIV) became the most common cause of severe immunodeficiency.

## Primary immunodeficiencies

PIDs are inherited disorders that primarily affect immune system cell function. Most are rare with prevalences less than 1 in 50,000 individuals. As would be expected, PIDs with less severe immunosuppressed phenotypes tend to be more common. Certain types of infections are associated with different classes of immunodeficiency (Table 17.2). Many PIDs lead to significant infections early in life. Deficiencies in multiple lymphocyte lineages tend to be devastating. Breastfeeding neonates with antibody deficiencies usually remain well for a time after birth due to passive transfer of maternal immunoglobulin during the first 6—9 months of life. Then the affected individual develops upper and lower respiratory tract infections including sinusitis, otitis, and pneumonia. Most of these infections are caused by bacteria such as *Streptococcus pneumoniae* and *Haemophilus influenzae* that have polysaccharide capsules, making them resistant to phagocytosis without effective antibody opsonization. If untreated, these recurrent infections cause significant destruction and scarring of lung tissue, resulting in bronchiectasis. Fortunately, treatment with pooled human immunoglobulin is effective in treating antibody deficiencies. More recently, it has been recognized that deficiency in one molecule (e.g., toll-like receptor-3 or TLR-3) may predispose individuals to only one or a few types of infections. We are just beginning to recognize these syndromes, and it is likely several more will be discovered in the near future.

**TABLE 17.2** Infections in immunodeficiencies.

| Host defense affected | Clinical example | History | Relevant pathogens |
|---|---|---|---|
| T-cells | AIDS | Disseminated infections<br>Opportunistic infections<br>Persistent viral infections | *Pneumocystis jiroveci*<br>*Cryptococcus neoformans*<br>Herpes viruses |
| B-cells | X-linked agammaglobulinemia | Recurrent respiratory infections<br>Chronic diarrhea<br>Aseptic meningitis | *Streptococcus pneumoniae*<br>*Haemophilus influenzae*<br>*Giardia lamblia*<br>*Enteroviruses* |
| Phagocytes | Chronic granulomatous disease | Gingivitis<br>Aphthous ulcers<br>Recurrent pyogenic infections | *Staphylococcus aureus*<br>*Burkholderia cepacia*<br>*Serratia marcescens*<br>*Aspergillus* spp. |
| Complement | Late complement component (C5, C6, C7, C8, or C9) deficiency | Recurrent bacteremia<br>Recurrent meningitis | *Neisseria* spp. |
| Pattern recognition receptor | TLR-3 defect | Herpes simplex encephalitis | HSV-1 |

## X-linked agammaglobulinemia (Bruton disease)

Patients with this condition present with recurrent respiratory tract infections, lack of tonsilar tissue, and a normal white blood cell count with normal lymphocyte percentage. When flow cytometry for surface membrane immunoglobulin (B-cell Receptor or BCR) is performed in these patients, few, if any, B-cells are found. Serum immunoglobulins are nonexistent. The defect is in Bruton's tyrosine kinase (BTK) gene, which is essential for B-cell development and survival. B-cells develop in the bone marrow from pluripotent hematopoietic stem cells, first becoming pro-B-cells and then pre-B-cells when the immunoglobulin H and L chain V region gene segments undergo rearrangement. During the next stage (immature B-cell), the BCR is expressed on the cell surface (sIgM), and these cells leave the bone marrow to undergo further maturation steps. These cells eventually develop into mature B-cells and then plasma cells, which, in the presence of antigen and T-cells, undergo antibody class switching and somatic hypermutation. BTK is involved in signal transduction at several stages of B-cell development. At the pre-B-cell stage, BTK signaling appears to be essential for directing L chain gene rearrangement events. Thus, defective BTK leads to arrest of B-cell differentiation at the pre-B-cell stage. The discovery of the gene involved in this defect has led to development of BTK inhibitors (e.g., ibrutinib) that have efficacy against chronic lymphocytic leukemia and immune complex—mediated diseases (e.g., lupus and rheumatoid arthritis).

## Hyper-IgM syndrome

Hyper-IgM syndrome, which is characterized by the presence of normal or elevated serum levels of IgM and low IgG and IgA, may be caused by one of at least 10 gene defects (Table 17.3). The most common (and the most clinically severe) of these is an X-linked deficiency in CD40 ligand (CD40L). This receptor is transiently expressed on activated T-cells and interacts with CD40 molecules constitutively expressed on the surface of B-cells and APCs. T-cell interaction with B-cells via CD40-CD40L interaction leads to immunoglobulin class switching and differentiation into plasma cells. In the absence of this signal, B-cells only produce IgM. Other T-cell/B-cell interactions can induce class switching, so some small amounts of IgG and IgA may be seen. However, the impaired production of IgG and IgA in CD40L deficiency leads to susceptibility to recurrent bacterial infections, particularly those involving the respiratory tract caused by encapsulated organisms. Importantly, T-cells also interact with APCs via CD40—CD40L. Thus, CD40L deficiency (as well as the more unusual autosomal recessive CD40 deficiency) leads to significant impairment in both antibody production and cell-mediated immunity, resulting in a clinical presentation that may include both infections with encapsulated bacteria and intracellular organisms. Other defects responsible for the hyper-IgM phenotype are either directly or indirectly involved in immunoglobulin class switching. Some do not involve T-cells and therefore do not induce a clinically mixed deficit.

**TABLE 17.3** Hyper IgM syndromes.

| Defective gene | Genetic transmission | T-cell defect | Somatic hypermutation |
|---|---|---|---|
| CD40L | X-linked | Yes | Decreased |
| CD40 | Autosomal recessive | Yes | Decreased |
| AID | Autosomal recessive | No | No |
| AID C terminal | Autosomal dominant | No | Yes |
| UNG | Autosomal recessive | No | Yes |
| AID-targeting | Autosomal recessive | No | Yes |
| DNA repair | Autosomal recessive | No | Yes |
| ATM | Autosomal recessive | Yes | Yes |
| MRE-11 | Autosomal recessive | Yes | ? |
| NB-S1 | Autosomal recessive | Yes | ? |

*AID*, activation-induced deaminase; *ATM*, ataxia telangiectasia mutated; *UNG*, uracil DNA glycosylase.

**TABLE 17.4** Genetic defects causing SCID.

| Defective gene | B-cell deficit | NK cell deficit |
|---|---|---|
| Common γ chain (X-linked) | No | Yes |
| Adenosine-deaminase (ADA) | Yes | Yes |
| CD3 δ-chain | No | No |
| CD3 ε-chain | No | No |
| CD45 | No | Yes |
| JAK3 | No | Yes |
| Artemis | Yes | No |
| IL-7 receptor α-chain | No | No |
| RAG1, RAG2 | Yes | No |

## Complement deficiency

Defects have been described for most of the components of complement. There is a propensity for individuals with C3 deficiency and with classical component defects (C1, C2, and C4) to develop systemic lupus erythematosus (SLE). While this is not well understood, classical pathway clearance of apoptotic cells and immune complexes may be important. C2 deficiency is relatively common, occurring in 1 in 20,000 individuals. These individuals have an increased propensity to develop infections, SLE, and myocardial infarctions. Defects in myelin basic protein are relatively common, and there is an association with infections in children. Finally, defects in the terminal components (C5-9) lead to a remarkable susceptibility ($\sim$5000-fold risk) to recurrent infections caused by pathogenic *Neisseria* spp., especially *Neisseria meningitidis*. Interestingly, mortality due to these infections is 10-fold lower than nondeficient individuals.

## Severe combined immunodeficiency

The severe combined immunodeficiency disorder (SCID) is characterized by the absence of T-cell differentiation, resulting in severely reduced or absent T-cells. At least 10 autosomal recessive or X-linked genetic defects have been described, which result in several possible phenotypes based on B-cell and NK cell development (Table 17.4). The most common cause of SCID is an X-linked deficiency in the common γ chain shared by receptors for several cytokines including IL-2 and IL-4. This form is characterized by complete absence of T-cells and NK cells. When untreated, SCID results in death within the first year of life from overwhelming opportunistic infections. Improved survival has been reported in infants receiving stem cell transplants. Using retroviral vectors, gene therapy has also been used successfully in some cases.

## Common variable immunodeficiency

Common variable immunodeficiency (CVID) is a heterogeneous syndrome, probably comprising several distinct genetic defects, characterized by recurrent infections and low antibody levels (IgG and IgA and/or IgM). The syndrome is usually diagnosed in the fourth decade of life and with a typical delay of >15 years from first symptoms to diagnosis. The estimated prevalence of CVID is about 1 in 25,000 individuals. The main clinical features are recurrent infections of the respiratory tract, chronic diarrhea, autoimmune disease, and malignancy. About 10 genetic defects have been described but these account for less than 15% of those that are diagnosed clinically, and most gene defects are currently unknown. The syndrome was first described in 1953 but it was not until 2003 that the first genetic defect was identified: a homozygous deletion in the gene for inducible co-stimulator (ICOS). This defect, which is transmitted as an autosomal recessive trait, accounts for about 2% of patients with CVID. The ICOS signaling pathway plays an important role in T-cell subset activation and in T-cell and B-cell interactions. Mutations in several molecules that form the BCR receptor complex, including CD19, CD21, and CD81, have been implicated in CVID. About 10% of CVID cases are caused by one of several mutations in the *t*ransmembrane *a*ctivator and calcium modulator and *c*yclophillin *i*nteractor (TACI) gene. TACI is a member of the TNF receptor family and is expressed on B-cells and activated T-cells. Ligands for TACI include *B*-cell *a*ctivating *f*actor of the TNF *family* (BAFF) and *a pr*oliferation *i*nducing *l*igand (APRIL). Both ligands bind other receptors on B-cells and T-cells. However, in response to ligation by APRIL, TACI mediates isotype switching to IgG and IgA. Similarly, BAFF mediates IgA switching. The penetrance of CVID phenotype in families carrying a mutant TACI gene is quite variable. Recently, a defect in BAFF receptor has also been identified as a cause of CVID.

## Chronic granulomatous disease

Phagocytes produce reactive oxygen species (including superoxide and hydrogen peroxide) to kill ingested pathogens. NADPH oxidase is a five-subunit enzyme that catalyzes the production of superoxide from oxygen for this purpose. Over 400 genetic defects in NADPH have been described that result in chronic granulomatous disease, which is characterized by recurrent, indolent bacterial and fungal infections caused by catalase positive organisms including *Staphylococcus aureus* and *Aspergillus* spp. Most mutations occur in the gp91phox gene located on the X chromosome. There are also autosomal recessive forms of the disease primarily involving p47phox. The lack of microbicidal oxygen species due to NADPH deficiency leads to a severe defect in intracellular killing of phagocytosed organisms. However, organisms that lack catalase produce peroxide as a byproduct of oxidative metabolism, which accumulates and leads to intracellular killing of these microbes. Tissue pathology is characterized by granulomas and lipid-filled histiocytes in liver, spleen, lymph nodes, and gut. The disease is treated with antibiotic prophylaxis, and chronic administration of IFN-γ may also be beneficial. More recently, gene therapy has been considered as a therapeutic option.

## X-linked proliferative disease (Duncan syndrome)

Caused by a mutation in the SH2D1A gene, this rare X-linked proliferative disease is manifested by dysregulated T-cell and NK cell proliferation and responses. SH2D1A encodes signaling *l*ymphocyte *a*ctivation *m*olecule (SLAM)-associated protein (SAP), a cytoplasmic adapter protein that binds the transmembrane protein SLAM and related molecules. SAP inhibits cell activation via SLAM. Defects in SAP lead to T-cell proliferation. Individuals with this disorder have a propensity to develop lymphoma and fulminant Epstein-Barr virus (EBV) infection.

## Hyper-IgE syndrome

Formerly known as Job syndrome, the *a*utosomal *d*ominant *hyper-IgE s*yndrome (AD-HIES) is a rare primary immunodeficiency caused by dominant negative mutations in the gene for *s*ignal *t*ransducer and *a*ctivator of *t*ranscription 3 (STAT3). Patients with AD-HIES have recurrent skin and sinopulmonary infections with *S. aureus*, eczema, mucosal infections with candida, abnormal dentition, and unique facial characteristics. As the name of the syndrome implies, serum IgE levels are typically very high and there is eosinophilia. STAT3 plays a critical role in signaling by IL-21 and IL-23, which are important mediators of the Th17 phenotype. Patients with AD-HIES have significant depletion of Th17 cells. A more severe form of HIES, with mortality often in the first decade of life, is caused by deficiency in *dedi*cator *of cyto*kinesis 8 (DOCK8). DOCK8 is highly expressed in the immune system and is a guanine nucleotide exchange factor that activates Rho-family GTPases (e.g., RAC). DOCK8-deficient patients have wide-ranging immune system effects (that are more severe compared to STAT3-deficiency) with defects in (1) T-cell function, (2) dendritic cell migration, and (3) the formation of memory B-cells. A distinguishing feature of DOCK8-deficiency is frequent viral infections of the skin.

## *Inherited susceptibility to herpes encephalitis*

The paradigm of a single gene lesion conferring vulnerability to multiple infections has been challenged by several recent discoveries. While over 80% of young adults are infected with herpes simplex virus type 1 (HSV-1), development of herpes simplex encephalitis (HSE) is quite rare (1/250,000 patient years). Jean Laurent Casanova and colleagues hypothesized that susceptibility to developing HSE was inherited as a monogenic trait. There was evidence that impaired interferon responses might predispose to HSE. Therefore, they screened otherwise healthy children with a history of HSE and found two unrelated children whose leukocytes had defective interferon production in response to HSV-1 antigens. Cells from these children had no evidence of increased susceptibility to infections. However, it was determined that they had impaired responses to agonists of TLR-7, TLR-8, TLR-9, and TLR-3. These responses were similar to those reported in mice lacking UNC-93B, an endoplasmic reticulum protein involved in activation by TLR-3, TLR-7, and TLR-9. It was subsequently determined that the patients lacked mRNA encoding UNC-93B. Each was homozygous for a different mutation in the UNC-93B gene: one had a 4-base deletion and one had a point mutation leading to alternative splicing. Previously, children with a deficiency in IL-1 *receptor*–*associated kinase*-4 had been described. These patients fail to signal through TLR-7, TLR-8, or TLR-9, and show a propensity to developing certain bacterial infections but not HSE. Therefore, Casanova and colleagues proposed that the HSE susceptibility of the UNC-93B-deficient children was due to impaired TLR-3 induction of interferon. One year later the same group identified two more otherwise healthy children with HSE and a single point mutation in the TLR-3 gene. TLR-3 is highly expressed in the central nervous system. The patients with TLR3 defects identified to date do not display susceptibility to other infections, including encephalitis caused by other viruses. Together, these data strongly suggest an important role for TLR-3 in protection against HSE and indicate that a single genetic defect may predispose to infection (or severe manifestations of infection) by a specific pathogen.

## *Inherited susceptibility to mycobacterial infections*

Susceptibility to infection with *Mycobacterium tuberculosis* and less virulent nontuberculous mycobacteria (NTM) and other intracellular infections such as listeriosis, salmonellosis, histoplasmosis, coccidioidomycosis, and others, has been demonstrated in patients with deficient Th1 responses. These include those with defects in IL-12, receptors for IFN-γ and IL-12, and STAT1 (which is involved in signaling via the IFN-γ receptor). These patients have generally been identified when presenting with disseminated NTM infection, which, besides infections caused by *Mycobacterium leprae* (leprosy), is essentially unheard of in an otherwise healthy immunocompetent individual.

## *Inherited susceptibility to mucocutaneous candidiasis*

Chronic mucocutaneous candidiasis (CMC) is characterized by recurrent or persistent symptomatic mucocutaneous infections caused by fungi of the genus *Candida*, affecting the nails, skin, and oral and genital mucosa. CMC can be seen in several conditions including newborns, those with T-cell deficiencies (e.g., AIDS, several primary immunodeficiencies, immunosuppressive drugs, and others), and those on broad spectrum antibiotics. Patients with AD-HIES and STAT3 mutations leading to significant Th17 deficiency often have CMC (along with several other features characteristic of this syndrome). Some individuals develop CMC without other significant underlying issues. This is a rare condition, occurring in about 1:100,000 individuals. Some of these patients may display other symptoms such as mild staphylococcal skin infections, recurrent mucocutaneous herpes virus disease, and milder autoimmune diseases (e.g., thyroid autoimmunity). Recently, inborn errors in the IL-17 axis (e.g., IL-17 and IL-17 receptor deficiencies) have been found to be responsible for at least some of these cases. In addition, high levels of neutralizing autoantibodies to IL-17A, IL-17F, and IL-22 are associated with CMC. Together this group of syndromes demonstrate the importance of Th17 immunity in host defenses against skin and mucous membrane infections, in particular those caused by *Candida* spp.

## Acquired immunodeficiencies

Exposure to a variety of factors such as infectious agents, immunosuppressive drugs, and environmental conditions are much more prevalent causes of immunodeficiency than genetic defects. Malnutrition is the most common cause of immunodeficiency worldwide. Metabolic diseases, such as diabetes, hepatic cirrhosis, and chronic kidney disease also lead to immunosuppression. Many viral and bacterial infections can result in transient immunosuppression. Infection with HIV can lead to a chronic and severe state of immunosuppression known as the Acquired Immune Deficiency Syndrome (AIDS).

## Acquired immune deficiency syndrome

It is currently estimated that over 40 million people are infected with HIV, with the majority in sub-Saharan Africa and South/Southeast Asia. HIV is a retrovirus that contains nine genes: *gag, pol, env, tat, rev, nef, vif, vpr,* and *vpu.* The gag gene product is split by the HIV protease into five structural proteins. The *pol* gene product is split into three enzymes: integrase, reverse transcriptase, and protease. The env gene product is cleaved to produce two envelope proteins, gp120 and gp41, which together constitute gp160. The other five genes encode regulatory proteins. The virus is shed into body fluids and the bloodstream. HIV infection is most typically spread via sexual contact, but transmission via contaminated hypodermic needles and blood products can also occur as can transmission from mother to infant.

HIV enters CD4+ T-cells through interaction of gp160 on the viral envelope and CD4 along with a co-receptor, either CCR5 or CXCR4, on the cell surface. Individuals expressing a mutant *CCR5* gene (*ccr5Δ32*) are protected from HIV infection. Once inside the cell, the virus uncoats and the reverse transcriptase, which is complexed to the viral RNA genome, transcribes viral RNA into double-stranded DNA. This is then transported to the cell nucleus where, with the help of virally encoded integrase, it is inserted into the cellular DNA. Thus, the virus establishes lifelong infection of the cell. Replication of the viral genome occurs along with cell replication.

Acute HIV infection is characterized by high viremia, immune activation, and CD4+ T-cell lymphopenia. The acute phase of infection lasts several weeks. Patients often develop nonspecific symptoms of fever, rash, headaches, and myalgias. Occasionally, opportunistic infections may occur in this period. The acute phase is followed by a period of clinical latency generally characterized by absence of significant symptoms. During this period, viral replication continues in the lymphoid tissue, resulting in lymphadenopathy, and there is increased susceptibility to certain infections such as tuberculosis. The latent period lasts several years. Higher viral loads predict shorter clinical latency. While initial host immune responses control viral infection somewhat, in almost all cases they inevitably fail. CD4+ T-cells continue to decline, and eventually the host becomes susceptible to opportunistic infections, including pneumocystis pneumonia, cryptococcal meningitis, disseminated cytomegalovirus, and NTM infections. If untreated, death occurs on average about 10 years after infection.

While the virus targets CD4+ T-cells, abnormalities in all parts of the immune system occur during HIV infection. There is profound disruption of lymphoid tissue architecture, resulting in an inability to mount responses against new antigens and severely impaired memory responses. CD4+ T-cells that are not killed directly are dysregulated. They have decreased IL-2 production and IL-2 receptor expression, resulting in diminished capacities to proliferate and differentiate. CD28 expression is also reduced. TCR Vβ repertoire can be significantly reduced in advanced HIV infection. CD8$^+$ T-cells also have decreased IL-2 receptor and CD28 expression. Cytotoxic lymphocyte activity is reduced as are production of cytokines and chemokines, including those that block HIV replication. B-cells are hyperactivated by gp41 in HIV infection. gp120 acts as a superantigen for B-cells carrying the V$_H$3 variable region. This leads to a gap in the B-cell antibody repertoire. B-cell dysregulation explains the propensity to developing infections with encapsulated bacteria. B-cells also express proinflammatory cytokines such as TNF and IL-6, which enhance HIV replication. Macrophages express CD4 and other HIV co-receptors and become infected with HIV. They serve as important reservoirs for HIV. Infection of macrophages also leads to functional abnormalities, including decreased IL-12 secretion, increased IL-10 production, decreased antigen uptake, and impaired chemotaxis. HIV infection of brain tissue macrophages or microglial cells plays a major role in HIV encephalopathy. Other immune cells shown to be dysregulated in HIV infection include NK cells and neutrophils.

Treatment with multiple drugs that target HIV enzymes, known as combined antiretroviral therapy (cART), is effective in reducing viremia and restoring normal CD4+ T-cell counts. cART has drastically reduced mortality rates in HIV infection. After 2−3 weeks, patients treated with cART may develop a severe inflammatory response to existing opportunistic infections known as the immune reconstitution inflammatory syndrome (IRIS). An HIV-infected man with acute myelogenous leukemia requiring bone marrow transplantation has apparently been cured of HIV infection. The so-called Berlin patient received a hematopoietic stem cell transplantation from an allogeneic donor homozygous for *ccr5Δ32*. Subsequently, he has remained free of HIV infection for over 8 years without cART. This remarkable success has stimulated the search for strategies to eradicate HIV or to induce long-term remission without requiring ongoing antiretroviral therapy.

## Cytokine autoantibodies

Falling somewhere between inherited immunodeficiencies, autoimmune diseases, and acquired immunodeficiencies as a cause of disordered immunity is the formation of autoantibodies that neutralize specific cytokines leading to susceptibility to certain infections. Antibodies to IFN-γ have been associated with disseminated NTM and other granulomatous infections. The presentation of these individuals mimics those with inborn errors of the IFN-$\gamma$ axis, except they present later in life. Many of the patients described with this condition are Asian, suggesting a genetic link. Confirming this suspicion, two HLA-II alleles were recently found to be associated with this phenotype. Production of autoantibodies to IL-17 and IL-22 have been described in patients with autoimmune polyendocrinopathy syndrome type 1. These patients have a propensity for developing CMC. Thus, there appear to be underlying genetic causes for the generation of these specific types of autoantibodies.

# Autoimmune diseases

T-lymphocytes and B-lymphocytes have incredible diversity in antigen recognition. They also carry a potent armamentarium, capable of destroying cells and causing damaging inflammation. While some pathogen-associated antigens are quite distinct from host molecules (PAMPs), there is often little fundamental difference between host antigens and those of pathogens. Therefore, lymphocytes that recognize host antigens may arise during the course of infection. To prevent self-destruction, the host needs to eliminate these autoreactive lymphocytes. Macfarlane Burnet originally formulated the clonal deletion theory, which proposes that all self-reactive lymphocytes (forbidden clones) are destroyed during development of the immune system. In fact, autoreactive T-cells and B-cells exist in many individuals that do not develop autoimmune disease. Autoimmunity occurs when T-cells or B-cells are activated and cause tissue destruction in the absence of ongoing infection.

Tolerance is the process that neutralizes these autoreactive T-cells and B-cells. Autoimmunity is the failure of tolerance mechanisms. B-cells may produce antibodies that recognize surface proteins, and these may directly cause disease by (1) initiating destruction of host cells or (2) mimicking receptor ligand, and causing hyperactivation. Antibodies that bind intracellular antigens, such as many of those formed in SLE, are generally believed to be secondary to the autoimmune process itself. Autoreactive B-cells may be deleted in the bone marrow or the lymph nodes and spleen. B-cells must receive a second signal (co-stimulation) following BCR ligation by antigen. Most often this second signal is delivered by T-cells. Without T-cell help, antigen-binding B-cells typically die via apoptosis.

Autoreactive T-cells are removed by two separate processes: central (thymus) and peripheral tolerance. During maturation, T-cells leave the bone marrow and travel to the thymus, where they encounter endogenous peptides complexed with MHC. If the receptors on a given T-cell bind these complexes with significant affinity, the cell is directed to die by apoptosis in a process called negative selection. Interestingly, moderate binding to self-antigens is necessary for T-cell survival, as a lack of binding to antigens presented in the thymus also triggers apoptosis. Not all self-antigens are presented in the thymus, and some autoreactive T-cells escape to the periphery. Similar to B-cells, presentation of antigens to T-cells in the absence of co-stimulation leads to deletion. Activated T-cells express Fas (CD95) on their cell surface. If they encounter Fas ligand, they will undergo apoptosis. Some tissues, such as the eye, constitutively express Fas ligand. When activated, T-cells expressing Fas enter the anterior chamber of the eye, encounter Fas ligand, and undergo apoptosis without causing tissue damage. Another molecule involved in peripheral tolerance is cytotoxic T-lymphocyte—associated protein (cytotoxic T-lymphocyte—associated protein (CTLA)-4 or CD152), which binds to B7-1 (CD80) on T-cells and B7-2 (CD86) on B-cells with higher affinity than the co-stimulatory molecule CD28, inhibiting T-cell activation. CTLA-4 polymorphism is associated with an increased predisposition to autoimmune diseases, including SLE, autoimmune thyroiditis, and type 1 diabetes. Another mechanism that plays a role in the development of autoimmunity is called immunological ignorance. Certain tissues, such as the central nervous system, are protected by barriers (like the blood brain barrier) that prevent entry of peripheral lymphocytes. Thus, the immune system remains largely ignorant of these sequestered antigens and may misidentify these as foreign upon initial exposure.

Recently, a new subset of T-cells has been described called Treg. These cells play an important role in controlling the magnitude and the quality of immune responses. Treg cells express CD4 and CD25 on their surface. They also produce a transcription factor, *f*orkhead *b*ox P3 (FOXP3), which is critical for the differentiation of thymic T-cells into Treg cells. FOXP3 interacts with other transcription factors to control expression of $\sim$700 gene products and leads to repression of IL-2 production and activation of CTLA-4 and CD25 expression. Treg cells mediate immunosuppression by several mechanisms. First, they may compete with specific naïve T-cells for antigen presented by APCs. Second, they may downregulate APCs via CTLA-4—dependent mechanisms. Finally, they may interact with effector T-cells to either kill them or inactivate them with immunosuppressive cytokines such as IL-10. The role of abnormal Treg cells in autoimmunity is still being established. Defects in certain genes specifically associated with Treg cells may lead to autoimmunity. In addition to CTLA-4, polymorphisms in IL-2 and CD25 are associated with susceptibility to autoimmune diseases. Mutations in *foxP3* lead to an immunodeficiency syndrome, IPEX (immune dysregulation, polyendocrinopathy, enteropathy, X-linked), associated with autoimmune diseases in endocrine organs. Environmental factors may

**TABLE 17.5** Autoimmune diseases.

| Name | Hypersensitivity[a] | Autoantibody[b] |
|---|---|---|
| Addison disease | | Anti-21-hydroxylase |
| Ankylosing spondylitis | | |
| Antiphospholipid antibody syndrome | | Anti-cardiolipin |
| Autoimmune hemolytic anemia | II | |
| Autoimmune hepatitis | | Anti-smooth muscle actin |
| Bullous pemphigoid | | Anti-bullous pemphigoid antigen |
| Celiac disease | IV | Anti-gliadin |
| Dermatomyositis | | Anti-Jo-1 |
| Diabetes mellitus type 1 | IV | Anti-insulin |
| Goodpasture syndrome | II | Anti-basement membrane collagen type IV |
| Graves disease | II | Anti-thyroid stimulating hormone receptor |
| Guillain-Barré syndrome | IV | Anti-ganglioside |
| Hashimoto thyroiditis | IV | Anti-thyroglobulin |
| Idiopathic thrombocytopenic purpura | II | Anti-platelet membrane glycoprotein IIb-III |
| Multiple sclerosis | IV | Anti-myelin basic protein |
| Myasthenia gravis | II | Anti-acetylcholine receptor |
| Pemphigus vulgaris | II | Anti-desmogein 3 |
| Pernicious anemia | II | Anti-intrinsic factor |
| Primary biliary cirrhosis | | Anti-mitochondrial |
| Rheumatoid arthritis | III | Rheumatoid factor |
| Sjögren syndrome | | Anti-SSB/La |
| Systemic lupus erythematosus | III | Anti-dsDNA |
| Temporal arteritis | IV | |
| Wegener granulomatosis | | Anti-neutrophil cytoplasmic |

[a]Hypersensitivity reaction associated with disease pathogenesis.
[b]A single relevant antibody is listed. Many syndromes are associated with more than one autoantibody.

also play a role in dysregulation of Treg cells. Compared to other T-cells, Treg cells have higher metabolic and proliferation rates and are therefore more susceptible to ionizing radiation and vitamin deficiencies. Other T-cells may also play a role in immunosuppression and tolerance, including CD8+, CD4 − CD8 − , and γ/δ T-cells.

Tolerance can be broken by several mechanisms. Infections are thought to be the main exogenous cause of autoimmunity. Infections can break ignorance by damaging barriers, leading to release of sequestered antigens. Superantigens, which stimulate polyclonal T-cell activation without the need for co-stimulation, are produced by certain microbes. Also, infection can produce significant inflammation with production of inflammatory cytokines and other co-stimulatory molecules, which may activate autoreactive lymphocytes (bystander activation). Another trigger of autoimmunity is a seemingly appropriate immune response to microbial antigens that mimic host antigens. For example, in the demyelinating disease Guillain-Barré syndrome which occasionally follows a bout of infectious gastroenteritis, antibody cross-reactivity has been demonstrated between human gangliosides and *Campylobacter jejuni* lipopolysaccharide. In fact, microbial antigens may evolve to resemble host antigens in order to evade the host immune response. Drugs, such as procainamide, represent another exogenous trigger of autoimmunity.

Several autoimmune diseases have been described (Table 17.5). These may be either systemic (such as SLE) or organ-specific (e.g., type 1 diabetes). Some of the more common autoimmune diseases are considered in the following sections with the understanding that the underlying mechanisms leading to host immune destruction in this diverse group of diseases are relatively similar.

## Systemic lupus erythematosus

SLE is a diverse systemic autoimmune syndrome with a significant range of symptoms and disease severity. The etiologies of SLE have not been clearly established. A viral illness often precedes the onset of SLE, and there is a temporal association with EBV infection and SLE. MHC genes (such as human leukocyte antigen (HLA)-A1, B8, and DR3) are linked to SLE. Deficiencies in the early complement component cascade (C1q, C2, or C4) are strongly associated with SLE. Several other genes may also be linked to SLE. The syndrome probably represents several distinct diseases that result in similar manifestations. There is a major propensity for females to develop this disease (9:1 vs. males), implicating a role for female hormones (or male hormones as protective). However, the reason for the gender-based susceptibility has not been established.

The hallmark of SLE is the development of autoantibodies, particularly those that bind double-stranded DNA (anti-dsDNA). Many other autoantibodies have been described in SLE and are associated with certain clinical manifestations. Antibodies to dsDNA, Sm antigen, and C1q are correlated with kidney disease. Antibodies to Ro and La antigens are associated with fetal heart problems, while antibodies to phospholipids are associated with thrombotic events and fetal loss in pregnancy. Tissue damage by autoantibodies has been most well studied with anti-dsDNA causing kidney damage. There are two possible mechanisms involved. First, anti-dsDNA may bind to fragments of DNA released by apoptotic cells (nucleosomes) in the bloodstream, forming immune complexes. These complexes are deposited in the glomerular basement membrane in the kidney and cause disease by Type III reactions. Second, anti-dsDNA may cross-react with another antigen expressed on kidney cells. One possible candidate for this antigen is α-actinin, which cross-links actin and is important in maintaining the function of renal podocytes.

Apoptotic cells express intracellular material in the form of blebs on the cell surface and complement may be important in clearing apoptotic debris. An intriguing hypothesis explaining the association of complement deficiencies and SLE is that defects in clearing apoptotic cells lead to continued exposure to self-antigens (such as nucleosomes, Sm, Ro, or La) resulting in SLE.

## Type 1 diabetes mellitus

Pancreatic β-cells produce insulin, which plays an essential role in glucose metabolism. In type 1 diabetes, cell-mediated destruction of β-cells occurs, leading to complete insulin deficiency. In contrast, in type 2 diabetes there is normal production of insulin, but cells do not respond appropriately (insulin resistance). Susceptibility to development of type 1 diabetes is linked to HLA-DR3/4 and DQ8 genes and is also associated with polymorphisms in the gene for CTLA-4. Potential exogenous triggers include viral infections (such as congenital rubella), chemicals (such as nitrosamines), and foods (including early exposure to cow milk proteins). Interaction with environmental triggers in an individual with genetic predisposition leads to infiltration of the pancreatic islets with CD4+ and CD8+ T-cells, B-cells, and macrophages, and the production of autoantibodies. Autoantibodies to β-cell antigens, insulin, and others (including anti-glutamic acid decarboxylase and anti-insulinoma-associated antigen) can precede the onset of type 1 diabetes by years. However, there is no direct evidence that these antibodies play a role in pathogenesis. Th1-activated autoreactive CD4+ T-cells together with cytotoxic CD8+ T-cells induce apoptosis of β-cells mediated via Fas-Fas ligand interaction and release of cytotoxic molecules. Clinical expression of disease occurs only after >90% of the β-cells are destroyed.

## Multiple sclerosis

Multiple sclerosis (MS) is a demyelinating disease presumed to be of autoimmune etiology. The disease presents with neurologic deficits and generally has a relapsing course followed by a progressive phase. The disease is more common in females (2:1 vs. males), and there appears to be a genetic predisposition. As with SLE and type 1 diabetes, there is an association with certain MHC, including HLA-DR2 and DQ6. The disease occurs more commonly in temperate climates, and relapses are often preceded by viral respiratory tract infections. However, the inciting factor is generally believed to be environmental. The hallmark of MS is the central nervous system inflammatory plaque containing a perivascular infiltration of myelin-laden macrophages and T-cells (both CD4+ and CD8+). These plaques tend to form in the white matter and involve both the myelin sheath and oligodendrocytes. There is often widespread axonal damage. Myelin-reactive T-cells can be isolated from individuals with and without MS. However, in the former, these T-cells are activated with a Th1 phenotype, whereas in the latter, these cells are naïve. There is increased antibody production in the central nervous system of MS patients and B-cells recovered from cerebrospinal fluid of MS patients display an activated phenotype.

## Celiac disease

Celiac disease occurs in genetically predisposed individuals following exposure to gluten. The disease results in diarrhea and malabsorption. It is unique among autoimmune diseases, as the environmental precipitant is known. Strict avoidance of the antigen, which is found in wheat, leads to remittance of symptoms in most subjects. Gliadin is the alcohol soluble fraction of gluten and is the primary antigen leading to an inflammatory reaction in the small intestine, characterized by chronic inflammatory infiltrate and villous atrophy. The enzyme, tissue transglutaminase, deamidases gliadin peptides in the lamina propria, increasing their immunogenicity. Gliadin-reactive T-cells recognizing antigen in the context of HLA-DQ2 or DQ8 lead to an inflammatory Th1 phenotype. While these HLA types are necessary for the development of celiac disease, they are not sufficient, and other genetic factors and environmental exposures also play a role in developing disease. IgA antibodies directed against gliadin, tissue transglutaminase, and connective tissue (such as anti-endomysial and anti-reticulin) are found in individuals with celiac disease. These patients are also susceptible to developing several types of cancer, most notably adenocarcinoma of the small intestine and enteropathy-associated T-cell lymphoma.

## The hygiene hypothesis

The hygiene hypothesis originally served to explain the increase in allergic disorders observed in industrialized countries over the past century. This theory has been attributed to David Strachan who noted an inverse relationship between the number of older siblings and the prevalence of hay fever and concluded that allergies could be attenuated by infections in early childhood. However, it was also noted that there was an increase in autoimmune diseases during this same period. This led to the idea that a reduction in exposure to infections early in life because of improved hygienic conditions might promote the development of chronic inflammatory disorders. In other words, when a vigorous immune system, which served as a survival advantage in a world abundant with microbes, is not stimulated appropriately but rather responds to environmental antigens, this results in increased allergic and autoimmune disease. Several epidemiologic studies support the hygiene hypothesis. The Karelia region in northeastern Europe, which is divided into a Finnish section and a Russian section, serves as a remarkable example. The people in this region are genetically related despite being divided by a political border. However, rates of autoimmunity and allergy are far higher on the more hygienic Finnish side, while infections are much more common on the Russian side, where public health measures are comparatively reduced.

In a variation to the hygiene hypothesis, Graham Rook proposed the so-called *old friends theory*, arguing that the important exposures are not influenza, measles, and other acute common childhood infections that have arisen rather recently in human history, but rather more ancient microbes that cause latent or mild chronic infections. The human immune system has evolved with these "old friends" and does not develop or function properly without them. Microbes that fit this description and have been reduced or almost eradicated in industrialized countries due to urbanization and public health measures, such as improved sanitation and control of food production, include helminthes and malaria. The observation that inflammatory bowel disease (IBD) is quite unusual in areas where helminthic infections are common, and the further finding that helminthic therapy is efficacious in murine models of colitis, has led researchers to test the efficacy of treating IBD in humans with ova from *Trichuris suis* (pig whipworm, which is unable to colonize humans very effectively), finding significant reduction in symptoms. Evidence from several murine models of autoimmune disease show that helminthic products in the gastrointestinal tract shift immune responses from an inflammatory state, characterized by Th1 and Th17 responses, to one that is more tolerogenic, characterized by increased Treg presence and IL-10 production. Helminthic therapy is now being studied in several autoimmune diseases in humans including MS, rheumatoid arthritis, type 1 diabetes, and SLE.

## Key concepts

- Upon re-exposure to environmental antigens called allergens, an inflammatory state can ensue, known as a hypersensitivity or allergic reaction. Four types of hypersensitivity states are classically recognized. Type I or immediate hypersensitivity is mediated by IgE and may occur as a local (e.g., asthma) or systemic (anaphylaxis) reaction. Type II hypersensitivity is mediated by IgM or IgG targeting cell membrane-associated antigens. The most common cause of type II reactions is medications, such as penicillin, which become associated with red blood cells or platelets leading to anemia and thrombocytopenia. Type III hypersensitivity occurs when antibodies bind antigen to form immune complexes that become deposited in tissues where they elicit inflammation. Examples of Type III reactions include serum sickness and the Arthus reaction. Type IV or delayed-type hypersensitivity (DTH) is mediated by T-cells. Examples of DTH include the tuberculin skin reaction and contact dermatitis from poison ivy.
- Individuals with antibody deficiencies often remain well for a time after birth due to passive transfer of immunoglobulin from breast milk during the first 6−9 months of life. Then they develop upper and lower respiratory tract infections including sinusitis, otitis, and pneumonia, primarily caused by encapsulated bacteria such as S. pneumoniae. If left untreated, these recurrent infections cause significant destruction and scarring of lung tissue, resulting in bronchiectasis.

- X-linked agammaglobulinemia is caused by a defect in Bruton's tyrosine kinase (Btk), which leads to arrest of B-cell differentiation at the Pre-B-cell stage. The syndrome is manifested by a complete lack of serum immunoglobulins and a lack of tonsilar tissue.
- The most common form of hyper-IgM syndrome is caused by an X-linked deficiency in CD40L, which is expressed on the surface of T-cells. In the absence of B-cell-T-cell interaction via CD40-CD40L, there is no immunoglobulin class switching and B-cells only produce IgM. In CD40L deficiency, there is significant impairment in both antibody production and cell-mediated immunity, resulting in a clinical presentation that may include infections with encapsulated bacteria and intracellular organisms.
- Defects in the terminal components of complement (C5-9) lead to a remarkable susceptibility to recurrent infections caused by pathogenic Neisseria spp.
- In chronic granulomatous disease (CGD), phagocytes are unable to produce reactive oxygen species to kill ingested pathogens due to genetic defects in the five subunit enzyme NADPH oxidase. The disease is characterized by recurrent, indolent infections caused by catalase positive organisms including S. aureus and Aspergillus spp.
- HIV is a retrovirus that enters CD4+ T-cells through interaction of gp160 on the viral envelope and CD4 along with a co-receptor, either CCR5 or CXCR4, on the cell surface. HIV infection leads to a progressive decline in CD4+ T-cells, leading to opportunistic infections including pneumocystis pneumonia, cryptococcal meningitis, disseminated CMV, and mycobacterial infections.
- The hallmark of SLE is the development of autoantibodies, particularly those that bind double-stranded DNA (anti-dsDNA). Deficiencies in the early complement component cascade (C1q, C2, or C4) are strongly associated with SLE.
- In type 1 diabetes, cell-mediated destruction of pancreatic b-cells occurs, leading to complete insulin deficiency.
- Celiac disease occurs in genetically predisposed individuals following exposure to wheat gluten. Th1-activated CD4 + T-cells recognizing the alcohol soluble fraction of gluten (gliadin) leads to an inflammatory reaction in the small intestine characterized by chronic inflammatory infiltrate and villous atrophy, resulting in diarrhea and malabsorption. Strict avoidance of the antigen leads to remittance of symptoms in most subjects.

## Suggested readings

[1] Casanova JL, Abel L. Inborn errors of immunity to infection: the rule rather than the exception. J Exp Med 2005;202:197−201.

[2] Abbas AK, Lichtman AH, Pillai S. Cellular and molecular immunology. New York: WB Saunders; 2014.

[3] Murphy KM, Weaver C. Janeway's immunobiology. New York: Garland Science; 2016.

[4] Gell PGH, Coombs RRA. Clinical aspects of immunology. Oxford: Blackwell; 1968.

[5] Johansson SG, Hourihane JO, Bousquet J, et al. A revised nomenclature for allergy. An EAACI position statement from the EAACI nomenclature task force. Allergy 2001;56:813−24.

[6] Rajan TV. The Gell-Coombs classification of hypersensitivity reactions: a re-interpretation. Trends Immunol 2003;24:376−9.

[7] Bruton OC. Agammaglobulinemia. Pediatrics 1952;9:722−8.

[8] Holland SM. Chronic granulomatous disease. Hematol Oncol Clin North Am 2013;27:89−99.

[9] Zhang SY, Jouanguy E, Ugolini S, et al. TLR3 deficiency in patients with herpes simplex encephalitis. Science 2007;317:1522−7.

[10] Bhaskaran K, Hamouda O, Sannes M, et al. Changes in the risk of death after HIV seroconversion compared with mortality in the general population. JAMA 2008;300:51−9.

[11] Kuritzkes DR. Hematopoietic stem cell transplantation for HIV cure. J Clin Invest 2016;126:432−7.

[12] Rahman A, Isenberg DA. Systemic lupus erythematosus. N Engl J Med 2008;358:929−39.

[13] Frohman EM, Racke MK, Raine CS. Multiple sclerosis − the plaque and its pathogenesis. N Engl J Med 2006;354:942−55.

[14] Lebwohl B, Ludvigsson JF, Green PH. Celiac disease and non-celiac gluten sensitivity. BMJ 2015;351:h4347.

[15] Versini M, Jeandel PY, Bashi T, et al. Unraveling the Hygiene hypothesis of helminthes and autoimmunity: origins, pathophysiology, and clinical applications. BMC Med 2015;13:81.

# Chapter 18

# Molecular basis of pulmonary disease

Dani S. Zander[1] and Carol F. Farver[2]

[1]Department of Pathology and Laboratory Medicine, University of Cincinnati, Cincinnati, OH, United States, [2]Director of Division of Education, Department of Pathology, University of Michigan Medical School, Ann Arbor, MI, United States

## Summary

Pulmonary pathology includes a large spectrum of both neoplastic and nonneoplastic diseases that affect the lung. Many of these diseases are the result of the unusual relationship of the lung with the outside world. Every breath that a human takes brings the outside world into the body in the form of infectious agents, organic and inorganic particles, and noxious agents of many types. Although the lung has many defense mechanisms to protect itself from these insults, these protections are not infallible and so lung pathology arises. In many cases, host cells and cellular products participate in the pathogenesis of the injury. Damage to the lung is particularly important given the role of the lung in the survival of the organism. Any impairment of lung function has widespread effects throughout the body, since all organs depend upon the lungs for oxygen. Pulmonary pathology encompasses adverse changes in the lung tissues and the mechanisms through which these occur. What follows is a review of pulmonary pathology and an overview of the current state of knowledge about the features and molecular pathogenesis of many key lung diseases.

## Neoplastic lung and pleural diseases

Lung cancer is a major cause of morbidity and mortality throughout the world. Estimates available from the *Surveillance, Epidemiology, and End Results* (SEER) program of the *National Cancer Institute* indicate that over 221,000 people in the United States were diagnosed with cancer of the lung and bronchus in 2015, and over 158,000 died due to this disease. However, in the past decades, incidence and mortality rates have begun to move in a more positive direction, particularly in men. These trends parallel changes in the prevalence of tobacco smoking, the most important risk factor for development of lung cancer.

Elucidation of the molecular pathogenesis of these neoplasms has progressed significantly, enabling development of new, molecularly targeted therapies and predictors of prognosis and therapeutic responsiveness. Recognition of precursor lesions for some types of lung cancers has been facilitated by our expanded understanding of early molecular changes involved in carcinogenesis.

The *World Health Organization* (WHO) classification scheme is the most widely used system for classification of lung and pleural neoplasms (Table 18.1). The most of the common malignant epithelial tumors can be grouped into the categories of non-small cell lung cancer (NSCLC) and small cell lung cancer (SCLC). NSCLCs account for 85%−90% of all lung cancers, and the two most common histologic types in this group are the adenocarcinomas (ACs) (about 40% of all lung cancers) and squamous cell carcinomas (SqCCs) (25%−30% of all lung cancers). Less common histologic types include large cell carcinomas (LCCs), adenosquamous carcinomas, sarcomatoid carcinomas, and other rarer types. SCLCs include cases of pure and combined small cell carcinoma. Neoplasms will usually declare themselves with clinical symptoms related to airway obstruction such as pneumonia or local invasion of adjacent structures such as chest, wall, nerves, superior vena cava, esophagus, or heart). SCLCs, known for early and widespread metastasis are particularly prone to being discovered through presentations with metastases to distant sites. Lung cancer are also discovered by pathophysiologic changes triggered by the release of soluble substances from neoplastic cells. These include endocrine syndromes with elaboration of hormones causing Cushing syndrome, syndrome of inappropriate antidiuretic hormone, hypercalcemia, carcinoid syndrome, gynecomastia, and others. Hypercoagulability may also occur with lung cancers, leading to manifestations such as venous thrombosis, nonbacterial thrombotic endocarditis, and disseminated

Essential Concepts in Molecular Pathology. DOI: https://doi.org/10.1016/B978-0-12-813257-9.00018-8

**TABLE 18.1** World health organization classification of tumors of the lung.

**Epithelial tumors**
Adenocarcinoma
   Lepidic adenocarcinoma
   Acinar adenocarcinoma
   Papillary adenocarcinoma
   Micropapillary adenocarcinoma
   Solid adenocarcinoma
   Invasive mucinous adenocarcinoma
      Mixed invasive mucinous and nonmucinous adenocarcinoma
   Colloid adenocarcinoma
   Fetal adenocarcinoma
   Enteric adenocarcinoma
   Minimally invasive adenocarcinoma
      Nonmucinous
      Mucinous
   Preinvasive lesions
      Atypical adenomatous hyperplasia
      Adenocarcinoma in situ
         Nonmucinous
         Mucinous
Squamous cell carcinoma
   Keratinizing squamous cell carcinoma
   Nonkeratinizing squamous cell carcinoma
   Basaloid squamous cell carcinoma
   Preinvasive lesion
      Squamous cell carcinoma in situ
   Neuroendocrine tumors
      Small cell carcinoma
         Combined small cell carcinoma
   Large cell neuroendocrine carcinoma
      Combined large cell neuroendocrine carcinoma
   Carcinoid tumors
      Typical carcinoid
      Atypical carcinoid
   Preinvasive lesion
      Diffuse idiopathic pulmonary neuroendocrine cell hyperplasia
Large cell carcinoma
Adenosquamous carcinoma
Pleomorphic carcinoma
Spindle cell carcinoma
Giant cell carcinoma
Carcinosarcoma
Pulmonary blastoma
Other and unclassified carcinomas
   Lymphoepithelioma-like carcinoma
   NUT carcinoma
Salivary gland-type tumors
   Mucoepidermoid carcinoma
   Adenoid cystic carcinoma
   Epithelial-myoepithelial carcinoma
   Pleomorphic adenoma
Papillomas
   Squamous cell papilloma
   Glandular papilloma
   Mixed squamous cell and glandular papilloma
Adenomas
   Sclerosing pneumocytoma
   Alveolar adenoma
   Papillary adenoma
   Mucinous cystadenoma
   Mucous gland adenoma

*(Continued)*

---

**TABLE 18.1** (Continued)

**Mesenchymal Tumors**
Pulmonary hamartoma
Chondroma
PEComatous tumors
   Lymphangioleiomyomatosis
  PEComa, benign
    Clear cell tumor
  PEComa, malignant
Congenital peribronchial myofibroblastic tumor
Diffuse pulmonary lymphangiomatosis
Inflammatory myofibroblastic tumor
Epithelioid hemangioendothelioma
Pleuropulmonary blastoma
Synovial sarcoma
Pulmonary artery intimal sarcoma
Pulmonary myxoid sarcoma with *EWSR1-CREB1* translocation
Myoepithelial tumors
  Myoepithelioma
  Myoepithelial carcinoma
**Lymphohistiocytic Tumors**
Extranodal marginal zone lymphoma of MALT
Diffuse large B-cell lymphoma
Lymphomatoid granulomatosis
Intravascular large B-cell lymphoma
Pulmonary Langerhans cell histiocytosis
Erdheim–Chester disease
**Tumors of Ectopic Origin**
Germ cell tumors
  Teratoma, mature
  Teratoma, immature
Intrapulmonary thymoma
Melanoma
Meningioma, NOS
**Metastatic tumors**

---

Reproduced with permission from Travis WD, Brambilla E, Burke AP, Marx A, Nicholson AG, editors. WHO classification of tumours of the lung, pleura, thymus and heart. Lyon: IARC; 2015.

intravascular coagulation. Hematologic changes may occur, such as anemia, granulocytosis, eosinophilia, and other abnormalities as do other paraneoplastic syndromes such as clubbing of the fingers, myasthenic syndromes, dermatomyositis/polymyositis, and transverse myelitis.

The American Joint Commission on Cancer's tumor, lymph nodes and metastasis staging system is the most widely used staging system. Overall, for lung cancers, the 5-year survival is 17.4%. An important factor leading to this relatively poor survival is the advanced stage at which many lung cancers are diagnosed. Information from the SEER database indicates that in 2010, 49.3% of patients were diagnosed with distant metastatic (stage IV) disease at the time their cancers were discovered.

Surgical resection is the preferred approach to treat localized NSCLCs, provided there is no medical contraindication to operative intervention. Lobectomy or more extensive resection (depending upon tumor extent) is usually recommended, unless other comorbid conditions preclude these procedures. However, more limited resections are now being pursued for certain localized cancers in some clinical scenarios. Intraoperative mediastinal lymph node sampling or dissection is also frequently recommended for accurate pathologic staging and determination of therapy. Subsets of patients also benefit from chemotherapy and/or radiotherapy. For more advanced NSCLC and for SCLC, chemotherapy and radiotherapy are the primary treatment modalities. Chemotherapeutic regimens can include conventional chemotherapeutic agents, molecularly targeted agents selected based upon genetic changes found in the cancer and/or immunotherapies.

## Common molecular genetic changes in lung cancers

Development of lung cancer occurs with multiple, complex, stepwise genetic and epigenetic changes involving allelic losses, chromosomal instability and imbalance, mutations in tumor suppressor genes (TSGs) and dominant oncogencs, epigenetic gcnc silencing through promoter hypermethylation, and aberrant expression of genes participating in control of cell proliferation and apoptosis. There are similarities as well as type-specific differences in the molecular alterations found in the histologic spectrum of NSCLCs and SCLCs (Fig. 18.1 and Table 18.2). Alterations include cell-cycle and other mutations, rearrangements, amplifications, and deletions. As our understanding of the molecular pathways involved in cancer initiation and progression continues to grow new molecularly targeted therapeutic agents and companion diagnostic tests predictive of therapeutic responsiveness to these agents will emerge (Fig. 18.2). These agents will also place greater emphasis upon the accurate histologic classification of cancers, upon which targeted therapies are selected. Collection of representative tissues for molecular testing has become critical for appropriate molecular testing. One challenge to successful therapy relates to the process of clonal evolution of cancer, in which progeny of a

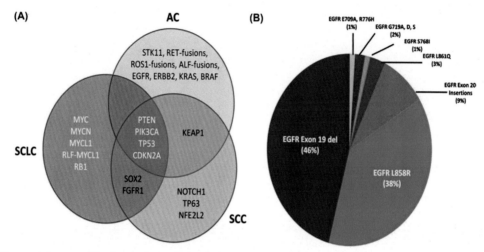

FIGURE 18.1 Gene mutations, amplifications, and gene fusions in lung cancer. (A) This Venn diagram displays the relationships of gene expression within the three common types of NSCLC. The patterns of genes depicted are relative and not absolute so as to depict trends within the cancer subtypes in terms of predominant genetic alterations. In particular the similarity of small cell lung cancer (SCLC) to squamous cell carcinoma (SCC) and adenocarcinoma (AC) is highlighted. (B) Within these genetic loci there are considerable variations in terms of the nature of the mutations present. As an example, *EGFR* mutations that are particularly common in AC encompass the illustrated mutations, amplifications, and deletions. *Reprinted with permission from Wood SL, Pernemalm M, Crosbie PA, Whetton AD. Molecular histology of lung cancer: From targets to treatments. Cancer Treat Rev 2015;41:361–75.*

**TABLE 18.2 Comparison of TCGA landmark articles for LUAD and LUSC.**

| Category | LUSC | LUAD |
|---|---|---|
| Publication year | 2012 | 2014 |
| No. of samples studied | 178 TCGA | 412 (230 TCGA) |
| Percentage of cases of past/present smoking | 96% | 81% |
| No mutations per megabase | 8.10 | 8.87 |
| Most frequently mutated genes | *TP53, CDKN2A, PTEN, PIK3CA, KEAP1* | *TP53, KRAS, EGFR, BRAF* |
| Notable amplifications | *SOX2, TP63,* chromosome 3q | *NKX2–1, TERT, MDM2, KRAS, EGFR, MET* |
| Notable deletions/loss of function | *FOXP1, NOTCH1, NOTCH2, ASCL4* | *CDKN2A, MET* exon 14 skipping |
| Possible cancer drivers for oncogene-negative tumors | *FAM123 B, HRAS, FBXW7, SMARCA4, NF1, SMAD4, EGFR* | *ERBB2, MET, NF1* |

*LUAD,* lung adenocarcinoma; *LUSC,* lung squamous cell carcinoma; *TCGA,* the Cancer Genome Atlas. Summarized from official TCGA publications for LUSC and LUAD.
Reprinted with permission from Chang JT, Lee YM, Huang RS. The impact of the cancer genome atlas on lung cancer. Transl Res 2015;166:568–85.

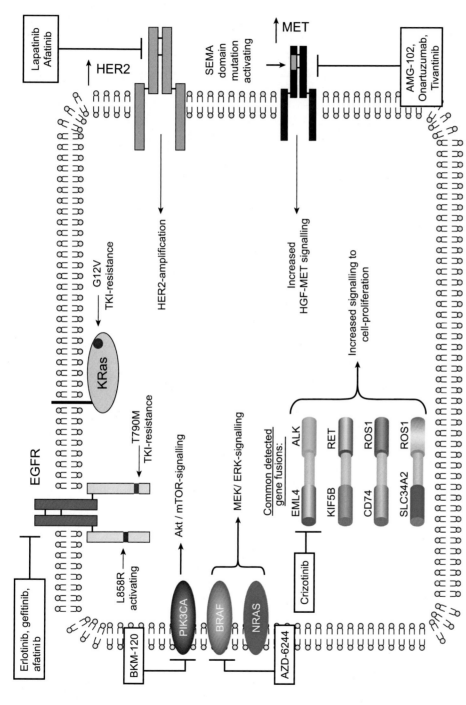

**FIGURE 18.2** Driver mutations identified in lung cancer: altered signaling events and personalized medicine. Molecular analysis of NSCLC by transcriptomic analysis (microarrays as well as RNA-seq) and DNA sequencing of lung cancer samples has identified numerous key driver mutations within lung cancer. While these alterations are depicted in this schematic as occurring within a single cell, it must be noted that they are not necessarily detected together and maybe mutually exclusive. The two most commonly detected mutational events in lung cancer are *EGFR* mutations (L858R and/or xon-19 deletions as activating mutations) and *KRAS* mutations (G12 V). *EGFR*-mutant tumors are currently targeted by the protein tyrosine kinase inhibitors (erlotinib, gefitinib, or afatinib). Resistance to EGFR-TKIs can occur via either *EGFR* gatekeeper mutations (such as T790 M) or *KRAS* mutations. Resistance to EGFR-TKIs can also be mediated by bypass pathway activation such as overexpression of the HER2 receptor or increased expression of the hepatocyte-growth factor (HGF) receptor. MET/HGF by-pass signaling occurs most commonly via MET receptor overexpression and/or the presence of activating mutations within the external SEMA-domain of the receptor. In addition to receptor overexpression and activating mutations, lung cancers also harbor numerous gene fusions that activate cellular signaling. Four commonly detected gene fusions are depicted of which the *EML4-ALK* fusion is one of the most studied and is the target for the therapeutic agent crizotinib. Less frequently detected cancer drivers in lung cancer include *PIK3CA* mutations and altered signaling via BRAF/NRAS. The common feature of all these lung cancer driver mutations is that they signal to increase cellular proliferation. In addition to these cell proliferative signals, lung cancers also harbor deletions/epigenetic silencing events for key tumor suppressor genes (these are omitted from the current figure). *Reprinted with permission from Wood SL, Pernemalm M, Crosbie PA, Whetton AD. Molecular histology of lung cancer: from targets to treatments. Cancer Treat Rev 2015;41:361–75.*

neoplastic cell acquire additional genomic alterations (branched evolution), generating subclones with differing genomic features (cancer heterogeneity) and sometimes differing responsiveness to drugs. Natural selection can drive expansion of drug-resistant subclones in tumors.

Driver mutations are essential for survival of cancer cells and play central roles in carcinogenesis. Knowledge about driver mutations has expanded significantly, particularly for ACs. Common driver mutations in ACs involve epidermal growth factor receptor (*EGFR*), *KRAS*, and anaplastic lymphoma kinase (*ALK*). The availability of *EGFR* tyrosine kinase inhibitors that are effective against cancers with *EGFR* mutations, and *ALK* inhibitors that are active against tumors with *ALK* rearrangements, has changed the therapeutic landscape for patients with these cancers in advanced stages. A related pathway, the phosphoinositide 3-kinase (PI3K)/AKT/mammalian target of rapamycin (mTOR) pathway, is frequently deregulated in pulmonary carcinogenesis. This pathway has been reported to mediate the effects of several tyrosine kinase receptors, including EGFR, c-Met, c-Kit, and IGF-IR, on proliferation and survival in NSCLC and SCLC. HER2/neu is another related receptor tyrosine kinase that is upregulated in approximately 20%−30% of NSCLCs, but unlike the situation with HER2/neu-positive breast cancers, treatment with anti-HER2/neu antibody does not yield comparable benefits for NSCLC when used alone or in combination with chemotherapy. Point mutations of RAS family oncogenes (most often at *KRAS* codons 12, 13, or 61) are detected in 20%−30% of lung ACs and 15%−50% of all NSCLCs. Although farnesyl transferase inhibitors prevent Ras signaling, these agents have not shown significant activity as single-agent therapy in untreated NSCLC or relapsed SCLC and targeted therapies effective against *KRAS*-mutated cancers are still actively sought. MYC family genes (*MYC*, *MYCN*, and *MYCL*), which play roles in cell cycle regulation, proliferation, and DNA synthesis, are more frequently activated in SCLCs than in NSCLCs, either by gene amplification or by transcriptional dysregulation.

In cancers, chromosomal regions harboring TSGs and oncogenes are often deleted or amplified, respectively. Allelic loss involving loci in 3p14-23 is a consistent feature of lung cancer pathogenesis. Allelic losses of 3p have been reported, often multiple and discontinuous, in 96% of the lung cancers studied and in 78% of the precursor lesions. Larger segments of allelic loss were noted in the majority of SCLCs (91%) and SqCCs (95%), compared to ACs (71%) and preneoplastic/preinvasive lesions. There was allelic loss in the 600-kb 3p21.3 deletion region in 77% of the lung cancers; 70% of the normal or preneoplastic/preinvasive lesions associated with lung cancers; and 49% of the normal, mildly abnormal, or preneoplastic/preinvasive lesions found in smokers without lung cancer, but no loss was seen in the samples from people who had never smoked. 8p21-23 deletions are also frequent and early events in the pathogenesis of lung carcinomas, and other common alterations include LOH at 13q, 17q, 18q, and 22p. Allelic losses that are more frequent in SqCCs than ACs include deletions at 17p13 (*TP53*), 13q14 (*RB*), 9p21 (p16$^{INK4a}$), 8p21-23, and several regions of 3p.

Inactivation of recessive oncogenes is believed to occur through a two-stage process. It has been suggested that the first allelic inactivation occurs often via a point mutation, and the second allele is later inactivated by a chromosomal deletion, translocation, or other alteration such as methylation of the gene promoter region. Inactivating mutations in the TSG *TP53*, which encodes the p53 protein, are the most frequent mutations in lung cancers. These mutations are found in up to 50% of NSCLCs and over 70% of SCLCs and are largely attributable to direct DNA damage from cigarette smoke carcinogens. *TP53* mutational patterns have a prevalence of G to T transversions in 30% of lung cancers in smokers versus only 12% of lung cancers in nonsmokers. p53 protein is a transcription factor and a key regulator of cell cycle progression. Cellular signals induced by DNA damage, oncogene expression, or other stimuli trigger p53-dependent responses, including initiating cell cycle arrest, apoptosis, differentiation, and DNA repair. Loss of p53 function in cancer cells results in inappropriate progression through the dysregulated cell cycle checkpoints and permits the inappropriate survival of genetically damaged cells.

The p16$^{INK4a}$-cyclin D1/CDK4/Rb pathway, which plays a central role in controlling the G1 to S phase transition of the cell cycle, is another important tumor suppressor pathway that is often disrupted in lung cancers. It interfaces with the p53 pathway through p14$^{ARF}$ and p21$^{Waf/Cip1}$. Thirty percent to 70% of NSCLCs contain mutations of p16$^{INK4a}$, including homozygous deletion, point mutations, or epigenetic alterations, leading to p16$^{INK4a}$ inactivation. Almost 90% of SCLCs and smaller numbers of NSCLCs display loss of Rb expression, and mutational mechanisms usually responsible include deletion, nonsense mutations, splicing abnormalities that lead to truncated Rb protein and p16$^{INK4a}$ leads to hypophosphorylation of the Rb protein, which causes arrest of cells in the G1 phase. The active, hypophosphorylated form of Rb regulates other cellular proteins including the transcription factors E2F1, E2F2, and E2F3, which are essential for progression through the G1/S phase transition. Loss of p16$^{INK4a}$ protein or increased complexes of cyclin D-CDK4/6 or cyclin E-CDK2 lead to hyperphosphorylation of Rb with resultant evasion of cell cycle arrest and progression into S phase. Cell cycle progression is inhibited by p21$^{Waf/Cip1}$ through its inhibition of the cyclin complexes. The 10%−30% of NSCLCs lacking detectable alterations in p16$^{INK4a}$ and Rb may have abnormalities of cyclin D1 and CDK4, which cause inactivation of the Rb pathway.

**TABLE 18.3** Principle microRNAs involved in the development or progression of lung cancer.

| microRNAs | Gene Targets | Biological Processes |
|---|---|---|
| Tumor Suppressor microRNAs With Downregulation in Lung Cancer | | |
| let-7 family | *RAS, HMGA2, CDK6, MYC, DICER1* | i.  Cell proliferation (*RAS, MYC, HMGA2*)<br>ii.  Cell cycle regulation (*CDK6*)<br>iii.  microRNA maturation (*DICER1*) |
| MiR-34 family | *MET, BCL2, PDGFRA, PDGFRB* | TRAIL-induced cell death and cell proliferation |
| MiR-200 family | *ZEB1, ZEB2, E-cadherin (CDH1), vimentin (VIM)* | Promotion of EMT and metastasis |
| Oncogenic microRNAs With Upregulation in Lung Cancer | | |
| MiR-21 | *PTEN, PDCD4, TPM1* | Apoptosis, cell proliferation, and migration |
| MiR-17-92 cluster | *E2F1, PTEN, HIF1A* | Cell proliferation and carcinogenesis |
| MiR-221/222 | *PTEN, TIMP3* | Apoptosis and cell migration |

*EMT*, epithelial-mesenchymal transition; *TRAIL*, TNF-related apoptosis-inducing ligand.
From Inamura K, Ishikawa Y. MicroRNA in lung cancer: novel biomarkers and potential tools for treatment. J Clin Med. 2016;5:36. Creative Commons Attribution 4.0 International license.

Epigenetic alterations (hypermethylation of the 5′ CpG island) of TSGs are also frequent occurrences during pulmonary carcinogenesis, and methylation profiles of NSCLCs show relationships to smoke exposure, histologic type, and geography. Methylation rates of *p16^INK4a* and *APC* and the mean methylation index (MI) (a reflection of the overall methylation status) in current or former smokers were significantly higher than in never smokers. The mean MI of cancers was highest in current smokers. Methylation rates of *APC*, *CDH13*, and *RARbeta* were significantly higher in ACs than in SqCCs. MicroRNAs are small noncoding RNAs that regulate gene expression post-transcriptionally by translational repression or degradation of target mRNAs. Accumulating evidence indicates that microRNAs play roles in the development and progression of lung cancer, and that they may be exploited as diagnostic and therapeutic tools. Table 18.3 lists the principle microRNAs that are involved in the development or progression of lung cancer. MicroRNA profiles appear to differ between lung cancer subtypes, and specific microRNA expression signatures may distinguish prognostically different groups of lung cancers. Circulating microRNAs show promise as biomarkers for early diagnosis of lung cancer and prognostic assessment, and modulation of microRNA activity represents a potential approach to therapy.

## Adenocarcinoma and its precursors

### Clinical and pathologic features

In the most recent version of the WHO classification scheme, AC is defined as "...a malignant epithelial tumor with glandular differentiation, mucin production, or pneumocyte marker expression...The tumors show an acinar, papillary, micropapillary, lepidic, or solid growth pattern, with either mucin or pneumocyte marker expression..." AC occurs primarily in smokers, but represents the most common type of lung cancer in people who have never smoked and in women. A small subset of these tumors arises in patients with localized scars or diffuse fibrosing lung diseases such as asbestosis and interstitial pneumonia associated with scleroderma. ACs usually arise in the periphery of the lung and are more likely to invade the pleura and chest wall than other histologic types of lung cancers. Radiologic studies may show one or more nodules, ground-glass opacities, or mixed solid and ground-glass lesions. On gross examination, the neoplasms are often solitary gray−white nodules or masses that pucker the overlying pleura and sometimes demonstrate necrosis or cavitation. Other presentations include a pattern of consolidation resembling pneumonia (usually lepidic adenocarcinoma) (Fig. 18.3), multiple nodules, diffuse interstitial widening due to lymphangitic spread, endobronchial lesions with submucosal infiltration, and diffuse visceral pleural infiltration and thickening resembling mesothelioma.

Common histologic patterns displayed by ACs include acinar (Fig. 18.4), papillary, micropapillary, lepidic (Figs. 18.5 and 18.6), and solid arrangements, and mixtures of these patterns are very frequent. ACs can exhibit differentiation toward Clara cells, type II pneumocytes, or mucin-containing cells (Fig. 18.7), sometimes having the morphology of goblet cells. In recent years, corresponding preinvasive lesions for AC have become better defined. Atypical adenomatous hyperplasia (AAH) has been recognized as a precursor lesion for peripheral pulmonary ACs, and is defined as "...*a small (usually ≤ 0.5 cm) localized proliferation of mildly to moderately atypical pneumocytes and/or Clara cells lining alveolar walls and sometimes respiratory bronchioles*..." (Fig. 18.8). AAH exists on a histologic

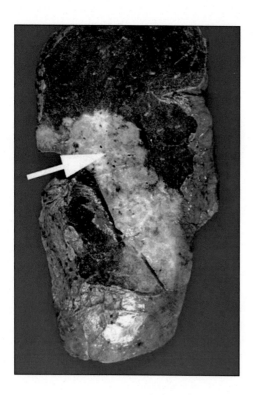

**FIGURE 18.3** Adenocarcinoma. The tan tumor (*arrow*) replaces a large portion of the normal lung parenchyma.

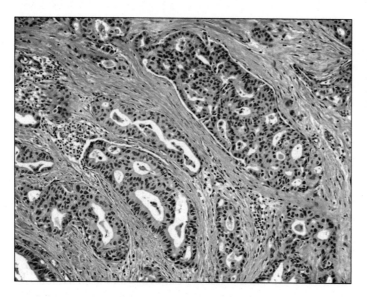

**FIGURE 18.4** Adenocarcinoma with acinar pattern. The tumor consists of abnormal glands, some showing cribriform architecture, in a desmoplastic stromal background. The desmoplastic stroma has a dense collagenized appearance and reflects the presence of invasion. The abnormal glandular structures are lined by columnar tumor cells with abundant cytoplasm and mildly pleomorphic nuclei.

continuum with adenocarcinoma in situ (AIS), which is defined as "...*a small (≤3 cm), localized adenocarcinoma with growth restricted to neoplastic cells along pre-existing alveolar structures (pure lepidic growth), and lacking stromal, vascular, or pleural invasion...*" (Fig. 18.6).

## Molecular pathogenesis

Data from *The Cancer Genome Atlas Research Network* showed high rates of somatic mutation (mean 8.9 mutations per megabase), with 18 statistically significant mutated genes in adenocarcinomas. In their series of 230 cases, mutations in *KRAS* (33%) were mutually exclusive with those in *EGFR* (14%). Other commonly mutated genes included *BRAF* (10%), *PIK3CA* (7%), *MET* (7%), and the small GTPase gene, *RIT1* (2%). TSGs with mutations included *STK11*

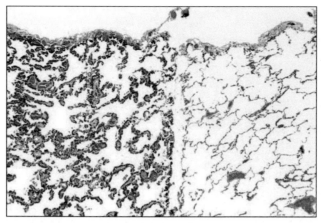

**FIGURE 18.5** Adenocarcinoma with lepidic pattern (left) compared with normal lung (right). The adenocarcinoma on the left displays a lepidic growth pattern, in which cancer cells extend along alveolar septa, maintaining the alveolar architecture of the lung. Notice that although open alveoli are present on both sides, the alveolar septa in the left portion are lined by cancer cells and have a thickened appearance, in contrast to the alveoli on the right, which have thin septa lined by flat pneumocytes.

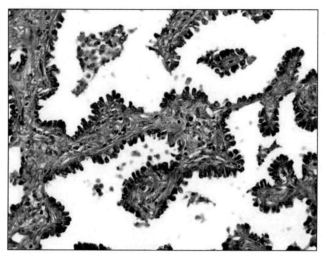

**FIGURE 18.6** Adenocarcinoma in situ. Columnar tumor cells with a hobnail appearance line thickened alveolar septa. The tumor cells have enlarged, hyperchromatic nuclei. These cells remain on the surface of the alveolar septa and do not invade the lung tissue.

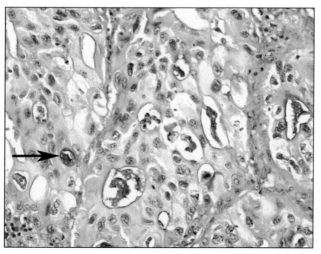

**FIGURE 18.7** Adenocarcinoma (mucicarmine stain). Intracytoplasmic (*arrow*) and luminal mucin stains dark pink. The production of mucin indicates glandular differentiation.

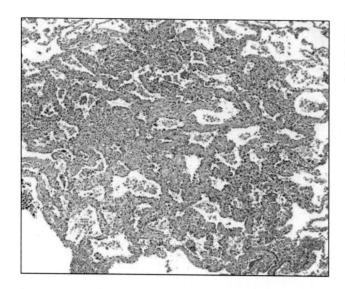

**FIGURE 18.8** Atypical adenomatous hyperplasia. This lesion, which has been defined as a precursor lesion for peripheral pulmonary adenocarcinomas, consists of a well-circumscribed nodule measuring several millimeters in diameter, in which alveolar septa are lined by mildly—moderately atypical cells.

(17%), *KEAP1* (17%), *NF1* (11%), *RB1* (4%), and *CDKN2A* (4%). Mutations in chromatin-modifying genes *SETD2* (9%), *ARID1A* (7%), and *SMARCA4* (6%), and the RNA splicing genes *RBM10* (8%) and *U2AF1* (3%) were also identified. Recurrent mutations in the *MGA* gene, which encodes a Max-interacting protein on the MYC pathway, were noted in 8% of samples, and loss-of-function (frameshift and nonsense) mutations were mutually exclusive with focal *MYC* amplification, suggesting a potential mechanism of MYC pathway activation. Mutational burdens in ACs of never-smokers are much lower (0.8—1 mutation per megabase) than those of smokers.

Driver gene alterations form the basis for molecularly targeted therapies. *EGFR* mutation (Fig. 18.9) and *ALK* rearrangement (Fig. 18.10) are the most important from a clinical perspective. Cancers with these alterations usually respond to specific targeted agents, offering an extended life expectancy to many individuals. Testing for *EGFR* mutations and *ALK* fusions is done to guide patient selection for therapy with an EGFR or ALK inhibitor, respectively, in patients with advanced-stage adenocarcinoma, and other testing for targetable oncogenes commonly includes *BRAF* mutation and *ROS1* rearrangement. Next generation sequencing has further extended the range of genetic alterations evaluated in many cancers. Although most cancers display one driver gene alteration, rare tumors have alterations in two driver genes. In addition, many patients also receive testing for alterations in *ROS1*, *RET*, *MET*, *BRAF*, and *HER2*. EGFR is a receptor tyrosine kinase whose activation by ligand binding leads to activation of cell signaling pathways such as Ras/mitogen-activated protein kinase (MAPK) and phosphatidylinositol-3-kinase, which in turn propagates signals for proliferation, blocking of apoptosis, differentiation, motility, invasion, and adhesion. Cancer-acquired mutations in the tyrosine kinase domain of EGFR, often associated with gene amplification, are associated with AC histology, never-smoker status, and female gender. The prevalence of *EGFR* mutation in AC patient populations varies with rates of 45% in Asian/Pacific, 24% in Caucasian, and 20% in African-American patients. *EGFR* mutations are frequently in-frame deletions in exon 19, single missense mutations in exon 21, or in-frame duplications/insertions in exon 20, and occasional missense mutations and double mutations can also be detected. *EGFR* mutation has an inverse correlation with methylation of the *p16^{INK4a}* gene and SPARC (secreted protein acidic and rich in cysteine), an extracellular Ca2+-binding glycoprotein associated with the regulation of cell adhesion and growth. *EGFR* status is an important predictor of response to EGFR kinase inhibitors: patients with *EGFR* mutations are most likely to have a significant response to EGFR tyrosine kinase inhibitor therapy. However, challenges remain in dealing with resistance. Mutations such as the threonine-790 to methionine (T790 M) point mutation can present major obstacles to responsiveness to EGFR tyrosine kinase inhibitors.

*KRAS* is a member of the Ras family of proteins, which function as signal transducers between cell membrane-based growth factor signaling and the MAPK pathways. *KRAS* mutations are associated with smoking, male gender, and poorly differentiated cancers, and are found in 20%—30% of lung ACs. Integrated analyses utilizing whole-exome and transcriptome sequencing have shown that components of the RTK/RAS/RAF pathway are usually affected (76% of cases) in AC. Development of effective targeted therapy against the *KRAS* mutant pathway has been difficult, but clinical trials of promising agents are underway. The echinoderm microtubule-associated protein like 4 and the anaplastic lymphoma kinase (*EML4-ALK*) fusion gene is found in 3%—7% of NSCLCs, particularly ACs, and is the best predictor of response to crizotinib, an ALK tyrosine kinase inhibitor. The *EML4-ALK* fusion gene results from an inversion in the short arm of chromosome two, fusing the N-terminal domain of *EML4* to the intracellular kinase domain of *ALK* (3′ gene region), producing a

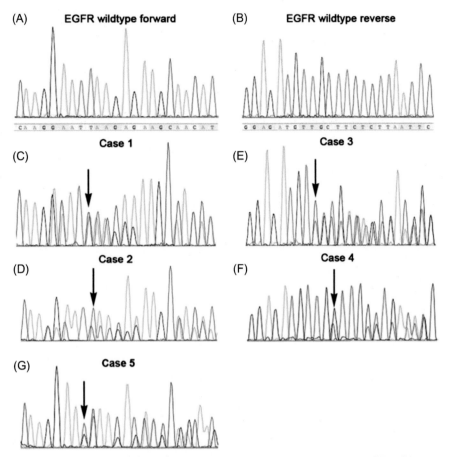

FIGURE 18.9 Epidermal growth factor receptor (EGFR) sequence analysis results for exon 19. (A and B) EGFR wild-type sequence by forward and reverse sequencing. (C—E) Sequence analysis results of cases 1, 2, 3, 4, and 5 by forward (C, D and G) or reverse (E and F) sequencing. Case 1: c.2238_2252del; case 2: c.2237 G > A and c.2238_2252del; case3: c.2238_2252del; case4: c.2238_2252del; and case5: c.2238_2252del (an *arrow* indicates the mutation for each case). *Reprinted with permission from Laack E, Simon R, Regier M, et al. Miliary never-smoking adenocarcinoma of the lung: Strong association with epidermal growth factor receptor exon 19 deletion. J Thorac Oncol 2011;6:199—202.*

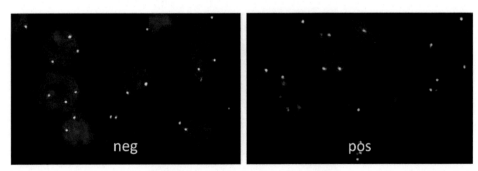

FIGURE 18.10 EML4-ALK fluorescence in situ hybridization (FISH). Representative FISH images showing normal fused signals (negative) and nuclei with multiple separated red signals (positive). *Reprinted with permission from Le Quesne J, Maurya M, Yancheva SG, et al. A comparison of immunohistochemical assays and FISH in detecting the ALK translocation in diagnostic histological and cytological lung tumor material. J Thorac Oncol 2014;9:769—74.*

constitutively active *ALK* tyrosine kinase. Patients tend to be younger and nonsmokers (70%—80%). *HER2* (also known as *EGFR2* or *ERBB2*), a member of the EGFR family of receptor tyrosine kinases, is mutated in less than 2% of NSCLC. The *HER2* mutations are in-frame insertions in exon 20 and are significantly more frequent in ACs (2.8%), never smokers (3.2%), Asian ethnicity (3.9%), and women (3.6%), similar to *EGFR* mutations.

Alterations in DNA methylation appear to be important epigenetic changes in cancer, contributing to chromosomal instability through global hypomethylation, and aberrant gene expression through alterations in the methylation levels at promoter CpG islands. Epigenetic differences exist between *EGFR*-mediated and *KRAS*-mediated tumorigenesis and may interact with the genetic changes. A study showed that the probability of having *EGFR* mutation was significantly lower among those with *p16INK4a* and *CDH13* methylation than in those without, and the methylation index was significantly lower in *EGFR* mutant cases than in wild-type. In contrast, *KRAS* mutation was significantly higher in *p16INK4a*-methylated cases than in unmethylated cases, and the methylation index was higher in *KRAS* mutant cases than in wild-type.

Limited data are available regarding molecular alterations in AAH, AIS, and minimally invasive adenocarcinoma (MIA). *EGFR* and *KRAS* mutations have been reported, each in up to about one-third of AAH lesions and in AIS lesions. AAH lesions contained an average of 2.2 mutations per lesion, with the most common mutations in *BRAF* and *ARID1B* (16% of AAHs each), and mutations also identified in *EGFR*, *MAML1*, *TP53*, and *KRAS*. Evaluation of AIS and MIA lesions found an average mutation rate of 6.2 and 10.8 mutations per patient, respectively. *EGFR* and *TP53* were the most frequently mutated genes in MIA patients,. Mutation frequencies in *EGFR* and *TP53* sequentially increased from early lesions to MIA, reflecting a transition to malignancy.

## Squamous cell carcinoma and its precursors

### Clinical and pathologic features

SqCC is defined as "…a malignant epithelial tumor that either shows keratinization and/or intercellular bridges, or is a morphologically undifferentiated non-small cell carcinoma that expresses immunohistochemical markers of squamous cell differentiation…" in the most recent WHO classification scheme. It is a common histologic type of NSCLC and closely linked to cigarette smoking. In most patients, this cancer arises in a mainstem, lobar, or segmental bronchus, producing a central mass on imaging studies. These cancers may have an endobronchial component that can cause airway obstruction, leading to postobstructive pneumonia, atelectasis, or bronchiectasis. Not infrequently, it is the pneumonia that prompts evaluation of the patient and leads to discovery of the cancer. Less often, SqCCs develop in the periphery of the lung.

A tan or gray mass that usually arises in a large bronchus and often includes an endobronchial component (Figs. 18.11 and 18.12). Partial or complete airway obstruction can be associated with changes of pneumonia,

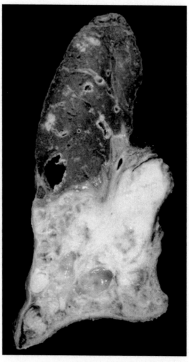

FIGURE 18.11 Invasive squamous cell carcinoma with postobstructive pneumonia and abscesses. This tan tumor lies in the central (perihilar) area of the lung and replaces the normal lung tissue. Distal to the tumor, the lung has extensive cystic changes reflecting abscesses and bronchiectasis, as well as a background of tan consolidation representing pneumonia.

FIGURE 18.12   Squamous cell carcinoma. The tumor has an endobronchial component (*arrow*) that partially obstructs the airway lumen and has a warty appearance.

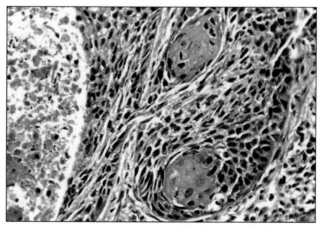

FIGURE 18.13   Invasive squamous cell carcinoma. This tumor consists of cells with hyperchromatic, pleomorphic nuclei, and eosinophilic cytoplasm. Two keratin pearls are present (center) and a portion of the tumor is necrotic (left).

bronchitis, abscess, bronchiectasis, or atelectasis. Necrosis and cavitation are very common in these tumors. Involvement of hilar lymph nodes by tan-gray tumor can be visible in some resected specimens. The key microscopic features of this tumor are its keratinization, sometimes with formation of keratin pearls, and intercellular bridges (Fig. 18.13). As in ACs, the degree of differentiation of this tumor ranges from very well differentiated cases with abundant keratinization and intercellular bridges and little cytoatypia, to very poorly differentiated cases, where keratinization and intercellular bridges can be quite inconspicuous and the tumor consists of sheets of large atypical cells with marked cytoatypia and frequent mitoses. Most cases fall more toward the middle of the spectrum. Invasiveness is reflected by the presence of irregular nests and sheets of cells that infiltrate through tissues, stimulating a fibroblastic response, or by cells inside vascular or lymphatic spaces.

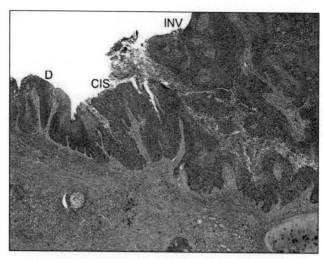

**FIGURE 18.14** Dysplasia, squamous cell carcinoma in situ, and invasive squamous cell carcinoma. The dysplastic squamous epithelium (D) demonstrates increased thickness of the basal layer with mild squamous atypia. The atypia is full thickness in the area of carcinoma in situ (CIS), and in this area the entire epithelium consists of similar appearing cells with increased nuclear:cytoplasmic ratios. Invasion (INV) into the underlying bronchial tissues is present as well.

Invasive SqCCs are often accompanied by SqCC in situ and dysplasia, their precursor lesions. These lesions arise in the bronchi and may be contiguous with the invasive tumor or exist as one or more separate foci. These precursor lesions may be observed without coexisting invasive carcinoma. Like SqCC, tobacco smoking is the main predisposing factor for SqCC in situ and dysplasia, though, these lesions are not invasive and do not extend through the basement membrane of the bronchial epithelium. Grossly, they may be invisible or appear as flat, tan or red discolorations of the bronchial mucosa, or tan wartlike excrescences. Microscopically, these lesions display a range of squamous changes that include alterations in the thickness of the bronchial epithelium, the maturational progress of squamous differentiation, cell proliferation, cell size, and nuclear characteristics (Fig. 18.14). As dysplasia increases, cell size, pleomorphism, and anisocytosis usually increase, and there is coarsening of the chromatin and appearance of nucleoli, nuclear angulations, and folding. In carcinoma in situ, although the epithelium may or may not be thickened and the cell size may be small, medium, or large, there is minimal or no maturation from the base to the superficial aspect, with atypical nuclear features present throughout the epithelium. Mitoses appear in the epithelium, initially in the lower third (mild dysplasia) and increasingly in the middle and upper thirds with higher grades of dysplasia and carcinoma in situ.

Basal cells in the bronchial epithelium are believed to represent the progenitor cells for invasive SqCC, and the sequence of changes leading to SqCC is believed to include basal cell hyperplasia, squamous metaplasia, squamous dysplasia, squamous cell carcinoma in situ, and invasive SqCC. Regression of lesions preceding invasive SqCC can occur, particularly the earlier lesions. However, severe dysplasia and carcinoma in situ are associated with a significantly increased probability of developing invasive SqCC in patients followed over time with surveillance bronchoscopy.

## Molecular pathogenesis

*The Cancer Genome Atlas Research Network* evaluated 178 pulmonary squamous cell carcinomas and revealed marked genomic complexity with a high overall mutation rate of 8.1 mutations/Mb. Statistically recurrent mutations were identified in 18 genes including mutation of *TP53* in nearly all specimens. Frequently altered pathways included *NFE2L2/KEAP1/CUL3* in 34%, squamous differentiation genes (*SOX2/TP63/NOTCH1* and others) in 44%, *PI3K/AKT* in 47%, and *CDKN2A/RB1* in 72% of cancers. *NFE2L2* and *KEAP1* code for proteins that regulate the cellular response to oxidative damage, chemo- and radiotherapy, and the alterations in this pathway had a pattern consistent with loss-of-function. Alterations in genes involved in squamous differentiation included overexpression and amplification of *SOX2* and *TP63*, loss-of-function mutations in *NOTCH1, NOTCH2*, and *ASCL4* and focal deletions in *FOXP1*. The TSG, *CDKN2A*, that encodes the INK4A/p16 and ARF/p14 proteins, was inactivated in 72% of cases by epigenetic silencing by methylation (21%), inactivating mutation (18%), exon 1β skipping (4%), and homozygous deletion (29%). Whole-transcriptome expression profiles yielded four subtypes designated as classical (36%), basal (25%), secretory (24%), and primitive (15%). Correlation of somatic mutations, copy number alterations, and gene expression signatures revealed significant subtype associations with alterations in the *TP53, PI3K, RB1*, and *NFE2L2/KEAP1* pathways.

On the other hand, *EGFR* and *KRAS* mutations and *ALK* fusions are rare in pulmonary SqCCs, and these cancers generally do not respond to targeted agents used for ACs with these alterations. However, potentially targetable oncogene alterations in SqCCs include *PIK3CA* mutations, *FGFR1* amplifications, *FGFR2/3* mutations, *FGFR3* rearrangements, and *DDR2* mutations.

## Large cell carcinoma

LCC is an undifferentiated NSCLC without light microscopic or immunohistochemical evidence of squamous or glandular differentiation, although squamous or glandular features may be detectable by ultrastructural examination. A diagnosis of LCC can only be made after the cancer has been thoroughly examined and cannot be made on biopsy or cytology samples. Previously defined histologic subtypes of LCC have been moved from this category to other categories, and the currently defined subtypes are based upon histology and immunohistochemical reactivities for TTF1, Napsin A, p63, p40, and CK5/6. Histologically, LCCs consist of sheets and nests of large cells with vesicular nuclei, prominent nucleoli, and moderate or abundant amounts of cytoplasm (Fig. 18.15).

Much of the early genetic data on LCC was derived from cases that would now be classified as AC or SqCC based upon current criteria. Cancers without immunohistochemical features justifying diagnosis as AC or SqCC (so-called marker-null LCC) exhibit *KRAS* or *EGFR* mutations in some cases, so molecular testing is recommended in this subgroup when targeted therapy is being considered.

## Neuroendocrine neoplasms and their precursors

### Clinical and pathologic features

The major categories of pulmonary neuroendocrine (NE) neoplasms include SCLC, large cell neuroendocrine carcinoma (LCNEC), typical carcinoid, and atypical carcinoid. SCLC and LCNEC are high-grade carcinomas, typical carcinoid is a low-grade malignant neoplasm, and atypical carcinoid occupies an intermediate position in the spectrum of biologic aggressiveness. In one large series, the 5-year and 10-year survival rates for typical carcinoid were 87% and 87%, 56% and 35% for atypical carcinoid, 27% and 9% for LCNEC, and 9% and 5% for SCLC, respectively. SCLC is characterized by early metastasis, and it is not uncommon for the metastasis to be discovered before the primary cancer. Almost all of these cancers can be detected as a lung mass or nodule on chest imaging typically prompted by a variety of clinical symptoms commonly including cough and shortness of breath.

Histologic features of these cancers include NE architectural features including organoid nesting, a trabecular arrangement, rosette formation, and palisading. These patterns are more prominent in carcinoids than in LCNECs and are often not visible in SCLCs, which usually consist of sheets and nests of cells. Typical carcinoids measure 5 mm or more, contain fewer than 2 mitoses per 2 mm$^2$, and lack necrosis (Fig. 18.16). Atypical carcinoids show 2−10 mitoses

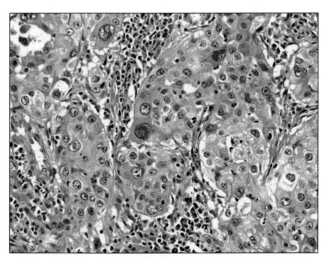

**FIGURE 18.15**  Large cell carcinoma. This poorly differentiated tumor displays large cell size with marked nuclear pleomorphism, large nucleoli, nuclear inclusions, and abundant eosinophilic cytoplasm. Evidence of squamous or glandular differentiation is not observed. The intervening stroma is inflamed.

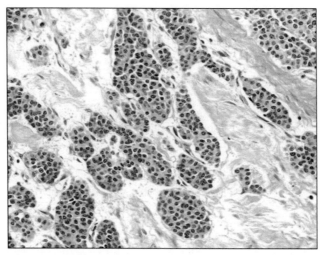

**FIGURE 18.16** Typical carcinoid. This tumor consists of nests of uniform tumor cells with round or ovoid nuclei, fine chromatin, and little nuclear cytoatypia. A moderate amount of cytoplasm is present. No necrosis or mitoses are observed. The stromal background is hyalinized.

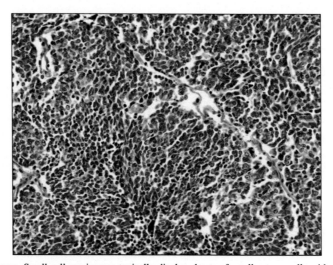

**FIGURE 18.17** Small cell carcinoma. Small cell carcinomas typically display sheets of small cancer cells with scant cytoplasm and nuclei demonstrating fine chromatin. Numerous mitoses and apoptotic cells are characteristic.

per 2 mm$^2$ and/or necrosis. SCLC have small, undifferentiated tumor cells with scant cytoplasm, finely granular chromatin, and absent or inconspicuous nucleoli (Fig. 18.17). Nuclear molding is characteristic, necrosis is common, and the mitotic rate is typically high, with a median of >80 mitoses per 2 mm$^2$. Combined SCLCs include a SCLC component accompanied by one or more histologic types of NSCLC. LCNECs consist of large tumor cells resembling those of LCC, with NE architectural patterns, necrosis, a high mitotic rate (average of 75 per 2 mm$^2$, but at least 11 per 2 mm$^2$) (Fig. 18.18), and NE differentiation reflected in immunohistochemical staining for one or more NE markers (chromogranin A, synaptophysin, or CD56 [neural cell adhesion molecule]) (Fig. 18.19) or the presence of neurosecretory granules on ultrastructural examination. Combined LCNECs are tumors that include an LCNEC component and a component of NSCLC.

Clinical differences also exist in the characteristics of patients with carcinoids compared to patients with SCLC and LCNEC. Patients with carcinoids are typically younger and less likely to smoke than those with SCLCs and LCNECs, the vast majority of whom have a current or previous history of tobacco smoking. Rare patients with carcinoids have the multiple endocrine neoplasia 1 (MEN1) syndrome, and carcinoid tumors can be associated with the preinvasive lesion, diffuse idiopathic pulmonary NE cell hyperplasia (DIPNECH). DIPNECH is a diffuse proliferation of single cells, small nodules (NE bodies), and linear proliferations of pulmonary NE cells that reside in the bronchial and/or bronchiolar epithelia and may be accompanied by extraluminal proliferations (tumorlets and carcinoids) (Fig. 18.20).

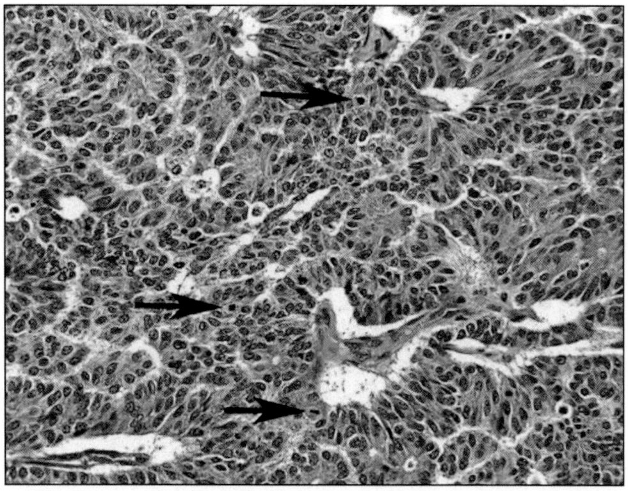

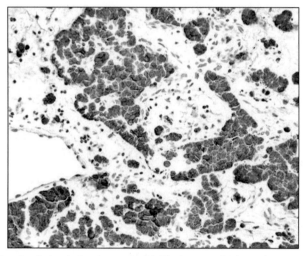

**FIGURE 18.18**  Large cell neuroendocrine carcinoma. The cancer cells form rosettes, a feature that is commonly observed in low-grade neuroendocrine tumors. Although necrosis is not present, mitoses are numerous (several indicated by *arrows*) and exceed 10 mitoses per 2 $mm^2$, justifying classification as a large cell neuroendocrine carcinoma.

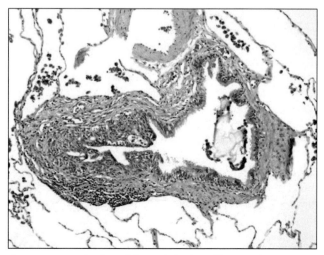

**FIGURE 18.19**  Carcinoid (immunohistochemical stain for synaptophysin). The brown-stained cytoplasm contains synaptophysin, a marker of neuroendocrine differentiation.

**FIGURE 18.20**  Diffuse idiopathic pulmonary neuroendocrine cell hyperplasia. A proliferation of neuroendocrine cells expands the epithelium of the left half of this bronchiole.

## Molecular pathogenesis

Protein markers of pulmonary NE tumors include chromogranin A, synaptophysin (Fig. 18.19), and N-CAM (CD56). These markers are expressed by all categories of NE tumors, with higher frequencies observed in the carcinoids and atypical carcinoids than in SCLC and LCNECs. Gastrin-releasing peptide, calcitonin, other peptide hormones, the insulinoma-associated 1 (*INSM1*) promotor, and the human achaete-scute homolog-1 (*hASH1*) gene have also been reported as overexpressed by these cancers. TTF-1 is expressed by 80%–90% of SCLCs, 30%–50% of LCNECs, and 0%–70% of carcinoids.

SCLC and many cases of LCNEC are driven by inactivating mutations in the *RB* and *TP53* genes. *TP53* inactivation causes genomic instability and leads to multiple sites of allelic imbalance, with losses on multiple chromosomes. SCLCs are aneuploid cancers with high frequencies of deletions on chromosomes 3p (including *ROBO1/DUTT1* [3p12.13], *FHIT* [3p14.2], *RASSF1* [3p21.3], β-*catenin* [3p21.3], *Fus1* [3p21.3], *SEMA3B* [3p21.3], *SEMA3F* [3p21.3], *VHL* [3p24.6], and *RARβ* [3p24.6]); 4q (including the proapoptotic gene *MAPK10* [4q21]); 5q; 10q (including the proapoptotic gene *TNFRSF6* [10q23]); 13q (location of the *Rb* gene); and 17p (*TP53*); and gains on 3q, 5p, 6p, 8q, 17q, 19q, and 20q. More than 90% of SCLCs and SqCCs demonstrate large, often discontinuous segments of allelic loss on chromosome 3p, in areas encompassing multiple candidate TSGs, including some of those listed previously. Atypical carcinoids show a higher frequency of LOH at 3p, 13q, 9p21, and 17p than typical carcinoids, but not as high as the high-grade NE tumors. Alterations compromising the p16^INK4a^/cyclin D1/Rb pathway of G1 arrest are consistent in high-grade pulmonary NE carcinomas, primarily through loss of Rb protein, but are less frequent in atypical carcinoids (59%) and are uncommon in typical carcinoids. Mutations in the *RB1* gene exist in many SCLCs, with associated loss of function of the gene product. The hypophosphorylated form of Rb protein functions as a cell cycle regulator for G1 arrest; cyclin D1 overexpression and P16^INK4a^ loss produce persistent hyperphosphorylation of Rb with consequent evasion of cell cycle arrest. Data also suggest that in SCLCs, overexpression of *MDM2* (a transcriptional target of p53) or p14^ARF^ loss leads to evasion of cell cycle arrest through the p53 and Rb pathway.

TSGs are inactivated in the majority of SCLCs, compared to >50% of NSCLCs, fewer atypical carcinoids, and virtually no typical carcinoids. Most of the *TP53* mutations in SCLCs are missense point mutations that result in a stabilized p53 mutant protein, which can be easily detected by immunohistochemistry. p53 protein overexpression occurs frequently in high-grade NE carcinomas, but is unusual in typical carcinoids and intermediate in atypical carcinoids. Dysregulation of p53 produces downstream effects on Bcl-2 and Bax. Antiapoptotic Bcl-2 predominates over proapoptotic Bax in the high-grade NE carcinomas, while the reverse is true for carcinoids. LCNECs resemble SCLCs in their high rates of *TP53* mutation and predominance of Bcl-2 expression over Bax expression.

Typical and atypical carcinoid tumors are often driven by mutations of the *MEN1* gene family on chromosome 11q13 and related chromatin modifier gene mutations, while these abnormalities occur with lower frequencies in SCLCs and LCNECs, supporting separate pathways of tumorigenesis. *MEN1* encodes for the nuclear protein menin, which is believed to play several roles in tumorigenesis by linking transcription factor function to histone-modification pathways, in part through interacting with the activator-protein-1 family transcription factor JunD, modifying it from an oncoprotein into a tumor suppressor protein. *MEN1* mutations were reported in 40% of carcinoids in patients without MEN1 disease, more often in atypical than typical carcinoids. About 20% of atypical carcinoids show mutation in *TP53* or *RB1*, with inactivation.

Integrative genome analyses have confirmed many of these findings and shed additional light on key somatic driver mutations of SCLC. SCLC have an extremely high mutation rate of $7.4 \pm 1$ protein-changing mutations per million base pairs, evidence for inactivation of *TP53* and *RB1*, and amplifications of *MYC* family members *MYCL1*, *MYCN*, and *MYC* occurred in 16% of cases. Recurrent mutations have been identified in the *CREBBP*, *EP300*, and *MLL* genes that encode histone modifiers, making this the second most frequently mutated class of genes in SCLC. Additional mutations in *PTEN, SLIT2,* and *EPHA7*, and focal amplifications of the *FGFR1* tyrosine kinase gene have also been observed. These cancers also contained the tobacco carcinogen–associated molecular signature, with a large number of $G \rightarrow T$ transversions caused by polycyclic aromatic hydrocarbons, often occurring at methylated CpG dinucleotides.

Next-generation sequencing technologies have revealed numerous mutated genes in SCLC, including genes encoding kinases, G protein-coupled receptors, and chromatin-modifying proteins. Several members of the SOX family of genes were mutated in SCLC, with *SOX2* amplification in about 27% of cases. Multiple fusion transcripts and a recurrent *RLF-MYCL1* fusion were found, In addition to known hotspots in *TP53*, *RB1*, *PIK3CA*, *CDKN2A*, and *PTEN*, new hotspot mutations included genes encoding Ras family regulators (*RAB37, RASGRF1,* and *RASGRF2*), chromatin-modifying enzymes or transcriptional regulators (*EP300, DMBX1, MLL2, MED12L, TRRAP,* and *RUNX1T1*), ionotropic glutamate receptor (*GRID1*), kinases (*STK38, LRRK2, PRKD3,* and *CDK14*), protein phosphatases (*PTPRD* and

*PPEF2*), and G protein−coupled receptors (*GPR55, GPR113*, and *GPR133*). Mutations clustering in particular gene families and pathways were found in the phosphatidylinositol 3-kinase (PI3K) pathway (*PIK3CA, AKT1−3, MTOR, RPS6KA2*, and *RPS6KA6*), the mediator complex (*MED12, MED12L, MED13, MED13 L, MED15, MED24, MED25, MED27*, and *MED29*), Notch and Hedgehog family members (*NOTCH1, NOTCH2, NOTCH3*, and *SMO*), glutamate receptor family members (*GRIA1, GRIA2, GRIA3, GRIA4, GRIND1, GRID2, GRM1−3, GRM5, GRM7*, and *GRM8*), SOX family members (*SOX3, SOX4, SOX5, SOX6, SOX9, SOX11, SOX14*, and *SOX17*) and DNA repair and/or checkpoint pathway genes (*ATM, ATR, CHEK1*, and *CHEK2*). *KRAS* mutation was not identified. Among the receptor tyrosine kinase genes, mutations were discovered in *FLT1, FLT4, KDR*, and *KIT* and members of the Ephrin family (*EPHA1-7* and *EPHB4*). Lastly, chromosome copy number analysis revealed recurrent copy gains (*MYC, SOX4*, and *KIT*) and losses (*RB1, RASSF1, FHIT, KIF2A*, and *CNTN3*).

Next-generation sequencing studies of 45 cases of LCNEC revealed three genomically distinct subsets with features of SCLC, NSCLC (predominantly AC), and occasionally highly proliferative carcinoid. Common genetic alterations included *TP53* (78%), *RB1* (38%), *STK11* (33%), *KEAP1* (31%), and *KRAS* (22%). Genomic profiles divided LCNEC into two major subsets and one minor subset: SCLC-like, with *TP53* + *RB1* co-mutation/loss and other SCLC-type alterations (*MYCL* amplification, *SOX2* amplification, *PTEN* mutation/loss, *FGFR1* amplification); NSCLC-like, characterized by a lack of co-altered *TP53* + *RB1* and almost universal occurrence of NSCLC- type mutations (*STK11, KRAS, KEAP1*); and carcinoid-like (n=2), showing *MEN1* mutations and low mutation burden. The NSCLC-like subset revealed low-level napsin A staining (AC-like), which was not observed in the SCLC-like group, while NE marker expression was similar in the two groups. However, the NSCLC-like LCNECs harbored more frequent mutations in *NOTCH* family genes (28%), implicated as key regulators of NE differentiation. Although no sensitizing *EGFR* mutations or *ALK* rearrangements were identified, at least one alteration that was potentially targetable by investigational agents was present in 67% of cases, more frequently in the NSCLC-like subset than the SCLC-like subset.

Pulmonary carcinoids were studied using gene copy number analysis, genome/exome and transcriptome sequencing. A mean somatic mutation rate of 0.4 mutations per megabase was determined and a smoking-related mutation signature was absent. Frequent mutations in chromatin-remodeling genes were observed including mutations of covalent histone modifiers and subunits of the SWI/SNF complex in 40% and 22.2% of the cases, respectively, with *MEN1, PSIP1*, and *ARID1A* being recurrently affected. In contrast to SCLC and LCNEC, *TP53* and *RB1* mutations were rare in typical carcinoids, suggesting that they develop via different molecular pathways than the high-grade pulmonary NE neoplasms.

Aberrant methylation of cytosine-guanine (CpG) islands in promoter regions of malignant cells is an important mechanism for silencing of TSGs (epigenetic inactivation). Methylation of DNA involves the transfer of a methyl group, by a DNA methyltransferase enzyme, to the cytosine of a CpG dinucleotide. *RASSF1A* is a potential TSG that undergoes epigenetic inactivation in virtually all SCLCs and a majority of NSCLCs through hypermethylation of its promoter region. NE tumors have lower frequencies of methylation of *p16, APC*, and *CDH13* (H-cadherin) than NSCLCs, and SCLCs have higher frequencies of methylation of *RASSF1A, CDH1* (E-cadherin), and *RARβ* than carcinoids. Promoter methylation of *CASP8*, which encodes the apoptosis-inducing cysteine protease caspase 8, was also found in 35% of SCLCs, 18% of carcinoids, and no NSCLCs, suggesting that *CASP8* may function as a TSG in NE lung tumors.

## Mesenchymal neoplasms

Mesenchymal neoplasms in the WHO classification scheme (Table 18.1) encompass a spectrum of malignant and benign proliferations that differentiate along multiple lineages. These tumors are much less common in the lung than are epithelial neoplasms and less information about their molecular pathogenesis is known... Pulmonary inflammatory myofibroblastic tumor (IMT) is composed of myofibroblastic cells, collagen, and inflammatory cells and primarily occurs in individuals less than 40 years of age and is the most common endobronchial mesenchymal lesion in childhood (Fig. 18.21). IMTs demonstrate clonal abnormalities with rearrangements of chromosome 2p23 and the *ALK* gene. The rearrangements involve fusion of tropomyosin (TPM) N-terminal coiled-coil domains to the ALK C-terminal kinase domain, producing two ALK fusion genes, *TPM4-ALK* and *TPM3-ALK*, which encode oncoproteins with constitutive kinase activity. Synovial sarcoma, usually a soft tissue malignancy, uncommonly arises in the pleura or the lung and often takes an aggressive course. Like their soft tissue counterparts, more than 90% of pulmonary and pleural synovial sarcomas demonstrate a chromosomal translocation t(X;18)(*SYT-SSX*) and detection of this translocation can be very helpful in confirming the diagnosis of synovial sarcoma in this unusual location. Pulmonary hamartomas are benign neoplasms consisting of mixtures of cartilage, fat, connective tissue, and smooth muscle, which present as coin lesions on chest radiographs and are excised in order to rule out a malignancy (Fig. 18.22). Many pulmonary hamartomas have

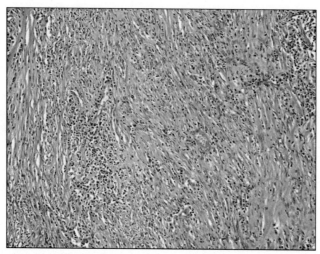

FIGURE 18.21 Inflammatory myofibroblastic tumor. The tumor consists of a proliferation of cytologically bland spindle cells in a background of collagen, with abundant lymphocytes and plasma cells.

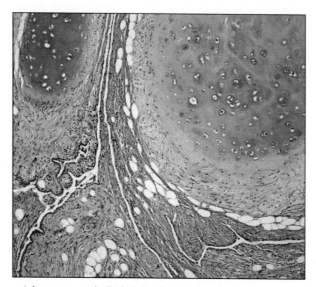

FIGURE 18.22 Pulmonary hamartoma. A hamartoma typically includes the components of mature cartilage, adipose tissue, and myxoid or fibrous tissue, all of which are shown here.

the translocation t(3;12)(q27-28;q14-15), representing a gene fusion of the high mobility group protein gene *HMGA2* and the *LPP* gene. HMG proteins are a family of nonhistone chromatin-associated proteins that serve an important role in regulating chromatin architecture and gene expression. Epithelioid hemangioendothelioma is an uncommon low-grade or intermediate-grade vascular tumor arising in the lung or in other anatomic sites, which is associated with a characteristic t(1;3)(p36.3;q25) translocation. Two genes are involved in this translocation: *WWTR1* (3q25), encoding a transcriptional coactivator expressed in endothelial cells, and *CAMTA1* (1p36), a DNA-binding transcriptional regulatory protein expressed during brain development. The *YAP1-TFE3* fusion has also been found in some epithelioid hemangioendotheliomas in young adults. Pulmonary myxoid sarcoma with *EWSR1-CREB1* translocation is a rare malignant tumor arising in the airways, and *EWSR1* gene rearrangements can also occur in pulmonary myoepithelial tumors.

Perivascular epithelioid cells, which have features of modified smooth muscle cells, give rise to lymphangioleiomyomatosis (LAM) and PEComas. LAM presents as an interstitial lung disease in women in their reproductive years, and PEComas typically form lung nodules and extrapulmonary nodules. Women with LAM develop worsening shortness of breath and can experience pneumothorax, pleural effusions, and pulmonary hemorrhages. LAM consist of spindle-shaped myoid cells with eosinophilic or clear cytoplasm, and often form bundles in the walls of cystically-enlarged air spaces. These cells coexpress muscle (smooth muscle actin) and melanocytic (HMB45, melan A, and microphthalmia transcription factor) markers, and often express estrogen and progesterone receptors.

LAM may occur as a sporadic disease or in association with tuberous sclerosis complex (TSC). TSC has been linked to a germline mutation affecting the TSGs TSC1 on chromosome 9q34 (which encodes hamartin) and TSC2 on chromosome 16p13 (which encodes tuberin). TSC-LAM is believed to occur as a result of a somatic mutation (second hit) occurring in addition to a germline mutation in TSC1 or TSC2 (first hit), and sporadic LAM is believed to develop via a two-hit model involving a somatic mutation and/or loss of heterozygosity in TSC2. Inactivating mutations in the TSC1 and TSC2 genes cause proliferation of LAM cells via constitutive activation of the mTOR pathway, a signaling pathway that participates in cell growth, proliferation, and metabolism, and is activated in many cancers.

Recently, LAM has been shown to be a low-grade metastasizing neoplasm with clonal cell that have identical TSC mutations in multiple lesions from different sites in individual patients, suggesting a common origin. LAM recurs in transplanted lungs, and the TSC mutations found in the recurrence match those in the host, suggesting that the pulmonary lesions are metastases. Women with pulmonary LAM have a high frequency of gynecologic LAM, and LAM involving pelvic and retroperitoneal lymph nodes, suggesting that LAM may originate in these regions. The mediastinum can also be involved and circulating LAM cells can be detected in many women with LAM. mTOR inhibitors have shown efficacy in stabilizing lung function, reducing chylous effusions and decreasing circulating LAM cells in affected individuals.

## Pleural malignant mesothelioma

### Clinical and pathologic features

Malignant mesothelioma (MM) is an uncommon, aggressive cancer arising from mesothelial cells on serosal surfaces, primarily the pleura and peritoneum, and less often the pericardium or tunica vaginalis. The most important risk factor for MM is exposure to the subset of asbestos fibers known as amphiboles (crocidolite and amosite). The incidence of this tumor in the United States peaked in the early to mid-1990s and is declining, likely related to decreases in the use of amphiboles since their peak period of use in the 1960s. These tumors are characterized by long (30–40 year) latency periods between asbestos exposure and clinical presentation of the tumor. Radiation, a nonasbestos fiber known as erionite, and potentially other processes associated with pleural scarring have also been implicated in the causation of smaller numbers of cases of malignant mesothelioma.

Pleural MM most commonly arises in males over the age of 60 years. Presenting features typically include a hemorrhagic pleural effusion associated with shortness of breath and chest wall pain. Weight loss and malaise are common. At presentation, patients usually have extensive involvement of the pleural surfaces and, with progression, the tumor invades the lung, chest wall, and diaphragm. Lymph node metastasis can cause superior vena caval obstruction and cardiac tamponade, subcutaneous nodules, and contralateral lung involvement can occur. From diagnosis, survival is usually about 12 months. Treatment may include surgery, chemotherapy, radiotherapy, immunotherapy, or other treatments, often in combination. The intent of therapy is to palliate and ideally to extend life.

Gross features of MM include pleural nodules, which grow and coalesce to fill the pleural cavity and form a thick, tan rind around the lung that can have a gelatinous consistency (Fig. 18.23). Extension along the interlobar fissures and invasion into the adjacent lung, diaphragm, and chest wall are characteristic. Further spread can occur into the pericardial cavity and around other mediastinal structures, and distant metastases can develop.

The major histologic categories of diffuse malignant mesotheliomas recognized by the WHO are epithelioid, sarcomatoid (including desmoplastic), and biphasic mesotheliomas. Epithelioid mesothelioma consists of round, ovoid, or polygonal cells with eosinophilic cytoplasm and round nuclei with little cytoatypia (Fig. 18.24). These cells often form sheets, tubulopapillary structures, or gland-like arrangements, and can have a myxoid appearance from production of large amounts of hyaluronate. Sarcomatoid mesothelioma is composed of malignant spindle cells occasionally accompanied by mature sarcomatous components (osteosarcoma, chondrosarcoma, others). Desmoplastic mesothelioma can be a diagnostic challenge. Helpful features for separating this tumor from organizing pleuritis include invasion of chest wall muscle or adipose tissue and necrosis. (Fig. 18.25). Biphasic mesotheliomas include both epithelioid and sarcomatoid elements, each comprising at least 10% of the tumor.

Pathologic diagnosis of MM has been greatly assisted by the expanded availability of antibodies for use in immunohistochemistry. Mesothelial differentiation can be supported by immunoreactivity for cytokeratin 5/6, calretinin

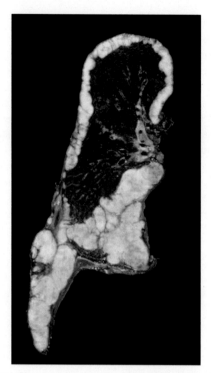

**FIGURE 18.23** Malignant mesothelioma. The tan/white tumor involves the entire pleura surrounding and compressing the underlying parenchyma, which appears congested but relatively unremarkable.

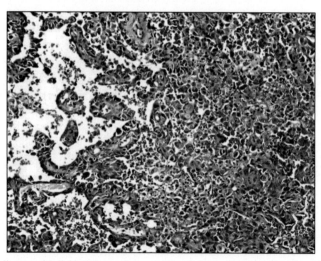

**FIGURE 18.24** Malignant mesothelioma, epithelioid. This neoplasm consists of sheets of polygonal cells with pleomorphic nuclei and also forms some papillary structures (left).

(Fig. 18.26), HBME-1, D2-40, and other antibodies. Additionally, in biopsies, although histologic differentiation of mesotheliomas from reactive mesothelial proliferations can represent a challenge, newer diagnostic approaches (*p16* FISH and BAP1 immunohistochemistry) can offer assistance with classification of the process. *p16* loss detected by FISH and loss of BAP1 expression as assessed by immunohistochemistry are observed in many MMs, but not in benign mesothelial disorders.

## Molecular pathogenesis

Exposure to asbestos fibers is believed to trigger the pathobiological changes leading to the majority of MMs. Currently, it is believed that asbestos causes genetic and cellular damage to mesothelial cells via reactive oxygen

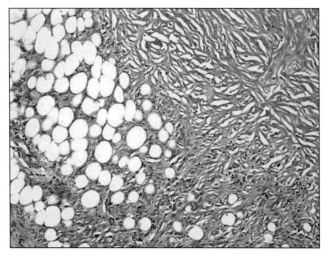

FIGURE 18.25 Malignant mesothelioma, desmoplastic. Abundant dense collagen is characteristic of this cancer and is shown in the upper right. Neoplastic cells are spindle shaped and relatively cytologically bland. The slitlike spaces observed in the dense collagen are another frequent feature. The tumor infiltrates adipose tissue, which is helpful in confirming that the cancer is a mesothelioma, as opposed to organizing pleuritis.

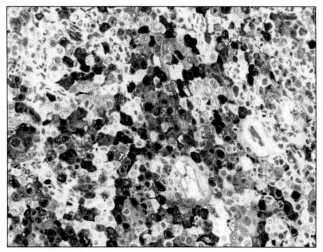

FIGURE 18.26 Malignant mesothelioma (immunohistochemical stain for calretinin). The cancer demonstrates cytoplasmic and nuclear staining (brown) for calretinin, which is expressed by many epithelioid malignant mesotheliomas.

species, disruption of the mitotic process, and possibly exposure of cells to adsorbed carcinogens. Accumulated data has revealed a number of important genetic alterations that are responsible for the development and progression of MM. Losses and gains of chromosomal regions are common in MMs. Frequent areas of loss are found on chromosome arms 1p, 3p, 4q, 6q, 9p, 13q, 14q, and 22q, and gains involve chromosome arms 1q, 5p, 7p, 8q, and 17q. The cyclin-dependent kinase inhibitor 2A/alternative reading frame (CDKN2A/ARF), neurofibromatosis type 2 (NF2), and *BRCA1-associated protein-1* (*BAP1*) genes are the most frequently mutated TSGs identified in MM. *CDKN2A/ARF* lies on chromosome 9p21.3 and *CDKN2A* encodes p16$^{INK4a}$ while ARF encodes p14$^{ARF}$. p16$^{INK4a}$ controls the cell cycle via the cyclin-dependent kinase 4/cyclin D retinoblastoma protein pathway, and p14$^{ARF}$ regulates p53 through inactivation of the human ortholog of mouse double minute 2, an upstream regulator of p53. As a consequence, the homozygous deletion of *CDKN2A/ARF*, found in about 70% of MMs, means that the tumor suppressing pathways of retinoblastoma and p53 are inactivated. Merlin, a tumor suppressor protein encoded by *NF2*, regulates multiple cell signaling pathways including the Hippo and mTOR, which regulate cell proliferation and growth. About 40%−50% of MMs harbor an inactivating mutation of *NF2*. BAP1 participates in histone modification and its inactivation induces the disturbance of global gene expression profiling. *BAP1*, which is on chromosome 3p21.1, is mutated in about

23%−63% of MMs. Targeted next-generation sequencing of advanced stage malignant pleural mesotheliomas revealed a complex mutational landscape with a higher number of genetic variations in the p53/DNA repair and phosphatidylinositol 3-kinase pathways. However, in MMs, unlike many other epithelial tumors, mutations in the *RB*, *EGFR*, and *RAS* genes are rare. Promoter methylation of TSGs is also found in MMs, suggesting that epigenetic inactivation of these genes (*E-cadherin, fragile histidine triad, retinoic acid receptor-β,* and *wnt inhibitory factor-1*) could contribute to tumor development and progression.

Receptor tyrosine kinase activation does occur and upregulates the Raf-MEK-extracellular signal-regulated kinase and PI3K-AKT pathways that are important for proliferation and/or survival of cells. The Wnt signal transduction pathway is also abnormally activated in MMs and appears to play a role in pathogenesis. Activation of the pathway leads to accumulation of β-catenin in the cytoplasm and its translocation to the nucleus. Interactions with TCF/LEF transcription factors promote expression of multiple genes including c-*myc* and *Cyclin D*. The mechanism of activation does not appear to involve mutations in the β-*catenin* gene, but may instead involve more upstream components of the pathway, such as the disheveled proteins. Evidence also suggests that the phosphatidylinositol 3-kinase (PI3-K/AKT) pathway is frequently activated in MMs, and that inhibition of this pathway can increase sensitivity to a chemotherapeutic agent.

# Non-neoplastic lung disease

## Chronic obstructive lung disease—emphysema/chronic bronchitis

### Clinical and pathologic features

The term chronic obstructive pulmonary disease (COPD) applies to emphysema, chronic bronchitis, and bronchiectasis, diseases in which airflow limitation is usually progressive, but unlike asthma, not fully reversible. The prevalence of COPD worldwide is estimated at 9%−10% in adults over the age of 40 years. The most common forms of the disease have similar clinical manifestations, including a progressive decline in lung function, a chronic cough and dyspnea. Emphysema and chronic bronchitis, result from cigarette smoking and usually exist together in most smokers. Chronic bronchitis is defined clinically as a persistent cough with sputum production for at least 3 months in at least two consecutive years without any other identifiable cause. Patients with chronic bronchitis typically have copious sputum with a prominent cough, more commonly get infections, and typically experience hypercapnia and severe hypoxemia, giving rise to the clinical moniker blue bloater. Emphysema is the destruction and permanent enlargement of the air spaces distal to the terminal bronchioles without obvious fibrosis. These patients have only a slight cough, while the overinflation of the lungs is severe, inspiring the term pink puffers.

The pathologic features of COPD are best understood if one considers the whole of COPD as a spectrum of pathology that consists of emphysematous tissue destruction, airway inflammation, remodeling, and obstruction. The pathologic features of chronic bronchitis include mucosal pathology that consists of epithelial inflammation, injury, regenerative epithelial changes of squamous and goblet cell metaplasia and the submucosa has smooth muscle hypertrophy and submucosal gland hyperplasia.

The pathology definition of emphysema is an abnormal, permanent enlargement of the airspaces distal to the terminal bronchioles accompanied by destruction of the alveolar walls without fibrosis. The four major pathologic patterns of emphysema are defined by the location of this destruction: centriacinar, panacinar, paraseptal, and irregular emphysema. The first two types are responsible for the overwhelming majority of the clinical disease. Centriacinar emphysema (sometimes referred to as centrilobular) represents 95% of the cases and is a result of destruction of alveoli at the proximal and central areas of the pulmonary acinus, including the respiratory bronchioles (Fig. 18.27). It predominantly affects the upper lobes (Fig. 18.28). Panacinar emphysema, usually associated with α1-antitrypsin (αAT) deficiency, results in a destruction of the entire pulmonary acinus from the proximal respiratory bronchioles to the distal area of the acinus and affects predominantly the lower lobes (Fig. 18.29). The remaining two types of emphysema, paraseptal and irregular, are rarely associated with clinical disease. In paraseptal emphysema, the damage is to the distal acinus and this proximity to the pleura may cause spontaneous pneumothoraces, typically in young, thin men. Irregular emphysema is tissue destruction and alveolar enlargement that occurs adjacent to scarring, secondary to the enhanced inflammation in the area. Though this is a common finding in a scarred lung, it is of little if any clinical significance to the patient.

Respiratory bronchiolitis refers to the inflammatory changes found in the distal airways of smokers with emphysema. These consist of pigmented macrophages filling the lumen and the peribronchiolar airspaces and chronic inflammation and fibrosis around the bronchioles (Fig. 18.30). The pigment represents the inhaled particulate matter of the cigarette smoke phagocytized by these cells. The macrophages release proteases, which destroy the elastic fibers in the surrounding area, resulting in the loss of elastic recoil and the obstructive symptoms.

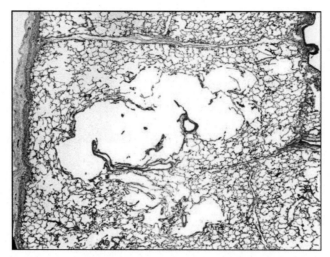

FIGURE 18.27 Centrilobular emphysema. Tissue destruction in the central area of the pulmonary lobule is demonstrated in this lung with mild centrilobular emphysema. The pattern of tissue destruction is in the area surrounding the small airway where pigmented macrophages release proteases in response to the cigarette smoke.

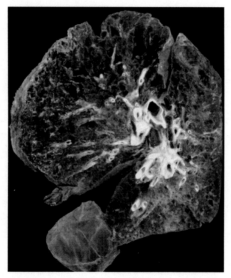

FIGURE 18.28 Centrilobular emphysema. This sagittal cut lung tissue section contains severe centrilobular emphysema with significant tissue destruction in the upper lobe and bulla forming in the upper and lower lobes.

## Molecular pathogenesis

COPD is a result of inflammation of the large airways that produces the airway remodeling characteristic of chronic bronchitis and the inflammation of the smaller airways that results in the destruction of the adjacent tissue and consequent emphysema. The predominant inflammatory cells involved in this process are the alveolar macrophages, neutrophils, and lymphocytes. The main theories of the pathogenesis of COPD support the interaction of airway inflammation with two main systems in the lung: the protease-antiprotease system and the oxidant-antioxidant system. These systems help to protect the lung from the many irritants that enter the lung via the large pulmonary surface area that interfaces with the environment.

The protease-antiprotease system, proteases are produced by a number of cells, including epithelial cells and inflammatory cells that degrade the underlying lung matrix. The most important proteases in the lung are the neutrophil elastases, part of the serine protease family, and the matrix metalloproteinases (MMPs) produced predominantly by macrophages. These proteases can be secreted in response to invasion by environmental irritants, most notably infectious agents such as bacteria. In this setting, their role is to enzymatically degrade the organism. Proteases can also be

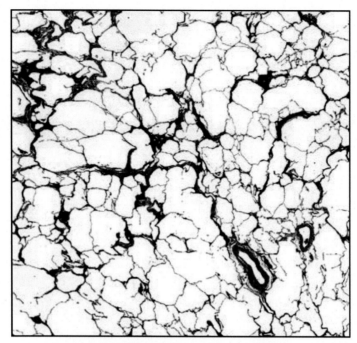

**FIGURE 18.29**  Panacinar emphysema. Tissue destruction in panacinar emphysema occurs throughout the lobule, producing a diffuse loss of alveolar walls unlike that of centrilobular emphysema with more irregular holes in the tissue.

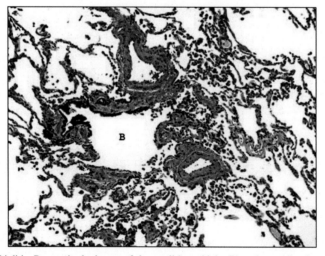

**FIGURE 18.30**  Respiratory bronchiolitis. Present in the lumen of the small bronchiole (B) and extending into the surrounding alveolar spaces are pigmented macrophages in a lung from a smoker. The pigment in these macrophages represents particulates from the cigarette smoke and stimulates the release of the proteases that are responsible for the tissue destruction in centrilobular emphysema.

secreted by both inflammatory and epithelial cells in a normal lung to repair and maintain the underlying lung matrix proteins. To protect the lung from unwanted destruction by these enzymes, the liver secretes antiproteases that circulate in the bloodstream to the lung and inhibit the action of the proteases. In addition, macrophages that secrete MMPs also secrete tissue inhibitors of metalloproteinases (TIMPs). A delicate balance of proteases and antiproteases is needed to maintain the integrity of the lung structure as an imbalance results in a relative excess of proteases (either by overproduction of proteases or underproduction of their inhibitors) leads to tissue destruction and the formation of emphysema.

In centriacinar emphysema, caused primarily by cigarette smoking, there is an overproduction of proteases primarily due to the stimulatory effect of chemicals in the smoke on the neutrophils and macrophages. Though the exact mechanism is not completely understood, most studies support that nicotine from the cigarette smoke acts as a chemoattractant, and ROS contained in the smoke stimulate an increased release of neutrophil elastases and MMPs from activated macrophages, leading to the destruction of the elastin in the alveolar spaces. This inflammatory cell activation may come about through the activation of the transcription factor NF-kB leading to TNFα production and the elastin peptides themselves may attract additional inflammatory cells to further increase the protease secretion and exacerbate the matrix destruction.

Pan-acinar emphysema is most commonly caused by a genetic deficiency of antiproteases, usually due to alpha-1-antitrypsin (αAT) deficiency, a condition that affects approximately 1 in 2000 to 1 in 5000 individuals. αAT deficiency is due to a defect in the gene that encodes the protein αAT, a glycoprotein produced by hepatocytes and the main inhibitor of neutrophil elastase. The affected gene, *SERPINA1*, is located on the long arm of chromosome 14 (14q31−32.3). Genetic mutations that occur have been categorized into four groups: base substitution, in-frame deletions, frame-shift mutations, and exon deletions. These mutations usually result in misfolding, polymerization, and retention of the aberrant protein within the hepatocytes, leading to decreased circulating levels. αAT deficiency, an autosomal codominant disease with approximately 120 variant alleles has phenotypes that are classified by a PI (for protease inhibitor) coding system with the names of the inherited alleles denoting the migration of the molecular in the isoelectric pH gradient from A for anodal variants to Z for slower migrating variants. PI*MM is for individuals homozygous for the normal M alleles, and PI*ZZ is for individuals homozygous for the Z allele. The Z allele is most commonly responsible for severe deficiency and disease. Z-type AAT molecules polymerize within the hepatocytes, stopping their secretion into the blood resulting in low serum αAT levels. Individuals who manifest the lung disease are usually homozygous for the alleles Z or S (ZZ and SS phenotype) or heterozygous for the 2M alleles (MZ or MS phenotype). αAT is the prototypic member of the serine protease inhibitor superfamily of proteins. Retention of these compounds in the liver leads to a loss of the natural antiprotease screen against neutrophil elastase in the lung and the loss of the antiinflammatory effects of AAT. An αAT concentration in plasma of less than 40% of normal confers a risk for emphysema. In individuals with the ZZ genotype, the activity of αAT is approximately one-fifth of normal.

The second system in the lung involved in the pathogenesis of emphysema is the oxidant−antioxidant system. The lung is protected from oxidative stress in the form of ROS by antioxidants produced by cells in the lung. ROS in the lung include oxygen ions, free radicals, and peroxides. The major antioxidants in the airways are enzymes including catalase, superoxide dismutase, glutathione peroxidase, glutathione S-transferase, xanthine oxidase, and thioredoxin, as well as nonenzymatic antioxidants including glutathione, ascorbate, urate, and bilirubin. The balance of oxidants and antioxidants in the lung prevents damage by ROS. However, cigarette smoke increases the production of ROS by neutrophils, eosinophils, macrophages, and epithelial cells. Further, ROS may induce a proinflammatory response that recruits more inflammatory cells to the lung. In animal models, cigarette smoke induces the expression of proinflammatory cytokines such as IL-6, IL-8, TNFα, and IL-1 from macrophages, epithelial cells, and fibroblasts, perhaps through activation of the transcription factor NF-kB. Epigenetic dysregulation may contribute to excessive activation of these proinflammatory cytokines and chemokines, especially in chronic smokers, leading to an erosion of the lung to repair this damage. The antioxidant transcription factor Nrf2 controls more than 100 of the genes involved in antioxidant defenses and may be the focus of future therapies to inhibit destructive effects of cigarette smoke.

## Bronchiectasis

### Clinical and pathologic features

Bronchiectasis represents the permanent remodeling and dilatation of the large airways of the lung most commonly due to chronic inflammation and recurrent pneumonia. This pathology dictates the clinical features of the disease, which include chronic cough with copious secretions and a history of recurrent pneumonia. The five major causes of bronchiectasis are infection, obstruction, impaired mucociliary defenses, impaired systemic immune defenses, and congenital. These may produce either a localized or diffuse form of the disease. Localized bronchiectasis is usually due to obstruction of airways by mass lesions or scars from previous injury or infection. Diffuse bronchiectasis can result from defects in systemic immune defenses in which either innate or adaptive immunity may be impaired. Diseases due to the former include chronic granulomatous disease, and diseases due to the latter include agammaglobulinemia/hypogammaglobulinemia and severe combined immune deficiencies. Defects in the mucociliary defense mechanism cause diffuse bronchiectasis. These include ciliary dyskinesias that result in cilia with aberrant ultrastructure and cystic fibrosis (CF). Congenital forms include Mounier-Kuhn syndrome and Williams-Campbell syndrome, the former causing enlargement of the trachea and major bronchi due to loss of bronchial cartilage, and the latter causing diffuse bronchiectasis of the major airways probably due to a genetic defect in the connective tissue.

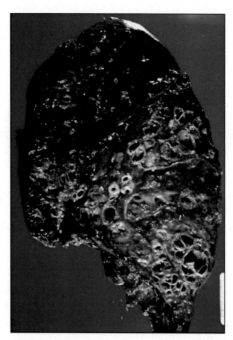

FIGURE 18.31    Cystic fibrosis. This sagittal cut section of a lung from a patient with cystic fibrosis demonstrates a diffuse bronchiectasis illustrated by enlarged, cystlike airways. This is the typical pathology for cystic fibrosis. The remainder of the lung contains some red areas of congestion.

The pathology of bronchiectasis is dilated airways containing infected secretions and mucous plugs localized either to a segment of the lung or diffusely involving the entire lung as in CF (Fig. 18.31). Bacteria may be found in these plugs, most notably *Pseudomonas aeruginosa*. Microscopic features include chronic inflammatory changes with ulceration of the mucosa and submucosa leading to destruction of the smooth muscle, and elastin in the airway wall.

*Molecular pathogenesis*

The pathogenic mechanism of bronchiectasis is complex and depends on the underlying etiology. In general, the initial damage to the bronchial epithelium is due to aberrant mucin (CF), dysfunctional cilia (ciliary dyskinesias), and ineffective immune surveillance (defects in innate and antibody-mediated immunity), leading to a cycle of tissue injury, repair, and remodeling that ultimately destroys the normal airway. Ciliary defects are found in primary ciliary dyskinesia, a genetically heterogeneous disorder, usually inherited as an autosomal recessive trait though cases of autosomal-dominant or X-linked inheritance have also been reported. These genetic defects produce immotile cilia with clinical manifestations in the lungs, sinuses, middle ear, male fertility, and organ lateralization. Over 250 proteins make up the axoneme of the cilia. Recent sequencing resulted in identification of primary ciliary dyskinesia-causing mutations in 30 genes. Of these, mutations in two genes, *DNAH5* (15%−21% prevalence) and *DNAI1* (2%−9% prevalence), which encode for proteins in the outer dynein arms, most frequently cause this disorder. In 30% of patients, the ciliary structure may be normal, requiring a genetic analysis for diagnosis. In these cases, hot spot screening of these two common genes may be the most efficient method of diagnosis.

Cystic fibrosis is an autosomal recessive genetic disorder that predominantly affects the mucin where there is a low volume of airway surface liquid causing sticky mucin that inhibits normal ciliary motion and effective mucociliary clearance of organisms. This is due to a defect in the CF transmembrane conductance regulator (*CFTR*) gene, located on chromosome 7 that encodes a cAMP-activated channel. This channel regulates the flow of chloride ions in and out of cells and intracellular vacuoles, helping to maintain the osmolality of the mucin. This protein is present predominantly on the apical membrane of the airway epithelial cells, though it is also involved in considerable subapical, intracellular trafficking, and recycling during the course of its maturation within these cells.

The genetic mutations in CF influence the CFTR trafficking in the distal compartments of the protein secretary pathway, and various genetic mutations produce different clinical phenotypes of the disease. Most mutations of the CFTR gene are missense alterations, and approximately 15% of the identified genetic variants are not associated with

**TABLE 18.4** Classes of CFTR mutations.

| Class | I Synthesis | II Processing | III Gating | IV Conductance | V Low Synthesis | VI Turnover |
|---|---|---|---|---|---|---|
| CFTR Defect | No functional CFTR protein | CFTR trafficking defect | Defective channel gating or regulation | Decreased channel conductance | Reduced synthesis of CFTR | Decreased CFTR stability |
| Type of mutation | Nonsense; frameshift; canonical splice | Missense; aminoacid deletion | Missense; aminoacid change | Missense; aminoacid change | Splicing defect; missense | Missense; aminoacid change |
| Cellular Defect | Unstable truncated RNA | Retension of misfolded protein at the endoplasmic reticulum and degradation in the proteasome | Defect in channel opening | Defects in the CFTR channel | Promoter or splicing defects | Reduced CFTR stability at the cell surface |

CFTR, cystic fibrosis transmembrane conductance regulator.
Adapted from Elborn JS. Cystic fibrosis. Lancet 2016;388:2519–31.

clinical disease. These mutations are divided into six classes. Classes I, II, and III mutations are associated with no residual CFTR function and these patients have a severe phenotype. Classes IV, V, and VI mutations have some residual function of the CFTR protein and have mild lung clinical symptoms and pancreatic insufficiency (Table 18.4).

# Interstitial lung diseases

## Idiopathic interstitial pneumonias—usual interstitial pneumonia

### Clinical and pathology features

The idiopathic interstitial pneumonias (IIPs) comprise a group of diffuse infiltrative pulmonary diseases with a similar clinical presentation characterized by dyspnea, restrictive physiology, and bilateral interstitial infiltrates on chest radiography. Pathologically, these diseases have characteristic patterns of tissue injury with chronic inflammation and varying amounts of fibrosis.

The pathologic classification of these diseases, originally defined by Liebow and Carrington in 1969, has undergone important revisions over the past 35 years with the latest revision by the *American Thoracic Society/European Respiratory Society* in 2002 and a subsequent update of this revision in 2013. The best known and most prevalent entity of the IIPs is idiopathic pulmonary fibrosis (IPF), with the current prevalence estimates of approximately 50 per 100,000, most of whom progress to respiratory failure and death within 5 years. IPF is known pathologically as usual interstitial pneumonia (UIP). UIP is a histologic pattern characterized by patchy areas of chronic lymphocytic inflammation with organizing and collagenous type fibrosis. Imaging studies usually reveal bilateral, basilar disease, with a reticular pattern. New anti-fibrolytic therapies have recently been introduced that stall disease progression but to date there is no evidence of regression or disease cure.

The pathology is characterized by a leading edge of chronic inflammation with fibroblastic foci that begin in different areas of the lung at different times. These processes produce a variegated pattern of fibrosis, usually referred to as a temporally heterogenous pattern of injury. Because it occurs predominantly in the periphery of the lung involving the subpleura and interlobular septae, the gross picture is one of more advanced peripheral and basilar disease (Fig. 18.32). The progression from inflammation to fibrosis includes interstitial widening, epithelial injury and sloughing, fibroblastic infiltration, and organizing fibrosis within the characteristic fibroblastic foci. Deposition of collagen by fibroblasts occurs in the latter stages of repair. This remodeling produces mucous-filled ectatic spaces giving rise to the gross picture of honeycomb spaces, which is seen in the advanced pathology (Fig. 18.32).

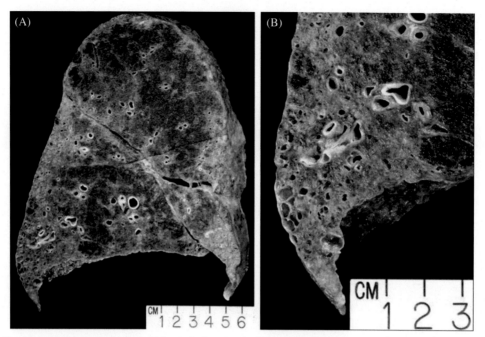

**FIGURE 18.32**    Usual interstitial pneumonia. A sagittal cut of a lung involved by usual interstitial pneumonia reveals the peripheral and basilar predominance of the dense, white fibrosis (A). A higher power view of the left lower lobe highlights the remodeled honeycomb spaces in the area of the lung with the end-stage disease (B).

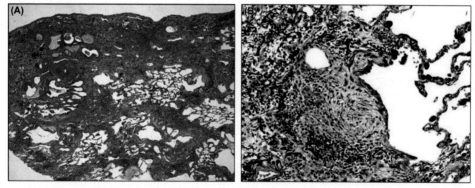

**FIGURE 18.33**    Usual interstitial pneumonia. The microscopic features of UIP lungs are characterized by inflammation and fibrosis that demonstrate the temporally heterogenous pattern of pathologic injury with normal, inflamed, and fibrotic areas of the lung, all seen at a single lower power view. (A) The leading edge of inflammation is represented by deposition of new collagen in fibroblastic foci. These consist of fibroblasts surrounded by collagen containing mucopolysaccharides highlighted in blue by this connective tissue stain (B).

## Molecular pathogenesis

Theories of the pathogenesis of IPF favor injury in the alveolar epithelial cells (AECs). The AECs consist of two populations: the type 1 pneumocytes and the type 2 pneumocytes. In normal lungs, type 1 pneumocytes line 95% of the alveolar wall, and type 2 pneumocytes line the remaining 5%. However, in lung injury, the type 1 cells, which are exquisitely fragile, undergo cell death, and the type 2 pneumocytes serve as progenitor cells to regenerate the alveolar epithelium. Though some studies have suggested that repopulation of the type 2 cells depends on circulating stem cells, this concept remains to be fully proven. According to current concepts, the injury and/or apoptosis of the AECs initiates a cascade of cellular events that produce the scarring in these lungs through stimulation of migration, proliferation, and activation of fibroblasts and myofibroblasts that leads to the characteristic fibroblastic foci of the UIP pathology and the deposition and accumulation of collagen and elastic fibers in the alveoli (Fig. 18.33). This unique pathology may be a result of the increased production of profibrotic factors such as transforming growth factor-$\alpha$ (TGF$\alpha$) and TGF$\beta$, fibroblastic growth factor-2, insulin-like growth factor-1, and platelet-derived growth factor. In support of this mechanism, is the finding that fibroblasts isolated from the lungs of IPF patients exhibit a profibrotic secretory phenotype.

Multiple factors, such as environmental particulates, drug or chemical exposures, and viruses, may trigger the initial injury to the AECs, but genetic factors also play a role. Approximately 2%–20% of patients with IPF have a family history of the disease with an inheritance pattern of autosomal dominance with variable penetrance. Genetic studies in a subset of patients with defined clinical syndromes with pulmonary fibrosis such as Hermansky-Pudlak syndrome and dyskeratosis congenita led to the association of the mutations in the *HPS* gene in patients with Hermansky-Pudlak syndrome and mutations in the *DKC, TERT*, and *TERC* genes in patients with dyskeratosis congenita.

In familial IPF, excluding these two syndromes, heterozygous mutations in the genes encoding the surfactant proteins A, B, and C have been associated with interstitial lung diseases. Surfactant protein A and B mutations, in general, are associated with neonatal lung disease. Mutations in the gene encoding surfactant protein C (*SFTPC*) have been associated with IPF in adults. Two types of mutations in this gene, skipping of exon 4 and deletion of the terminal 37 amino acids, and a Q188L missense mutation of the C-terminal portion were found. Additional mutations in the *SPC* gene associated with pulmonary fibrosis have been found in approximately 1% of the sporadic cases of IPF. In addition, in sporadic IPF patients, whole-genome linkage analysis has identified chromosome 10q22 as an affected area, where genes for surfactant proteins A1, A2, and D reside. In general, the mutations in the genes for both the SFA2 and SFC proteins result in the production of abnormal protein structures that are retained in the endoplasmic reticulum (ER) and fail to form oligomers. How this leads to lung fibrosis is not clear, but some suggest that the ER stress may result in AEC injury, inhibiting the ability of these cells to maintain the epithelial integrity, which leads to the abnormal proliferation of fibroblasts and myofibroblasts in these lungs.

In addition to surfactant protein-related genes, there is evidence that mutations in age-related genes may predispose toward pulmonary fibrosis including genes for telomerase reverse transcriptase (*TERT*) and telomerase RNA (*TERC*). Telomeres are specialized nucleoprotein structures that contain 100 to 10,000 repeat sequences of TTAGGG that protect the chromosomal ends. During normal cell division, these repeat sequences may be lost resulting in chromosomal shortening. TERT and TERC are enzymes that catalyze the addition of these repeat DNA sequences in the telomere region, protecting the chromosomes from loss of material during mitoses. Mutations in these genes lead to short telomeres, telomere dysfunction, and limits in tissue renewal. Sequencing of *TERT* and *TERC* genes in both familial and sporadic pulmonary fibrosis patients reveals mutations in *TERT* and *TERC* in 15% of familial pulmonary fibrosis cases and in rare cases of sporadic IPF. In addition, familial IPF patients have been found to have mutations in genes encoding other telomerase-related genes including *DKC1, TINF2*, and *RTEL1*. Although the mechanisms through which telomerase pathway mutations lead to lung fibrosis are unclear, it has been speculated that the loss of function in these enzymes disrupts AEC repair mechanisms, and similar to mutations in genes that control the expression of the surfactant proteins by the AECs (*SPA2, SPC*), result in the AEC injury that leads to the fibrosis.

## Surfactant dysfunction diseases

The alveolar surface of the lung is lined by type I and type II alveolar epithelial cells (AECs) that directly interact with the air inhaled into the alveolar sacs. In normal hosts, surfactant is essential to maintaining the low surface tension needed for proper alveolar inflation and gas exchange at this liquid-air interface. Surfactant is composed of phospholipids; phosphatidylcholine (PC); phosphatidylglycerol (PG); and surfactant proteins A, B, C, and D. The critical role of maintaining the proper composition and amount of surfactant in the alveoli is performed by the type 2 AECs and alveolar macrophages (AMs). The type 2 AECs synthesize surfactant in the endoplasmic reticulum and Golgi, store it as lamellar bodies, and then secrete it into the airways. Type 2 AECs and AMs control the catabolism of surfactant, regulating the amount present in the alveolar sacs. However, when this homeostasis is disturbed, accumulation of surfactant in the alveolar sacs may occur causing progressive respiratory insufficiency. Two examples of these diseases are pulmonary alveolar proteinosis (PAP), which occurs predominantly in adults, and hereditary surfactant protein disorders, which occur predominantly in children.

### Pulmonary alveolar proteinosis

*Clinical and pathologic features*

PAP, also known as alveolar proteinosis, lipoproteinosis, or perhaps most accurately phospholipoproteinosis, is a rare disease of the lungs characterized by accumulation of surfactant in the alveolar spaces. PAP takes three forms clinically: (1) congenital (hereditary) (2%), (2) secondary (linked mainly to cancers of the blood or systemic inflammatory diseases) (5%–10%), and (3) autoimmune (acquired) (88%–93%). Congenital PAP is a disease of children and the

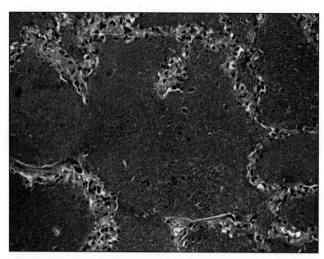

**FIGURE 18.34**    Pulmonary alveolar proteinosis. The microscopic features of this disease reveal a periodic acid-schiff positive surfactantlike substance filling the alveoli that otherwise show only a minimum of inflammatory changes.

presentation depends upon the mutated gene. Secondary and autoimmune PAP arises in previously healthy adults with the median age at diagnosis of approximately 40 years and a male-to-female ratio of 2.7:1. The clinical presentation usually includes an insidious onset of slowly progressive dyspnea, a dry cough, and other symptoms of respiratory distress, including fatigue and clubbing. However, almost one-third of patients are asymptomatic and are found clinically by abnormal chest X-rays.

Chest imaging studies in both the autoimmune and secondary forms most commonly show fine, diffuse, feathery nodular infiltrates, centered in the hilar areas, sparing the peripheral regions. On chest computerized tomographs, the infiltrates may have a geometric-type shape, sometimes referred to as crazy paving.

The most prominent microscopic feature of both the autoimmune and secondary PAP is the filling of the alveoli with finely granular period acid-Schiff-positive diastase-resistant (PASD) acellular material (Fig. 18.34). The material consists of phospholipids (90%); surfactant proteins A, B, C, and D (10%); and carbohydrate (<1%). AMs with prominent foamy cytoplasm are commonly seen, while alveolar septa are remarkably normal in appearance. In some alveolar spaces there are denser, more solid clumps of PASD-positive material. Definitive pathologic differences between the autoimmune and secondary forms of PAP have not been well documented.

### Molecular pathogenesis

The etiologies of the three forms of PAP have focused on the abnormal accumulation of the surfactant-like material within the alveolar spaces with most evidence suggesting that the clearance of surfactant by the AM is decreased.

The mechanisms by which this happens differ in each clinical form of PAP. In autoimmune PAP, there are circulating autoantibodies to GM-CSF that bind to the GM-CSF receptor, depressing the effect of GM-CSF on the AMs. The knowledge that most patients with autoimmune PAP have circulating autoantibodies to GM-CSF has led to a therapeutic replacement of GM-CSF in these patients, resulting in AM differentiation and appropriate surfactant clearance. The mechanism of secondary PAP is poorly understood. Autoantibodies have been reported in patients with secondary PAP. However, the main mechanism by which surfactant accumulates in the secondary form of PAP is presumed to be by reduced alveolar macrophage number and function. Finally, the congenital or hereditary form of PAP is caused by mutations in the genes that encode the GM-CSF receptor. Normal activity of GM-CSF requires signaling via the dimerization of its receptors, consisting of an α-chain and common β-chain that activates JAK/STAT5 signaling to enhance macrophage differentiation. Mutations in *CSF2RA* and *CSF2RB* genes encoding the GM-CSF receptor α and β chains, respectively, result in loss of GM-CSF receptor function.

Finally, it is important to note that GM-CSF activates its receptor via transcription factors such as STAT3, PU.1, and PPAR-γ. These control a number of AM functions such as microbial killing, adaptive immunity, innate immunity, phagocytosis, and proinflammatory cytokine signaling among others. Therefore, PAP patients with decreased levels of GM-CSF activity may also have varying degrees of immune suppression making them susceptible to a myriad of infections.

## Hereditary disorders of surfactant dysfunction

### Clinical and pathologic features

Surfactant dysfunction disorders represent a heterogenous group of inherited disorders caused by genetic defects in the surfactant protein B (*SFTPB*, chromosome 2p12-p11.2), surfactant protein C (*SFTPC*, chromosome 8p21), and adenosine triphosphate (APT)-binding cassette transporter subfamily A member 3 (*ABCA3*, chromosome 16p13.3). Defects in *SFTPB* and *ABCA3* have an autosomal recessive inheritance pattern, and defects in *SFTPC* have an autosomal dominant pattern.

Diseases due to *SFTPB* deficiency usually present at birth and affected babies die within weeks or months reflecting the lack of effective therapy. Diseases due to *ABCA3* or *SFTPC* deficiency may present within a week of birth or years later. The former has a poor prognosis, but the latter has a more variable prognosis with some patients surviving into adulthood. In general, the pathology of all of these diseases range from a PAP-like pattern to a chronic pneumonitis of infancy (CPI) pattern with alveolar walls expanded by chronic inflammation and a cuboidal alveolar epithelium.

### Molecular pathogenesis

The SP-B gene (*SFTPB*) is approximately 10 kb in length and is located on chromosome 2. There are over 30 recessive loss-of-function mutations associated with the *SFTPB* gene. These mutations include deletion, termination, missense, and substitution mutations that interfere with the synthesis of pro-SP-B or produce aberrant pro-SP-B that cannot be fully processed and is inactive. In most forms of hereditary SP-B deficiency, active SP-B cannot be isolated from the surfactant. Because active SP-B is required for both intracellular and extracellular surfactant homeostasis, surfactant replacement in these infants is not effective.

SP-C protein deficiency is due to a defect in the *SFTPC* gene localized to human chromosome 8. The most common mutation is a threonine substitution for isoleucine in codon 73 (I73T), found in 25% of the cases, including both sporadic and inherited disease. This mutation leads to a misfolding of the SP-C protein, inhibiting its progression through the intracellular secretory pathway, usually within the Golgi apparatus or the endoplasmic reticulum. Infants with documented mutated proSP-C protein where the larger primary translation product from which SP-C is proteolytically cleaved can have either respiratory distress syndrome or CPI. In older individuals, pathologic patterns observed in the lungs with these mutations include UIP.

The ABCA3 protein is a member of the family of ATP-dependent transporters, which includes the CFTR, and is expressed in epithelial cells. Mutation in *ABCA3* gene results in severe respiratory failure that is refractory to surfactant replacement. Surfactant from these patients is deficient in both PG and PC and has little or no activity. The presence of abnormal lamellar bodies within the type 2 cells by ultrastructural analysis suggests a disruption in the normal surfactant synthesis and packaging in this disease.

## Pulmonary vascular diseases

### Pulmonary hypertension

### Clinical and pathologic features

Pulmonary hypertension consists of a group of distinct diseases whose pathology is characterized by abnormal destruction, repair, remodeling, and proliferation of all compartments of the pulmonary vascular tree, including arteries, arterioles, capillaries, and veins. The classification of these diseases has undergone a number of revisions. The most recent revision (in 2003) groups these diseases based on both their pathologic and clinical characteristics. There are five major disease categories in the current classification system: (1) pulmonary arterial hypertension (PAH), (2) pulmonary hypertension with left heart disease, (3) pulmonary hypertension associated with lung disease and/or hypoxemia, (4) pulmonary hypertension due to chronic thrombotic and/or embolic disease, and (5) miscellaneous causes, including sarcoidosis, histiocytosis X, and lymphangioleiomyomatosis. The clinical course of most patients with pulmonary hypertension begins with exertional dyspnea and progresses through chest pain, syncope, increased mean pulmonary artery pressures, and eventually right heart failure. Current therapies rarely prevent progression of the disease, and lung transplantation provides the only hope for long-term survival.

The major group of this classification, PAH, can be subdivided into heritable PAH, idiopathic PAH, PAH associated with other conditions (such as connective tissue diseases, HIV, congenital heart disease), and PAH secondary to drugs and toxins (such as anorexigens, cocaine, and amphetamines). In these diseases, the primary pathology is localized

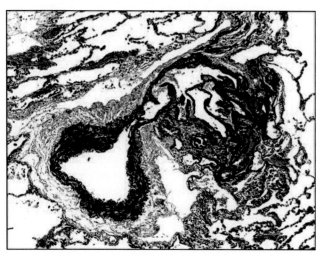

**FIGURE 18.35**   Pulmonary hypertension. A plexogenic lesion in a lung from a patient with idiopathic pulmonary hypertension reveals slitlike spaces (upper right corner) emerging from a pulmonary artery. These remodeling vascular spaces represent the irreversible damage done to these vessels in this disease.

predominantly in the small pulmonary arteries and arterioles. However, two other diseases in this group (pulmonary veno-occlusive disease and pulmonary capillary hemangiomatosis) involve predominantly other components of the pulmonary vasculature, the veins, and the capillaries, respectively. The pathologic changes seen in the pulmonary vessels of these patients primarily reflect injury to and repair of the endothelium. Early pathologic changes include medial hypertrophy and intimal fibrosis that narrows and obliterates the vessel lumen. These are followed by remodeling and revascularization, producing a proliferation of abnormal endothelial-lined spaces. These structures are known as plexogenic lesions and are the pathognomonic feature of PAH (Fig. 18.35). In the most severe pathologic lesions, these abnormal vascular structures become dilated or angiomatoid-like and may develop features of a necrotizing vasculitis with transmural inflammation and fibrinoid necrosis.

## Molecular pathogenesis

The heritable form of PAH, with a 2:1 female-to-male prevalence, has an autosomal dominance inheritance pattern with low penetrance. The genetic basis for this has been found to be germline mutations in the gene encoding the bone morphogenetic protein receptor type 2 (*BMPR2*), a member of the TGFβ superfamily of proteins that play a role in the growth and regulation of many cells, including those of the pulmonary vasculature. These mutations account for approximately 80% of heritable PAH. The mechanism by which a single mutation to the *BMPR2* gene induces vascular smooth muscle proliferation and decreased apoptosis is not completely understood. However, it is known that in the vascular smooth muscle cells of the lung, TGFβ signaling causes a proliferation of smooth muscle in pulmonary arterioles, while BMPR2 signaling causes an inhibition of the proliferation of these cells, favoring an apoptotic environment. The BMPR2 signaling occurs through an activation of a receptor complex (BMPR1 and BMPR2) that leads to phosphorylation and activation of a number cytoplasmic mediators, most notably the Smad proteins (Mothers against decapentaplegic). These Smad proteins, especially the Smad 1, Smad 5, and Smad 8 complex with Smad 4, translocate to the nucleus where they target gene transcription that induces an antiproliferative effect in the cell. In heritable PAH, the *BMRPR2* gene mutation may lead to insufficient protein product and subsequent decreased protein function, in this case decreased BMPR2 receptor function, decreased Smad protein activation, and decreased antiproliferative effects in the vascular smooth muscle cells. The imbalance between the proproliferative effects of TGFβ and the antiproliferative effects of the BMPs results in the formation of the vascular lesions of PAH (Fig. 18.36).

Only 30% of patients with the mutation develop clinical disease. Suggesting that genes confer susceptibility but a second hit is required to develop the clinical disease, such as modifier genes or environmental triggers that produce a proinflammatory response. In addition, though *BMPR2* mutations have been found in both the heritable and the idiopathic form of PAH, they are present in only 30% of all PAH patients, suggesting that further research is needed to uncover additional etiologic agents

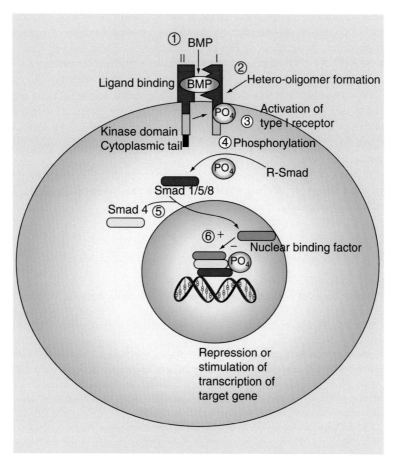

**FIGURE 18.36** Bone morphogenetic protein signaling pathway. 1 and 2: BMPR1 and BMPR2 are present on most cell surfaces as homodimers or hetero-oligomers. With ligand (bone morphogenetic proteins; BMP) binding, a complex of ligand, two type I receptors and two type II receptors, is formed. 3: After ligand, two type I receptors and two type II receptors phosphorylate the type I receptor in its juxtamembrane domain. 4: The activated type I receptor then phosphorylates a receptor-regulated Smad (R-Smad); thus, the type I receptors determine the specificity of the signal. 5: Once activated by phosphorylation, the R-Smads interact with the common mediator Smad 4 to form hetero-oligomers that are translocated to the nucleus. 6: In the nucleus, the Smad complex interacts with transcription factors and binds to DNA to induce or suppress transcription of target genes. *Reprinted with permission from Runo JR, Loyd JE. Primary pulmonary hypertension. Lancet 2003;316:1533—44.*

# Key concepts

- Driver mutations promote survival of cancer cells and play central roles in carcinogenesis. Common driver mutations in adenocarcinomas involve the epidermal growth factor receptor (EGFR), KRAS, and anaplastic lymphoma kinase (ALK). MYC family genes (MYC, MYCN, and MYCL), which play roles in cell cycle regulation, proliferation, and DNA synthesis, are more frequently activated in small cell lung cancers. TSG abnormalities involving TP53, RB, p16INK4a, and new candidate TSGs on the short arm of chromosome 3 (DUTT1, FHIT, RASFF1A, FUS-1, BAP-1) are common in lung cancers. Typical and atypical carcinoid tumors are often driven by mutations of the MEN1 gene family on chromosome 11q13 and related chromatin modifier gene mutations. The most common genetic abnormality in MM is a deletion in 9p21, the locus of the TSG CDKN2A encoding the tumor suppressors p16INK4a and p14ARF, which participate in the p53 and Rb pathways and inhibit cell cycle progression.
- Targeted therapeutic agents in use and under investigation for treatment of lung cancers include EGFR tyrosine kinase inhibitors, multi-kinase inhibitors, Ras/Raf/MEK pathway inhibitors, HER2 inhibitors, tumor suppressor gene therapies, and others.
- Precursor lesions have been defined for ACs, SqCCs, and carcinoids, and manifest histologic and molecular alterations that overlap with the corresponding neoplasms.
- The two main categories of asthma are atopic (allergic) and non-atopic (non-allergic). Both are associated with a Type 1 hypersensitivity response with a predominance of Type 2 helper cells with CD4+ phenotype.

- Emphysema, the loss of elastic fibers in the lung, is the result of an imbalance of elastases, which destroy elastic fibers and anti-elastases, which inhibit this destruction. Smoking causes an excess of elastase production and a centrilobular pattern of emphysema. A decrease in the circulating level of one anti-elastase (aAT) is the result of a genetic mutation on the SERPINA1 gene and results in a panacinar pattern of emphysema.
- Usual interstitial pneumonitis (UIP) is the pathologic pattern most associated with the clinical disease idiopathic pulmonary fibrosis (IPF). It is most likely the result of injury to the alveolar epithelial cells (AEC) that activates fibroblasts and myofibroblasts to deposit collagen and elastic fibers in the form of fibroblastic foci in these lungs.
- Lymphangioleiomyomatosis is a low grade metastasizing neoplasm characterized by abnormal smooth muscle cells in the lung, a result of genetic mutations in either the TSC1 or TSC2 gene causing a loss of inhibition in the mTOR pathway and uncontrolled cell growth.
- Pulmonary arterial hypertension (PAH) is the result of a germline mutation in the bone morphogenetic protein receptor type 3 (BMPR2) in most forms of familial PAH and some forms of sporadic PAH. However, a minority of patients with the mutation develop clinical disease.

# Suggested readings

## Neoplastic lung diseases

[1] SEER Cancer Statistics Factsheets: Lung and Bronchus Cancer. National Cancer Institute. Bethesda, MD, http://seer.cancer.gov/statfacts/html/lungb.html.

[2] Travis WD, Brambilla E, Burke AP, Marx A, Nicholson AG, editors. WHO classification of tumours of the lung, pleura, thymus and heart. Lyon: IARC; 2015.

[3] Swanton C, Govindan R. Clinical implications of genomic discoveries in lung cancer. N Engl J Med 2016;374:1864—73.

[4] Cancer Genome Atlas Research Network. Comprehensive molecular profiling of lung adenocarcinoma. Nature 2014;511:543—50.

[5] Lindeman NI, Cagle PT, Aisner DL, et al. Updated molecular testing guideline for the selection of lung cancer patients for treatment with targeted tyrosine kinase inhibitors: guideline from the College of American Pathologists, the International Association for the Study of Lung Cancer, and the Association for Molecular Pathology. J Mol Diagn 2018;20:129—59.

[6] Weir BA, Woo MS, Getz G, et al. Characterizing the cancer genome in lung adenocarcinoma. Nature 2007;450:893—8.

[7] Izumchenko E, Chang X, Brait M, et al. Targeted sequencing reveals clonal genetic changes in the progression of early lung neoplasms and paired circulating DNA. Nat Commun 2015;6:8258.

[8] Cancer Genome Atlas Research Network. Comprehensive genomic characterization of squamous cell lung cancers. Nature 2012;489:519—25.

[9] Dreijerink KM, Hoppener JW, Timmers HM, Lips CJ. Mechanisms of disease: multiple endocrine neoplasia type 1-relation to chromatin modifications and transcription regulation. Nat Clin Pract Endocrinol Metab 2006;2:562—70.

[10] Peifer M, Fernandez-Cuesta L, Sos ML, et al. Integrative genome analyses identify key somatic driver mutations of small-cell lung cancer. Nat Genet 2012;44:1104—10.

[11] Pleasance ED, Stephens PJ, O'Meara S, et al. A small-cell lung cancer genome with complex signatures of tobacco exposure. Nature 2010;463:184—90.

[12] Rekhtman N, Pietanza MC, Hellmann MD, et al. Next-generation sequencing of pulmonary large cell neuroendocrine carcinoma reveals small cell carcinoma-like and non-small cell carcinoma-like subsets. Clin Cancer Res 2016;22:3618—29.

[13] Fernandez-Cuesta L, Peifer M, Lu X, et al. Frequent mutations in chromatin-remodelling genes in pulmonary carcinoids. Nat Commun 2014;5:3518.

[14] Begueret H, Galateau-Salle F, Guillou L, et al. Primary intrathoracic synovial sarcoma: a clinicopathologic study of 40t(X;18)-positive cases from the French Sarcoma Group and the Mesopath Group. Am J Surg Pathol 2005;29:339—46.

[15] Flucke U, Vogels RJ, de Saint Aubain Somerhausen N, et al. Epithelioid hemangioendothelioma: clinicopathologic, immunohistochemical, and molecular genetic analysis of 39 cases. Diagn Pathol 2014;9:131.

[16] Fujita A, Ando K, Kobayashi E, et al. Detection of low-prevalence somatic TSC2 mutations in sporadic pulmonary lymphangioleiomyomatosis tissues by deep sequencing. Hum Genet 2016;135:61—8.

[17] McCormack FX, Travis WD, Colby TV, Henske EP, Moss J. Lymphangioleiomyomatosis: calling it what it is: a low-grade, destructive, metastasizing neoplasm. Am J Respir Crit Care Med 2012;186:1210—12.

[18] Galateau-Salle F, Churg A, Roggli V, Travis WD. The 2015 world health organization classification of tumors of the pleura: advances since the 2004 classification. J Thorac Oncol 2016;11:142—54.

[19] Sekido Y. Molecular pathogenesis of malignant mesothelioma. Carcinogenesis 2013;34:1413—19.

[20] Jean D, Daubriac J, Le Pimpec-Barthes F, Galateau-Salle F, Jaurand MC. Molecular changes in mesothelioma with an impact on prognosis and treatment. Arch Pathol Lab Med 2012;136:277—93.

## Non-neoplastic lung diseases

[21] Global initiative for chronic obstructive lung disease (GOLD): global strategy for the diagnosis MOCOPD NHLBI/who workshop report. 2003.

[22] Wright JL, Churg A. Advances in the pathology of COPD. Histopath 2006;10:1—9.

[23] Taraseviciene-Stewart L, Voelkel NF. Molecular pathogenesis of emphysema. J Clin Invest 2008;118:394—402.

[24] Stoller JK, Aboussouan LS. A review of alph-1-antitrypsin deficiency. Am J Respir Crit Care Med 2012;185:246—59.

[25] Tuder RM, Petrache I. Pathogenesis of chronic obstructive pulmonary disease. J Clin Invest 2012;122:2749—55.

[26] Tilley AE, Walters MS, Shaykhiev R, et al. Cilia dysfunction in lung disease. Ann Rev Physiol 2015;77:379—406.

[27] Elborn JS. Cystic fibrosis. Lancet 2016;388:2519—31.

[28] Travis WD, Costabel U, Hansell DM, et al. An official American Thoracic Society/European Respiratory Society statement: update of the international multidisciplinary classification of the idiopathic interstitial pneumonias. Am J Respir Crit Care Med 2013;188:733—42.

[29] Wolters PJ, Collard HR, Jones KD. Pathogenesis of idiopathic pulmonary fibrosis. Ann Rev Pathol Mech Dis 2014;9:157—79.

[30] Katzenstein AL, Myers JL. Idiopathic pulmonary fibrosis: clinical relevance of pathologic classification. Am J Respir Crit Care Med 1998;157:1301—15.

[31] Farver CF. Pathology of advanced interstitial diseases: pulmonary fibrosis, sarcoidosis, histiocytosis X, autoimmune pulmonary disease, lymphangioleiomyomatosis. In: Maurer JR, editor. Non-neoplastic advanced lung disease. New York: Marcel Dekker, Inc; 2003. p. 29—58.

[32] Steele MP, Schwartz DA. Molecular mechanisms of progressive idiopathic pulmonary fibrosis. Ann Rev Med 2013;64:265—76.

[33] Kropski JA, Blackwell TS, Loyd JE. Emerging genetic studies offer new insights into the fundamental mechanisms of pulmonary fibrosis. Eur Respir J 2015;445:1539—41.

[34] Ben-Dov I, Segel MJ. Autoimmune pulmonary alveolar proteinosis: clinical course and diagnositic criteria. Autoimmun Rev 2014;13:513—17.

[35] Carey B, Trapnell BC. The molecular basis of pulmonary alveolar proteinosis. Clin Imuunol 2010;135:223—35.

[36] Whitsett JA, Weaver TE. Alveolar development and disease. Am J Respir Cell Mol Biol 2015;53:1—7.

[37] Whitsett JA, Wert SE, Xu Y. Genetic disorders of surfactant homeostasis. Biol Neonate 2005;87:283—7.

[38] Simonneau G, Galie N, Rubin HL, et al. Clinical classification of pulmonary hypertension. J Am Coll Cardiol 2004;43:5S—12S.

[39] Chin KM, Rubin LJ. Pulmonary arterial hypertension. J Am Coll Cardiol 2008;51:1527—38.

[40] Austin ED, Loyd JE. Genetics and mediators in pulmonary arterial hypertension. Clin Chest Med 2007;28:43—57.

[41] Humbert M. Update in pulmonary arterial hypertension. Am J Respir Crit Care Med 2008;177:574—9.

[42] Michardo RD, Southage L, Eichstaedt CA, et al. Pulmonary arterial hypertension: a current perspective on established and emerging molecular genetic defects. Hum Mutat 2015;36:1113—27.

Chapter 19

# Molecular basis of diseases of the gastrointestinal tract

Antonia R. Sepulveda[1] and Armando J. Del Portillo[2]

[1]Department of Pathology, George Washington University School of Medicine and Health Sciences, Washington, DC, United States, [2]Department of Pathology and Cell Biology, Columbia University College of Physicians and Surgeons, New York, NY, United States

**Abstract**

Much of the progress in the understanding of gastrointestinal disorders has continued to center on the molecular underpinning of gastrointestinal neoplasia in the 21st century. First, the development of cancer in the setting of inflammatory conditions is well represented by the association of *Helicobacter pylori* with gastric cancer and of inflammatory bowel diseases with colorectal cancer. Second, the development of cancer in patients with hereditary predisposition syndromes has shed light not only in the mechanisms of hereditary neoplasia, but has also led to major progress in the understanding of the molecular basis of the more common forms of sporadic cancer. The molecular characterization of the steps of gastrointestinal neoplastic development and progression has led to advances in disease diagnosis and treatment and has opened the opportunity for development of more targeted approaches to cancer prevention, surveillance, and novel therapeutics. This chapter focuses on the disease processes that most clearly illustrate the concepts and advances in molecular pathology of the gastrointestinal tract. It includes neoplastic diseases associated with a background of chronic inflammation, well-characterized gastrointestinal hereditary cancer syndromes, and the so-called sporadic cancers of the gastrointestinal tract, primarily reviewing gastric and colonic carcinogenesis.

## Introduction

Our understanding of gastrointestinal disorders is critically centered on the molecular underpinnings of gastrointestinal neoplasia. The development of cancer in the setting of inflammatory conditions is well represented by the associations of *Helicobacter pylori* with gastric cancer, reflux esophagitis and esophageal adenocarcinoma, and inflammatory bowel diseases and colorectal cancer (CRC). The development of cancer in patients with hereditary predisposition syndromes has shed light not only in the mechanisms of hereditary neoplasia, but has also led to major progress in the understanding of the molecular basis of more common forms of sporadic cancer. Molecular characterization of the steps required for gastrointestinal neoplasia development and progression has led to advances in disease diagnosis and treatment and development of more targeted approaches to cancer prevention, surveillance, and precision therapy.

In this chapter, we will focus on the disease processes that most clearly illustrate the concepts and advances in molecular pathology of the gastrointestinal tract. This includes neoplastic diseases associated with a background of chronic inflammation, well-characterized gastrointestinal hereditary cancer syndromes, and sporadic cancers of the gastrointestinal tract, primarily reviewing gastric and colorectal carcinogenesis.

## Gastric cancer

Gastric carcinoma is the fifth most frequent cancer worldwide, with highest rates in Asia, countries in Eastern Europe, and areas of Central and South America, while it is less frequent in Western countries. Overall, gastric cancer represents the third most common cause of death from cancer ($\sim$700,000/year). In the United States, >24,000 new cases of stomach cancer were estimated in 2015. Most gastric cancers develop as sporadic cancers without a well-defined hereditary predisposition. A small proportion of gastric cancers arise as a consequence of a hereditary predisposition caused by specific inherited germ line mutations in critical cancer-related genes.

Essential Concepts in Molecular Pathology. DOI: https://doi.org/10.1016/B978-0-12-813257-9.00019-X

## Nonhereditary gastric cancer

### Risk factors

Most nonhereditary gastric cancers, also referred to as sporadic cancers, arise in a background of chronic gastritis, which is most commonly caused by *H. pylori* infection of the stomach. There may be clustering of gastric cancer within some affected families. Family relatives of patients with gastric carcinoma have an increased risk for gastric carcinoma of about three-fold. Patients with both a positive family history and infection with a CagA+ *H. pylori* strain were reported to have a greater than eight-fold risk of gastric carcinoma as compared to others without these risk factors.

The pathogenesis of gastric cancer is multifactorial, resulting from the interactions of host genetic susceptibility factors; environmental exogenous factors that have carcinogenic activity, such as dietary elements and smoking; and complex damaging effects of chronic gastritis. The diets that have been implicated in increased risk of gastric cancer include those enriched for salted, smoked, pickled, and preserved foods (rich in salt, nitrite, and preformed N-nitroso compounds), and diets with reduced vegetables and vitamin intake.

Chronic infection of the stomach by *H. pylori* is the most common form of chronic gastritis. Therefore, since most gastric cancers develop in a background of chronic gastritis, *H. pylori* is the most significant known risk factor for the development of gastric cancer. *H. pylori* was first implicated as the causal agent of most cases of chronic gastritis and ulcers in the seminal studies of Warren and Marshall. *H. pylori* was classified as a human carcinogen based on a strong epidemiological association of *H. pylori* gastritis and gastric cancer. Overall, the risk of gastric cancer in patients with *H. pylori* gastritis is six-fold higher than that of the population without *H. pylori* gastritis. The risk of gastric cancer increases exponentially with increasing grade of gastric atrophy and intestinal metaplasia and has been reported to reach 90-fold higher risk in patients with severe multifocal atrophic gastritis affecting the antrum and corpus of the stomach compared to individuals with normal, non-infected stomachs. However, even non-atrophic gastritis, as compared to healthy non-*H. pylori*−infected individuals, raises the gastric cancer risk to approximately two-fold.

Additional evidence supporting the association of *H. pylori* and gastric cancer includes (1) prospective studies demonstrating gastric cancer development in 2.9% of infected patients over a period of about 8 years, and in 8.4% of patients with extensive atrophic gastritis and intestinal metaplasia during a 10-year surveillance; (2) animal models that develop gastric cancer associated with *H. pylori* infection, including Mongolian gerbils and mice; and (3) eradication of *H. pylori* infection in patients with early gastric cancer resulted in the decreased appearance of new cancers. Furthermore, *H. pylori* eradication reduced the incidence of gastric cancer in patients without atrophy and intestinal metaplasia, suggesting that eradication may contribute to prevention of progression from gastritis to gastric cancer.

### Stepwise progression of H. pylori *gastritis to gastric carcinoma: histologic changes of the gastric mucosa*

The chronicity associated with *H. pylori* gastritis is critical to the carcinogenic potential of *H. pylori* infection. *H. pylori* is generally acquired during childhood and persists throughout life unless the patient undergoes eradication treatment. Gastric cancer develops several decades after acquisition of the infection, in a sequence of mucosal damage with development of specific histological alterations.

*H. pylori* infection of the stomach activates both humoral and cellular inflammatory responses within the gastric mucosa involving dendritic cells, macrophages, mast cells, recruitment and expansion of T-lymphocytes and B-lymphocytes, and neutrophils. Despite a continuous inflammatory response, *H. pylori* organisms are able to evade the host immune mechanisms and persist in the mucosa, causing chronic gastritis.

Histologically, the progression of *H. pylori*-associated chronic gastritis to gastric cancer starts with chronic gastritis, which leads to progressive damage of gastric glands resulting in atrophy (atrophic gastritis). There is patchy replacement of normal gastric glands by intestinal metaplasia over the entire stomach, but usually it is more severe in the antrum than in the body/fundus. Later, dysplasia and carcinoma may develop in some patients (Fig. 19.1).

The potential role of bone marrow−derived stem cells in chronic gastritis and *H. pylori*-associated neoplastic progression has been proposed based on studies in animal models. The current hypothesis is that *H. pylori*-associated inflammation and glandular atrophy create an abnormal microenvironment in the gastric mucosa that favors engraftment of bone marrow−derived stem cells into the inflamed gastric epithelium. It is postulated that engrafted bone marrow−derived stem cells do not follow a normal differentiation pathway and undergo uncontrolled replication, progressive loss of differentiation, and neoplastic behavior. However, the potential role of bone marrow−derived stem cells in human disease remains unclear.

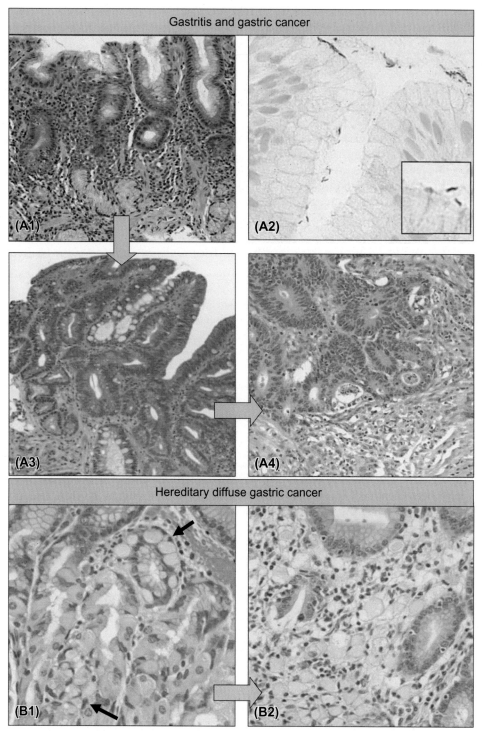

**FIGURE 19.1** Gastric carcinogenesis: stepwise progression of *H. pylori*-associated gastric cancer (Panels A1–A4), and hereditary diffuse gastric cancer (Panels B1 and B2). Panel A1: chronic active gastritis involving the mucosa of the gastric antrum (H&E stain, original magnification 10×); Panel A2: immunohistochemical stain highlights *H. pylori* organisms with typical S and comma shapes, seen at higher magnification in the inset. *H. pylori* organisms typically appear attached or adjacent to the gastric surface and foveolar epithelium (original magnification 40×); panel A3: gastric mucosa with intestinal metaplasia and low-grade dysplasia/adenoma (H&E stain, original magnification 10×); panel A4: gastric carcinoma of intestinal type (moderately differentiated adenocarcinoma) (H&E stain, original magnification 10×). Gastric mucosa of patient with hereditary diffuse gastric cancer with in situ signet ring cell carcinoma (*arrow*) (B1) and invasive signet ring cell carcinoma expanding the lamina propria between the gastric glands (H&E stain, original magnification 20×). *Panels B1 and B2 Courtesy of Dr. Adrian Gologan, Jewish General Hospital, McGill University.*

Stomach cancers are classified according to the World Health Organization guidelines based on their grade of differentiation into well-differentiated, moderately differentiated, and poorly differentiated adenocarcinomas. In addition, gastric adenocarcinomas can be categorized into intestinal and diffuse types (Lauren classification), based on the morphologic features on hematoxylin and eosin (H&E)-stained tissue sections.

Gastric cancers arising on the inflammatory background of *H. pylori*-associated chronic gastritis are most commonly intestinal-type adenocarcinoma, which are predominantly well-differentiated to moderately differentiated adenocarcinomas, but diffuse-type cancers, which are poorly differentiated or are signet ring cell carcinomas, also occur in the sequence of *H. pylori* gastritis (Fig. 19.1).

Progression to gastric cancer is higher in patients with extensive forms of atrophic gastritis with intestinal metaplasia involving large areas of the stomach, including the gastric body and fundus. This pattern of gastritis has been described as pangastritis or multifocal atrophic gastritis. Extensive gastritis involving the gastric body and fundus results in hypochlorhydria, allowing for bacterial overgrowth and increased carcinogenic activity in the stomach through the conversion of nitrites to carcinogenic nitroso-N compounds. *H. pylori*-associated pangastritis is frequently seen in the family relatives of gastric cancer patients, which may contribute to gastric cancer clustering in some families.

### Molecular mechanisms underlying gastric epithelial neoplasia associated with H. pylori infection

How *H. pylori* gastritis promotes gastric carcinogenesis involves an interplay of mechanisms that include (1) a longstanding inflammation in the mucosa, with increased oxidative damage of gastric epithelium, (2) a number of alterations of epithelial and inflammatory cells induced by *H. pylori* organisms and by released bacterial products, (3) inefficient host response to the induced damage, and (4) mechanisms of response related to host genetic susceptibility, which may mediate the variable levels of damage in different individuals.

*H. pylori* bacterial products and factors released by activated or injured epithelial and inflammatory cells both contribute to persistent chronic inflammatory response involving the activation of innate and acquired immune responses in the infected gastric mucosa. This chronic and continuously active mucosal inflammatory infiltrate is in part responsible for potential damage to the epithelium, through the release of oxygen radicals and the production of chemokines that may alter the normal regulation of molecular signaling in epithelial cells. *H. pylori* organisms and bacterial products, including the *H. pylori* virulence factors (CagA and VacA), may directly alter gastric epithelial cells, as well as inflammatory cells, leading to alterations of signaling pathways, gene transcription, and genomic modifications. These changes lead to modifications in cell behavior, such as increased apoptosis and proliferation, as well as increased rates of mutagenesis. A number of genetic susceptibility factors that increase the risk of gastric cancer development in *H. pylori*-infected patients have been identified. Interleukin-1 (IL-1) gene polymorphisms, in IL-1beta and IL-1RN (receptor antagonist), have been shown to increase the risk of gastric cancer and gastric atrophy in *H. pylori*-infected patients. Individuals with the IL-1B-31*C or IL-1B-511*T and the IL-1RN*2/*2 genotypes are at increased risk of hypochlorhydria, gastric atrophy, and gastric cancer in response to *H. pylori* infection. A gene polymorphism that may affect the function of OGG1, a protein involved in the repair of mutations induced by oxidative stress, was reported frequently in patients with intestinal metaplasia and gastric cancer, suggesting that deficient OGG1 function may contribute to increased mutagenesis during gastric carcinogenesis.

### Mechenisms and spectrum of epigenetic changes, mutagenesis, and gene expression changes of the gastric epithelium induced by H. pylori infection

Epigenetic modification and mutagenesis precede cancer development and accompany neoplastic progression during gastric carcinogenesis. Both of these changes have been shown to occur in *H. pylori* gastritis and subsequent preneoplastic and neoplastic mucosal lesions. The combined effects of epigenetic modifications, mutagenesis, and functional gene expression changes in gastric epithelial cells in response to inflammatory mediators and bacterial virulence factors result in abnormal gene expression and function in the various stages of progression to gastric cancer (gastritis, intestinal metaplasia, dysplasia, and cancer). The effects of *H. pylori* on gastric epithelial cells are likely to occur primarily during the phase of gastritis and intestinal metaplasia, while additional molecular events associated with neoplastic progression from dysplasia to invasive cancer may be independent of *H. pylori*. However, the background inflammatory milieu associated with ongoing chronic infection may influence the mechanisms of neoplastic progression.

During gastric carcinogenesis a number of genes are regulated by CpG methylation of the promoter regions at CpG sites, with potential promoter inactivation. Many genes are regulated by promoter CpG methylation in the transitions from gastritis to intestinal metaplasia, dysplasia, and gastric cancer (*MLH1, CDKN2A, MGMT, E-cadherin, RASSF1,* and others). The mechanisms that regulate CpG methylation and gene silencing during *H. pylori*-associated gastritis and resulting gastric mucosal lesions are not well understood. For example, pro-inflammatory IL-1β polymorphisms were associated with CpG island methylation of target genes, and CpG methylation of the *E-cadherin* promoter was induced in cells treated with IL-1β. These data suggest that components of the inflammatory cascade induced by *H. pylori* may contribute to orchestrate the epigenetic response in *H. pylori*-associated carcinogenesis.

Mutations are likely to accumulate during *H. pylori* chronic gastritis because of increased damaging factors in the mucosa and also because of overall deficiency of some DNA repair functions. DNA damage during *H. pylori* gastritis is caused primarily by reactive oxygen species (ROS) and reactive nitrogen species. Additionally, when mucosal atrophy develops, the resulting reduced acid levels may allow the overgrowth of other bacteria and activation of environmental carcinogens with mutagenic activity.

Increased cyclooxygenase (COX2) has been reported in *H. pylori*-associated gastritis and may contribute to increased mutagenesis through oxidative stress. Further, with reduced levels of oxygen radical scavengers, such as glutathione and glutathione-S-transferase, relatively higher levels of oxygen radicals may accumulate in the mucosa of *H. pylori*-infected patients. DNA 8-hydroxydeoxyguanosine (8OHdG) can be used as a marker for oxidative DNA damage. Increased mutagenesis may affect progenitor cells such as LGR5 progenitor cells in response to *H. pylori* infection. The gastric mucosa with *H. pylori* gastritis and preneoplastic lesions (intestinal metaplasia and atrophy) contain increased levels of 8OHdG, and the levels of 8OHdG in the gastric mucosa significantly decrease after eradication of *H. pylori*, supporting the role of active infection in the accumulation of mutations. Indeed, the Lgr5+ epithelial stem cell pool is expanded in *H. pylori*-associated gastritis in the antrum of patients with gastric cancer and the Lgr5+ epithelial cells may be more susceptible to DNA damage than Lgr5-negative epithelial cells. Mutations associated with oxidative damage include point mutations in genes involved in gastric carcinogenesis such as *TP53*.

Accumulation of mutations during *H. pylori* gastritis may be enhanced because of a relatively deficient DNA repair system, resulting in persistence of ROS-induced mutations and uncorrected DNA sequence replication errors that are transmitted to future epithelial cell generations. Several DNA repair systems are required for correction of DNA damage occurring during *H. pylori* gastritis: (1) the DNA mismatch repair (dMMR) system, which repairs DNA replication-associated sequence errors, and (2) several other proteins that primarily repair DNA lesions induced by oxidative and nitrosative stress, including MGMT and OGG1 glycosylase. The dMMR system functions through the action of MutS proteins (MSH2, MSH3, and MSH6) and MutL proteins (MLH1, PMS1, PMS2, and MLH3). dMMR deficiency leads to frameshift mutagenesis, which can generate mutations in the coding region of genes, as well as in repetitive regions known as short tandem repeats (STRs) or microsatellite regions. The mutations in microsatellite regions result in microsatellite instability (MSI), which can be used as surrogate markers of dMMR deficiency. High levels of MSI (MSI-H), defined as instability in greater than 30% of the microsatellite markers in a panel of five microsatellite loci, correlate well with loss of dMMR function. Several studies have reported a role of dMMR deficiency in accumulation of mutations during *H. pylori* infection. In experimental conditions, when gastric epithelial cells are co-cultured with *H. pylori* organisms, the levels of dMMR proteins, including MSH2 and MLH1, are greatly reduced, and both point mutations and frameshift/microsatellite type mutations accumulate in the *H. pylori*-exposed cells. Microsatellite instability was reported in 13% cases of chronic gastritis, 20% of intestinal metaplasias, 25% of dysplasias, and 38% of gastric cancers, indicating a stepwise accumulation of MSI during gastric carcinogenesis. Microsatellite instability has been detected in intestinal metaplasia from patients with gastric cancer in several studies, indicating that MSI can occur in preneoplastic gastric mucosa. Several studies reported that patients with MSI+ cancers showed a significantly higher frequency of active *H. pylori* infection, which supports the notion that *H. pylori* infection can underlie dMMR deficiency and MSI in the various steps of gastric carcinogenesis.

Deficient function of other genes involved in the repair of oxidative stress—induced mutations may contribute to mutagenesis during *H. pylori* gastritis. Repair of 8-OHdG is accomplished by DNA repair proteins including a polymorphic glycosylase (OGG1). This protein may be less efficient in carriers of a gene polymorphism that was reported frequently in patients with intestinal metaplasia and gastric cancer. O6-methylguanine-DNA methyltransferase (MGMT) function includes the repair of O6-alkylG DNA adducts. In the absence of functional MGMT, these adducts mispair with T during DNA replication, resulting in G-to-A mutations. MGMT-promoter methylation has been reported in a subset of cases of *H. pylori* gastritis and in various stages of gastric carcinogenesis, suggesting a possible role for this DNA repair protein in gastric carcinogenesis.

Gene expression analysis with microarrays has revealed expression signatures associated with *H. pylori* gastritis and gene expression induced by the effect of *H. pylori* organisms in gastric epithelial cells, confirming the increased expression of some genes by epithelial cells in response to *H. pylori* (such as IL-8) and expanding the knowledge of signaling pathways involved in *H. pylori* pathogenesis.

## *Mutational, epigenetic, gene expression, and microrna patterns of gastric intestinal metaplasia, dysplasia/adenoma, and cancer*

Intestinal metaplasia, dysplasia, and carcinoma represent cell populations with a clonal origin that manifest epigenetic and genetic alterations incurred by the non-neoplastic epithelium, as well as additional events that occur during neoplastic progression. Mutational events during gastric neoplastic development and progression include MSI-type mutations, chromosomal instability manifesting as loss of heterozygosity (LOH) and gene amplifications, and point mutations of cancer-related genes. High level of MSI is associated with loss of expression and promoter hypermethylation of *MLH1* in gastric adenomas and cancer. MSI has been reported in 17%−59% of gastric cancers from various studies and in 22% of gastric cancers from *The Cancer Genome Atlas* (TCGA). Gastric cancers with MSI-H may harbor frameshift mutations in the coding regions of cancer-related genes, such as *BAX, IGFRII, TGFβRII, MSH3*, and *MSH6*. Point mutations and LOH at multiple gene loci have been detected in intestinal metaplasia, adenomas, and gastric carcinomas. Mutations of *TP53* and adenomatous polyposis coli (*APC*) genes have been reported in intestinal metaplasia and gastric dysplasia. *TP53* mutations in exons 5−8 resulting in G:C to A:T transitions are frequent in gastric carcinogenesis. *APC* mutations, including stop-codon and frameshift mutations, were reported in 46% of gastric adenomas and 5q allelic loss in 33% of informative cases of gastric adenoma and in 45% of carcinomas. *KRAS* mutations in codon 12 were reported in 14% of biopsies with atrophic gastritis and in less than 10% of adenomas, dysplasia, and carcinomas.

Gene regulation through epigenetic modification occurs at multiple steps of gastric carcinogenesis. Variations of CpG methylation during the steps of disease progression from *H. pylori* gastritis to intestinal metaplasia and gastric cancer have been observed for some genes, but not for others. Genes that play a role in cell cycle progression, DNA repair, cell adhesion, and a number of tumor suppressor genes may be regulated by epigenetic mechanisms through promoter methylation. In gastric carcinogenesis, CpG island methylation occurs in genes such as *MLH1, p14, p15, p16, E-cadherin, RUNX3*, thrombospondin-1 (*THBS1*), tissue inhibitor of metalloproteinase 3 (*TIMP-3*), *COX-2*, and *MGMT*.

Genome-wide gene expression analysis with microarrays has yielded extensive information on gene expression of the pathological lesions associated with gastric carcinogenesis. One study reported a signature of diffuse-type cancers that exhibited altered expression of genes related to cell-matrix interaction and extracellular-matrix components, whereas intestinal-type cancers had a pattern of enhancement of cell growth. Another study reported several combinations of genes that could discriminate between normal and cancer samples, and intestinal metaplasia cases were characterized by a gene expression signature resembling that of adenocarcinoma. The lack of reproducibility of studies using microarrays for gene expression has limited the impact of this technology to cancer applications. Further, many of the observed expression signatures await additional studies to confirm their significance in carcinogenesis and will require integration of expression data using powerful computational methods.

MicroRNAs (miRNAs) are small noncoding RNAs that have been shown to regulate gene expression and may be aberrantly expressed in cancer. Available data indicate that most gastric cancers show overexpression of *miR21*, while *miR218−2* is downregulated. The functional implications of these findings remain unclear and await further studies.

## *The Cancer Genome Atlas Research Network—molecular characterization of gastric adenocarcinomas*

*The Cancer Genome Atlas Research Network* (TCGA) classified gastric adenocarcinomas into four distinct subtypes based on molecular features: (1) Epstein-Barr Virus (EBV)-associated cancers, (2) MSI-H cancers, (3) genomically-stable cancers (GS), and (4) cancers with chromosomal instability (CIN). The observed frequency for each subtype was 9% EBV, 22% MSI, 20% GS, and 50% CIN. The distribution within the stomach of each subtype varied slightly, with EBV-associated cancers occurring more often in the body/fundus, and CIN cancers are overrepresented among tumors of the cardia. Of note, no differences were found in survival or in the frequencies of each subtype in East Asian or Western patients.

## Genomic alterations in EBV-associated gastric adenocarcinomas

EBV-associated gastric adenocarcinomas have a prominent CpG island methylator phenotype (CIMP) distinct from the MSI subgroup. For example, hypermethylation of the *CDKN2A* promoter is more prevalent in EBV-type cancers, and *MLH1* promoter hypermethylation is more prevalent in MSI-type tumors. In EBV-associated cancers, amplification of a 9p locus is enriched, a locus that contains therapeutic targets such as *JAK2, PD-L1*, and *PD-L2*. EBV-associated cancers also frequently contain mutations in *PIK3CA, ARID1A*, and *BCOR*, but unlike other subtypes lack mutations in *TP53*.

## MSI cancers

MSI gastric adenocarcinomas have a hypermutated phenotype due to loss of MMR proteins, similar to MSI colonic adenocarcinomas. MSI gastric adenocarcinomas show frequent mutations in *PTEN* and *PIK3CA*, as well as in receptor tyrosine kinase (RTK)-RAS pathway proteins, such as *KRAS/NRAS*, *EGFR*, *ERBB2*, *ERBB3*, *JAK2*, *FGFR2*, and *MET*, among others. However, these cancers have far fewer genetic amplifications in these targetable genes as compared to the other molecular subtypes. Unlike most sporadic colorectal MSI carcinomas, gastric MSI carcinomas do not contain *BRAF* V600 E mutations.

## GS cancers

GS cancers, as the name implies, harbor very few mutations. Almost 40% of these cancers contain somatic mutations in *CDH1*, the gene mutated in the germline in patients with hereditary diffuse gastric cancer (HDGC). No pathogenic germline mutations in this gene were found in the cohort analyzed, emphasizing the importance of this gene in sporadic gastric cancers. Therefore, this pathway to carcinogenesis does not progress by selection from a pool of mutations as the result of persistent injury. Instead, it appears that mutation of *CDH1* rapidly primes the cells for neoplastic transformation, as is corroborated by the high frequency of adenocarcinomas arising in patients with germline mutations in *CDH1*. Additional recurrent alterations found in GS cancers include mutations in *ARID1A* and *RHOA*, and fusion of *CLDN18* and *ARHGAP26*. Interestingly, this fusion was found to be mutually exclusive with *RHOA* mutation.

## CIN Cancers

The most common molecular subtype of gastric adenocarcinomas described in the TCGA report are CIN cancers, and most of these cancers (71%) have mutations in *TP53*. CIN cancers have mutations in RTK-*RAS* and *PI3K* pathway genes and more frequent gene amplifications, especially when compared to MSI cancers. The most frequently amplified RTK-*RAS* pathway gene in CIN cancers is *ERBB2*, which is the target of trastuzumab.

The data from the TCGA Research Network highlights the variable molecular landscape seen in gastric cancers and points to the multiple molecular pathways underlying carcinogenesis in the stomach.

## Familial gastric cancer

### Genetic and molecular basis of familial and hereditary gastric cancer

Familial gastric cancer may be inherited as an autosomal dominant disease, occurring as the main tumor in HDGC or as one of the tumor types in various cancer predisposing syndromes. Familial gastric cancer is likely to be due to multiple factors and may also occur as family clustering. Familial gastric cancer represents less than 10% of all stomach cancers. HDGC associated with germline mutations in the *E-cadherin* gene account for 30%−40% of known cases of hereditary diffuse gastric cancer. In up to 70% of the cases of familial gastric cancer, the underlying genetic defect is unknown. Gastric carcinomas represent one of the tumors occurring in the following hereditary cancer syndromes: (1) germ-line mutations in the *E-cadherin* gene (*CDH1*) underlie some but not all HDGC families (of note, germ-line mutation in the *E-cadherin* gene does not underlie hereditary intestinal-type gastric cancer); (2) Li-Fraumeni syndrome, associated with germline *TP53* mutations; (3) hereditary nonpolyposis colon cancer (HNPCC), with most HNPCC-associated gastric cancers representing the intestinal-type; (4) gastric cancer may also occur in Peutz-Jeghers syndrome (PJS), where hamartomatous polyps in the stomach occur in ∼24% of patients, but the overall risk of gastric cancer is small; (5) patients with familial adenomatous polyposis (FAP) frequently develop fundic gland polyps in the stomach and may develop gastric adenomatous polyps in about 10% of individuals with FAP, but the risk of gastric cancer is small. Gastric cancer occurs in 5.7% of families with the *BRCA2* 6174delT mutation.

### Hereditary diffuse gastric cancer: genetic basis

The *CDH1 gene*, which encodes the protein E-cadherin, is the only gene known to be associated with HDGC. Mutations in other genes may account for susceptibility to HDGC, but the evidence is limited. The human *CDH1* gene consists of 16 exons that span 100 kb. The identified mutations are scattered throughout the gene and are truncating mutations, caused by frameshift mutations, exon/intron splice site mutations, point mutations, and missense mutations.

## *Hereditary diffuse gastric cancer: molecular mechanisms, clinical and pathologic features*
### Natural history and pathologic features

The average age of diagnosis of HDGC is 38 years (ranging from 14 to 69 years), with most cases occurring before the age of 40 years. The lifetime risk of gastric cancer, by age 80 years is 67% for men and 83% for women. In addition to gastric cancer, women also have a 39% risk for lobular breast cancer. Histologically the adenocarcinomas in patients with HDGC are characteristically poorly differentiated carcinomas with signet ring morphology (signet ring cell carcinomas). Tumor foci are initially confined mostly in the superficial zone of the gastric mucosa and appear to arise in the lower proliferative zone of the gastric foveolae, and may be multifocal. In situ signet ring cell carcinoma, characterized by disorderly oriented signet ring cells within glands or foveolar epithelium may be observed (Fig. 19.1). Grossly, the cancer extends through the gastric layers and gastric wall with the development of linitis plastica. These cancers do not arise in a background of intestinal metaplasia and gastric atrophy or dysplastic preneoplastic lesions. E-cadherin immunohistochemistry may show reduced to absent expression in both the in situ and invasive areas of the tumor.

### Molecular mechanisms and pathologic correlates

In HDGC with known germline mutations, E-cadherin loss-of-function caused by pathologic mutations underlie the development of cancer. In HGDC, loss of E-cadherin expression is also associated with transcriptional downregulation of the wild-type *CDH1* allele by promoter hypermethylation. However, up to 70% of cases of familial gastric cancer of diffuse-type do not have a well-defined genetic defect. E-cadherin is a member of the cadherin family of transmembrane glycoproteins, characterized by five extracellular domains and a cytoplasmic domain. E-cadherin has been shown to be essential for establishing and maintaining the polarized differentiated organization of epithelia. It plays important roles in signal transduction, gene expression, differentiation, and cell motility. The activity of E-cadherin in cell adhesion is dependent upon its association with the actin cytoskeleton through the interaction with the catenin ($\alpha$-catenin, $\beta$-catenin, and $\gamma$-catenin) family of proteins. Loss of E-cadherin expression is seen in most diffuse gastric cancers, including sporadic-type gastric carcinomas of diffuse-type, and in lobular breast cancer. The E-cadherin/catenin complex is important to suppress invasion/metastasis and cell proliferation. Somatic mutation of E-cadherin is associated with increased activation of epidermal growth factor receptor (EGFR) followed by enhanced recruitment of the downstream signal transduction and activation of RAS. The activation of EGFR by E-cadherin mutants in the extracellular domain explains the enhanced motility of cancer cells in the presence of an extracellular mutation of E-cadherin.

### Hereditary diffuse gastric cancer: genetic testing and clinical management

Criteria for consideration of *CDH1* molecular genetic testing in individuals with diffuse gastric cancer have been recommended. DNA/sequence analysis of the *CDH1* gene is the current test recommended for confirmation of the diagnosis. The clinical management of individuals who have an identified *CDH1* cancer−associated mutation varies from intense surveillance for early detection of gastric cancer or prophylactic gastrectomy.

# Colorectal cancer

Worldwide CRC ranks fourth in frequency in men and third in women, with $\sim$ 1 million cases annually. Men and women are similarly affected. In the United States there were $\sim$ 135,000 cases of CRC in 2016. Most CRCs do not develop in association with a hereditary cancer syndrome and are known as sporadic cancers. Several hereditary colon cancer syndromes have been characterized (Table 19.1). The most frequent is Lynch syndrome or HNPCC, representing 3%−4% of CRCs. FAP represents about 1%, and the remaining cancer syndromes (Table 19.1) are responsible for less than 1% of CRCs.

## Sporadic colon cancer

Colon cancers generally arise from precursor lesions of the colonic epithelium histologically described as dysplasia or adenoma, with progression to high-grade dysplasia (HGD) (previously referred to as carcinoma in situ) and invasive adenocarcinomas (Fig. 19.2). There are two main types of CRC precursor lesions: the most common are conventional adenomas or adenomatous lesions and the second most common are serrated polyps. The epithelium that constitutes conventional colonic adenomas displays cytologic features of dysplasia, but in polyps that do not arise in inflammatory bowel disease, these cytologic changes are routinely called adenomatous features, and the polyps are diagnosed as adenomas. Conventional adenomas are further classified into tubular, tubulovillous, or villous adenomas. In the initial

**TABLE 19.1** Gastrointestinal hereditary cancer syndromes and other polyposis syndromes.

| Syndrome | Inheritance | Key Clinical Features | Gene | Gene Product Function |
|---|---|---|---|---|
| Familial adenomatous polyposis and variants Gardner | Autosomal Dominant | >100 colonic adenomas, near 100% lifetime risk for colorectal carcinoma FAP plus CHRPE, osteomas, desmoid tumors | APC | Growth inhibitory:β-catenin sequestration; targeting of β-catenin for destruction |
| Turcot[a] Attenuated FAP | | FAP plus medulloblastoma >15 but <100 colonic adenomas | | |
| MYH-associated polyposis | Autosomal recessive | FAP-like presentation; no APC mutation identifiable | MYH | DNA damage repair |
| Hereditary nonpolyposis colorectal cancer (HNPCC; Lynch syndrome) and variants | Autosomal dominant | 70%−85% lifetime risk for colorectal carcinoma; predisposition for extracolonic malignancy including endometrial | MSH2 MLH1 MSH6 PMS2 MLH3 | DNA mismatch repair |
| Muir-Torre | | HNPCC plus sebaceous tumors and extracolonic malignancies | | |
| Peutz-Jeghers | Autosomal dominant | Hamartomatous gastrointestinal polyps; predisposition for multiple extracolonic malignancies | STK11 | Serine/threonine kinase |
| Cowden and[b]BRR syndrome | Autosomal dominant | Hamartomatous gastrointestinal polyps | PTEN | Protein phosphatase |
| Juvenile Polyposis | Autosomal dominant | Hamartomatous gastrointestinal polyposis; increased risk for colorectal carcinoma | SMAD4 BMPR1AENG | TGF-β signaling |

*APC*, adenomatous polyposis coli; *CHRPE*, Congenital Hypertrophy of the Retinal Pigment Epithelium; *FAP*, familial adenomatous polyposis.
[a]*Two-thirds of Turcot syndrome occur in patients with APC gene mutation and one-third in DNA mismatch repair gene mutation.*
[b]*BRR: Bannayan-Ruvalcaba-Riley syndrome.*

steps, conventional adenomas consist of epithelium with low-grade dysplasia, which may progress to HGD and invasive adenocarcinomas (Fig. 19.2). Serrated precursor lesions are histologically of two types: the sessile serrated adenoma/polyp (SSA/P) and traditional serrated adenoma (Fig. 19.3). SSA/P lesions are diagnosed based on their crypt architectural features, consisting of basal dilated and irregular crypts, without cytologic dysplasia. However, later they may develop cytologic dysplasia. Traditional serrated adenomas consist of serrated epithelium as well as adenomatous features.

The molecular pathways of colon cancer development include a stepwise acquisition of mutations, epigenetic changes, and alterations of gene expression, that lead to uncontrolled cell division and to invasive neoplastic behavior. The molecular changes underlying colorectal neoplasia correlate partially with histopathological variants. At the genomic level, the main molecular pathways of CRC development characterized to date include (1) the chromosomal instability pathway (CIN), (2) the microsatellite instability pathway (MSI), and (3) the CIMP pathway.

Most colon cancers progress through a CIN pathway characterized by genomic alterations that include aneusomy and gains and losses of chromosomal regions. The molecular mechanisms underlying CIN are poorly understood. CIN appears to result from deregulation of the DNA replication and mitotic-spindle checkpoints. Mutation of the mitotic checkpoint regulators *BUB1* and *BUBR1* and amplification of *STK15* are seen in a subset of CIN-colon cancers. In CRCs that follow the CIN pathway, the dominant genomic abnormality is inactivation of tumor suppressor genes, such as *APC* (chromosome 5q), *TP53* (chromosome 17p), *DCC* (deleted in colon cancer), *SMAD2*, and *SMAD4* (chromosome 18q). *APC* gene mutations occur early in colorectal neoplasia. *APC* mutations are detected in aberrant crypt foci, the lesion that precedes the development of adenomas, and have been detected in 50% of sporadic adenomas and 80% of sporadic colon cancers. Inactivation of *APC* results in activation of Wnt/beta-catenin signaling, which in turn can induce chromosomal instability in colon cancer. *DCC* gene loss occurs late in neoplastic progression, with frequent deletion in carcinomas (73%) and in high-grade adenomas (47%). The tumor suppressor *TP53* is also involved in the later steps of colon carcinogenesis. The TP53 protein has DNA-binding activity; contains a transcription activation domain; and regulates target genes that mediate cell cycle arrest, apoptosis, and DNA repair. In the sporadic colorectal

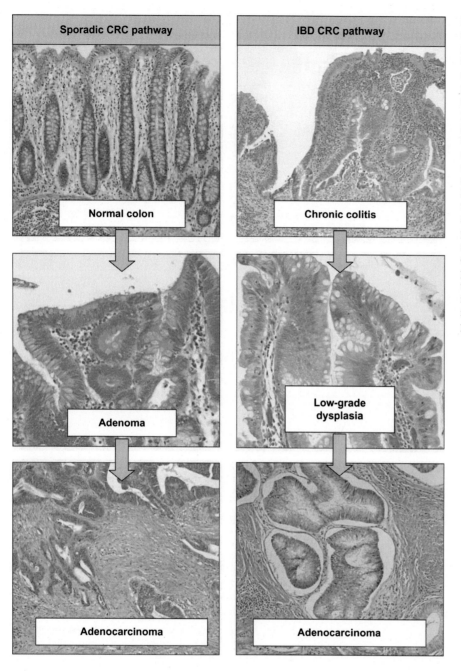

**FIGURE 19.2** Stepwise progression of colorectal neoplasia: progression of neoplasia in sporadic colorectal cancer and in inflammatory bowel disease-associated colitis. In the sporadic colorectal cancer pathway, adenomas characterized in the early stages by low-grade epithelial dysplasia precede the development of high-grade dysplasia, which may then progress to invasive adenocarcinoma. In IBD-associated neoplasia, the background colonic mucosa reveals variable degrees of chronic colitis, and eventually foci of low-grade dysplasia develop, which in turn may progress to high-grade dysplasia and invasive adenocarcinoma. The morphologic features of the neoplastic lesions significantly overlap between sporadic colorectal cancer and IBD-associated neoplasia, but the inflammatory environment that characterizes chronic colitis dictates a number of different molecular mechanisms of neoplastic development and progression.

carcinogenesis pathway, *TP53* gene mutations occur in adenomas with high-grade dysplasia (50%) and carcinomas (75%). The most common mutations are missense point mutations of one allele followed by deletion of the second allele, resulting in LOH of the *TP53* gene locus on chromosome 17p.

The activation of oncogenes contributes to both the development and the progression of neoplasia. The *RAS* family of oncogenes is frequently activated in CRC. Activation of *RAS* following activation of EGFR leads to signaling through several pathways including the Raf/MAPK and PI3K/Akt pathways. Activating *RAS* mutations in CRC affect *KRAS* and *NRAS* codons 12 and 13 of exon 2, 59 and 61 of exon 3, and 117 and 146 of exon 4. Overall, at least 50% of all CRCs carry *RAS* mutations. *KRAS* mutations have been observed in the first stages of colonic neoplasia in aberrant crypt foci. Sporadic, FAP-associated, and CIMP colorectal neoplasms have the highest rates of activating *KRAS* mutations (50%–80%), while *KRAS* mutations are rare in both sporadic MSI-H and HNPCC-associated cancers.

The MSI pathway occurs in ∼15% of sporadic CRCs. The MSI pathway was first characterized in HNPCC or Lynch syndrome. Cancers arising through the microsatellite instability pathway are characterized by a deficiency of

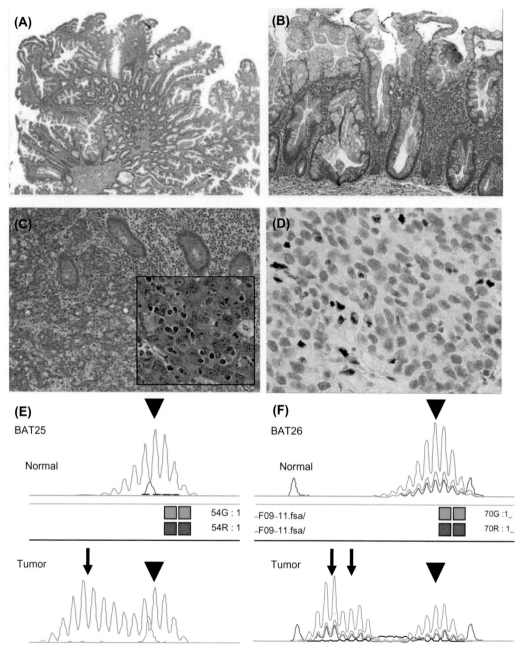

**FIGURE 19.3** Colorectal cancer pathways: histopathology and molecular correlates. Panels A and B: The serrated pathway. Adenocarcinomas that develop through the serrated pathway arise from serrated polyps that include traditional serrated adenomas (A) and sessile serrated adenomas (B) (H&E stain, original magnification 5× and 10×, for panels A and B, respectively). Traditional serrated adenomas show both serrated architecture and dysplasia similar to that seen in adenomatous mucosa, whereas sessile serrated adenomas reveal architectural abnormalities but no evidence of classic dysplasia. Panels C and D: the microsatellite instability pathway. Poorly differentiated colonic adenocarcinoma with prominent intratumoral lymphocytes, best seen in the inset (C) (H&E stain, original magnification 10×). By immunohistochemistry, the cancer cells are negative for MSH2, while the surrounding lymphocytes and stromal cells show preserved expression of MSH2 protein in the nonneoplastic cell nuclei (D) (Immunohistochemistry, original magnification 10×). (E and F) The presence of microsatellite instability in the tumor DNA is demonstrated by microsatellite instability at the microsatellite markers *BAT25* (E) and *BAT26* (F) characterized by the appearance of new PCR amplification peaks of smaller size (*tailed arrows*) as compared to nonneoplastic DNA from the same patient (*arrow tip*).

dMMR proteins. The main dMMR proteins are MLH1, MSH2, MSH6, and PMS2. Loss-of-function of one of the main dMMR proteins results in high levels of mutagenesis, and this molecular phenotype is known as MSI. Through the use of microsatellite loci panels, cancers can be classified by their levels of MSI. Cancers with MSI-H show loss of expression of major dMMR proteins in tumor cell nuclei. In sporadic colorectal carcinomas, MSI is usually caused by loss of expression of MLH1, secondary to *MLH1* promoter hypermethylation, while in HNPCC the underlying defect is caused by inherited mutations of one of the *dMMR* genes. The deficiency of dMMR may lead to accumulation of secondary mutations in cancer-related genes. In both sporadic and familial MSI-H colon cancers, mutational inactivation of the *TGFβRII* gene by MSI occurs in 80% of tested cancers and is frequently a late event, occurring at the transition from adenoma to cancer. Studies comparing HNPCC cancers and sporadic MSI colon cancers showed that HNPCC tumors are more frequently characterized by aberrant nuclear β-catenin. Aberrant *TP53* expression, 5q LOH, and *KRAS* mutation is infrequent in both MSI-cancer groups. Despite sharing a similar underlying pathway (dMMR deficiency), there are differences between sporadic CRC with MSI and HNPCC-associated tumors. Sporadic MSI-H cancers are more frequently poorly differentiated, mucinous, and proximally located than HNPCC tumors. In sporadic MSI-H cancers, contiguous adenomas are frequently serrated lesions, while conventional adenomas are usually seen in HNPCC. Lymphocytic infiltration is common in both types of tumors (Fig. 19.3).

Epigenetic regulation is important in cancer development and progression. The CIMP pathway is characterized by widespread CpG methylation in neoplasms. Extensive CpG methylation in the CIMP pathway may result in inactivation of important genes involved in tumorigenesis. For example, deficiency of the DNA repair protein MGMT in sporadic CRC may be due to inactivation by promoter CpG hypermethylation (39%−42%), and hypermethylation can be detected in 49% of adenomas. The promoter hypermethylation of *MLH1* triggers the MSI pathway, accelerating neoplastic progression.

The CIMP pathway underlies MSI in most cases of sporadic CRCs associated with *MLH1* hypermethylation, and is tightly associated with *BRAF* mutation in sporadic CRC. However, only about one-third of CIMP+ tumors are MSI-H. Thus, CIMP appears to be independent of microsatellite status. Several studies have confirmed that CIMP+ tumors are associated with *BRAF* mutations, wild-type *TP53*, inactive WNT/β-catenin, and low-level of genomic instability of the CIN-type.

CIMP is frequent in serrated pathway lesions. *BRAF* mutations are also associated with serrated colorectal lesions, while *KRAS* mutations are associated with hyperplastic polyps, in particular goblet cell-rich hyperplastic polyps, and conventional adenomas. Sessile serrated adenomas have frequent *BRAF* mutation (78%) and have rare *KRAS* mutations (11%).

## Molecular mechanisms of neoplastic progression in inflammatory bowel disease

### Natural history of neoplasia in inflammatory bowel disease

Inflammatory bowel diseases (IBD) include both ulcerative colitis (UC) and Crohn disease. Patients with IBD have an increased risk of dysplasia and CRC, associated with longstanding chronic colitis. Colorectal adenocarcinomas in IBD develop from foci of low-grade dysplasia, which may progress to HGD and ultimately invasive adenocarcinoma (Fig. 19.2). The risk of colon cancer for patients with IBD was reported to increase by 0.5%−1.0% every year after 8−10 years of diagnosis. CRC in IBD has an incidence 20-fold higher and is detected on average in patients 20 years younger than the non-IBD population.

### Molecular mechanisms of colorectal cancer development and progression in IBD-associated colitis

In IBD-associated colitis, there is a continued inflammatory environment of the mucosa (Fig. 19.2) with damage of the colonic epithelium associated with increased cell proliferation and deregulation of apoptosis. Driving factors in the inflammation-associated cancer pathway include the increased oxidative damage with increased mutagenesis caused by the heightened inflammatory infiltration of the mucosa, resulting in a stepwise progression of neoplastic lesions. The molecular players of IBD-associated carcinogenesis are similar to those seen in sporadic colorectal carcinogenesis, but the timing of occurrence of molecular events is different. During chronic colitis, there is activation of NF-kB in the epithelium. NF-kB activates the expression of COX2, several pro-inflammatory cytokines (including IL-1, TNFα, IL-12p40, and IL-23p19), anti-apoptotic factor inhibitor of apoptosis protein, and B-cell leukemia/lymphoma (Bcl-xL). Prostaglandins and cytokines such as IL-6 are released in the inflammatory milieu and activate intracellular serine-threonine kinase Akt signaling, with inhibition of pro-apoptotic factors TP53, BAD, and FoxO1, and increased cell survival.

In colitis-associated cancers, genetic instability includes CIN and MSI, similar to sporadic colon cancers, occurring in 85% and 15% of the cases, respectively. In UC-associated colon cancers, *APC* mutations or LOH of the *APC* locus are seen in 14%−33% of cancers. *APC* mutations occur late in UC-associated neoplasia, in the transition from HGD to carcinoma. This is in contrast to sporadic colon carcinogenesis, where the *APC* gene is mutated in the majority (80%)

of cancers and occurs early in the neoplastic process. This significant difference in the timing of genetic events in IBD-associated versus sporadic colon carcinogenesis is likely due to the inflammatory environment in IBD. Since other molecular mechanisms, such as the activation of NF-kB in colonic epithelial cells by inflammatory-cell released cytokines, promote cell proliferation and inhibit apoptosis, *APC* inactivation may not be necessary to drive the early steps of colitis-associated carcinogenesis.

The tumor suppressor gene *TP53* is mutated at a high rate in both sporadic and UC colon cancers, but in contrast to sporadic cancers, in UC *TP53* mutations occur earlier in the colitis-associated carcinogenesis pathway. Mutations in *TP53* were detected in 19% of biopsies without dysplasia, with the frequency increasing with higher grades of dysplasia. This is in contrast to sporadic CRC in which *TP53* mutations and LOH are associated with the progression from high-grade adenoma to cancer. Early TP53 loss-of-function in UC-associated carcinogenesis contributes to the rapid progression to colon cancer observed in these patients. Epithelial cells with deficient TP53 function have a survival advantage. In the normal mucosa, where there is limited DNA damage, TP53 induces *p21* gene expression and delays cell cycle progression in S-phase to allow for repair of the damaged DNA. If DNA damage is more extensive, TP53 can induce apoptosis, preventing DNA replication and persistence of significant mutations in daughter cells. In the inflamed mucosa, early mutation of *TP53* is associated with inflammation-related oxidative damage. The reduced DNA repair, also caused by the inflammatory environment, together with the increased mutation burden and the reduced ability to remove cells with significant mutations, results in clonal selection, expansion of neoplastic cells, and cancer development.

In MSI+ UC colon cancers, the cause of MSI-H is loss of MLH1 protein expression associated with *MLH1* promoter hypermethylation. However, MSI in the non-neoplastic mucosa is significantly more frequent in the colitic mucosa as compared to the background mucosa of sporadic colon cancers. This may be explained by reduced expression and function of DNA repair enzymes induced by oxidative stress free-radicals. These data point to a role of deficient dMMR without complete loss of gene expression underlying mutagenesis in the epithelium before overt neoplasia becomes histologically apparent. In contrast to sporadic colon cancers, in UC-associated colon cancers with MSI, *TGFβRII* mutations are much less common, observed in only 17% of these cancers, and typically occur early in the neoplastic process.

Another mechanism that contributes to neoplastic development and progression in UC-associated CRC is epigenetic gene regulation, in particular CpG island hypermethylation. Methyl-binding proteins recognize hypermethylated sites and recruit histone deacetylases, leading to histone deacetylation, chromatin condensation, and inactivation of genes. Several genes have been reported to be hypermethylated in dysplasia and/or cancer in UC, including genes that have been reported as targets of CpG island methylation in sporadic CRC such as *MLH1* and *CDKN2A/p16INK4a*. Increased methylation in HGD of patients with UC versus controls without UC, was reported for estrogen receptor (*ESR1*), *MYOD*, p16 exon 1, and *CSPG2*, and hypermethylation of three of these genes had similar methylation in the colitic mucosa of patients with HGD. The *p16* methylation levels averaged 2% in the mucosa of controls, 3% in UC patients who had mucosa without dysplasia, 8% in the normal appearing epithelium of patients with HGD/CA, and 9% in the dysplastic epithelium (HGD/CA). Additionally, methylation was present not just in the neoplastic mucosa but also in the non-neoplastic-appearing epithelium from UC patients with HGD/cancer, suggesting that the increased levels of methylation are widespread in the inflammation-afflicted colon and occur early in the process of tumorigenesis, preceding the histological appearance of dysplasia.

Methylation of the hyperplastic polyposis gene 1 (*HPP1*) was observed in 50% of UC adenocarcinomas and in 40% of dysplasias. In contrast, there was no methylation in non-neoplastic UC mucosa. Methylation of the *CDH1* promoter was detected in 93% of the patients with dysplastic biopsy samples, in contrast to only 6% of the patients without dysplasia and by immunohistochemistry areas of dysplasia displayed reduced E-cadherin expression. In IBD, promoter hypermethylation of the gene for DNA repair protein O6-methylguanine-DNA methyltransferase (*MGMT*) was detected in 16.7% adenocarcinomas and in 3.7% of mucosal samples with mild inflammation. Notably, *MGMT* is more frequently methylated in sporadic adenomas and carcinomas than in IBD. Extensive methylation characteristic of the CIMP was observed in 17% of the UC-related cancers, while global DNA methylation measured with an LINE-1 assay was seen in 58% of UC-associated cancers.

Activation of the Raf/MEK/ERK (MAPK) kinase pathway either through *RAS* or *BRAF* mutation was detected in 27% of all UC-related cancers. Non-dysplastic UC mucosa of patients with UC cancer did not show *BRAF* mutations, indicating that *BRAF* mutations are not an initiating event in UC-related carcinogenesis, but are associated with MMR deficiency through *MLH1* promoter hypermethylation in advanced lesions. Conversely, *KRAS* mutations may occur in dysplasia, but also in villous regeneration and active colitis. *KRAS* mutations are inversely correlated with *BRAF* mutations in UC cancers, similar to sporadic CRCs.

## IBD-associated neoplasia: diagnosis and clinical management

The most significant predictor of the risk of malignancy in IBD is the presence of dysplasia on colonic biopsies. Colonoscopy with biopsies to rule out dysplasia is currently used during follow-up of patients with IBD. The pathological diagnosis of dysplasia relies on a classification based on morphological criteria established in the early 1980s, and includes the following categories: (1) negative for dysplasia, (2) indefinite/indeterminate for dysplasia, and (3) positive for low-grade dysplasia, HGD, or invasive cancer.

The identification of definitive dysplasia places a patient with chronic IBD in a higher risk group that may require frequent repeat colonoscopies or colonic resection. Through use of currently available histological methods, it is often difficult to determine whether a lesion represents dysplasia or a reactive process, and a diagnosis indicating indefinite/indeterminate for dysplasia is rendered, requiring additional colonoscopy. The *identification* of molecular markers of dysplasia for a more accurate screening of dysplasia in IBD patients remains a high priority.

## Hereditary nonpolyposis colorectal cancer

### Hereditary nonpolyposis colorectal cancer: genetic and molecular basis

HNPCC is an autosomal dominant cancer predisposition syndrome, characterized by early onset of CRC and tumors from other organs, including endometrium, ovary, urothelium, stomach, brain, and sebaceous glands. HNPCC with underlying dMMR gene mutations and characteristic dMMR/MSI+ cancer represent the HNPCC/Lynch syndrome, while a group of patients with clinical features of HNPCC, but lacking dMMR gene mutations, represent a separate group of cancers. HNPCC/Lynch syndrome is estimated to represent 4%–6% of all CRC cases. At the molecular level, HNPCC/Lynch syndrome patients inherit germline mutations in one of the *dMMR* genes, leading to defects in the corresponding dMMR proteins. Mutations are found most commonly in the *MLH1* and *MSH2* and less frequently in the *MSH6* and *PMS2* genes.

HNPCC/Lynch syndrome is subdivided into Lynch syndrome I, characterized by colon cancer susceptibility, and Lynch syndrome II, which shows all the features of Lynch syndrome I, but patients are also at increased risk for carcinoma of the endometrium, ovary, and other sites. The spectrum of tumors that occur in patients with HNPCC is referred to as HNPCC cancers and include carcinomas of the colon and rectum, small bowel, stomach, biliary tract, pancreas, endometrium, ovary, urinary bladder, ureter, and renal pelvis. Sebaceous gland adenomas and keratoacanthomas are part of the Muir-Torre syndrome, and tumors of the brain, usually glioblastomas, are seen in the Turcot syndrome (Table 19.1).

The *dMMR* genes encode the MutS proteins (MSH2, MSH3, and MSH6) and the MutL proteins (MLH1, PMS1, PMS2, and MLH3). Functional dMMR requires the formation of MutS and MutL complexes between the different MMR proteins. The MSH2 protein interacts with either MSH6 (forming the MutS-alpha complex) or with MSH3 (forming the MutS-beta complex). MutS-alpha and MutS-beta complexes bind to post-DNA replication mismatched nucleotide sequences. The MutS-alpha complex is required to repair base-base mispairs and small insertion or deletion mispairs, while the MutS-beta complex is primarily involved in the correction of insertion or deletion mispairs. MutL heterodimers bind MutS-alpha or MutS-beta. MutL heterodimers include MutL-alpha (MLH1 and PMS2), which appears to be responsible for most of the dMMR. The MutL-beta heterodimer (MLH1 and PMS1) does not appear to be significantly involved in dMMR.

Known germline mutations of *dMMR* genes in HNPCC affect the coding region of *MLH1* ($\sim$40%), *MSH2* ($\sim$40%), *MSH6* ($\sim$10%), and *PMS2* ($\sim$5%). Another mechanism underlying HNPCC/Lynch syndrome, in up to 6.3% of all Lynch syndrome cases, involves deletions in the 3′ end of *EPCAM* that result in epigenetic silencing of the adjacent *MSH2* gene by promoter hypermethylation.

### Hereditary nonpolyposis colorectal cancer: natural history, clinical and pathologic features, and molecular mechanisms

The average age of presentation of CRC in HNPCC patients is 45 years of age. Tumors are often multiple or associated with other synchronous or metachronous neoplasms of the HNPCC-cancer spectrum. In addition to CRCs, patients may present with tumors of the Muir-Torre syndrome or with the Turcot syndrome. In HNPCC/Lynch syndrome patients, the lifetime risk of CRC is up to 80%, and up to 60% for endometrial carcinoma. CRCs are located in the proximal colon in two-thirds of cases and have detectable MSI in more than 90% of the cases.

A few histopathologic features are characteristically associated with MSI-H colorectal adenocarcinomas that may suggest the possibility of HNPCC (Fig. 19.3). These features are not specific to HNPCC cancers, in that they are also seen frequently in sporadic MSI-H tumors and in some tumors that are microsatellite stable (MSS). Three major histopathologic groups of MSI-H cancers can be recognized: (1) poorly differentiated adenocarcinomas (also described as medullary type cancers), (2) mucinous adenocarcinomas and carcinomas with signet ring cell features, and (3) well-differentiated to moderately differentiated adenocarcinomas. The presence of prominent tumor-infiltrating lymphocytes (TILs) is the most predictive finding of dMMR/MSI-H status (Fig. 19.3). TILs may be particularly numerous in poorly differentiated cancers, but also occur in the other morphologic types of HNPCC-associated cancers. TILs are CD3+ T-cells and most are CD8+ cytotoxic T-lymphocytes. Peritumoral lymphocytic inflammation and lymphoid aggregates forming a Crohn-like reaction are also frequent in MSI-H carcinomas.

Patients with HNPCC develop adenomas more frequently and at an earlier age than non-carriers of dMMR gene mutations. In HNPCC, patients develop one or few colonic adenomas of conventional type (tubular adenomas and tubulovillous adenomas). The progression from adenoma to invasive carcinoma occurs rapidly, in many patients occurring in <3 years, in contrast to a mean of 15 years in patients without HNPCC. Detection of colonic adenomas in HNPCC mutation carriers occurs at an average age of 42–43 years (range 24–62 years). Compared to sporadic adenomas, HNPCC adenomas are more frequently proximal in the colon, and more frequently show HGD. In HNPCC, a significant association was found between MSI-H and HGD in adenomas, with associated loss of either MLH1 or MSH2. Based on these findings it was recommended that immunohistochemical (IHC) staining/MSI testing of large adenomas with HGD in young patients (younger than 50 years) may be performed to help identify patients with suspected HNPCC.

The molecular mechanisms that underlie the neoplastic development and progression in HNPCC are those of the MSI pathway. The deficient or absent dMMR protein(s) in HNPCC/Lynch syndrome is dictated by the gene that carries a germline mutation. A second genetic hit is responsible for loss-of-function of the second allele. Epigenetic silencing through CpG methylation of *MLH1*, which is the underlying mechanism of MSI in sporadic colon cancer, is rare in HNPCC tumors. In addition, the patients with unambiguous germline mutation in *dMMR* genes do not appear to carry *BRAF*-activating mutations in their tumors. As in sporadic MSI carcinomas, loss of dMMR may lead to accumulation of mutations in cancer-related genes, such as the *TGFβRII* gene. HNPCC tumors show frequent aberrant nuclear β-catenin, but aberrant *TP53* expression, 5q LOH, and *KRAS* mutation is uncommon, similar to sporadic MSI-cancers.

## Hereditary nonpolyposis colorectal cancer: molecular diagnosis, clinical management, and genetic counseling

To identify patients with HNPCC, criteria known as the Amsterdam criteria were established in the early 1990s, with later revisions. Because the Amsterdam criteria do not include some patients with known germline MMR gene mutations, another set of guidelines, known as the Bethesda guidelines, was established to help decide whether or not a patient should undergo further molecular testing to rule out HNPCC. Through use of the Amsterdam II Criteria, patients are diagnosed with HNPCC when the following criteria are present: (1) the family includes three or more relatives with an HNPCC-associated cancer, (2) one affected patient is a first-degree relative of the other two, (3) two or more successive generations are affected, (4) cancer in one or more affected relatives is diagnosed before the age of 50 years, (5) familial adenomatous polyposis is excluded in any cases of CRC, and (6) tumors are verified by pathological examination. Alternatively, one of the following criteria (Modified Amsterdam) needs to be met: (1) very small families, which cannot be further expanded, can be considered to have HNPCC with only two CRCs in first-degree relatives if at least two generations have the cancer and at least one case of CRC was diagnosed by the age of 55 years; (2) in families with two first-degree relatives affected by CRC, the presence of a third relative with an unusual early onset neoplasm or endometrial cancer is sufficient; or (3) if an individual is diagnosed before the age of 40 years and does not have a family history that fulfills the preceding criteria (Amsterdam II and Modified Amsterdam criteria), that individual is still considered as having HNPCC.

If an individual has a family history that is suggestive of HNPCC but does not fulfill the Amsterdam, modified Amsterdam, or young age at onset criteria, that individual is considered to be HNPCC variant, or familial CRC X. A large group of patients representing 60% of all cases that meet Amsterdam I or Amsterdam II criteria for HNPCC do not have characteristic features of MMR deficiency. Compared to the MSI-HNPCC patients, the MSS HNPCC patient's age at diagnosis is 6 years higher on average, and most CRCs appear on the left side of the colon. The underlying genetic defect for these tumors is not yet known.

Bethesda criteria recommend testing patients to rule out HNPCC if there is one of the following criteria: (1) patient is diagnosed with CRC before the age of 50 years; (2) presence of synchronous or metachronous colorectal or other HNPCC-related tumors (stomach, urinary bladder, ureter and renal pelvis, biliary tract, glioblastoma, sebaceous gland adenomas, keratoacanthomas, or small bowel), regardless of age; (3) CRCs with an MSI-H morphology (presence of TILs, Crohn-like lymphocytic reaction, mucinous or signet ring cell differentiation, or medullary growth pattern) that was diagnosed before the age of 60 years; (4) CRC patient with one or more first-degree relatives with CRC or other HNPCC-related tumors, and one of the cancers must have been diagnosed before the age of 50 years; or (5) CRC patient with two or more relatives with CRC or other HNPCC-related tumors, regardless of age. Bethesda guidelines increased identification of MSI+ colon cancers compared to previous guidelines. The effect of setting the age cut-off to <50 years and inclusion of histopathologic features in the selection of tumors that should be tested for MSI and dMMR protein immunohistochemistry significantly increased the detection rate of potential HNPCC-associated cancers. With the Bethesda guidelines, when selected for age (CRC in patients less than 50 years of age) and histologic features suggestive of MSI-tumor in patients less than 60 years of age, MSI-H tumors were detected in ~5% of CRC cases meeting Bethesda guidelines. Algorithms have been devised to determine whether a patient has sporadic MSI+ cancer or HNPCC/Lynch syndrome. The EGAPP (*Evaluation of Genomic Applications in Practice and Prevention Working Group*) recommended that all newly diagnosed CRC patients should be screened for HNPCC/Lynch syndrome to reduce the morbidity and mortality from CRC in their at-risk unaffected relatives. Testing of CRC tissues to help determine whether the tumor represents HNPCC/Lynch syndrome, includes testing for MMR status, with MSI DNA tests and/or IHC analysis of tumors for MSH2 and MLH1, MSH6, and PMS2 dMMR proteins.

The MSI test is based on the evaluation of instability in small DNA segments that consist of repetitive nucleotides of generally 100−200 base pairs in length, called microsatellite regions or STRs. The nucleotide sequences within the repetitive elements include mononucleotide repeats of Adenine (A)n or cytosine-adenine (CA)n dinucleotide repeats. During DNA replication, these repetitive sequences are susceptible to variations in length because of DNA strand slippage. Such changes in length of the nucleotide repeat are termed MSI. In cells with deficient dMMR, as occurs in patients with HNPCC/Lynch syndrome, these mutations are not repaired and persist in the DNA of future cell generations in the tumor tissue. MSI can be detected in the tumor DNA using PCR approaches. Tissue sections from cancer tissue used for routine pathologic diagnosis embedded in paraffin are adequate for the MSI test. The set of microsatellite markers recommended by an NCI consensus group consists of two mononucleotide repeat markers (*BAT25* and *BAT26*) and three dinucleotide repeat markers (*D2S123, D5S346,* and *D17S250*). The results of the MSI test using the NCI panel of five microsatellite markers are reported as MSI-H, MSI-Low (MSI-L), and MSS. MSI-H cancers show MSI in at least two of the five markers, MSI-L cancers show MSI in only one marker, and no instability is detected in any of the five markers in MSS cancers. Fig. 19.3 illustrates a CRC with loss of expression of MSH2 in the cancer cells and associated MSI-H identified by microsatellite instability at both the *BAT25* and *BAT26* markers (Fig. 19.3). Alternative microsatellite marker panels are available. If the cancer tissue reveals MSI-H and/or there is loss of expression of one of the DNA repair proteins by immunohistochemistry, germline testing should be performed for the gene encoding the deficient protein, after appropriate genetic counseling of the patient. If tissue testing is not feasible, or if there is sufficient clinical evidence of HNPCC, it is acceptable to proceed directly to germline analysis of the *MSH2* and/or *MLH1* genes. Germline mutation analyses are done with DNA extracted from the patient's blood samples and can be performed by several methods, including DNA sequencing, with recent widespread use of next-generation sequencing and methods to detect large deletions and gene duplications including array comparative genomic hybridization and multiplex ligation-dependent probe amplification.

If no loss of expression of *MSH2* or *MLH1* is seen in MSI-H tumors or if the tumor is MSI-L or MSS, but there is suspicion of HNPCC, evaluation of other MMR genes, in particular *MSH6* and *PMS2*, should be performed, first by IHC stains, followed by germline mutational analyses. Identification of a germline mutation in index cancer patients is important because it confirms a diagnosis of HNPCC, and the identified mutation may be used to screen at-risk relatives who may be mutation carriers. If the tumor tissue reveals loss of expression of MLH1 by immunohistochemistry, but no mutation in the *dMMR* genes underlying HNPCC/Lynch syndrome are found, two other tests (*MLH1* methylation and *BRAF* mutation tests) may help discriminate between a sporadic MSI-tumor and HNPCC tumor with undetected *MLH1* mutation. Through use of quantitative methylation analyses, HNPCC patients showed no or low level of *MLH1* promoter methylation, in contrast to high levels of methylation (greater than a cutoff value of 18% methylation) in sporadic MSI cancers. In addition, none of the patients with unambiguous germline mutation in *dMMR* genes demonstrated *BRAF* mutation. Therefore, adding *BRAF* mutation and *MLH1* methylation tests in the algorithm for testing of MSI-H colon tumors with loss of expression of MLH1 protein in the tumor can help determine whether a tumor is likely to be a sporadic or an HNPCC-associated tumor.

After a germline mutation is identified or the patient is diagnosed with HNPCC, at-risk relatives should be referred for genetic counseling and tested if they wish. If no MMR gene mutation is found in a proband with an MSI-H tumor and/or a clinical history of HNPCC, the genetic test result is non-informative. The patients and the relatives at risk should be counseled as if HNPCC was confirmed and high-risk surveillance should be performed.

In families that meet strict clinical criteria for HNPCC, germline mutations in *MSH2* and *MLH1* have been found in 45%−70% of the families, and germline mutations in these two genes account for 95% of HNPCC cases with an identified mutation. The reported data show that despite extensive testing there is still a significant number of families without an identified germline mutation that accounts for HNPCC.

I ndividuals diagnosed as carriers of dMMR gene mutations seen in the HNPCC/Lynch syndrome are recommended to have colonoscopic surveillance starting at early age. Colonoscopy is recommended every 1−2 years beginning at ages 20−25 years (age 30 years for those with *MSH6* mutations) or 10 years younger than the youngest age of the person diagnosed in the family. Although there is limited evidence regarding efficacy, the following are also recommended annually: (1) endometrial sampling and transvaginal ultrasound of the uterus and ovaries (ages 30−35 years), (2) urinalysis with cytology (ages 25−35 years), and (3) history, examination, review of systems, education, and genetic counseling regarding Lynch syndrome (age 21 years). For individuals who will undergo surgical resection of a colon cancer, subtotal colectomy is favored. Further, evidence supports the efficacy of prophylactic hysterectomy and oophorectomy. Recently, a multi-society task force developed guidelines to assist health care providers with the appropriate provision of genetic testing and management of patients at risk for and affected with Lynch syndrome.

## Familial adenomatous polyposis and variants

### Familial adenomatous polyposis: genetic basis

FAP is a cancer predisposition syndrome characterized by numerous adenomatous colorectal polyps, with virtually universal progression to colorectal carcinoma at an early age. It accounts for <1% of all CRC cases in the United States and affects 1 in 8000−10,000 individuals. The majority of cases of FAP are caused by germline mutations in the *APC* gene on chromosome 5q and is transmitted in an autosomal dominant fashion, although up to a third of cases may present as de novo germline mutations. FAP patients, particularly some cases presenting with attenuated FAP (AFAP), carry bi-allelic mutations in the *MUTYH* gene.

FAP is one of the best characterized cancer predisposition syndromes, as the elucidation of the molecular basis of the syndrome was paramount in the formulation and advancement of the multistep model of carcinogenesis. The first verified case of adenomatous polyposis was reported in 1881, and familial cases of adenomatous polyposis were the subject of scientific inquiry throughout the 1900s, leading to registries and early intervention with colectomy as early as the 1920s. A series of major breakthroughs both in regard to FAP and as related to the understanding of carcinogenesis started with the association of the syndrome with a deletion on chromosome 5q21−22, and the genetic defect was subsequently identified to reside within the *APC* gene at that locus.

### Familial adenomatous polyposis: natural history and clinical, molecular, and pathologic features

Patients affected by FAP have a nearly 100% lifetime risk for the development of colorectal carcinoma in the absence of aggressive treatment, which often includes prophylactic colectomy. In addition, patients have a 90% lifetime risk for the development of upper gastrointestinal tract polyps, with a 50% risk of developing advanced duodenal polyposis by the age of 70 years. The lifetime risks for development of gastric adenocarcinoma (<1%) and duodenal and ampullary carcinoma (5%−10%) are considerably higher than those of the general population. FAP mutation carriers have increased risk of fundic gland polyps involving the stomach. Among the other associated lesions, congenital hypertrophy of retinal pigmented epithelium (CHRPE) is found in 70%−80%, and desmoid tumors are found in 15% of FAP patients. Thyroid carcinoma, both papillary and follicular type, is estimated to occur in 1%−2% of FAP patients. In young children with a family history of FAP, the relative risk of hepatoblastoma is increased compared to the general population.

Clinically, the common manifestation of FAP and its variants is the presence of numerous (sometimes in excess of 1000) adenomatous polyps distributed throughout the colon and rectum. The characteristic gross findings, combined with the histologic findings and predictable progression to colorectal adenocarcinoma, have historically made FAP a robust model system to better understand carcinogenesis development and progression. There are several variants of FAP, including Gardner Syndrome, Turcot Syndrome, and AFAP, summarized in Table 19.1. Gardner syndrome is characterized by multiple extracolonic manifestations including osteomas, desmoid tumors, dental abnormalities, ophthalmologic abnormalities (including CHRPE), and cutaneous cysts. Some suggest that some degree of these extraintestinal manifestations may be identified with close scrutiny in typical FAP patients. Turcot syndrome is the association of the colorectal polyposis and brain tumors, most commonly medulloblastoma. AFAP demonstrates a reduction in the number of colonic polyps, usually falling short of the 100 polyps necessary for a diagnosis of FAP, but with sufficient colonic polyposis (frequently over 15) to raise suspicion for an underlying polyposis syndrome. FAP and its variants may present clinically in several manners, including through identification of colorectal polyposis or colorectal carcinoma at an unusually early age, the presence of extracolonic manifestations, or as part of screening in the setting of a known family history.

One of the most striking features of FAP and its variants (except AFAP) is the presence of hundreds, perhaps even thousands, of polyps throughout the colon and rectum, leading to a carpet appearance of the colorectal mucosa (Fig. 19.4). The polyps are frequently sessile and appear as early as late childhood/early adolescence, requiring that endoscopic screening in familial cases begin early in life. The histologic features of polyps in FAP are essentially indistinguishable from those seen in sporadic adenomas. However, it is common to identify lesions at various stages in the dysplasia-adenoma-carcinoma sequence, further underscoring the multistep nature of carcinogenesis (Fig. 19.4).

The molecular changes in FAP have been extensively characterized and serve as an excellent demonstration of the relationship between the understanding of cellular signaling pathways and their effect on disease process. The detailed examination of the molecular changes of the lesions at various stages in FAP was a cornerstone of the current understanding of carcinogenesis as a multistep process involving gradual increases in tumor size, disorganization, and accumulation of genetic changes. The majority of cases of FAP and its variants are attributed to germline mutation in the *APC* gene, located at the 5q21−22 chromosome locus. The protein product of the *APC* gene serves as a key mediator in the Wnt pathway of signal transduction for cellular growth and proliferation (Fig. 19.4). Wnt binding to cellular receptors initiates a series of downstream signals that result in increased transcription of cell growth and proliferation-associated genes, through the effect of the protein β-catenin. β-catenin is a transcriptional regulator, which must be localized to the nucleus in order to impart its effect on transcription. In the absence of a Wnt-mediated growth signal, APC serves as part of a complex that destabilizes β-catenin through phosphorylation, thereby targeting it for destruction by the proteasome and preventing its nuclear localization and transcriptional effects (Fig. 19.4A). In the presence of Wnt-mediated growth signal, the APC-β-catenin protein complex is disrupted, and phosphorylation cannot occur, resulting in increased stability and nuclear localization of β-catenin (Fig. 19.4B). As is common in many examples of neoplasia, the disease state is one which mimics the activated state, and mutations in APC commonly cause a disruption of the protein complex, which destabilizes β-catenin, resulting in constitutive activation of the Wnt pathway (Fig. 19.4C).

## Familial adenomatous polyposis and related syndromes: molecular diagnosis, clinical management, and genetic counseling

As the majority of cases of FAP have been attributed to mutation in *APC*, there has been extensive investigation of the function of its protein product. In addition to its role in FAP, the understanding of the role of *APC* bears special relevance, as over 70% of non-syndromic CRC are found to have somatic mutation of *APC*. APC codes for a 312 kDa protein that is expressed in a wide range of tissues and is thought to participate in several cellular functions including Wnt-mediated signaling, cell adhesion, cell migration, and chromosomal segregation. Within the APC protein, there are multiple domains which are responsible for these varied functions, including (1) an oligomerization domain, (2) an armadillo domain region, which is thought to be involved in binding of APC to proteins related to cell morphology and motility, (3) β-catenin-binding domain, (4) axin-binding domain, and (5) a microtubule-binding domain (Fig. 19.5). The majority of germline mutations associated with FAP are either frameshift or nonsense mutations, which lead to a truncated protein product, thereby disrupting the interaction with β-catenin, leading to its stabilization and subsequent downstream signaling. Two hotspots have been identified for germline mutations in *APC*, at codons 1061 and 1309, which account for 17% and 11% of all germline APC mutations, respectively (Fig. 19.5). The region between codons 1286 and 1513 is termed the mutation cluster region (MCR) to reflect the observation that this region encompasses many of the identified *APC* mutations.

Both within and beyond the MCR, there is some degree of association between the location of the germline *APC* mutation and the clinical phenotype. Mutations within the MCR are associated with a profuse polyposis, with the development of over 5000 colorectal polyps. Attenuated polyposis is seen in the settings of mutations in the 5′-end of the *APC* gene (exons 4 and 5), the alternatively spliced form of exon 9, or the 3′-distal end of the gene. An intermediate phenotype is observed with mutations between codon 157 and 1249 and 1465 and 1595. Similarly, genotype-phenotype correlations for extraintestinal manifestations have been identified. Desmoid tumors and upper gastrointestinal lesions are clinically the most pressing extraintestinal manifestations, as there is significant morbidity and mortality associated with them. The association of FAP with desmoid tumor formation has been correlated to mutations downstream of codon 1400. An unequivocal correlation between upper gastrointestinal tumors and specific mutations has not been established, although there are data to suggest that some specific mutations are associated with a higher rate of these lesions, including mutations beyond codon 1395, those beyond codon 934, and within codons 564−1465. CHRPE is associated with mutations falling between codons 311 and 1444 (Fig. 19.5).

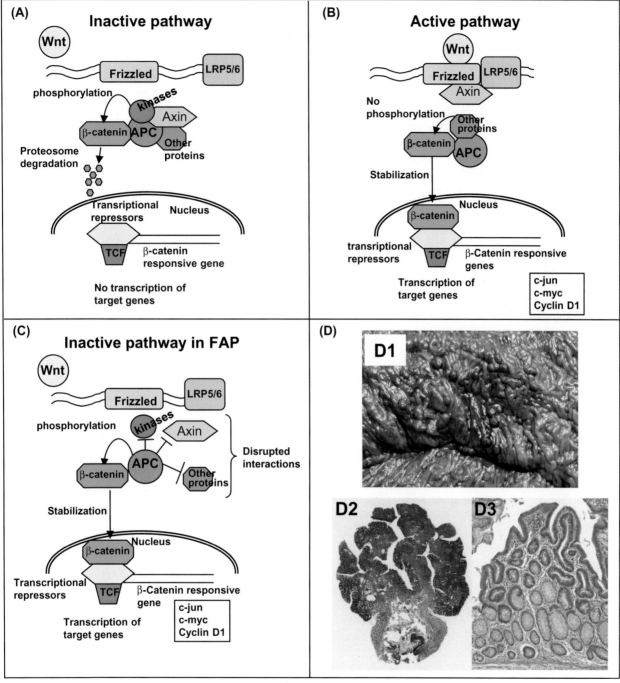

**FIGURE 19.4**   The Wnt pathway and familial adenomatous polyposis. (A) Signaling of the Wnt pathway is mediated through the Frizzled family of receptors, and a co-receptor LRP5 or 6. In the absence of ligand, the pathway is inactive through the negative regulation of the downstream effector β-catenin. When present in sufficient quantity in the nucleus, β-catenin stimulates transcription of target genes. Lack of signaling through the Frizzled receptor results in sequestration of β-catenin in a multiprotein complex including APC, Axin, and other proteins, which exert a negative effect both through the cytoplasmic sequestration of β-catenin and by targeting it for destruction by the proteasome through phosphorylation. (B) In the presence of Wnt ligand, the Frizzled receptor and LRP5/6 form a complex, which results in the recruitment and sequestration of Axin at the cell surface, thereby inhibiting the kinase activity of the APC complex. This results in the stabilization and nuclear translocation of β-catenin, thus resulting in transcription of target genes. (C) In the setting of an *APC* gene mutation, the multiprotein APC/Axin/β-catenin complex is disrupted, most commonly due to truncation of the APC protein in domains responsible for protein-protein interaction. This results in stabilization and nuclear translocation of β-catenin, with the net effect of constitutively active transcription of growth-promoting genes otherwise under tight regulation. (D) Typical appearance of polyps in FAP, which are indistinguishable from spontaneous non-syndromic polyps, but are numerous and may eventually cover most of the surface of the colon with a carpet appearance. Tubular adenoma (D2) and adenomatous colonic mucosa (D3) at an early stage before the development of larger adenomas and adenocarcinomas (H&E stain, original magnification 5×). *(D1) From the files of the Department of Pathology and Cell Biology, Columbia University College of Physicians and Surgeons.*

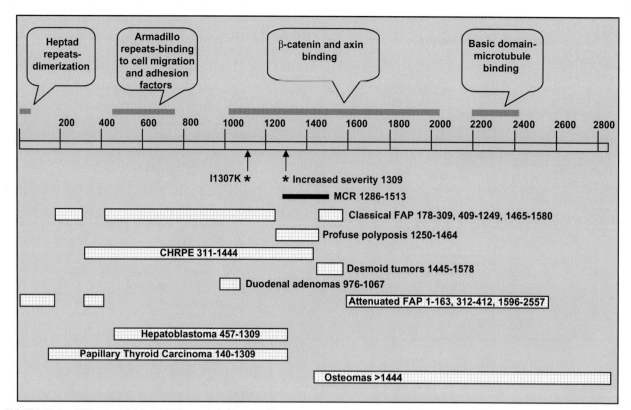

**FIGURE 19.5** APC gene mutation and phenotype correlation in FAP.
The figure illustrates the *APC* gene functional domains and mutation-phenotype correlations. The APC protein product consists of 2843 amino acids with multiple functional domains, including a dimerization domain, microtubule domain, and binding sites for β-catenin and axin (*pictured as gray boxes*). Mutations within certain codon ranges may correlate to a clinical phenotype as depicted in the yellow boxes. Germline mutations generally fall within the entire spectrum of depicted mutational sites, while somatic mutations tend to cluster within the mutation cluster region. Of note, a profuse polyposis phenotype is seen in patients in whom a germline mutation between codons 1250 and 1464 is seen, with the clinical presentation of >5000 polyps. Mutation at codon 1309 is associated with profuse polyposis and earlier onset of disease. I1307 K represents a mutation common in the Ashkenazi Jewish population. *Adapted from Hofgartner WT, Thorp M, Ramus MW, et al. Gastric adenocarcinoma associated with fundic gland polyps in a patient with attenuated familial adenomatous polyposis. Am J Gastroenterol 1999;94:2275—81; Ahnen DJ. The genetic basis of colorectal cancer risk. Adv Intern Med 1996;41:531—52; Michils G, Tejpar S, Thoelen R et al. Large deletions of the APC gene in 15% of mutation-negative patients with classical polyposis (FAP): a Belgian study. Hum Mutat 2005;25:125—34*

Well-established methodologies are in place for screening patients in whom a diagnosis of FAP is a consideration. Recommendations for indications and approach have been suggested and were reported by the *American College of Gastroenterology*. FAP screening is indicated in individuals who have a personal history of >10 cumulative colorectal adenomas, a family history of one of the adenomatous polyposis syndromes, or a history of adenomas and FAP-type extracolonic manifestations. Genetic testing of patients with suspected adenomatous polyposis syndromes should include *APC* and *MUTYH* gene mutation analysis.

In de novo cases or in cases in which a family mutation is unknown, there are several high-throughput approaches to conduct the mutational analysis. Sequencing of the entire coding region is the gold standard for diagnosis, although other methods such as protein truncation tests and mutation scanning approaches have been used. In some settings, initial targeting of common mutational hotspots is the preferred method for initial screening, followed by more extensive sequence analysis if the common mutations are not identified. Once a mutation is identified, targeted genetic testing can be carried out for potentially affected family members. Additionally, there is a growing use of such targeted genetic testing as part of preimplantation genetic diagnosis.

In a subset of patients, a discrete *APC* mutation is not identified using first-line approaches to diagnosis. In some cases, adjusting the approach to screen individual alleles of *APC* yields a diagnostic mutation, or in some cases large exonic or entire gene deletions. However, in other cases, these approaches do not identify a discrete *APC* mutation. An alternative target for genetic testing has recently been identified as *MUTYH*, in which bi-allelic mutations are detected in a significant minority of *APC* mutation-negative cases of polyposis and is correlated to cases of AFAP. There have been a number of approaches for identifying the underlying molecular changes in *APC* mutation-negative cases not resolved by the evaluation of individual alleles or *MUTYH* gene, including examination of epigenetic regulation of the *APC* gene, evaluation of genes encoding other proteins involved in the β-catenin pathway (such as Axin), evaluation of allelic mRNA ratios, and evaluation of somatic *APC* mosaicism. Germline hypermethylation of the *APC* gene has been shown not to be a significant cause of FAP in *APC* mutation-negative cases. One study has shown unbalanced *APC* allelic mRNA expression in cases in which no other discrete mutation was identified. Interestingly, in the tumor specimens from these cases, there was loss of the remaining wild-type allele, indicating that the reduced dosage of *APC* and unbalanced *APC* allelic mRNA expression may create a functional haploinsufficiency that engenders the same predisposition to colorectal carcinogenesis. The mechanism of unbalanced allelic mRNA expression is unclear. Mutations in *AXIN2* have also been rarely identified in *APC* mutation-negative FAP.

Surveillance of those affected by FAP begins early in life with annual sigmoidoscopy or colonoscopy (beginning at age 10−12 years), followed by prophylactic colectomy, usually by the time the patient reaches his or her early 20 s. Additionally, regular endoscopy with full visualization of the stomach, duodenum, and peri-ampullary region has been recommended. However, the optimal timing of these screening evaluations is not well established and is generally managed based on the severity of upper gastrointestinal disease burden. Additional recommendations are based on extracolonic manifestations that have the propensity for morbidity and mortality. These include annual palpation of the thyroid, with some advocating a low threshold for referral for ultrasound examination and serum alpha-fetoprotein levels and abdominal palpation every 6 months in young children in FAP families to detect hepatoblastoma.

## Key concepts

- Gastric carcinoma is the fifth most frequent cancer worldwide and the third most common cause of death from cancer. Most gastric cancers develop as sporadic cancers, without a well-defined hereditary predisposition. The majority of sporadic gastric cancers arises in a background of chronic gastritis, which is most commonly caused by *H. pylori* infection of the stomach.
- The pathogenesis of gastric cancer is multifactorial, resulting from the interactions of host genetic susceptibility factors, environmental exogenous factors that have carcinogenic activity such as dietary elements (salted, smoked, pickled, and preserved foods) and smoking, and the complex damaging effects of chronic gastritis related to *H. pylori* infection.
- The Cancer Genome Atlas (TCGA) Research Network characterized gastric adenocarcinomas into 4 distinct subtypes based on molecular features: Epstein Barr Virus (EBV)-associated tumors, microsatellite instability-high (MSI) tumors, genomically stable tumors (GS), and chromosomal instability (CIN) tumors. Molecular testing for genomic alterations is necessary for personalized targeted therapies of patients with advanced gastric cancers.
- Colorectal cancer is one of the most frequent types of cancers worldwide with approximately 1 million cases annually. Most colorectal cancers do not develop in association with a hereditary cancer syndrome. However, several hereditary colon cancer syndromes have been characterized, including Lynch syndrome (HNPCC), familial adenomatous polyposis (FAP), and several others.
- The molecular pathways of colon cancer development include a stepwise acquisition of mutations, epigenetic changes, and alterations of gene expression, resulting in uncontrolled cell division, and manifestation of invasive neoplastic behavior. The major molecular pathways of colorectal cancer development include (1) the chromosomal instability pathway (CIN), (2) the microsatellite instability pathway (MSI), and (3) the CpG island methylator pathway (CIMP).
- A comprehensive molecular characterization of human colon and rectal cancer has been reported by the Cancer Genome Atlas Research Network. Molecular testing for genomic alterations is necessary for personalized targeted therapies of patients with advanced colorectal cancers.
- HNPCC is an autosomal dominant cancer predisposition syndrome, characterized by early onset CRC (accounting for 4%−6% of CRC) and tumors in other organs (including endometrium, ovary, urothelium, stomach, brain, and sebaceous glands). HNPCC patients inherit germline mutations in one of the DNA mismatch repair genes (most commonly in the *MLH1* and *MSH2*), leading to defects in the corresponding DNA mismatch repair.
- FAP is a cancer predisposition syndrome characterized by numerous adenomatous colorectal polyps, with uniform progression to colorectal carcinoma at an early age, accounting for more than 1% of all CRC cases in the United States. The majority of FAP cases are related to autosomal dominant inheritance of germline mutations in the APC gene on chromosome 5q. Clinically, the common manifestation of FAP is the presence of numerous (sometimes >1000) adenomatous polyps distributed throughout the colon and rectum.

# Suggested readings

[1] Correa P. Human gastric carcinogenesis: a multistep and multifactorial process--first American Cancer Society Award Lecture on Cancer Epidemiology and Prevention. Cancer Res 1992;52:6735−40.

[2] Sepulveda AR. Helicobacter, inflammation, and gastric cancer. Curr Pathobiol Rep 2013;1:9−18.

[3] Uehara T, Ma D, Yao Y, et al. H. pylori infection is associated with DNA damage of Lgr5-positive epithelial stem cells in the stomach of patients with gastric cancer. Dig Dis Sci 2013;58:140−9.

[4] Cancer Genome Atlas Research Network. Comprehensive molecular characterization of gastric adenocarcinoma. Nature 2014;513:202−9.

[5] Barber M, Fitzgerald RC, Caldas C. Familial gastric cancer − aetiology and pathogenesis. Best Pract Res Clin Gastroenterol 2006;20:721−34.

[6] Vogelstein B, Fearon ER, Hamilton SR, et al. Genetic alterations during colorectal-tumor development. N Engl J Med 1988;319:525−32.

[7] Peltomaki P. Role of DNA mismatch repair defects in the pathogenesis of human cancer. J Clin Oncol 2003;21:1174−9.

[8] Cancer Genome Atlas Research Network. Comprehensive molecular characterization of human colon and rectal cancer. Nature 2012;487:330−7.

[9] Peltomaki P. Role of DNA mismatch repair defects in the pathogenesis of human cancer. J Clin Oncol 2003;21:1174−9.

[10] Sepulveda AR, Hamilton SR, Allegra CJ, et al. Molecular biomarkers for the evaluation of colorectal Cancer: guideline from the American Society for Clinical Pathology, College of American Pathologists, Association for Molecular Pathology, and American Society of Clinical Oncology. J Mol Diagn 2017;19:187−225.

[11] Giardiello FM, Allen JI, Axilbund JE, et al. Guidelines on genetic evaluation and management of Lynch syndrome: a consensus statement by the US Multi-Society Task Force on colorectal cancer. Gastroenterology 2014;147:502−26.

[12] Lipton L, Tomlinson I. The genetics of FAP and FAP-like syndromes. Fam Cancer 2006;5:221−6.

[13] Syngal S, Brand RE, Church JM, et al. ACG clinical guideline: genetic testing and management of hereditary gastrointestinal cancer syndromes. Am J Gastroenterol 2015;110:223−62.

Chapter 20

# Molecular basis of liver disease

Satdarshan P.S. Monga[1,2] and Jaideep Behari[2]

[1]Division of Experimental Pathology, Department of Pathology, Pittsburgh Liver Research Center, School of Medicine, University of Pittsburgh, Pittsburgh, PA, United States, [2]Division of Gastroenterology, Hepatology, and Nutrition, Department of Medicine, Pittsburgh Liver Research Center, School of Medicine, University of Pittsburgh, Pittsburgh, PA, United States

**Summary**

Liver diseases are a cause of global morbidity and mortality. While the predominance of specific liver diseases varies with geographical location, the diversity of hepatic diseases affecting the underdeveloped, developing, and developed countries is high, with diseases ranging from infectious diseases of the liver to neoplasia and obesity-related illnesses. This chapter will discuss progress made toward an improved understanding of the cellular and molecular basis of various aspects of liver pathophysiology, including the processes of hepatic development, metabolic zonation, and liver regeneration. Equally relevant to this review is recent progress toward elucidation of basic mechanisms that lead from hepatic injury and inflammation to hepatic fibrosis and cirrhosis. Furthermore, there has been a notable improvement in our understanding of cellular and molecular aberrations in liver diseases such as alcoholic liver disease, non-alcoholic fatty liver disease, and both benign and malignant tumors of the liver, in the hope of improving diagnostic, prognostic, and therapeutic tools. In this chapter, we concisely discuss fundamental concepts in hepatic pathobiology, with emphasis on cellular and molecular mechanisms, summarizing the biological and translational significance of such concepts.

## Molecular basis of liver development

Embryonic liver development is a function of precisely regulated spatiotemporal signals throughout the development process (Fig. 20.1). During gastrulation and early somitogenesis, the endoderm is patterned into the foregut, midgut, and hindgut. In the anterior endoderm, the suppression of Wnts and Fibroblast growth factor 4 (FGF4) results in foregut fate. The next major step is that of hepatic competence, brought about by expression of pioneer transcription factors FOXA and GATA, that occupy the albumin (*Alb*) gene enhancer prior to liver specification and have the ability to bind and open compacted chromatin. In addition, BMP signaling plays a positive role in hepatic competence, in part, by enhancing histone acetylation, which can influence FOXA1 binding. Liver specification or hepatic induction occurs next, under the effect of FGF, BMP, and Wnt, leading to endodermal cells giving rise to hepatoblasts. These hepatic-inducing signals arise from the cardiac mesoderm and septum transversum mesenchyme (STM) in mice, and the lateral plate mesoderm (LPM) in zebrafish. FGF ligands, FGF1 and FGF2, but not FGF8, could replace the cardiac mesoderm to induce *Alb* expression from the foregut endoderm in mouse explant cultures. The mitogen-activated protein kinase (MAPK) pathway appears to be the signaling molecule downstream of FGFs in liver specification. *Bmp2* and *Bmp4* were shown to be expressed in the STM when liver specification occurs in mice. In vivo studies showed a missing liver bud in the *Bmp4* mutant mice. An unbiased forward genetic screen in zebrafish led to the identification of *wnt2bb* mutants, which display very small or lack of liver buds. Prox1 expression is greatly reduced in this mutant, and overexpression of the dominant-negative Tcf blocked liver formation, suggesting the positive role of Wnt/β-catenin signaling in liver specification. Foxa3-cre-driven β-catenin deletion is evident at E9.5 in hepatoblasts and did not affect the hepatoblast compartment at this stage.

This stage is followed by the phase of embryonic liver growth characterized by the expansion and proliferation of the resident cells within the hepatic bud. Several transcription factors including Hex, Gata6, and Prox1 are the earliest known mediators of this phase. Once the hepatic program is fully expressed, liver growth continues and progresses to hepatic morphogenesis. The epithelial cells at this stage are now considered the hepatoblasts, or bipotential progenitors, which means that they are capable of giving rise to both major lineages of the liver, the hepatocytes, and the biliary

Essential Concepts in Molecular Pathology. DOI: https://doi.org/10.1016/B978-0-12-813257-9.00020-6

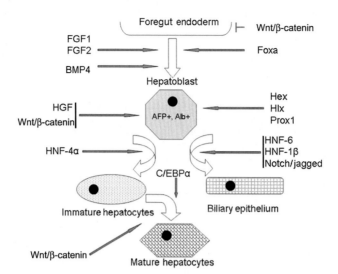

**FIGURE 20.1** Summary of molecular signaling during liver development in the mouse. *AFP*, α-fetoprotein; *Alb*, albumin; *BMP*, bone morphogenic protein; *C/EBP*α, CCAAT enhancer−binding protein-alpha; *FGF*, fibroblast growth factor; *HGF*, hepatocyte growth factor; *HNF*, hepatocyte nuclear factor.

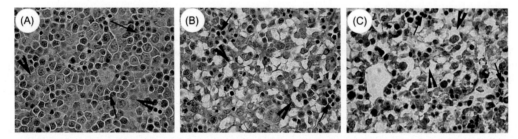

**FIGURE 20.2** Histology of the developing mouse liver. (A) several hematopoietic cells (*arrows*) are seen interspersed among hepatoblasts (*arrowheads*), which display large nuclei and scanty cytoplasm in a histological section from E14 liver (H&E (hematoxylin and eosin) stain). (B) Fewer hematopoietic cells (*arrows*) are observed among the hepatocytes (*arrowheads*), which begin to display cuboidal morphology, large clear cytoplasm, and polarity in a histological section from E17 liver (H&E stain). (C) Similar cuboidal morphology of hepatocytes (*arrowheads*) is seen in a histological section from E19 mouse liver (H&E stain).

epithelial cells. Hepatoblasts undergo expansion under the control of BMP, FGF, hepatocyte growth factor (HGF), Wnt, and RA signaling that activate PI3 kinase, β-catenin, smad, MAPK, and Jun N-terminal kinase (JNK) signaling. Additional factors such as NF-κB, c-jun, XBP1, K-ras, and others are also important during hepatic morphogenesis.

The final stage is characterized by the differentiation of hepatoblasts to mature functional cell types—the hepatocytes and the biliary epithelial cells. As cells mature to hepatocytes, they acquire cuboidal morphology and show clear cytoplasm due to accumulation of glycogen (Fig. 20.2). Maturing hepatocytes begin to express genes that are associated with functions including various cytochrome P450s and enzymes with metabolic and synthetic functions. Various factors such as oncostatin M, glucocorticoids, HGF, Wnt, and others regulate hepatocyte maturation. Oncostatin M is released from developing hematopoietic compartment within the liver and is an important mediator of hepatocyte differentiation. Additionally, HGF promotes hepatic morphogenesis having effects on hepatoblast expansion along with hepatocyte maturation. Wnt signaling through β-catenin was shown to promote hepatocyte maturation. The major transcription factors that are expressed in hepatocytes as they emerge from hepatoblasts include FoxA1/2, HNF1α and HNF1β, HNF4α, and HNF6. The downstream effectors of these various signaling cascades may not only directly induce target gene expression responsible for hepatocyte functions but also induce indirectly through induction of hepatocyte-enriched transcription factor expression. In fact, HNF4α has been shown to bind to promoters of nearly half of the genes associated with hepatocyte polarity, junctional integrity, and metabolic and xenobiotic functions.

Several signaling molecules are implicated in lineage determination of hepatoblasts for cholangiocyte differentiation. There exists a gradient of activin/TGF-β with the highest exposure limited to the cells around the portal mesenchyme. This has been shown to be important for differentiation of hepatoblasts to cholangiocytes. Similarly, Jagged1 expression occurs in the periportal mesenchyme and biliary cells, whereas Notch2 is present in biliary cells. The function of Notch signaling in biliary differentiation is also substantiated by the phenotype observed on hepatic inactivation of RBP-Jκ, a transcriptional effector of Notch signaling, which results in diminished biliary differentiation of hepatoblasts. Notch signaling regulates the expression of Hes1, which is required later for biliary tubulogenesis. It should be noted that even before the discovery of temporal expression and regulation of Notch signaling during intrahepatic bile duct development, human Alagille syndrome, which is associated with paucity of bile ducts, was identified to be due to mutations in Jagged-1 (JAG1) or Notch-2 (NOTCH2) resulting in deficient Notch signaling. Some of the key transcription factors that allow for biliary differentiation include Sox9, whereas expression of Onecut transcription factors HNF6 and OC-2 counteracts the activin/TGF-β signaling away from the portal mesenchyme where hepatoblasts differentiate into hepatocytes and acquire distinct hepatocyte-enriched transcription factors such as HNF4α. It is relevant to note that in response to mesenchyme signals, the hepatoblasts located adjacent to the portal vein not only upregulate biliary factors such as Sox9, HNF6, OC2, and HNF1β but also concurrently downregulate hepatocyte-enriched transcription factors such as HNF4α. In addition, there are important roles for HNF6 and OC2 in modulating hepatoblast differentiation to hepatocytes and cholangiocytes. An important role of C/EBPα in repressing gene expression associated with both HNF6 and HNF1β has been shown in hepatocytes in the liver parenchyma, whereas suppression of its expression in periportal hepatoblasts allows for a higher expression of both HNF6 and HNF1β. It has also been reported that adjacent to the single layer of primary cholangiocytes derived from hepatoblasts, a second layer of cholangiocytes appears later in hepatic development but only at specific locations. This layer is derived from undifferentiated hepatoblasts that exist toward the hepatocyte parenchyma in developing livers. A luminal space between the two cell layers has been observed and as these appose to form asymmetric ducts with lumen, the hepatoblasts undergo rapid differentiation to cholangiocytes. Ductular morphogenesis follows next and is under the control of Notch signaling, and Hes1 plays an important role in this process.

## Molecular basis of metabolic zonation in the liver

In an adult liver, hepatocytes acquire location-specific functions within a hepatic lobule, a process termed as metabolic zonation. This term was initially coined by Jungermann who pointed out the distinct properties of cells within the liver lobule based on their location. The hepatocytes located in close proximity to the hepatic inflow around the portal triad are referred to as being in the periportal zone (zone 1), whereas those around central vein are located in the centrizonal or pericentral zone (zone 3). The functions of hepatocytes in periportal, midzonal, or pericentral zone are distinct. Zonation of the hepatic lobule requires expression of specific genes in pericentral versus periportal hepatocytes for optimum hepatic function in regulating metabolism.

β-Catenin signaling was discovered to be active in the pericentral area, and many of the genes downstream of this signaling pathway are highly expressed in the hepatocytes located in this zone. Hepatic tissues of mice with liver-specific knockout of β-catenin do not express *Glul* (encodes glutamine synthetase (GS)), *Cyp2e1* (encodes cytochrome p450 2e1), or *Cyp1a2*. Expression of GS, cytochrome p450 2e1 (Cyp2e1), Cyp1a2, and certain glutathione S-transferases is under the control of β-catenin. More recent studies have shown that knockout of Wnt co-receptors low-density lipoprotein—related protein 5 and 6 (LRP5/6) from mouse hepatocytes prevents expression of these genes, showing that β-catenin is regulated by WNT signaling in mouse liver. The source of Wnt proteins seems to be endothelial cells that may be lining the central veins. Wnt2 and Wnt9b may be the specific Wnt proteins from these cells that regulate basal β-catenin activity in these hepatocytes in an adult liver. Recently, R-Spondin-Lgr has been shown to contribute to activation of pericentral gene expression in partnership with the Wnt-β-catenin signaling.

How is β-catenin activity restricted to pericentral hepatocytes? Because Wnt proteins are acylated, which in turn is essential for their biologic activity, this posttranslational modification renders them insoluble. In this state, Wnts can only diffuse to limited distances that restrict their activity to pericentral hepatocytes. Furthermore, it is feasible that there are two kinds of β-catenin—TCF targets, namely, low threshold and high threshold. It is likely that β-catenin targets expressed in pericentral hepatocytes are high threshold meaning that very high levels of Wnt proteins are required to stabilize and activate β-catenin in sufficient quantity to induce their expression. Furthermore, β-catenin activity in zones distant from the central vein is maintained at low levels because of basally higher expression of the APC gene product, which is responsible for β-catenin degradation.

How is the dynamic process of metabolic zonation regulated? Although HNF4α is considered to be a liver-enriched transcription factor, it regulates expression of genes in periportal hepatocytes. Loss of HNF4α from hepatocytes increases expression of pericentral genes by the periportal hepatocytes. Therefore, there appears to be a mechanism whereby HNF4α suppresses gene activation by β-catenin and TCF4. The consensus sequence of the WNT-responsive element in gene promoters is similar to that of the HNF4α-responsive element. Hence, HNF4α can bind to the TCF4-binding site and vice versa. The presence of HNF4α prevents β-catenin-dependent transcription, whereas the presence of β-catenin prevents HNF4α-dependent transcription. This could be the basis of overall metabolic zonation in the liver. Further studies are needed to elucidate these interactions and their regulation, but the functional interplay between Wnt/β-catenin signaling pathway and IINF4α signaling may regulate the process of metabolic zonation.

Another mechanism of zonation may be the reciprocal relationship between Wnt and Yap signaling. β-Catenin activity is restricted to the pericentral hepatocytes while Yap signaling may be driving expression of target genes in the periportal zone, in addition to being active in cholangiocytes. The authors induced endogenous Yap activation through deletion of *Mst1/Mst2* deletion that led to conversion of hepatocytes in the periportal area to duct-like cells, but intriguingly led to diminution of pericentral gene expression program. Furthermore, acute loss of YAP led to expansion of the GS-positive hepatocytes to multiple layers around central vein. This suggests a push and pull mechanism exists between the two signaling pathways that may critical for maintaining liver zonation.

## Molecular basis of liver regeneration

The liver is a unique organ with an innate ability to regenerate, which is important because of its strategic location and function. After absorption from intestines, both nutrients and toxins along with bacterial products make their way to the liver by way of the portal vein. The liver is vital to functions such as detoxification, metabolism, and synthesis, which contribute to overall homeostasis. The ability of liver to regenerate may suggest the criticality of the need for a minimal functional hepatic mass to perform these vital functions. One of the best models to study the ability of liver to regenerate is after surgical resection of the two-thirds of liver mass, which triggers the process of regeneration. Liver regeneration is enabled by concurrent activation of multiple signaling pathways based on evidence from many animal studies. This ensures proliferation and expansion of all liver cell types, enabling restoration of lost hepatic mass. Various pathways that form the basis of initiation, continuation, and termination of the regeneration process have been identified which has implications in hepatic regenerative medicine including stem cell differentiation, tissue engineering, cell therapy, and in the setting of liver transplantation. Dysregulation of such pathways is also often seen in aberrant growth in benign and malignant liver tumors.

Partial hepatectomy triggers a sequence of events that proceeds in an orderly fashion to restore the lost mass within 7 days in rats, 10−14 days in mice, and 8−15 days in humans. Following these periods, the liver lobules become larger and the thickness of hepatocyte plates is doubled as compared with the prehepatectomy livers. However, over several weeks there is gradual lobular and cellular reorganization, leading to an unremarkable and indistinguishable liver histology compared to a normal liver. It should be noted that all cell types within the liver undergo replication after surgical resection, albeit at specific times. Peak hepatocyte proliferation is evident at 24 h and 40−48 h after hepatectomy in rats and mice, respectively. There is a 24−48 h lag in the proliferation of biliary epithelial cells, stellate cells, macrophages, and endothelial cells.

Immediately following the hepatectomy, there is a dramatic change in the hemodynamics of the liver especially because relatively same volume of blood continues to traverse through a reduced tissue mass, which may induce shear stress on sinusoidal endothelial cells. Although very few studies conclusively address a role of these changes in initiation of regeneration, there are reports of the impact of such changes on activation of HGF, a major initiator of regeneration. Key metabolic changes including hypogyclemia also occur early after hepatectomy. It's correction via dextrose supplementation affects hepatocyte proliferation. Similarly, microsteatosis is noted during early liver regeneration. Suppression of hepatocellular fat accumulation by either leptin supplementation or performing hepatectomies on hepatocyte-specific knockouts of the glucocorticoid receptor, are associated with impaired hepatocellular proliferation in mice.

More than 95% of hepatocytes will undergo cell proliferation during the process of regeneration. The earliest known signals involve both the growth factors and cytokines. Although it is difficult to layout the exact chronology of events, it is important to say that many events are concomitant, concurrently ensuring proliferation and maintaining liver function. The earliest events observed include activation of urokinase plasminogen activator (uPA) enabling activation of plasminogen to plasmin, which induces matrix remodeling that leads to activation of HGF from the bound hepatic matrix. HGF is a known hepatocyte mitogen that acts through its receptor c-Met, a tyrosine kinase, and is a master effector of hepatocyte proliferation and survival both in vitro and in vivo. Similarly, EGF, which is continually present in the portal circulation, also promotes hepatocyte proliferation.

Other factors activated during liver regeneration include the Wnt/β-catenin pathway, Notch/Jagged pathway, norepinephrine, serotonin, and TGF-α. These factors work in autocrine and paracrine fashion, and cell sources of the various factors include the hepatocytes, Kupffer cells, stellate cells, and sinusoidal endothelial cells. For example, an important role of Wnt/β-catenin signaling during liver regeneration has been established. It is now known that β-catenin signaling in hepatocytes during the process of liver regeneration is Wnt dependent, and these Wnt proteins originate from sinusoidal endothelial cells and macrophages. Concurrently, tumor necrosis factor-α (TNF-α) and interleukin-6 (IL-6) are being released from the Kupffer cells and have been shown to be important in normal liver regeneration through genetic studies. These pathways are known to act through NF-κB and Stat3 activation. Bile acids have also been shown to play a vital role in normal liver regeneration through activation of transcription factors such as FoxM1b and c-myc, necessary for cell cycle transition as well, and decreased hepatocyte proliferation after partial hepatectomy was observed in animals depleted for bile acids with the use of cholestyramine or in the FXR null mice. The eventual goal of these changes is to initiate and cell cycle in hepatocytes with a successful G1 to S phase transition dependent on key cyclins such as cyclin A, cyclin D, and cyclin E to ensure DNA synthesis and mitosis. Additional growth factors such as Yap, FGF, PDGF, and insulin are also important in liver regeneration and might contribute to normal regeneration process.

In liver regeneration, while the hepatocytes are the main cell type undergoing proliferation, division of nonparenchymal occurs at specific times as well. Peak proliferation of the biliary epithelial cells occurs at 48 h, Kupffer and stellate cells at 72 h, and sinusoidal endothelial cells at 96 h. Several key growth factors are mitogenic for these cell types and work in an autocrine as well as paracrine manner, because several of these factors are produced by one or more cell types within the regenerating liver. Factors such as HGF, PDGF, vascular endothelial growth factor (VEGF), FGFs, and angiopoietins are important for Kupffer cells and endothelial cell proliferation and homeostasis during regeneration. Neurotrophins and nerve growth factors have been shown to be important in regeneration and survival of stellate cells during regeneration.

Liver regeneration is terminated after it reaches its pre-hepatectomy mass. The liver mass is restored after hepatectomy at 14 days in mice and 7 days in rats. It may initially exceed its original mass, when transient apoptosis gets initiated. Based on the mitoinhibitory action of TGF-β on the hepatocytes, it has been suggested to be a terminator of regeneration. Recent sets of studies have also shown an important role of extracellular matrix (ECM) in regulating liver growth at baseline and after regeneration. Integrin-linked kinase (ILK), which is a cell−ECM−adhesion component, is implicated in cell−ECM signaling via integrins. Deletion of ILK from hepatocytes led to increased proliferation, which was associated with increased Yap and β-catenin activation. After this transient proliferation of all epithelial components, proliferation subsided and final liver to body weight ratio in livers with ILK-deficient hepatocytes stabilized at around double that of littermate controls. A follow-up study showed that when these mice are subjected to partial hepatectomy, there was a termination defect in liver regeneration such that the mice with ILK-deficient hepatocytes regenerated to 58% larger liver than their original pre-hepatectomy weight.

Multiple cytokines and growth factors are activated in responses to partial hepatectomy. No single signal transduction pathway's suppression leads to complete abrogation of liver regeneration after partial hepatectomy in rodents. Inhibition of EGFR, TGF-α, Jagged/Notch, and Wnt/β-catenin; of cytokines, such as TNF-α and IL-6; and of additional modulators, such as bile acids, norepinephrine, serotonin and components of hepatic biomatrix, and components of the complement, all show varying range of decrease or delay of regeneration. Interference with the HGF/Met signaling pathway appears to cause the most profound but still not complete interference with the regenerative process because of redundancy with EGF. In addition, when all else fails, additional cell types such as the oval cells can be called on to restore hepatocytes. These cells are the facultative hepatic progenitors or transient amplifying progenitor cells that originate from the two mature epithelial cells of the liver based on the repair needed. Thus, during various forms of hepatic injury, first molecular signaling, and then cellular redundancy, ensures repair and eventually the health of this indispensable Promethean organ.

## Liver stem cells in liver health and disease

Despite its capacity to regenerate and the presence of redundant molecular signaling that allows for regeneration to occur even under most extraordinary circumstances such as in absence of key components of various mitogenic pathways, there is sometimes a need for a progenitor cell activation in the liver. These facultative adult progenitors (or oval cells) exhibit oval nuclei, small shape, and high nuclear to cytoplasmic ratio. Oval cells proliferate and expand and have been shown to be precursors of both hepatocytes and biliary epithelial cells, based on the type of cell injury. It should be emphasized that unlike gut, hematopoiesis, and skin, the liver does not depend on an active progenitor cell compartment for homeostasis. Adult hepatocytes and cholangiocytes are capable of replacing slowly dying cells over the lifespan of the organ.

The basal presence of these facultative stem cells or oval cells or transiently amplifying hepatic progenitors in a normal adult liver remains debated. However, on certain specific kinds of injury, there is a clear appearance and activation of hepatic progenitors in adults, which can then give rise to mature epithelial cells of the liver. The appearance of these cells is almost always periportal. Label-retaining studies have shown oval cells to be a subset of biliary epithelial cells residing in proximal biliary branches and intralobular ducts within the space of Disse. In fact, several studies have shown that cholangiocytes express many markers that have traditionally been associated with progenitor cell markers, even at baseline. The prerequisite to expansion of these cells is that an injury to the hepatocytes should coexist with the inability of mature hepatocytes to proliferate. This creates a selection and proliferative pressure on oval cells, which eventually differentiate into hepatocytes. This scenario is classically attained in rats by the use of acetylaminofluorine (AAF), which cross-links DNA in hepatocytes (rendering them unable to proliferate), followed by surgical partial hepatectomy that leads to appearance of oval cells in periportal area (Fig. 20.3). The origin of oval cells in this model has been quite convincingly shown to be from cholangiocytes. Administration of biliary toxin methylene dianiline in the AAF−partial hepatectomy model led to a substantial reduction in the appearance of oval cells. Furthermore, oval cell activation occurred both in response to periportal or pericentral injury. The evidence that oval cells give rise to hepatocytes in vivo was shown in this model by labeling of oval cells with the use of tritiated thymidine to follow their differentiation into hepatocytes. Does oval cell activation occur in mice?

Unfortunately, AAF does not work in mice because they lack the enzyme required for metabolic activation of this toxin. In mice, alternatives used include administration of diets containing 3,5-diethoxycarbonyl-1,4-dihydrocollidine (DDC) or the choline-deficient, ethionine-supplemented (CDE) diet. However, DDC primarily causes biliary injury due to formation of porphyrin plugs in bile ducts, which leads to extravasation of bile into the parenchyma, and hence the injury is also primarily periportal. The response is in the form of atypical ductular proliferation or reaction, which has sometimes been equated with an oval cell response (Fig. 20.3). However, fate-tracing studies show that this injury model, like other cholestatic injury models such as bile duct ligation (BDL), lacks repair of hepatocytes from the atypical ductular reaction or oval cells. Similarly, repair in the livers of CDE diet (which causes a notable hepatocyte injury) was also shown to be non-oval cell dependent. In fact, the major difference between AAF/partial hepatectomy model in rats and models such as DDC, CDE, and BDL in mice is that, while hepatocytes proliferation is completely inhibited in the former, it continues uninterrupted in the latter group. Thus, there is a lack of selective pressure on the cholangiocytes to transdifferentiate to hepatocytes when uninjured hepatocytes continue to replicate and replace lost mass to maintain liver function and enable survival. This hypothesis was directly tested where hepatocytes deficient in $Mdm2$, which have high levels of expression of p21, a known mitoinhibitor. When these mice were subjected to CDE diet, labeled cholangiocyte-derived hepatocytes were clearly observed as a mechanism of liver repair. Several additional studies have now shown this phenomenon. Although it is quite clear that under appropriate conditions biliary epithelial cells (or oval cells) do give rise to hepatocytes and more biliary epithelial cells, the molecular pathways underlying these reprogramming events remain an enigma. Elucidating such pathways could be exploited to promote repair in innovative ways.

DDC primarily causes a biliary injury that also leads to periportal hepatocyte injury. Hence the phenotype observed is one that of atypical ductular proliferation along with appearance of periportal intermediate hepatocytes that often

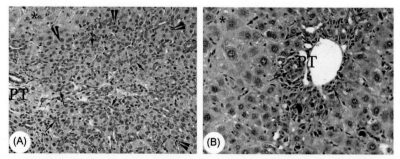

**FIGURE 20.3** Histology demonstrating the activation of oval cells or facultative adult liver stem cells in experimental models. (A) Several cells with oval shape, smaller size, and higher cytoplasmic to nuclear ratio (*arrows*) are observed around the portal triads (PTs) in rat livers after acetylaminofluorine/PHx treatment (H&E (hematoxylin and eosin) stain). Several intermediate hepatocytes (*arrowhead*) are evident as these oval cells undergo progressive differentiation. The * denotes normal hepatocytes adjoining the region of ongoing oval cell activation, expansion, and differentiation. (B) Mouse liver after 2 weeks of 3,5-diethoxycarbonyl-1,4-dihydrocollidine administration displays atypical ductular hyperplasia (*arrows*), equivalent to oval cell activation, and (*) adjoining normal hepatocytes (H&E stain).

express biliary markers (Fig. 20.3). Are these polygonal intermediate hepatocytes representative of differentiation of atypical ductules toward hepatocytes or dedifferentiation of hepatocytes toward biliary cells? Fate-tracing studies have shown that it is likely that these cells are hepatocytes transdifferentiating toward biliary epithelial cells. Michalopoulos and colleagues first put forth this concept when they observed biliary epithelial cells derived from hepatocytes when using in vitro rolling bottle cultures of hepatocytes. In a follow-up study, they showed such event also occurs in rats in vivo based on the observation that a subset of bile ducts was derived from transplanted chimeric hepatocytes. Our group showed that when $\beta$-catenin conditional knockout mice (lacking $\beta$-catenin in 100% cholangiocytes and more than 95% hepatocytes) were subjected to chronic DDC diet, islands of $\beta$-catenin-positive hepatocytes emerged and expanded. Intriguingly, only after expansion of such hepatocyte islands, did some biliary structures appear that were $\beta$-catenin-positive, suggesting the derivation of cholangiocytes from hepatocytes via this lineage trace. More elegant fate-tracing studies have now clearly shown that these events commonly occur after chronic biliary damage secondary to DDC and CDE, although this phenomenon repairs only a small fraction of bile ducts. While a more comprehensive molecular basis of this repair needs to be investigated for therapeutic exploitation, roles of Notch and Yap signaling in hepatocyte to biliary differentiation have been suggested. At the same time, hepatocytes expressing stable $\beta$-catenin after chronic DDC diet showed a notable improvement in cholestasis as evident by lower alkaline phosphatase levels, which coincided with enhanced expression of biliary markers in almost all hepatocytes. Mores studies will again be of essence to elucidate the exact mechanism.

Various markers have been applied to detect these oval cells. By histology, these cells are smaller than the hepatocytes and possess high nuclear to cytoplasmic ratio and typically seen in periportal region. These cells are concomitantly positive for biliary (CK-19, CK-7, A6, OV6), hepatocyte (Hepar-1, albumin), and fetal hepatocyte markers ($\alpha$-fetoprotein). However, at any given time, only a subset of these cells are positive for all markers possibly reflecting different stages of differentiation of these cells or heterogeneity in these cells. Reactive ducts are also positive for neural cell adhesion molecule or vascular cell adhesion molecule. Additional surface markers for the oval cells have been identified in rodent studies, which give an advantage in cell-sorting studies. The six unique markers include CD133, claudin-7, cadherin 22, mucin-1, Ros1 (oncogene v-ros), and $\gamma$-aminobutyrate, type A receptor $\pi$ (Gabrp). More recently, markers such as Lgr5, Trop2, and Foxl1 have been shown to be of relevance as more specific markers of oval cells and more studies will be required to demonstrate their true function and regulation.

Progenitor cell activation has been seen in patients after various forms of hepatic injury. This is an attempt by the diseased liver to restore the lost cell type to maintain hepatic functions. An acute ductular reaction is observed in the setting of submassive necrosis due to hepatitis, drugs, alcohol, or cholestatic disease, which is subtle during the early stages. These cells expand in numbers and show hepatocytic differentiation with passage of time as revealed by limited studies from serial liver biopsies from these patients. Hepatic progenitor cell activation has been observed in alcoholic and non-alcoholic fatty liver disease (ALD and NAFLD) patients as well. ALD and NAFLD are associated with increased lipid peroxidation, generation of reactive oxygen species (ROS), and additional features of elevated oxidative stress, which are known inhibitors of hepatocyte proliferation. A positive correlation between the oval cell response and the stage of hepatic fibrosis and the fatty liver disease has been identified. In viral hepatitis, oval cell activation is also observed. This is classically observed in the periportal region and is typically proportional to the extent of inflammatory infiltrate. Moderate to severe inflammation is often associated with greater oval cell response that is comprised of intermediate hepatocytes. Based on these findings, a paracrine mechanism of oval cell activation triggered by the inflammatory cells has been proposed. These scenarios also bring into perspective the progenitor or oval cell origin of a subset of hepatocellular cancers (HCCs). HCC occurs more frequently in the backgrounds of cirrhosis observed in ALD, NAFLD, and hepatitis than non-diseased liver. It is feasible that the oval cell activation that occurs in certain liver pathologies, while providing a distinct advantage of maintenance of hepatic function, might also serve as a basis for neoplastic transformation in the ideal microenvironmental milieu. However, conclusive studies to this end are still missing.

## Molecular basis of hepatocyte death

Hepatocyte death is a common hallmark of many forms of liver pathology. This is often seen as diffuse or zonal hepatocyte death and dropout. In many cases, hepatocyte death occurs due to death receptor activation, which leads to hepatocyte apoptosis and ensuing liver injury. Such a mechanism of liver injury has been identified in hepatitis, inflammatory hepatitis, ALD, ischemia−reperfusion injury, and cholestatic liver disease. Hepatocytes express the typical death receptors that belong to the TNF-$\alpha$ receptor superfamily. These include (1) Fas, which is also called Apo 1/CD95 or Tnf receptor superfamily6 (Tnfrsf6), (2) TNF-receptor 1 (TNF-R1), which is also called CD120a or Tnfrsf10a, and (3)

TRAIL-R1 or Tnfrsf10A and TRAIL-R2 or Tnfrsf10B. The ligands for these receptors are the Fas ligand (FasL), which is also called CD178 or Tnfsf6, TNFα or Tnfsf2, and TRAIL or Apo 2 L or Tnfsf10, respectively. The former two subgroups of death receptors and their ligands are known to be associated with various liver diseases, whereas TRAIL/TRAIL-R have not yet been identified as a mechanism of hepatic injury pertinent to a specific liver disease.

## Fas-activation-induced liver injury

This mode of death is known to be associated with liver diseases such as viral hepatitis, inflammatory hepatitis, Wilson disease, cholestasis, and ALD. FasL, present on inflammatory cells or Fas-activating agonistic antibodies such as Jo-2 injection (in experimental models), leads to Fas activation resulting in massive hemorrhagic liver injury with extensive hepatocyte apoptosis and necrosis (Fig. 20.4). Most mice die within 4—6 h after Jo-2 injection. On activation, homotrimeric association of Fas receptors occurs, which recruits Fas-associated death domain protein (FADD) adaptor molecule and initiator caspase-8. This complex or the death-inducing signaling complex (DISC) requires mitochondrial involvement and cytochrome $c$ release, which is inhibited by anti-death Bcl-2-family proteins (Bcl-2 and Bcl-x$_L$). Cytochrome $c$ stimulates the activation of caspase-9 and then caspase-3, inducing cell death. Interestingly, prodeath Bcl-2 family proteins (Bid, Bax, and Bak) are also needed for hepatocyte death induced by Fas. In fact, Bid is cleaved by caspase-8 after Fas activation and cleaved Bid translocates to mitochondria. This in turn activates Bax or Bak in the mitochondria to stimulate the release of apoptotic factors such as cytochrome $c$. At the same time, Bid also induces mitochondrial release of Smac/DIABLO, which inactivates the inhibitors of apoptosis (IAP). The role of IAPs is at the level of caspase-3 activation, which occurs in two steps. The first step is caspase-8-induced severance of larger subunit of caspase-3. The second step is the removal of the prodomain by its autocatalytic activity, which is essential for caspase-3 activation and inhibited by IAPs. While caspase-8 can directly activate caspase-3, bypassing the mitochondrial involvement, the execution of the entire pathway ensures the process of death in the hepatocytes. Thus, the relative expression and activity levels of caspase-8, Bid, and other modulators of this pathway (such as IAPs and Smac/DIABLO) determine whether Fas-induced hepatocyte apoptosis will or will not utilize mitochondria to induce cell death.

A recent discovery unveils another important regulatory step in the Fas-mediated cell death. Under normal circumstances, Fas receptor was shown to be sequestered with c-Met (the HGF receptor) in hepatocytes. This makes Fas receptor unavailable to the Fas ligand. On HGF stimulation, this complex becomes destabilized and hepatocytes become more sensitive to Fas-agonistic antibody. Recently, lack of this Fas antagonism by Met was identified in fatty liver disease. While this explains how HGF could be prodeath at high doses or in combination with other death signals, a

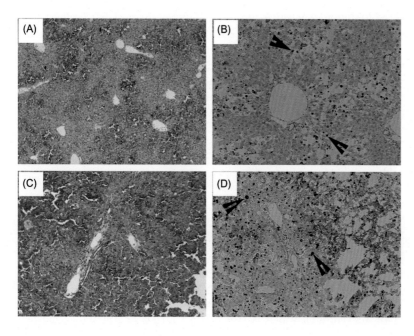

FIGURE 20.4 Histology and terminal deoxynucleotidyl transferase dUTP nick end labeling (TUNEL) immunohistochemistry exhibiting apoptotic cell death after Fas-mediated and tumor necrosis factor-α—mediated liver injury. (A) H&E (hematoxylin and eosin) stained histologic section shows massive cell death 6 h after Jo-2 antibody administration. (B) Several TUNEL-positive apoptotic nuclei (*arrowheads*) are evident in the same liver. (C) H&E stained histologic section shows massive cell death 7 h after D-galactosamine/lipopolysaccharide administration in mice. (D) Several TUNEL-positive apoptotic nuclei (*arrowheads*) are evident in the same liver.

paradoxical effect of HGF on promoting cell survival is also observed, albeit at lower doses. This effect is observed with other growth promoting factors such as TGFα and is mediated by elevated expression of Bcl-$x_L$ that inhibits mitochondrial release of cytochrome $c$ and inhibits Bid-induced release of Smac/DIABLO.

## TNF-α-induced liver injury

This mode of hepatocyte death is commonly observed in ischemia–perfusion liver injury and ALD. In mice, TNF-α is induced by bacterial toxin administration such as lipopolysaccharide (LPS, 25–50 µg/kg). Because it was identified that LPS alone initiates NF-κB-mediated protective mechanisms, an inhibitor of transcription (D-galactosamine) or translation (cycloheximide) is used before the LPS injection, for successful execution of cell death. This induces massive hepatocyte death due to apoptosis as seen by terminal deoxynucleotidyl transferase dUTP nick end labeling immunohistochemistry (Fig. 20.4). TNF-α binds to the TNF-α-R1 on hepatocytes to induce receptor trimerization and DISC formation. DISC is composed of TNF-α-R1 and TNFR-associated death domain (TRADD), which can recruit FADD and caspase-8 via an unknown mechanism, to further activate caspase-3. The significance of additional mechanisms is relatively unknown in liver biology. However, one additional mechanism that deserves mention and is relevant to liver injury is that TNF-α-R1 engagement also induces cathepsin B release from lysozyme, which induces cytochrome $c$ release from mitochondria. However, how much effect of TNF-α on apoptosis is mediated via the mitochondrial pathway components including Bcl-2, Bcl-$x_L$, and Bid in TNF-α-mediated apoptosis remains debated, but it is believed that some effects are via this intrinsic pathway.

One of the important effects of TNF-α stimulation is the unique and concomitant activation of the NF-κB pathway as a protective mechanism. How this occurs is not fully understood, but is relevant as an ongoing protective mode for maintaining hepatic homeostasis. TNF-α-R1 engagement leads to the recruitment of TRADD, which can further induce recruitment of TNF associated factor 2 (TRAF2) and receptor-interacting protein (RIP). This complex recruits and activates IKK complex containing I-κB kinase 1 (IKKα) and IKKβ, which leads to the phosphorylation of I-κB and its degradation. This allows for the release and activation of NF-κB p50 and p65 dimers, inducing the phosphorylation of p65, and their nuclear translocation to direct transcriptional activation of cytoprotective genes such as IAP, iNOS, Bcl-$x_L$, and others. Several additional regulators of NF-κB exist that regulate its transcriptional activity. These include glycogen synthase kinase-3 (GSK3) and TRAF2-associated kinase (T2K). GSK3 knockout embryos die at around E13–E15 stage due to massive liver cell death related to TNF-α toxicity.

# Molecular basis of non-alcoholic fatty liver disease

Accumulation of fat in hepatocytes (called steatosis) in the absence of significant alcohol intake is called non-alcoholic fatty liver disease (NAFLD). The term NAFLD encompasses a spectrum of liver diseases ranging from simple steatosis (non-alcoholic fatty liver or NAFL) to steatosis with inflammation and hepatocyte injury called non-alcoholic steatohepatitis (NASH) (Fig. 20.5). The latter condition, which occurs in ~20% of NAFLD patients, can progress to fibrosis, cirrhosis, and HCC. A classic pericellular fibrosis precedes development of advanced fibrosis and cirrhosis in NAFLD (Fig. 20.5).

NAFLD has been associated with many drugs, genetic defects in metabolism, and abnormalities in nutritional states. However, it is most commonly associated with the metabolic syndrome. The metabolic syndrome is a group of related clinical features linked to visceral obesity, including insulin resistance (IR), dyslipidemia, and hypertension. NAFLD is strongly associated with metabolic syndrome and is considered the liver manifestation of this condition. NASH is now the most common chronic liver disease and expected to become the most common indication for liver transplantation in the United States.

The pathogenesis of NAFLD remains incompletely understood. However, several common themes have recently emerged and, conceptually, its pathogenesis can be thought of as an interplay between (1) IR and adipose–tissue axis, (2) diet, gut permeability, and the gut–liver axis, and (3) hepatic lipotoxicity and mechanisms leading to hepatocyte injury, inflammation, and cell death (Fig. 20.6).

### Diet, insulin resistance, and the adipose–tissue–liver axis

#### Role of fructose in non-alcoholic fatty liver disease

Fructose is a sweetener used in many processed foods and commercial soft drinks. Use of fructose is associated with increased risk of NAFLD and hepatic fibrosis. Several mechanisms have been postulated for the negative effects of

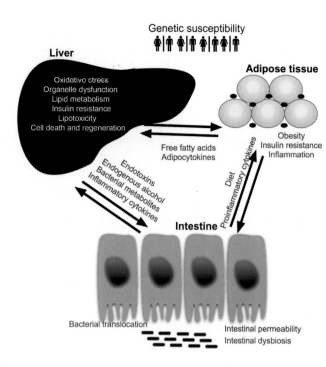

Genetic susceptibility

**Liver**

Oxidative stress
Organelle dysfunction
Lipid metabolism
Insulin resistance
Lipotoxicity
Cell death and regeneration

**Adipose tissue**

Obesity
Insulin resistance
Inflammation

Free fatty acids
Adipocytokines

Endotoxins
Endogenous alcohol
Bacterial metabolites
Inflammatory cytokines

Diet
Proinflammatory cytokines

**Intestine**

Bacterial translocation

Intestinal permeability
Intestinal dysbiosis

**FIGURE 20.5** A complex molecular interplay between the liver, intestines, and adipose tissue underlies the pathogenesis of nonalcoholic fatty liver disease. In genetically susceptible individuals, several factors (multiple parallel hits) lead to the development of steatohepatitis. intestinal dysbiosis, increased intestinal permeability, gut-derived metabolites, and inflammatory signals lead to disruption of hepatic and adipose tissue metabolic function. Obesity and adipose tissue inflammation and insulin resistance promote free fatty acid and adipocytokine release from adipose tissue. Increased hepatic exposure to free fatty acids and other lipid species and inflammatory signals, along with changes in lipid and carbohydrate homeostasis, lead to hepatic steatosis and create conditions that favor hepatic lipotoxicity. Oxidative and endoplasmic reticulum stress, dysregulation of lipid metabolic pathways, and organelle dysfunction lead to hepatocyte injury and death (steatohepatitis). Chronic liver injury and inadequate or defective repair and regenerative pathways favor progression to fibrosis and cirrhosis.

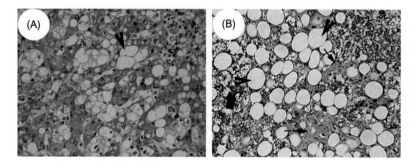

**FIGURE 20.6** Fatty liver disease along with hepatic fibrosis. (A) Patient with nonalcoholic fatty liver disease shows significant macrovesicular (*arrowhead*) and microvesicular steatosis (*blue arrowhead*) in a representative H&E (hematoxylin and eosin) stained histological section. (B) Masson trichrome staining of liver from a mouse on the methionine- and choline-deficient diet for 4 weeks reveals hepatic fibrosis (*arrows*) along with macrovesicular steatosis (*arrowheads*).

fructose in NAFLD pathogenesis. Fructose induces increased intestinal permeability, microbial translocation, induction of Toll-like receptors (TLRs) and TLR-dependent pathways, de novo lipogenesis, and inhibition of fatty acid β-oxidation in NAFLD. It may also promote gut dysbiosis, which contributes to the subsequent development of NASH.

## Insulin resistance

The liver, muscles, and adipose are important insulin-responsive tissues that participate in systemic energy regulation. IR is strongly associated with the development of NAFLD. Normally, insulin acts on the insulin receptor of myocytes to tyrosine-phosphorylate the insulin receptor substrate (IRS). IRS, in turn, activates phosphatidyl-inositol 3-kinase and protein kinase B and results in translocation of the glucose transporter to the plasma membrane with resultant rapid uptake of glucose from the blood into the myocyte. The net effect is decreased blood glucose and therefore decreased insulin secretion from the pancreas. In the setting of obesity, fat-laden myocytes become resistant to the signaling effects of insulin. This inability of skeletal muscle to take up glucose from the circulation after insulin stimulation leads to elevated blood glucose levels and increased insulin secretion from the pancreas, with metabolic consequences in the liver.

Besides skeletal muscle, IR also causes increased blood glucose via decreased insulin action in the liver. In the setting of IR, insulin is unable to suppress hepatic glucose production, which, along with decreased myocyte glucose uptake, leads to high blood glucose levels and high circulating insulin levels. The net effect of these changes is increased hepatic lipogenesis and triglyceride accumulation, which is mediated by the synergistic action of two transcription factors: carbohydrate response element−binding protein (ChREBP) and sterol regulatory element−binding protein-1c (SREBP-1c). In the setting of IR and hyperglycemia, ChREBP mediates the conversion of glucose to fatty acids by upregulating both glycolysis and lipogenesis. Besides glucose, insulin can also regulate de novo hepatic fatty acid synthesis via SREBP-1c, a member of the β-HLH-Zip family of transcription factors. A third process that favors hepatic steatosis results from IR in the adipose tissue.

### Adipose tissue inflammation and adipocytokines

Under conditions of positive energy balance, adipose tissue stores triglycerides. In the fat-laden adipocytes in obese individuals, IR causes defective insulin-mediated inhibition of hormone-sensitive lipase, increased lipolysis, and increased free fatty acid (FFA) release into the circulation. Increased FFA uptake into the liver from the circulation also favors the development of hepatic steatosis and inflammation. Adipose tissue inflammation and adipocytokine production promote fatty acid release from adipocytes. Obesity is associated with macrophage accumulation in adipose tissue and a state of low-grade inflammation, which influences development of IR. Adipose tissue inflammation leads to secretion of cytokines and chemokines, including TNF-α, IL-6, and CC-chemokine ligand-2 (CCL2).

## Intestinal permeability, gut dysbiosis, and the gut−liver axis

### Intestinal permeability and the gut−liver axis

A critical role for the gut−liver axis in NAFLD pathogenesis has been proposed, which is likely mediated by changes in intestinal microbiota, altered intestinal permeability, increased endotoxin levels, upregulated expression of hepatic TLRs (TLR-2, TLR-4, TLR-5, and TLR-9), and increased hepatic inflammation. Two strains of obese mice, leptin-deficient *ob/ob* and leptin receptor−deficient *db/db* mice, display increased intestinal permeability, altered distribution of tight junction proteins occludin and zonula occludens-1 in the intestinal mucosa, and higher circulating levels of inflammatory cytokines and portal endotoxemia compared with lean control mice. A metaanalysis of clinical studies revealed that NAFLD patients, and particularly NASH patients, have increased risk of increased intestinal permeability compared with healthy controls. Further evidence of the gut−liver link emerged from an elegant study using multiple mouse models that demonstrated that NLRP6 and NLRP3 inflammasomes and the effector protein IL-18 negatively regulate development of metabolic syndrome and NAFLD progression. Mice with inflammasome deficiency demonstrated changes in gut microbiota, increased TLR-4 and TLR-9 agonists in the portal circulation, and higher TNF-α expression, ultimately leading to exacerbation of NASH.

### Gut-derived hormones

The enteroendocrine cells of the gut secrete a large number of peptide hormones with critical roles in maintaining gut barrier integrity and effects on insulin action, as well as hepatic metabolism and systemic energy balance. Glucagon-like peptide (GLP-1), an intestinal hormone secreted after nutrient ingestion increases glucose-stimulated insulin secretion by pancreatic β cells. GLP-1 receptor (GLP-1R) is expressed in rodent and primary human hepatocytes. GLP-1 ameliorates NAFLD through several mechanisms that include decreased hepatic lipogenesis via activation of AMP-activated protein kinase, inhibiting c-JNK, attenuation of hepatic IR, and stimulation of hepatic lipid oxidation. In addition to favorable metabolic effects, activation of the GLP-1R has been shown to suppress hepatic oxidative stress and production of inflammatory cytokines and transcription factors. GLP-1 analog therapy was shown to be beneficial for NASH in preliminary human studies and human studies have been proposed.

### Intestinal dysbiosis

Several studies have demonstrated altered gut microbiota in obesity and NAFLD. SCFAs are the end products of dietary fiber fermentation by intestinal anaerobic bacteria. SCFAs have been shown to confer many beneficial metabolic effects, such as the regulation of cholesterol, glucose, and fatty acid metabolism. Choline in the diet is essential for very low-density lipoprotein synthesis and export, and its dietary deficiency is associated with the development of hepatic steatosis. The development of hepatic steatosis on a choline-deficient diet is associated with changes in gut Gammaproteobacteria and Erysipelotrichi. Gut microbiota also regulates systemic energy balance, and NAFLD development is via bile acid metabolism. Normally, bile acids are synthesized in the liver, and the gut microbiota converts them into secondary bile acids. Gut microbiota regulates secondary bile acid metabolism and inhibits bile acid synthesis in the liver via FXR inhibition in the distal small intestine.

## Common pathways of hepatocyte lipotoxicity, injury, repair and cell death

### Lipotoxicity and factors modifying lipotoxicity

Lipotoxicity plays an important role in the pathogenesis of NAFLD. Several mechanisms for hepatic lipotoxicity in NASH have been postulated such as the generation of ROS during mitochondrial and peroxisomal fatty acid oxidation. Another mechanism is via changes in stress response and metabolic signaling pathways involving transcription factors such as TLRs, which are innate immune receptors. Lipotoxicity can activate inflammatory pathways, endoplasmic reticulum (ER) stress, and impair autophagy. Several modifying factors such as accumulation of iron in the liver can increase oxidative stress and increase susceptibility to lipotoxicity.

### Inflammation and signaling pathways of inflammation

The NF-κB pathway has been extensively studied for its role in steatohepatitis and is upregulated in NASH patients and in animal models of the disease. However, its role in pathogenesis of NASH is complex, and while activation of NF-κB signaling in liver cells induces inflammation and steatosis, its inactivation also causes steatohepatitis and liver cancer in a mouse model. Deletion of the JNK1 isoform of c-JNK, a mediator of TNF-induced apoptosis, is protective in a mouse model of steatohepatitis, suggesting that JNK signaling is important in the pathogenesis of NASH. Proinflammatory cytokines, TNF-α and IL-6, are increased in NASH patients. In animal models of steatohepatitis, decreasing TNF-α expression causes decreased hepatic steatosis, cell injury, and inflammation, suggesting a role of this cytokine in the pathogenesis of NASH.

### Oxidative stress and lipid peroxidation

Serum markers of oxidative stress and markers of oxidative damage for DNA are also increased in NASH, whereas antioxidant factors such as glutathione S-transferase and catalase are reduced. Polyunsaturated fatty acids in the cell can undergo peroxidation by ROS, resulting in the formation of malondialdehyde and trans-4-hydroxy-2-nonenal by-products. Lipid peroxides contribute to liver injury by increasing production of TNF-α, increasing the influx of inflammatory cells, impairing protein and DNA synthesis, and depleting levels of protective cellular antioxidants such as glutathione.

### Mitochondrial dysfunction

Ultrastructural studies have demonstrated the presence of hepatocyte megamitochondria with paracrystalline inclusions in patients with NASH. Functional differences in mitochondria include impaired ability to synthesize ATP after a fructose challenge, which causes transient liver ATP depletion, lower expression levels of mitochondrial DNA—encoded proteins, and lower activity of complexes of the mitochondrial respiratory chain in patients with NASH. Mice heterozygous for the mitochondrial trifunctional protein exhibit defective β-oxidation and age-dependent development of hepatic steatosis and systemic IR. In turn, the defective mitochondrial fatty acid β-oxidation causes hepatic IR that is independent of high-fat diet—induced obesity. Furthermore, studies with rat suggest that development of mitochondrial dysfunction in the liver precedes development of fatty liver and IR.

### Endoplasmic reticulum stress and the unfolded protein response

The UPR is activated in patients with NAFLD. Activation of UPR leads to hepatic IR, upregulation of triglyceride genes, and increased lipogenesis. Inactivation of ATF6, a transcription factor downstream of XBP1 in the UPR, results in hepatic steatosis, and its inactivation is protective against steatosis via changes in lipoprotein export, carbohydrate metabolism, and β-oxidation of fatty acids. ER stress and the UPR appear to be critical in hepatic liver metabolism and play a central role in NAFLD pathogenesis.

### Autophagy

Inhibition of macroautophagy by genetic or pharmacological approaches increases triglyceride content in hepatocytes due to decreased lipolysis and fatty acid β-oxidation. Knockout mice with hepatocyte-specific defect in macroautophagy demonstrate striking increases in hepatic triglyceride and cholesterol levels. Hyperinsulinemia and IR are associated with decreased autophagic function in the liver. Autophagy regulates organelle function and insulin signaling, and defective hepatic autophagy contributes to ER stress and IR, providing a potential molecular link between obesity, IR, and NAFLD.

## Hepatic progenitor cell activation

Pediatric patients with biopsy-proven NAFLD demonstrate expansion of the hepatic progenitor cell compartment. The hepatic progenitor cells express adiponectin, resistin, and GLP-1, which are associated with histological severity of NAFLD. In chronic liver disease, where hepatocyte proliferation is less effective, progenitor cell activation in the form of ductular reaction can occur, which can also be sources of pro-fibrogenic and pro-inflammatory cytokines, thus contributing to disease processes.

## Extracellular vesicles

Extracellular vesicles (EV) are small membrane-bound particles released from stressed and injured cells and are involved in intercellular communication. There is a correlation between increased abundance of circulating EVs and diet-induced development of NAFLD in experimental rodent models. These findings have been replicated in human subjects with NASH, with circulating plasma EV abundance correlating with progression of NAFLD. Hepatocyte-derived microparticles, a type of EV, promote angiogenesis and liver damage in steatohepatitis. Additionally, EVs derived from adipose tissue stimulate or inhibit insulin signaling in hepatocytes depending on their adipokine content, suggesting a potential role in regulating systemic IR.

## MicroRNA

miRNAs have been implicated in regulation of hepatic triglyceride homeostasis and pathogenesis of NAFLD. The most abundant miRNA in the liver, miR-122, is a key regulator of hepatic cholesterol and fatty acid metabolism. Experimental inhibition of miR-122 with antisense oligonucleotide in a murine model reduced plasma cholesterol levels, decreased hepatic fatty acid and cholesterol synthesis, and increased hepatic fatty acid oxidation. miR-122 inhibition also improved liver steatosis in a diet-induced obesity model. Besides miR-122, other miRNAs have also been implicated in the pathogenesis of NAFLD.

## Nuclear receptors

Dysregulation of signaling pathways and metabolic processes controlled by several nuclear receptors of the NR1 subfamily have been implicated in the pathogenesis of NAFLD. Peroxisome proliferator−activated receptor (PPAR) subfamily members are major regulators of lipid metabolism. Activation of PPAR-α decreases hepatic steatosis and inflammation. PPAR-β/δ (NR1C2) is expressed in the muscle and activated by FFAs and eicosanoids, and its activation also decreases hepatic steatosis and inflammation, as well as IR. PPAR-γ activation ameliorates hepatic steatosis and IR, and treatment with the PPAR-γ agonist, pioglitazone, has been shown to benefit patients with biopsy-proven NASH. The pregnane X receptor (PXR, NR1I2) is expressed in the liver and intestines and both ligand-dependent activation and disruption of PXR, results in hepatic steatosis. Activation of Constitutive androstane receptor (CAR; NR1I3) decreases hepatic steatosis, diabetes, and IR. Activation of the liver X receptor (LXR) variants, LXRα (NR1H3) and LXRβ (NR1H2), increases fatty acid synthesis via increased SREBP-1c expression. Activation of Farnesoid X receptor α (FXRα; NR1H4) reduces hepatic fat and fibrosis in animal models of NAFLD. Obeticholic acid, an FXR ligand, improves histological features of NASH and is under clinical investigation as a treatment for NASH.

## Cell death

Patients with NASH have increased hepatocyte FasL expression that can trigger apoptosis. Cytokeratin 18, an intermediate filament protein associated with apoptotic cells, is increased in serum from NASH patients and has been proposed as a biomarker for NASH. In addition to hepatocyte apoptosis, adipocyte apoptosis also plays a critical role in NAFLD pathogenesis by initiating macrophage infiltration into adipose tissue, leading to IR, increased release of FFAs from adipose tissue, increased hepatic FFA uptake, and development of NAFLD.

## Genetic factors

Several genome-wide association studies have identified multiple risk alleles for NAFLD development. A *PNPLA3* genetic variant *PNPLA3* (rs738409[G], encoding Ile148Met) is associated with increased susceptibility to hepatic steatosis, NASH, and fibrosis. Interestingly, this increased risk is not only independent of the risk of obesity and the metabolic syndrome but also confers increased risk of progressive liver disease with other risk factors such as viral hepatitis and ALD. A second polymorphism, rs58542926 within the *TM6SF2* gene, is also associated with increased risk of

NAFLD. This variant confers increased risk of NAFDL/NASH and fibrosis but decreased cardiovascular risk. Together, these studies raise the prospect of precision medicine in the field of NAFLD with therapy directed at subpopulations at differential risk of NAFLD-associated adverse outcomes.

# Molecular basis of alcoholic liver disease

Alcoholic fatty liver develops in 90% of individuals that consume over 60 g/day of alcohol, and 5%−15% of patients may progress to alcohol-related fibrosis and cirrhosis. The histological spectrum of alcohol-induced liver disease spans steatosis, steatohepatitis, fibrosis, and cirrhosis. ALD shares many histological features as well as molecular and physiologic pathogenic mechanisms with NAFLD, and the salient features of the molecular basis of ALD are summarized here.

## Alcohol metabolism and generation of oxidative stress

There are four pathways of ethanol metabolism, of which two are major oxidative pathways and two minor non-oxidative pathways. The major oxidative pathways include alcohol dehydrogenase (ADH) and aldehyde dehydrogenase (ALDH) 1, which is present in the cytosol, and ALDH2, which is located in the mitochondria. Initial oxidation of ethanol occurs through the cytosolic ADH isoenzymes that produce acetaldehyde, which is then converted into acetate, and these reactions result in the reduction of nicotinamide adenine dinucleotide to nicotinamide adenine dinucleotide (reduced form) (NADH). Excessive ADH-mediated hepatic NADH generation can lead to redox imbalance and in turn in inhibition of the Krebs cycle and fatty acid oxidation to promote hepatic steatosis. However, with chronic ethanol consumption, the redox state is normalized but alterations in lipid metabolism regulatory genes appear occurs.

Ethanol is also metabolized by the microsomal ethanol-oxidizing system, of which cytochrome P450 2E1 (Cyp2E1), which is induced by chronic ethanol consumption. Induction of Cyp2E1 plays a role in alcohol-induced liver injury through the production of ROS lead to exacerbation of oxidative stress.

## Common pathways of hepatocyte injury in alcoholic liver disease

Mitochondrial dysfunction has been implicated in the pathogenesis of ALD. In a murine model, chronic ethanol-mediated liver injury was associated with increased sensitivity of the hepatic mitochondrial permeability transition pore induction. ER stress also plays an important role in the pathogenesis of ALD. Feeding mice ethanol through the intragastric route induces UPR/ER stress genes. The mechanism of ER stress involves hyperhomocysteinemia, which occurs due to methionine synthase downregulation. Ethanol ingestion-induced ER stress can, in turn, activate SREBP1c and SREBP2, which promote hepatic steatosis through cholesterol and triglyceride accumulation in hepatocytes.

Autophagy has been proposed as a protective mechanism in ALD as a mechanism for removal of damaged cellular organelles such as mitochondrial and lipid droplets. Ethanol administration induces macroautophagy in a process that requires ethanol metabolism, generation of ROS, and inhibition of mammalian target of rapamycin signaling. Experimental suppression of macroautophagy exacerbates hepatocyte apoptosis and liver injury. The macroautophagy appears to be selective for lipid droplets and damaged mitochondria in hepatocytes and pharmacologic upregulation of macroautophagy ameliorates alcohol-induced acute and chronic hepatotoxicity, while inhibition of autophagy by treatment with chloroquine exacerbates steatosis and liver injury.

### *Apoptosis, necrosis, and Necroapoptosis*

Alcohol-induced production of ROS, which induces lipid peroxidation and depletion of glutathione, can cause hepatocyte death via apoptosis, necrosis, or necroapoptosis. Both organelle-based (intrinsic pathway) and death receptor−induced (extrinsic pathway) apoptosis can occur in ALD. Hepatocyte apoptosis is increased human alcoholic hepatitis. Furthermore, the severity of ALD correlates with the magnitude of hepatocytes apoptosis. Similar to apoptosis, necrosis is also a regulated process, involving receptor-interacting protein kinase 1 (RIP1) and RIP3. Necroapoptosis is programmed necrosis initiated by death receptors similar to the extrinsic apoptotic pathway. Increased RIP3 expression is detected in liver biopsies of patients with ALD and ethanol-feeding results in RIP3-induced hepatocyte necroapoptosis in murine livers. This process appears to be CYP2E1 dependent as CYP2E1-deficient mice do not induce RIP3.

## *Hypoxia*

Increased oxygen consumption associated with ethanol metabolism can induce hypoxia in zone 3 (pericentral) hepatocytes. Ethanol feeding is also associated with upregulation of hypoxia-inducible factors (HIFs) as an adaptive response to hypoxia. Induction of HIF1a ameliorates alcoholic steatosis in mice. However, mice with liver-specific *Hif1b* disruption are resistant to alcohol-induced liver injury, suggesting a complex role of HIFs in ALD.

## *Innate and adaptive immune responses*

Alcohol exposure significantly alters stress signaling, and innate and adaptive immune responses. Hepatocytes that suffer injury release danger-associated molecular patterns (DAMPs), which impact Kupffer cells, sinusoidal endothelial cells, and hepatic stellate cells (HSCs). High mobility group box 1 (HMGB1) is a DAMP that recruits HSCs and endothelial cells to sites of ethanol-induced hepatocyte injury.

Alcohol intake activates the innate immune system in the liver. Alcohol induces circulating levels of gut-derived endotoxin. In addition, ethanol exposure sensitizes hepatic Kupffer cells to TLR4-mediated signaling, resulting in increased expression of inflammatory mediators such as cytokines and chemokines via M1 polarization. Compared with healthy individuals and stable, abstinent alcoholic cirrhosis patients, cytotoxic T lymphocytes and natural killer cells from patients with severe alcoholic hepatitis demonstrate impaired cytotoxic function and reduced activation. Dendritic cells are antigen-presenting cells. Dendritic cells from ethanol-fed rats secrete higher levels of cytokines, such as IL1b and IL10, and lower levels of TNF-α, IFN-γ, and IL-12, suggesting suppressed dendritic cell phenotype.

## *Gut—liver axis in alcoholic liver disease*

ALD is also associated with increased intestinal permeability, compromise of the intestinal tight junctions, and gut dysbiosis. Both acute and chronic ingestion of alcohol increase intestinal permeability. Treatment with probiotics in a rat model of alcoholic steatohepatitis was found to ameliorate gut leakiness and alcohol-induced liver injury. The increased gut permeability promotes translocation of endotoxin and increased risk of liver injury. Alcohol consumption is also associated with alteration of the colonic microbiome in a subset of alcoholic individuals and correlates with endotoxemia. Mice humanized with human intestinal microbiota from a patient with severe alcoholic hepatitis develop more severe liver inflammation and injury compared with mice treated with intestinal microbiota from an alcoholic patient without alcoholic hepatitis.

## *Genetic factors predisposing to alcoholic liver disease*

Given the many similarities in the pathogenesis of NAFLD and ALD, it is not surprising that risk loci are also shared for the two conditions. In a GWAS for alcohol-related cirrhosis in individuals of European descent, risk loci in three genes with role in lipid turnover were identified, including rs738409 in PNPLA3, and new variants in TM6SF2 and MBOAT7.

# Molecular basis of hepatic fibrosis and cirrhosis

Any chronic liver injury can lead to liver fibrosis, which can advance to cirrhosis. The process of fibrosis entails excessive deposition of ECM in the liver, especially collagen, which can be detected by specific stains such as Sirius red or Masson trichrome (Fig. 20.7). As it advances, the hepatic architecture is disrupted leading to hemodynamic alterations. Functional hepatocyte mass is reduced leading to hepatic insufficiency and hepatic failure. Eventually, cirrhosis is also a major risk factor for the development of HCC.

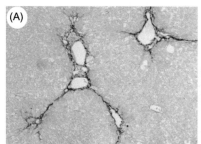
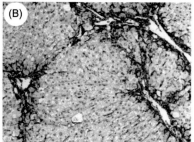

**FIGURE 20.7** Hepatic fibrosis in a mouse model of repeated carbon tetrachloride injection. (A) Sirius red staining identifies collagen deposition that is in the form of bridging fibrosis. (B) Immunohistochemistry for α-smooth muscle actin depicts activated stellate cells that are the predominant source of collagen production.

The fundamental cell type involved in the process of hepatic fibrosis is the HSC. Chronic insult to the liver leads to the differentiation of HSCs to activated myofibroblasts, which can be detected by staining for α-smooth muscle actin (Fig. 20.7). Mesothelial cells have been shown to contribute to HSC and myofibroblast pool after liver injury. Exploiting the presence of enzymes of retinol metabolism in HSC, a new transgenic mouse expressing cre-recombinase under lecithin retinol acyltransferase, HSCs were shown to be the major source of activated myofibroblasts after a number of pathological insults. The process of HSC activation entails increased proliferation, migration and contractility. Additionally, activated myofibroblasts overexpress several profibrotic genes whose products are deposited in hepatic parenchyma. These include the products of ECM genes such as type I collagen (α1 and α2), type III collagen, laminin, fibronectin, and proteoglycans such as decorin, hyaluronan, heparin sulfate, and chondroitin sulfate. In addition to HSCs, portal fibroblasts have also been implicated as an important source of activated myofibroblasts. These cells, which reside normally in periportal area, are particularly relevant cell source of myofibroblasts in cholestatic liver injury.

Quiescent HSCs are typically characterized by the presence of perinuclear lipid droplets containing vitamin A or retinoid. These droplets are lost following activation of HSC, when the retinoid is released as retinol. Multiple cellular mechanisms lead to the activation of HSCs. In ALD HSC activation is mediated through increased oxidative stress due effect of CYP2E1-mediated ROS as well as of acetaldehyde on HSCs. Around 20%–40% of NASH patients progress to hepatic fibrosis, secondary to elevated ROS secondary to PPAR-α-mediated increased FFA oxidation.

There has also been a growing interest in gut microbiome and loss of intestinal barrier in hepatic fibrosis. Mice, lacking TLR4 and whose gut has been sterilized, fail to develop fibrosis. Qualitative and quantitative changes in intestinal microbiome have been shown to contribute to hepatic fibrosis. Carbon tetrachloride administration to mice to induce hepatic fibrosis produces differential overgrowth of *Firmicutes* and *Actinobacteria* but appeared to be a result rather than a cause of liver fibrosis. In the BDL model of hepatic fibrosis, there is bacterial overgrowth after surgery and precedes development of fibrosis. Simultaneous administration of high fat diet induced a greater magnitude dysbiosis. High-fat diet alone increased *Clostridium* cluster XI and enhanced metabolism of primary bile acids to deoxycholic acid, which produced greater ROS, DNA damage, and development of HCC. In yet another model of fibrosis-driven HCC, diethylnitrosamine administration is followed by repeated injections of carbon tetrachloride to produce liver injury and disease. In this model, sterilization of the gut led to suppression of fibrosis-associated HCC in a TLR4-independent manner.

Because quiescent HSC stores lipid and HSC activation is associated with loss of fat droplets, it was speculated that inhibition of autophagy may prevent HSC activation. Indeed, HSCs that lacked autophagy capability failed to lose stored fat droplets and in turn remained quiescent following toxic insults.

Several pathways have been identified in HSC activation including PDGF and TGF-β/Smad signaling. PDGF is a potent mitogen for HSC, and its cognate receptor PDGFR expression goes up during HSC activation. PDGF activation has also been shown to stimulate PI3 kinase, which induces HSC proliferation. In addition, recruitment of Ras to PDGF receptor also leads to ERK activation via sequential activation of Raf-1, MAPK-1/2, and ERK-1 and ERK-2. Nuclear ERK can regulate target genes responsible for proliferation of HSCs. Additionally, JNK activation has been shown to positively regulate HSC proliferation. TGF-β is a potent mediator of hepatic fibrosis and produced by liver macrophages and HSCs. TGF-β signaling leads to increased expression of type I and III collagens, decorin, elastin, and others. Increased expression of KLF6, a transcription factor and a tumor suppressor, which acts as a chaperone for collagen, is also evident during HSC activation. Connective tissue growth factor (CTGF) is upregulated by TGF-β during HSC activation in chronic viral hepatitis and BDL. Decrease in PPAR-γ has been associated with HSC activation. IL-17, which is mainly produced by the Th17 cells, can activate NF-κB and STAT3 in Kupffer cells and HSCs yielding high expression of profibrotic genes in HSCs. Mice lacking IL-17 or IL-17RA are protected from fibrosis following carbon tetrachloride exposure or BDL. IL-22 has an anti-fibrotic effect by inducing HSC senescence via STAT3 and p53. IL-33 promotes hepatic fibrosis. IL-33 levels are elevated in patients with cirrhosis, along with its receptor ST2. Elevated levels are also seen in rodents with hepatic fibrosis secondary to a variety of insults. The proximal event seems to be the release of IL-33 from injured hepatocytes, which acts on innate immune lymphoid cells of the liver (ILC2), which promotes TGF-β signaling in HSCs.

During the early process of hepatic fibrosis, increased fibrogenesis is counterbalanced by factors that negatively regulate ECM deposition. However, as the chronic insult to the liver continues, the process of fibrogenesis, observed as continued activation of myofibroblasts derived from HSC and perivascular fibroblasts, exceeds fibrinolysis. It remains a conundrum to be able to successfully predict the risk of developing cirrhosis in patients with similar liver pathology. The most promising advances that are being reported include the identification of genetic polymorphisms that predict the risk of progression of liver fibrosis. Studies reporting SNPs as predictors of progression of fibrosis are beginning to appear. Recently, cytokine SNPs were identified that successfully predict disease progression. Similarly, SNPs in the DDX5 gene successfully predicted fibrosis progression in hepatitis C patients.

While liver fibrosis was originally thought to be irreversible, several studies have now shown that rectification of the primary cause of hepatic injury can lead to regression of hepatic fibrosis. This process appears to be driven by apoptosis of activated myofibroblasts. Using lineage-tracing experiments, it was shown that removal of fibrosis-inciting injury in mice led to regression of fibrosis that corresponded with return of around half of the activated myofibroblasts to quiescent HSC. It is now accepted that if the primary insult is removed, and fibrosis has not progressed beyond a specific stage, fibrosis is reversible.

# Molecular basis of hepatic neoplasms

## Hepatocellular adenoma

Hepatic or hepatocellular adenomas (HCAs) are benign liver tumors that occur in greater frequency in females and are observed as benign proliferation of hepatocytes in an otherwise normal liver. These monoclonal tumors have been classically associated with the use of estrogen-containing oral contraceptives or androgen-containing steroid anabolic drugs. Such tumors are usually observed as solitary masses and are asymptomatic. Additionally, glycogen storage diseases, especially type I (von Gierke) and type III, are known risk factors for the development of HA. However, in these circumstances, HCAs frequently occur as multiple lesions with a greater propensity to undergo malignant transformation. Macroscopically, HAs present as solitary, yellowish masses (due to lipid accumulation), display a pseudoencapsulated appearance because of the compression of adjacent hepatic tissue, and can range from 0.5 to 15 cm in diameter. Some of the histological features include areas of fatty deposits and hemorrhage, cordlike or plate-like arrangements of larger hepatocytes containing excessive glycogen and fat, sinusoidal dilatation (the result of the effects of arterial pressure as these tumors lack a portal venous supply), presence of few and non-functioning Kupffer cells, and classical absence of portal triads and interlobular bile ducts (Fig. 20.8). The extensive hypervascularity and lack of a true capsule make this tumor prone to hemorrhage. The tumors can regress in size following discontinuation of inciting factors (such as oral contraceptives). At other times, they continue to exhibit a stable or progressive disease. However, the risk of associated hemorrhage and neoplastic transformation is always present. In view of these risks, surgical resection is often recommended. Several molecular classes of this tumor type have now been identified.

Biallelic inactivating mutations in HNF1α or TCF1 genes have been identified in around 30%−40% of HCA. These tumors display marked steatosis and excess glycogen accumulation. HCA with HNF1α inactivation displays an extremely low risk of malignant transformation and is almost exclusive to the tumors displaying activation of β-catenin and IL-6 (inflammatory). Around 10%−15% HCAs display Wnt/β-catenin activation secondary to mutations in exon-3 of the *CTNNB1* gene. Mutant β-catenin may occur in combination with other mutations in inflammatory HCA. In fact, around half of the β-catenin-active HCAs are inflammatory. Recently, β-catenin gene mutations have also been

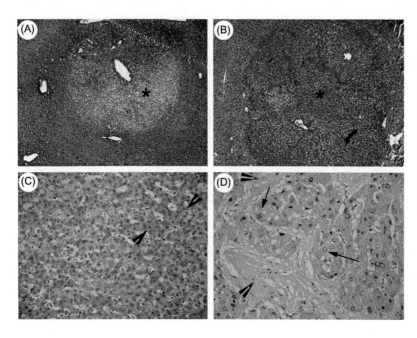

FIGURE 20.8 Histology of hepatic adenoma and hepatocellular cancer (HCC) in mouse model of chemical carcinogenesis and HCC in patients. (A) A hepatic adenoma (*) is evident in this H&E (hematoxylin and eosin) stained histological section of liver from a mouse 6 months after diethylnitrosamine (DEN) injection. (B) HCC (*) is evident in this H&E stained histological section of liver from a mouse 9 months after DEN injection. (C) H&E stained histological liver section shows the abnormal trabecular pattern (*arrowhead*) associated with in HCC in a human patient. (D) H&E stained histological liver section showing the fibrolamellar variant of HCC in a human patient with lamellar fibrosis (*arrowheads*) pattern surrounding large cancer cells (*arrows*).

identified outside of exon 3 in HCAs. Specifically, these mutations affected codon 335 (exon 7) and 387 (exon 8). These HCAs were initially categorized as either unclassified or inflammatory group. These mutations lead to β-catenin activation albeit to a lesser extent than classical exon 3 mutations, both in vitro and in vivo. Overall mutations in β-catenin occur more frequently in HCAs that develop in males. Histologically, these HCAs exhibit cholestasis and cell dysplasia and show frequent cytological abnormalities and pseudoglandular formation. The β-catenin target gene GS is upregulated at RNA and protein levels in β-catenin-mutated HCA. Because detection of nuclear β-catenin by staining is often challenging, combining it with GS staining enables diagnosis of β-catenin-mutated HCA with a greater sensitivity. It has been shown that β-catenin-mutated HCAs have a greater propensity for malignant transformation. Another group of HCAs are the inflammatory type. These display activation of the JAK/STAT pathway. Histologically, these tumors are characterized by polymorphic inflammatory infiltrates. This type of HCA shows mutations in Fyn-related kinase, IL6ST (coding for gp130), STAT3, and GNAS complex locus. A subset of this HCA also exhibit *CTNNB1* mutation, irrespective of the molecular driver, and poses an increased risk of malignant transformation.

## Hepatoblastoma

HB is a rare primary tumor of the liver (incidence is 1:1,000,000 births) but is the most frequent liver tumor in children under the age of 3 years. HBs are composed of hepatoblasts. HB can be fetal, embryonal, mixed (embryonal and fetal), macrotrabecular, and small cell undifferentiated subtypes (Fig. 20.9). Mixed epithelial—mesenchymal HBs, in addition to containing any of these epithelial cell types, also contain mesenchymal components such as myofibroblastic, chondroid, and osteoid tissues. In addition, teratomatous HBs contain derivatives from all three germ layers. Most HBs are sporadic, but can be component of the Beckwith—Wiedemann syndrome (BWS) or Familial Adenomatous Polyposis.

HB typically displayed lower rate of chromosomal imbalances with a small subset showing none versus HCCs showing multiple chromosomal aberrations ranging from 5 to 16 per tumor. The most relevant molecular aberration in HB is activation of Wnt/β-catenin signaling. Sequence analysis of β-catenin gene (*CTNNB1*) has revealed missense mutations or interstitial deletions in up to 90% of HBs. In addition to mutations in *CTNNB1*, mutations in *AXIN* and *APC* have also been reported in HB. These events lead to nuclear and/or cytoplasmic accumulation of β-catenin in HB and coincide with upregulation of several targets of the Wnt pathway (Fig. 20.9).

Although there is compelling data that shows Wnt/β-catenin activation in HB, its exact contribution to tumorigenesis in this disease is not clear. Various target genes of Wnt signaling such as c-Myc, cyclin-D1, GS, EGFR, Axin-2, and others have been reported in various histological subtypes of HB. As reviewed by Armengol et al., nuclear β-catenin is

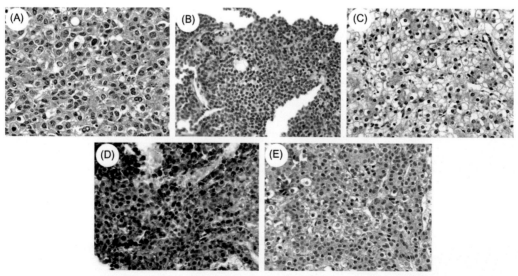

FIGURE 20.9 Histology and immunohistochemistry for β-catenin in pediatric hepatoblastoma. (A) H&E (hematoxylin and eosin) stained histological section displaying embryonal pattern of hepatoblastoma. (B) H&E stained histological section displaying small cell embryonal hepatoblastoma. (C) H&E stained histological section displaying fetal hepatoblastoma. (D) Nuclear and cytoplasmic localization of β-catenin in an embryonal hepatoblastoma (immunohistochemistry (IHC)). (E) Nuclear localization of β-catenin along with some membranous staining in a fetal hepatoblastoma (IHC).

evident in embryonal HB (or in embryonal areas of an HB) and coincides with lack of GS (a Wnt target in mature hepatocytes), whereas fetal HB (or fetal component of HB) showed membranous, cytoplasmic, and nuclear β-catenin and coincided with GS positivity.

A recent study demonstrated just that in around 80% of all HB, β-catenin was nuclear as was the Yes-associated protein 1 (YAP), a component of the Hippo signaling pathway. Although the mechanism of nuclear localization and activation of Yap remains undetermined, it was nuclear in a predominant subset of HB irrespective of histological subtype. Also in HB cell lines only, β-catenin and Yap were found to associate with each other and regulate cell proliferation and survival synergistically. This was unique to HB and was not observed in HCC cells. Using sleeping beauty (SB) transposon/transposase, we expressed deletion-mutant β-catenin along with point-mutant active Yap in a small number of hepatocytes in the livers of young mice through hydrodynamic tail vein injection. Co-expression of both genes led to generation of HB in mice. The tumors displayed upregulation of various Wnt and Yap targets. Availability of this model will enable investigation of modulation of β-catenin and Yap pathways or additional downstream signaling for therapeutic benefit.

## Hepatocellular cancer

HCC is the most common primary neoplasm of the liver accounting for 85% of all primary malignant tumors. HCC is a sexually dimorphic disease with greater incidence in men. In fact, HCC is the fifth most common cancer type in men and the ninth most frequent cancer in women. HCC is the second most common cause of death from cancer worldwide. The incidence rates and death rates associated with HCC are increasing in many parts of the world, including North America, Latin America, and central Europe, making it prudent to understand molecular and cellular mechanisms of HCC for improved understanding and therapies.

Advanced fibrosis and cirrhosis secondary to chronic liver diseases or chronic hepatitis of any etiology (infection, metabolic changes, toxins such as alcohol or aflatoxin) remain the most important risk factors for HCC. In fact, repeated cycles of hepatocyte necrosis, followed by hepatocyte proliferation due to the innate regenerative capacity of the liver, can be important initiating events of dysplasia. Hepatocyte death also evokes inflammation and a wound-healing response in the form of activation of stellate cells or portal fibroblasts, which by themselves can create a microenvironment that may play a role in tumorigenesis. As the process continues, there is formation of regenerating nodules containing proliferating hepatocytes and clearly defined by collagen borders. However, due to inflammation, activated myofibroblasts, and other yet-uncharacterized elements, there is persistent oxidative stress and the free radicals incite DNA damage in proliferating hepatocytes. Accumulating aberrations in the form of loss of tumor suppressor genes and activating mutations in oncogenes can suddenly provide a growth and survival advantage to a group of cells leading to their outgrowth. This sets the stage for progressive tumorigenesis, from dysplastic nodules to frank, progressive and even metastatic HCC. Thus, some of the regenerating nodules evolve into low-grade and high-grade dysplastic nodules, which then lead to HCC. This common basis of HCC initiation and development is very distinct from other less common forms (such as the minor subsets of HCAs), which proceed to HCC (Fig. 20.8). Histologically, HCC may be well differentiated to poorly differentiated depending on degree of nuclear atypia, anaplasia, and nucleolar prominence, and the tumor cells usually arrange in a trabecular pattern composed of uneven layers of hepatocytes with occasional mitotic figures (Fig. 20.8). Thus, overall, there is a progression of the preneoplastic events into neoplasia, and clearly understanding the molecular basis of this evolution will have strong clinical implications.

The possibility of a cancer stem cell origin of some of the HCC is also being entertained. Cancer stem cells are formed by mutations in the normal existing stem cells or a progenitor cell within a tissue. It has also been suggested that mature cells (such as hepatocytes) can dedifferentiate into progenitor-like cells and become targets for mutational events. Dysplastic lesions within liver have been shown to have pre-neoplastic precursors. In addition, a subset of HCCs exhibit mixed hepatocholangiocarcinoma phenotype that might represent bipotential stem cell origin of these tumors. Likewise, some early HCCs have been shown to harbor smaller proliferating oval cells or stem cells. In fact, the pathways that are known to regulate stem cell renewal (such as the Wnt, Hedgehog, and Notch) have been shown to play a role in oval cell activation, as well as in a subset of HCCs.

During the process of hepatocarcinogenesis, several alterations ranging from point mutations in individual genes to gain or loss of chromosome arms have been reported. Several candidate genes include *MYC* (8q), *Cyclin-A2* (4q), *Cyclin-D1* (11q), *Rb1* (13q), *AXIN1* (16p), *p53* (17p), *IGFR-II/M6PR* (6q), *p16* (9p), *E-Cadherin* (16q), *SOCS* (16p), and *PTEN* (10q). A recent study performed exome sequencing on 243 liver tumors and recorded their unique molecular signatures. The most frequent mutation affected the promoter region of *TERT* promoter in around 60% of all HCCs that led to increased telomerase expression. Next common mutations were those affecting *CTNNB1* and *TP53* each of which were evident in around 35% and 25% of HCC cases. Several additional mutations occurred consistently albeit at lower frequencies including those affecting *ARID1A*, *ALB*, *AXIN1*, *APOB*, *CDKN2A*, *ARID2*, *RPS6SKA3*, and others.

The relevance of telomerase activation in human HCC must be emphasized. In fact, telomerase activation is evident in more than 90% of all human HCC cases albeit due to differing mechanisms, including promoter mutations, gene amplification, and HBV insertion within the *TERT* promoter. Frequently, *TERT* promoter mutations coexist with mutations in *CTNNB1*. Mutations that lead to activation of β-catenin are usually present in exon 3, exon 7/8, or in components of β-catenin degradation complex components such as *AXIN1*. Mutations in genes that regulate the redox state of a cell including *NRF2* and *KEAP1* have also been reported.

Eventually, activation of the following signaling pathways alone or in combination has been shown in HCC in multiple studies: Wnt/β-catenin, Notch, TGF-β, IGF2, AKT/mTOR, RAS/MAPK, and Met. Various preclinical and clinical studies have shown some advantages of targeting these pathways in HCC, although several studies are preliminary at this stage. Similarly, several of these pathways (excluding Wnt signaling) activate RTKs and the signal is transduced via Ras/MAPK, PI3 kinase, or Jak/Stat pathways. Success of Sorafenib in HCC is attributable to identification of such molecular aberrations in HCC.

Generation of animal models to replicate human disease will be instrumental to further delineate the biology of disease and develop improved therapies. There have been significant issues with tumor xenograft models. SB transposon/transposase models have gained popularity to generate efficient animal models of human disease and use a reductionist approach. For example, examining the human HCC landscape, we identified around 10% of all HCCs to demonstrate Met activation or overexpression along with *CTNNB1* mutations. Using SB and hydrodynamic tail vein injections, we generated an animal model that co-expressed Met and mutant β-catenin gene and led to HCC development, which showed high resemblance to human HCC at a molecular level. Such approaches may eventually be critical to understand the cooperative or individual roles of key pathways in HCC. Other animal models that replicate tumor microenvironment and a cancer field effect to study HCC are also invaluable. These include mice that are given chemical carcinogen such as diethylnitrosamine followed by repeated insults through administration of fibrosis causing agents such as carbon tetrachloride. In summary, such animal models may be instrumental in better understanding of the disease process and hence may allow development of improved therapies for HCC.

## Fibrolamellar hepatocellular cancer

One of the uncommon variants of HCC that deserves mention is fibrolamellar HCC (FL-HCC). This neoplasm usually occurs in younger patients (5−35 years) and in a non-cirrhotic hepatic background. The histology is characterized by lamellar pattern of fibrosis surrounding larger tumor cells that contain abundant granular cytoplasm and prominent nucleoli (Fig. 20.8). Although initially thought to have better prognosis, FL-HCC appears to be equally aggressive with a 45% 5-year survival. In fact the improved prognosis appears to be due to absence of cirrhosis in FL-HCC as compared with existing cirrhosis in the conventional HCC cases. The molecular basis of this variant of HCC remains largely obscure. Because of rarity of these tumors, lesser studies are available that have characterized molecular and genetic aberrations in FL-HCC. The most important study in molecular basis of FL-HCC has come more recently. A chimeric transcript that is expressed in tumors but not in adjacent normal liver was observed in 100% of FL-HCC. This was a result of a ∼400-kilobase deletion on chromosome 19 that led to an anomalous fusion between two genes DNAJB1 (a member of the heat shock 40 protein family) and PRKACA (which encodes cAMP-dependent protein kinase A catalytic subunit alpha). The fusion protein when expressed in cells showed clear kinase activity. While the mechanism and exact significance of these results is not completely clear, based on the presence of this genetic alteration in 100% of FL-HCC, it is likely that this fusion plays a major role in the pathogenesis of this tumor type.

# Key concepts

- Liver development entails precise and spatiotemporal modulation of several signaling pathways that can sometimes play seemingly opposite roles in same cells.
- Several signaling pathways are modulated in tandem to ensure liver regeneration after surgical resection, which enable regain of pre-surgical mass and function.
- There exists redundancy in various key pathways, such that elimination of key component of a signaling pathway is quickly compensated by other mechanisms that enable execution of liver regeneration. Only HGF and EGF combined loss is powerful enough to block liver regeneration completely.

- Hepatocytes and cholangiocytes act as facultative progenitors for each other when hepatocytes or cholangiocytes are unable to sufficiently repair respective cell type via replication.
- Both NASH and ALD are multifactorial diseases with multiple genetic and other disease modifiers being identified. These two diseases remain challenging to treat and are the focus of intense ongoing research and discovery.
- Hepatic fibrosis is reversible up to a point if the insult to the primary cells in the liver is removed, which has been specifically visible following advent of new HCV treatments. Several subsets of cases that received direct acting anti-virals and attained sustained viral response showed reversal of fibrosis. However, others that had progressed to cirrhosis or late stage fibrosis become autonomous to underlying disease and continued to progress.
- Hepatocellular cancer has shown gradual increase in incidence due to prevalence of fibrosis and cirrhosis in turn due to underlying hepatic diseases. GWAS studies reveal key molecular perturbations that could have therapeutic implications and become the basis of precision medicine in this disease of poor prognosis.

## Suggested readings

[1] Zaret KS, Watts J, Xu E, et al. Pioneer factors, genetic competence, and inductive signaling: programming liver and pancreas progenitors from the endoderm. Cold Spring Harb Symp Quant Biol 2008;73:119—26.
[2] Planas-Paz L, Orsini V, Boulter L, et al. The RSPO-LGR4/5-ZNRF3/RNF43 module controls liver zonation and size. Nat Cell Biol 2016;18:467—79.
[3] Monga SP. Beta-catenin signaling and roles in liver homeostasis, injury, and tumorigenesis. Gastroenterology 2015;148:1294—310.
[4] Michalopoulos GK. Principles of liver regeneration and growth homeostasis. Compr Physiol 2013;3:485—513.
[5] Michalopoulos GK. Advances in liver regeneration. Expert Rev Gastroenterol Hepatol 2014;8:897—907.
[6] Yanger K, Zong Y, Maggs LR, et al. Robust cellular reprogramming occurs spontaneously during liver regeneration. Genes Dev 2013;27:719—24.
[7] Yimlamai D, Christodoulou C, Galli GG, et al. Hippo pathway activity influences liver cell fate. Cell 2014;157:1324—38.
[8] Singh R, Kaushik S, Wang Y, et al. Autophagy regulates lipid metabolism. Nature 2009;458:1131—5.
[9] Evans RM, Mangelsdorf DJ. Nuclear receptors, RXR, and the big bang. Cell 2014;157:255—66.
[10] Kozlitina J, Smagris E, Stender S, et al. Exome-wide association study identifies a TM6SF2 variant that confers susceptibility to nonalcoholic fatty liver disease. Nat Genet 2014;46:352—6.
[11] Nagy LE, Ding WX, Cresci G, et al. Linking pathogenic mechanisms of alcoholic liver disease with clinical phenotypes. Gastroenterology 2016;150:1756—68.
[12] Buch S, Stickel F, Trepo E, et al. A genome-wide association study confirms PNPLA3 and identifies TM6SF2 and MBOAT7 as risk loci for alcohol-related cirrhosis. Nat Genet 2015;47:1443—8.
[13] Seki E, De Minicis S, Osterreicher CH, et al. TLR4 enhances TGF-beta signaling and hepatic fibrosis. Nat Med 2007;13:1324—32.
[14] Yoshimoto S, Loo TM, Atarashi K, et al. Obesity-induced gut microbial metabolite promotes liver cancer through senescence secretome. Nature 2013;499:97—101.
[15] Kisseleva T, Cong M, Paik Y, et al. Myofibroblasts revert to an inactive phenotype during regression of liver fibrosis. Proc Natl Acad Sci USA 2012;109:9448—53.
[16] Schulze K, Imbeaud S, Letouze E, et al. Exome sequencing of hepatocellular carcinomas identifies new mutational signatures and potential therapeutic targets. Nat Genet 2015;47:505—11.
[17] Honeyman JN, Simon EP, Robine N, et al. Detection of a recurrent DNAJB1-PRKACA chimeric transcript in fibrolamellar hepatocellular carcinoma. Science 2014;343:1010—14.
[18] Lu WY, Bird TG, Boulter L, et al. Hepatic progenitor cells of biliary origin with liver repopulation capacity. Nat Cell Biol 2015;17:971—83.
[19] Russell JO, Lu WY, Okabe H, et al. Hepatocyte-specific β-catenin deletion during severe liver injury provokes cholangiocytes to differentiate into hepatocytes. Hepatology 2019;69:742—59.
[20] Raven A, Lu WY, Man TY, et al. Cholangiocytes act as facultative liver stem cells during impaired hepatocyte regeneration. Nature 2017;547:350—4.
[21] Mederacke I, Hsu CC, Troeger JS, et al. Fate tracing reveals hepatic stellate cells as dominant contributors to liver fibrosis independent of its aetiology. Nat Commun 2013;4:2823.

Chapter 21

# Molecular basis of diseases of the exocrine pancreas

Matthias Sendler[1], Julia Mayerle[2] and Markus M. Lerch[1]

[1]Department of Medicine A, University Medicine Greifswald, Greifswald, Germany, [2]Department of Medicine II, Ludwigs-Maximilian-University Munich, Munich, Germany

**Summary**

Pancreatitis has been characterized as self-digestion of the pancreas by its own proteases. A premature activation of digestive enzymes within pancreatic acinar cells results in cellular injury and the activation of the immune system. The serine protease trypsinogen plays a crucial role in this process. Trypsinogen is activated by the lysosomal hydrolase cathepsin B within acinar cells and this defines the onset of the disease. This event is accompanied by a local and systemic immune response. The significant role of trypsinogen is underlined by genetic data; different mutations within *PRSS1*, the cationic trypsinogen, or genes that are related to the activation, degradation, or inhibition of trypsin are associated with an increased risk of developing chronic pancreatitis.

## Introduction

Pancreatitis has been characterized as self-digestion of the pancreas by its own proteases. A premature activation of digestive enzymes within pancreatic acinar cells results in cellular injury and the activation of the immune system. The serine protease trypsinogen plays a crucial role in this process. Trypsinogen is activated by the lysosomal hydrolase cathepsin B within acinar cells and this defines the onset of disease. This event is accompanied by a local and systemic immune response. The significant role of trypsinogen is underlined by genetic data; different mutations within *PRSS1*, the cationic trypsinogen, or genes that are related to the activation, degradation, or inhibition of trypsin are associated with an increased risk of developing chronic pancreatitis.

## Acute pancreatitis

Acute pancreatitis presents clinically as a sudden inflammatory disorder of the pancreas, which is associated with premature intracellular activation of pancreatic proteases leading to self-destruction of acinar cells and to autodigestion of the organ. Necrotic cell debris causes a systemic inflammatory reaction, which, in due course, can lead to multiorgan failure. The incidence of acute pancreatitis differs regionally from 20 to 120 cases per 100,000. Acute pancreatitis varies considerably in severity and can be categorized into two forms of the disease. The majority of cases (85%) present with a mild form classified as edematous pancreatitis, with absent or only mild and transient organ failure. In a minority of cases (15%), pancreatitis runs a severe course accompanied by extended ($>48$ h) multiorgan failure. This latter form of pancreatitis is referred to as severe necrotizing pancreatitis. Severe necrotizing pancreatitis is associated with a high mortality (10%−20%) and may lead to long-term complications such as the formation of pancreatic pseudocysts or impairment of exocrine and endocrine function of the pancreatic gland.

In many patients ($\sim 50\%$), the underlying cause of acute pancreatitis is the migration of a gallstone obstructing the pancreatic duct at the papilla of Vater. Another disease-triggering factor is alcohol abuse. In 25%−40% of acute pancreatitis, increased alcohol consumption by the patient is regarded as the cause of the disease. Removal of the underlying disease-causing agents results in complete regeneration of the pancreas and preserved exocrine and endocrine function in the majority of cases. Recurrent attacks of the disease can originate from chronic alcohol abuse, repeated gallstone passage, genetic predisposition, sphincter dysfunction, metabolic disorders, or pancreatic duct strictures, all of

*Essential Concepts in Molecular Pathology.* DOI: https://doi.org/10.1016/B978-0-12-813257-9.00021-8

which can contribute to the development of chronic pancreatitis. In the remaining cases (10%−20%), no apparent clinical cause or etiology of the disease can be identified and these cases are referred to as idiopathic pancreatitis. It became clear during the last decade that previously unknown genetic factors play a major role in these cases. The initial cellular mechanism causing acute pancreatitis is probably independent of the underlying etiology of the disease.

## Early events in acute pancreatitis and the role of protease activation

The pancreas is known as the enzyme factory of the human organism producing and secreting large amounts of potentially hazardous digestive enzymes, many of which are synthesized as proenzymes, known as zymogens. Under physiological conditions these enzymes are secreted in response to hormonal stimulation. Activation of these proenzymes requires hydrolytic cleavage of their activation peptide by proteases. After entering the small intestine, the pancreatic zymogen trypsinogen is first activated to trypsin by an intestinal protease, enterokinase (enteropeptidase). Trypsin then proteolytically processes other pancreatic enzymes to their active forms. Hence, under physiological conditions, pancreatic proteases remain inactive during synthesis, intracellular transport, secretion from acinar cells, and transit through the pancreatic duct. They are activated only when reaching the lumen and brush border of the small intestine (Fig. 21.1).

More than a century ago, Hans Chiari proposed autodigestion of the exocrine pancreatic tissue by proteolytic enzymes as the underlying pathophysiological mechanism for the development of pancreatitis and today this theory is well accepted. Nevertheless, this theory suggests premature intracellular activation of zymogens in the absence of enterokinase and escape of all physiological defense mechanisms, such as the synthesis of endogenous protease inhibitors and the storage of proteases in the membrane-confined compartment of zymogen granules.

Data from various animal models suggest that after the initial insult a variety of pathophysiological factors determine the disease onset. These include (1) a block in secretion, (2) the co-localization of zymogens with lysosomal enzymes, (3) the activation of trypsinogen and other zymogens, and (4) acinar cell injury. In vitro and in vivo studies have demonstrated the importance of premature zymogen activation in the pathogenesis of pancreatitis since the inception of the hypothesis by Chiari. The activation of trypsinogen and other pancreatic zymogens can be demonstrated in

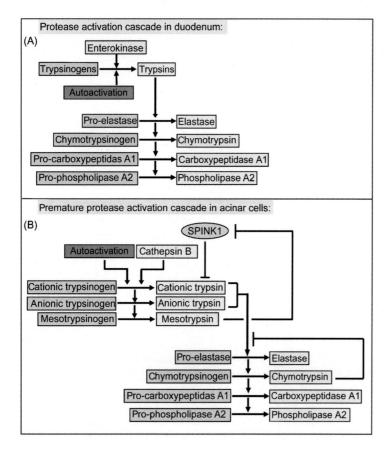

FIGURE 21.1 Activation of pancreatic proteases under normal conditions and in disease. Protease activation cascade in the duodenum (A), enterokinase activates trypsinogen by proteolytic cleavage of the trypsin activation peptide and further autoactivation contributes to this process. Trypsin then activates other digestive proenzymes in a cascade-like fashion. Within acinar cells, premature activation of trypsinogen by cathepsin B is involved in the onset of pancreatitis (B). Intracellular trypsin may activate other digestive proenzymes, in spite of presence of the trypsin inhibitor SPINK1 and the trypsin-degrading enzyme chymotrypsin C.

pancreatic homogenates from experimental animals and this zymogen activation appears to be an early event. Trypsin activity is detected as early as 10 min after supramaximal stimulation with the cholecystokinin analogue caerulein in rats and increases over time. The activation of trypsinogen requires the hydrolytic cleavage of a 7−10 amino acid pro-peptide called TAP (trypsin activation peptide) at the N-terminus of trypsinogen. Localization of immunoactive TAP in the pancreas after caerulein-induced pancreatitis in rats showed trypsinogen activation in the secretory compartment of acinar cells. Furthermore, TAP was detected in serum and urine of patients, and the amount of TAP seems to correlate with the severity of the disease. In addition to the increased activity of trypsin in the early phase of acute experimental pancreatitis, elastase activity also increases. In addition to the activation peptide of trypsinogen (TAP), the activation peptide of carboxypeptidase A1 (PCA1) can be identified in serum at an early stage of pancreatitis. Premature activation of these proenzymes leads to the development of necrosis and to autodigestion of pancreatic tissues. Recent studies defining the localization of early activation of proenzymes suggest that this process, which leads to pancreatitis and pancreatic tissues necrosis, originate in acinar cells (Fig. 21.1).

In conclusion, premature intracellular activation of zymogens to active proteases in the secretory compartment of acinar cells results in acinar cell death by either apoptosis or necrosis and contributes to the onset of pancreatitis. As a result of injury the acinar cells release chemokines and cytokines, which initiate the later events in pancreatitis, including recruitment of inflammatory cells into the tissue. Trypsin seems to be the key enzyme in the process of activating other digestive proenzymes prematurely, and one of the crucial questions in understanding the pathophysiology of acute pancreatitis is to identify the mechanisms, which prematurely activate trypsinogen inside acinar cells. However, it must be noted that the term trypsin, as defined by the cleavage of specific synthetic or protein substrates, comprises a group of enzymes whose individual role in the initial activation cascade may differ considerably.

## The mechanism of zymogen activation

One hypothesis for the initiation of the premature activation of trypsinogen suggests that during the early stage of acute pancreatitis, pancreatic digestive zymogens become colocalized with lysosomal hydrolases. Recent data show that the lysosomal cysteine proteinase cathepsin B plays an important role for the activation of trypsinogen. Many years ago in vitro data could demonstrate an activation of trypsinogen by cathepsin B. Most lysosomal hydrolases are synthesized as inactive proenzymes, but in contrast to digestive zymogens, they are activated by posttranslational processing in the cell. During protein sorting in the Golgi system, lysosomal hydrolases are sorted into pre-lysosomes, in contrast to zymogens that are packed into condensing vacuoles. The sorting of lysosomal hydrolases depends on a mannose-6-phosphate-dependent pathway, which leads to a separation of lysosomal hydrolases from other secretory proteins and to the formation of prelysosomal vacuoles. However, this sorting is incomplete, and under physiological conditions a significant fraction of hydrolases enter the secretory pathway. It has been suggested that these incorrectly sorted hydrolases play a role in the regulation of zymogen secretion. In acute pancreatitis the separation of digestive zymogens and lysosomal hydrolases is impaired. This leads to further co-localization of lysosomal hydrolases and zymogens within cytoplasmic vacuoles of acinar cells. This co-localization could also be shown by electron microscopy, as well as in subcellular fractions isolated by density gradient centrifugation. The redistribution of cathepsin B from the lysosome-enriched fraction was noted within 15 min of the start pancreatitis induction, and trypsinogen activation was observed in parallel. There are two main theories trying to explain the co-localization of cysteine and serine proteases: (1) fusion of lysosomes and zymogene granules or (2) damaged sorting of zymogens and hydrolases in the process of vacuole maturation.

Final evidence that cathepsin B is involved in activation of trypsinogen during caerulein-induced experimental pancreatitis comes from experiments in cathepsin B knockout mice. In these animals, trypsin activity after induction of experimental pancreatitis was reduced to less than 20% compared with wild-type animals, and the severity of the disease was markedly ameliorated. These data showed unequivocally the importance of cathepsin B for the pathogenesis of acute pancreatitis (Fig. 21.1).

The cathepsin B theory implies one further critical point: Trypsinogen is expressed and stored in the presence of different potent intrapancreatic trypsin inhibitors. To activate trypsinogen, cathepsin B needs to override these defensive mechanisms to start the premature intracellular activation cascade. Recently it has become clear that cathepsin B activates cationic and anionic trypsinogen, as well as mesotrypsinogen. Mesotrypsin, the third trypsin isoform expressed in the human pancreas, is resistant to trypsin inhibitors such as SPINK1 or soybean trypsin inhibitor. Moreover, mesotrypsinogen is able to degrade trypsin inhibitors. Under physiological conditions, mesotrypsin is activated in the duodenum by enterokinase, where it degrades exogenous trypsin inhibitors to ensure normal tryptic digestion. In vitro mesotrypsin rapidly inactivates trypsin inhibitors such as SPINK1 by proteolytic cleavage. Therefore, activation of trypsins by cathepsins might not only trigger a proteolytic cascade but also involve the removal of trypsin inhibitors such as SPINK1 via the activation of mesotrypsin.

New data from T7 trypsinogen knockout mice suggest a critical role of trypsinogen activation for disease severity analogous in $CTSB^{-/-}$ mouse experiments. Activation of an immune response, as well as endoplasmic reticulum (ER) stress during pancreatitis, is independent of the presence of trypsinogen. In contrast to acute pancreatitis, chronic pancreatitis was not influenced by the absence of trypsinogen or cathepsin B. These results have challenged the role of the protease cascade in the development of pancreatitis and are in contrast to the observation that the only autosomal dominant variety of hereditary pancreatitis is invariably associated with mutations in the cationic trypsinogen (*PRSS*) gene.

## The degradation of active trypsin

During the early phase of pancreatitis, trypsinogen and other zymogens are rapidly activated, whereas later in the disease course their activity declines to physiological levels suggesting degradation of the active enzymes. This phenomenon has been termed autolysis or autodegradation. Because this process self-limits autoactivation of trypsinogen, it is regarded as a safety mechanism to counteract premature zymogen activation.

One theory to possibly explain how uncontrolled trypsinogen activation can be antagonized involves the existence of a serine protease that is capable of trypsin degradation. In 1988, Heinrich Rinderknecht discovered an enzyme that rapidly degrades active cationic and anionic trypsin and named this protease enzyme Y. Recent in vitro data suggests that the autodegradation of trypsin is a very slow process and that most of trypsin degradation is not mediated by trypsin itself but by another enzyme. Chymotrypsin C (CTRC) has the capability to proteolytically cleave cationic trypsin at Leu(81)-Glu(82) in the $Ca^{2+}$-binding loop. This leads to rapid autodegradation and catalytic inactivation of trypsin by cleavage at the Arg(122)-Val(123) position. Thus, CTRC has the capability to induce trypsin-mediated trypsin autodegradation during pancreatitis and is most likely identical to the enzyme Y that Rinderknecht proposed nearly 20 years ago. However, CTRC also has the ability to induce trypsin-mediated trypsinogen autoactivation by proteolytic cleavage at the Phe(18)-Asp (19) position of cationic trypsinogen. The balance between autoactivation and autodegradation of cationic trypsin mediated by CTRC is regulated via the $Ca^{2+}$ concentration. In the presence of $Ca^{2+}$ (1 mM), degradation of trypsin is blocked, and autoactivation of trypsinogen is induced. Under physiological conditions, such high calcium concentrations are effective in the duodenum and trypsinogen can be readily activated to promote digestion. In the absence of high $Ca^{2+}$ concentrations, CTRC degrades active trypsin and protects against premature activation of trypsin (Fig. 21.2).

## Calcium signaling

Under physiological conditions, pancreatic acinar cells maintain a $Ca^{2+}$ gradient across the plasma membrane with low intracellular concentration and high extracellular concentration of calcium. In response to hormonal stimulation, $Ca^{2+}$ is released from intracellular stores and taken up from the extracellular space to regulate signal-secretion coupling. In pancreatic acinar cells, acetylcholine (ACh) and cholecystokinin (CCK) regulate the secretion of digestive enzymes via the generation of repetitive local cytosolic $Ca^{2+}$ signals. In response to secretagogue stimulation with ACh or CCK, $Ca^{2+}$ is

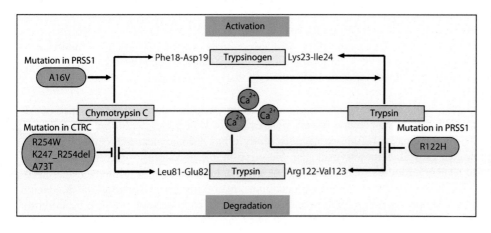

FIGURE 21.2 Chymotrypsin C has different functions in the processing of trypsin or trypsinogen. The major role in pancreatitis is degrading active trypsin. This function is disturbed by mutations within the *CTRC* gene or by high levels of $Ca^{2+}$. The activation of trypsinogen to trypsin is mediated by chymotrypsin C. Trypsin itself has the capability to autoactive or to self-degrade. The R122H mutation results in the decreased autolysis of active trypsin.

initially released from intracellular stores near the apical pole of acinar cells. This induces the fusion of zymogen granules with the apical plasma membrane and the activation of $Ca^{2+}$-dependent $Cl^-$ channels in the apical membrane. The pattern of intracellular calcium signal in response to secretagogues stimulation is dependent on the neurotransmitter or hormone concentration. ACh at physiological concentrations elicits repetitive calcium spikes and oscillations, which are restricted to the secretory pole of the cell. High concentrations of CCK lead to short-lasting spikes followed by longer calcium transients that spread to the entire cell. Each oscillation is associated with a burst of exocytotic activity and the release of zymogen into the duct lumen. In contrast, supramaximal stimulation of acinar cells induces a completely different pattern of calcium signals. Instead of oscillatory activity observed with physiological doses of CCK, there is a much larger rise followed by a sustained elevation associated to a block of enzyme secretion and premature intracellular protease activation. Calcium is released from the ER in response to stimulation. The ER is located in the basolateral aspect of the acinar cell with extensions into the apical region enriched with zymogen granules. The whole ER is filled with $Ca^{2+}$, but $Ca^{2+}$ is released only at the apical pole in response to CCK or ACh due to the higher density of $Ca^{2+}$ release channels at the apical pole of the ER. Two types of calcium channels are expressed in the ER of acinar cells, inositol triphosphate receptors ($IP_3$) and ryanodine receptors, and both are required for apical $Ca^{2+}$ peaks. ACh activates phospholipase C and initiates the $Ca^{2+}$ release via the intracellular messenger $IP_3$. In contrast, CCK does not activate PLC but increases the intracellular concentration of nicotinic acid adenine dinucleotide phosphate in a dose-dependent manner. The higher density of $Ca^{2+}$ channels in the apical region of the ER explains the initiation of calcium signals in the granule part of the cytoplasm. The apical, zymogen granule-enriched part of the acinar cell is surrounded by a barrier of mitochondria, which absorb released calcium, preventing higher $Ca^{2+}$ concentrations from expanding beyond the apical part of acinar cells. The spatially limited release of $Ca^{2+}$ at the apical pole prevents an unregulated chain reaction across gap junctions, which would affect neighboring cells. The mitochondrial $Ca^{2+}$ uptake further leads to an increase in metabolism and generation of ATP. ATP is required for the reuptake of $Ca^{2+}$ in the ER via the sarcoplasmic ER calcium ATPase and for exocytosis across the apical membrane. Thus, $Ca^{2+}$ homeostasis plays a crucial role in maintaining signal-secretion coupling in pancreatic acinar cells (Fig. 21.3).

Elevated calcium concentrations in the extracellular compartment or within acinar cells are known to be a risk factor for the development of acute pancreatitis. Disturbances in the calcium homeostasis of pancreatic acinar cells occur early in the secretagogue-induced model of pancreatitis. An attenuation of the cytosolic calcium elevation in acinar cells by the calcium chelator BAPTA-AM prevents zymogen activation, and therefore gives evidence that calcium is essential for zymogen activation. In the absence of extracellular calcium, the activation of trypsinogen induced by supramaximal doses of caerulein is attenuated, suggesting that the initial and transient rise in calcium caused by the release of calcium from the internal stores is not sufficient to permit trypsinogen activation. In contrast, interference with high calcium plateaus by the natural calcium antagonist magnesium or a calcium chelator in vivo abolishes trypsinogen activation as well as pancreatitis.

## Inflammation: cause and consequence of acinar cell damage

In addition to intracellular activation of zymogens, the immune reaction also plays an important role for the severity of pancreatitis. Activation of the immune system is an initial step during manifestation of pancreatitis. Acinar cells respond to CCK stimulation with an activation of NFκB, a major transcription factor for the activation of the immune system. This leads to expression of a plurality of immune effective factors such as TNF-α, MCP-1, and IL-6. Cytokine and chemokine release from acinar cells function to recruit neutrophil granulocytes and monocytes to the organ. The activation of the immune system is independent of bacterial pathogens and pancreatitis. Hence, pancreatitis is primarily a sterile disease. Bacterial infections (such as infectious pancreatic necrosis) represent later events during disease progression. Bacterial translocation from the gut is responsible for infectious necrosis and is associated with an increased mortality of disease. The time course of the immune response during severe acute pancreatitis can be divided into an initial systemic inflammatory response syndrome (SIRS) and a later compensatory anti-inflammatory response syndrome (CARS). The initial immune reaction is characterized by a pro-inflammatory cytokine storm (SIRS), followed by an anti-inflammatory down regulation of the immune response characterized by anti-inflammatory cytokines (CARS). During CARS, the capability of the immune system to respond to bacteria is severely limited. As a consequence, bacteria cross the intestinal barrier, which leads to infectious necrosis. The initial pro-inflammatory reaction can also lead to a severe course of disease with multi-organ failure. Systemic damage such as lung injury is a common consequence of acute pancreatitis and is regulated by the pro-inflammatory cytokine IL-6. Animal experiments underline the importance of immune reaction during pancreatitis. Transgenic mice overexpressing IL-1β under a pancreas-specific promoter

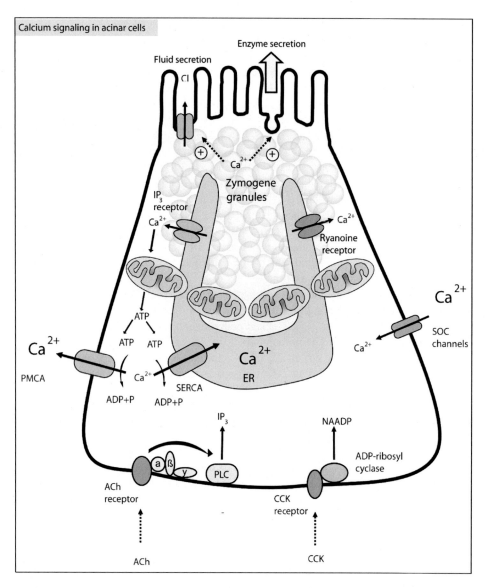

FIGURE 21.3 Calcium signaling in acinar cells. Intracellular $Ca^{2+}$ homeostasis regulates the secretion of zymogens and fluid. Acetylcholine (ACh) regulates apical $Ca^{2+}$ influx from intracellular stores (endoplasmic reticulum (ER)) via the second messenger inositol-3-phosphate. Cholecystokinin (CCK) leads to the production of nicotinic acid adenine dinucleotide phosphate (NAADP), which interacts with ryanodine receptor in the ER membrane and regulates the apical $Ca^{2+}$ influx. The plasma membrane calcium ATPase (PMCA) or the sarcoplasmic endoplasmic reticulum calcium ATPase (SERCA) regulates cytoplasmic $Ca^{2+}$ decrease. A mitochondrial barrier inhibits a global $Ca^{2+}$ increase by absorbing free $Ca^{2+}$ from the apical zymogen-enriched part of the acinar cell. Interferences in the calcium homeostasis lead to a global $Ca^{2+}$ increase in the cytoplasm, which results in premature activation of zymogens.

develop a chronic form of pancreatitis with excessive infiltration of immune cells. This model illustrates impressively the role of infiltrating immune cells on the acute and chronic forms of pancreatitis.

Different subpopulations of leukocytes (cells of the innate immune system) transmigrate during the first hours after onset of the disease directly into the pancreas. Neutrophil granulocytes are the first population of immune cells transmigrating to the pancreas. These cells mediate pancreatic damage by producing reactive oxygen species or by degrading tissue architecture via release of enzymes such as neutrophil elastase. Depletion of neutrophil granulocytes attenuates the severity of pancreatitis in animal models. Neutrophils are able to form extracellular traps (NETs) in response to bacterial infection. This cellular suicide is accompanied by an excessive cytokine release and recruits other immune cells to the site of infection. New data show that during pancreatitis (which is primarily sterile inflammation), NET formation plays an important role in the recruitment of immune cells to the pancreas and that this is directly related to pancreatic damage (Fig. 21.4).

Similar to neutrophils, monocytes are cells of the innate immune system and transmigrate at an early disease stage into the pancreas. Recruitment and activation of monocytes is affected by the acinar cells itself during secretion of MCP-1 and conversely by the release of DAMPs (damage-associated molecular patterns). DAMPs are extracellular DNA, histones, or free ATP. The inflammasome pathway becomes activated by various DAMPs. Monocytes increase pancreatic damage by secretion of TNF-α. Animal models demonstrate an important role of TNF-α for the induction of regulated acinar cell necrosis via the RIP1/RIP3 (receptor interacting protein) complex formation. Deletion of RIP3,

The role of innate immune response in acute pancreatitis

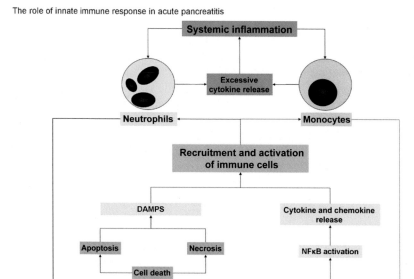

FIGURE 21.4   The role of innate immune response in acute pancreatitis. A pathophysiological stimulus induces acinar cell damage, which is accompanied by activation of the transcription factor NFκB, which results in secretion of pro-inflammatory cytokines, and induction of cell death pathways. Both the release of cytokines and the release of damage-associated molecular patterns (DAMPs) lead to the recruitment of neutrophils and macrophages to the pancreas. Both cell types increase pancreatic damage by secretion of TNF-α or the production of reactive oxygen species (ROS). This cycle results ultimately in an excessive cytokine storm, which is responsible for systemic inflammation associated with multiple complications and organ failure. *NETs*, neutrophil extracellular traps; *PMN*, polymorphonuclear neutrophil.

as well as treatment with necrostatin, an inhibitor of the necroptosis pathway, leads to markedly reduced severity of disease with decreased pancreatic necrosis. Pro-inflammatory macrophages are responsible for pancreatic necrosis during the acute phase of the disease (Fig. 21.4). By changing their polarization from pro-inflammatory to anti-inflammatory phenotype (M1 to M2 macrophages), cytokine secretion is also altered dramatically. Instead of pro-inflammatory mediators such as TNF-α, macrophages begin to secrete anti-inflammatory cytokines such as IL-10. M2 macrophages are the major source of TGF-β, which activates pancreatic stellate cells and stimulates them to secrete extracellular matrix proteins such as collagens. This is an essential part of fibrosis development during chronic pancreatitis.

## Chronic and hereditary pancreatitis

Chronic pancreatitis is clinically defined as recurrent bouts of a sterile inflammatory disease characterized by persistent and often progressive and irreversible morphological changes, typically causing pain and permanent impairment of pancreatic function. Chronic pancreatitis histologically represents a transformation of focal necrosis into perilobular and intralobular fibrosis of the parenchyma, pancreatic duct obstruction by pancreatic stones and tissue calcification, and the development of pseudocysts. In the course of the disease, progressive loss of endocrine and exocrine function can be observed. It should be noted that the clinical distinction between the pathophysiology associated with acute and chronic pancreatitis is becoming less distinct, and similar or identical onset mechanisms may play roles in both varieties. Such a mechanism is the premature and intracellular activation of digestive proteases. Much of our present knowledge about this process has been generated since a genetic basis for pancreatitis was first reported in 1996. Hereditary pancreatitis represents a genetic disorder associated with mutations in the cationic trypsinogen gene and presents with a disease penetrance of up to 80%. Patients with hereditary pancreatitis suffer from recurrent episodes of pancreatitis, which progress in the majority of cases to chronic pancreatitis. The disease usually begins in early childhood, but onset can vary from infancy to the sixth decade of life.

### Mutations within the *PRSS1* gene

Hereditary pancreatitis is associated with genetic mutations in the cationic trypsinogen gene, suggesting that mutations in a digestive protease such as trypsin can cause the disease. Hereditary pancreatitis follows, for the most part, an autosomal dominant trait with an 80% disease penetrance. The gene coding for cationic trypsinogen (*PRSS1*) is approximately 3.6 kb long, located on chromosome 7, and structured in 5 exons. The precursor of cationic trypsinogen is a 247-amino-acid protein, the first 15 amino acids of which represent the signal sequence, the next 8 amino acids function as the activation peptide, and the remaining 224 amino acids form the backbone and catalytic center of the digestive enzyme. Exactly one century after Chiari proposed his theory on autodigestion of the pancreas as a pathogenic mechanism of pancreatitis, Whitcomb et al. (1996) reported a mutation in exon 3 of the cationic trypsinogen gene (*PRSS1*) on chromosome 7 (7q36) that was strongly associated with hereditary chronic pancreatitis. This single point mutation causes an arginine-(R)-(CGC) to histidine-(H)-(CAC) substitution, at position 22 of the cationic trypsinogen gene (pR122H). This amino acid alteration is located in the hydrolysis site of trypsin and can prevent the autodegradation of active trypsin (Fig. 21.2). Once trypsin has been activated intracellularly, the R122H mutation would interfere with the elimination with active trypsin by autodegradation. This conclusion was derived from in vitro data using recombinant R122H mutant trypsinogen. In the same *Escherichia coli*—based system Sahin-Toth and coworkers showed that the R122H mutation leads to an increase in trypsogen autoactivation. Therefore, the R122H mutation represents a dual gain of function mutation, which facilitates intracellular trypsin activity and results in a higher stability of R122H-trypsin. The direct pathogenic role of the R122H mutations for the development of pancreatitis was confirmed by the group of Bar-Sagi. These investigators generated a transgenic mouse in which the expression of the murine *PRSS1* mutant R122H (R122H_mPRSS1) was targeted to pancreatic acinar cells by fusion to the elastase promoter. Pancreata from transgenic mice displayed early-onset acinar cell injury and inflammatory cell infiltration. With progressing age, the transgenic mice developed areas of pancreatic fibrosis and displayed acinar cell dedifferentiation. Interestingly, no increased trypsin activity was found in these animals under experimental supramaximal stimulation.

Shortly after the identification of the R122H mutation, a second mutation was reported in kindreds with hereditary pancreatitis. The R122C mutation is a single amino acid change affecting the same codon as the R122H mutation. In contrast to the R122H mutation, the R122C mutation causes a decreased trypsinogen auto-activation, and biochemical studies demonstrated a 60%—70% reduced activation in the Cys-122 trypsinogen mutant induced by either enterokinase or cathepsin B activation. The amino acid alteration of the R122C mutation leads to altered cysteine—disulfide bonds and consequently a misfolded protein structure, which accounts for the reduced catalytic activity.

Since the initial discovery, several other mutations (64 until today) in the trypsinogen gene have been reported, but the R122H mutation is still by far the most common. In addition to the R122H mutation, four other mutations in different regions of the *PRSS1* gene associated with hereditary pancreatitis have been biochemically characterized: A16V, D22G, K23R, E79K, and N29I. It has been suggested that these mutations may have different structural effects on the activation and activity of trypsinogen (Table 21.1).

Recent data from the group of Sahin-Toth have identified CTRC, a trypsin degrading enzyme, as a key factor in the pathogenesis of pancreatitis associated with *PRSS* mutation. The most frequent mutations simply render cationic trypsin more resistant to degradation by CTRC and in this context confer a gain of function. At present the underlying mechanisms which best explain why trypsin (PRSS1) mutations cause pancreatitis are (1) resistance of the mutant protein to degradation by CTRC, i.e. a defective defense mechanism (2) an increased propensity to autoactivate, and (3) misfolding of the mutant protein which leads to significant endoplasmic reticulum stress.

## Mutations within the *PRSS2* gene

The fact that mutations in the *PRSS1* gene encoding for cationic trypsinogen are associated with hereditary pancreatitis suggests that genetic alterations of the anionic trypsinogen gene (*PRSS2*) might also be associated with chronic pancreatitis. The E79K (c.237 G > A) mutation in the cationic trypsinogen gene, while reducing auto-activation by 80%—90%, leads to a two-fold increase in activation of anionic trypsinogen and suggests a potential role of *PRSS2*. Genetic analysis of the *PRSS2* gene in 2466 chronic pancreatitis patients and 6459 healthy individuals revealed an increased rate of a rare mutation in the anionic trypsinogen gene in control subjects. A variant of codon 191 (G191R) was present in 220/6459 (3.4%) controls but in only 32/2466 (1.3%) affected individuals. Biochemical analysis of the recombinant G191R protein showed a complete loss of tryptic function after enterokinase or trypsin activation, as well as rapid autolytic proteolysis of the mutant protein. This was a first report of a loss of trypsin function (in *PRSS2*) that has a protective effect on the onset of pancreatitis.

## Mutations in the chymotrypsin C gene

Because it was known that chymotrypsin can degrade all human trypsin enzymes and trypsinogen isoforms with high specificity, the chymotrypsin C gene (*CTRC*) was sequenced in a German cohort suffering from idiopathic and

**TABLE 21.1** Common mutations associated with pancreatitis mutations within the *PRSS1* gene or genes inactivating active trypsin (e.g., SPINK1 or CTRC) are burdened with different phenotypes, but all result in an increased risk for developing chronic pancreatitis. Mutations within the *CFTR* gene are frequently associated with pancreatitis. It has been suggested that these mutations have an effect on the rate of enzyme secretion from acinar cells, which leads to an accumulation of enzymes and subsequently to premature activation of zymogens.

**Most common mutations associated with pancreatitis**

|  | Gene | Mutation | Comments |
|---|---|---|---|
| pancreatic proteases | *PRSS1* | R122H | Increased autoactivation and decreased autolysis of cationic trypsin (most common mutation) |
|  |  | R122C | Decreased autoactivation of trypsinogen, decreased autolysis of trypsin and decreased trypsin activity |
|  |  | N29I | Increased autoactivation of trypsinogen |
|  |  | A16V | Increased autoactivation of trypsinogen |
|  |  | D22G | Increased autoactivation of trypsinogen |
|  |  | K23R | Increased autoactivation of trypsinogen |
|  |  | E79K | Increased activation of anionic trypsinogen, decreased autoactivation of trypsinogen |
|  | *PRSS2* | G191R | Loss of trypsin activity, increased protein degradation (protective mutation) |
|  | *CTRC* | R254W | Decreased activity of chymotrypsin C |
|  |  | K247_R254del | Loss of function of chymotrypsin C |
|  |  | A37T | Decreased trypsin degradation of chymotrypsin C |
|  | *CPA1* | Various | Loss of function of carboxypeptidase A1, causes endoplasmatic reticulum stress by misfolding protein |
| trypsin inhibitor | *SPINK1* | N34S | Damaged inhibition of active trypsin |
|  |  | R65Q | 60% loss of protein expression |
|  |  | G48E | Nearly complete loss of protein expression |
|  |  | D50E | Nearly complete loss of protein expression |
|  |  | Y54H | Nearly complete loss of protein expression |
|  |  | R67C | Nearly complete loss of protein expression |
|  |  | L14P & L14R | Rapid intracellular degradation of SPINK1 |
| other | *CEL* | CEL-HYB | Protein accumulation causes endoplasmatic reticulum stress |
|  | *CFTR* | Various >1000 | Decreased secretion ratio of acinar cells and pancreatic duct cells |
|  | *FUT2* |  | Blood group B and FUT-2 secretion status is a new risk factor for chronic pancreatitis |
|  | *CLDN2* |  | Unknown patho-mechanismus |

hereditary pancreatitis. Two variants in the *CTRC* gene were found in association with hereditary and idiopathic chronic pancreatitis. The c.760 C > T (p.R254W) variant occurred with a frequency of 2.1% (19/901) in affected individuals compared with 0.6% (18/2804) in healthy controls, whereas the c738_761del24 (p.K247_R254del) variant occurs with a frequency of 1.2% compared with 0.1% in controls. In a confirmative cohort of different ethnic background, the authors detected a third mutation in affected patients with a frequency of 5.6% (4/71) compared with 0% (0/84) in control individuals. This mutation leads to an amino acid exchange at the position 73 (c.217 G > A, p.A73T). The suggested pathogenic mechanism of *CTRC* mutations is based on lowered enzyme activity in the p.R254W variant and a total loss of function in the deletion mutation (p.K247_R254del) and the p.A73T variant (Fig. 21.2). Hence, CTRC is an enzyme that can counteract the disease-causing effect of trypsin, and loss of function mutations impairs its protective role in pancreatitis by allowing prematurely activated trypsin to escape its degradation.

## Mutations in serine protease inhibitor kazal type 1

It was found that mutations in the *SPINK1* gene (the pancreatic secretory trypsin inhibitor or PSTI, OMIM 167790) are associated with idiopathic chronic pancreatitis in children. *SPINK1* mutations are frequently detected in cohorts of patients who do not have a family history and exhibit none of the classical risk factors for chronic pancreatitis. The most common mutation (N34S) is located in exon 2 of the *SPINK1* gene, and asparagine (AAT) is substituted by serine (AGT). Homozygous and heterozygous N34S mutations were detected in 10%−20% of patients with pancreatitis, compared with 1%−2% of healthy controls, suggesting *SPINK1* as a disease-modifying factor. Structural modeling of SPINK1 predicted that the N34S region near the 41-lysine residue functions as the trypsin binding pocket of SPINK1 and that the N34S mutation changes the structure of the trypsin binding pocket resulting in decreased inhibitory capacity. In contrast to the computer-modeled prediction, in vitro experiments using recombinant N34S SPINK1 and wild-type SPINK1 demonstrated identical trypsin inhibitory activities. Recently, two novel SPINK1 variants in the exon that affects the secretory signal peptide have been reported. Seven missense mutations occurring within the mature peptide of PSTI associated with chronic pancreatitis were analyzed for expression level. The N34S and the P55S mutations do not result in a change of PSTI activity nor in a change in expression. The R65Q involves the substitution of a positively charged amino acid for an uncharged amino acid, resulting in a 60% reduction of protein expression. G48E, D50E, Y54H, and R67C, all of which affect strictly conserved amino acid residues, cause nearly complete loss of PSTI expression. As the authors had excluded the possibility that the reduced protein expression resulted from reduced transcription of unstable mRNA, they concluded that these missense mutations probably cause intracellular retention of mutant proteins. The discovery of *SPINK1* mutations in humans provides additional support for a role of active trypsin in the development of pancreatitis, although the most common variety, N34S is not defective in its inhibitory capacity towards trypsin. SPINK1 is believed to form the first line of defense in inhibiting prematurely activated trypsinogen in the pancreas.

## Mutations in genes independent of trypsin processing

Beside genes that are directly linked to trypsinogen processing, such as *CTRC* and *SPINK1*, mutations in genes of other pancreatic enzymes are also associated with chronic forms of pancreatitis. Various mutations in the gene of carboxypeptidase A1 (*CPA1*), but not in carboxypeptidase A2 or B1 (*CPA2/CPB1*), have been found to be associated with chronic pancreatitis, although this report remains unconfimed. The nonfunctional character of the mutations suggests ER stress by misfolding protein structure to be responsible for pancreatitis. Variants in the gene encoding for pancreatic carboxylester lipase (*CEL*) produce disease through a similar mechanism involving ER stress. The fusion of the *CEL* gene with the pseudogene *CELP* results in the hybrid *CEL-HYP* allele. The CEL-HYP fusion protein displays impaired secretion and intracellular accumulation of the protein, which leads to cell stress. ER stress is a common phenomenon in the progression of chronic pancreatitis. These genetic data underline a new disease causing mechanism that is independent of the classical protease−anti-protease balance.

Another protease-independent risk factor of chronic pancreatitis is the ABO blood type B. Recent studies show a significant association of SNPs in the blood group specifying transferases A/B (ABO), fucosyltransferase-2 (FUT2), and chymotrypsinogen B2 with elevated serum lipase levels. The FUT-2 non-secretor status, as well as ABO blood type B, is associated not only with high serum lipase activity but also with chronic pancreatitis. The mechanism governing these findings is not well understood but must involve glycosylation events of pancreatic proteins.

## *CFTR* mutations—a new cause of chronic pancreatitis

Cystic fibrosis is an autosomal recessive disorder with an estimated incidence of 1:2500 characterized by pancreatic exocrine insufficiency and chronic pulmonary disease. The extent of pancreatic involvement varies between a complete loss of exocrine and endocrine function to nearly unimpaired pancreatic function. In 1996, the first reports of mutations in the cystic fibrosis gene in patients with hereditary chronic pancreatitis emerged. Analysis of larger cohorts revealed recurrent episodes of pancreatitis in 1%−2% of patients with cystic fibrosis and normal exocrine function and rarely in patients with exocrine insufficiency. Compared with an unaffected population (*CFTR* mutation rate ∼15%), patients who suffer from idiopathic pancreatitis carry mutations in the cystic fibrosis conductance regulator gene (*CFTR*) in

The role of ductal system during secretion

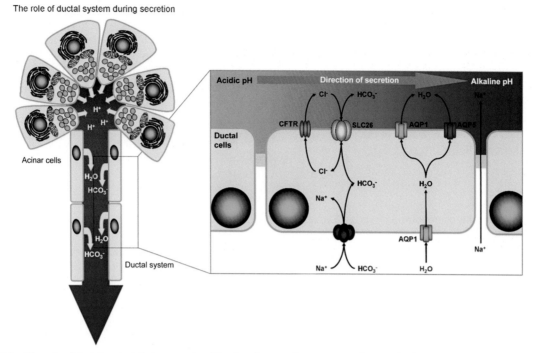

**FIGURE 21.5** The role of ductal system during secretion. The ductal system is necessary to increase luminal pH by secretion of bicarbonate and support transportation by secreting fluid via water channels (AQP1 and AQP5). The chloride transporter CFTR secretes $Cl^-$, which is needed by the SLC26 antiporter system to secrete $HCO_3^-$ into the ductal system. Defects in this system such as a decreased function or expression of CFTR (more than 1000 mutations are known) are associated with chronic pancreatitis.

16.7%−25.9%. CFTR is a chloride channel, regulated by $3',5'$-cAMP and phosphorylation, and is essential for the control of ion transport in epithelial cells. The transport of $Cl^-$ by CFTR into the pancreatic duct is necessary for bicarbonate secretion by the $HCO_3^-/Cl^-$ antiporter SLC26. The anion exchanger from type SLC26 regulates the $HCO_3^-$ secretion of pancreatic ductal cells. Fluid secretion is closely related to the secretion of anions into the ductal lumen and is regulated by osmotic gradient. $H_2O$ secretion is controlled by aquaporins, which are expressed in pancreatic ductal system (Fig. 21.5). Hence, CFTR is directly linked to pancreatic fluid secretion. Evidence from *CFTR* knockout mice shows diminished pancreatic fluid secretion under physiological conditions. New data suggest that alcohol, a major risk factor for development of chronic pancreatitis, influence the expression level and the subcellular localization of CFTR in pancreatic ductal cells.

The level of executable protein function (as well as the protein expression level) determines the type and the severity of the disease phenotype. Today more than 2000 mutations within the *CFTR* gene are known and several of them have been reported in direct association with chronic pancreatitis. For heterozygous carriers of *CFTR* mutations, the risk of developing pancreatitis is about twofold compared to unaffected individuals. Reduced levels of functional CFTR are a clear risk factor for chronic pancreatitis.

## Perspectives

Recent advances in cell biology and molecular biology techniques have enabled investigators to address the intracellular pathophysiology and genetics underlying pancreatic diseases in a much more direct manner than was previously considered possible. Studies that have employed these techniques have changed our knowledge about the disease onset. Pancreatitis has long been considered an autodigestive disorder in which the pancreas is destroyed by its own digestive proteases. Under physiological conditions, pancreatic proteases are synthesized as inactive precursor zymogens and stored by acinar cells in zymogen granules. Independent of the pathological stimulus that triggers the disease the pathophysiological events that eventually lead to tissue destruction begin within the acinar cells and involve premature intracellular activation of proteases. Cell injury subsequently induces a systemic inflammatory response. Much of our present understanding of the underlying pathogenic mechanisms comes from genetic studies, which support a crucial role of trypsinogen activation. Different mutations within the *PRSS1* gene (like the pR122H mutation), or in genes coding for endogenous inhibitors of active trypsin (such as *SPINK1*) or in trypsin-degrading enzymes (such as *CTRC*), have been found in association with different varieties of pancreatitis. New genetic data from genome-wide association studies suggest that different pathogenic mechanisms can also cause pancreatitis; ER stress−related mutations affecting *CEL* or *CPA1* are independent from trypsinogen activation as well as from mutations in *CFTR*. This demonstrates the complexity of the disease in which a variety of factors can drive the onset and development of the disease. Genetic factors determine the risk of developing pancreatitis but do not necessarily determine the severity of disease course. The immune system seems to play a crucial role for the outcome of pancreatitis and may offer a therapeutic target. Nevertheless, the molecular mechanisms that regulate the balance between proteases and antiproteases, as well as the role of individual digestive enzymes in the proteolytic cascades that precedes cell injury still need to be defined by further experimental studies.

## Key concepts

- Pancreatitis has long been considered an autodigestive disorder in which the pancreas is destroyed by its own digestive proteases.
- Under physiological conditions, pancreatic proteases are synthesized as inactive precursor zymogen and stored by acinar cells in zymogen granules. Independent of a pathological stimulus that triggers the disease, the pathophysiological events that eventually lead to tissue destruction begin within the acinar cells and involve premature intracellular activation of pancreatic zymogens. This results in acinar cell necrosis followed by a systemic inflammatory response.
- The intracellular rise of $Ca^{2+}$ concentration is a prerequisite for the activation of proteases at the apical pol of acinar cells.
- Pancreatitis is a primary sterile inflammation. Macrophages and neutrophils migrate into the damaged organ and contribute to severity of the disease by releasing pro-inflammatory cytokines.
- Genetic studies on patients suffering from hereditary and idiopathic pancreatitis suggest that the serine protease trypsin, which is activated by the lysosomal cysteine proteinase cathepsin B, contributes significantly to the development of acute pancreatitis.
- The inhibition of active trypsin by PSTI, as well as the degradation of active trypsin mediated by chymotrypsin C in a calcium-dependent manner, is a protective mechanism of the pancreas against pancreatitis.
- The R122H mutation within the PRSS1 gene (cationic trypsinogen) is the most frequent mutation associated with an autosomal dominant variety of hereditary pancreatitis with an 80% penetrance. This gain of function mutation leads to increased autoactivation of trypsinogen and decrease degradation by chymotrypsin C. Since the initial discovery of this mutation, several other mutations within trypsinogen have been reported in association with chronic or hereditary pancreatitis.
- The degradation of active trypsin is a protective mechanism against pancreatitis. Different mutations within the chymotrypsin C gene, a trypsin degrading enzyme that leads to a loss of function of trypsin, are associated with the development of pancreatitis.
- Mutations within the pancreatic secretory trypsin inhibitor (SPINK-1) results in a increased risk to develop pancreatitis. A change in the trypsin binding pocket or other structural changes can result in a decreased inhibitory capacity of SPINK-1. For the most common SPINK-1 mutation, N34S, this has not been shown and it remains unknown why this variant causes pancreatitis.
- Different mutations within the CFTR gene confer an increased risk for the development of chronic pancreatitis in patients with sufficient exocrine pancreatic function.

## Suggested readings

[1] Hofbauer B, Saluja AK, Lerch MM, et al. Intra-acinar cell activation of trypsinogen during caerulein-induced pancreatitis in rats. Am J Physiol 1998;275:G352−62.

[2] Halangk W, Lerch MM, Brandt-Nedelev B, et al. Role of cathepsin B in intracellular trypsinogen activation and the onset of acute pancreatitis. J Clin Invest 2000;106:773−81.

[3] Dawra R, Sah RP, Dudeja V, et al. intra-acinar trypsinogen activation mediates early stages of pancreatic injury but not inflammation in mice with acute pancreatitis. Gastroenterology 2011;141:2210−17.

[4] Whitcomb DC, Gorry MC, Preston RA, et al. Hereditary pancreatitis is caused by a mutation in the cationic trypsinogen gene. Nat Genet 1996;14:141−5.

[5] Rosendahl J, Witt H, Szmola R, et al. Chymotrypsin C (CTRC) variants that diminish activity or secretion are associated with chronic pancreatitis. Nat Genet 2008;40:78−82.

[6] Krüger B, Albrecht E, Lerch MM. The role of intracellular calcium signaling in premature protease activation and the onset of pancreatitis. Am J Pathol 2000;157:43−50.

[7] Petersen OH. Ca2+ signalling and Ca2+-activated ion channels in exocrine acinar cells. Cell Calcium 2005;38:171−200.

[8] Sendler M, Dummer A, Weiss FU, et al. Tumour necrosis factor α secretion induces protease activation and acinar cell necrosis in acute experimental pancreatitis in mice. Gut 2013;62:430−9.

[9] Gukovsky I, Gukovskaya AS, Blinman TA, Zaninovic V, Pandol SJ. Early NF-kappaB activation is associated with hormone-induced pancreatitis. Am J Physiol 1998;275:G1402−14.

[10] Gukovskaya AS, Vaquero E, Zaninovic V, et al. Neutrophils and NADPH oxidase mediate intrapancreatic trypsin activation in murine experimental acute pancreatitis. Gastroenterology 2002;122:974−84.

[11] Sendler M, Weiss FU, Golchert J, et al. Cathepsin B-mediated activation of trypsinogen in endocytosing macrophages increases severity of pancreatitis in mice. Gastroenterology 2018;154 704−718.e10.

[12] He S, Wang L, Miao L, et al. Receptor interacting protein kinase-3 determines cellular necrotic response to TNF-alpha. Cell 2009;137:1100−11.

[13] Xue J, Sharma V, Hsieh MH, et al. Alternatively activated macrophages promote pancreatic fibrosis in chronic pancreatitis. Nat Commun 2015;6:7158.

[14] Szabó A, Sahin-Tóth M. Increased activation of hereditary pancreatitis-associated human cationic trypsinogen mutants in presence of chymotrypsin C. J Biol Chem 2012;287:20701−10.

[15] Witt H, Sahin-Toth M, Landt O, et al. A degradation-sensitive anionic trypsinogen (PRSS2) variant protects against chronic pancreatitis. Nat Genet 2006;38:668−73.

[16] Witt H, Luck W, Hennies HC, et al. Mutations in the gene encoding the serine protease inhibitor, Kazal type 1 are associated with chronic pancreatitis. Nat Genet 2000;25:213−16.

[17] Witt H, Beer S, Rosendahl J, et al. Variants in CPA1 are strongly associated with early onset chronic pancreatitis. Nat Genet 2013;45:1216−2120.

[18] Fjeld K, Weiss FU, Lasher D, et al. A recombined allele of the lipase gene CEL and its pseudogene CELP confers susceptibility to chronic pancreatitis. Nat Genet 2015;47:518−22.

[19] Weiss FU, Schurmann C, Guenther A, et al. Fucosyltransferase 2 (FUT2) non-secretor status and blood group B are associated with elevated serum lipase activity in asymptomatic subjects, and an increased risk for chronic pancreatitis: a genetic association study. Gut 2015;64:646−56.

[20] Cohn JA, Friedman KJ, Noone PG, et al. Relation between mutations of the cystic fibrosis gene and idiopathic pancreatitis. N Engl J Med 1998;339:653−8.

[21] Maléth J, Balazs A, Pallagi P, et al. Alcohol disrupts levels and function of the cystic fibrosis transmembrane conductance regulator to promote development of pancreatitis. Gastroenterology 2015;148 427−439.e16.

Chapter 22

# Molecular basis of diseases of the endocrine system

Alan L.-Y. Pang[1] and Wai-Yee Chan[2]

[1]TGD Life Company Limited, Hong Kong Science Park, Sha Tin, Hong Kong, [2]School of Biomedical Sciences, Faculty of Medicine, The Chinese University of Hong Kong, Sha Tin, Hong Kong

**Summary**

It has long been recognized that endocrine disorders are caused by too much or too little hormone. However, the etiology of hormonal excess or deficiency was not known until the identification of gene mutations that affect hormonal action, from hormone synthesis to receptor response, and the development of hypothalamic—pituitary—target organ axis. In this chapter, we review the molecular basis of endocrine disorders based on examples from several well-established hormonal systems and the associated organs, including the pituitary, thyroid gland, parathyroid gland, adrenal gland, and gonad. We also summarize the more common gene mutations that have been shown to be the cause of or associated with endocrine disorders of the above systems.

## Introduction

It has long been recognized that endocrine disorders are caused by too much or too little hormone. However, the etiology of hormonal excess or deficiency was not known until the advent of molecular biology made it possible to elucidate many of the steps involved in hormone synthesis, hormone function, and target response, furthering our understanding of the molecular basis of some of the causes of endocrine disorders. The endocrine system is extremely complex and affects all human activities. In this chapter, we direct our attention to several well-established hormonal systems and the more common gene mutations that are known to be the cause of or associated with endocrine disorders.

## The pituitary gland

The pituitary gland is located at the base of the brain. Despite its small size, it is one of the most important organs of the body. The pituitary gland functions as a relay between the hypothalamus and target organs by producing, storing, and releasing hormones that affect different target organs in the regulation of basic physiological functions (such as growth, stress response, reproduction, metabolism, and lactation). Anatomically, the pituitary gland is composed of two compartments. The anterior pituitary, the largest part of the gland, is composed of distinct types of hormone-secreting cells: growth hormone (GH) by the somatotrophs, thyroid-stimulating hormone (TSH) by the thyrotrophs, adrenocorticotropin (ACTH) by the corticotrophs, follicle-stimulating hormone (FSH) and luteinizing hormone (LH) by the gonadotrophs, prolactin (PR) by the lactotrophs, and melanocyte-stimulating hormone by the melanotrophs. The posterior pituitary stores and releases hormones (such as antidiuretic hormone and oxytocin) produced by the hypothalamus.

The function of the pituitary gland is regulated by the hypothalamus. Secretion of anterior pituitary hormones (except PR) is stimulated or suppressed by specific hypothalamic releasing or inhibiting factors, respectively. The release of these factors is regulated by feedback mechanisms from hormones produced by the target organs as exemplified by the hypothalamic—pituitary—thyroid axis shown in Fig. 22.1. The same feedback mechanism also acts on the pituitary to fine-tune the production of pituitary hormones. At the molecular level, secretion of a specific pituitary hormone is triggered by binding of the respective hypothalamic releasing hormone to the corresponding membrane receptors on specific hormone-producing cells in the anterior pituitary. The pituitary hormone is released into the

Essential Concepts in Molecular Pathology. DOI: https://doi.org/10.1016/B978-0-12-813257-9.00022-X

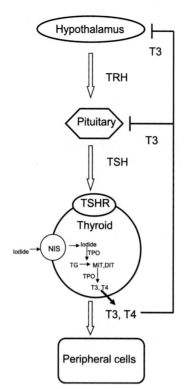

**FIGURE 22.1** The hypothalamic−pituitary−thyroid axis. Hypothalamic thyroid-stimulating hormone releasing hormone (TRH) stimulates pituitary thyroid-stimulating hormone (TSH) secretion. TSH binds to TSH receptors (TSHRs) in the thyroid gland (target organ) to trigger thyroid hormone (T3 and T4) secretion. Thyroid hormones are released into bloodstream and elicit their physiological functions in peripheral cells through receptor-mediated mechanism. Meanwhile, thyroid hormones inhibit further hypothalamic TRH and pituitary TSH secretion through negative feedback. A stea-dy circulating level of thyroid hormones is achieved. The example shown reflects the principal mode of regulation of the hypothalamic−pituitary−target organ axis. ↓ indicates stimulation, ⊢ indicates inhibition. *DIT*, diiodotyrosyl; *MIT*, monoiodotyrosyl; *NIS*, sodium iodide symporter; *TG*, thyroglobulin; *TPO*, thyroid peroxidase.

bloodstream, and its binding to the cell surface receptors in target organs triggers hormone secretion to carry out the rel-evant physiological functions.

Pituitary disorders are caused by either hyposecretion (hypopituitarism) or hypersecretion. Hypopituitarism is the deficiency of a single or multiple pituitary hormones. Target organs usually become atrophic and function abnormally due to the loss of their stimulating factors from the pituitary. Hypopituitarism can be acquired by physical means, including traumatic brain injury, tumors that destroy the pituitary or physically interfere with hormone secretion, vascu-lar lesions, and radiation therapy to the head or neck. In contrast, congenital hypopituitarism is caused mainly by abnor-mal pituitary development. Mutations in genes encoding transcription factors controlling the formation of nascent pituitary (Rathke pouch), cellular proliferation, and differentiation of cell lineages have been identified. These mutants generally result in combined pituitary hormone deficiency (CPHD), which is characterized by pituitary malformation and a concomitant or sequential loss of multiple anterior pituitary hormones. Isolated pituitary hormone deficiency (IPHD) is caused by mutations in transcription factors controlling the differentiation of a particular anterior pituitary cell line or mutations affecting the expression or function of individual hormones or their receptors. Pituitary hyperse-cretion disorders are usually caused by ACTH (Cushing syndrome) and GH (gigantism or acromegaly).

## Combined pituitary hormone deficiency

A number of genes have been shown to cause CPHD. Among these are the homeobox expressed in embryonic stem cells 1 (*HESX1*), LIM homeobox protein 3 and 4 (*LHX3/LHX4*), prophet of pit-1 (*PROP1*), POU domain class 1, tran-scription factor 1 (*POU1f1*), and zinc finger protein Gli2 (*GLI2*).

### Homeobox expressed in embryonic stem cells 1

*HESX1* is a paired-like homeobox transcriptional repressor essential to the early determination and differentiation of the pituitary. It is one of the earliest markers of the pituitary primordium. *HESX1* is localized on chromosome 3p14.3. Both autosomal recessive and autosomal dominant inheritance have been suggested in HESX1 deficiency. Mutations in *HESX1* are associated with CPHD and IPHD as well as familial cases of septooptic dysplasia.

Homozygous and heterozygous mutations (including mis-sense, insertion, and deletion mutations) of *HESX1* have been identified. Heterozygotes mostly demonstrate incomplete penetrance and milder phenotypes compared with homozygotes. The mutated proteins display diminished DNA binding capacity or impaired recruitment of corepressor molecules. In two cases (g.1684delG and p.E149K), the ability of HESX1 mutants to interact with PROP1 (an opposing transcription factor to HESX1) is altered. The change in balance between HESX1 on PROP1 is believed to disrupt the normal timing of PROP1-dependent pituitary program, which consequently leads to hypopituitarism.

## LIM homeobox protein 3 and 4

*LHX3* and *LHX4* are LIM homeodomain transcription factors important to early pituitary development and maintenance of mature anterior pituitary cells. The two genes are located on chromosome 9q34.3 and 1q25.2, respectively. Missense and non-sense mutations, small deletion with frameshift, and partial and complete gene deletion of *LHX3* have been identified. Inheritance of LHX3 deficiency syndrome follows an autosomal recessive pattern. Accordingly, all individuals with these mutations are homozygous and have CPHD but in most cases display normal ACTH levels. A hypoplastic pituitary is commonly observed. Except for gene deletion, the described mutations cause a change in amino acid residues or removal of the LIM domains and/or homeodomains. When translated, the mutant proteins display compromised DNA binding ability and variable loss of gene transactivation activity.

The inheritance pattern of LHX4 deficiency syndrome is autosomal dominant. The first identified mutation is derived from a familial case with complex disease phenotype including CPHD associated with a hypoplastic pituitary. The patients display an intronic G-to-C transversion at the splice-acceptor site preceding exon 5, which abolishes normal *LHX4* splicing and potentially generates two LHX4 mutant proteins that possess altered DNA binding capacity. The other mutations are located in the coding region. All mutations are associated with CPHD and/or IPHD, and affected patients show anterior pituitary hypoplasia. The LHX4 mutant proteins show reduced or complete loss of DNA binding and transactivation properties on target gene promoters.

## Prophet of PIT-1

*PROP1* is a paired-like transcription factor restricted to the developing anterior pituitary. PROP1 is required for the determination of somatotrophs, lactotrophs, and thyrotrophs and differentiation of gonadotrophs. As reflected by its name, PROP1 expression precedes and is required for *PIT-1* (now called *POU1F1*) expression. In humans, *PROP1* mutations are the leading causes of CPHD, accounting for 30%–50% of familial cases. The majority of patients have a hypoplastic or normal anterior pituitary. However, a hyperplastic pituitary with subsequent involution has also been reported.

*PROP1* is located on chromosome 5q35.3. Most mutations of *PROP1* are found in homozygous and compound heterozygous patients. However, three of the mutations, namely, IVS2−2 A > T, c.301_302delAG, and p.R120C, are identified in heterozygotes even though the inheritance of the disorder caused by *PROP1* mutation follows an autosomal recessive mode. The c.301_302delAG (also known as c.296_297delGA) mutation, which leads to p.S109X, is the most frequently encountered mutation. With the exception of p.W194X mutation, all *PROP1* mutations affect the DNA-binding homeodomain, which leads to a reduced or abolished DNA binding and/or gene transactivation activity of the transcription factor.

## POU domain, class 1, transcription factor 1

*POU1F1*, also known as *PIT-1*, encodes a POU domain protein essential to the terminal differentiation and expansion of somatotrophs, lactotrophs, and thyrotrophs. It functions as a transcription factor that regulates the transcription of itself and other pituitary hormones and their receptors, including *GH*, *PRL*, TSH beta subunit (*TSHβ*), TSH receptor (*TSHR*), and growth hormone−releasing hormone receptor (*GHRHR*).

*POU1F1* is located on chromosome 3p11. Mutations of *POU1F1* have been shown to be responsible for GH, PRL, and TSH deficiencies. Both autosomal recessive and autosomal dominant inheritance have been suggested, and both normal and hypoplastic anterior pituitaries have been observed. Various types of gene mutation (mis-sense, non-sense, frameshift, and splicing mutation, as well as gene deletion) have been found in homozygous or compound heterozygous patients, with six distinct mutations (p.P14L, p.P24L, p.K145X, p.Q167K, p.K216E, and p.R271W) found only in heterozygotes. Compound heterozygous patients appear to have a more severe disease phenotype. The *POU1F1* mutations are mainly found in the POU-specific and POU-homeodomains, which are important interfaces for high-affinity DNA binding on *GH* and *PRL* genes and protein−protein interaction with other transcription factors. Consequently,

DNA binding and/or gene transactivation capacity is impaired. In some cases, the mutant proteins act as dominant inhibitors of gene transcription.

## Zinc finger protein GLI2

GLI family of transcription factors is implicated as the mediators of Sonic hedgehog (Shh) signals in vertebrates. In humans, Shh signaling is associated with the forebrain defect holoprosencephaly. *GLI2* gene is localized on chromosome 2q14. *GLI2* deficiency phenotype follows an autosomal dominant inheritance pattern. Four heterozygous mutations (c.2274del1, p.W113X, p.R168X, and IVS5+1 G > A) were identified in 7 of 390 holoprosencephaly patients who also displayed malformed anterior pituitary and pan-hypopituitarism. The mutant GLI2 proteins exhibit either loss-of-function or dominant negative activity.

## GHRH-GH-IGF1 axis

The GHRH-GH-IGF1 axis plays a central role in the regulation of somatic growth. Consequently, genetic lesions affecting any component of the axis usually lead to GH deficiency (GHD). The hormones and receptors along the axis represent the hot spots for GHD.

### Growth hormone–releasing hormone receptor

GHRHR is a G protein–coupled receptor (GPCR) expressed specifically in somatotrophs. It is essential to the synthesis and release of GH in response to GHRH and the expansion of somatotrophs during the final stage of pituitary development. *GHRHR* is localized on chromosome 7p14. Biallelic mutation in *GHRHR* is a frequent cause of isolated growth hormone deficiency (IGHD) type 1b. At least 14 different inactivating mutations are found in different parts of the gene, including the promoter, introns, or exons. The mutant receptors are unable to bind to their ligands or elicit cyclic AMP (cAMP) responses after GHRH treatment. These mutations are identified in homozygous and compound heterozygous patients in accordance with an autosomal recessive mode of inheritance. However, heterozygous mutant alleles have also been reported. In vitro studies show that mutated GHRHR displays dominant negative effect on wild-type GHRHR functions. Patients with homozygous mutations tend to display a more severe GHD, but the overall disease phenotype is variable.

### Growth hormone

Human GH is a 191–amino-acid (~22 kDa) single-chain polypeptide hormone synthesized and secreted by somatotrophs in the anterior pituitary. It is the major player in many physiological processes related to growth and metabolism. The human GH gene *GH1* resides on chromosome 17q24.2. This chromosomal region contains a cluster of *GH*-like genes arising from gene duplication (namely *CSHP*, *CSH1*, *GH2*, and *CSH2*). The highly homologous nature of the intergenic and coding regions makes the *GH* locus susceptible to gene deletions arising from homologous recombination.

Both autosomal dominant and autosomal recessive inheritance have been suggested for IGHD type 1a, the most severe form of IGHD. Nonetheless, the severity of disease phenotype due to *GH1* mutation varies, with heterozygous mutations giving a milder phenotype than homozygotes. The disorder typically results from large gene deletion or inactivating mutations of *GH1*. The most frequent *GH1* mutations involve the donor splice site of intron 3, which leads to the skipping of exon 3. Mutations resulting in exon 3 skipping were also identified in exon or intron splice enhancer elements. The exon skipping event leads to an internal deletion of amino acid residues 32–71 and the production of a smaller (17.5 kDa) GH, which inhibits normal GH secretion in a dominant negative manner. Five mis-sense mutations, namely, p.C53S (homozygous mutation), p.R103C, p.D112G, p.D138G, and p.I179M (heterozygous mutations), lead to the production of a biologically inactive GH that can counteract GH action, impair GHR binding and signaling, or prevent productive GHR dimerization.

### Growth hormone receptor

*GHR* is localized on chromosome 5p12-13 and encodes a single transmembrane domain (TMD) glycoprotein belonging to the cytokine receptor superfamily. In mature form, *GHR* encodes 620 amino acid residues, with the first 246 residues (encoded by exons 3–7) constituting the extracellular domain (ECD) for GH binding and receptor dimerization, the middle 24 residues representing the TMD (exon 8), and the last 350 residues (exon 9 and 10) forming the intracellular domain for GH signaling. GHR is predominantly expressed in liver. The binding of GH triggers receptor dimerization and activation of downstream signaling events leading to IGF1 production. The absence of GHR activity causes an autosomal recessive disorder called Larons syndrome. Affected patients display clinical features similar to IGHD but are characterized by a low level of IGF1 and an increased level of GH. Over 50 *GHR* mutations, which affect the coding region and splicing of *GHR* transcripts, have been identified. A vast majority of them impact the ECD of GHR. The mutated GHRs display reduced affinity to GH. In other cases, *GHR* mutations lead to defective homodimer formation and defective GH binding. The intracellular domain of GHR is lost in certain splice site mutants. The truncated receptor, which carries only the ECD, is unable to anchor to the cell membrane. Although they can bind GH, these GHR mutants fail to activate downstream signaling processes, and they can deplete the binding sites from functional GHRs in a dominant negative manner.

### Insulin-like growth factor 1

IGF1, originally called somatomedin C, is a 70-amino-acid polypeptide hormone encoded by a gene localized on chromosome 12q22-23. IGF1 is the major mediator of prenatal and postnatal growth. It is produced primarily in liver and serves as an endocrine (as well as paracrine and autocrine) hormone mediating the action of GH in peripheral tissues such as muscle, cartilage, bone, kidney, nerves, skin, lungs, and the liver itself.

IGF1 deficiency follows an autosomal recessive mode of transmission. It can be the result of molecular defects that affect any of the upstream components of the GHRH-GH-IGF1 axis and *IGF1* itself. Despite its rarity, several *IGF1* mutations have been described. Homozygous deletion of exons 4 and 5 as well as complete heterozygous deletion of *IGF1* lead to IGF1 insufficiency in patients. In another patient with low circulating IGF level, a homozygous T-to-A transversion was found at the 3′-UTR of *IGF1* so that the E-domain of IGF1 precursor, which is essential to IGF maturation, is altered. In two homozygous mis-sense mutations, the V44M mutant shows a severely reduced affinity to IGF1 receptor. The other mutant, which has a homozygous Arg-to-Gln substitution at the C domain, displays a partial loss of affinity for the receptor. In addition, frameshift mutation owing to an insertion of a 4-nucleotide duplication as well as splicing mutation (c.402+1 G > C) have been reported (www.growthgenetics.com).

## Growth hormone hypersecretion

Excessive production of GH causes IGF1 overproduction leading to gigantism in children and acromegaly in adults. In general, the patients display an abnormally increased somatic growth and distorted body proportions, which is in contrast to the dwarfism phenotype displayed by patients with GHD. Almost all GH hypersecretion syndromes are caused by benign pituitary GH-secreting adenomas, either as isolated disorders or associated with other genetic conditions such as multiple endocrine neoplasia (MEN1), McCune Albright syndrome, neurofibromatosis, or Carney complex. In rare cases, it can be attributed to a hypothalamic tumor or carcinoid that ectopically secretes GHRH. Activating mutations of the gene encoding G protein subunit $G_s\alpha$ and inactivating mutations of *GHR*, which lead to overactivation of GHRHR and impaired negative feedback on GH production, respectively, are identified in GH-secreting pituitary tumors. An inactivating mutation of the gene encoding aryl hydrocarbon receptor-interacting protein may also predispose the formation of pituitary adenoma.

## The thyroid gland

The thyroid, one of the largest endocrine glands, is composed of two types of secretory elements: the follicular and parafollicular cells. The former are responsible for thyroid hormone production, the latter elaborate calcitonin. The factors that control thyroidal embryogenesis and account for the majority of cases of thyroid agenesis or dysgenesis are presently poorly understood.

Although the thyroid is capable of some independent function, thyroid hormone production is regulated by the hypothalamic−pituitary−thyroid axis (Fig. 22.1). Hypothalamic tripeptide TSH-releasing hormone (TRH) stimulates pituitary TSH production. TSH in turn stimulates the thyroid follicular cells to secrete prohormone thyroxine (T4) and its active form triiodothyronine (T3). Because thyroid hormone has a profound influence on metabolic processes, it plays a key role in normal growth and development.

## Hypothyroidism

Impaired activity of any component of the hypothalamic—pituitary—thyroid axis will result in decreased thyroid function. The common causes are faulty embryogenesis, inflammation (most often autoimmune thyroiditis: Hashimoto thyroiditis) or a reduction in functioning thyroid tissue due to prior surgery, radiation damage, and cancer.

Congenital hypothyroidism is the most common congenital endocrine disorder with an incidence of about 1 in 3500 births. The majority of cases (85%) are non-goitrous and are due to faulty embryogenesis. A small number are goitrous, the result of enzyme defects in hormone biosynthesis (dyshormonogenesis). Clinically, congenital hypothyroidism often presents within the first few days of life. Hypothyroidism in the adult is most often seen in middle age. Females are eight times more likely to develop this disease than males.

The genes responsible for the development of thyroid follicular cells include thyroid transcription factor 1 (*TIF1*, also known as *TITF1*, *NKX21*, or *T/EBP*), thyroid transcription factor 2 (*TTF2*, also known as *TITF2*, *FOXE1*, or *FKHL15*), paired box transcription factor 8 (*PAX8*), *TSH*, and *TSHR*. Mutation of these genes accounts for about 5% of hypothyroidism patients. *TITF1*, *TITF2*, and *PAX8* mutations give rise to non-syndromic congenital hypothyroidism, whereas mutations of *TSH* and *TSHR* cause syndromic congenital hypothyroidism. Mutations in sodium iodide symporter (*NIS*), pendrin (*PDS*, also known as *SLC264A*), thyroid peroxidase (*TPO*), thyroid oxidase 2 (*THOX2*, also known as dual oxidase DUOX2, *LNOX2*), and thyroglobulin (*TG*) affect organification of iodide and have been linked to congenital hypothyroidism.

Etiologically, three types of congenital hypothyroidism can be distinguished: (1) central congenital hypothyroidism, including hypothalamic (tertiary) and pituitary (secondary) hypothyroidism, (2) primary hypothyroidism (impairment of the thyroid gland itself), and (3) peripheral hypothyroidism (resistance to thyroid hormone).

### Central congenital hypothyroidism

Central hypothyroidism may be caused by hypothalamic or pituitary diseases leading to deficiency of TSH or TRH. TRH receptor mutation is very rare. The majority of cases of central congenital hypothyroidism are caused by *TSH* mutations.

**Thyroid-stimulating hormone**. TSH is a heterodimeric glycohormone sharing a common α chain with FSH, LH, and choriogonadotropin (CG). The gene encoding the α chain is located on chromosome 6q14.3. The gene encoding β chain of TSH is localized on chromosome 1p22. Transmission of congenital isolated TSH deficiency follows an autosomal recessive pattern. Patients with homozygous and compound heterozygous mutations in TSHβ have been identified in different ethnic populations. The most common mutation is a single base deletion in exon 3, which results in the substitution of cysteine-105 for valine and a frameshift with the appearance of a premature stop codon at position 114 (p. C105fs114X). Cysteine-105 is a mutation hot spot in TSHβ. Different mutations have different effects on TSHβ. Nonsense mutations lead to a truncated TSHβ subunit, whereas other mutations may cause a change in conformation. Thus, mutated TSHβ may be immunologically active but biologically inactive, depending on how significant the mutation is on the structure of the protein.

**TSH receptor-inactivating mutations**. The effects of TSH on thyroid follicular cells are mediated by TSHR, a cell surface receptor which, together with the receptor for FSH (FSHR) and the receptor for LH (LHR) and CG, forms a subfamily of the GPCR family. This subfamily of glycohormone receptors is characterized by the presence of a large amino-terminal ECD involved in hormone-binding specificity. The signal of hormone binding is transduced by a TMD containing 7 α-helices. The intracytoplasmic C-terminal tail (ICD) of the receptor is coupled to $G_s\alpha$. Binding of hormone to the receptor causes activation of $G_s\alpha$ and the adenylyl cyclase cascade, resulting in augmented synthesis of cAMP. The receptor is also coupled to $G_q$ and activates the inositol phosphate cascade. TSHR is located on chromosome 14q31 and is composed of 10 exons.

A subgroup of patients with familial congenital hypothyroidism due to TSH unresponsiveness has homozygous or compound heterozygous loss-of-function mutations in *TSHR*. Mutations of *FSHR* are distributed throughout the receptor gene sequence with no obvious mutation hot spot. Depending on where the mutation is located, the effects can be defective receptor synthesis due to premature truncation, accelerated mRNA or protein decay, abnormal protein structure, abnormal trafficking, subnormal expression on cell membrane, ineffective ligand binding, and altered signaling. All mutant TSHRs are associated with defective cAMP response to TSH stimulation. Monoallelic heterozygous inactivating mutations have been reported. In these cases, the other TSHR allele is probably inactivated by unrecognized mutations in the intronic, 5'-noncoding region, or 3'-noncoding region, because there is no evidence to support a dominant negative effect of an inactivating *TSHR* mutation.

Patients with mutated TSHR with residual activities are less severely affected than those with activity void mutated TSHR. There is also good correlation between phenotype of resistant patients observed in vivo with receptor function measured in vitro.

## Thyroid dyshormonogenesis

Iodide is accrued from the blood into thyroidal cells through NIS. It is attached to tyrosyl residue of TG through the action of TPO in the presence of hydrogen peroxide ($H_2O_2$). Two iodinated TG couple with diiodotyrosyl (DIT) to form T4; one DIT and one TG coupled with monoiodotyrosyl (MIT) to form T3. The coupling reaction is also catalyzed by TPO. TG molecules that contain T4 and T3 on complete hydrolysis in lysosomes yield the iodothyroxines, T4 and T3. Disorders resulting in congenital primary hypothyroidism have been identified in all major steps in thyroid hormonogenesis.

**NIS.** A homozygous or compound heterozygous mutation of NIS causes an iodide transport defect and an uncommon form of congenital hypothyroidism characterized by a variable degree of hypothyroidism and goiter. NIS is localized on chromosome 19p12−13.2 with 15 exons. Mis-sense and non-sense mutations, splicing mutation leading to shifting of reading frame and in-frame deletion, of NIS have been reported. The majority of the patients are homozygous, and about half of them are products of consanguineous marriage. These mutations cause inactivation of NIS resulting in abnormal uptake of iodide by the thyroid.

**Pendrin.** Pendred syndrome is an autosomal recessive disorder first described in 1896 and characterized by the triad of deafness, goiter, and a partial organification defect. While many patients with Pendred syndrome are euthyroid, others have subclinical or overt hypothyroidism. The variability in clinical expression may be influenced by dietary iodide intake.

Pendred syndrome was mapped to chromosome 7q22−31. The candidate gene *PDS* (also known as *SLC26A4*) was cloned and mutations were identified. *PDS* is a member of the solute carrier family 26A. It encodes a 780-amino-acid protein pendrin, which is predominantly expressed in thyroid, inner ear, and kidney. More than 140 different mutations of *PDS* have been described (http://www.healthcare.uiowa.edu/labs/pendredandbor/) in patients with classical Pendred syndrome. The majority of *PDS* mutations are mis-sense mutations, and some of these mutant proteins appear to be retained in the endoplasmic reticulum. Other forms of mutations include non-sense mutation, splicing mutation, partial duplication, as well as insertion and deletion leading to shifting of reading frame. Individuals with Pendred syndrome from consanguineous families are usually homozygous for *PDS* mutations, whereas sporadic cases typically harbor compound heterozygous mutations.

At the thyroid level, the role of pendrin has not been defined. In thyroid follicular cells, pendrin is inserted into the apical membrane and acts as an iodide transporter for apical iodide efflux. Abnormality of pendrin affects iodide transport and may lead to iodide organification defects. This is often mild or absent, leading to the hypothesis that an as-yet-undefined mechanism may compensate for the lack of pendrin.

**Thyroid peroxidase.** TPO is a key enzyme in thyroid hormone biosynthesis. It catalyzes both iodination and coupling of iodotyrosine residues in TG. Human TPO is located on chromosome 2p25, with 17 exons. The mRNA is about 3 kb in length, encoding a thyroid-specific glycosylated hemoprotein of 110 kDa bound at the apical membrane of thyrocytes. The majority of patients with congenital hypothyroidism have defects in the synthesis or iodination of TG that are attributable to TPO deficiency. Subnormal or absence of TPO activities result in thyroid iodide organification and give rise to goitrous congenital hypothyroidism. Total iodide organification defect (TIOD) is inherited in an autosomal recessive mode.

*TPO* mutations are one of the most frequent causes of thyroid dyshormonogenesis. Mutations of *TPO* have also been identified in follicular thyroid carcinoma and adenoma. Various forms of mutation of *TPO* have been identified in patients with congenital goitrous hypothyroidism (Table 22.1), including mis-sense and non-sense mutation, frameshift due to insertion or deletion, and splicing mutations due to deletion of intronic sequences or mutation of nucleotide at intron−exon junctions. Among the 17 exons of *TPO*, mutations have mostly been found in exons 8 and 9. These two exons encode part of the catalytic site of *TPO*, and any mutation will make the enzyme susceptible to inactivation. Mutations affecting protein folding, such as the c.1429_1449 deletion that removes seven amino acids, will disturb the structure of the enzyme. The fact that TPO is glycosylated suggests that any mutation affecting a potential glycosylation site (N-X-S/T), such as the mis-sense mutation at nucleotide 391, which causes the replacement of Ser131 by Pro and disturbs the glycosylation site at Asn129, may alter TPO activity. Exons 15−17 encode the membrane-spanning region and the cytoplasmic tail of TPO. Mutations affecting the membrane spanning region will disturb the insertion of TPO

**TABLE 22.1** Partial list of inactivating mutations of thyroid peroxidase (*TPO*).

| Exon | Mutation and nucleotide position | Amino acid position and effect on protein |
| --- | --- | --- |
| 2 | c.51dup20bp | fsX92 |
| 3 | c.157G > C | p.A53P |
| 3 | c.215delA | p.Q72fsX86 |
| 4 | c.349G > C | GG/gt to GC/gt |
| 5 | c.387delC | p.N129fsX208 |
| 5 | c.391T > C | p.S131P |
| 6 | c.523C > T | p.R175X |
| 7 | c.718G > A | p.D240N |
| 8 | c.843delC | p.Q252fsX309 |
| 8 | c.875C > T | p.S292F |
| 8 | c.920A > C | p.N307T |
| 8 | c.976G > A | p.A326T |
| 8 | c.1132G > A | p.E378K |
| 8 | c.1152G > T | p.Q384D |
| 8 | c.1159G > A | p.G387R |
| 8 | c.1183_1186insGGCC | p.N396fsX472 |
| 8 | c.1242G > T | p.Q384D |
| 8 | c.1274A > G | p.N425S |
| 8 | c.1297G > A | p.V433M |
| 9 | c.1335delC | p.A445, fsX472 |
| 9 | c.1339A > T | p.I447F |
| 9 | c.1357T > G | p.Y453D |
| 9 | c.1373T > C | p.L458P |
| 9 | c.1429_1449del | p.A477_N483del |
| 9 | c.1477G > A | p.G453S |
| 9 | c.1496C > T | p.P499L |
| 9 | c.1496delC | p.P499fsX501 |
| 9 | c.1567G > A | p.G453S |
| 9 | c.1581G > T | p.W527C |
| 9 | c.1597G > T | p.G533C |
| 9 | exon/intron 9+1G > T | |
| 10 | c.1618C > T | p.R540X |
| 10 | c.1690C > A | p.L564I |
| 10 | c.1708C > T | p.R540X |
| 10 | c.1718_1723del | p.D574_L575del |
| 10 | c.1768G > A, AG/gt to AA/gt | |
| 10 | intron 10, GA+1, AG/gt to AG/at | |
| 11 | c.1955insT | p.F653fsX668 |

*(Continued)*

**TABLE 22.1** (Continued)

| Exon | Mutation and nucleotide position | Amino acid position and effect on protein |
|---|---|---|
| 11 | c.1978C > G | p.Q660E |
| 11 | c.1993C > T | p.R655W |
| 11 | c.1994G > A | p.R665Q |
| 11 | c.1999G > A | p.G667S |
| 12 | c.2077C > T | p.R693W |
| 12 | c.2153_2154delTT | p.fsF718X |
| 13 | c.2243delT | p.fsX831 |
| 13 | c.2268insT | p.E756fsX756 |
| 13 | c.2311G > A | p.G711R |
| 13 | c.2386G > T | p.D796Y |
| 14 | c.2395G > A | p.E799K |
| 14 | c.2413delC | p.fsX831 |
| 14 | c.2415insC | p.fsX879 |
| 14 | c.2422T > C | p.C808R |
| 14 | c.2422delT | p.C808fsX831 |
| 14 | c.2512T > A | p.C838S |
| 15 | c.2579G > A | p.G860R |
| 15 | c.2647C > T | p.P883S |
| 15 | Intron 15-exon 16 boundary, g.10delccacaggaca | |
| 16 | c.2722delCGGGCCGCAG | p.R908fxX969 |
| 16 | c.2748G > A | AG/gt to AA/gt |

The first base of the coding sequence is assigned nucleotide number 1, and the first amino acid of the encoded polypeptide is denoted as residue number 1.

into plasma membrane. Frameshift mutations in these exons cause premature truncation of the protein, resulting in improper membrane anchoring. A heterozygous 10 base pair deletion at the intron 15-exon 16 boundary presumably causes splicing error and affects proper membrane insertion of the protein.

Most cases of congenital hypothyroidism are caused by homozygous or compound heterozygous mutations of *TPO*. Carrier parents of affected individuals have normal thyroid function. There are examples of congenital hypothyroidism due to uniparental disomy for chromosome 2p or monoallelic expression of the mutant allele. In these cases, the genetic defect is confined to the affected patient and is not inherited. TIOD is often associated with inactivating mutations in *TPO*. Mutant *TPO* alleles with residual activities are associated with milder thyroid hormone insufficiency or partial iodide organification defects (PIOD). There are also individuals with PIOD in whom only one mutated allele of *TPO* was identified. In such cases, the other *TPO* allele may have a mutation located in intronic sequences or in a distal region of the gene to affect transcription. Another possibility could be an inactivating mutation in thyroid oxidase 2 (*THOX2*), which has been described in patients with iodide organification defects.

**Thyroid oxidase 2**. THOX2 generates $H_2O_2$ that is used by TPO in the oxidation of iodine prior to iodination of the tyrosyl residues of TG. Human THOX2 is located on chromosome 15q15.3 and is only 16 kb apart from the highly homologous THOX1 gene. Even though both THOX genes are expressed in thyroid glands, THOX2 is preferentially expressed in thyroid while THOX1 is preferentially expressed in airway epithelium and skin. THOX2 has 34 exons. The mRNA encodes a polypeptide of 1548 amino acids, 26 of which comprise a signal peptide. The mature protein is glycosylated. The N-terminal segment of THOX2 is homologous to peroxidases, whereas the C-terminal domain is highly homologous to NADPH oxidases. Thus, THOX2 is also known as dual oxidase 2 (DUOX2).

**TABLE 22.2** Partial list of mutations of thyroid oxidase 2.

| Mutation | Genotype | Perchlorate discharge | CH phenotype |
|---|---|---|---|
| c.108 G > C, p.Q36H/ | Comp heterozygous | 46%, PIOD[a] | Permanent, mild |
| c.2895_2898delGTTC, p.S965fsX994 | | | |
| c.602insG, skip exon 5, p.fsX254 | Heterozygous | n.d.[b] | n.d.[b] |
| c.602insG, skip exon 5, p.fsX254/ | Comp heterozygous | n.d.[b] | n.d.[b] |
| c.1516 G > A, p.D506N | | | |
| c.1126 C > T, p.R376W/ | Comp heterozygous | 28%, PIOD[a] | Permanent, mild |
| c.2524 C > T, p.R842X | | | |
| c.1253delG, p.G418fsX482/ | | | |
| g.IVS19−2 A > C | Comp heterozygous | 68%, PIOD[a] | Permanent |
| c.1300 C > T, p.R434X | Homozygous | 100%, TIOD[c] | Permanent, severe |
| c.2056 C > T, p.Q686X | Heterozygous | 66%, PIOD[a] | Transient, mild |
| c.2101 C > T, p.R701X | Heterozygous | 41%, PIOD[a] | Transient, mild |
| c.2895_2898delGTTC, p.S965fsX994 | Heterozygous | 40% PIOD[a] | Transient, mild |
| | | 20%−63% | n.d.[b] |
| c.3329 G > A, p.R1110Q | Homozygous | 72.8%, PIOD[a] | Permanent, mild |
| p.Q1026X | Heterozygous | 20%−63%, PIOD[a] | n.d.[b] |

CH indicates congenital hypothyroidism. Numbering system is the same as in Table 22.1.
[a]*Partial iodide organification defect (PIOD) as defined by the perchlorate discharge test.*
[b]*n.d., not determined.*
[c]*Total iodide organification defect (TIOD) as defined by the perchlorate discharge test.*

Defects in THOX2 result in a lack or shortage of $H_2O_2$ and consequently hypothyroidism. Various mutations in *THOX2* have been identified (Table 22.2), which include mis-sense and non-sense mutations, splicing mutations leading to exon skipping, and small deletions leading to frameshift and premature truncation of polypeptide. In one patient, genomic DNA sequencing reveals a G-insertion at position 602 in exon 5. This insertion predicts shifting of the reading frame producing a stop at codon 300. However, sequencing of the cDNA shows that in the mutant allele, exon 5 is skipped. This abnormal splicing leads to a shift of reading frame beginning at codon 172 and produces a premature stop codon at position 254.

All *THOX2* mutations were identified as congenital hypothyroidism in childhood. The exception is the p.R1110Q mutation, which was found in a homozygous adult woman whose thyroid function remained almost normal until the age of 44 years. Homozygous and compound heterozygous inactivating mutations in *THOX2* lead to complete disruption of thyroid hormone synthesis and are associated with severe and permanent congenital hypothyroidism. Compound heterozygotes with one or both mutant alleles displaying residual activity have permanent but mild (subclinical) hypothyroidism, whereas heterozygotes are associated with milder, transient hypothyroidism (Table 22.2). It is unusual that a genetic defect results in a transient phenotype at a younger age. This can be due to a changing requirement of thyroid hormone with age.

**Thyroglobulin.** TG serves as iodine storage and is the most abundant protein in the thyroid. Some of its tyrosine residues are iodinated by the action of TPO, and the iodinated tyrosines are coupled to form T3 and T4. TG is produced by thyroid follicular cells and secreted into the follicular lumen. Mutations of gene-encoding factors responsible for the development and growth of thyroid follicular cells, such as TTF1, TTF2, PAX8, and TSHR, cause dysgenetic congenital hypothyroidism. A rarer cause of congenital goitrous hypothyroidism is mutation of TG resulting in structural defects of the protein.

TG deficiency is transmitted in an autosomal recessive mode. Affected individuals are either compound heterozygous or homozygous. The patient phenotypes range from mild to severe goitrous hypothyroidism. No clear phenotype−genotype correlation has been observed. TG is encoded by a gene, which spans 270 kb on chromosome 8q24.2−24.3, and composed of 48 exons. The polypeptide carries 2768 amino acids, of which the first 19 amino acids constitute a signal peptide. The majority of *TG* inactivating mutations are mis-sense mutation. Non-sense mutations, single nucleotide insertion or deletion, and splice site mutations have also been reported. These mutations are found in exons and introns. Four mutations (p.R277X, p.C1058R, p.C1245R, and p.C1977S) are the most frequently reported. Most mutations alter the folding, stability, or intracellular trafficking of the mature protein. Interestingly, some nonsense mutations, such as p.R277X, cause the production of a smaller peptide that is sufficient for the synthesis of T4 in the N-terminal domain and results in a milder phenotype.

## Resistance to thyroid hormone

Resistance to thyroid hormone is characterized by reduced clinical and biochemical manifestations of thyroid hormone action relative to the circulating hormone levels. Patients have high circulating free thyroid hormones, reduced target tissue responsiveness, and an inappropriately normal or elevated TSH.

The thyroid hormones, T3 and T4, exert their target organ effects via nuclear thyroid hormone receptors (TR). After entering target cells, T4 undergoes deiodination to T3, which enters the nucleus and binds to TR. On T3 binding, the conformation of TR changes, which causes the release of corepressors and recruitment of coactivators resulting in the activation or repression of target gene transcription.

### Thyroid hormone receptor

There are two TR genes, *TRα* and *TRβ*, located on chromosomes 17 and 3, respectively. Both genes produce two isoforms, *TRα1* and *TRα2* by alternative splicing, and *TRβ1* and *TRβ2* by utilization of different transcription start sites. TR binds as a monomer, or with greater affinity as homodimers or heterodimers, to thyroid hormone response elements (TREs) in target genes to regulate gene expression. The TR protein has a central DNA-binding domain. The carboxy-terminal portion of TR contains the T3 ligand—binding domain, the transactivation domain, and the dimerization domain. Unlike the other TRs, TRα2 does not bind thyroid hormone. TRs are expressed at different levels in different tissues and at different stages of development. The biological effects of different TRs are compensatory to some degree. However, some thyroid hormone effects are TR isoform specific.

About 85% of the cases of thyroid hormone resistance are caused by *TRβ* mutation. Over a hundred different mutations of *TRβ* have been identified in subjects showing thyroid hormone resistance syndrome, in which p.R338W is the most common one. The vast majority of the mutations cluster within three CpG-rich hot spots, namely, amino acids 234—282 (hinge domain), amino acids 310—353, and amino acids 429—460 (ligand-binding domain). Most mutated receptors display reduced T3 binding or abnormal interaction with one of the cofactors involved in thyroid hormone action. Somatic *TRβ* mutations have been identified in TSH-producing pituitary adenomas.

Dimerization of TRs results in dominant negative effects of mutant TRβs. Besides impairing gene transactivation, the mutant receptors also interfere with the function of wild-type receptor by dimerizing with it, resulting in resistance to thyroid hormone. On the other hand, individuals expressing a single wild-type *TRβ* allele are normal despite the fact they lost the other allele due to deletion. Mutations in the hinge domain of TRβ do not interfere with T3 binding and translocation into the nucleus but cause impairment of transactivation function and resistance to thyroid hormone. Differences in the degree of hormonal resistance in different tissues are due to the absolute and relative expression levels of TRβ and TRα.

Once thought to be absent, several *TRα* mutations have recently been identified. Interestingly, most of them (including mis-sense, non-sense, and frameshift mutations) are de novo mutations. The mutant TRαs display reduced gene transactivation activity and a dominant negative effect on the wild-type receptor.

### Thyroid hormone cell transporter

Thyroid hormone action requires transport of the hormone from the extracellular compartments across the cell membrane, which is mediated by the protein encoded by monocarboxylate transporter gene 8 (*MCT8*). Clinical features of patients with *MCT8* mutations are those of Allan—Herndon—Dudley syndrome (AHDS). These patients have unusual combination of high concentration of active thyroid hormone and low level of an inactive metabolite (reverse T3). The severity of the disease is variable among families.

Human *MCT8* (also known as *SLC16A2*; solute carrier family 16, member 2) is located on chromosome Xq13.2. It belongs to the family of *SLC16* genes, the products of which catalyze the proton-linked transport of monocarboxylates, such as lactate, pyruvate, and ketone bodies. *MCT8* encodes two putative proteins of 613 and 539 amino acids, respectively, by translation from two in-frame start sites. However, it is not known whether there are two human MCT8 proteins expressed in vivo. MCT8 contains 12 TMDs, with both the amino-terminal and carboxyl-terminal peptides located within the cell. In the pituitary, it is expressed in folliculostellate cells rather than TSH-producing cells.

Various types of mutation of *MCT8* have been reported, ranging from mis-sense and non-sense mutations to deletion of a single base pair, a triplet codon, an entire exon, and several exons. There is also insertion of a triplet codon as well as insertion of a single nucleotide resulting in shifting of reading frame and production of truncated protein. Besides these mutations, there is a family with a mis-sense mutation affecting the putative translational start codon (c.1A > T, p.M1L). This mutation is present in the patient as well as his healthy brother whose T3 level is normal, indicating that it is not pathogenic. This finding implies that another methionine at amino acid position 75 can serve as the translational start of a functional MCT8 protein. In another family, a single nucleotide deletion in the second to last codon results in a frameshift and bypassing of the natural stop codon until another stop is encountered 195 nucleotide downstream. This mutation leads to an extension of the MCT8 protein by 65 amino acids. In a female patient with complete AHDS clinical features, *MCT8* was disrupted at the X-breakpoint of a de novo translocation t(X:9)(q13.2;p24). A complete loss of MCT8 expression was observed in cultured fibroblasts from the patient.

The majority of the mutations result in complete loss of MCT8 transporter function, which can be due to reduced protein expression, impaired trafficking to plasma membrane, or reduced substrate affinity. Mis-sense *MCT8* mutations (such as p.S194F, p.V235M, p.R271H, p.L434W, and p.L568P) show residual activities when compared with wild-type MCT8 protein. Individuals with these mutations may have a milder phenotype. There is no clear-cut phenotype—genotype correlation with *MCT8* mutations. The only consistent feature among mutated MCT8 is elevated level of serum T3.

## Familial non-autoimmune hyperthyroidism

Patients with a family history of thyrotoxicosis with early disease onset typically have thyromegaly, absence of typical signs of autoimmune hyperthyroidism, and recurrence after medical treatment. A number of germ-line constitutive-activating mutations of the *TSHR* have been identified in patients with non-autoimmune hyperthyroidism (Table 22.3). The inheritance of the trait follows an autosomal dominant pattern. The majority of these mutations are located in exon 10 of *TSHR*, encoding the transmembrane and intracellular domains. In approximately 70% of the activating mutations, a purine is exchanged for a pyrimidine nucleotide, in particular, guanine for cytosine or adenosine for thymidine. Aside from germ-line activating mutations, a number of sporadic mutations have also been identified in patients with non-autoimmune hyperthyroidism (Table 22.3). Germ-line *TSHR* mutations usually present later in life with milder clinical manifestation, whereas sporadic *TSHR* mutations cause severe thyrotoxicosis and goiter with neonatal or infancy onset. Mis-sense mutations in one codon resulting in different amino acid exchanges have been observed for several gain-of-function mutations, including codons 281, 505, 568, 597, 631. Mutation hot spots include p.Ala623 and p.Phe631. Biological activities of the mutant TSHRs differ in terms of adenylyl cyclase and inositol phosphate pathway activation. Most mutations cause only constitutive activation of the cAMP pathway, whereas a few mutants activate both cAMP and phospholipase C cascades.

The expressivity of the activating mutations may be affected by environmental factors such as iodine uptake. Even among cases with the same mutation of *TSHR*, phenotypic expression can vary with respect to the age of onset and severity of hyperthyroidism and thyroid growth. At least in one report, there is apparent anticipation of disease onset across generations, presumably due to increased iodine supplementation in the area where the affected family resides.

Although less frequently, somatic activating *TSHR* mutations have been identified as the cause of solitary toxic adenomas and multinodular goiter and thyroid carcinomas. About 80% of toxic thyroid adenomas (hot nodules) exhibit *TSHR* gain-of-function mutations. An overlap of somatic and germ-line phenotype has only been observed in 3 (p. M453T, p.L629F, and p.P639S) of 23 known germ-line activating mutations. This has been ascribed to the fact that hereditary mutations cause a less severely affected phenotype, with a marginal effect on reproductive fitness. Among the 10 sporadic mutations, only 2 occur as somatic mutations in toxic adenomas.

## The parathyroid gland

Histologically, the parathyroid glands comprise two types of cells: chief cells and oxyphil cells. Each cell type may appear in clusters or occasionally mixed. Chief cells have a pale staining cytoplasm with centrally placed nuclei. The cytoplasm contains elongated mitochondria and two sets of granules: secretory granules that take a silver stain and others that stain for glycogen. The oxyphil cells are fewer in number, larger than the chief cells, and their cytoplasm has more mitochondria. The oxyphil cells do not appear until after puberty and their function is unclear.

Parathyroid hormone (PTH) is secreted by the chief cells of the parathyroid glands. It is encoded by a gene located on chromosome 11p15.3. It has three exons encoding a mature peptide of 84 amino acids and a signal peptide of 25 amino acids. The hormone has two main targets: bones and kidneys. In bone, PTH activates lysosomal enzymes in osteoclasts to

**TABLE 22.3** Partial list of activating mutations of *TSHR* in familial Nonautoimmune hyperthyroidism.

| Amino acid position | Germ line | Sporadic | Somatic |
|---|---|---|---|
| p.R183K | X | | |
| p.S281N | X | | |
| p.S281I | | X | |
| P.N372T | X | | |
| p.A428V | | X | |
| p.G431S | X | | |
| p.M453T | X | | X |
| p.M463V | X | | |
| p.A485V | X | | |
| p.S505R | X | | |
| p.S505N | X | X | |
| p.V509A | X | | |
| p.L512Q | X | X | |
| p.I568V | X | | |
| p.I568T | | X | X |
| p.V597L | | X | |
| p.V597F | X | X | |
| p.D617Y | X | | |
| p.A623V | X | | |
| p.M626I | X | | |
| p.L629F | X | | X |
| p.F631L | | X | X |
| p.F631S | X | | |
| p.T632I | | X | |
| p.M637R | X | | |
| p.P639S | X | | X |
| p.N650Y | X | | |
| p.N670S | X | | |
| p.C672Y | X | | |

Numbering system is the same as in Table 22.1.

mobilize calcium. In kidney, it promotes calcium resorption and phosphate and bicarbonate excretion by activating the cAMP signaling pathway. The synthesis and secretion of PTH is regulated by serum-ionized calcium concentration.

## Calcium homeostasis

The calcium level in blood is kept within a very narrow range (between 2.2 and 2.55 mM). About 30% of serum calcium is bound to protein, mainly albumin, 10% is chelated (or complexed), mostly to citrate, and the remaining 60% is free or ionized. In humans, the regulation of the ionized fraction of serum calcium depends on the interaction of PTH and the active form of vitamin D-1,25-dihydroxycholecalciferol ($1,25(OH)_2D$).

Vitamin D is a prohormone that requires hydroxylation for its conversion to the active form. Vitamin D from food (ergocalciferol, $D_2$) is absorbed in the proximal gastrointestinal tract and is also produced in the skin by exposure to ultraviolet light (cholecalciferol, $D_3$). It is carried by a vitamin D–binding protein to the liver where it is hydroxylated to form 25-hydroxy-vitamin D. The 25-hydroxy-vitamin D is then further hydroxylated in kidneys to $1,25(OH)_2D$ under the influence of PTH to become the physiologically active hormone.

Decreasing serum-ionized calcium, acting via a transmembrane G protein–coupled calcium-sensing receptor, stimulates PTH secretion. PTH in turn acts on osteoclasts to release calcium from bones and inhibits calcium excretion from kidneys. PTH, hypocalcemia, and hypophosphatemia augment the formation of $1,25(OH)_2D$, which stimulates absorption of calcium from the gut and aids PTH in mobilizing calcium from bone, while at the same time inhibiting 24-hydroxylation. The reverse occurs when calcium levels rise, and PTH secretion is inhibited. 24-hydroxylation is induced, whereas 21-hydroxylation is inhibited, resulting in a shift to the inactive metabolite of vitamin D. Movement of calcium from bone and gut into the extracellular fluid compartment ceases, so does renal reabsorption, thereby restoring serum calcium to its normal physiological concentration. Any disturbance in this regulation may result in hypocalcemia or hypercalcemia. Symptoms and signs may vary from mild, asymptomatic incidental findings to serious life-threatening disorders depending on the duration, severity, and rapidity of onset.

## Hypoparathyroidism

Hypoparathyroidism is characterized by levels of PTH insufficient to maintain normal serum calcium concentration. The most common cause of hypoparathyroidism is injury to the parathyroid glands during head and neck surgery or due to autoimmunity (either isolated or as a polyglandular syndrome). Familial forms of congenital and acquired hypoparathyroidism are relatively rare. Both familial isolated hypoparathyroidism and familial hypoparathyroidism accompanied by abnormalities in multiple organ systems have been described.

### Familial isolated hypoparathyroidism

Several forms of familial isolated hypoparathyroidism with autosomal dominant, autosomal recessive, and X-linked recessive inheritance have been described. Mutations affecting the signal peptide of pre-pro-PTH have been reported in families with hypoparathyroidism following an autosomal recessive or autosomal dominant inheritance pattern. A patient with a single base substitution resulting in the replacement of Cys18 by Arg, which affects the processing of the mutant pre-pro-PTH, manifested autosomal dominant hypoparathyroidism. In another family with autosomal recessive hypoparathyroidism, a single base substitution gave rise to the replacement of Ser23 by Pro. The mutant pre-pro-PTH cannot be cleaved, resulting in the degradation of mutant peptide in endoplasmic reticulum. A third mutation that occurs at the exon 2-intron 2 boundary results in the skipping of exon 2 and splicing of exon 1 to exon 3. This causes the loss of the initiation codon and signal peptide and gives rise to autosomal recessive isolated hypoparathyroidism.

Glial cells missing (GCM) belongs to a small family of key regulators of parathyroid gland development. There are two forms: *GCMA* and *GCMB*. Five different homozygous mutations of *GCMB* (two mis-sense mutations, p.R47L and p.G63S; an intragenic microdeletion removing exons 1–4, and two single nucleotide deletions leading to frameshift, c.1389delT and c.1399delC) have been identified in patients with isolated hypoparathyroidism. These mutations inactivate GCMB function by either disrupting the DNA-binding domain or removing the transactivation domain of the protein. The heterozygous carriers of these mutations are asymptomatic, indicating this disorder follows an autosomal recessive mode of inheritance.

Aside from mutations of *PTH* and *GCMB*, X-linked recessive hypoparathyroidism due to mutation of a gene localized to chromosome Xq26-q27 has been reported in two multigenerational kindreds. However, the candidate gene has not yet been confirmed.

### Familial hypoparathyroidism as part of a complex congenital defect

Hypoparathyroidism may be part of a polyglandular autoimmune disorder or associated with multiple organ abnormalities. It is likely the abnormalities are not intrinsic to the parathyroid gland but consequential to other abnormalities. Hypoparathyroidism occurs in HAM syndrome (hypoparathyroidism, Addison disease, mucocutaneous candidiasis; also known as polyglandular failure type 1 syndrome or autoimmune polyendocrinopathy–candidiasis–ectodermal dystrophy). This syndrome is caused by mutations of *AIRE1* (autoimmune regulator type 1). Cells destined to become parathyroid glands emerge from the third and fourth pharyngeal pouches. Failure of development of the derivatives of these two pouches results in absence or hypoplasia of the parathyroids and thymus and gives rise to DiGeorge syndrome

(DGS). Most cases of DGS are sporadic, but autosomal dominant inheritance of DGS has been observed. The vast majority of cases display imbalanced translocation or microdeletion of 22q11.2. The broad spectrum of abnormalities in DGS patients and the large genomic region in the deletion suggest the possibility of involvement of multiple genes. Similarly, abnormality of *PTH* is excluded in familial hypoparathyroidism, sensorineural deafness, and renal dysplasia (HDR syndrome), which is inherited in an autosomal dominant mode. Deletion analyzes of HDR patients indicate the absence of *GATA3* is responsible for HDR. GATA3 is a member of the GATA-binding family of transcription factors, encoded by a gene with six exons located on chromosome 10q14. The protein has 444 amino acids with two transactivation domains and two zinc finger domains. GATA3 is expressed in the developing parathyroid gland, inner ear, kidney, thymus, and central nervous system. Thus, *GATA3* mutations will affect parathyroid development. Various forms of mutations of *GATA3* have been identified, including mis-sense and non-sense mutations, in-frame deletion, insertions or deletions leading to premature stop codons, donor splice site mutation, and whole gene deletion. These mutations affect DNA-binding, mRNA and protein stability, or protein synthesis. The pathogenic mechanism responsible for HDR syndrome is believed to be GATA3 haploinsufficiency. Other diseases with hypoparathyroidism but with no genetic abnormality of *PTH* include several mitochondrial disorders, Kenny–Caffey and Sanjad–Sakati syndrome, Barakat syndrome, and Blomstrand disease.

## Hyperparathyroidism

Primary hyperparathyroidism is a genetically heterogeneous disorder that occurs most commonly in the 60+ age group and is more common in females. It results when PTH is secreted in excess due to a benign parathyroid adenoma, parathyroid hyperplasia, multiple endocrine neoplasia (MEN1 or MEN2A), or cancer. Elevated serum calcium together with low serum phosphorus is suggestive of hyperparathyroidism and becomes diagnostic when supported by increased urinary calcium excretion and elevated PTH level.

### *Familial isolated hyperparathyroidism*

Familial isolated hyperparathyroidism (FIHP) is a rare disorder characterized by single or multiple glandular parathyroid lesions without other syndromes or tumors. FIHP has been reported in 16 familial hyperparathyroidism patients from 14 families. The mode of transmission is autosomal dominant. Most FIHP kindreds have unknown genetic background. Studies of 36 kindreds reveal that a few families are associated with multiple endocrine neoplasia gene 1 (*MEN1*), with DNA markers near the hyperparathyroidism–jaw tumor (HPT-JT) syndrome locus at 1q25–q32, or inactivating calcium-sensing receptor (*CaSR*) mutations.

### *Multiple endocrine neoplasia type 1*

MEN type 1 (MEN1) is an autosomal dominant disorder with a high degree of penetrance. Over 95% of patients develop clinical manifestations by the fifth decade. The most common feature of MEN1 is parathyroid tumors. Tumors of the pancreas and anterior pituitary also occur in a significant, albeit smaller percentage of patients. MEN1 is caused by germ-line mutations of *MEN1* that is located on chromosome 11q13. The gene has 10 exons and encodes the 610-amino-acid protein menin, a tumor suppressor and nuclear protein that participates in transcriptional regulations, genome stability, cell division, and proliferation. MEN1 tumors frequently display loss of heterozygosity (LOH) of the *MEN1* locus. Somatic abnormalities of *MEN1* have been reported in MEN1 and non-MEN1 endocrine tumors.

There are a total of 565 different *MEN1* mutations among the 1336 mutations reported, with 459 being germ-line and 167 being somatic (61 mutations occur both as germ-line and somatic), scattered throughout *MEN1*. Among these mutations, 41% are frameshift deletions or insertions, 23% are non-sense mutations, 20% are mis-sense mutations, 9% are splicing mutations, 6% are in-frame deletions or insertions, and 1% represent whole or partial gene deletion. Somatic mis-sense mutations occur more frequently than germ-line mis-sense mutations. The majority of these mutations ($>70\%$) are predicted to give rise to truncated or destabilized proteins. Four mutations, namely, c.249_252delGTCT (deletion at codons 83–84), c.1546_1547insC (insertion at codon 516), c.1378C > T (R460X), and c.628_631delACAG (deletion at codons 210–211), account for 12.3% of all mutations and are potential mutation hot spots. No phenotype–genotype correlations have been observed in MEN1. In fact, similar *MEN1* mutations have been observed in FIHP and MEN1 patients.

## *Multiple endocrine neoplasia type 2*

MEN type 2 (MEN2) describes the association of medullary thyroid carcinoma (MTC), pheochromocytomas, and parathyroid tumors. There are three clinical variants of MEN2, the most common of which is MEN2A, which is also known as Sipple syndrome. MEN2A is inherited in an autosomal dominant manner. All MEN2A patients develop MTC, 50% develop bilateral or unilateral pheochromocytoma, and 20%−30% develop primary hyperparathyroidism. Primary hyperparathyroidism can occur by the age of 70 in up to 70% of patients. MEN2A is caused by activating point mutations of the *RET* proto-oncogene. *RET* encodes a tyrosine kinase receptor with cadherin-like and cysteine-rich ECDs, and a tyrosine kinase intracellular domain. In 95% of patients, MEN2A is associated with mutations of the cysteine-rich ECD, and mis-sense mutation in codon 634 (Cys to Arg) accounts for 85% of MEN2A mutations.

The second clinical variant is MEN2B. These patients exhibit a relative paucity of hyperparathyroidism and development of marfanoid habitus, mucosal neuromas, and intestinal ganglioneuromatosis. Of these patients, 95% present with mutation in codon 918 (Met to Thr) of the intracellular tyrosine kinase domain of *RET* proto-oncogene. The third clinical variant of MEN2 is MTC-only. This variant is also associated with mis-sense mutation of RET in the cysteine-rich ECD, and most mutations are identified in codon 618. The precise mechanism of the genotype−phenotype relationship in the three clinical variants is unknown.

## *Hyperparathyroidism-jaw tumor syndrome*

HPT-JT syndrome is an autosomal dominant disorder characterized by the development of parathyroid adenomas and carcinomas and fibro-osseous jaw tumors. The gene causing the syndrome, *HRPT2* (also known as *CDC73*), is located on chromosome 1q25. It consists of 17 exons and encodes a ubiquitously expressed 531-amino-acid protein called parafibromin. *HRPT2* is a tumor suppressor gene. Its inactivation is directly involved in the predisposition to HPT-TJ syndrome and in the development of some sporadic parathyroid tumors. To date, more than 10 different heterozygous inactivating mutations that predict truncation of parafibromin have been reported.

## Calcium-sensing receptor and related disorders

Extracellular ionized calcium concentration regulates the synthesis and secretion of PTH as well as parathyroid cell proliferation. The action of calcium is mediated by the CaSR. Human *CaSR* is located on chromosome 3q13.3−21. It is composed of 6 exons encoding a protein of 1078 amino acids. CaSR is a member of family CII of the GPCR superfamily. The protein has a large ECD with 612 amino acids, a TMD of 250 amino acids containing 7 transmembrane helices, and an intracellular domain of 216 amino acids. The receptor is heavily glycosylated, which is important for normal cell membrane expression. CaSR forms homodimers via intramolecular disulfide linkages within the ECD. The ECD is also responsible for calcium binding. The TMD is responsible for transducing the signal of calcium binding, whereas the intracellular domain interacts with $G\alpha_i$ or $G\alpha_o$ and activates different signal transducing pathways. The critical pathway(s) through which CaSR mediates its biological effects has not been defined.

CaSR is highly expressed in parathyroid cells, C cells of the thyroid, and kidneys. Parathyroid cells are capable of recognizing small perturbation in serum calcium and respond by altering the secretion of PTH. Mutations of *CaSR* can result in loss of function or gain of function of the receptor. Heterozygous mutations of the *CaSR* are the cause of a growing number of disorders of calcium metabolism, which typically manifest as asymptomatic hypercalcemia or hypocalcemia, with relative or absolute hypercalciuria or hypocalciuria. On the other hand, homozygous or compound heterozygous inactivating mutations produce a severe and sometimes lethal disease if left untreated.

### *Disorders due to loss-of-function mutations of calcium-sensing receptor*

There are two hypercalcemic disorders caused by inactivating mutations in *CaSR*: (1) familial hypocalciuric hypercalcemia (FHH) and (2) neonatal severe primary hyperparathyroidism.

**Familial hypocalciuric hypercalcemia.** A disorder that must be recognized as distinct from primary hyperparathyroidism is FHH, also known as familial benign hypercalcemia. In the majority of cases the cause appears to be an autosomal dominant loss-of-function mutation in CaSR. The mutation in FHH reduces the receptor sensitivity to calcium, which leads to defective triggering of PTH release in parathyroid glands. In the kidney, the defect causes an increase in tubular calcium and magnesium resorption. The net effect is hypercalcemia, hypocalciuria, and frequently hypermagnesemia.

About two-thirds of the FHH kindreds studied have unique heterozygous mutations. The inactivating mutations include 89 mis-sense mutations, 8 non-sense mutations, 7 insertions and/or deletions, including an *Alu* element insertion, and 2 splicing mutations. These mutations are described in the *CaSR* mutation database (http://www.casrdb. mcgill.ca). Some of these mutations cause a more severe hypercalcemia by inhibiting the wild-type CaSR in mutant wild-type heterodimers. There are ~30% of FHH patients without an identifiable mutation. Linkage analyzes have linked FHH in some of these patients to the long and short arms of chromosome 19.

**Neonatal severe primary hyperparathyroidism**. Neonatal severe primary hyperparathyroidism (NSHPT) represents the most severe expression of FHH. Symptoms manifest very early in life with severe hypercalcemia, bone demineralization, and failure to thrive. Most NSHPT patients have either homozygous or compound heterozygous CaSR inactivating mutations. There are also three cases of NSHPT in which de novo mutation was found in the ECD of CaSR. The patients have only one copy of CaSR mutated, and no CaSR mutations are found in the parents. There are also asymptomatic patients with homozygous CaSR inactivating mutations. Thus, the severity of symptom may be affected by factors other than mutant gene dosage.

### Disorders due to gain-of-function mutation of the calcium-sensing receptor

There are two hypocalcemic disorders associated with gain of CaSR function due to activating mutations of the receptor. These mutated receptors are more sensitive to extracellular calcium required for PTH secretion and cause reduced renal calcium resorption. The two disorders are autosomal dominant hypocalcemia (ADH) and Bartter syndrome type V.

**Autosomal dominant hypocalcemia**. ADH is a familial form of isolated hypoparathyroidism characterized by hypocalcemia, hyperphosphatemia, and normal to hypoparathyroidism. Inheritance of the disorder follows an autosomal dominant mode. The patients are generally asymptomatic. A significant fraction of cases of idiopathic hypoparathyroidism may in fact be ADH.

More than 80% of the reported ADH kindreds have CaSR mutations. There are more than 40 activating mutations of CaSR reported in the literature. These mutations produce a gain of CaSR function when expressed in vitro. The majority of the ADH mutations are mis-sense mutations within the ECD and TMD of CaSR. In addition, a deletion in the intracellular domain, p.S895_V1075del, has also been described in an ADH family. The mechanism of CaSR activation by these mutations is not known. Almost every ADH family has its own unique mis-sense heterozygous CaSR mutation. Most ADH patients are heterozygous. The only deletion-activating mutation occurs in a homozygous patient in an ADH family. However, there is no apparent difference in the severity of the phenotype between heterozygous and homozygous patients.

**Bartter syndrome type V**. In addition to hypocalcemia, patients with Bartter syndrome type V have hypercalciuria, hypomagnesemia, potassium wasting, hypokalemia, metabolic alkalosis, elevated renin and aldosterone levels, and low blood pressure. Four different activating mutations of CaSR (p.K29E, p.L125P, p.C131W, and p.A843E) have been identified in patients with Bartter syndrome type V. Functional analyzes show that these CaSR mutations result in more severe receptor activation when compared with the other activating mutations.

## The adrenal gland

The adrenal glands are two crescent-shaped structures located at the superior pole of each kidney. Each gland is composed of two separate endocrine tissues, the inner medulla and the outer cortex, which have different embryonic origins. The adrenal medulla is responsible for the production of catecholamines, epinephrine, and norepinephrine, which are involved in fight-or-flight responses. The adrenal cortex produces a number of steroid hormones that regulate diverse physiological functions. Despite the difference in origin and physiology, the hormones produced by the medulla and cortex often act in a concerted manner.

The adrenal cortex is composed of three distinct cell layers: (1) the zona glomerulosa, (2) fasciculata, and (3) reticularis. The outermost layer (zona glomerulosa) secretes mineralocorticoids, the most important of which is aldosterone. Aldosterone primarily affects the distal renal tubular sodium–potassium exchange mechanism to promote sodium resorption, thereby regulating circulatory volume and having a major impact on cardiac function. The synthesis and release of aldosterone is influenced by angiotensin II, derived from angiotensin I under the control of renin. Renin is made by the juxtaglomerular cells of the renal cortex in response to variations in renal blood flow. The production of aldosterone is, to a lesser extent, influenced by adrenocorticotropic hormone (ACTH).

The middle layer of the adrenal cortex (zona fasciculata) secretes glucocorticoids. The most important glucocorticoid is cortisol. Cortisol maintains blood glucose level and blood pressure and modulates stress as well as inflammatory responses. Its action affects almost every organ and tissue in the body. Secretion of cortisol is regulated by the hypothalamic—pituitary—adrenal axis. The hypothalamic corticotropin-releasing hormone (CRH) stimulates ACTH secretion by the pituitary corticotrophs through ligand—receptor-mediated mechanism. ACTH triggers cortisol production in adrenocortical cells by binding to melanocortin receptor 2 (MC2R, which is the ACTH receptor). A series of enzymatic reactions is initiated to mediate the uptake of cholesterol and the biosynthesis of cortisol. The circulating cortisol feeds back to the pituitary and hypothalamus to suppress further ACTH secretion.

The innermost layer of the adrenal cortex (zona reticularis) produces adrenal androgens. The production of adrenal androgens is also regulated by ACTH. In the fetal adrenal cortex (during weeks 7—12 of gestation), androgens are secreted and regulate the differentiation of male external genitalia. However, in adults the contribution of androgens from adrenal glands is quantitatively insignificant.

## Congenital primary adrenal insufficiency

As with other endocrine glands, both hypofunction and hyperfunction of the adrenal are associated with significant disorders. Adrenal insufficiency generally refers to an inadequate cortisol production by adrenals. Primary adrenal insufficiency is caused by defects of the adrenals themselves. The most frequent cause is the destruction of adrenal cortex due to autoimmunity. Congenital primary adrenal insufficiency is caused by inactivating mutations of genes responsible for normal adrenal development or cortisol production. Depending on the severity, a deficiency of cortisol alone or with other adrenal steroids occurs.

### Adrenal hypoplasia congenita

Adrenal hypoplasia congenita (AHC) is the underdevelopment of adrenal glands that typically results in adrenal insufficiency during early infancy. Two forms of AHC are known: (1) an autosomal recessive miniature adult form and (2) an X-linked cytomegalic form. Patients affected by the latter display low or absent glucocorticoids, mineralocorticoids, and androgens and do not respond to ACTH stimulation. The adrenals are atrophic and structurally disorganized, with no normal zona formation in cortex. Impaired sexual development, owing to hypogonadotropic hypogonadism, manifests in affected males who survive childhood.

The cytomegalic form of AHC is an X-linked disorder caused by the deletion or inactivating mutation of *NR0B1*, which encodes DAX1 (dosage-sensitive sex reversal—AHC critical region on the X chromosome gene 1). As a result, females are unaffected carriers and males are affected by *NR0B1* mutation. DAX1 is indispensable to the development of adrenal glands, gonads, and kidneys. It is an orphan nuclear receptor that interacts with SF1 (steroidogenic factor 1; encoded by *NR5A1*). Binding of DAX1 inhibits SF1-mediated transactivation of genes involved in the development of hypothalamic—pituitary—adrenal—gonad axis and biosynthesis of steroid hormones. Numerous mutations of *NR0B1*, which affect the DNA-binding or ligand-binding domain of DAX1, have been described. The genetic lesions (mostly mis-sense, non-sense, and frameshift mutations) are translated into truncated DAX1 proteins that lose the ability to repress SF1 transactivation activity. Phenotypic heterogeneity, in terms of severity and onset of disorder, is observed in patients with different or even the same *DAX1* mutations, suggesting that other genetic, epigenetic, or environmental factors may be involved in AHC. Alternatively, the presence of residual activity in *DAX1* mutants (positional effect of mutations) may determine the extent of phenotypic variation. The expression of a novel *NR0B1* splice variant in the adrenal glands is also suspected to influence AHC phenotype.

Similar to DAX1, SF1 is also an orphan nuclear receptor essential to normal adrenogonadal development, besides its pivotal role in steroidogenesis. However, inactivating mutation of *NR5A1* does not always associate with adrenal insufficiency. The cause of such variability is still unknown.

### Congenital adrenal hyperplasia

Congenital adrenal hyperplasia (CAH) is one of the most common autosomal recessive disorders in humans. It is the result of inborn metabolic errors in adrenal steroid biosynthesis (Fig. 22.2). Cholesterol is transported from the cytoplasm into mitochondria in adrenocortical cells by steroidogenic acute regulatory protein (StAR) and is first converted to pregnenolone, the common precursor of all adrenal steroids, by rate-limiting cholesterol side-chain cleavage enzyme P450scc (encoded by *CYP11A1*). Adrenal steroids are produced under the action of various cytochrome P450s, with some of them participating in more than one steroidogenic pathway. Depending on the enzymatic defect, an impaired production of glucocorticoids (and mineralocorticoids or androgens) may occur in utero. The feedback to the pituitary is undermined such that ACTH production becomes excessive. Consequently, the adrenals are overstimulated and

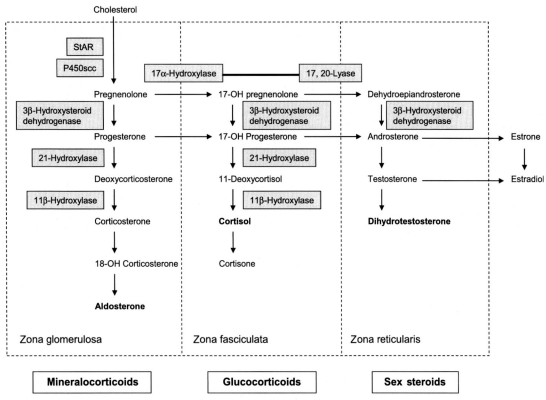

**FIGURE 22.2**   Pathways of adrenal steroidogenesis. The three classes of adrenal steroids are shown in bold. Enzymes or proteins that cause congenital adrenal hyperplasia when defective are shown in *gray boxes*. the 17α-hydroxylase and 17,20-lyase activities are encoded by the same enzyme (CYP17A1). *StAR*, steroidogenic acute regulatory protein.

hyperplastic. The overproduced steroid precursors are shunted to the androgen biosynthetic pathway, leading to androgen overproduction. Genital ambiguity in newborn females (owing to excessive fetal exposure in utero) and precocious pseudopuberty in both sexes are commonly observed in CAH patients. In severe cases, salt-wasting CAH occurs with life-threatening vomiting and dehydration within the first few weeks of life.

**21-Hydroxylase deficiency**. The most frequent CAH variant is 21-hydroxylase deficiency (21-OHD), which is caused by inactivating mutations or deletion of *CYP21A2* and accounts for ~95% of classical CAH. In many cases the term CAH refers to 21-OHD. The failure of conversion of progesterone and 17-hydroxyprogesterone leads to their accumulation and shunting to the androgen biosynthetic pathway. Over 100 mutations, most of which are mis-sense mutations, have been described. Based on the severity of enzyme defect, three clinical forms of 21-OHD CAH are recognized: the most severe classical salt-wasting form (due to aldosterone insufficiency and androgen excess), the moderately severe classical simple-virilizing form (due to androgen excess), and the least severe non-classical form.

**11β-Hydroxylase deficiency**. The second most common cause of CAH is 11β-hydroxylase deficiency (11β-OHD). The enzyme 11β-hydroxylase (encoded by *CYP11B1*) catalyzes the conversion of deoxycorticosterone and 11-deoxycortisol to corticosterone and cortisol, respectively (Fig. 22.2). Similar to 21-OHD, the enzyme defect leads to virilizing CAH due to the accumulation of immediate steroid precursors. The accumulation of deoxycorticosterone also causes salt retention and hypertension. More than 50 inactivating *CYB11B1* mutations have been identified. Most of these mutations completely abolish enzymatic activity.

A variant form of CAH related to 11β-OHD is caused by chimeric gene formation due to unequal crossing-over. *CYP11B2*, which is >95% identical to and is located ~40 kb upstream of *CYP11B1*, encodes an aldosterone synthase that is exclusively expressed in zona glomerulosa to catalyze the conversion of deoxycorticosterone to corticosterone. The high degree of homology renders the two isoforms susceptible to crossing-over during DNA replication, with regulatory sequence from one fused to the coding region of the other. *CYP11B2* promoter is unable to drive the production of chimeric CYP11B2/B1 protein in zona fasciculata because the promoter is inactive in this region of the adrenal. Thus, carriers of this mutation display 11β-OHD. Conversely, the *CYP11B1* promoter-driven expression of *CYP11B1/B2* chimera, which retains aldosterone synthase activity, leads to mineralocorticoid overproduction and results in an autosomal dominant disorder called glucocorticoid-suppressible hyperaldosteronism (or familial hyperaldosterone type 1). A rare case of salt-wasting CAII with 11-OHD due to homozygous internal deletion of the *CYP11B2/B1* chimera has also been reported.

**Other less common steroidogenic enzyme deficiency in CAH.** The key enzyme in steroid hormone biosynthesis is 3β-hydroxysteroid dehydrogenase (3β-HSD). It catalyzes the conversion of progesterone, 17-OH progesterone, and androstenedione from their respective $\Delta^5$ steroid precursors. As a result, inactivation of 3β-HSD leads to incomplete genital development and impaired aldosterone synthesis due to a deficiency of all classes of adrenal steroids. In humans, *HSD3B2* is the isoform expressed specifically in the adrenals and gonads. Inactivating *HSD3B2* mutations, mostly mis-sense mutations, have been reported. Aside from reduced and complete loss of enzymatic activity, some of the mutated enzymes display impaired stability.

*CYP17A1* encodes an enzyme with dual (17α-hydroxylase and 17,20-lyase) activities, mediating the biosynthesis of cortisol and sex steroid precursors (Fig. 22.2). Inactivating mutations of the gene frequently produce combined enzyme deficiencies, displaying hypertension, hypokalemia, and sexual infantilism. Despite its rarity in causing CAH, more than 70 inactivating mutations of *CYP17A1* have been identified. Most of the reported mutations alter the enzyme structure or generate truncated products and completely abolish both enzymatic activities. Isolated 17,20-lyase deficiency also occurs when lyase activity is selectively impaired by mutations of amino acid residues (such as p.R347 or p.R358) that are crucial to the interaction between CYP17A1 and its redox partners (P450 oxidoreductase and cytochrome *b*5) in the P450-electron donor complex.

Although uncommon, lipoid congenital adrenal hyperplasia (LCAH) represents the most severe form of CAH. Life-threatening mineralocorticoid and glucocorticoid deficiencies are common in infants and children. Male infants are under-virilized, and puberty is delayed in both sexes. The enlarged adrenals are filled with lipid globules. Similar damage is observed in the gonads. LCAH results from defective cholesterol-to-pregnenolone conversion, the first step in steroidogenesis, mediated by P450scc. Only several *CYP11A1* mutations have been described. Most of the genetic lesions actually stem from defective cholesterol transport into the mitochondria by inactivating mutation of *StAR*. The lack of substrate leads to adrenal steroid deficiency and absence of feedback for ACTH suppression. The elevated level of ACTH stimulates excessive cholesterol uptake and adrenal cell growth, resulting in adrenal hyperplasia. More than 60 inactivating mutations of *StAR*, most of which are mis-sense, non-sense, and frameshift mutations, are known. In all cases, the activity of the mutated StARs, in terms of ligand binding and cholesterol-to-pregnenolone conversion, is severely impaired or totally lost.

## Adrenocorticotropic hormone resistance syndromes

ACTH resistance syndromes represent a group of disorders that lead to unresponsiveness to ACTH in the production of glucocorticoid by adrenal cortex. The disorders can occur in patients with familial glucocorticoid deficiency (FGD) or triple A syndrome.

**Familial glucocorticoid deficiency.** FGD is a rare autosomal recessive disease characterized by cortisol deficiency without mineralocorticoid deficiency as well as pituitary structural defects. Specifically, FGD patients show an extremely high plasma ACTH level, which indicates a resistance to ACTH action. They almost always develop hyper-pigmentation, and adrenarche is absent in children with FGD.

FGD is caused by defects of *MC2R*. Based on gene mutations, FGD is subdivided into two types. FGD type 1 is caused by inactivating mutations of *MC2R*. The mutated receptors are completely inactive or display subnormal cortical response to ACTH stimulation. At least 39 *MC2R* mutations have been reported, most of them are mis-sense mutations in the coding region. Nucleotide substitution at the promoter region is also associated with a lower cortisol response to ACTH stimulation and results in FGD in conjunction with a frameshift mutation in the other allele. FGD type 2 is caused by inactivating mutations of the MC2R accessory protein (MRAP). MRAP assists the trafficking and expression of MC2R at cell surface. A dozen of *MRAP* mutations have been identified; almost all of them lead to the production of severely truncated proteins and loss of MC2R-interacting TMD. Genetic defects of *MC2R* and *MRAP* account for 45% of FGD cases, suggesting that more genes may be involved in the etiology of FGD.

**Triple A syndrome.** Triple A syndrome is a complex disorder characterized by adrenal failure, alacrima, and achalasia. Not all patients exhibit adrenal failure, but those with adrenal failure ∼80% will have isolated glucocorticoid deficiency. Mutations of the causative gene, AAAS, are identified in patients with the syndrome. However, defects in AAAS do not account for all cases of triple A syndrome, and the expression pattern of the gene correlates poorly to pathology.

## Secondary adrenal insufficiency

Secondary adrenal insufficiency can be attributed to a lack of ACTH. The adrenal glands become atrophic and cortisol, but not aldosterone, and production is extremely low or undetectable. In acquired cases, ACTH deficiency may result from pituitary tumor or other physical lesions that prevent ACTH secretion. Congenital cases are caused by disorders of the hypothalamic–pituitary axial components that control ACTH production.

ACTH secretion is regulated by the hypothalamic CRH. The 41-amino-acid neuropeptide binds to CRH receptor (CRHR) on pituitary corticotrophs to stimulate ACTH production through enzymatic cleavage of its precursor proopiomelanocortin (POMC) under the action of prohormone convertase 1 (PC-1). Malfunctioning of any of these genes can lead to ACTH deficiency. POMC deficiency syndrome, owing to recessive inactivating mutations of *POMC*, was described in seven patients, and two cases of compound heterozygous mutations of *PC-1* were reported. No *CRH* or *CRHR* mutation is known. The small number of mutations reflects the rarity of these genetic defects in the etiology of ACTH deficiency. In contrast, mutations of *TPIT*, which encodes a T-box transcription factor important to POMC expression and the terminal differentiation of pituitary POMC-expressing cells, cause adrenal insufficiency more frequently.

## Generalized glucocorticoid resistance/insensitivity

Generalized glucocorticoid resistance is a rare genetic condition characterized by generalized partial target tissue insensitivity to glucocorticoid. Patients show increased levels of ACTH and cortisol, resistance of the hypothalamic–pituitary–adrenal axis to dexamethasone suppression, but no clinical sign of hypercortisolism. Production of mineralocorticoids and androgens is increased as a result of excessive ACTH, which is reflected by symptoms such as hypertension, hypokalemic alkalosis, ambiguous genitalia, and gonadotropin-independent precocious puberty. The disorder is caused by the inactivation of glucocorticoid receptor-$\alpha$ isoform (GR$\alpha$), which is the classical GR that functions as a ligand-dependent transcription factor. Almost all known mutations are heterozygous, implying that a complete loss of the receptor is incompatible with life. Besides one case of a small deletion at the exon–intron 6 boundary, the remaining nine *GR*$\alpha$ mutations are mis-sense mutations. All mutations impair one or multiple actions of glucocorticoid, ranging from defective nucleocytoplasmic shuttling, reduction of gene transactivation activity and affinity for ligands, to partial or complete loss of interaction with coactivators. In some cases, the mutated receptors display a dominant negative effect on wild-type receptors.

## Hypercortisolism (Cushing syndrome)

Excessive hormone production by the adrenal cortex may result from abnormal pituitary ACTH stimulation, pituitary tumors, ectopic ACTH produced by a neoplasm, or pathology within the adrenals themselves. The clinical picture resulting from hypersecretion of cortisol, androgens, and deoxycorticosterone is variable, depending on the etiology. The manifestations, nevertheless, are those of glucocorticoid excess.

Cushing syndrome is caused by excessive circulating cortisol, which affects 10–15 in every million people at age 20–50 years. The affected individuals are characterized by moon faces and buffalo hump, with central obesity, high blood pressure and glucose levels, and gonadal dysfunction. The cause of Cushing syndrome is mostly iatrogenic, resulting from prolonged steroid ingestion (such as from glucocorticoid drug hormone medication). Cushing syndrome is primarily caused by ACTH-secreting pituitary adenoma (also known as Cushing disease), which accounts for ∼70% of all cases of ACTH-dependent Cushing syndrome. Other causes include adrenal hyperplasia or neoplasia (the second most common genetic cause) and ectopic ACTH production by other tumors (including small cell lung cancer and carcinoid tumors). All defects result in adrenal gland overgrowth and cortisol overproduction. The elevated level of cortisol promotes protein catabolism, wasting of muscles, thinning of skin, and conversion to fat without weight gain, which leads to the characteristic appearance.

ACTH-independent Cushing syndrome constitutes the remaining subgroup of the disorder. It is well known to be caused by adrenocortical tumors or inherent endocrine tumor-forming diseases such as MEN1 and primary pigmented nodular adrenocortical disease. An activating mutation of *MC2R* (p.F278C) was shown to be associated with the disorder and adrenal hyperactivity. Other endocrine factors emerge to play a role in disease etiology. For instance, LHR was shown to be present in adrenocortical cells, and LH/hCG is involved in adrenal hyperfunctions.

# Puberty

Puberty refers to the physical changes by which a child becomes an adult capable of reproduction. The maturation of the reproductive system is very complex, involves almost the entire endocrine system, and occurs in a phasic manner.

During fetal development, neuroendocrine cells appear in the rostral forebrain, from where they migrate to an area in the hypothalamus to become the gonadotropin-releasing hormone (GnRH) pulse generator. During infancy, gonadotropin secretion is inhibited centrally by a sensitive negative feedback control, which keeps the reproductive system quiescent.

The onset of puberty is contingent on maturation of the central pulse generator and disinhibition and reactivation of the hypothalamic–pituitary–gonadal axis. Temporal and developmental changes in GnRH pulse frequency have differential effects on FSH and LH. Slow GnRH pulse frequencies preferentially stimulate FSH synthesis and release, whereas higher frequencies preferentially stimulate LH production. The pulsatile LH secretion promotes growth and maturation of the gonads (ovaries and testes), sex hormone production, and development of secondary sex characteristics. As puberty progresses, there is an increase in the amplitude and frequency of LH pulses, leading eventually to adult steroid production and gametogenesis. The sex hormones stimulate growth as well as function and transformation of the brain, bones, muscle, and skin in addition to the reproductive organs. Growth accelerates at the onset of puberty and stops at its completion.

Before puberty, the body differences between the sexes are mainly confined to the genitalia. During puberty, major differences of size, form, shape, composition, and function develop. Breast budding in girls and testicular enlargement in boys are usually the first signs of puberty. The time of onset and subsequent course of hormonal and physical changes during puberty are highly variable and influenced by genetic makeup, nutrition, and general health. In addition to the striking changes brought about by activation of the pituitary–gonadal axis, secretion of androgens by the adrenal gland also increases. However, gonadal and adrenal androgen production (adrenarche) is not causally related temporally, and discordance in the two is seen in a number of situations. Adrenarche occurs in response to rising adrenal androgen secretion, usually precedes the pubertal rise in gonadotropins and sex steroids by about 2 years, and is associated with differentiation and growth of the adrenal zona reticularis. The major adrenal androgens are androstenedione and epiandrosterone. They are weak androgens, which in the female, directly or by peripheral conversion, account for about 50% of serum testosterone and are largely responsible for female sexual hair development. Unlike the major gonadal sex steroids, adrenal androgens do not play an important role in the adolescent growth spurt.

Any disturbance in the integrity of the brain, the hypothalamic–pituitary axis, or the peripheral endocrine system may interfere with normal puberty, whether it is developmental, inflammatory, genetic, neoplastic, or functional. Absence of one or more trophic or pituitary hormones interferes with hormonal effectiveness or end organ response, which may be on a genetic or acquired basis. Lack of signs of puberty in a female 13 years of age or a male 14 years of age becomes a concern and deserves evaluation.

## Delayed puberty

The most common cause of failure of pubertal development in the female is gonadal dysgenesis, a disorder of the X chromosome. Short stature and a variety of congenital anomalies associated with elevated serum gonadotropins readily confirm the diagnosis. The exact chromosomal abnormality is established by the karyotype. Primary amenorrhea after some pubertal development may indicate an abnormality of Müllerian development.

The most common cause of delay in the onset of puberty in boys is constitutional. It is often diagnosed in retrospect if boys initiate puberty spontaneously before the age of 18 years. There is frequently a positive family history. Short stature, a younger appearance, and a delayed bone age are common findings. In the absence of signs of puberty on physical examination, a serum testosterone level >50 mcg/dL foreshadows testicular hormone production and suggests cautious inaction and follow-up. If serum testosterone is < 50 mcg/dL, evaluation of the functional integrity of hypothalamic–pituitary–gonadal axis is required to ascertain whether the delay is a temporary or a permanent feature. If hypogonadotropic hypogonadism is present at 18 years of age or older, the diagnosis is isolated hypogonadotropic hypogonadism (IHH). IHH combined with a history of anosmia/hyposmia suggests Kallmann syndrome. Otherwise, pituitary deficiency (idiopathic or acquired), pituitary stalk interruption syndrome, hypothyroidism, and a variety of genetic syndromes need to be considered.

### Hypogonadotropic hypogonadism

IHH describes the condition, which is the consequence of defects in the pulsatile release of gonadotropins or the deficiency of gonadotropin action. A number of conditions are associated with the Mendelian forms, including developmental defects of the hypothalamus or abnormal pituitary gonadotropin secretion. Structural defects of LH and FSH affect the action of gonadotropins. Mutations of three genes (*KAL1*, *GNRHR*, and *FGF1*) account for most of the known cases of IHH.

**X-linked Kallmann syndrome and KAL1**. When associated with anosmia or hyposmia, IHH is referred to as Kallmann syndrome. IHH is caused by GnRH deficiency due to failure of embryonic migration of GnRH-synthesizing neurons, whereas anosmia is related to hypoplasia or aplasia of the olfactory bulbs. The phenotype associated with Kallmann syndrome mutations varies significantly, indicating the potential influence of modifying genes and other factors. There are two forms of Kallmann syndrome: (1) an X-linked form caused by mutations of *KAL1* and (2) an autosomal dominant form caused by mutations of *KAL2*.

*KAL1* is located on Xp22.3. It is composed of 14 exons encoding an ~100 kD extracellular matrix glycoprotein, anosmin-1, which shares homology with neural cell adhesion molecule. Anosmin-1 has been suggested to provide a scaffold to direct neuronal migration of both GnRH and olfactory neurons to their proper embryonic destination. Mutations in *KAL1* account for approximately 10%−15% of anosmic male patients in sporadic cases but may comprise 30%−60% of familial cases. A variety of mutational changes have been observed with *KAL1*. Genomic deletions, including microdeletion of chromosomal region containing the *KAL1* locus, deletion of the entire *KAL1* gene, and intragenic deletion of multiple exons, have been reported. A recent study found that 12% of male Kallmann syndrome patients have intragenic *KAL1* deletions. There are also deletions of nucleotides, resulting in frameshift and occurrence of premature stop codon. Another mutation shows the replacement of four nucleotides and the creation of an in-frame stop codon (c.1651_1654delinsAGCT resulting in p.P551_E552delinsSX). In one patient, a deletion mutation causes a splicing error. The deletion of nine nucleotides starting at codon 400 (Asn) removes seven bases from exon 8 and the two most 5′-nucleotides of intron 8 (p.N400_IVS8delAACAACAgt). This deletion potentially changes the splicing of intron 8 and creates a different amino acid sequence downstream of residue Asn400. There are also splicing mutations reported that affect IVS1, IVS4, IVS6, and IVS12 and insertions that lead to frameshift and early truncation of the encoded protein (c.1166_1167insA, c.570_571insA). Mis-sense and non-sense mutations have also been reported. These mutations are distributed throughout *KAL1*. The majority of the mutations result in the formation of truncated proteins. Some mutations affect the formation of disulfide bond in the N-terminal cysteine-rich region (WAP domain), whereas other mutations affect the structure of the four fibronectin type III (FNIII) domains of anosmin-1. The WAP and the first FNIII domain are highly conserved across many species and are believed to be essential for the binding of anosmin-1 with heparin sulfate and its other ligands.

**Autosomal dominant Kallmann syndrome and KAL2/FGFR1**. The autosomal dominant form of Kallmann syndrome is caused by an inactivating mutation of *KAL2/FGFR1* that is localized on chromosome 8p11.23-p11.22. *KAL2/FGFR1* contains 18 exons and encodes fibroblast growth factor receptor 1 (FGFR1). FGFR1 is a single spanning transmembrane receptor. The protein has three immunoglobulin-like domains, a heparin-binding domain, and two tyrosine kinase domains. FGFR1 is expressed in GnRH neurons, and FGF signaling is involved in GnRH neuron specification, migration, and axon targeting. Anosmin-1 is a ligand for FGFR1 and induces neurite outgrowth and cytoskeletal changes through an FGFR1-dependent mechanism.

Mutations of *KAL2/FGFR1* occur in approximately 7%−10% of patients (male and female) with autosomal dominant Kallmann syndrome. Although most *KAL2/FGFR1* mutations are identified in Kallmann patients, mutations of this gene in normosmic IHH patients have also been reported. Mutations of *KAL2/FGFR2* show reduced penetrance and variable expressivity.

More than 40 inactivating mutations of *KAL2/FGFR1* leading to IHH with or without anosmia have been described. About 70% of them are mis-sense mutations, 18% are non-sense mutations, 9% are frameshift deletion/insertions, and two cases are splicing mutations. A number of the mis-sense mutations cluster in the first immunoglobulin domain of FGFR1, suggesting that this domain is important to FGFR1 function. The non-sense mutations cluster at the C-terminal tyrosine kinase domain and result in truncated receptors lacking the autophosphorylated tyrosine residues, which likely impede the signaling activity of FGFR1.

**Normosmic IHH and GnRH Receptor (GnRHR)**. About 50% of familial cases of normosmic IHH are associated with loss-of-function *GnRHR* mutations. *GnRHR* is located on chromosome 4q21.2 with an open reading frame of 981 nucleotides and encodes a protein of 327 amino acids. It is a member of the GPCR family. GnRHR is expressed on the cell surface of pituitary gonadotrophs. The ligand for GnRHR is GnRH-1, a decapeptide that is derived from a 92-amino-acid pre-pro-protein. It is released in a pulsatile manner in the preoptic area of the hypothalamus and delivered to the anterior pituitary gland. There, it binds and activates GnRHR, resulting in the synthesis and release of gonadotropins (LH and FSH). Although it is rare, several GnRH mutations causing IHH have been identified.

Currently, 21 loss-of-function mutations of *GnRHR* have been identified in patients with IHH. All are mis-sense mutations with the exception of a non-sense mutation and a splicing mutation. These mutations are transmitted in an autosomal recessive mode. Most patients are compound heterozygotes. The two most prevalent mutations are p.Q106R (32%) and p.R262Q (15%). The former mutation reduces GnRH binding affinity, and the latter mutation reduces signal transduction. Slightly more than half of the *GnRHR* mutations interfere with ligand binding, whereas the remainder affects signal transduction.

There is no report of *GnRHR* mutations in anosmic or hyposmic IHH patients. *GnRHR* mutations are present in 1%−4.6% of all IHH patients, whereas it is 6%−11% in autosomal recessive IHH families. In most cases the phenotype correlates with the functional alterations of the GnRHR in vitro. Patients with complete inactivating GnRHR variants on both alleles present with severe hypogonadism, whereas patients homozygous for a partially inactivating GnRHR variant present with partial hypogonadism.

More recently, another GPCR, *GPR54* (also known as *KISS1R*), has been found to be associated with autosomal recessive IHH. *GPR54* is localized on chromosome 19p13.3 and is composed of 1191 nucleotides encoding a protein of 396 amino acids. At least five mis-sense mutations, one non-sense mutation, a 155-bp deletion, and one insertion of *GRP54* have been described in homozygous and compound heterozygous IHH patients. Impaired signaling capacity of the mutated receptor has been observed. GPR54 is the receptor for kisspeptins, which are potent stimulators of LH and FSH secretion. Kisspeptin-induced GPR54 signaling is also believed to be a major control point for GnRH release. Phenotypic expression of *GPR54* inactivating mutation does not differ from that of *GnRHR* mutation.

**Isolated gonadotropin deficiency.** LH and FSH are the main regulators of gonadal steroid secretion, pubertal maturation, and fertility. A small number of hypogonadotropic hypogonadism patients have been found to carry loss-of-function mutations of the hormone-specific β subunit of LH and FSH. Both LH and FSH are heterodimers with an α subunit that is shared among the glycohormones. The β subunit of each hormone gives it its specificity. *LHβ* is located on chromosome 19q13.32, and FSHβ is on 11p13. Loss-of-function mutation of the β subunit renders the hormone inactive, giving rise to hypogonadotropic hypogonadism.

At present there are seven individuals, four women and three men, who are hypogonadal due to *FSHβ* mutation. Four different mutations have been identified: p.C51 G, p.C82R, p.Y76X, and p.V61delTGfs87X. These mutations interfere with subunit dimerization and render the hormone inactive. All four women with homozygous inactivating *FSHβ* mutations have sexual infantilism and infertility because of a lack of follicular maturation, primary amenorrhea, low estrogen production, undetectable serum FSH, and increased LH. The men with homozygous inactivating mutations are all normally masculinized, with normal to delayed puberty, but azoospermic.

Mutation of *LHβ* is rare. Only two mis-sense mutations (p.G36D and p.Q54R), two splicing mutations (IVS2 +1 G > C and IVS2+1 G > T), and two deletions (c.28delTTGCTGCTGCTG and c.88delCACCCCATC) of *LHβ* have been described in seven men and three women. The mis-sense mutations either interfere with the formation of active heterodimers or affect the interaction with LHR. The splicing mutation produces a truncated or unstable LHβ subunit and abrogates LH secretion. Male homozygotes are normally masculinized at birth but lack postnatal sexual differentiation, are hypogonadal with delayed pubertal development, absence of mature Leydig cells, spermatogenic arrest, and absence of circulating LH and testosterone. The only woman with the homozygous splicing mutation has normal pubertal development, secondary amenorrhea, and infertility. All heterozygotes are fertile and have normal basal gonadotropin and sex steroid levels.

## Hypergonadotropic hypogonadism

Hypergonadotropic hypogonadism and infertility/subfertility in both sexes are the result of gonadotropin resistance caused by inactivating mutations in receptors of the two gonadotropins, LH and FSH. Both LHR and FSHR are members of GPCR family. The ECDs of the receptors convey ligand specificity and are different among the receptors, whereas the TMD is highly homologous between species and among the glycohormone receptors.

**Leydig cell hypoplasia (LCH)—inactivating mutations of LHR.** LHR is shared between LH and CG, thus the name LH/CG receptor. CG exerts its effect during early embryogenesis to induce Leydig cell maturation. LH also promotes steroidogenesis by Leydig cells, especially around the period of puberty. Inactivating mutations of LHR are recessive in nature. In males with homozygous or compound heterozygous inactivating mutations, a loss of LHR function causes resistance to LH stimulation, resulting in failure of testicular Leydig cell differentiation. This gives rise to LCH. A number of features distinguish LCH from other forms of male pseudohermaphroditism. LCH patients are genetic males with a 46, XY karyotype. The hormonal profile of them shows elevated serum LH level, normal to elevated FSH level, and low testosterone level, which is unresponsive to CG stimulation. Clinical presentation of LCH is variable, ranging from hypergonadotropic hypogonadism with microphallus and hypoplastic male external genitalia to a form of male pseudo-hermaphroditism with female external genitalia. In between are patients with variable degree of masculinization of the external genitalia. LCH patients show no development of either male or female secondary sexual characteristics at puberty. In females, LHR inactivation causes hypergonadotropic hypogonadism and primary amenorrhea with subnormal follicular development and ovulation, and infertility.

*LHR* is located on human chromosome 2p21. It has 11 exons. The first 10 exons encode the ECD, whereas the last exon encodes a small portion of the ECD, the TMD, and the ICD. More than two dozens of inactivating mutations, distributed throughout *LHR*, have been identified in LCH patients. The majority of them are single base substitutions leading to either mis-sense or non-sense mutations. All single base substitutions affect highly conserved amino acids. Both in-frame and out-of-frame loss-of-function insertion mutations have been identified. A 33-bp in-frame insertion has been found as heterozygous as well as homozygous mutation in kindreds with male pseudohermaphrodites. The insertion occurs between amino acid residues 18 and 19, immediately upstream of the signal peptide cleavage site. Exon deletions and minor deletion have also been identified. The majority of the inactivating mutations are found in homozygous patients. There are a number of LCH kindreds in which *LHR* mutations have not been identified, indicating that inactivating mutations are very heterogeneous or that LCH is caused by mutation of LHR as well as other gene(s).

Effect of the different mutations on LHR activity is variable. Depending on the location of the mutation in the receptor protein, it can cause diminished hormone binding, reduced surface expression, abnormal trafficking, or reduced coupling efficiency, all of which result in reduction or abolition of signal transduction triggered by hormone binding. Clinical presentation of LCH patients can be correlated with the amount of residual activity of the mutated receptor. Mutated LHRs of patients with the most severe phenotype, i.e., male pseudohermaphroditism, have zero or minimal signal transduction activity. On the other hand, patients with male hypogonadism have mutated LHRs with reduced, but not abolished, signal transduction.

**Ovarian dysgenesis—inactivating mutation of FSHR**. FSHR mediates the action of FSH. In females, FSH function is essential for ovarian follicular maturation. In male, FSH regulates Sertoli cell proliferation before and at puberty and participates in the regulation of spermatogenesis. Loss-of-function mutations of FSHR are expected to be found in connection with hypergonadotropic hypogonadism associated with retarded follicular maturation and anovulatory infertility in women and small testicles and impaired spermatogenesis in normally masculinized males.

*FSHR* is located on chromosome 2p, next to *LHR*. With an exceptional case where a compounded heterozygosity of mis-sense mutation and partial deletion of exons 8 and 9 of *FSHR* is reported, inactivating mutations of *FSHR* are mostly mis-sense mutations in the ECD and TMD. The inheritance of these mutations follows an autosomal recessive mode. The most frequently detected mutation is p.A198V. Female patients homozygous with this mutation present with ovarian dysgenesis that includes hypergonadotropic hypogonadism, primary or early onset secondary amenorrhea, variable pubertal development, hypoplastic ovaries with impaired follicle growth, and high gonadotropin and low estrogen levels. Compound heterozygous female patients with totally or partially inactivating *FSHR* mutations have less prominent phenotypes. The male phenotype of inactivating *FSHR* mutations is less assuring. The five male subjects with homozygous p.A198V mutation are normally masculinized, with moderately or slightly decreased testicular volume, normal plasma testosterone, normal to elevated LH but high FSH levels, and variable spermatogenic failure. In some cases, fertility is maintained, suggesting that FSH action is not compulsory for spermatogenesis.

There is good correlation between the phenotype and the degree of receptor inactivation, as well as the site of mutation and its functional consequences. All mutations in the ECD cause a defect in ligand binding and targeting of FSHR to the cell membrane. In the TMD, the mutations have minimal effect on ligand binding but impair signal transduction to various extents.

# Precocious puberty

Pubertal development is considered precocious if it occurs before 7 or 8 years of age in girls, or 9 years in boys. It is more common in girls and is mostly idiopathic. In contrast, precocious puberty in boys is due to some underlying pathology in half of the cases.

Precocious pubertal development that results from premature activation of the hypothalamic—pituitary—gonadal axis, and hence gonadotropin dependent, is termed central precocious puberty (CPP). Precocious sexual development that is not physiological, not gonadotropin dependent, but the result of abnormal sex hormone secretion, is termed sexual precocity or gonadotropin-independent precocious puberty (also called pseudopuberty or PPP). CPP is not uncommon in otherwise normal healthy girls.

Clinically, the earliest sign of puberty in girls is breast enlargement (thelarche) in response to rising estrogen secretion, which usually precedes but may accompany or follow the growth of pubic hair. The earliest sign of puberty in boys is an increase in the size of testes and in testosterone production, followed by penile and pubic hair growth. Precocious thelarche and adrenarche may need to be considered but are readily differentiated from true precocious puberty.

## Gonadotropin-dependent precocious puberty

A study of 156 children with idiopathic CPP indicates 27.5% of it to be familial. Segregation analysis reveals autosomal dominant transmission with incomplete sex-dependent penetrance. Studies of a girl with CPP show a heterozygous single base substitution leading to the replacement of Arg386 by Pro (c. G1157C) in the carboxy-terminal tail of *GPR54*. This mutation leads to prolonged activation of intracellular signaling pathways in response to kisspeptin. The kisspeptin—GPR54 signaling complex has been proposed as a gatekeeper of pubertal activation of GnRH neurons and the reproductive axis. Whether mutations affecting the kisspeptin—GPR54 duet are the genetic causes of gonadotropin-dependent precocious puberty need further study.

## Gonadotropin-independent precocious puberty

**Familial male-limited precocious puberty (FMPP)—activating mutations of LHR.** Gain-of-function mutations, resulting in constitutive activation of LHR, cause luteinizing hormone—releasing hormone (LHRH)—independent isosexual precocious puberty in boys. Constitutive activity of LHR leads to stimulation of testicular Leydig cells in the fetal and prepubertal period in the absence of the hormone, resulting in autonomous production of testosterone and pubertal development at a very young age. This autosomal dominant condition is termed FMPP or testotoxicosis. Signs of puberty usually appear by 2—3 years of age in boys with FMPP. These patients have pubertal to adult levels of testosterone, whereas the basal and LHRH-stimulated levels of gonadotropins are prepubertal. There is also lack of a pubertal pattern of LH pulsatility. Activating mutations of *LHR* have no apparent effect on female carriers.

All activating mutations of the *LHR* identified in FMPP patients are single base substitutions. Seventeen activating mutations in exon 11 of *LHR* have been identified among kindreds with FMPP. These mutations affect 14 amino acids. Half of the mutations are located in transmembrane helix VI (transmembrane VI), which represent 67.5% of all mutations identified in FMPP patients. The most frequently mutated amino acid is Asp578, with >55% of all activating mutations affecting this amino acid. The most common mutation is the c. T1733G transition, which results in the replacement of Asp578 by Gly. This mutation represents >51% of all mutations identified in FMPP patients so far. With the exception of the replacement of Ile542 by Leu in transmembrane V, all mutations result in the substitution of amino acids, which are conserved among the glycohormone receptors.

*LHR* is prone to mutation. Among the kindreds of FMPP with confirmed molecular diagnoses, over 25% are caused by new mutations. There is no difference in clinical manifestation between familial and new mutations. Rare mutations occur more often in patients of non-Caucasian ethnic background. Detection of *LHR* mutations is unsuccessful in about 18% of patients diagnosed to have FMPP.

All FMPP mutations reside in the TMD. Agonist affinity of the mutated receptors is largely unchanged, whereas cell surface expression is either the same or reduced when compared with the wild-type receptor. All FMPP mutations have been shown to confer constitutive activity to the mutated LHR by in vitro expression studies. There is no consensus on the phenotype—genotype correlation in FMPP. However, mutations that give the highest basal level of cAMP in in vitro assays are associated with an earlier age of pubertal development.

Besides the germ-line mutations found in FMPP patients, a somatic activating mutation of *LHR* (p.D578H) has also been identified in tumor tissue of a number of patients with testicular neoplasia. Even though a couple of FMPP patients developed testicular neoplasia, the p.D578H mutation has never been found as a germ-line mutation in any FMPP patient. Its presence is confined to patients with testicular neoplasia.

**Activating mutation of FSHR.** *FSHR* is not prone to mutation. So far, there are only two gain-of-function *FSHR* mutations identified in males. In one case, the patient had been previously hypophysectomized but maintained normal spermatogenesis in spite of undetectable gonadotropins. He has a heterozygous activating mutation p.D567G in the third intracytoplasmic loop of FSHR. In vitro expression of the mutated receptor shows elevated basal activity of the receptor. In the other case, the asymptomatic man exhibits normal spermatogenesis but suppressed serum FSH. He has a heterozygous mutation p.N431I in the first extracellular loop of FSHR.

**Ovarian hyperstimulation syndrome (OHSS).** OHSS is an iatrogenic complication of ovulation induction therapy. In its most severe form, this syndrome involves massive ovarian enlargement and the formation of multiple ovarian cysts and can be fatal. OHSS can arise spontaneously during pregnancy owing to a broadening of the specificity of FSHR for hCG at high concentration of the hormone. So far, eight mis-sense mutations affecting six amino acids (p.S128Y, p. T449I, p.T449A, p.V514A, p.I545T, p.D567N, p.D567G, and p.A575V) have been identified in OHSS patients. All except one of the mutations are found in the TMD of *FSHR*. The transmission of these mutations follows an autosomal dominant mode. Patients analyzed so far are heterozygous for these mutations. These mutations relax the ligand specificity of the receptor in such a way that it also binds and becomes activated by hCG. Response of the receptor to TSH was also found for mutations that are located in the TMD. It is possible the promiscuous stimulation of follicles during the first trimester of pregnancy by hCG results in excessive follicular recruitment observed in this disorder.

## Acknowledgments

This work was supported in part by the School of Biomedical Sciences, Faculty of Medicine, The Chinese University of Hong Kong, and TGD Life Company Limited.

## Key concepts

- The pituitary acts as a relay between the hypothalamus and target organs (the hypothalamus-pituitary target organ axis) in the regulation of normal physiology by hormone secretion. The production of pituitary hormones, and their releasing factors from the hypothalamus, is subject to negative feedback regulation from the target organs. The action of pituitary hormones and their release are mediated through the ligand-receptor binding mechanism. Inactivating mutations of the genes encoding pituitary hormones or their receptors lead to pituitary hormone deficiency phenotypes. Activating mutations of hormone receptors, or the loss of negative feedback regulation on pituitary hormone secretion, can lead to hypersecretion of pituitary hormones and/or overstimulation of target organs.

- The major hormone-secreting cells in the pituitary are located at the anterior pituitary. Genes that are essential to the development of pituitary hormone-secreting cells have been identified, including HESX1, LHX3/4, PROP1, POU1F1, and GLI2. Loss-of-function mutations of these genes lead to abnormal pituitary development and are the major causes of congenital hypopituitarism.

- The majority of cases of congenital hypothyroidism is nongoitrous and is due to faulty embryogenesis (athyreosis, agenesis, or dysembryogenesis). A small number are goitrous, the result of enzyme defects in hormone biosynthesis (dyshormonogenesis). Etiologically, three types of congenital hypothyroidism can be distinguished: central congenital hypothyroidism including hypothalamic (tertiary) and pituitary (secondary), primary (impairment of the thyroid gland itself), and peripheral (resistance to thyroid hormone). The majority of cases of central congenital hypothyroidism are caused by TSH mutations. A subgroup of patients with familial congenital hypothyroidism is the result of loss-of-function mutations in TSHR. Constitutive activating mutations of TSHR give rise to familial nonautoimmune hyperthyroidism. Disorders resulting in congenital primary hypothyroidism have been identified in all major steps in thyroid hormonogenesis.

- In humans, the regulation of the ionized fraction of serum calcium depends on the interaction of PTH and the active form of vitamin D-1,25-dihydroxycholecalciferol [1,25(OH)2D]. Familial isolated hypoparathyroidism can be caused by mutations of PTH and Glial cells missing (GCM). On the other hand, primary hyperparathyroidism is genetically heterogeneous. Mutations of the Calcium-Sensing Receptor (CaSR) can result in loss-of-function or gain-of-function of the receptor. Disorders due to loss-of-function mutations of CaSR include familial hypocalciuric hypercalcemia and neonatal severe primary hyperparathyroidism, whereas disorders due to gain-of-function mutations of the CaSR include Autosomal Dominant Hypocalcemia (ADH) and Bartter syndrome type V.

- The adrenal cortex of adrenal glands produces mineralocorticoids and glucocorticoids for homeostasis, and adrenal androgens for fetal external genitalia development. The cause of congenital primary adrenal insufficiency can be attributed to defective development of the adrenal glands (Adrenal Hypoplasia Congenita; AHC) or inborn errors in adrenal steroid biosynthesis (Congenital Adrenal Hyperplasia; CAH). The cytomegalic form of AHC is an X-linked disorder caused by inactivating mutations of NR0B1 (DAX1). CAH is caused most frequently by inactivating mutations of CYP21A2 and CYP11B1, which encode enzymes that regulate the biosynthesis of mineralocorticoids and glucocorticoids. Excessive adrenal hormone production results in hypercortisolism, which is caused mostly by ACTH-secreting pituitary adenoma (Cushing's disease).

- Mutations in three genes (KAL1, GNRHR, and FGF1) account for most of the known cases of Isolated Hypogonadotropic Hypogonadism (IHH). A small number of hypogonadotropic hypogonadism patients have been found to carry loss of function mutations of the hormone-specific β subunit of LH and FSH. On the other hand, hypergonadotropic hypogonadism and infertility/subfertility in both sexes are the result of gonadotropin resistance caused by inactivating mutations in receptors of LH and FSH; inactivating mutations of the LHR cause Leydig Cell Hypoplasia (LCH); those of FSHR cause Ovarian Dysgenesis. Activating mutations of LHR are responsible for the development of familial male-limited precocious puberty (FMPP).

# Suggested readings

[1] Mullis PE. Genetics of growth hormone deficiency. Endocrinol Metab Clin North Am 2007;36:17—36.

[2] Cohen LE, Radovick S. Molecular basis of combined pituitary hormone deficiencies. Endocr Rev 2002;23:431—42.

[3] Ayuk J, Sheppard MC. Growth hormone and its disorders. Postgrad Med J 2006;82:24—30.

[4] Winter WE, Signorino MR. Review: molecular thyroidology. Ann Clin Lab Sci 2001;31:221—44.

[5] Glaser B. Pendred syndrome. Pediatr Endocrinol Rev 2003;1:199—204.

[6] Rivolta CM, Targovnik HM. Molecular advances in thyroglobulin disorders. Clin Chim Acta 2006;374:8—24.

[7] Refetoff S, Dumitrescu AM. Syndromes of reduced sensitivity to thyroid hormone: genetic defects in hormone receptors, cell transporters and deiodination. Best Pract Res Clin Endocrinol Metab 2007;21:277—305.

[8] Visser WE, Friesema EC, Jansen J, et al. Thyroid hormone transport in and out of cells. Trends Endocrinol Metab 2008;19:50—6.

[9] Duprez L, Parma J, Van Sande J, et al. TSH receptor mutations and thyroid disease. Trends Endocrinol Metab 1998;9:133—40.

[10] Thakker RV. Genetics of endocrine and metabolic disorders: parathyroid. Rev Endocr Metab Disord 2004;5:37—51.

[11] Ferris RL, Simental Jr. AA. Molecular biology of primary hyperparathyroidism. Otolaryngol Clin North Am 2004;37:819—31.

[12] Egbuna OI, Brown EM. Hypercalcaemic and hypocalcemic conditions due to calcium-sensing receptor mutations. Best Pract Res Clin Rheumatol 2008;22:129—48.

[13] Layman LC. Hypogonadotropic hypogonadism. Endocrinol Metab Clin North Am 2007;36:283—96.

[14] Bhagavath B, Layman LC. The genetics of hypogonadotropic hypogonadism. Semin Reprod Med 2007;25:272—86.

[15] Cadman SM, Kim SH, Hu Y, et al. Molecular pathogenesis of Kallmann's syndrome. Horm Res 2007;67:231—42.

[16] de Roux N. GnRH receptor and GPR54 inactivation in isolated gonadotropic deficiency. Best Pract Res Clin Endocrinol Metab 2006;20:515—28.

[17] Huhtaniemi IT, Themmen AP. Mutations in human gonadotropin and gonadotropin-receptor genes. Endocrine 2005;26:207—17.

[18] Huhtaniemi I, Alevizaki M. Gonadotrophin resistance. Best Pract Res Clin Endocrinol Metab 2006;20:561—76.

[19] Turcu AF, Auchus RJ. The next 150 years of congenital adrenal hyperplasia. J Steroid Biochem Mol Biol 2015;153:63—71.

[20] Reynaud R, Saveanu A, Barlier A, et al. Pituitary hormone deficiencies due to transcription factor gene alterations. Growth Horm IGF Res 2004;14:442—8.

[21] Zhu X, Lin CR, Prefontaine GG, et al. Genetic control of pituitary development and hypopituitarism. Curr Opin Genet Dev 2005;15:332—40.

[22] Coya R, Vela A, Perez de Nanclares G, et al. Panhypopituitarism: genetic versus acquired etiological factors. J Pediatr Endocrinol Metab 2007;20:27—36.

# Chapter 23

# Molecular basis of gynecologic diseases

**Samuel C. Mok[1], Kwong-Kwok Wong[1], Karen H. Lu[1], Karl Munger[2] and Zoltan Nagymanyoki[3]**

[1]*Department of Gynecologic Oncology, University of Texas MD Anderson Cancer Center, Houston, TX, United States,* [2]*Department of Developmental, Molecular and Chemical Biology, Tufts University School of Medicine, Boston, MA, United States,* [3]*Department of Pathology, West Pacific Medical Laboratory, Santa Fe Springs, CA, United States*

**Summary**

Gynecologic diseases in general are diseases that involved the female reproductive track. These diseases include benign and malignant tumors, pregnancy-related diseases, infection, and endocrine diseases. Among them, malignant tumors are the most common cause of death. In recent years, the causes of some of these diseases have been elucidated. For example, human papillomavirus infection has been shown to be one of the major etiological factors associated with cervical cancer. Inactivation of the *BRCA1* tumor suppressor gene has been implicated in hereditary ovarian cancer. In spite of these findings, the molecular basis of most gynecologic diseases remains largely unknown. In this chapter, we will focus on benign and malignant tumors of the female reproductive organs, as well as pregnancy-related diseases, which are relatively well understood on a molecular basis.

## Introduction

Gynecologic diseases, in general, are diseases that involved the female reproductive track. These diseases include benign and malignant tumors, pregnancy-related diseases, infection, and endocrine diseases. Among them, malignant tumors are the most common cause of death. In recent years, the causes of some of these diseases have been elucidated. For example, human papillomavirus (HPV) infection has been shown to be one of the major etiological factors associated with cervical cancer. Inactivation of the *BRCA1* tumor suppressor gene has been implicated in hereditary ovarian cancer. In spite of these findings, the molecular basis of most gynecologic diseases remains largely unknown. In this chapter, we will focus on benign and malignant tumors of the female reproductive organs, as well as pregnancy-related diseases, which are relatively well understood on a molecular basis.

## Benign and malignant tumors of the female reproductive tract

### Cervix

Infections of the genital mucosa with HPVs represent the most common virus-associated sexually transmitted disease, and at age 50 years, $\sim 80\%$ of all females will have acquired a genital HPV infection sometime during their life. It has been estimated that in the United States there are $>79$ million people affected by genital HPV infection, with an estimated annual incidence of 14 million new infections. Genital HPV infections are particularly prevalent in sexually active younger individuals. Even though most of these infections are transient and may not cause any overt clinical symptoms or disease, the total annual cost of preventing and treating HPV-associated disease exceeds $8 billion in the United States alone.

### *Human papillomaviruses associated with cervical lesions and cancer*

HPVs are members of the Papillomaviridae family. They have a tropism for squamous epithelial cells and cause the formation of generally benign hyperplastic lesions that are commonly referred to as papillomas or warts. Papillomaviruses contain closed circular double-stranded DNA genomes of $\sim 8000$ base pairs that are packaged into $\sim 55$-nm non-enveloped, icosahedral capsids, consisting of three regions. The early (E) region encompasses up to eight open reading

Essential Concepts in Molecular Pathology. DOI: https://doi.org/10.1016/B978-0-12-813257-9.00023-1

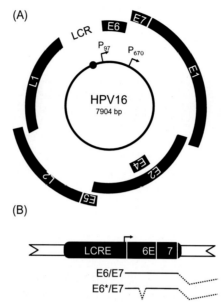

**FIGURE 23.1** The human papillomavirus (HPV) genome. (A) Schematic representation of the HPV16 genome. The double-stranded circular DNA genome is represented by the central circle. Early (E) and late (L) genes are all encoded in one of the two DNA strands in each of the three possible reading frames. The long control region (LCR) does not have extensive coding potential but contains the viral origin of replication (designated by the *black circle*) as well as the major early promoter, $P_{97}$. The differentiation-specific late promoter, $P_{670}$, is contained within E7. See text for details. (B) Representation of the minimal HPV16 genome fragment that is consistently retained after integration into a host chromosome. The major E6/E7 transcripts are shown underneath.

frames (ORFs) designated E followed by a numeral, with the lowest number designating the longest ORF, and the late (L) region encodes the major and minor capsid proteins, L1 and L2, respectively. Papillomavirus gene transcription is unidirectional, and early and late ORFs are encoded on the same DNA strand. A third region, referred to as the long control region (LCR), does not have significant coding capacity but contains multiple regulatory DNA sequences that control viral genome replication and transcription (Fig. 23.1).

Approximately 40 HPV types infect mucosal epithelia, and these viruses are further classified as low risk or high risk depending on the propensity for malignant progression of the lesions that they cause. Low-risk HPVs (such as HPV6 and HPV11) cause genital warts, whereas high-risk HPVs (such as HPV16 and HPV18) cause intraepithelial neoplasia that can progress to frank carcinoma. Harald zur Hausen's group discovered the association of HPVs with anogenital tract lesions and isolated HPVs from genital warts. Using these sequences as hybridization probes under low stringency conditions, they succeeded in detecting HPV sequences in cervical carcinomas.

The most abundant high-risk HPV is HPV16, and it is detected in ~50% of all cervical carcinomas. Based on an extensive review of epidemiological, medical, and basic science data, the *World Health Organization* (WHO) has classified HPV types 16, 18, 31, 33, 35, 39, 45, 51, 52, 56, 58, 59, and 66 as "...*carcinogenic to humans*..." The following sections are focused on a review of these mucosal high-risk HPVs and their contributions to cervical lesions and cancers.

## Detection of human papillomavirus—associated lesions

Papanicolaou tests (also known as Pap tests) are named after their inventor, Georgios Papanicolaou, and serve to detect HPV-associated lesions in the cervix. On implementation, this relatively inexpensive cytology test dramatically reduced the incidence and mortality rates of cervical cancer. The test involves collecting exfoliated epithelial cells from the outer opening of the cervix and either directly smearing the cells on a microscope slide, or immediately preserving and storing the cells in fixative liquid medium followed by automated processing into a monolayer. In either format, cells are stained and examined for cytological abnormalities. Liquid-based monolayer cytology has a reduced rate of false positivity presumably due to standardized specimen preparation and immediate fixation of the sampled cells. Although nuclear features are currently used for diagnosis, attempts are underway to assess HPV DNA positivity and/or expression of high-risk HPV-specific biomarkers. The most useful biomarker for high-risk HPV-associated cervical lesions is p16$^{INK4A}$, an inhibitor of CDK4/CDK6 cyclin D complexes. High-level p16$^{INK4A}$ expression in high-risk HPV-associated precancerous lesions and cancers represents a cellular response to oncogenic stress caused by expression of the E7 protein. In normal cells, p16$^{INK4A}$ induces senescence, which is mediated by the RB1 tumor suppressor. Because RB1 is inactivated by E7 in high-risk HPV-associated lesions, cells proliferate despite high-level p16$^{INK4A}$ expression. Hence, concurrent staining of cytology specimens for p16$^{INK4A}$ and the proliferation marker KI67 allows sensitive and specific detection of high-risk HPV-associated lesions and cancer.

The *American Cancer Society* (ACS) and the *American College of Obstetricians and Gynecologists* (ACOG) recommend that females aged 21–65 years should receive cervical cancer screening. Until age 30 years this should involve cytology screening every 3 years. Women aged 30–65 years should preferably be screened by combined cytology/HPV DNA testing every 5 years. In 2014, the FDA approved the PCR-based cobas HPV test with greatly enhanced sensitivity and specificity over the previously used nucleic acid–based hybridization technologies, and this has reignited the discussion of whether HPV molecular testing should entirely replace cytology-based screening.

## Diagnosis and treatment

Genital warts are frequent and clinically obvious manifestations of low-risk mucosal HPV infections. Even though they generally do not undergo malignant progression and sometimes regress spontaneously, patients generally insist on their removal. Because no HPV-specific antivirals currently exist, standard therapeutic modalities include surgical excision, laser therapy, cryotherapy, topical administration of various caustic chemicals, or immunomodulatory agents. In very rare cases, low-risk HPV genital tract infections can also cause serious disease. The giant condyloma of Buschke–Lowenstein can arise in patients who cannot adequately control and/or clear a low-risk HPV infection. Although these are slow-growing tumors, they are highly destructive to adjacent normal tissue, which eventually can form local and distant metastases.

Lesions caused by high-risk HPV infections are detected by Pap tests. The procedure involves application of an acetic acid–based solution to the cervix, whereupon lesions appear as white masses on colposcopic evaluation. When lesions are detected, a biopsy is performed and the tissue is examined histologically. Lesions are classified as cervical intraepithelial neoplasia (CIN), carcinoma in situ (CIS), or invasive cervical carcinoma. Treatments for CIN include cryotherapy, laser ablation, or loop electrosurgical excision, whereas carcinomas are treated by surgery and/or chemotherapy. Given that HPV-associated cancers express non–cell-derived viral antigens, there have been many attempts to develop therapeutic HPV vaccines. The clinical introduction of immunological checkpoint inhibitors has greatly galvanized these efforts.

## Prevention and vaccines

Condoms reduce, but do not negate, the risk of HPV infection. Preclinical studies in a mouse model have shown that the polysaccharide carrageenan, which is used as a lubricant, greatly inhibits HPV transmission. A clinical study with the microbicide candidate Carraguard has suggested some efficacy in inhibiting HPV transmission.

The currently available prophylactic HPV vaccines consist of recombinant HPV L1 proteins that self-assemble into virus-like particles (VLPs). Gardasil was developed by Merck and is a quadrivalent formulation that contains VLPs of the most prevalent low-risk (HPV6 and HPV11) and high-risk HPVs (HPV16 and HPV18). Cervarix is produced by GlaxoSmithKline (GSK) and is a bivalent vaccine targeting HPV16 and HPV18. More recently, Gardasil 9, a nonavalent vaccine that targets HPV16, HPV18, HPV6, HPV11, and other high-risk HPVs (HPV 31, 33, 45, 52, and 58), has been developed. These vaccines are administered as three doses over the course of 6 months and are highly efficacious in providing type-specific protection from new infections with vaccine HPV types. The Gardasil vaccines are approved in females aged 9 through 26 years for prevention of HPV-associated cervical, vulvar, vaginal, and anal precancers and cancers, as well as genital warts, and in males aged 9–26 years (Gardasil) or aged 9–15 years (Gardasil 9) for prevention of HPV-associated precancerous anal lesions and cancers, as well as genital warts. Cervarix is approved for females aged 16–25 years for cervical cancer prevention. These vaccines have the potential to dramatically reduce the burden of HPV-associated cancers and genital warts. Because these prophylactic vaccines trigger humoral immune responses and inhibit viral entry, they are not predicted to affect HPV infections that are already active at the time of vaccination. Given that cervical cancer generally develops decades after the initial infection, it has been estimated that incidence and mortality rates of cervical cancer will not decrease for 25–40 years. Hence, recommendations for cervical cancer screening remain unchanged for vaccinated individuals. The high cost of the current vaccines and the need for cold storage greatly impedes their extensive use in some of the countries with the highest incidence of cervical cancer, and there are many efforts to develop alternative, less costly, and more stable vaccine preparations.

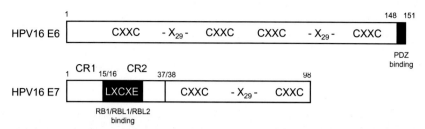

**FIGURE 23.2**   HPV16 E6 (top) and E7 (bottom) oncoproteins. The position of the binding site for cellular PDZ proteins on E6 is indicated by a *black box*. The amino terminal domains of HPV16 E7 that are similar to a portion of conserved region (CR1) and to CR2 of the adenovirus E1A protein are indicated. The CR2 domain contains the LXCXE core RB1 binding site, which is indicated by a *black box* that is not to scale.

## Biological and biochemical activities of high-risk HPV E6 oncoproteins

The HPV16 E6 are related in sequence, but not in structure, to the carboxyl terminal domain of E7 (Fig. 23.2). The best-known activity of high-risk HPV E6 proteins is their ability to associate with the TP53 tumor suppressor and the ubiquitin ligase UBE3A (E6-AP), which targets TP53 for ubiquitin-mediated proteasomal degradation. UBE3A interacts through an LXXLL (L, leucine; X, any amino acid) motif with a hydrophobic/basic groove of E6. Different E6 proteins selectively associate with specific LXXLL-containing cellular proteins. LXXLL protein binding causes a conformational change in E6, which, in the case of UBE3A binding to high-risk HPV E6 proteins, is critical for TP53 interaction. TP53 senses cellular stress and triggers G1 growth arrest and/or apoptosis (Fig. 23.3). TP53 inactivation by E6 has evolved to abrogate a cytostatic/cytotoxic TP53 response to aberrant S-phase induction by E7. HPV E6 expression also causes telomerase activation, which contributes to cellular immortalization of primary human epithelial cells (Fig. 23.3).

High-risk HPV E6 proteins contain a short peptide sequence (S/T)-X-V-I-L (S, serine; T, threonine; V, valine; I, isoleucine; L, leucine, X, any amino acid) at their carboxyl termini, which mediates association with cellular PDZ domain−containing proteins (Fig. 23.2). The integrity of the PDZ binding sequence on E6 is important for the HPV viral life cycle, as well as the transforming activities of E6. Because PDZ protein and UBE3A binding by E6 are not mutually exclusive, E6-associated PDZ proteins can also be targeted for ubiquitination by UBE3A.

## Contributions of human papillomavirus oncoproteins to induction of genomic instability

Cervical cancers generally develop years or decades after the initial infection, and these tumors have suffered a multitude of genomic aberrations. The ability of the high-risk HPV E6 and E7 oncoproteins to activate telomerase activity and to inhibit the TP53 and RB1 tumor suppressors is sufficient to lead to extended uncontrolled proliferation and cellular immortalization, but acquisition of additional host genome mutations is necessary for malignant progression (Fig. 23.3). A defining biological activity of high-risk HPV E6/E7 proteins is their ability to subvert genomic integrity. Hence, high-risk HPV E6/E7 oncoproteins not only contribute to initiation but also actively promote malignant progression.

HPV16 E7 expression in primary human epithelial cells causes several types of mitotic abnormalities. These include supernumerary centrosomes, lagging chromosomes and anaphase bridges. Because multipolar mitoses are a histopathological hallmark of high-risk HPV-associated cervical lesions, induction of supernumerary centrosomes by E7 has been studied in greatest detail. Induction of supernumerary centrosomes by HPV16 E7 is strictly dependent on CDK2 activity. HPV16 enhances CDK2 activity through increased expression of the CDK2 catalytic subunits, cyclins E and A, secondary to E2F activation, and by inactivation of CDK2 inhibitors, CDKN1A and CDKNA2. In addition, HPV16 E7 deregulates expression of polo-like kinase 4, a critical regulator of centriole synthesis. As a consequence, E7 expression triggers synthesis of multiple daughter centrioles from a single maternal centriole template during S-phase. Expression of HPV16 in primary cells effectively induces centrosome overduplication, whereas E6 co-expression is necessary for induction of multipolar mitoses. HPV16 E7 expression also causes a higher incidence of DNA double-strand breaks. These can trigger breakage fusion bridge cycles, which result in chromosomal translocations.

The lagging chromosomes that have been observed in HPV16 E7-expressing cells arise as a consequence of a prometaphase delay. This defect may be caused by multiple mechanisms. E7 interacts with the nuclear mitotic apparatus protein 1 (NUMA1), which causes dynein delocalization from mitotic spindles. Moreover, E7 interferes with the degradation of the EMI1 inhibitor of the anaphase-promoting factor (APC), thereby dysregulating APC activity.

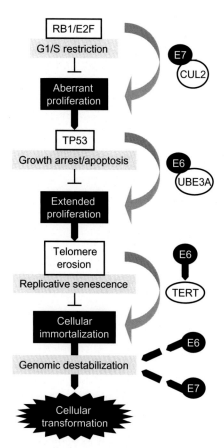

**FIGURE 23.3** Schematic depiction of some of the major biochemical and biological activities of high-risk HPV E6/E7 proteins and how they may cooperate in the development of cervical disease and cancer.

The ability of the HPV16 E6 protein to contribute to genomic destabilization is based on TP53 inactivation. As a consequence, the DNA damage−induced G1/S checkpoint is not functional and HPV16 E6 expressing cells also exhibit mitotic checkpoint defects. HPV16 E6-mediated TP53 degradation causes subversion of a postmitotic checkpoint that is triggered when cells reenter a G1-like state after a failed mitosis. Such cells contain a tetraploid complement of chromosomes, rather than the normal diploid chromosome number, and contain two centrosomes in G1 rather than one. TP53-defective cells will disregard this checkpoint, reenter S-phase, reduplicate centrosomes, and may eventually undergo tetrapolar mitosis, which can lead to aneuploid progeny. Consistent with this model, HPV16 E6-expressing cells show evidence of genomic instability.

## Uterine corpus

The uterine corpus represents the second common site for malignancy of the female reproductive track. These neoplasms can be divided into epithelial, mesenchymal, and trophoblastic tumors.

### Epithelial cancers

Uterine cancer is the most common gynecologic cancer in the United States. In 2015, the ACS estimated 54,870 new cases of uterine cancer, as compared with 21,290 new cases of ovarian cancer and 12,900 new cases of cervix cancer. Most uterine cancers are adenocarcinomas and develop from the endometrium, the inner lining of the uterus. Therefore, they are referred to as endometrial cancers. Risk factors for the development of endometrial cancer include obesity, unopposed estrogen use, polycystic ovary syndrome (PCOS), insulin resistance and diabetes, and estrogen-secreting ovarian tumors. Given the spectrum of risk factors, the development of endometrial cancer is strongly associated with an excess of systemic estrogen and a lack of progesterone. This hyperestrogenic state is presumed to result in endometrial hyperproliferation leading to endometrial cancer. Over the last 15 years there has been increased research into understanding the molecular pathogenesis of endometrial cancer.

Endometrial cancer can be broadly divided into type I and type II categories, based on risk factors, natural history, and molecular features. Women with type I endometrial cancers exhibit classic risk factors, have tumors with low-grade endometrioid histology, frequently display concurrent complex atypical hyperplasia (CAH), and typically present with early-stage disease. Molecular features important in type I endometrial cancers include presence of estrogen (ER) and progesterone (PR) receptors, microsatellite instability (MSI), *PTEN* mutations, *KRAS* mutations, and *CTNNB1* (encoding β-catenin) mutations. Women with type II endometrial cancers are typically older, non-obese, have tumors with serous and clear cell histologies, as well as high-grade endometrioid histology, and may present with more advanced stage disease. In general, type II tumors have a poorer clinical prognosis, in part due to their propensity to metastasize even with minimal myometrial invasion. The molecular alterations commonly seen in type II cancers include *p53* mutations and chromosomal aneuploidy. Although not all tumors fall neatly into these categories, the classification is helpful to broadly define endometrial cancers. A number of investigators are working toward developing more specific and rational targeted therapies. Most recently a large-scale next-generation sequencing effort led by the *National Cancer Institute* (The Cancer Genome Atlas (TCGA)) classified endometrial cancers into four categories: (1) *POLE* ultra-mutated, (2) MSI hypermutated, (3) copy number low, and (4) copy number high. This section will focus on defining the molecular changes that have been described for endometrial hyperplasia/neoplasia, as well as for the different types of endometrial cancers, with an emphasis on clinical relevance.

**Endometrial hyperplasia.** Endometrial hyperplasia is a proliferation of the endometrial glands without evidence of frank invasion. According to WHO, there are four categories of endometrial hyperplasia: (1) simple hyperplasia, (2) simple hyperplasia with atypia, (3) complex hyperplasia, and (4) complex hyperplasia with atypia. Complex hyperplasia with atypia, or complex atypical hyperplasia, is considered a precursor lesion to endometrial cancer. Simple hyperplasia is characterized by proliferating glands of irregular size, separated by abundant stroma. Cytologically, there is no atypia. Simple hyperplasia with atypia is characterized by mild architectural changes in the proliferating glands and cytologic atypia but is rarely found in clinical practice. Complex hyperplasia is characterized by more densely crowded and irregular glands; however, intervening stroma is present. Cytologically, there is no atypia. CAH is characterized by densely crowded and irregular glands with intervening stroma, as well as cytologic atypia based on enlarged nuclei, irregular nuclear membranes, and loss of cellular polarity. From a clinical standpoint, only CAH is associated with a substantial risk of developing endometrial cancer. One study estimated the risk of progression to endometrial cancer to be 1% for simple hyperplasia, 3% for complex hyperplasia, and 28.6% for CAH. Another study estimated that complex hyperplasia without atypia had a 2% risk of progression and CAH had a 51.8% risk of progression. In a nested case−control study of 138 patients with endometrial hyperplasia, simple hyperplasia and complex hyperplasia without atypia was found to have minimal (relative risk of 2 and 2.8, respectively) risk of progression to cancer. However, CAH had a substantial (relative risk of 14) risk of progression. Recently, a *Gynecologic Oncology Group* study (GOG 167) demonstrated that in 289 cases of community-diagnosed CAH, 123 (43%) cases had a concurrent grade I endometrial cancer. Of the 123 cases of endometrial cancer, 38 (31%) had some degree of invasion into the myometrium. In addition, there was substantial discrepancy between the community-diagnosed CAH and subsequent pathologic review by gynecologic pathologists, with both underdiagnosis and overdiagnosis. Finally, even when a panel of expert gynecologic pathologists reviewed cases of CAH, there was only modest reproducibility with a kappa value of 0.28. The current clinical management of CAH is simple hysterectomy and bilateral salpingoooophorectomy. In young women who desire future fertility, progestins including megestrol acetate have been used to successfully reverse CAH. An initial dilatation and curettage with careful pathologic evaluation is necessary before initiating treatment. In addition, an MRI to rule out an invasive process can also be helpful. Close surveillance with endometrial sampling every 3 months is recommended.

Risk factors for CAH and type I endometrial cancers are the same, including obesity, diabetes and insulin resistance, unopposed estrogen use, and PCOS. Certain, but not all, molecular features are shared. *PTEN* mutations, *KRAS* mutations, and MSI have all been identified in complex endometrial hyperplasia and likely represent early molecular alterations in the pathogenesis of endometrioid endometrial cancer. A novel estrogen-regulated gene (*EIG121*) shows a similar increase in expression in CAH and grade I endometrial cancers. *IGFI-R* is also increased and activated in both CAH and grade I endometrial cancers.

**Endometrial intraepithelial neoplasia (EIN).** EIN is a monoclonal premalignant endometrial glandular lesion that precedes the development of endometrioid-type endometrial adenocarcinoma. EIN is associated with a 45-fold increased risk of developing endometrial adenocarcinoma. There is substantial overlap between EIN and CAH, such as frequent *PTEN* and

*KRAS* mutations, and many lesions could be diagnosed as both EIN and CAH. However, applying the stricter diagnostic criteria of EIN (gland crowding with >50% gland to stroma ratio, cytologic alteration compared with background normal endometrial glands, and exclusion of mimics) results in a more clinically useful model of precancerous neoplasia with an ~15% risk of concurrent endometrioid endometrial carcinoma and 40% risk of progression to cancer within 18 months.

**Endometrioid endometrial cancer.** Endometrioid endometrial cancer accounts for ~70%−80% of newly diagnosed cases of endometrial cancer, and these cancers are considered type I. Risk factors for the development of endometrioid endometrial cancer are associated with an excess of estrogen and a lack of progesterone. The most common genetic changes in endometrioid endometrial cancers include mutations in *PTEN*, MSI, mutations in *KRAS* and *CTNNB1*. Recently, mutations were found in the *POLE* gene resulting in an ultra-mutation phenotype in about 10% of endometrial cancers. A new mechanism of epigenetic alterations has recently been described involving mutations in two of the SWI/SNF chromatin remodeling complex genes, *ARID1A* and *ARID5B*. Somatic mutations in the *PTEN* tumor suppressor gene are the most common genetic defect in endometrioid endometrial cancers and occur in ~40%−50% of cancer case. *PTEN* is a tumor suppressor gene located on chromosome 10. It encodes a lipid phosphatase that acts to negatively regulate AKT. The importance of *PTEN* inactivation in endometrial carcinogenesis has been demonstrated in a *PTEN* heterozygous mouse model. In this model, 100% of the heterozygote mice will develop endometrial hyperplasia by 26 weeks of age, and ~20% will develop endometrial carcinoma. Further, in this mouse model, phosphorylation of AKT followed by activation of ERα was demonstrated in the mouse endometrium. Decrease of endometrial ERα levels and activity reduced the development of endometrial hyperplasia and cancer. In humans, germ-line *PTEN* mutations are the underlying genetic defect in individuals with Cowden syndrome, which is a hereditary syndrome characterized by skin and gastrointestinal hamartomas, and increased risk of breast and thyroid cancers. Women with Cowden syndrome are at increased risk for endometrial cancer.

In addition to *PTEN* mutations, mutation and inactivation of other components of the PIK3CA/AKT/mTOR pathway have been described. In one study, m utations in the oncogene *PIK3CA* (encoding phosphatidylinositol-3-kinase) have been found in 39% of endometrial carcinomas, but only 7% of CAH cases. In another study, a 36% rate of mutations in *PIK3CA* and found that there was a high frequency of tumors with both *PIK3CA* and *PTEN* mutations. Our group described loss of TSC2 and LKB1 expression in 13% and 21% of endometrial cancers, with the subsequent activation of mTOR. A heterozygous *LKB1* mouse model has been described recently, which develops highly invasive endometrial adenocarcinomas. In humans, germ-line *LKB1* mutations are responsible for Peutz−Jeghers syndrome. However, women with Peutz−Jeghers syndrome are not at increased risk for endometrial cancer. Clearly, somatic abnormalities in the components of the PTEN/AKT/TSC2/MTOR pathway have been identified in a substantial number of endometrial cancers. Currently, clinical trials are either underway or recently completed examining mTOR inhibitors, including CCI-779 and RAD-001. Further investigation of whether specific alterations in the pathway correlates to response to therapy will be necessary.

Approximately 30% of sporadic endometrioid endometrial cancers demonstrate MSI. MSI identifies tumors that are prone to DNA replication repair errors. Microsatellites are well-defined short segments of repetitive DNA (example: CACACA) scattered throughout the genome. Tumors that demonstrate gain or loss of these repeat elements at specific microsatellite loci, when compared with normal tissue, are considered to have MSI. MSI occurs in ~30%−40% of all endometrioid endometrial cancers, but rarely in type II endometrial cancers. In addition, MSI also occurs in ~20% of all colon cancers.

The mechanism of MSI is due to either a somatic hypermethylation or silencing of the *MLH1* promoter or an inherited defect in one of the mismatch repair genes (*MSH1*, *MSH2*, *MSH6*, *PMS2*). An inherited defect in one of the mismatch repair genes is the cause of Lynch syndrome, a hereditary cancer predisposition syndrome. Individuals with Lynch syndrome are at increased risk for colon and endometrial cancer, as well as cancers of the stomach, ovary, small bowel, and ureters. Women with Lynch syndrome have a 40%−60% lifetime risk of endometrial cancer, which equals or exceeds their risk of colon cancer. Although the risk of endometrial cancer is high in these women, overall Lynch syndrome only accounts for ~2%−3% of all endometrial cancers. Endometrial cancers that develop in women with Lynch syndrome almost uniformly demonstrate MSI.

Somatic hypermethylation of *MLH1* and MSI occurs in ~30% of all endometrioid endometrial cancers and is an early event in disease pathogenesis. MSI has been identified in CAH lesions. It is presumed that MSI specifically targets tumor suppressor genes, resulting in the development of cancer. Reported target genes include *FAS*, *BAX*, *IGF*, *IGF2R* (encoding insulin-like growth factor 2 receptor), and TGFβRII (encoding transforming growth factor beta receptor type 2). Studies focusing on the clinical significance of MSI have been mixed. Although some investigators have found an association between MSI and tumors with a more aggressive clinical course, a more recent study examining only endometrioid endometrial cancers found no difference in clinical outcome between tumors demonstrating MSI and those without.

The MSI assay, as well as immunohistochemistry for MSH1, MSH2, MSH6, and PMS2, can be very useful as a screening tool to identify women with endometrial cancer as having Lynch syndrome. Although collecting and interpreting a family history is helpful, these molecular tools can be useful in targeting certain populations that may be at higher risk for Lynch syndrome, such as women under the age of 50 years.

*KRAS* mutations have been identified in 20%−30% of endometrioid endometrial cancers. There is a higher frequency of *KRAS* mutations in cancers that demonstrate MSI. Mutations in *β-catenin* have been seen in ∼20%−30% of endometrioid endometrial cancers. These mutations occur in exon 3 and result in stabilization of the β-catenin protein as well as nuclear accumulation. Nuclear β-catenin plays an important role in transcriptional activation. One study demonstrated that *PTEN*, MSI, and *KRAS* mutations frequently coexist. However, *β-catenin* mutations do not typically occur in conjunction with these other abnormalities. *β-Catenin* mutations have been identified in CAH, suggesting that this is an early step in the pathogenesis of endometrial cancer.

**Type II endometrial cancers.** Type II endometrial cancers have a more aggressive clinical course and include poorly differentiated endometrioid tumors as well as papillary serous and clear cell endometrial cancers. Patients with stage I papillary serous endometrial cancer have a 5-year overall survival of 74%, significantly lower than the 90% 5-year overall survival for women with endometrioid endometrial cancer. The average age of diagnosis of patients with papillary serous endometrial cancers is 68 years, and risk factors typically associated with type I endometrial cancers are not present. In addition, the genetic alterations seen in type I endometrial cancers are not frequently found in papillary serous endometrial cancers. Microarray studies examining type I versus type II cancers have identified distinct gene signatures associated with type I and type II cancers. *P53*, *PIK3CA*, and *PPP2R1A* mutations are particularly common in uterine papillary serous carcinomas.

**Molecular classification of endometrial cancers.** A large-scale sequencing effort by TCGA classified endometrial cancers into four categories: (1) *POLE* ultra-mutated, (2) MSI hypermutated, (3) copy number low, and (4) copy number high. Besides mutation analysis, somatic copy number alterations (SCNAs) were assessed and carcinomas were found to cluster in four distinct groups based on their SCNAs. Cluster 1 tumors were nearly devoid of broad SCNAs, averaging less than 0.5% genome alteration. Copy number clusters 2 and 3 consisted mainly of endometrioid tumors (all grades), distinguished by more frequent 1q amplification in cluster 3 than in cluster 2. Cluster 4 is characterized by a very high degree of SCNAs and consisted mostly of serous carcinomas. SCNA clusters correlated well with progression-free survival.

Taking both the mutation panel and the copy number alterations into account, investigators created an algorithm to classify endometrial carcinomas into four distinct categories. If POLE mutation was present (10%) and the tumor showed an ultramutated phenotype, the tumor was categorized as POLE ultramutated. These cases showed a nearly 100% progression-free survival. If POLE was not mutated, but MSI genes were affected (28%), the tumor was categorized into MSI hypermutated. These tumors showed 75%−80% progression-free survival. If mutations in POLE or MSI genes were not present, tumors were categorized into copy number low (38%) or high (25%) groups based on their copy number alteration profiles. These last two groups showed progression-free survival of ∼75% and 50%, respectively.

**Mesenchymal tumors.** Endometrial mesenchymal tumors are derived from the mesenchyme of the corpus and are composed of cells resembling those of the proliferative phase endometrial stroma. Among them, uterine fibroids are the most common mesenchymal tumors of the female reproductive tract. Uterine fibroids represent a group of benign, smooth muscle tumors of the uterus. Recent studies have shown that the lifetime risk of fibroids in a woman above the age of 45 years is more than 60%, with higher incidence in African-Americans than in Caucasians. The natural history of uterine fibroids remains largely unknown, and the molecular basis of these diseases remains to be determined. Recent molecular and cytogenetic studies have revealed genetic heterogeneity in various histological types of uterine fibroids. Chromosome 7q22 deletions are common in uterine leiomyoma, the most common type of uterine fibroid. Several candidate genes, including *ORC5L* and *LHFPL3*, have been identified, but their roles in the pathogenesis of the disease have not been elucidated. Loss of a portion of chromosome 1 is common in the cellular form of uterine leiomyomata.

Endometrial stromal sarcoma is the most common malignant mesenchymal tumor of the uterus. They often arise from endometriosis. Genetic studies suggest that abnormalities of chromosomes 1, 7, and 11 may play a role in uterine sarcoma initiation or progression. Fusion of two zinc fingers (*JAZF1* and *JJAZ1*) by translocation t(7;17) has been described.

## Ovary and fallopian tube

Multiple benign and malignant diseases have been identified in the ovary and the fallopian tube. The most common ones are benign and malignant tumors of the ovary. Although molecular studies in all these diseases have identified changes in multiple genes, the etiology of most of these diseases remains largely unknown.

## Borderline tumors

Borderline tumors account for 15%–20% of epithelial ovarian tumors and were first recognized by Howard Taylor in 1929. He described a group of women of reproductive age with large ovarian tumors whose course was rather indolent. In the early 1970s, the *Federation of Gynecologists and Obstetricians* defined these semimalignant tumors as borderline ovarian tumors (BOTs). Later on, at the 2003 WHO workshop, the term low malignant potential became an accepted synonym for BOTs.

Borderline tumors with different cell types (serous, mucinous, endometrioid, clear cell, transitional, and mixed epithelial cells) have been reported. However, the serous and mucinous are the most common types. Mutational analyzes have identified several gene mutations in borderline tumors (Table 23.1). Both *BRAF* and *KRAS* mutations are very common in borderline serous tumor. However, *BRAF* mutation is rare in borderline mucinous tumors. Moreover, the

**TABLE 23.1** Common somatic mutations in human sporadic ovarian tumors.

| | Gene name | Number of samples screened | Number of positive samples | Percent mutated (%) |
|---|---|---|---|---|
| Ovarian adenoma | | | | |
| Serous | BRAF | 142 | 14 | 10 |
| | KRAS | 179 | 11 | 6 |
| | TP53 | 87 | 1 | 1 |
| Mucinous | CDKN2A | 20 | 7 | 35 |
| | KRAS | 146 | 50 | 34 |
| | TP53 | 20 | 2 | 10 |
| Borderline tumors | | | | |
| Serous | BRAF | 557 | 206 | 37 |
| | KRAS | 370 | 77 | 21 |
| | TP53 | 41 | 4 | 10 |
| | ERBB2 | 105 | 7 | 7 |
| | PIK3CA | 56 | 2 | 4 |
| Mucinous | CDKN2A | 32 | 18 | 56 |
| | KRAS | 233 | 100 | 43 |
| | TP53 | 26 | 2 | 8 |
| | BRAF | 163 | 5 | 3 |
| Endometrioid | CTNNB1 | 10 | 8 | 80 |
| | PTEN | 8 | 1 | 12 |
| | KRAS | 11 | 1 | 9 |
| Carcinomas | | | | |
| Serous | TP53 | 1253 | 828 | 66 |
| | KRAS | 1866 | 95 | 5 |
| | BRCA1 | 1056 | 43 | 4 |
| | BRCA2 | 844 | 22 | 3 |
| | CDKN2A | 895 | 18 | 2 |
| | PIK3CA | 1176 | 26 | 2 |
| | BRAF | 1668 | 23 | 1 |
| | ARID1A | 806 | 6 | 1 |

*(Continued)*

**TABLE 23.1** (Continued)

|  | Gene name | Number of samples screened | Number of positive samples | Percent mutated (%) |
|---|---|---|---|---|
| Mucinous | TP53 | 49 | 27 | 55 |
|  | KRAS | 245 | 100 | 41 |
|  | CDKN2A | 58 | 9 | 16 |
|  | BRAF | 148 | 15 | 10 |
|  | PTEN | 36 | 3 | 8 |
|  | PIK3CA | 73 | 6 | 8 |
| Clear cell | ARID1A | 175 | 86 | 49 |
|  | PIK3CA | 457 | 155 | 34 |
|  | TERT | 320 | 51 | 16 |
|  | TP53 | 86 | 11 | 13 |
|  | BRCA2 | 16 | 2 | 13 |
|  | CDKN2A | 77 | 8 | 10 |
|  | KRAS | 460 | 31 | 7 |
|  | BRAF | 353 | 4 | 1 |
| Endometrioid | TP53 | 134 | 86 | 64 |
|  | ARID1A | 39 | 12 | 31 |
|  | CTNNB1 | 295 | 78 | 26 |
|  | PIK3CA | 238 | 49 | 21 |
|  | PTEN | 170 | 29 | 17 |
|  | KRAS | 395 | 43 | 11 |
|  | BRAF | 263 | 9 | 3 |

frequency of *KRAS* mutation is higher in borderline mucinous tumor than serous tumors. On the other hand, *CTNNB1* and *PTEN* mutations have been found in borderline endometrioid tumors. The difference in mutation spectrum indicates different pathogenic pathways for these histological subtypes.

The progression of serous BOT (SBOT) to low-grade serous carcinoma is supported by clinical, pathological, and molecular evidence; however, low-grade serous carcinoma may also develop de novo. Studying genetic changes in different types of ovarian tumors provides insight into the molecular pathogenesis for ovarian cancer. Mutations in *KRAS* have been found in 63% of mucinous BOTs and 75% of invasive mucinous ovarian cancers. These data suggest that *KRAS* mutations are involved in the development of mucinous BOTs and support the notion that mucinous BOTs may represent a phase of development along the pathologic continuum between benign and malignant mucinous tumors. A mucinous cystadenoma would give rise to a mucinous BOT, a subset of which may progress, giving rise to invasive low-grade or high-grade mucinous carcinomas. On the other hand, a serous cystadenoma would give rise to a serous BOT, which may progress, giving rise to invasive low-grade serous carcinoma but rarely to high-grade serous carcinoma.

## Malignant tumors

Ovarian cancer is a general term that represents a diversity of cancers that are believed to originate in the ovary. Over 20 microscopically distinct types can be identified, which can be classified into three major groups: (1) epithelial cancers, (2) germ cell tumors, and (3) specialized stromal cell cancers. These three groups correspond to the three distinct cell types of different functions in the normal ovary: (1) the epithelial covering the ovary or the fallopian tube fimbriae may give rise to the epithelial ovarian cancers (EOCs), (2) the germ cells may give rise to the germ cell tumors, and (3) the steroid-producing cells may give rise to the specialized stromal cell cancers.

**Epithelial ovarian tumors**. The majority of malignant ovarian tumors in adult women represent EOC. Based on the histology of the neoplastic cells, EOCs are classified into different catagories—serous, mucinous, endometrioid, clear cell, transitional, squamous, mixed, and undifferentiated. From the Sanger Center's Catalogues of Somatic Mutations in Cancer (COSMIC) database, the most frequently identified mutations for each histological subtype are listed in Table 23.1. In addition to these mutations, hundreds of rare mutations have also been identified (http://www.sanger.ac. uk/genetics/CGP/cosmic/). Each histological subtype has a unique mutation profile.

Genetic analysis, mouse models, and other high-throughput methods have been exploited to understand the pathogenic pathways in ovarian cancer. The cell origin of EOC has been controversial. For many years, it is hypothesized that most ovarian tumors develop from ovarian inclusion cysts arising from the ovarian surface epithelial (OSE) cell. Thus, the majority of high-grade serous carcinomas derive de novo from ovarian cysts or ovarian endosalpingiosis. However, recent evidence suggests that EOC of the serous subtype may also derive from the fallopian tube epithelium. Extensive examination of prophylactic ovaries and fallopian tubes from women with *BRCA1/2* mutations has identified serous tubal intraepithelial carcinoma (STIC) in high-risk patients with *BRCA1/2* mutations. STIC strongly expresses p53 protein, and *TP53* mutations have been identified in STIC, termed p53 signature. Because almost all high-grade serous ovarian tumors have *TP53* mutations, STIC may be the precursor of high-grade serous cancer. Alternatively, based on similar high frequency of *ARID1A* mutations in both endometrioid and clear cell cancers, these tumors may arise from endometriosis and are distinct entities from high-grade serous tumor.

Loss-of-heterozygosity (LOH) studies have been widely used to identify minimally deleted regions where tumor suppressor genes may reside. BOTs showed significantly lower LOH rates (0%–18%) than invasive tumors at all loci screened, suggesting that the LOH rate at autosomes is less important in the development of BOTs than more advanced tumors. Based on these and other results, the importance of LOH at the AR locus in BOTs and invasive cancers remains undetermined. However, several studies have found differences in LOH rates on other chromosomes. LOH at the p73 locus on 1p36 was found in both high-grade and low-grade ovarian and surface serous carcinomas, but not in borderline ovarian cancers. In one study, LOH rates at 3p25, 6q25.1–26, and 7q31.3 were significantly higher in high-grade serous carcinomas than in low-grade serous carcinomas, mucinous carcinomas, and borderline cancers. In another study, LOH rate at a 9-cM region on 6q23–24 were significantly higher in papillary serous carcinomas of the peritoneum than in serous epithelial ovarian carcinomas. In addition, multiple minimally deleted regions have been identified on chromosomes 11 and 17. LOH rates at a 4-cM region on chromosome 11p15.1 and an 11-cM region on chromosome 11p15.5 were found only in serous invasive cancers, and the LOH rates at 11p15.1 and 11p15.5 were significantly higher in high-grade than low-grade serous tumors. Similarly, significantly higher LOH rates were identified at the *TP53* locus on 17p13.1 and the *NF1* locus on 17q11.1 in high-grade serous carcinomas than in low-grade and borderline serous cancers and all mucinous neoplasms. LOH at the region between *THRA1* and D17S1327, including the *BRCA1* locus on 17q21, was found exclusively in high-grade serous tumors. In general, the fact that LOH rates at multiple chromosomal sites were significantly higher in serous than in mucinous cancer subtypes suggests that these neoplasms have different pathogenic pathways. Because tumor suppressor genes are typically located in chromosomal regions displaying LOH, further analysis of genes located in the minimal deleted regions may provide insights into genes that may be important for the pathogenesis of serous and mucinous ovarian cancers. Studying genetic changes in endometriosis, endometrioid, and clear cell ovarian cancer also provides insights into pathogenic pathways in the development of these tumor types. Endometriosis is highly associated with endometrioid and clear cell carcinomas (28% and 49%, respectively), in contrast to its very low frequency of association with serous and mucinous carcinomas of the ovary (3% and 4%, respectively). This fact furnishes strong evidence that endometriosis is a precancerous lesion for both endometrioid and clear cell carcinomas. One study showed that endometrioid but not serous or mucinous epithelial ovarian tumors had frequent *PTEN/MMAC* mutations. In addition, the loss of PTEN immunoreactivity has been reported in a significantly higher percentage of clear cell and endometrioid ovarian cancers than cancers of other histologic types. The expression of oncogenic *KRAS* or conditional *PTEN* deletion within the OSE induces preneoplastic ovarian lesions with an endometrioid glandular morphology. Furthermore, the combination of these mutations in the ovary led to the induction of invasive and widely metastatic endometrioid ovarian adenocarcinomas. These data further suggest that tumors with different histologic subtypes may arise through distinct developmental pathways.

**Germ cell cancer**. The germ cell cancers commonly occur in young women and can be very aggressive. Fortunately, chemotherapy is usually effective in preventing these neoplasms from recurring. The pathogenesis of these cancers is not well understood. Several subtypes exist, which include dysgerminoma (DG), yolk sac tumor, embryonal carcinoma, polyembryoma, choriocarcinoma, immature teratoma, and mixed germ cell tumors. A recent study identified *c-KIT* mutation and the presence of Y-chromosome material in DG. Furthermore, DNA copy number changes were detected by comparative genomic hybridization in DGs. The most common changes in DGs were gains from chromosome arms 1p (33%), 6p (33%), 12p (67%), 12q (75%), 15q (42%), 20q (50%), 21q (67%), and 22q (58%). DGs are also characterized by frequent *KIT* mutations (26%). However, overexpression or mutation of *p53* was not observed in ovarian germ cell tumors.

**Stromal cancer**. The specialized stromal cell cancers (granulose cell tumors, theca cell tumors, and Sertoli–Leydig cell tumors) are uncommon. One interesting characteristic of these neoplasms is that they can produce hormones. Granulosa and theca cell tumors are frequently mixed and can produce estrogen, which will result in premature sexual development and short stature if the tumors develop in young girls. *FOXL2* mutation is found in almost all granulosa cell tumors of the ovary. *FOXL2* encodes a transcription factor involved in the breakdown of fats, steroid hormones, and removal of reactive oxygen species in the ovaries. Sertoli–Leydig cell tumors produce male hormones, which will cause defemininization. Subsequently, male pattern baldness, deep voice, excessive hair growth, and enlargement of the clitoris will develop in patients with Sertoli–Leydig cell tumors. Recurrent somatic *DICER1* mutations have been identified in 60% (26/43) of Sertoli–Leydig cell tumors. DICER1 is an endoribonuclease in the RNase III family for processing microRNAs. Overexpression of the *BCL2* gene in Sertoli–Leydig cell tumor of the ovary has been reported. Fortunately, these specialized stromal cell cancers are usually not aggressive cancers and involve only one ovary.

**Cancer of the fallopian tube**. The fallopian tubes are the passageways that connect the ovaries and the uterus. Fallopian tube cancer is very rare. It accounts for less than 1% of all cancers of the female reproductive organs. Only 1500–2000 cases have been reported worldwide, primarily in postmenopausal women. There is some evidence that women who inherit a mutation in the *BRCA1* gene (a gene already linked to breast and ovarian cancer) seem to have an increased risk of developing fallopian tube cancer. In one recent analysis of several hundred women who were carriers of the *BRCA1* gene mutation, the incidence of fallopian tube cancer was increased more than 100-fold. Likewise, a substantial proportion of women with the diagnosis of fallopian tube cancer test positively for either the *BRCA1* or *BRCA2* gene mutation. Recent recommendations suggest that any woman with a diagnosis of fallopian tube cancer be tested for the BRCA mutations. Based on the pathological study of fallopian tubes from *BRCA1* carriers, a majority of serous ovarian tumors developed from the secretory fimbrial cells of the tube.

## Vagina and vulva

Vaginal intraepithelial neoplasia and squamous cell carcinoma are the most common forms of vaginal cancer. Persistent infection with high-risk HPV (e.g., types 16 and 18) is a major etiological factor for both diseases. Overexpression of p53 and Ki67 has been identified in 19% and 75% of vaginal cancers, respectively.

Similar to vaginal malignancies, squamous cell carcinoma is the most common form of neoplasia in the vulva. Precursor lesions include vulvar intraepithelial neoplasia (VIN), the simplex (differentiated) type of VIN, lichen sclerosis, and chromic granulomatous. Although vaginal carcinomas and VIN are associated with HPV (usually of types 16 and 11), the simplex form of VIN and other precursor lesions are not HPV-associated. Genetic studies identified both *TP53* and *PTEN* mutations in VIN, as well as in carcinomas, suggesting that these molecular changes are early events in the pathogenesis of vulvar cancers. Chromosome losses affecting 3p, 4p, 5q, 8p, 22q, Xp, and 10q and chromosome gains affecting 3q, 8p, and 11q have been reported. In addition, high frequencies of allelic imbalances have been found in both HPV-positive and HPV-negative lesions at multiple chromosomal loci: 1q, 2q, 3p, 5q, 8p, 8q, 10p, 10q, 11p, 11q, 15q, 17p, 18q, 21q, and 22q.

## Disorders related to pregnancy

Conception, maintenance of the pregnancy, and delivering the baby are the most important functions of the female reproductive tract. Proper implantation of the embryo is a crucial step in healthy pregnancy. Failure of this process can compromise the life of the fetus and the mother. During implantation, trophoblast cells basically invade into the endometrium in a controlled fashion and transform the spiral arteries to supply the growing fetus with oxygen and nutrition. Failure of this process results in fetal hypoxia and growth retardation, which might lead to complications such as abortion, preeclampsia, and preterm delivery. On the other hand, uncontrolled invasion can lead to deep implantation (placenta acreta, increta, percreta), or in gestational trophoblastic diseases (GTDs), it can result in persistent disease, metastases, and development of neoplasia.

Trophoblast invasion is regulated by at least three factors: (1) the endometrial environment (including adhesion molecules and vascularization at the implantation site), (2) the maternal immune system, and (3) the invasive and proliferative potentials of the trophoblast cells. Complications following normal conception are mainly due to the failure of the maternal endometrium or immune system, whereas following genetically abnormal conceptions, complications are most often the result of the abnormal trophoblastic cell function. Preeclampsia and GTDs can lead to maternal death. This fact has put these diseases into the focus of clinical and molecular researchers. The molecules described here may also play roles in other implantation-related diseases, such as unsuccessful in vitro fertilization cycles, recurrent spontaneous abortions, preterm deliveries, and others.

In GTDs, the problem is the overly aggressive behavior of trophoblastic cells. Trophoblast cells in GTD are hyperplastic and display more aggressive behavior than in normal pregnancy. Trophoblast cells might invade into the myometrium and persist after evacuation, which without adequate chemotherapy can compromise the life of the mother. Trophoblast hyperplasia in GTD also can result in true gestational trophoblastic neoplasia (GTN).

In preeclampsia, hypertension and vascular dysregulation are indirect results of inadequate placentation due to different combinations of factors. To date, the primary reason why trophoblast cells fail to invade and transform the spiral arteries is still unknown, but mechanisms and factors leading to the classic symptoms after fetal stress have been identified, and clinical trials have initiated to control these factors and prevent preeclampsia.

## Gestational trophoblastic diseases

GTD is a broad term covering all pregnancy-related disorders in which the trophoblast cells demonstrate abnormal differentiation and hyperproliferation. Molar pregnancy is the most common GTD, characterized by focal (partial mole) or extensive (complete mole) trophoblast cell proliferation on the surface of the chorionic villi. GTN, such as gestational choriocarcinoma, can arise from molar pregnancies (5%−20%) and on very rare occasions from normal pregnancies (0.01%).

Partial molar pregnancy is a result of the union of an ovum and two sperm. Therefore, the genetic content is imbalanced with overrepresentation of the paternal genetic material. Because the maternal chromosomes are present, the embryo develops to a certain point. The complete mole is the result of a union between an empty ovum and generally two sperm. Rarely, conception is accomplished with one sperm and the genetic content of the sperm doubles in the ovum. The maternal chromosomes are not present; therefore no embryo develops. One study described familial recurrent hydatidiform molar pregnancies where balanced biparental genomic contribution could be found. In these cases, gene mutation is suspected on the long arm of chromosome 19. The *NALRP7* gene has been proposed to be the causative gene in these rare familial molar pregnancies.

### Epigenetic changes in trophoblastic diseases

Because of unbalanced conception, gene expression in molar pregnancy greatly differs from normal pregnancy. Gene expression differences can partly be explained by the lack or excess of the chromosome sets, but more importantly epigenetic parental imprinting has substantial role in the modification of the expression profile. Maternally imprinted genes are normally expressed from one set of paternal chromosomes. In molar pregnancy, these genes are expressed from two or more sets of paternal chromosomes. Furthermore, paternally imprinted genes are normally expressed only from the maternal allele, but in complete moles these genes are not expressed at all due to the lack of maternal chromosomes in complete GTDs. Complete molar pregnancy, as a unique uniparental tissue, is in the focus of epigenetic studies of placental differentiation. Several imprinted genes ($P57^{kip2}$, IGF-2, HASH, HYMAI, P19, LIT-1) have been investigated in complete mole, and paternally imprinted $p57^{kip2}$ is now used as a diagnostic tool to differentiate complete mole from partial mole and hydropic pregnancy by detecting maternal gene expression. Table 23.2 lists known hydatidiform mole-associated parentally imprinted genes and their functions.

### Immunology and trophoblastic diseases

As partial and complete moles are foreign tissues in the maternal uterus, vigorous maternal immune response would be expected. However, trophoblast cells express several immunosuppressive factors to attenuate the maternal immune response and protect the semiallogeneic fetus.

Trophoblast cells, unlike other human cells, lack class I major histocompatibility complexes (MHCs). Instead, they express atypical MHCs such as human leukocyte antigens (HLA) G, E, and F. The atypical HLAs do not present antigens to the immune cells, but they are still able to inactivate the natural killer cells. Therefore, they protect the semiallogeneic fetus from immune cell attack. The expression of immunosuppressive factors shows significant differences between normal and molar trophoblast cells.

**TABLE 23.2 Imprinted genes in embryonic development associated with hydatidiform moles.**

| Gene | Transcribed from | Function |
|---|---|---|
| p57$^{kip2}$ | Maternal allele | Negatively regulates cell proliferation by binding G1 cyclin complexes [211] |
| IGF-2 | Paternal allele | Increases the expression of passive amino acid transporters on trophoblast cells, increases trophoblast invasiveness [212] |
| HYMAI | Maternal allele | Associated with transient neonatal diabetes mellitus [213] |
| H19 | Maternal allele | Necessary for muscular differentiation, associated to Wilms tumor [212,214] |
| LIT-1 | Maternal allele | LIT-1 is required for p57$^{kip2}$ expression and specific deletion results in BWS [215] |

BWS, Beckwith–Wiedemann syndrome; HYMAI, hydatidiform mole–associated imprinted gene; IGF-2, insulin-like growth factor 2; LIT-1, long QT intronic transcript.

**TABLE 23.3 Immunosuppressive molecules secreted by trophoblast cells.**

| | Increases the expression | Ligand | Immune cell or system affected | Function |
|---|---|---|---|---|
| HLA molecules (C, G, E, F) | IL-10 | KIR | NK | Inhibition |
| Soluble HLA-G | IL-10 | KIR | Tc cell | Apoptosis |
| Fas L | Th2 cytokines and hCG | Fas | Tc cell | Apoptosis |
| Cytokines (TGFβ, IL-10, IL-4, IL-6) | Unknown | Cytokine receptors | Th cells, Tc cells, NK cells Trophoblast cells | Suppression HLA-G, HLA-E, HLA-F overexpression |
| Indoleamine 2,3-dioxygenase | γ-INF | L-Tryptophan | T cell Complement system | Inhibition of T cell maturation by tryptophan depletion Complement inhibition |
| DAF, MCP, CD59 | Unknown | Multiple | Complement system | MAC inhibition |
| PI-9 | Unknown | Granzyme B | Tc cell, NK cell killing | Granzyme B inhibitor |
| hCG | Unknown | hCG receptor | Trophoblast Immune cells | Fas L overexpression hCG → progesterone → PIBF |

DAF, decay accelerating factor; Fas L, Fas ligand; HLA, human leukocyte antigen; IL, interleukin; INF, interferon; KIR, killer inhibitor receptor; MAC, membrane attack complex; MCP, membrane cofactor protein; NK, natural killer; PIBF, progesterone-induced blocking factor; Tc, cytotoxic T cells; Th, helper T cells.
Adapted From Hayes MP, Wang H, Espinal-Witter R, et al. PIK3CA and PTEN mutations in uterine endometrioid carcinoma and complex atypical hyperplasia. Clin Cancer Res 2006;12:5932–35.

Table 23.3 lists the molecules that play roles in the development of systematic and local gestational immunosuppression. In GTDs, the hyperproliferation of the trophoblast cells results in an elevated level of immunosuppressive factors in the mother. Furthermore, some of the immunosuppressive factors in GTD and GTN are not only elevated because of the higher amount of hyperplastic trophoblast cells, but also due to overexpression of these factors by the trophoblast cells. For example, soluble and membrane-bound HLA-G was shown to be overexpressed in molar pregnancies and GTN. Furthermore, molar villous fluid effectively suppresses cytotoxic T cells in vitro. Shaarawy et al. found that the soluble IL-2 receptor level is significantly higher in the serum of mothers who subsequently developed persistent GTD.

## Oncogenes and tumor suppressor gene alterations in trophoblastic diseases
As molar pregnancies and especially GTN demonstrate excessive trophoblast cell proliferation, numerous studies were undertaken to characterize the expression of tumor suppressor genes and oncogenes. One study demonstrated that both

complete mole and choriocarcinoma were characterized by overexpression of *p53*, *p21*, and *RB* tumor suppressor genes and *c-myc*, *c-erbB-2*, and *bcl-2* oncogenes, whereas partial mole and normal placenta generally did not strongly express these molecules. p53 and RB molecules were mainly found in the nuclei of cytotrophoblastic cells, whereas p21 could be seen in syncytiotrophoblast cells. *DOC-2/hDab2* tumor suppressor gene expression was significantly stronger in normal placenta and partial mole compared with complete mole and choriocarcinoma.

In a study that compared the hydatidiform mole gene expression profile to normal placenta gene expression using microarray methods, 508 differentially expressed genes were identified. Most of these genes encoded proteins involved in MAP-RAS kinase, Wnt, and Jak-STAT5 pathways. Some other genes, such as *Versican*, might have a role in cell migration or drug interactions. These findings provide important insight into the development of the GTN. However, to date it is not clear which gene(s) drive development of GTD or neoplasia.

## Diagnosis of molar pregnancies

Beyond classic morphologic analysis by microscopic examination of H&E-stained tissue specimens, molecular studies of short tandem repeat (STR) polymorphisms now offer a robust and reliable method for identifying molar pregnancies. STR polymorphisms discriminate maternal and paternal alleles if maternal blood is available for comparison. High-resolution semiquantitative capillary electrophoresis determines the relative paternal to maternal allele ratios enabling diagnosis of complete and partial hydatidiform moles.

## Detection of trophoblastic diseases

Gestational trophoblastic neoplasms are typically chemosensitive malignancies. Human choriogonadotropin (hCG) follow-up is employed to diagnose persistent disease or GTN before administration of chemotherapy. Several studies have attempted to identify a reliable molecular or histologic marker that predicts persistent disease at the time of evacuation, but to date none seems to be sufficiently sensitive and specific to alter current prognostic scoring and hCG follow-up in clinical practice. However, some studies have produced promising results toward improved diagnosis and prognostication of GTN. Cole et al. recently demonstrated that hyperglycosylated hCG (hCG-H) appears to reliably identify active trophoblastic malignancy, but hCG-H alone is unable to predict persistence at the time of primary evacuation. Serum IL-2 and soluble IL-2 receptor levels at the time of evacuation are associated with clinical outcome and persistence. Immunostaining for IL-1, EGFR, c-erbB-3, and antiapoptotic MCL-1 on trophoblast cells significantly correlates with the development of persistent postmolar GTN. Further studies are needed to find a reliable marker or marker panel to predict persistent GTDs.

## Key concepts

- High-risk HPVs contribute to the genesis of almost all human cervical carcinomas and have also been associated with a number of other anogenital malignancies including vulvar, anal and penile carcinomas.
- HPV associated cervical cancers are quite unique in that they represent the only human solid tumor for which the initiating carcinogenic agents has been identified at a molecular level. The fact that the high-risk HPV E6 and E7 oncoproteins contribute to initiation as well as progression and that they are necessary for the maintenance of the transformed phenotype of cervical cancer cells suggests that these proteins and/or the processes that they regulate should provide targets for intervention.
- Uterine cancer is the most common gynecologic cancer in the United States. Most uterine cancers are adenocarcinomas and develop from the endometrium, the inner lining of the uterus.
- Risk factors for the development of endometrial cancer include obesity, unopposed estrogen use, polycystic ovarian syndrome, insulin resistance and diabetes, and estrogen secreting ovarian tumors.
- Endometrial cancer can be broadly divided into Type I and Type II categories, based on risk factors, natural history and molecular features.
- The majority of malignant ovarian tumors in adult women are epithelial ovarian cancer, which can be classified into serous, mucinous, endometrioid, clear cell, transitional, squamous, mixed and undifferentiated. They all have different pathogenetic pathways.
- High-grade serous ovarian carcinomas can derive from both the ovarian surface epithelium and form the fallopian tube epithelium.
- *BRCA1* and *BRCA2* are the key genes involved in the development of familial ovarian cancer.
- Low-grade and high-grade serous ovarian cancer are developed through a two-tier system.

- Gestational trophoblastic disease is a broad term covering all pregnancy related disorders, in which the trophoblast cells demonstrate abnormal differentiation and hyperproliferation. Molar pregnancies are the most common gestational trophoblastic diseases, which are characterized by focal (partial mole) or extensive (complete mole) trophoblast cell proliferation on the surface of the chorionic villi.

## Suggested readings

[1]  Schiffman M, Castle PE, Jeronimo J, Rodriguez AC, Wacholder S. Human papillomavirus and cervical cancer. Lancet 2007;370:890−907.

[2]  Lee C, Laimins LA. The differentiation-dependent life cycle of human papillomaviruses in keratinocytes. In: Garcea RL, DiMaio D, editors. The Papillomaviruses. New York: Springer; 2007. p. 45−68.

[3]  Munger K, Baldwin A, Edwards KM, et al. Mechanisms of human papillomavirus-induced oncogenesis. J Virol 2004;78:11451−60.

[4]  Barr E, Tamms G. Quadrivalent human papillomavirus vaccine. Clin Infect Dis 2007;45 609-7.

[5]  Cancer Genome Atlas Research Network. Integrated genomic characterization of endometrial carcinoma. Nature 2013;497:67−73.

[6]  Brinton LA, Berman ML, Mortel R, et al. Reproductive, menstrual, and medical risk factors for endometrial cancer: results from a case-control study. Am J Obstet Gynecol 1992;167:1317−25.

[7]  Trimble CL, Kauderer J, Zaino R, et al. Concurrent endometrial carcinoma in women with a biopsy diagnosis of atypical endometrial hyperplasia: a gynecologic oncology group study. Cancer 2006;106:812−19.

[8]  Levine RL, Cargile CB, Blazes MS, et al. PTEN mutations and microsatellite instability in complex atypical hyperplasia, a precursor lesion to uterine endometrioid carcinoma. Cancer Res 1998;58:3254−8.

[9]  Risinger JI, Hayes K, Maxwell GL, et al. PTEN mutation in endometrial cancers is associated with favorable clinical and pathologic characteristics. Clin Cancer Res 1998;4:3005−10.

[10]  Lu KH, Wu W, Dave B, et al. Loss of tuberous sclerosis complex-2 function and activation of Mammalian target of rapamycin signaling in endometrial carcinoma. Clin Cancer Res 2008;14:2543−50.

[11]  MacDonald ND, Salvesen HB, Ryan A, et al. Frequency and prognostic impact of microsatellite instability in a large population-based study of endometrial carcinomas. Cancer Res 2000;60:1750−2.

[12]  Risinger JI, Maxwell GL, Chandramouli GV, et al. Microarray analysis reveals distinct gene expression profiles among different histologic types of endometrial cancer. Cancer Res 2003;63:6−11.

[13]  Bell DA, Longacre TA, Prat J, et al. Serous borderline (low malignant potential, atypical proliferative) ovarian tumors: workshop perspectives. Hum Pathol 2004;35:934−48.

[14]  Bonome T, Lee JY, Park DC, et al. Expression profiling of serous low malignant potential, low-grade, and high-grade tumors of the ovary. Cancer Res 2005;65:10602−12.

[15]  Shih Ie M, Kurman RJ. Ovarian tumorigenesis: a proposed model based on morphological and molecular genetic analysis. Am J Pathol 2004;164:1511−18.

[16]  Cancer Genome Atlas Research Network. Integrated genomic analyses of ovarian carcinoma. Nature 2011;474:609−15.

[17]  Hashiguchi Y, Tsuda H, Inoue T, Berkowitz RS, Mok SC. PTEN expression in clear cell adenocarcinoma of the ovary. Gynecol Oncol 2006;101:71−5.

[18]  Jarboe E, Folkins A, Nucci MR, et al. Serous carcinogenesis in the fallopian tube: a descriptive classification. Int J Gynecol Pathol 2008;27:1−9.

[19]  Genest DR, Dorfman DM, Castrillon DH. Ploidy and imprinting in hydatidiform moles. Complementary use of flow cytometry and immunohistochemistry of the imprinted gene product p57KIP2 to assist molar classification. J Reprod Med 2002;47:342−6.

[20]  Kato HD, Terao Y, Ogawa M, et al. Growth-associated gene expression profiles by microarray analysis of trophoblast of molar pregnancies and normal villi. Int J Gynecol Pathol 2002;21:255−60.

[21]  Feltmate CM, Batorfi J, Fulop V, et al. Human chorionic gonadotropin follow-up in patients with molar pregnancy: a time for reevaluation. Obstet Gynecol 2003;101:732−6.

Chapter 24

# Molecular basis of kidney disease

Roderick J. Tan[1], Sheldon I. Bastacky[2] and Youhua Liu[2]

[1]Renal-Electrolyte Division, Department of Medicine, University of Pittsburgh School of Medicine, Pittsburgh, PA, United States, [2]Department of Pathology, University of Pittsburgh School of Medicine, Pittsburgh, PA, United States

## Summary

The kidney is a complex organ populated by a variety of unique, specialized cells that function in a coordinated fashion to maintain homeostasis in the body. Diseases affecting the kidney are numerous, and their clinical manifestations are dependent on specific cellular, molecular, and immunological abnormalities. Current research has revealed much about the molecular pathology underlying diabetic nephropathy, membranous nephropathy, focal segmental glomerulosclerosis, IgA nephropathy, acute tubular necrosis/acute kidney injury, and interstitial nephritis, among many others. The final common pathway of all progressive kidney diseases leading to renal failure is tubulointerstitial fibrosis, characterized by the replacement of normal kidney with scar tissue. Inflammation, myofibroblast activation, matrix accumulation, and tubular and vascular atrophy and dysfunction are all involved in this fibrotic response. This chapter correlates common clinical symptoms and histopathologic findings with our current molecular understanding of kidney disease.

## Introduction

The kidney performs innumerable functions for maintenance of body fluid homeostasis. The basic functional unit of the kidney is the nephron, and each nephron consists of a glomerulus and a tubule. The glomerulus performs blood filtration, with the filtrate then traveling through the renal tubules, where additional absorption and secretion occurs. The final product is urine that is eliminated from the body. Glomeruli consist of endothelial cells, specialized epithelial cells called podocytes, and supporting mesangial cells, while tubules are comprised of their own unique epithelium. Surrounding the tubules is the microvasculature, while the renal interstitium provides scaffolding. Pathologic assessment is particularly important in classification and treatment, and new data over the past decade has greatly enhanced our understanding of the molecular pathogenesis of disease. This chapter will focus on clinical, pathological, and molecular aspects of a variety of renal diseases, with a specific focus on fibrosis as the common endpoint of all kidney disorders.

## Clinical manifestations

The ultimate clinical manifestation of progressive chronic kidney disease (CKD) is a decrease in renal function. Uremia occurs when toxins are no longer adequately eliminated and manifests as nausea, anorexia, fatigue, and cognitive deficits. The presence of severe symptoms and/or life-threatening sequelae indicates end-stage renal disease (ESRD) and the need for dialysis or renal transplantation to sustain life. The kidney is also responsible for control of blood pressure, volume status, red blood cell mass, and bone and mineral metabolism. Therefore, it is common for CKD patients to have hypertension, edema, anemia, vitamin D deficiency, and bone disease. The correction of these derangements is a key principle of CKD care.

Once CKD has developed, there are no effective therapies to reverse the loss of function. Renin—angiotensin system (RAS) blockade with angiotensin-converting enzyme inhibitors (ACEIs) and angiotensin receptor blockers (ARBs) are non-specific therapies that can prolong renal function but are not a cure. Improved understanding of kidney disease has the promise to provide better therapies, highlighting the intrinsic value of molecular pathology to this field.

Essential Concepts in Molecular Pathology. DOI: https://doi.org/10.1016/B978-0-12-813257-9.00024-3

# Diagnosis of renal disease

## Serum measurements of renal function

Renal function is commonly based on elimination of serum creatinine in the urine. Most estimations of the glomerular filtration rate (GFR) are derived from formulas utilizing the serum creatinine. However, creatinine is generated by muscle and is therefore contingent on a person's muscle mass, leading to inaccuracies using available formulas. The serum marker cystatin C is gaining favor as it is not subject to the same inaccuracies.

## Urinalysis

The presence of protein in the urine (proteinuria) is an important disease marker incorporated into formal CKD definitions. The protein is usually largely comprised of albumin (e.g., albuminuria) and generally indicates glomerular disease. It is known that proteinuria itself can cause intrarenal inflammation and fibrosis. Proteinuria is therefore both a cause and effect of CKD, and higher levels generally portend a greater risk of CKD progression.

Several renal diseases also lead to the presence of blood (hematuria) or inflammatory cells (pyuria) in the urine. Glomerular leak of red blood cells is a sign of severe injury with a high likelihood of causing ESRD. The presence of white blood cells indicates inflammatory conditions such as infection or interstitial nephritis. These different cells can become trapped in urinary proteins and identified by microscopic examination of urine as red blood cell casts and white blood cell casts, respectively. Granular casts are consistent with acute tubular injuries.

## Renal biopsy

A biopsy is required for definitive diagnosis of most kidney diseases. The biopsy provides a wealth of information regarding the cause, severity, and overall prognosis of the patient. The goal of the biopsy is to obtain both glomeruli and the surrounding tubules, which are both necessary for diagnosis (Fig. 24.1A). Complete specimen evaluation includes light microscopy, immunofluorescence, and electron microscopy (EM).

## Biomarkers for diagnosis and prognosis

Surrogate urinary or serum biomarkers for kidney disease have been discovered to replace or enhance results of kidney biopsy. For example, urinary proteomics has also identified insulin-like growth factor-binding protein 7 (IGFBP7) and tissue inhibitor of metalloproteinases-2 (TIMP-2) as novel acute kidney injury (AKI) biomarkers. Urinary matrix metalloproteinase 7 (MMP-7) is also proposed as a valuable biomarker for predicting the severity of AKI after cardiac surgery in patients. The presence of phospholipase $A_2$ receptor in immune complexes or its autoantibody in the serum

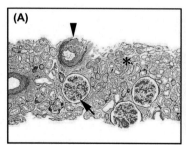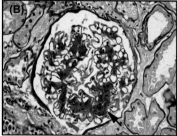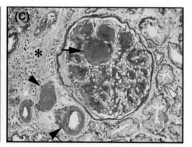

FIGURE 24.1 Diabetic nephropathy. (A) Low power view of a renal core biopsy from a diabetic patient showing the major components of the cortex: glomerulus (*arrow*), vasculature (*arrowhead*), and renal tubules (*asterisk*). Note in this early stage of disease the relatively close apposition of tubules with minimal visible interstitium. Periodic acid Schiff (PAS) stain. (B) In this enlarged image of the glomerulus, mesangial expansion can be appreciated, a pathologic finding in diabetic nephropathy (DN) (*arrow*). This mesangial expansion is non-proliferative and is not accompanied by an increase in cellularity (compare to IgA nephropathy, Fig. 24.4). (C) Nodular glomerulosclerosis occurs in severe disease (Kimmelstiel−Wilson nodule, *arrow*). In DN, both the afferent and efferent arterioles demonstrate hyalinosis, with strongly PAS-positive material composed of plasma proteins leading to thickening of the vessel walls (arrowheads). Note in this specimen the presence of more advanced interstitial expansion/fibrosis (*asterisk*) creating a widened space between tubules. There is also increased wrinkling of the tubular basement membranes (*yellow arrow*). PAS stain. For comparison with normal tubules, see Fig. 24.10A.

is a marker for idiopathic membranous nephropathy. Tissue transglutaminase, α-smooth muscle actin (α-SMA), platelet-derived growth factor receptor-β (PDGFR-β) and other proteins could theoretically be utilized in diagnosis and prognosis.

Genomic and proteomic techniques have great potential in kidney disease management. Gene expression profiling can differentiate inflammatory versus fibrotic lesions, and stable versus progressive disease. Similarly, laser microdissection of biopsy tissue followed by mass spectrometry can identify specific renal disorders, including amyloidosis. Biomarker discoveries are summarized in Table 24.1.

## Specific glomerular and tubular diseases

Although a comprehensive description of all renal diseases is beyond the scope of this chapter, it is useful to highlight recent molecular discoveries that have transformed our understanding and treatment of disease. Several broad categories of renal disease exist. Nephrotic syndrome is characterized by significant proteinuria, usually greater than 3.5 g per day. Nephritic syndromes are characterized by hematuria, dysmorphic red blood cells, and red blood cell casts on urinalysis. Both nephrotic and nephritic syndromes generally involve the glomerulus. Tubulointerstitial diseases, on the other

**TABLE 24.1** Biomarkers for kidney injury.

| Markers | Role |
|---|---|
| **Serum** | |
| Creatinine | Most widely used serum biomarker for estimating GFR |
| Cystatin C | Endogenously produced protein being investigated for GFR estimation, more accurate than creatinine |
| NGAL | NGAL is present in higher levels in serum in acute and chronic kidney injury |
| Asymmetric dimethylarginine | Marker of endothelial function (inhibits nitric oxide synthase), higher levels associated with disease progression and death |
| **Urine** | |
| PLA$_2$R autoantibody | Specific for idiopathic membranous nephropathy |
| Creatinine | Most widely used marker for determining renal clearance |
| Protein/albumin | Non-specific marker of kidney injury, degree of proteinuria correlates with progression |
| NGAL | Marker of tubular epithelial cell damage and both acute and chronic kidney injury |
| IGFBP7 and TIMP-2 | Marker of acute kidney injury and cell cycle arrest |
| L-FABP | Correlates with proteinuria and predicts progression |
| Uromodulin | Decreased excretion in diabetic nephropathy |
| Collagen fragments | Decreased excretion in CKD (hypothesized to be due to decreased protease activity) |
| MMP-7 | Urinary marker of pathologic Wnt/β-catenin signaling |
| α1-Antitrypsin | Urinary marker of glomerular diseases with nephrotic syndrome |
| RBP, β-trace protein, β2-microglobulin | Each protein individually correlated with renal function |
| Inulin, iothalamate, iohexol, DTPA, EDTA | Exogenously given filtration markers to estimate GFR. Inulin is the ideal filtration marker and performs better than creatinine. |
| **Tissue/Biopsy** | |
| α-SMA | Predicts disease progression, marker for myofibroblasts |
| Tissue transglutaminase | Positive immunohistochemical staining is correlated with worsened fibrosis |
| MMP-7, MMP-9, IL-8, β4-integrin, urokinase receptor | Gene subset that predicted disease progression (from kidney biopsy samples) |
| PLA$_2$R | Specific for idiopathic membranous nephropathy |

*DTPA*, diethylenetriaminepentaacetic acid; *EDTA*, ethylenediaminetetraacetic acid; *GFR*, glomerular filtration rate; *L-FABP*, liver-type fatty acid–binding protein; *MMP*, matrix metalloproteinase; *NGAL*, neutrophil gelatinase associated lipocalin; *PLA2R*, phospholipase A2 receptor; *RBP*, retinol-binding protein; *SMA*, smooth muscle actin.

hand, affect the tubular compartment and the corresponding interstitium preferentially, rather than glomeruli. Cystic diseases are characterized by the formation of pathologic cysts in the tubulointerstitial compartment.

## Diabetic nephropathy

### Clinical and pathologic features

DN is of particular importance as it accounts for approximately 50% of ESRD cases in the developed world. Both type 1 and type 2 diabetes can cause DN, and the majority of cases are characterized by albuminuria. An array of pathologic features accompanies DN (Fig. 24.1). Among these are mesangial expansion and/or nodular sclerosis in the glomeruli, progression to global glomerulosclerosis, and increased basement membrane thickness on EM. Hyalinosis of both the afferent and efferent arterioles helps to differentiate DN from other entities such as renovascular disease, which affects only the afferent arteriole. Interstitial fibrosis and tubular atrophy develop as the disease progresses and are discussed later.

### Molecular pathogenesis

Glomerular hyperfiltration plays a major role and is the basis for its only specific therapy. The hyperfiltration is largely mediated by tubuloglomerular feedback, in which the macula densa in the early distal tubule elaborates soluble mediators such as adenosine and nitric oxide to modulate constriction of the afferent arteriole in the glomerulus. In DN there is an increase in filtration of glucose at the level of the glomerulus, coupled with increased proximal tubule sodium reabsorption mediated by sodium–glucose cotransporters. This leads to a decrease in distal tubule delivery of sodium and chloride that activates tubuloglomerular feedback leading to hyperfiltration, which causes glomerular damage. Reduction in hyperfiltration with the use of ACEIs and ARBs has been a mainstay of DN treatment since 1993.

Hyperglycemia also has direct effects on the glomerulus. The mesangium is disproportionately affected by hyperglycemia, and exposure to high glucose is detrimental to mesangial cells. Hyperglycemia also leads to the production of advanced glycation end products, or AGEs, which cause oxidative stress, inflammation, and other injury. Treatment of these effects by enhancing the antioxidant and anti-inflammatory nuclear factor erythroid 2–related factor-2 (Nrf2) pathway is protective experimentally but has been limited in clinical trials by adverse side effects. Podocytes express insulin receptors that increase glucose uptake, and insulin resistance is directly detrimental to the cell. Upregulated mammalian target of rapamycin complex 1 (mTORC) causes mislocalization of podocyte slit diaphragm proteins and endoplasmic reticulum (ER) stress.

Soluble or circulating factors have been identified as potential mediators of DN. Both preclinical and clinical data suggest that soluble endothelin-1, produced locally in the kidney or arriving via the circulation, causes harmful glomerular hyperfiltration and podocyte injury, primarily via the endothelin-A receptor. Activated protein C (APC), vascular endothelial growth factor (VEGF), and bone morphogenetic protein-7 (BMP-7) could also be future targets for therapy.

## Membranous nephropathy

### Clinical and pathologic features

Idiopathic membranous nephropathy (IMN) is a common cause of primary nephrotic syndrome in adults. Traditionally, IMN had been treated with non-specific immunosuppression and control of proteinuria with RAS blockade. Fortunately, with the understanding of IMN as an antibody-mediated autoimmune disease, treatment is now evolving with anti-B-cell therapy being a viable and effective treatment.

IMN is characterized pathologically by glomerular capillary wall thickening, and the deposition of sub-epithelial electron dense deposits on EM. These deposits are immune complexes (ICs) comprised of immunoglobulin and/or complement. Growth of GBM around these deposits can produce characteristic "spikes" on silver stains. Newly developed immunofluorescence stains and serum tests for phospholipase $A_2$ receptor autoantibodies have aided in the diagnosis of IMN (Fig. 24.2).

### Molecular pathogenesis

IC formation is the main pathogenic mechanism of IMN. IC deposition leads to the activation of the membrane attack complex comprised of C5b-C9, which normally leads to lysis of targeted cells. However, in podocytes the attack complex leads to activation rather than lysis, resulting in oxidative stress, protease activation, extracellular matrix (ECM) deposition, and TGF-β expression. In addition, podocyte apoptosis, cell cycle arrest, and foot process effacement all occur.

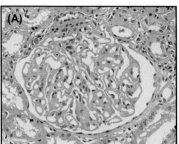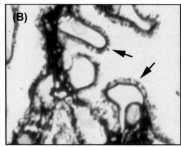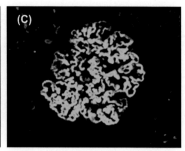

**FIGURE 24.2**  Idiopathic membranous nephropathy (IMN). (A) In IMN, the glomerulus is characterized by thickened capillary loops. Periodic acid Schiff stain. (B) Higher magnification of glomerular capillary loops with a silver stain reveals typical spikes (*arrows*) where glomerular basement membrane extends between immune complex deposits (which are invisible on this stain). (C) Immunofluorescence staining for the $PLA_2R$ autoantigen (green color) is highly specific for IMN and has become a helpful biomarker for diagnosis.

A breakthrough occurred in 2009, when M-type phospholipase A2 receptor ($PLA_2R$) was identified as the autoantigen present in 70% of IMN patients. $PLA_2R$ is normally expressed in podocytes and appears to be the site of *in situ* IC formation. Specificity of testing was impressive—patients with other nephrotic syndromes as well as secondary forms of membranous lacked this antibody entirely. Furthermore, autoantibody levels were associated with disease activity. This discovery rapidly led to the incorporation of $PLA_2R$ in diagnosis of IMN and as a marker of disease activity. The $PLA_2R$ antigen can now be detected in ICs by immunofluorescence of renal biopsy samples, and the autoantibody can also be detected in serum. Anti-B-cell therapy with rituximab has successfully treated IMN in clinical trials, with the disappearance of $PLA_2R$ antibodies associated with remission. This rapid improvement in diagnosis and care of IMN is a direct reflection of improved molecular understanding of this disease.

## Focal segmental glomerulosclerosis

### Clinical and pathologic features

Primary focal segmental glomerulosclerosis (FSGS) is another major cause of nephrotic syndrome. As its name suggests, FSGS is characterized by unique segmental sclerosis, affecting only a part of each glomerulus. The disease is also characteristically patchy (focal), affecting a portion of total glomeruli (Fig. 24.3). This pattern of injury can occur in nearly any CKD, making the diagnosis of primary FSGS challenging, but all are related to podocyte injury. FSGS can also occur as sequelae caused by viral infections (including HIV), medications, and obesity. Treatment typically consists of RAS blockade as well as steroids and other immunosuppression.

### Molecular pathogenesis

Genome-wide studies have greatly increased our understanding of FSGS. Because there is a strong prevalence of FSGS in African populations, an admixture scan was performed to identify disease-causing genes. Initially, studies found that the *MYH9* gene locus on chromosome 22 conferred an increased risk of ESRD. These findings were later refined to implicate the nearby *APOL1* gene, which is in linkage disequilibrium with *MYH9*. The presence of two risk alleles of *APOL1* (G1 and G2) conferred an odds ratio of 10.5 for FSGS development and a 7.3 odds ratio of hypertensive ESRD. Interestingly, APOL1 is a lytic factor for *Trypanosoma brucei*, the pathogen responsible for African sleeping sickness. The increased incidence of these alleles in African populations may be explained by the protective effects of the G1 and G2 *APOL1* gene variants in *Trypanosoma* infection. Patients with *APOL1* risk alleles also exhibit increased FSGS lesions and worse prognosis. In a more global sense, *APOL1* is also associated with progression of CKD and incidence of ESRD, regardless of the cause. The exact mechanism by which *APOL1* leads to increased kidney injury is unclear but may involve activation of stress-activated protein kinases, intracellular potassium depletion, endosomal trafficking and autophagic flux.

Circulating factors causing FSGS have been implicated due to observations of recurrence after renal transplant and reports of successful treatment with plasmapheresis in some patients. Most studies have focused on the soluble urokinase-type plasminogen activator receptor, or suPAR. suPAR is a result of cleavage of cell surface−associated urokinase-type plasminogen activator receptor (uPAR). suPAR is upregulated in FSGS and could cause proteinuria in animal models via interactions with β3-integrin on podocytes, causing foot process effacement. However, subsequent studies were unable to replicate either result. There was also evidence that suPAR levels increase non-specifically when

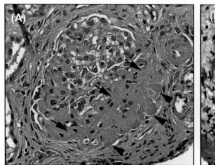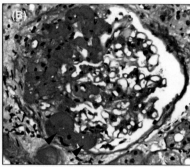

**FIGURE 24.3** Focal segmental glomerulosclerosis (FSGS). (A,B) In FSGS, there is segmental sclerosis affecting only a portion of the glomerulus (*arrows*). This is usually accompanied by hyalinosis, or the accumulation of eosinophilic material, in injured segments (*arrowheads*). Periodic acid Schiff stain.

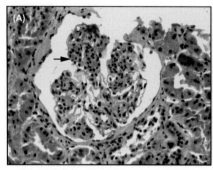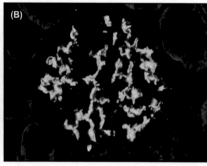

**FIGURE 24.4** IgA nephropathy (IGAN). (A) As in diabetic nephropathy, glomeruli in IGAN are characterized by expansion of the mesangial space (*arrowhead*). However, IGAN lesions are also characterized by increased cellularity. Hematoxylin and eosin stain. (B) Immunofluorescence shows a predominance of IgA staining.

GFR declines, which could have led to alternative erroneous conclusions in earlier studies. These data clearly indicate the need for additional studies to understand suPAR biology and its contribution to FSGS and CKD.

Still other mechanisms may play a role in FSGS. B7-1, also known as CD80, is found on podocytes and is upregulated in glomerular disease. Its role appears to be the disruption of β1-integrin leading to a more pathogenic motile phenotype in the podocyte. Treatment with the anti-B7-1 drug abatacept was capable of inducing disease remission in a small number of patients. Another mechanism of podocyte injury appears to be imbalance in the activities of the transient receptor potential cation channels 5 and 6 (TRPC5 and TRPC6) which affect the actin cytoskeleton.

## IgA nephropathy

### Clinical and pathologic features

IgA nephropathy (IGAN) is one of the most common glomerular diseases worldwide and a major cause of ESRD. More commonly it is characterized by a nephritic syndrome with hematuria but can also lead to nephrotic-range proteinuria. Many patients will report occurrence or worsening of symptoms at the time of mucosal upper respiratory or gastrointestinal infections, perhaps as a result of increases in IgA production. Asian populations have a much higher prevalence, as high as 16% in some countries. Characteristic pathology includes mesangial expansion and detection of IgA by immunofluorescence (Fig. 24.4). EM shows mesangial immune complex deposits.

### Pathologic mechanisms

The molecular cause of IGAN is still not understood. There is now evidence for galactose-deficient IgA antibodies in patients, but its presence alone is not enough to cause disease. Instead, IGAN appears to require autoantibodies to exposed glycans in these galactose-deficient IgA. It was proposed that since viruses and bacteria contain similar residues recognized by these anti-glycan antibodies, this would be another explanation for the worsening of IGAN with infections. Since IGAN is more common in regions with endemic worm and parasite infections, it has also been proposed that selective pressure to adapt to these infections has also led to IGAN. These hypotheses, while interesting, remain to be proven.

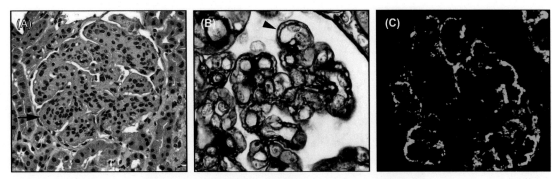

**FIGURE 24.5** Membranoproliferative glomerulonephritis (MPGN). (A) MPGN glomeruli have a lobular appearance with endocapillary proliferative lesions with increased cellularity in both the mesangium and within capillaries (*arrow*). Hematoxylin and eosin stain. (B) Tram track lesions or double contours are characteristic of MPGN and represent duplication of the glomerular basement membrane (*arrowhead*). Silver stain. (C) Immunofluorescent staining pattern for MPGN is shown. Positive staining for both immunoglobulins and complement is seen in immune complex——mediated disease. In complement-mediated disease, only complement will stain positive.

## Membranoproliferative glomerulonephritis

### Clinical and pathologic features

Membranoproliferative glomerulonephritis (MPGN) remains a challenging entity to treat. Its clinical manifestations can range from nephrotic to nephritic, and it can progress slowly or very rapidly. Low complement levels are characteristic serologic findings.

Light microscopic examination reveals increased numbers of mesangial cells and endocapillary proliferation leading to a lobular appearance, as well as double contours in the glomerular capillaries (Fig. 24.5). Traditional categorization relied on electron microscopic evidence of immune deposits in sub-endothelial and/or sub-epithelial locations. However, in keeping with an improved understanding of the disease, newer classification schemes focus on IC-mediated or complement-mediated disease with corresponding immunofluorescence staining of antibody and/or complement, respectively.

### Molecular pathogenesis

IC-mediated MPGN is often related to infection, autoimmune diseases, and monoclonal gammopathy. Viral hepatitis, systemic lupus erythematosus, and monoclonal diseases are all associated with MPGN and are treated by addressing the underlying disease. Complement-mediated disease is characterized by abnormal activation of the complement alternative pathway leading to glomerular deposition of pathway components. A variety of mutations and autoantibodies have been identified and are used in clinical diagnosis. Eculizumab, a monoclonal antibody targeting C5 to prevent terminal complement pathway activation, is now being used to treat this type of MPGN.

## Polycystic kidney disease

Autosomal dominant polycystic kidney disease is the most common genetic disorder affecting the kidney. In this disease, innumerable cysts are generated that ultimately lead to renal dysfunction and ESRD in around half of afflicted patients (Fig. 24.6). The pathophysiology appears related to two genes, polycystin-1 (PKD1) and polycycstin-2 (PKD2), which are expressed on primary cilia that sense flow. Mutations in these genes appear to cause derangement of calcium second messengers, G protein signaling, mTOR, Wnt, and several other pathways leading to cyst formation. Therapies aimed at these pathways are under clinical testing.

## Acute kidney injury (acute tubular necrosis)

### Clinical and pathological features

AKI is defined by a rapid loss of renal function, occurs in more than 20% of hospitalized patients in the developed world, and is associated with nearly 25% mortality. Most AKI results from tubular injury, or acute tubular necrosis (ATN), as a result of impairment of renal perfusion (for a number of reasons, including circulatory collapse, sepsis, volume depletion, and cardiac surgery) as well as exposure to toxins or severe glomerular injury. ATN injury is reversible

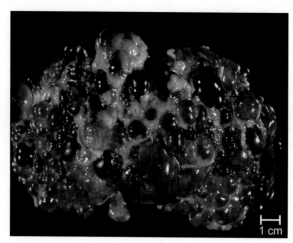

**FIGURE 24.6** Polycystic kidney disease. Innumerable fluid-filled cysts involve the kidney, which can become significantly enlarged. Bar equals 1 cm.

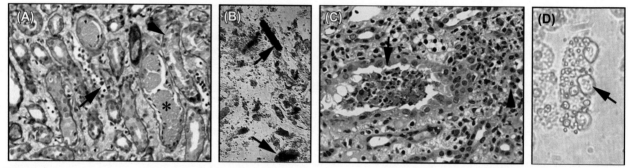

**FIGURE 24.7** Acute tubular necrosis and acute interstitial nephritis. (A) Tubular cells are damaged in acute tubular necrosis (ATN) and are sloughed into the tubular lumens (*arrow*). Tubules also lose their normal brush borders (*arrowhead*). As a result of the injury, cast formation occurs in tubular lumens (*asterisk*). Hematoxylin and eosin (H&E) stain. (B) Microscopic assessment of sediment from a fresh urine sample from a patient with ATN showing granular casts (*arrows*) that correspond to the casts seen in (A). Thus, urinalysis can be very helpful for diagnosis of ATN in the absence of a kidney biopsy. (C) In acute interstitial nephritis (AIN), eosinophils and other inflammatory cells infiltrate the interstitium (*arrowhead*) and can also form casts in tubular lumens (*arrow*). H&E stain. (D) Microscopic urinalysis in AIN can reveal the presence of white blood cell casts (leukocyte at *arrow*).

to some extent but in some cases can lead to CKD and ESRD in the short or long term. Patients require supportive care, including dialysis in severe AKI. However, there is no proven therapy in humans to reverse the course of disease.

Although its name suggests tubular cell death, the degree of clinical deterioration often does not correlate with findings of frank necrosis. More common findings include tubular cell detachment and sloughing into the lumen, deterioration of normal brush borders of proximal tubules, and the presence of tubular casts comprised of protein and cellular debris (Fig. 24.7). The casts can be excreted into the urine and take on a granular appearance. These granular casts can be extremely helpful in diagnosis (Fig. 24.7B). Interstitial edema, inflammation, and congestion of peritubular capillaries are also found on histology.

## Molecular pathogenesis

Tubular cell death appears to be a combination of apoptosis and necrosis. Apoptosis is initiated by a number of pathways including cellular and ER stressors as well as by TNF-$\alpha$ and Fas ligand. Bcl-2 proteins and mitochondria play major roles, with newer data also showing the phenomenon of mitochondrial fission. Necrosis is characterized by a more haphazard release of cell contents compared with apoptosis. Unfortunately, preclinical data has not supported a strong role for blockade of apoptosis or necrosis in AKI treatment.

Tubular cells have an enormous capacity for repair, but maladaptive repair can still occur. AKI-affected tubular cells become arrested in G2/M phase after AKI and upregulate c-Jun NH2-terminal kinase (JNK) signaling and pro-fibrotic cytokines such as TGF-β. More recently it was found that histone deacetylase inhibitors may ameliorate G2/M arrest and improve outcomes.

A rapid influx of a variety of inflammatory cells occurs after AKI. Tubular necrosis releases damage-associated molecular pattern proteins (DAMPs) that activate local and infiltrating cells to secrete an array of chemokines and cytokines. Initial migration is characterized by pro-inflammatory neutrophils, M1 macrophages, and pathologic B lymphocytes and T lymphocytes. However, after initial waves there is an attempt at repair, as M1 macrophages transform into reparative M2 forms. Interestingly, an initial AKI insult can lead to long-term population of the kidney with activated and effector memory T-cells, which can play a role in further injury.

The endothelium plays a unique role in AKI as well. Vascular congestion and reduced blood flow worsen hypoxia and are related to a pro-thrombotic phenotype due to decreased thrombomodulin and APC. Barrier integrity is broken due to changes including endothelial apoptosis and alterations in cytoskeleton and adherens junctions, leading to increased inflammation and edema. Finally, loss of blood vessel density (i.e., microvascular rarefaction) has been described and may predispose to future injury as well as fibrotic change. Oxidative stress is increased in AKI, particularly in areas of hypoperfusion.

## Acute interstitial nephritis

Acute interstitial nephritis (AIN) causes AKI and can manifest as fever, rash and eosinophilia. Identification of white blood cell casts in the urine can be helpful (Fig. 24.7D). Drugs are the major etiology of AIN, with antibiotics, nonsteroidal anti-inflammatory drugs, and proton pump inhibitors being the most common. Tuberculosis and IgG4-related autoimmune kidney disease are rarer causes. Steroids are the mainstay of therapy, along with removal of underlying causes. The pathogenesis of AIN is thought to be allergic in nature, with timing and manifestation reminiscent of a type IV (delayed-type) hypersensitivity response. It is possible that drugs bind to tubular basement membranes and act as haptens to increase immunogenicity and lead to autoantigen stimulation of the immune system.

Renal biopsy reveals interstitial inflammation including lymphocytes, monocytes, and eosinophils (Fig. 24.7C). Tubulitis is also present and indicates inflammation involving the renal tubules. In IgG4-related disease, the infiltrate is rich in IgG4-producing plasma cells.

## Tubulointerstitial fibrosis

### Tubulointerstitial fibrosis as a final common pathway of CKD

Tubulointerstitial fibrosis (TIF) refers to the replacement of normal renal tubules and adjacent interstitium with excessive ECM and represents the final outcome of all progressive CKD leading to ESRD, regardless of inciting cause. TIF is associated with poor prognosis and is incorporated in pathologic grading classifications for many renal diseases. The presence of extensive areas of fibrosis is considered to be irreversible and may diminish enthusiasm for aggressive renal-specific therapies that carry significant side effects.

The main components of the normal tubulointerstitial compartment are the tubular epithelium, the vasculature, and the interstitium. In progressive renal injury, the tubules become atrophic, usually adhering to one of three patterns of injury: classic, endocrine, and thyroidization. Classic injury is characterized by thickened, wrinkled tubular basement membranes (Fig. 24.8C). In endocrine injury, tubular cells are plump and cuboidal with very small lumens, resembling endocrine histology in other tissues. Thyroidization injury appears as dilated tubules filled with proteinaceous fluid and lined by thin epithelial cells (Fig. 24.9A).

With tubular atrophy comes interstitial expansion, largely through the accumulation of excess ECM. Numerous studies have identified increases in structural matrix proteins including collagens, laminin, fibronectin, decorin, and heparin sulfate. Matrix accumulation can be identified histologically with stains such as picrosirius red and Masson's trichrome stain. With the picrosirius red stain, collagen fibers stain red with light microscopy and exhibit birefringence under polarized light (Fig. 24.8A and B). With the Masson's trichrome stain, collagen is identified through blue coloration (Fig. 24.8D). Fibroblasts and myofibroblasts are the main contributors to matrix production, but other resident cells can also be induced to produce matrix.

Renal vasculature changes can have varying morphologies. Aging, diabetes, and hypertensive changes are characterized by thickened arteries/arterioles with hyperplastic tunica media and intimal fibrosis (Fig. 24.9B). Hyaline arteriosclerosis is also seen in DN and is characterized by extrusion of eosinophilic proteinaceous plasma material

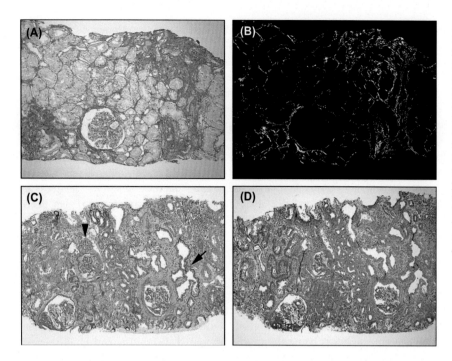

**FIGURE 24.8** Renal fibrosis is revealed with different histologic stains. (A) Picrosirius red stains collagen fibers red under normal light microscopy, but also non-specifically stains other non-collagen components. (B) Polarization of the picrosirius image in (A) shows birefringence, which is highly specific for the collagen fibers only. Note the large areas of increased fibrosis in this diseased kidney. (C) Periodic acid Schiff staining highlights basement membranes and shows the wrinkled tubular basement membranes of atrophic tubules (*arrow*). Also shown are discrete foci of inflammatory cells (*arrowhead*), which likely contribute to injury. (D) Masson's trichrome stain of a serial section of same specimen in (C) facilitates identification of collagen with dark blue staining.

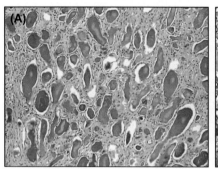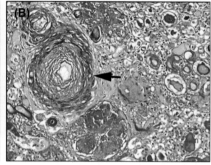

**FIGURE 24.9** Tubular and vascular changes associated with renal fibrosis. (A) Thyroidization type of tubular atrophy. Tubules are dilated and filled with proteinaceous debris and bears a resemblance to thyroid histology. (B) Vascular damage in renal fibrosis. Blood vessel showing marked intimal fibrosis (*arrow*) in a patient with end-stage diabetic kidney disease. Both images use Periodic acid Schiff stain.

into vessel walls. Another feature is loss of blood vessels, a phenomenon known as microvascular rarefaction, which likely contributes to tissue hypoxia.

## Molecular mechanisms

Decades of research have led to vastly improved understanding of the molecular pathways leading to the development of TIF. A variety of growth factors, cytokines, and intracellular and intercellular signaling pathways, as well as virtually all resident cell types and infiltrating cells from the circulation are involved in TIF progression. Conceptually, TIF occurs via priming, activation, execution, and progression stages.

### Priming: inflammation generates a pro-fibrotic environment

Ample evidence supports a role for inflammation in the fibrotic response. General anti-inflammatory interventions, including specific antagonists of C−C chemokine receptor types 1 and 2, TNF-α, and the IL-1 receptor, have been shown to reduce the development of fibrosis. Acute injury appears to be partially mediated by infiltrating pro-inflammatory M1-type macrophages. As injury progresses, anti-inflammatory M2 macrophages emerge and promote tubular repair, but can also promote fibrosis through elaboration of TGF-β and connective tissue growth factor (CTGF). Mice in which lymphocytes are ablated are protected from renal fibrosis, and this was reversed by reconstitution of the T-cell population. Acute injuries also lead to infiltration of long-lasting effector memory lymphocytes that may predispose to further injury. A regulatory T-lymphocyte population conversely promotes repair in the tubular epithelium.

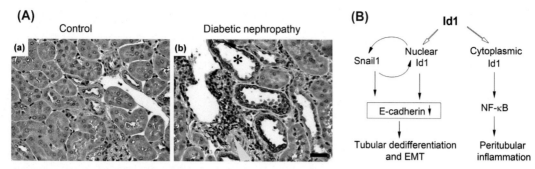

**FIGURE 24.10** Inhibitor of differentiation-1 (Id1) has roles in both tubular epithelial cell dedifferentiation and peritubular inflammation. (A) Immunohistological staining demonstrates upregulated Id1 in dilated tubules (*) in murine diabetic nephropathy. Notably, Id1 was upregulated in a region proximal to inflammation (*yellow arrowhead*). (a) Control kidney. (b) Diabetic kidney. Scale bar, 25 μm. (B) Diagram detailing the dual roles of Id1. Id1 shuttles between the nucleus and cytoplasm, with nuclear Id1 inducing dedifferentiation of tubular epithelium through downregulation of E-cadherin or upregulation of Snail signaling. Meanwhile, cytoplasmic Id1 increases nuclear factor-kappa B (NF-κB) signaling leading to chemokine production and peritubular inflammation. *EMT*, epithelial-to-mesenchymal transition. *Reproduced with permission Li Y, Wen X, Liu Y. Tubular cell dedifferentiation and peritubular inflammation are coupled by the transcription regulator Id1 in renal fibrogenesis. Kidney Int 2012;81:880–91.*

Kidney dendritic cells (DCs) have an increased capacity for lymphocyte activation after injury, and DC ablation attenuates disease. Cytokines from inflammatory cells likely render fibroblasts and tubular epithelia more likely to enter a fibrogenic pathway. *In vitro* culture of tubular cells with conditioned medium from activated leukocytes leads to epithelial-to-mesenchymal transition (EMT, see below). Direct cell-to-cell interactions also play a role.

Much attention has been given to NF-κB as a central mediator of progressive renal disease. Although well-known for promoting inflammation, it has been implicated in pro-fibrotic pathways as well. The cytokine TNF-α can activate the NF-κB pathway in epithelial cells and lead to prolonged Snail1 signaling. Snail1 serves as a pro-fibrotic transcription factor associated with EMT and fibroblast migration. Meanwhile, development of renal fibrosis was decreased when NF-κB activation was blocked specifically in fibroblasts *in vivo*.

An interesting dual role for inhibitor of differentiation-1 (Id1) has also been described in tubular epithelium (Fig. 24.10). Originally viewed merely as an antagonist of basic helix−loop−helix transcription factors, Id1 can shuttle between the nucleus and cytoplasm. Cytoplasmic Id1 upregulates NF-κB signaling and RANTES expression and facilitates renal inflammation. Meanwhile, nuclear Id1 directly affects fibrogenesis by decreasing E-cadherin and ZO-1 expression, a step in epithelial dedifferentiation.

## Activation: recruitment of matrix-producing cells

The development of TIF is inherent on the activation of fibroblasts that will ultimately produce the fibrotic matrix. This is a process that appears closely linked to inflammation. Although other cell types in the kidney can produce ECM, it is the activated fibroblast, or myofibroblast, which is the primary source of fibronectin and collagen in renal fibrosis. Current evidence implicates at least five different originator cells that can become myofibroblasts in diseased kidneys: (1) interstitial fibroblasts, (2) tubular epithelial cells, (3) endothelial cells, (4) pericytes, and (5) circulating fibrocytes (Fig. 24.11). This has been one of the most controversial and active areas of research in renal fibrosis.

**Interstitial fibroblasts**. Resident fibroblasts are present in the interstitium of the kidney and provide the framework for renal structures. Much of this skeleton is provided by fibroblasts as well as actin-containing cell processes that interconnect capillary and epithelial basement membranes. Utilizing their abundant endoplasmic reticulum, the fibroblast is capable of secreting the ECM components necessary for the normal structure of the kidney.

On the other hand, disease states are characterized by the appearance of myofibroblasts. These activated fibroblasts greatly upregulate matrix production and develop a contractile phenotype with expression of α-smooth muscle actin (α-SMA). Proliferative capacity is also increased and driven by soluble factors including TGF-β, PDGF, CTGF, Wnt ligands, sonic hedgehog, and tissue plasminogen activator (tPA). This dramatic phenotypic change is a major impetus for fibrosis development.

**Epithelial cells**. In epithelial to mesenchymal transition (EMT), epithelial markers such as E-cadherin and zona occludens-1 (ZO-1) are downregulated and coupled with upregulation or de novo expression of mesenchymal markers such as vimentin, α-SMA, and transcription regulators Snail1, Snail2/Slug, and β-catenin. The *in vitro* occurrence of EMT has been clearly demonstrated and widely accepted, but the *in vivo* correlation is less certain. Tubular cell cultures are induced to undergo EMT via exposure to TGF-β1. Fate mapping experiments initially supported EMT, as tubular

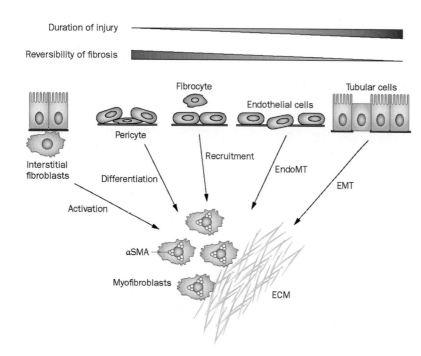

Duration of injury

Reversibility of fibrosis

Fibrocyte

Tubular cells

Endothelial cells

Pericyte

Interstitial fibroblasts

Recruitment

Differentiation

EndoMT

EMT

Activation

αSMA

Myofibroblasts

ECM

**FIGURE 24.11** Multiple origins of myofibroblasts. Current evidence points to at least five sources of activated myofibroblasts: (1) activation of interstitial fibroblasts, (2) differentiation of vascular pericytes, (3) infiltration of circulating fibrocytes, (4) capillary endothelial-to-mesenchymal transition (EndoMT), and (5) tubular epithelial-to-mesenchymal transition (EMT). The relative contributions of each source to the myofibroblast pool remains a highly controversial area of active research. It is likely that effects on endothelium and tubular cells represent more severe or prolonged injury which is less likely to be reversible. Myofibroblasts possess increased levels of α-smooth muscle actin (α-SMA) as well as enhanced extracellular matrix (ECM) secretory capacity. *Reproduced with permission Liu Y. Cellular and molecular mechanisms of renal fibrosis. Nat Rev Nephrol 2011;7:684—96.*

epithelium contributed 36% of the interstitial fibroblasts in a mouse model of obstructive injury. However, in humans, there is only circumstantial evidence that epithelial cells upregulate expression of mesenchymal markers. More recent animal studies now directly contradict the earlier fate mapping experiments and suggest other sources of fibroblasts.

Recently, a pair of studies have refined our understanding of its role in fibrosis. In these studies, the specific deletion of *Snail* or *Twist1* in renal tubular cells led to a dramatic reduction in the development of fibrosis. This deletion was associated with reduction in the time spent in G2/M cell cycle arrest, which was previously implicated in impaired tubular repair after injury. In addition, blocking Snail1 activation reduced—and even reversed—previously established injury. In these studies, a full EMT was not observed. Rather, it appeared that a partial phenotypic change was enough to explain the effects on fibrosis. As such, partial EMT may be a better explanation for tubular involvement in TIF and will be a promising target for future therapies (Fig. 24.10).

**Endothelial cells.** A process similar to EMT has been described in endothelial cells, with upregulation of α-SMA and concomitant downregulation of the CD31 and cadherin normally present in these cells. Such an endothelial-to-mesenchymal transition (EndoMT) was demonstrated in mouse models of unilateral ureteral obstruction, DN, and Alport syndrome utilizing fate mapping techniques. Specific ablation of the TGF-β type II receptor in endothelial cells led to reduction of EndoMT, supporting a role for pro-fibrotic TGF-β in this process.

**Vascular pericytes.** Pericytes are support cells for the peritubular capillaries that supply renal tubules. In newer studies contradicting earlier experiments showing EMT, fate mapping experiments suggested that it is pericytes (not epithelial cells) that differentiate into myofibroblasts in experimental kidney injury models. These pericytes also seem to have properties of mesenchymal stem cells. In addition to contributing to the myofibroblast pool, the differentiation of pericytes may lead to microvascular destabilization and rarefaction, leading to further injury.

**Circulating fibrocytes.** Cells known as fibrocytes circulate in blood and share the phenotype of both fibroblasts and blood monocytes. They are derived from the bone marrow and express CD45, which is present on all hematopoietic cells. Fibrocytes elaborate both pro-inflammatory cytokines and collagen types I and III and vimentin. Fibrocytes can migrate into injured kidney and may contribute to up to 20% of the myofibroblasts.

Regardless of the actual source of myofibroblasts, the end effect is an activated cell that can produce excess matrix leading to scar formation. Myofibroblasts can be distinguished from normal fibroblasts by their morphology and unique combination of cell markers; however, no individual marker is completely specific (Table 24.2).

## Execution: assembly and activation of the fibrogenic matrisome

The actual production of ECM by activated fibroblasts/myofibroblasts depends on a number of molecular interactions between the cell and its external environment. The term matrisome has been proposed to describe the multicomponent,

**TABLE 24.2** Markers for renal interstitial fibroblasts and myofibroblasts.

| Markers | Fibroblasts | Myofibroblasts | Function | Other cell types |
|---|---|---|---|---|
| α-SMA | − | ++ | Stress fiber, cell contractility | Vascular smooth muscle cells |
| Fsp1 | + | + | Actin-binding protein, cell motility | CD45$^+$ leukocytes |
| Vimentin | − | ++ | Intermediate filament protein, cytoskeleton | Glomerular podocytes |
| Nestin | − | + | Intermediate filament protein, cytoskeleton | Glomerular podocytes |
| CD73 | ++ | + | Ecto-5′-nucleotidase, conversion of 5′-AMP to adenosine | Proximal tubular cells, mesangial cells, T-cells |
| PDGFR-β | + | + | Cell proliferation, matrix production | Vascular smooth muscle cells, pericytes |

*AMP*, adenosine monophosphate; *Fsp1*, fibroblast-specific protein 1; *PDGFR*, platelet-derived growth factor receptor; *SMA*, smooth muscle actin.

integrin-associated, molecular machinery that relays, integrates, and executes pro-fibrotic signals leading to the transcription of matrix and other fibrogenic genes.

**Matrix production**. The production of ECM depends on a balance of pro-fibrotic and anti-fibrotic signals. Mediating fibrogenesis are TGF-β, PDGF, CTGF, fibroblast growth factor-2 (FGF-2), and angiotensin II. By binding to plasma membrane receptors, these cytokines activate intracellular signaling cascades resulting in transcriptional activation of collagen and fibronectin genes. On the other hand, anti-fibrotic cytokines such as HGF and bone morphogenic protein-7 (BMP-7) interfere with signaling.

The TGF-β family of proteins, of which the most abundant is TGF-β1, has been shown to result in the increased production of matrix components from fibroblasts, myofibroblasts, and epithelial cells and may also decrease matrix degradation. TGF-β1 can cause EMT *in vitro* and leads to expression of plasminogen activator inhibitor-1 (PAI-1), which itself has pro-fibrotic influence. These effects are mediated via TGF-β receptors on the cell surface, leading to intracellular activation of Smad transcription factors that translocate to the nucleus to initiate target gene expression, including collagens.

Once thought to be active only during kidney development, the Wnt/β-catenin signaling pathway becomes dramatically reactivated during TIF. The Wnt family consists of at least 19 different soluble factors that bind to frizzled receptors. This leads to the nuclear translocation of the transcription regulator β-catenin and upregulation of target genes including Snail1, PAI-1 and the RAS genes. Wnt/β-catenin pathway inhibition ameliorates fibrosis in animal models. Recent data also implicates another developmental growth factor, sonic hedgehog (Shh). Studies now show that renal injuries lead to Shh upregulation in tubules. Fibroblasts respond to this Shh by increasing proliferation and transforming into myofibroblasts (Fig. 24.10). Shh blockade leads to decreased TIF. The Gli proteins are transcriptional effectors of Shh and are markers for myofibroblast progenitors. It appears that tubular-derived Shh and fibroblast-derived Wnts may play a role in intercellular crosstalk to promote fibrosis development.

Numerous other factors are involved in either promoting or inhibiting fibrotic responses, with integration of their signals within cells determining the actual final response. PDGF family members have been shown to be pro-fibrotic. CTGF promotes EMT and ECM accumulation and appears to be necessary for persistence of fibrosis. Both PDGF and CTGF appear to act in concert with TGF-β to increase fibrotic responses. On the other hand, HGF acts to antagonize TGF-β and Smad signaling, and BMP-7 also exerts anti-fibrotic effects by similar mechanisms. Finally, tubular injuries lead to impairment of fatty acid oxidation (FAO) which induces tubular cell apoptosis and dedifferentiation via TGF-β (Fig. 24.12).

Physical interactions between fibroblasts and the ECM can also activate fibrogenic signaling pathways. Integrin transmembrane proteins connect ECM to the intracellular cytoskeleton. On binding to ECM, the integrins activate focal adhesion kinase (FAK) and integrin-linked kinase (ILK), which in turn activates the β-catenin and Snail1 pathways. ILK also recruits PINCH, leading to EMT. ILK, integrin, and their related proteins are core components of the matrisome and serve to integrate many diverse signals. TGF-β1 upregulates β1-integrin, ILK, and PINCH1, angiotensin II induces both ILK and β1-integrin, and CTGF upregulates ILK. Meanwhile, PDGF regulates matrisome signaling via interactions with PINCH through an adaptor protein Nck. tPA influences β1-integrin and ILK via the LDL receptor–related protein 1 (LRP1) and is known to induce collagen production. The result of all of these signals appears to be the formation of an ILK–PINCH–parvin complex, the disruption of which can blunt matrix production. Similarly, ILK inhibition and its subsequent effects on Snail1 and β-catenin also lead to prevention of renal fibrosis. These pathways are summarized in Fig. 24.13.

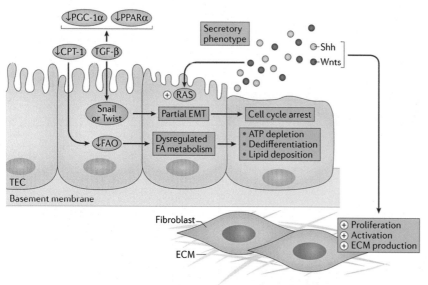

FIGURE 24.12 Epithelial cell pathways leading to renal fibrosis. TGF-β induces the expression of Snail1 and Twist1, leading to a partial EMT phenotype, followed by cell cycle arrest. These cells are markedly pro-fibrotic and lead to increased fibroblast activation. In addition, TGF-β leads to defective FAO, leading to tubular dedifferentiation and/or cell death. Tubular injury also leads to the elaboration of tubule-derived Shh and Wnts that in turn stimulate fibroblast activation. Wnts also upregulate pro-fibrotic RAS genes. *CPT-1*, carnitine palmitoyltransferase 1; *EMT*, epithelial-to-mesenchymal transition; *FAO*, fatty acid oxidation; *PGC-1α*, PPARγ coactivator-1α; *PPARα*, peroxisome proliferator-activated receptor-α; *RAS*, renin—angiotensin system; *Shh*, sonic hedgehog; *TGF-β*, tissue growth factor-β. *Reproduced with permission Zhou D, Liu Y. Renal fibrosis in 2015: understanding the mechanisms of kidney fibrosis. Nat Rev Nephrol 2016;12:68—70.*

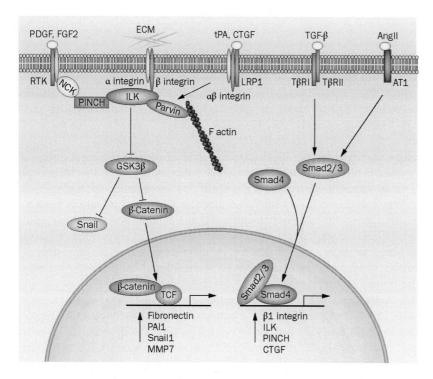

FIGURE 24.13 Molecular integration of fibrogenic signals leading to matrix production. A complex (or matrisome) containing PINCH, integrin-linked kinase (ILK), and cell surface integrins coordinate numerous signals leading to pro-fibrotic gene expression. Via Smad signaling, transforming growth factor-β1 (TGF-β1) is responsible for upregulation of expression of β1-integrin, ILK, and PINCH. Angiotensin II contributes to this via direct activation of Smad3 signaling and also affects TGF-β1 signaling. Platelet-derived growth factor (PDGF) and fibroblast growth factor-2 (FGF-2) can affect the matrisome via an adaptor protein Nck. Connective tissue growth factor (CTGF) and tissue plasminogen activator (tPA) both activate the matrisome through recruitment of β1-integrin. Matrisome activation leads to inhibition of GSK-3β, which in turn activates β-catenin and Snail1 leading to transcription of fibrogenic genes. *Reproduced with permission Liu Y. Cellular and molecular mechanisms of renal fibrosis. Nat Rev Nephrol 2011;7:684—96.*

**Matrix deposition**. Accumulation of excess ECM is a coordinated event with the initial steps being the production of fibronectin. Early in the process, a relatively weak skeleton of fibronectin and fibrillar collagen is produced, and it has been proposed that this stage is reversible. Next, the collagen becomes cross-linked, and basement membrane proteins, glycoproteins, and proteoglycans are incorporated. Secreted protein acidic and rich in cysteine (SPARC) and tissue transglutaminase contribute to the process. Mature fibrosis is resistant to degradation. Traditional views of matrix metalloproteinases (MMPs) indicate that they would be responsible for degrading excess matrix and that decreased expression or increased inhibition by tissue inhibitors of metalloproteinases (TIMPs) would lead to fibrosis. However, MMPs also participate in growth factor regulation, EMT, cell survival and apoptosis, inflammation, and basement membrane integrity, and as such potentially could be pro-fibrotic. It is likely that the exact microenvironment in which a particular MMP is active will greatly determine its overall effects.

*Progression: damage above and beyond scar formation*

**Tubular dysfunction and atrophy.** A number of cellular changes take place in tubular epithelia involving many of the same pathways active in the execution stage. Partial EMT, driven by NF-κB, ILK, HIF-1, Wnts, and TGF-β, leads to a loss of tubular transporters critical for normal tubular function. Severe injury can lead to cell death and tubular atrophy, while G2/M cell cycle arrest impairs the ability of remaining cells to repopulate the tubule and restore barrier function. Arrested cells are also pro-fibrotic and promote TIF formation. The end result is tubular dysfunction and tubular atrophy which would greatly impair renal function.

**Microvascular rarefaction.** Microvascular rarefaction refers to the loss of peritubular capillaries in renal fibrosis and has many possible underlying mechanisms. Just as partial EMT may lead to tubular atrophy, EndoMT and pericyte transformation into myofibroblasts may lead to loss of the vasculature. A downstream effect of microvascular rarefaction is hypoxia, a process worsened by decreased oxygen diffusion across fibrotic regions as well as the high metabolic activity of renal tubules. This would predispose the kidney to further hypoxic injury and progressive CKD.

# Conclusions

Primary renal diseases exhibit significant heterogeneity with a variety of pathways leading to organ dysfunction and injury. There is now improved understanding of the cellular and molecular mechanisms causing kidney disease. Furthermore, careful interrogation of the pathways leading to renal fibrosis, the final common pathway of all progressive CKD, is providing hope for the development of anti-fibrotic therapies in the future. Remaining challenges include (1) biomarker discovery for earlier detection of TIF and to differentiate progressive versus non-progressive CKD; (2) elucidation of the origins, as well as the mechanism of activation, of the matrix-producing cells; (3) dissection of the molecular machinery responsible for matrix production, assembly, and deposition; and (4) development of therapies that effectively target molecular pathways of renal fibrosis without affecting normal homeostasis. Integration of new findings in histological analysis, molecular diagnostics, and experimental biology is generating optimism that new discoveries will eventually be able to improve diagnosis and treatment of CKD patients.

# Key concepts

- Modern molecular pathology is utilizing genomics, proteomics, and biomarker identification to supplement traditional serum creatinine, urinalysis, and renal biopsy.
- Diabetic nephropathy remains a major cause of kidney failure and results from glomerular hyperfiltration and hyperglycemia, among other factors.
- Membranous nephropathy diagnosis and treatment has been revolutionized by identification of the PLA$_2$R autoantigen.
- FSGS susceptibility may be driven by *APOL1* risk alleles that arose as protection against endemic *Trypanosoma* infections. Circulating factors also contribute.
- Acute kidney injury/acute tubular necrosis involves inflammation, tubular cell dysfunction and death, cell cycle arrest, and microvascular rarefaction.
- Tubulointerstitial fibrosis is a final common pathway for all kidney diseases and involves the replacement of normal kidney tissue with fibrotic scar. Inflammation, activation of matrix-producing cells, generation of extracellular matrix, and damage to normal cells all contribute to this process.

# Suggested readings

[1] Johnson SA, Spurney RF. Twenty years after ACEIs and ARBs: emerging treatment strategies for diabetic nephropathy. Am J Physiol Ren Physiol 2015;309:F807−20.
[2] Boor P, Ostendorf T, Floege J. Renal fibrosis: novel insights into mechanisms and therapeutic targets. Nat Rev Nephrol 2010;6:643−56.
[3] Bomback AS. Management of membranous nephropathy in the PLA$_2$R era. Clin J Am Soc Nephrol 2018;13:784−6.
[4] Fogo AB. Causes and pathogenesis of focal segmental glomerulosclerosis. Nat Rev Nephrol 2015;11:76−87.
[5] Genovese G, Friedman DJ, Ross MD, et al. Association of trypanolytic ApoL1 variants with kidney disease in African Americans. Science 2010;329:841−5.
[6] Lai KN. Pathogenesis of IgA nephropathy. Nat Rev Nephrol 2012;8:275−83.
[7] Tan RJ, Zhou D, Zhou L, Liu Y. Wnt/beta-catenin signaling and kidney fibrosis. Kidney Int Suppl 2014;4:84−90.
[8] Liu Y. Cellular and molecular mechanisms of renal fibrosis. Nat Rev Nephrol 2011;7:684−96.

[9]  Suzuki H, Fan R, Zhang Z, et al. Aberrantly glycosylated IgA1 in IgA nephropathy patients is recognized by IgG antibodies with restricted heterogeneity. J Clin Invest 2009;119:1668−77.

[10]  Sethi S, Fervenza FC. Membranoproliferative glomerulonephritis−a new look at an old entity. N Engl J Med 2012;366:1119−31.

[11]  Rabb H, Griffin MD, McKay DB, et al. Inflammation in AKI: current understanding, key questions, and knowledge gaps. J Am Soc Nephrol 2016;27:371−9.

[12]  Ferenbach DA, Bonventre JV. Mechanisms of maladaptive repair after AKI leading to accelerated kidney ageing and CKD. Nat Rev Nephrol 2015;11:264−76.

[13]  Molitoris BA. Therapeutic translation in acute kidney injury: the epithelial/endothelial axis. J Clin Invest 2014;124:2355−63.

[14]  Yang L, Besschetnova TY, Brooks CR, Shah JV, Bonventre JV. Epithelial cell cycle arrest in G2/M mediates kidney fibrosis after injury. Nat Med 2010;16(535−543):531 pp. following 143.

# Chapter 25

# Molecular pathogenesis of prostate cancer

William B. Coleman

*American Society for Investigative Pathology, Rockville, MD, United States*

**Summary**

The prostate is a male-specific hormone-responsive gland that is anatomically located in the retroperitoneal space and is physically associated with the urethra and neck of the bladder. The anatomy of the prostate gland reflects four biologically distinct zones: (1) the peripheral zone, (2) the central zone, (3) the transitional zone, and (4) the periurethral zone. Histologically, the prostate is a glandular tissue with a basal layer of low cuboidal epithelial cells covered by a layer of columnar secretory cells, with abundant fibromuscular stroma separating individual glands. Androgens regulate the growth and survival of the cells composing the prostatic tissue, and elimination of androgens (via castration) leads to atrophy of the prostate. Various prostate pathologies tend to occur in specific anatomic regions of the gland (zones). For example, the majority of hyperplastic proliferative lesions occur in the transitional zone, and the majority of prostate cancers occur in the peripheral zone. The prostate is affected by three major forms of pathology: (1) inflammation (or prostatitis), (2) benign nodular hyperplasia (or benign prostatic hyperplasia (BPH)), and (3) malignant prostate cancer. Prostatitis may develop secondary to bacterial infection (acute or chronic infection) or in the absence of recurrent infection (chronic abacterial prostatitis) and is often a granulomatous lesion. BPH occurs commonly among older men ($>50$ years old) and results from hyperplasia of prostate epithelial and stromal cells, presenting as discrete nodules in the periurethral region. With progressive enlargement of these nodules, obstruction of the urethra can occur resulting in difficulty in urination and an inability to efficiently empty the bladder. In-depth discussion of the pathogenesis of prostatitis and BPH is beyond the scope of this chapter. Rather, this chapter will focus on the molecular pathogenesis of prostate cancer.

## Introduction

The prostate is a male-specific hormone-responsive gland that is anatomically located in the retroperitoneal space and is physically associated with the urethra and neck of the bladder. The anatomy of the prostate gland reflects four biologically distinct zones: (1) the peripheral zone, (2) the central zone, (3) the transitional zone, and (4) the periurethral zone. Histologically, the prostate is a glandular tissue with a basal layer of low cuboidal epithelial cells covered by a layer of columnar secretory cells, with abundant fibromuscular stroma separating individual glands. Androgens regulate the growth and survival of the cells composing the prostatic tissue, and elimination of androgens (via castration) leads to atrophy of the prostate. Various prostate pathologies tend to occur in specific anatomic regions of the gland (zones). For example, the majority of hyperplastic proliferative lesions occur in the transitional zone, and the majority of prostate cancers occur in the peripheral zone.

The prostate is affected by three major forms of pathology: (1) inflammation (or prostatitis), (2) benign nodular hyperplasia (or benign prostatic hyperplasia or BPH), and (3) malignant prostate cancer. Prostatitis may develop secondary to bacterial infection (acute or chronic infection) or in the absence of recurrent infection (chronic abacterial prostatitis) and is often a granulomatous lesion. BPH occurs commonly among older men ($>50$ years old) and results from hyperplasia of prostate epithelial and stromal cells, presenting as discrete nodules in the periurethral region. With progressive enlargement of these nodules, obstruction of the urethra can occur resulting in difficulty in urination and an inability to efficiently empty the bladder. In-depth discussion of the pathogenesis of prostatitis and BPH is beyond the scope of this chapter. Rather, this chapter will focus on the molecular pathogenesis of prostate cancer.

**Essential Concepts in Molecular Pathology. DOI: https://doi.org/10.1016/B978-0-12-813257-9.00025-5**

# Incidence and etiology of prostate cancer

Risk factors for development of prostate cancer include age, race, inherited genes, and environmental factors (such as diet). In the United States in 2017, an estimated 161,360 new cases of prostate cancer were diagnosed, representing 19% of all cancers in men. During that same year, 26,730 deaths associated with prostate cancer occurred in the United States, representing 8% of all cancer-associated deaths among men. The incidence rate for prostate cancer has increased over the last several decades at a rate of approximately 4% per year, and this increase is paralleled by an increase in the mortality rate. These increases are likely related to the increasing average age of the male population, increased reporting, and increased screening of older men. Detection of prostate cancer can be achieved through the application of the digital rectal examination, screening based on detection of the prostate-specific antigen (PSA) in serum, and using ultrasonography. Elevations of PSA are associated with both BPH and malignant prostate cancer, but serum levels of this protein are markedly elevated in patients with prostate cancer. The serum PSA assay is extremely sensitive, but it lacks specificity, which limits its usefulness as a definitive diagnostic for prostate cancer. However, when used in combination with the digital rectal exam and ultrasonography, PSA increases the ability to detect occult prostate cancer.

Cancer of the prostate provides a dramatic example of the age-dependent cancer development. Cancers at this site occur with negligible frequency in men that are less than 55 years of age, and the vast majority of cases (56%) occur in men over the age of 65. Almost all prostate cancer cases (97%) occur in men older than age 50. In addition, prostate cancer occurs with greatly varied frequency among men of different races and ethnicity. In the United States, African-American men exhibit a significantly higher incidence of this cancer than Caucasian men, whereas American Indians demonstrate the lowest incidence of all groups. The underlying mechanisms that account for these dramatic differences in prostate cancer incidence are not readily apparent. However, difference in the levels of circulating testosterone among men from these different groups is one factor suggested to contribute to the observed variations in prostate cancer occurrence. The incidence of prostate cancer also varies widely from one region to another. In 2008, approximately 903,500 new cases of prostate cancer were diagnosed worldwide. Men living in developed countries are affected by prostate cancer more often than men living in other parts of the world. In 2008, prostate cancer affected 648,400 men in developed countries (ranked first for cancer incidence in this cohort) compared with 255,000 men in developing countries (ranked sixth for cancer incidence in this cohort). It is notable that the incidence of prostate cancer among men from the various regions of Africa is significantly lower than the rates for African-American men living in the United States.

Although the precise molecular mechanisms underlying prostate carcinogenesis and progression are currently unknown, histological studies indicate a multistage developmental progression of neoplastic lesions. The existence of premalignant lesions has been demonstrated in the prostate based on shared histological and molecular features with adenocarcinoma, as well as prevalence and severity. Ample evidence supports the hypothesis that prostate cancer can progress from intraepithelial neoplasia to invasive carcinoma, and ultimately metastasis and androgen-independent lethal disease. Prostatic intraepithelial neoplasia (PIN) is a term currently used to describe the closest precursor lesion to carcinoma. Histologically, the transition from normal epithelium to PIN involves nuclear atypia, epithelial cell crowding, and some component of basal cell loss. Autopsy studies suggest that PIN lesions precede the appearance of carcinoma. High-grade PIN lesions are often spatially associated with carcinoma and may exhibit molecular alterations similar to those found in prostate tumor cells. The cells of origin of premalignant lesions of the prostate and prostate cancer remain controversial, but greater understanding is emerging. Likewise, the role of prostate cancer stem cells in the development of prostate cancers and premalignant lesions continues to be investigated.

Glandular atrophy is frequently observed in the prostate tissue of aging men. Histologically, luminal spaces appear dilated with flattened epithelial linings. Atrophic lesions in the prostate consist of a number of histological variants, and some of these have also been proposed to be potential precursors to prostate cancer based in part on their frequent occurrence in proximity to malignant lesions. Proliferative inflammatory atrophy (PIA) is a term used to describe a range of morphologies from simple atrophy to postatrophic hyperplasia. PIA lesions are highly proliferative and often associated with inflammation. Epithelial cells within PIA foci express both luminal and basal markers. The lesions likely result from cellular injury initiated by inflammation and/or carcinogen insult, and at times show evidence of transition to high-grade PIN or rarely to invasive cancer. Given the intermediate phenotype of these cells and the characteristic expression of stress-associated genes, PIA lesions are thought to reflect the morphological manifestation of prostate epithelial damage and regeneration. Some genetic and epigenetic alterations observed in PIN and prostatic adenocarcinoma have also been found in PIA lesions, albeit to a lesser degree. These findings suggest a potential link between a subset of atrophic lesions and adenocarcinoma. A multistep progression model of prostate cancer development has been proposed. In this model, normal prostatic epithelium undergoes focal atrophy, which leads to PIN, followed by progression to invasive cancer. In this model, ongoing injury to the prostatic epithelium secondary to inflammation and/or carcinogen exposures, results in cell damage and death leading to tissue regeneration, which manifests morphologically as prostatic atrophy. Atrophic prostate cells acquire and accumulate somatic genetic alterations during self-renewal, including methylation of the *GSTP1* promoter, telomere shortening, and activation of MYC, leading to neoplastic transformation.

Prostate cancer arises from the accumulation of genetic and epigenetic alterations. Cytogenetic analyzes demonstrate the prevalence of chromosomal aberrations associated with prostate tumorigenesis, and many genes that map to deleted or amplified regions have been investigated for their roles in disease progression. In recent years, genome-wide profiling of prostate cancer has led to the identification of a number of biomarkers and pathways that are altered during prostate carcinogenesis, some of which may represent potential molecular targets for therapy. Despite the advancing knowledge of genes altered in prostate cancer, the precise molecular pathways, their combinatorial relationships, and the temporal order of molecular events occurring in the development of preneoplastic lesions and adenocarcinoma are still being discovered and refined.

## Genetic contributions to prostate cancer risk

Prostate cancer is known to have a hereditary component, and numerous investigators have contributed to the identification of familial prostate cancer genes. A limited set of germline polymorphisms and mutations is associated with increased prostate cancer risk. Based on linkage analysis, the inherited susceptibility locus *HPC1* (hereditary prostate cancer 1), which encodes Ribonuclease L (RNASEL), was mapped to human chromosome region 1. Germline mutations in the macrophage scavenger receptor 1 (*MSR1*) locus have also been linked to prostate cancer risk. The *MSR1* gene is located at chromosomal region 8p22 and is expressed in infiltrating macrophages. Many other loci have been implicated as well, and further characterization of these genes in the etiology of prostate cancer is warranted. Several groups have employed genome-wide association studies (GWAS) to show a number of single nucleotide polymorphisms at novel loci related to prostate cancer risk, several of which map to chromosome 8q24. More recently, whole genome sequencing and whole exon sequencing have been applied to prostate cancer. These studies have identified a number of signaling pathways that are activated in subsets of prostate cancer.

## Somatic alterations in gene expression

Acquired somatic gene alterations in prostate cancer cells include cytosine methylation alterations within CpG dinucleotides, point mutations, deletions, amplifications, and telomere shortening. Inactivation of classic tumor suppressor genes (for example, *TP53* and *RB1*) has been found in primary prostate cancers and cancer cell lines. However, these alterations are far more common in advanced hormone-refractory and/or metastatic cancers. As with other epithelial cancers, cytogenetic studies using fluorescence in situ hybridization (FISH) and comparative genomic hybridization have identified chromosomal regions frequently gained and lost in prostate cancer. The most common chromosomal abnormalities are losses at 8p, 10q, 13q, and 16q and gains at 7p, 7q, 8q, and Xq. Recent developments based on prostate cancer cytogenetics and molecular characteristics have identified novel biomarkers that may be useful in replacements for PSA testing in prostate cancer detection.

### MYC

MYC protein functions as a nuclear transcription factor that impacts a wide range of cellular processes including cell cycle progression, metabolism, ribosome biogenesis, and protein synthesis. The *MYC* oncogene maps to chromosome 8q24, and this region is amplified in a number of human cancers. The complexity of MYC-regulated transcriptional networks has been intensely studied since the late 1990s, yet the precise role of the *MYC* oncogene during neoplastic transformation and its direct molecular targets in prostate tumorigenesis remain largely unknown. Initial reports of increased *MYC* gene expression in PIN lesions and ~30% of invasive cancers, combined with the correlation of 8q24 amplification with high Gleason grade and metastatic carcinoma, suggested that alterations in MYC were associated with advanced disease. This was in contrast to increased levels of *MYC* mRNA detected in most prostate cancers, including low-grade cases. However, evidence suggests that MYC upregulation at the protein level is an early and common event in primary prostate cancer cases. Using an improved antibody for immunohistochemical analyzes on tissue microarrays, nuclear MYC staining was increased in the luminal epithelial cells of PIN, PIA, and malignant lesions compared to benign tissues. Interestingly, FISH analysis revealed a positive correlation between gain of 8q24 and Gleason grade, but not overall MYC protein levels. These findings suggest a key role for MYC upregulation in the initiation of prostate tumorigenesis that is likely independent of gene amplification.

## NKX3.1

*NKX3.1* is one of the several candidate tumor suppressor genes located on chromosome 8p. The *NKX3.1* gene encodes a homeodomain transcription factor that is the earliest known marker of prostate epithelium during embryogenesis. Its expression persists in the epithelial cells of the adult gland and is required for maintenance of ductal morphology and the regulation of cell proliferation. Human *NKX3.1* maps within the minimal deletion interval of chromosomal region 8p21. Loss of heterozygosity at 8p21 has been observed in 63% of high-grade PIN lesions and ~70% of prostate cancers, although this percentage may be significantly less. Methylation of CpG dinucleotides upstream of the *NKX3.1* transcriptional start site and the existence of germline variants within the homeodomain have been reported. However, the order of occurrence relative to 8p21 loss and prostate cancer initiation is unknown. Nevertheless, the evidence suggests that haploinsufficiency of *NKX3.1* plays a role in prostate cancer development.

Several studies have analyzed NKX3.1 expression by immunohistochemistry in human prostate cancers. In studies of PIN and prostate adenocarcinoma, both decreased intensity and loss of NKX3.1 protein staining compared to benign tissue have been reported. Decreased expression is correlated with high Gleason score, advanced tumor stage, the presence of metastatic lesions, and hormone-refractory disease. Immunohistochemical analyzes with a novel NKX3.1 antibody revealed a dramatic decrease in the level of NKX3.1 in PIA lesions. As NKX3.1 is thought to regulate cell proliferation, loss of expression in atrophic epithelial cells may contribute to increased proliferative capacity and amplify genetic changes that occur in epithelial cells within these regenerative lesions. Consistent with the previous reports, NKX3.1 staining intensity was significantly diminished in PIN lesions and prostate adenocarcinoma. Diminished NKX3.1 protein expression correlated with 8p loss in high-grade tumors, but not PIN or PIA. Intriguingly, the levels of *NKX3.1* mRNA and protein were found to be discordant in most prostate cancer cases examined, confirming previous reports and suggesting that multiple mechanisms lead to sporadic loss of NKX3.1 during prostate cancer development and progression.

Although the precise transcriptional targets of NKX3.1 are largely unknown, microarray analyzes suggest that lack of *NKX3.1* results in increased oxidative DNA damage by controlling expression of antioxidant enzymes. A limited set of proteins has been identified to physically interact with NKX3.1, including serum response factor and prostate-derived Ets transcription factor. The serine-threonine kinase CK2 and the ubiquitin ligase TOPORS regulate the stability of *NKX3.1* in prostate cancer cells in vitro. The functional significance of these interactions in precursor and cancer lesions is not yet clear.

## PTEN

The phosphatase and tensin homolog (*PTEN*) gene is a well-characterized tumor suppressor that maps to chromosomal region 10q23 and acts as a negative regulator of the phosphatidylinositol 3-kinase/AKT (PI3K/AKT) signaling, which functions in cell survival. PTEN inhibits growth factor signals sent through PI3 kinase by dephosphorylating the PI3K product, phosphatidylinositol 3,4,5-trisphosphate (PIP3). Loss of PTEN expression results in the downstream activation (phosphorylation) of AKT, an inhibitor of apoptosis and promoter of cell proliferation. Aberrant PTEN expression has been implicated in numerous cancers, including metastatic prostate cancers, which at times exhibit homozygous deletion of *PTEN*. However, the majority of primary prostate cancer cases with genetic alterations in *PTEN* harbor loss of heterozygosity at the *PTEN* locus without mutations in the remaining allele.

## Androgen receptor

The role of androgen receptor (AR) is central to prostate pathobiology, as the prostate is dependent on androgens for normal growth and maintenance. AR is a nuclear steroid hormone receptor with high expression in luminal epithelial cells and little expression in basal epithelial cells. In the absence of ligand, AR is inactive and bound to heat shock chaperone proteins. On binding of the active form of testosterone (dihydrotestosterone, DHT), AR is released and translocates to the nucleus where it physically associates with cofactors to regulate target gene transcription. AR activity is essential for the development of prostate cancer, and AR expression is evident in high-grade PIN and most adenocarcinoma lesions. Androgen ablation, through the use of antiandrogens, castration, or gonadotropin superagonists is the mainline therapy for advanced, metastatic prostate cancer. Although the majority of patients respond to this treatment, it eventually fails and the tumors become androgen-independent. As prostate cancer progresses from androgen-dependence to a hormone-refractory state, AR is often amplified at the Xq12 region or mutated to respond to a range of ligands for androgen-independent activation and cancer growth. In addition, AR has been reported to engage in cross-talk with other mitogenic signaling pathways such as PI3K/AKT and MAPK as a means of enhancing androgen-independent tumor progression.

## TMPRSS2-ETS gene fusions

Ets-related gene-1 (ERG) is an oncogenic transcription factor upregulated in the majority of prostate cancers. Evidence of recurrent chromosomal rearrangements in prostate cancer was first reported by Tomlins and colleagues. By the use of a bioinformatic approach to analyze DNA microarray studies, gene fusions between the promoter/enhancer region of the androgen-responsive *TMPRSS2* gene and members of the *ETS* family (including *ERG* and *ETV1*) were found in ~90% of prostate cancers with known overexpression of *ERG*. This discovery has fueled efforts to characterize the functional implications of aberrant chromosomal fusions on tumor progression and clinical outcome. Presently, there are conflicting data regarding the presence of *TMPRSS-Erg* gene fusions and clinical outcomes. Although initial studies suggested these fusions were not present in PIN lesions, it has become clear that some PIN lesions indeed harbor fusions, and this suggests that gene fusion may lead to neoplastic transformation itself and not specifically to the invasive phenotype. These studies demonstrate that upregulation of *ERG* may contribute to transformation of prostate epithelial cells but is insufficient to initiate prostate cancer.

## p27$^{kip1}$

The cyclin-dependent kinase inhibitor p27$^{kip1}$ is a candidate tumor suppressor encoded by the *CDKN1B* gene. To prevent cell cycle progression, p27$^{kip1}$ binds to and inhibits cyclin E/CDK2 and cyclin A/CDK2 complexes. Although mutations are rare, loss of p27$^{kip1}$ expression results in hyperplasia and malignancy in many organs, including the prostate. In normal and benign prostate tissues, p27$^{kip1}$ is expressed at high levels in most luminal epithelial cells and much more variably in basal cells. However, p27$^{kip1}$ expression is decreased in most high-grade PIN and malignant lesions. Several studies have shown decreased p27$^{kip1}$ to correlate positively with increased cell proliferation, PSA relapse, high tumor grade, and advanced cancer stage. The molecular mechanism by which p27 protein is decreased in prostate cancer has not been clarified, although posttranscriptional regulation of p27$^{kip1}$ is evident in prostate cancer cases where *p27$^{kip1}$* mRNA is expressed at high levels, but there is a lack of protein expression.

## Telomeres

Telomeres are specialized structures composed of repeat DNA sequences at the ends of chromosomes that are complexed with binding proteins and required for maintenance of chromosomal integrity. In cells lacking sufficient levels of the enzyme telomerase, telomeres progressively shorten with each cell division as a result of the end-replication problem and/or oxidant stress. The enzyme telomerase can add new repeat sequences to the ends of chromosomes, which stabilizes the telomeres and ensures proper telomere length. Excessive shortening can lead to improper segregation of chromosomes during cell division, genomic instability, and the initiation of tumorigenesis. The majority of high-grade PIN and prostate cancer cases have abnormally short telomeres exclusively in the luminal cells. This supports the notion that cells in the luminal compartment may be the target of neoplastic transformation. Oxidative damage can result in telomere shortening, and this is consistent with the proposed inflammation-oxidative stress model of prostate cancer progression.

## MicroRNAs

MicroRNAs (miRNAs) are small noncoding RNA molecules that negatively regulate gene expression by interfering with translation. They are initially generated from primary transcripts (pri-mRNAs) and processed by the RNase III endonucleases Drosha and Dicer to produce mature miRNA molecules. In the cytoplasm, the mature miRNA associates with the RNA-induced silencing complex (RISC) and binds to the 3′ UTR of its target mRNA, leading to degradation or transcriptional silencing. Since their discovery, miRNAs have been shown to play key roles in development, and there is increasing evidence of their widespread dysregulation in cancer. A limited number of studies have reported a predominant decrease in the levels of miRNAs in prostate carcinoma, including let-7, miR-26a, miR-99, and miR-125-a-b. MicroRNAs that are consistently dysregulated in prostate tumorigenesis have a strong relationship with disease progression. Recent studies have identified microRNA signatures associated with the multistage development and progression of prostate cancer, from premalignant lesions in prostate cancer metastasis. Hence, microRNAs and microRNA signatures represent potentially valuable biomarkers for development of prostate cancer and prediction of disease course.

## Epigenetics

Epigenetic alterations in prostate carcinogenesis include changes in chromosome structure through abnormal deoxycytidine methylation of wild-type DNA sequences and histone modifications (acetylation, methylation). Epigenetic events occur earlier in prostate cancer progression and more consistently than recurring genetic changes. However, the mechanisms through which these changes arise are poorly understood. Silencing of genes secondary to aberrant DNA methylation may occur as a result of altered DNA methyltransferase (DNMT) activity. DNMTs establish and maintain the patterns of methylation in the genome by catalyzing the transfer of methyl groups to deoxycytidine in CpG dinucleotides. Several genes that are silenced via DNA methylation-dependent epigenetic alterations have been identified.

### GSTP1

Glutathione S-transferases are enzymes responsible for the detoxification of reactive chemical species through conjugation to reduced glutathione. The *GSTP1* gene, encoding the pi class glutathione S-transferase, was the first hypermethylated gene to be characterized in prostate cancer. GSTP1 protects prostate epithelial cells from carcinogen-associated and/or oxidative stress−induced DNA damage, and loss of GSTP1 renders prostate cells unprotected from such genomic insults. As an example, in LNCaP prostate cancer cells (which are devoid of GSTP1 as a result of epigenetic gene silencing) restoration of GSTP1 function affords protection against metabolic activation of the dietary heterocyclic amine carcinogen 2-amino-1-methyl-6-phenylimidazo [4,5-β] pyridine (PhIP), known to cause prostate cancer when ingested to rats. The loss of enzymatic defenses against reactive chemical species encountered as part of dietary exposures or arising endogenously associated with epigenetic silencing of GSTP1 provides a plausible mechanistic explanation for the marked impact of the diet and of inflammatory processes in the pathogenesis of human prostate cancer.

The most common genomic DNA mark accompanying epigenetic gene silencing in cancer cells is an accumulation of 5-meC bases in CpG dinucleotides clustered into CpG islands encompassing transcriptional regulatory regions. Hypermethylation of these CpG island sequences directs the formation of repressive heterochromatin that prevents loading of RNA polymerase and transcription of hnRNA that can be processed for translation into protein. For *GSTP1*, the 5-meCpG dinucleotides begin to appear in the gene promoter region of rare cells in PIA lesions, with more dense CpG island methylation changes emerging as PIA lesions progress to PIN and then cancer. The proliferative expansion of cells with hypermethylated *GSTP1* CpG island sequences as PIA lesions progress hints at some sort of selective growth advantage, although how loss of GSTP1 can be selected during prostatic carcinogenesis has not been established. In addition to *GSTP1*, many other critical genes undergo epigenetic silencing during the pathogenesis of prostate cancer. The mechanisms by which epigenetic defects arise in prostate cancer cells, or in other human cancer cells, have not been completely discerned. However, the consistent appearance of such changes in inflammatory precancerous lesions and conditions, including PIA lesions, inflammatory bowel disease, chronic active hepatitis, and others, supports the contribution of inflammatory processes to some sort of epigenetic or DNA methylation catastrophe.

### APC

The adenomatous polyposis coli (APC) protein is a component of the Wnt/β-catenin signaling pathway that negatively regulates cell growth. APC complexes with glycogen synthase kinase-3 (GSK3) and axin. This complex is responsible for targeting free cytosolic β-catenin for ubiquitin-mediated degradation. The pathway is activated by the binding of the Wnt protein to the frizzled family of seven transmembrane receptors and LRP5/6, followed by downregulation of GSK3β, which allows accumulation of β-catenin and subsequent translocation of β-catenin to the nucleus. Once inside the nucleus, β-catenin is able to activate transcription of Wnt target genes.

Indirect support of the concept that APC inactivation may be mechanistically tied to prostate cancer progression comes from studies showing frequent hypermethylation of its promoter region and that the extent of methylation correlates with stage, grade, and biochemical recurrence. Activating mutations occur in approximately 5% of prostate cancers, and aberrant nuclear localization of β-catenin appears to occur only somewhat more frequently. The latter finding suggests that if APC is functioning as a tumor suppressor in prostate adenocarcinoma, then its primary role in the prostate may not relate to nuclear translocation of β-catenin. It is not well established if APC is involved in prostate carcinogenesis and cancer progression or not. However, methylation of the *APC* promoter could become a useful biomarker in prostate cancer diagnosis because this methylation event may be detectable in bodily fluids such as urine or blood.

## Key concepts

- Prostatic adenocarcinomas arise from precursor lesions termed proliferative inflammatory atrophy (PIA) and prostatic intraepithelial neoplasia (PIN).
- PIA lesions, which are characterized by epithelial damage, regeneration, and inflammatory cell infiltration, appear in response to a variety of pro-carcinogenic stresses.
- Heredity contributes significantly to the risk of prostate cancer development. Inherited prostate
- cancer susceptibility genes/loci discovered to date include RNASEL, encoding an enzyme that acts as a defense against viral infection, MSR1, encoding a receptor for bacteria on macrophages, and several sites at 8q24.
- Consistent alterations in the expression of key genes, with overexpression of MYC and AR, and underexpression of NKX3.1, PTEN, p27, GSTP1, and APC, accompany the pathogenesis of prostate cancer.
- Shortening of telomere sequences ubiquitously appears early during prostate cancer development.
- Somatic epigenetic alterations, with hypermethylation and transcriptional silencing of several key genes, are the most abundant and earliest genome abnormalities, evident in PIA and PIN as well as in prostate cancer.
- Acquired genetic defects in prostate cancer include translocations and deletions that give rise to fusion transcripts between androgen-regulated genes, such as TMPRSS2, and genes encoding ETS family transcription factors.

## Suggested readings

[1] Aaron L, Franco OE, Hayward SW. Review of prostate anatomy and embryology and the etiology of benign prostatic hyperplasia. Urol Clin North Am 2016;43:279—88.

[2] Siegel RL, Miller KD, Jemal A. Cancer statistics. CA Cancer J Clin 2017;67:7—30.

[3] Jemal A, Bray F, Center MM, et al. Global cancer statistics. CA Cancer J Clin 2011;61:69—90.

[4] Packer JR, Maitland NJ. The molecular and cellular origin of human prostate cancer. Biochim Biophys Acta 2016;1863:1238—60.

[5] Barbieri CE, Tomlins SA. The prostate cancer genome: perspectives and potential. Urol Oncol 2014;32:53.e15—22.

[6] Eeles R, Goh C, Castro E, et al. The genetic epidemiology of prostate cancer and its clinical implications. Nat Rev Urol 2014;11:18—31.

[7] Fonseka LN, Kallen ME, Serrato-Guillen A, et al. Cytogenetics and molecular genetics of prostate cancer: a comprehensive update. J Assoc Genet Technol 2015;41:100—11.

[8] Leite KR, Tomiyama A, Reis ST, et al. MicroRNA expression profiles in the progression of prostate cancer — from high-grade prostate intraepithelial neoplasia to metastasis. Urol Oncol 2013;31:796—801.

[9] Keil KP, Vezina CM. DNA methylation as a dynamic regulator of development and disease processes: spotlight on the prostate. Epigenomics 2015;7:413—25.

[10] Rodrigues DN, Boysen G, Sumanasuriya S, et al. The molecular underpinnings of prostate cancer: impacts on management and pathology practice. J Pathol 2017;241:173—82.

[11] Hoang DT, Iczkowski KA, Kilari D, et al. Androgen receptor-dependent and -independent mechanisms driving prostate cancer progression: opportunities for therapeutic targeting from multiple angles. Oncotarget 2017;8:3724—45.

# Chapter 26

# Molecular biology of breast cancer

Mariana Brandão, Philippe Aftimos, Hatem A. Azim, Jr. and Christos Sotiriou
*Institut Jules Bordet, Université Libre de Bruxelles, Brussels, Belgium*

## Summary

Breast cancer has been increasingly recognized as a heterogeneous disease with distinct subtypes being identified with the advances of molecular techniques and diagnostics. Although subtyping mainly depended on immunohistochemical markers such as the estrogen receptor, the progesterone receptor, and the human epidermal growth factor receptor 2 (HER2), novel subtypes within the three historical subtypes have been defined using gene expression profiling and next-generation sequencing. This new classification has both prognostic and predictive value not only in the advanced setting but also in the treatment of early breast cancer. Furthermore, the discovery of genomic markers of sensitivity and resistance has led to the approval of multiple targeted therapies, whereas others are still in development. The use of biomarkers will also serve treatment deescalation because the first results of large studies testing genomic signatures in the early setting favor the use of gene expression profiling in treatment decisions. Finally, the discovery of immune markers is paving the way for new immune signatures. Indeed, tumor-infiltrating lymphocytes are starting to be used as prognostic and predictive markers while novel immunotherapy agents are being tested both in the advanced and early setting.

## Introduction

In recent years, breast cancer has increasingly been recognized as a biologically heterogeneous disease, thanks to advances in gene expression profiling and massively parallel sequencing. Many of these differences are attributable to the heterogeneity that exists at the molecular level, and as a result the discovery and development of molecular markers have recently taken center stage. Biomarkers associated with resistance to endocrine or other targeted therapies are increasingly being incorporated into treatment algorithms. Furthermore, stroma and immune cells are being recognized as prognostic and predictive factors. These and other new developments hold promise for advancing the concept of precision oncology in breast cancer, giving us the tools with which to unravel the mysteries behind metastasis and to improve cure rates in early disease.

## Histopathological classification

Breast cancer has traditionally been classified according to histopathological subtypes and anatomical staging (TNM classification). For many years, this classification was the only tool to assess relapse risk after local therapy and to inform decision-making about adjuvant systemic therapy. Afterwards, estrogen receptor (ER), progesterone receptor (PR), and human epidermal growth factor receptor 2 (HER2) expression status have emerged as vital immunohistochemical markers both for prognosis and for guiding decision-making about endocrine and targeted therapy. Nowadays, newer paradigms are evolving to further refine breast cancer characterization at the histopathological level, including the heterogeneity of multifocal disease, infiltrating lymphocytes, the role of tumor microenvironment, and many others.

### Histopathological features of breast cancer

Histopathological features create the foundation of traditional breast cancer pathology. They consist of (1) histological type, (2) grade, and (3) vascular invasion and provide clinicians with essential information for prognosis and treatment decision-making.

Essential Concepts in Molecular Pathology. DOI: https://doi.org/10.1016/B978-0-12-813257-9.00026-7

## Histological type

The definition of breast cancer comprises invasive carcinoma and *in situ* carcinoma (which is noninvasive). The fourth edition of the WHO Classification of Tumors of the Breast defines 21 subtypes of invasive breast carcinoma, the most frequent being what was formerly called invasive ductal carcinoma, now known as invasive carcinoma of no special type (NST) or invasive ductal carcinoma NST. Today, after demonstration that most breast cancers arise from the same location—the terminal duct lobular unit—the histomorphological differences between ductal and lobular carcinomas are regarded as manifestations of their distinct molecular profiles, also responsible for distinct clinical courses, despite similar survival rates.

**Lobular Neoplasia.** Lobular neoplasias are composed of noninvasive lesions, such as atypical lobular hyperplasia and lobular carcinoma *in situ*, and invasive lobular carcinoma (ILC). They represent a continuum of diseases, ranging from risk factors for cancer to cancers with the potential for metastases. ILC accounts for about 14% of invasive breast cancers and is represented by diverse subtypes, the classic one and its variants, which have worse prognosis than the classic form. Lobular neoplasia (*in situ* lobular neoplasia and ILC) is characterized by a population of small aberrant cells with small nuclei, individual private acini, and a lack of cohesion between cells. The distinctive molecular feature of lobular neoplasia is the loss of E-cadherin (*CDH1*) that can be demonstrated by immunohistochemistry (IHC). Because ductal carcinomas can also be E-cadherin-negative, it should not be considered pathognomonic for lobular neoplasia. Other molecular characteristics of lobular carcinomas are epidermal growth factor receptor 1 (EGFR-1) and HER2 receptor negativity, but positivity for antibody 34b E12 (recognizing cytokeratins 1, 5, 10, and 14), ER, and PR. These markers, while helpful, are also not pathognomonic, with the pleomorphic variant often described as ER-negative, PR-negative, and HER2-positive. Therefore, such molecular markers still need to be correlated with the classic histopathological findings. A further refinement of the ILC subtypes has recently become possible through the application of omics technologies, which have uncovered molecular alterations potentially driving the disease.

**Ductal Carcinoma *in situ*.** DCIS is characterized by the proliferation of malignant cells within the ducts without invasion of the surrounding stromal tissue. The European Organisation for Research and Treatment of Cancer (EORTC) grading system, based on cytonuclear patterns, groups DCIS on low, intermediate, and high grade. Most ductal carcinomas are diffusely positive for luminal cell markers (CK8, CK18, CK19) but negative for basal cell markers (CK5/6 and CK14). By contrast, benign ductal hyperplasia may show a mosaic-staining pattern for any of these markers, indicating a heterogeneous underlying cell population. DCIS is multifocal in 30% of the cases, often in the same breast. In 2%–6% of cases, axillary lymph node invasion is found at pathological examination, and this is thought to occur as a result of an unidentified invasive component. In the era of widespread mammographic screening, DCIS is more frequently detected and represents 25% of screen-detected breast cancers.

The University of Southern California/Van Nuys prognostic index incorporating four independent predictors of local recurrence (tumor size, margin width, pathologic classification, and age) is commonly used to determine which patients require radiation therapy in addition to excision. However, our understanding of the biology of DCIS remains crude, and further clinical trials are necessary to optimize local and systemic treatment modalities.

**Invasive Ductal Carcinoma.** Invasive carcinoma of NST, formerly called invasive ductal carcinoma, is the most common invasive carcinoma of the breast, representing roughly 70% of all cases. ILC is the second most common histopathological type representing 10%–20% of all breast malignancies. The other 10%–20% consists mostly of medullary, mucoid, papillary, and inflammatory carcinoma (occurring in 1%–2% of all cases). Inflammatory carcinomas typically mimic the clinical presentation of benign inflammatory disease although, histologically, they are characterized by extensive invasion of the lymphatic vessels within the dermis. Other types of rare invasive breast cancer include adenoid cystic carcinoma, apocrine carcinoma, and carcinoma with squamous metaplasia. An in-depth characterization of these special types might allow novel therapeutic targets to be identified.

Often the distinction between invasive and noninvasive carcinoma is difficult to ascertain because carcinoma cells can be found in both types of lesions and there are no reliable markers that differentiate between invasive and noninvasive cells. E-cadherin is often helpful in distinguishing malignant disease, because it is mostly positive in ductal carcinomas but negative in lobular carcinomas. Invasive ductal carcinoma may express a variety of markers that have been extensively evaluated.

## Vascular invasion

The term vascular invasion refers to the invasion of lymphatic and blood vessels with cancer cells. It is assessed in hematoxylin–eosin stained slides. Lymphovascular invasion is routinely included in the pathological evaluation and reporting of all breast cancers; however, its interpretation is often difficult. It is a poor prognostic factor and predicts for increased local failure and reduced overall survival (OS). Lymphovascular invasion by cancer cells is found in ~15% of invasive ductal carcinoma of the breast and is present in 10% of cancers without axillary lymph node invasion.

### Histological grading

Histological grading is the classic and simplest evaluation method for prognostication in breast cancer patients, requiring only hematoxylin−eosin staining. It typically consists of three factors: tubule formation, mitotic count, and nuclear atypia. These factors are all important in identifying patients at high risk of recurrence; however, the relative importance of each feature is unclear. In addition, intratumor heterogeneity and interobserver variation add to the difficulty of accurate prognostication based on grading.

## Biomarkers

Molecular analyzes of cancer have led to the discovery of oncogenes and tumor suppressor genes. In practice, pathologists look for protein products (biomarkers) of the genes mainly through IHC or immunocytochemistry. Biomarkers are potentially helpful in distinguishing between various histopathological types of breast cancer, as well as in assessing prognosis and predicting response to specific systemic therapy.

## The estrogen receptor

ERs are members of a large family of nuclear transcriptional regulators that are activated by steroid hormones, such as estrogen. ERs exist as two isoforms, $\alpha$ and $\beta$, which are encoded by two different genes. Although both isoforms are expressed in the normal mammary gland, it appears that only ER$\alpha$ is critical for normal gland development. Nonetheless, there is growing evidence that ER$\beta$ may antagonize the function of ER$\alpha$ and that high levels of ER$\beta$ are associated with a more favorable response to tamoxifen treatment.

Steroid hormones, particularly estrogen, play a major role in the development of breast cancer and approximately 75% of all breast cancers express the ER (ER+). The actions of estrogen and the ER are complex, with binding followed by dimerization and translocation to estrogen-responsive elements in the promoter region of genes and subsequent modulation of transcription. Patients with ER-negative (ER −) cancers derive no benefit from endocrine therapy, whereas for ER+ cancers, estrogen signaling blockade is an important therapeutic strategy that can lead to improved outcomes, including increasing cure rates in early breast cancer, improving response rates (RRs) and disease control in advanced disease, and reducing breast cancer incidence.

### Testing for estrogen receptor

The introduction of routine ER testing in the early 1990s changed the way in which endocrine therapy is prescribed, particularly in preventing unnecessary tamoxifen-related toxicity for patients with ER − disease. Yet, the treatment of individual patients with ER+ breast cancer is less clear-cut, largely because of the ambiguity created by varying methodologies and cutoffs employed by different laboratories for ER testing. Indeed, responses have been demonstrated in cancers with as few as 1%−10% of ER+ cells by IHC. It is therefore more important clinically to distinguish between ER-absent disease and measurable, but low-level ER expression, variably reported as either ER+ or ER −, depending on which cutoff values are used.

Quantitative ER as a continuous variable has been demonstrated to be proportionally correlated to the response to endocrine therapy in metastatic breast cancer. In the adjuvant setting, a large meta-analysis has shown a similar relationship between ER expression and tamoxifen efficacy, and in the neoadjuvant setting, for ER expression and response to both tamoxifen and letrozole. Nevertheless, quantitative ER is still an imprecise predictive tool, because even breast cancers in the highest strata of ER expression can be endocrine resistant (*de novo* resistance). Furthermore, a significant proportion of patients with ER+ advanced disease who initially respond to endocrine therapy will eventually fail treatment (acquired resistance). Although this resistance is not fully understood, several mechanisms have been hypothesized to be responsible for the development of resistance, including (1) loss of ER, (2) selection of clones with ER mutations, (3) deregulation of cell cycle components including ER regulatory proteins, and (4) crosstalk between ER and other growth factor receptor pathways. Given the limitations of ER assessment, the development of other molecular tools, reflecting the complex biology of these cancers, is needed to improve treatment for patients with ER+ disease.

## Resistance to endocrine therapy

Feedback loops with growth factor signaling pathways may play a vital role in resistance to endocrine therapy. Aberrant signaling through the phosphatidylinositol 3-kinase (PI3K)/Akt/mammalian target of rapamycin (mTOR) signaling pathway is one mechanism of endocrine resistance acting via phosphorylation of the function domain 1 of the ER by S6 kinase 1, a substrate of mTOR complex 1. Targeting this pathway has been shown to have important clinical implications, as demonstrated by the improvement in progression-free survival by the addition of a mTORC1 inhibitor (everolimus) or a PI3K-alpha inhibitor (alpelisib or taselisib) to endocrine therapy, in ER+ metastatic breast cancer patients after progression or relapse on a nonsteroidal aromatase inhibitor. Only patients with *PIK3CA*-mutated tumors derived benefit from PI3K-alpha inhibitors, but unfortunately, no biomarker is available yet to identify which patients derive superior benefit from everolimus.

Dysregulation of the cell cycle is a second important pathway targeted for drug development that has led to the regulatory approval of a new class of drugs. The cyclin-dependent kinases (CDKs) 4/6 play a role in the regulation of the G1/S transition through action on the phosphorylation of pRb. Pivotal trials showed that the addition of CDK4/6 inhibitors (palbociclib, ribociclib or abemaciclib) to endocrine therapy, both in endocrine-sensitive and endocrine-resistant metastatic breast cancer patients, led to gains in progression-free survival.

## The progesterone receptor

The *PR* gene is an estrogen-regulated gene. PR mediates the effect of progesterone in the development of the mammary gland and breast cancer. Among breast cancers that express ER, more than half also express PR. It has been hypothesized that PR levels in breast cancer may be a marker of an intact ER signal transduction pathway and that PR levels may therefore add independent predictive information. Emerging laboratory and clinical data also suggest that among ER+ cancers, PR status may predict differential sensitivity to antiestrogen therapy.

Regarding the nature of ER+PR − cancers, some studies have shown that this phenotype may evolve through loss of PR, whereas other studies suggest that they reflect a distinct molecular origin, with unique epidemiologic risk factors when compared with ER+PR+ disease. Patients with ER+PR- cancers appear to derive less benefit from adjuvant tamoxifen than patients with ER+PR+ cancers, but may have increased responsiveness to aromatase inhibitors, although this is still controversial. Loss of PR may also be a negative prognostic factor, independently of quantitative ER levels.

## The HER2 receptor

HER2 has emerged as an important molecular target in the treatment of breast cancer. Much of the recent success with anti-HER2 therapy has been a direct result of being able to properly select the right patient subpopulation for targeted treatment.

HER2 belongs to the human epidermal growth factor receptor family of tyrosine kinases consisting of EGFR (HER1), HER2, HER3, and HER4. All of these receptors have an extracellular ligand-binding region, a single membrane-spanning region, and a cytoplasmic tyrosine−kinase-containing domain, except HER3, which does not possess an intracellular kinase domain. Ligand binding to the extracellular region results in homodimer and heterodimer activation of the cytoplasmic kinase domain and phosphorylation of a specific tyrosine kinase. This leads to the activation of various intracellular signaling pathways, such as the mitogen-activated protein kinase (MAPK) and the PI3K-AKT pathways involved in cell proliferation and survival. HER signaling can become dysregulated via a number of mechanisms, including (1) overexpression of a ligand, (2) overexpression of the normal HER receptor, (3) overexpression of a constitutively activated mutation of the HER receptor, or (4) defective HER receptor internalization, recycling, or degradation.

Early studies suggested that as many as 30% of breast cancers demonstrate HER2 overexpression, which is associated with a more aggressive phenotype and a poorer disease-free survival (DFS). Most importantly, HER2 status was found to be predictive of benefit from anti-HER2 therapies, such as trastuzumab.

## Identifying the HER2 receptor

The accurate determination of HER2 status is vitally important because it is such a useful marker for therapeutic decision-making, both in the early and advanced settings. Two different methods of determining HER2 are routinely available: (1) IHC and (2) fluorescence or silver *in situ* hybridization (FISH/SISH). IHC is a semiquantitative method that identifies HER2 receptor expression on the cell surface using a grading system (0, 1+, 2+, and 3+), where an IHC result of 3+ is regarded as HER2 overexpression. It is the most widely used technique. While fairly easy to perform at relatively low cost, results can vary depending on different fixation protocols, assay methods, scoring systems, and the selected antibodies. SISH/FISH are quantitative methods measuring the number of copies of the HER2 gene present in each tumor cell and are reported as either positive or negative. They are highly reproducible but comparatively more time-consuming and expensive.

Concordance between IHC and FISH has been extensively studied. In a study conducted on 2963 samples using FISH as the standard method, the positive predictive value of an IHC 3+ result was 91.6%, and the negative predictive value of an IHC 0 or 1+ result was 97.2%. Nonetheless, FISH had a significantly higher failure rate (5% versus 0.08%), was more costly, and required more time for testing and interpretation than IHC. Another problem with HER2 testing is the poor reproducibility between laboratories, even when the same technique is used.

In recognizing the importance of accurate HER2 testing, guidelines from the American Society of Clinical Oncology (ASCO)/CAP provide recommendations for HER2 evaluation using an algorithm for positive, equivocal, and negative results, based on IHC and/or FISH/SISH.

## Targeting HER2

Multiple therapeutic strategies have been developed to target the HER2 receptor: (1) monoclonal antibodies, (2) small molecule kinase inhibitors, (3) antibody–drug conjugates, and (4) immunotherapy combinations.

Trastuzumab is a recombinant, humanized anti-HER2 monoclonal antibody and was the first clinically active anti-HER2 therapy to be developed. Trastuzumab exerts its action through several mechanisms, including (1) induction of receptor downregulation/degradation, (2) prevention of HER2 ectodomain cleavage, (3) inhibition of HER2 kinase signal transduction via antibody-dependent cell-mediated cytotoxicity, and (4) inhibition of angiogenesis.

In metastatic breast cancer, trastuzumab monotherapy produces a median duration of disease control of 9 months. Preclinical studies have also shown additive or synergistic interactions between trastuzumab and multiple cytotoxic agents, including platinum analogs, taxanes, anthracyclines, vinorelbine, gemcitabine, capecitabine, and cyclophosphamide. The use of trastuzumab combined with chemotherapy can further increase RRs, time to progression, and OS. Furthermore, continuing trastuzumab beyond progression is now standard of care.

In the adjuvant setting, trastuzumab also produces significant benefit in reducing recurrence and mortality. Despite differences in patient population and trial design, remarkably consistent results were reported across trials: a 33%–58% reduction in recurrence rate and a 30% reduction in mortality. This degree of benefit in early breast cancer is the largest reported since the introduction of tamoxifen for ER+ disease.

There are many other novel drugs developed for use following failure of trastuzumab, either because of *de novo* or acquired resistance. The second class of anti-HER2 agents to reach the clinic was the pan-HER tyrosine kinase inhibitors. Lapatinib is effective in combination with trastuzumab or with chemotherapy in advanced breast cancer patients, who are trastuzumab-refractory. More recently, it was shown that adjuvant neratinib improves invasive-DFS when taken sequentially after trastuzumab. Novel anti-HER2 antibodies have also been developed such as pertuzumab, a humanized monoclonal antibody that binds HER2 at a different epitope of the HER2 extracellular domain (subdomain II), preventing HER2 from dimerizing with other ligand-activated HER receptors, most notably HER3. This has led to double-HER2 blockade strategies. Although the combination of lapatinib and trastuzumab was approved in the metastatic setting, it failed in the adjuvant setting. The combination of chemotherapy, trastuzumab, and pertuzumab produced an unprecedented OS benefit and is now the standard of care in the first-line metastatic setting. This double-HER2 blockade also improves pathologic complete response (pCR) rates in the neoadjuvant setting and it is now standard of care as well. Furthermore, it was demonstrated that adjuvant double-HER2 blockade modestly improves invasive-DFS. T-DM1 is an antibody–drug conjugate, incorporating the HER2-targeted antitumor properties of trastuzumab with the cytotoxic activity of the microtubule-inhibitory agent DM1 (derivative of maytansine). It improves OS in advanced breast cancer patients with trastuzumab-refractory disease. Recently, it has been shown that adjuvant T-DM1 improves invasive-DFS, as compared to trastuzumab, in patients who fail to achieve pCR after neoadjuvant therapy.

Despite many attempts to refine patient selection for anti-HER2-targeted agents, HER2 overexpression and/or amplification remain the sole predictive biomarker for anti-HER2 therapy. ER status is also emerging as a prognostic and predictive marker within HER2+ disease, but it is not yet firmly established.

## The androgen receptor

Androgen receptor (AR) is widely expressed by IHC in all subtypes of breast cancer: 88% of ER+, 59% of HER2+, and 32% of triple-negative breast cancers (TNBC). Similar to ER and PR, AR expression is more commonly found in well-differentiated, relatively indolent cancers. TNBC remains the most promising field for antiandrogen therapy with the identification of a molecular subgroup characterized by high expression of AR at the mRNA level, which closely resembles the gene expression pattern in ER+ breast cancers.

## Proliferative and stromal biomarkers

A high proliferative rate is associated with poor breast cancer survival in untreated patients, but it is also associated with favorable response to chemotherapy. Although many proliferative biomarkers are under investigation, none of these are routinely used in clinical practice. The main obstacles lie in the (1) poor standardization of detection methods, (2) vaguely defined cutoff values, and (3) requirement of fresh-frozen tissue in some cases.

### Ki67

Ki67 is a nuclear antigen present in mid G1, S, G2, and the entire M phase of the cell cycle based on IHC. Although it has been extensively studied, its precise cellular function is still unknown. Many studies have shown that overexpression of Ki67 correlates with poor metastases-free survival and OS. Yet, as with all biomarkers with continuous values, the cutoff value differentiating high Ki67 from low Ki67 expression is somewhat arbitrary. It has been suggested that the optimal cutoff value for Ki67 is 15%. Nevertheless, this cutoff is constantly challenged and a 20%−29% fork has been proposed recently. Routine use of this marker in clinical practice is not recommended given its poor reproducibility.

### Tumor-infiltrating lymphocytes

TILs had been suspected for years to be a biomarker predicting pCR and benefit from specific treatments. Extensive studies have been performed on the immune component of the tumor microenvironment, and a meta-analysis from the German Breast Group demonstrated an independent association between the percentage of stromal TILs as a continuous variable and pCR and survival. The quantification of TILs has led to the definition of lymphocyte-predominant breast cancer (LPBC) with a definition ranging from 50% to 60%. The incidence of LPBC varies from 4% to 28% and is associated with a good prognosis. Nonetheless, the predictive and prognostic values of TILs are not homogeneous among all breast cancer subtypes of breast cancer and are particularly associated with higher pCR rates in triple-negative and HER2+ breast cancers.

## Multifocal breast cancer

Multifocal breast cancer is defined as multiple simultaneous ipsilateral and synchronous breast cancer lesions, provided they are macroscopically distinct and measurable using current traditional pathological and clinical tools. Even though multifocal breast cancer has been associated with a number of more aggressive features, including an increased rate of axillary lymph node metastases, larger tumors, and possible adverse patient outcome, a review of the published literature yields a contradictory split between the expressed points of view. Current CAP guidelines recommend characterizing only the largest lesion unless the grade and/or histology are different and treatment strategies do not differ. However, reports show a difference in biology that needs to be addressed in the future. Indeed, one study showed that mismatches in ER status were present in 4.4% of cases, PR in 15.9%, histological grade in 18.6%, Ki67 in 15%, and HER2 in 9.7%. This may lead to different treatment in 12.4% of patients compared with what would have been done if only one lesion had been investigated.

On the molecular level, targeted gene sequencing on 171 tumors from 36 patients with multifocal breast cancer revealed that as many as 33% of patients (12 patients) did not share any substitutions or indels in the genetic makeup of their cancers, with inter-lesion heterogeneity observed for oncogenic mutation(s) in genes such as *PIK3CA*, *TP53*, *GATA3*, and *PTEN*. Genomically heterogeneous lesions tended to be further apart in the mammary gland than homogeneous lesions. Although it might not be feasible to molecularly characterize all multifocal breast cancer lesions in daily practice today, it remains vital that pathologists analyze all identified lesions to avoid situations in which patients are withheld potentially life-saving therapies, such as hormonal therapy or anti-HER2 therapy.

# Gene expression profiling

Through their ability to interrogate thousands of genes simultaneously, microarray studies have enabled the comprehensive molecular and genetic profiling of cancers. Not only have these studies changed the way in which we have traditionally classified breast cancer, but the results of these studies have also yielded molecular signatures with the potential to have a significant impact on clinical care by providing a molecular basis for treatment tailoring.

## Microarray technology

Gene expression profiling, using microarray technology, relies on the accurate binding, or hybridization, of DNA strands with their precise complementary copies, where one sequence is bound onto a solid-state substrate. These are hybridized to probes of fluorescent cDNAs or genomic sequences from normal or cancer tissue. Through analysis of the intensity of the fluorescence on the microarray chip, a direct comparison of the expression of all genes in normal and cancer cells can be made.

## Molecular classification of breast cancer

One of the most important discoveries stemming directly from microarray studies has been the reclassification of breast cancer into molecular subtypes. This new classification has not only furthered our understanding of cancer biology, but it has also altered the way that physicians and investigators conceptually regard breast cancer—not as one disease, but a collection of several biologically different ones. Four main molecular classes of breast cancer have been consistently distinguished by gene expression profiling. These subtypes are (1) basal-like, (2) HER2-enriched (HER2-E), (3) luminal A, and (4) luminal B breast cancers.

In the basal-like subtype, there is a high expression of basal cytokeratins (CK5/6 and CK17) and proliferation-related genes, as well as laminin and fatty acid−binding protein 7. In the HER2-E subtype, there is a high expression of genes in the erbb2 amplicon, such as GRB7. The luminal cancers are ER+. Luminal A is characterized by a higher expression of ER, GATA3, and X-box-binding protein trefoil factor 3, hepatocyte nuclear factor 3 alpha, and LIV-1. Luminal B cancers are generally characterized by a lower expression of luminal-specific genes.

Beyond differing gene expression profiles, these molecular subtypes have distinct clinical outcomes and responses to therapy. The basal-like and HER2-E subtypes are more aggressive, having a higher proportion of *TP53* mutations, and a markedly higher likelihood of being grade III ($P < .0001$ and $P=.0002$) than luminal A cancers. Despite the poorer prognosis, they tend to respond better to chemotherapy, including higher pCR rates after neoadjuvant therapy. By contrast, fewer than 20% of luminal tumors have mutations in *TP53*, and they are often grade I. They tend to be more sensitive to endocrine therapy, less responsive to conventional chemotherapy, and demonstrate better overall clinical outcomes.

This classification is not optimal, and the basal-like subtype is frequently misrepresented as the triple-negative subtype. Analysis of gene expression profiles of 587 TNBC cases has led to the discovery of 6 different subtypes for TNBC, which were later updated to 4 subtypes (TNBC type-4): basal-like 1 (BL1), basal-like 2 (BL2), mesenchymal (M), and luminal androgen receptor (LAR). BL1 has higher genomic instability, M is enriched in EGFR and Notch signaling pathways, and LAR is enriched in hormonally regulated pathways and in *PIK3CA* mutations, despite being triple-negative. More importantly, these subtypes exhibit differential sensitivity to anticancer drugs such as cisplatin or PARP inhibitors for BL1, EGFR inhibitors for M, and androgen antagonists or PI3K inhibitors for LAR.

## Gene expression signatures to refine prognostication

Traditional prognostic factors based on clinical and pathological variables are unable to fully capture the heterogeneity of breast cancer, as they are being overcome by molecular factors.

Using the top-down approach, where gene expression data are correlated with clinical outcome without a prior biological assumption, several multigene signatures were developed in the early 2000s. MammaPrint is a 70-gene prognostic signature, including mainly genes involved in cell cycle, invasion, metastasis, angiogenesis, and signal transduction. Multiple studies have shown it to be a strong predictor for distant metastases-free survival, independent of adjuvant treatment, tumor size, histological grade, and age. Improvements in RNA processing have enabled microarray diagnostics to be feasible using formalin-fixed paraffin-embedded (FFPE) tissue.

Using the top-down approach, investigators developed a RS based on 21 genes that appeared to accurately predict the likelihood of distant recurrence in tamoxifen-treated patients with node-negative, ER-positive breast cancer. A final panel of 16 cancer-related genes and 5 reference genes forms the basis for the Oncotype DX breast cancer assay.

EndoPredict (EP) was also developed using the top-down approach and provides a dual score: the EP score consisting of eight cancer-related genes of interest and three normalization genes; and the EP clin score integrating clinical parameters (tumor size and lymph node status). Validation was performed on samples from two large trials (ABCSG-6 and ABCSG-8), which randomized patients with ER+/HER2- cancers to receive 5 years of endocrine therapy (tamoxifen alone or tamoxifen and an aromatase inhibitor sequentially). Continuous EP was an independent predictor of distant recurrence in multivariate analysis.

By using a different, hypothesis-driven approach (the bottom-up approach), another study examined whether gene expression patterns associated with histologic grade could improve prognostic capabilities, especially within the class of intermediate-grade cancers. Accounting for 30%−60% of all breast cancers, these intermediate-grade tumors display the largest heterogeneity in both phenotype and outcome. Of the unique 97 genes that formed the gene expression grade index (GGI), most were associated with cell cycle progression and differentiation. These genes were differentially expressed between low-grade and high-grade breast cancers, without a distinct gene expression pattern to distinguish the intermediate group. Instead, the intermediate-grade cancers showed expression patterns and clinical outcomes matching those of either low-grade or high-grade cases. The GGI, therefore, could potentially improve treatment decision-making for patients confronted with problematic intermediate-grade cancers by reclassifying them into two distinct and clinically relevant subtypes, especially within the ER-positive group.

The Genomic Grade Assay was created as a polymerase chain reaction (PCR)-based assay using six reporter genes and four reference genes. The concordance of the Genomic Grade Assay and GGI was assessed in 44 paired frozen and FFPE samples and then was validated in 336 samples of node-negative, ER+ breast cancer. Increasing Genomic Grade Assay score was significantly associated with lower disease recurrence−free interval and added independent prognostic information to clinicopathological prognostic factors.

The Breast Cancer Index (BCI) was developed as a continuous risk index using a dichotomous index combining two gene expression assays, HOXB13:IL17BR (H:I) and five-gene molecular grade index (MGI). This was performed analyzing tumors from patients treated in the randomized Stockholm trial conducted from 1976 to 1990 to examine the efficacy of adjuvant tamoxifen compared with no adjuvant treatment among postmenopausal women. The dichotomous H:I+MGI was significantly associated with distant recurrence and breast cancer death.

The PAM50 (ROR-S) was developed as an RT-PCR test that could make an intrinsic subtype diagnosis. The normal-like class (for quality control) was represented using true normal samples taken from reduction mammoplasty or grossly uninvolved tissue. Prototypic cancer samples were identified from an expanded intrinsic gene set comprising genes found in four previous microarray studies. Of 189 individuals, 122 breast cancers profiled by quantitative real-time PCR (qRT-PCR) and microarray had significant clusters representing the intrinsic luminal A, luminal B, HER2E, basal-like, and normal-like subtypes, and 50 genes were selected to distinguish between them. PAM50 was evaluated in two studies involving more than 1400 patients and, although it was associated with risk of relapse in the multivariate model, when using c-index it failed to show an added value to Adjuvant! Online in node-positive or node-negative disease.

There is significant potential for gene expression profiles to aid in the tailoring treatment for individual breast cancer patients. Prognostic signatures can differentiate subpopulations based on risk of relapse, and these signatures may be most useful in identifying patients at low risk of recurrence who could potentially be spared adjuvant chemotherapy. Landmark prospective trials, such as TAILORx (testing Oncotype Dx) and MINDACT (testing MammaPrint), have recently been reported.

The large, international, prospective randomized MINDACT trial assessed the added value of MammaPrint to the commonly used clinicopathological criteria for selecting patients with node-negative or 1−3 node-positive breast cancer for adjuvant chemotherapy. Results have shown that the primary endpoint was met: the 5-year distant-metastasis free-survival in the primary test population (high-clinical/low-genomic risk patients assigned not to receive chemotherapy) was 94.7% with a 95% CI ranging from 92.5% to 96.2%, thus not crossing the non-inferiority lower boundary.

The TAILORx trial is also a large, randomized study designed to evaluate whether women with node-negative, ER +/HER2- breast cancer need chemotherapy based on the RS. Patients with an RS < 11 (low risk) were given only hormonal therapy. An RS > 25 (high risk) meant that patients received chemotherapy in addition to hormone therapy. Patients with an RS of 11−25 (intermediate risk) were randomly assigned to receive hormone therapy or chemotherapy followed by hormone therapy. It was shown that low-risk patients had a 5-year invasive DFS-rate of 93.8% (95% confidence interval [CI], 92.4−94.9), and a 5-year OS rate of 98.0% (95% CI, 97.1−98.6). In the intermediate-risk group,

**TABLE 26.1** Summary of genomic tests.

|  | Oncotype Dx | MammaPrint | Genomic grade assay | Prosigna (PAM50) | Breast cancer index | EndoPredict |
|---|---|---|---|---|---|---|
| Provider | Genomic Health | Agendia | Ipsogen | Nanostring Technologies | Biotheranostics | Sividon diagnostics |
| Assay | 21-Gene recurrence score | 70-Gene assay | 6 reporter and 3 reference genes | 50-Gene assay | 2-Gene ratio HOXB13 to IL17R and molecular grade index | 11-Gene assay |
| Tissue sample | FFPE | Fresh, frozen, or FFPE | FFPE | FFPE | FFPE | FFPE |
| Prospective validation | TAILORx (positive) + RxPONDER (ongoing) | MINDACT (positive) | ASTER 70s (Ongoing) | OPTIMA trial (ongoing) | None | None |
| Technique | qRT-PCR | DNA microarray | qRT-PCR | qRT-PCR | qRT-PCR | qRT-PCR |

after a median follow-up of 90 months, non-inferiority of endocrine treatment alone was demonstrated (iDFS rate: 83.3% with endocrine treatment vs. 84.3% with chemotherapy plus endocrine treatment, HR 1.08; 95% CI, 0.94–1.24). No OS difference was observed between both groups (93.9% with endocrine treatment vs. 93.8% with chemotherapy plus endocrine treatment).

The RxPONDER trial is also testing Oncotype DX, but in ER+/HER2 − early breast cancer patients with 1−3 positive lymph nodes and an RS score ≤ 25. Patients are randomized to endocrine therapy alone or chemotherapy followed by endocrine therapy.

These trials show that gene expression signatures can be trusted for adjuvant treatment assignment in this subpopulation of patients with early breast cancer. Gene expression signatures (Table 26.1) have now been integrated in the recommendation guidelines for the treatment of early breast cancer both in Europe and the United States, albeit at different levels of evidence.

## Signatures of immune response

Seven *in silico* gene expression modules recapitulating key biologic processes in cancer, namely tumor invasion/metastasis, impairment of immune response, sustained angiogenesis, evasion of apoptosis, self-sufficiency in growth signals, and ER and HER2 signaling were defined in a meta-analysis of gene expression and clinicopathologic profiles of 2100 breast cancer patients. While proliferation and histologic grade determined the prognosis of ER+ breast cancer, markers of immune response were associated with prognosis in HER2+ and TNBCs. Indeed, only the immune response was significantly associated with relapse-free survival in TNBC, and tumor invasion and immune response showed significant correlation with clinical outcomes in HER2+ breast cancer. These findings have potential therapeutic implications, given that a pooled transcriptome analysis of 996 breast cancers, all treated with anthracycline-based or anthracycline and taxane-based neoadjuvant chemotherapy, showed that high scores of immune gene expression signatures were associated with an increased rate of pCR. Although the association was seen across all breast cancer subtypes, multivariate analysis showed statistical significance only in HER2+ and TNBC.

## Massively parallel sequencing

The advent of next-generation sequencing (NGS), initially used to study whole genomes, has evolved to address defined regions of the genome, such as single genes associated with cancer. Two preparatory approaches allow the exploration of specific regions of the genome: (1) PCR, and (2) hybrid capture.

## Polymerase chain reaction (PCR)

Multiple primer pairs in a mixture are combined with genomic DNA of interest in a multiplex approach to preserve precious DNA. The use of multiplex primer pairs couples the high throughput of NGS platforms, and the fact that each sequence read represents a single DNA product in the mixture because of the nature of the sequencing platforms. Following this first step, a library is created using the resulting fragments with platform-specific adaptors ligated to their ends.

## Hybrid capture

DNA fragments are hybridized from a whole-genome library to complementary sequences that have been synthesized and combined into a mixture of probes designed with high specificity for the matching regions in the genome. Covalently linked biotin moieties enable a secondary capture by mixing the probe library complexes with streptavidin-coated magnetic beads. The targeted regions of the genome are selectively captured from solution by applying a magnetic field, whereas most of the remainder of the genome is washed away in the supernatant. Subsequent denaturation releases the captured library fragments from the beads into solution, ready for post-capture amplification, quantitation, and sequencing. Exome sequencing is performed when the probes are designed to capture essentially all of the known coding exons in a genome.

The use of NGS has allowed different international initiatives such as The Cancer Genome Atlas (TCGA) or the International Cancer Genome Consortium (ICGC) to define the genomic landscape of early-stage breast cancer.

## The molecular landscape of breast cancer and novel classification

DNA sequencing, DNA methylation, microRNA expression, and proteomics were applied by TCGA to characterize more thoroughly the molecular background of breast cancer. The TCGA analysis made it possible to identify genes implicated in breast cancer: *PIK3CA, PTEN, AKT1, TP53, GATA3, CDH1, RB1, MLL3, MAP3K1,* and *CDKN1B*. It also led to the identification of novel genes such as *TBX3, RUNX1, CBFB, AFF2, PIK3R1, PTPN22, PTPRD, NF1, SF3B1,* and *CCND3*.

mRNA expression profiling made it possible to correlate the frequently altered genes with breast cancer subtypes. Significantly mutated genes were considerably more diverse and recurrent within luminal A and luminal B cancers than within basal-like and HER2E subtypes. However, overall mutation rate was lowest in luminal A and highest in basal-like and HER2E subtypes. Recurrently mutated genes across the different subtypes are detailed in Table 26.2. Normal tissue DNA was used to test a selected number of genes, and 47 of 507 patients whose cancers were sequenced were shown to contain deleterious germ-line variants, representing nine different genes (*ATM, BRCA1, BRCA2, BRIP1, CHEK2, NBN, PTEN, RAD51C,* and *TP53*). Ten percent of sporadic breast cancers may have a strong germ-line contribution.

In another study, an integrated analysis of copy number and gene expression identified 10 different molecular subtypes with distinct clinical outcomes (labeled as IntClust 1 to IntClust10), which split many of the intrinsic subtypes. A subgroup of high-risk ER+ cancers was clustered in IntClust 2 and was enriched with *CCND1* amplifications. IntClust 3 was marked by low genomic instability and comprised luminal A cancers and tumors with a good prognosis (lobular and tubular carcinomas). Luminal B cancers appeared in a cluster with an intermediate prognosis (IntClust 1). HER2-amplified cancers, including the HER2E subtype and the luminal subtype, appeared in the same cluster (IntClust 5).

TABLE 26.2 Highlights of molecular alterations across breast cancer subtypes.

| Subtypes | Luminal A | Luminal B | Basal-like | HER2-enriched |
|---|---|---|---|---|
| Mutations | PIK3CA (49%); TP53 (12%); GATA3 (14%); MAP3K1 (14%) | TP53 (32%); PIK3CA (32%); MAP3K1 (5%) | TP53 (84%); PIK3CA (7%) | TP53 (75%); PIK3CA (42%); PIK3R1 (8%) |
| Copy number variations | Most diploid; many with quiet genomes; 1q, 8q, 8p11 gain; 8p, 16q loss; 11q13.3 amp (24%) | Most aneuploid; many with focal amp; 1q, 8q, 8p11 gain; 8p, 16q loss; 11q13.3 amp (51%); 8p11.23 amp (28%) | Most aneuploid; high genomic instability; 1q, 10p gain; 8p, 5q loss; MYC focal gain (40%) | Most aneuploid; high genomic instability; 1q, 8q gain; 8p loss; 17q12 focal ERRB2 amp (71%) |

This table was adapted from Stephens PJ, Tarpey PS, Davies H, et al. The landscape of cancer genes and mutational processes in breast cancer. Nature 2012;486:400–04.

This preliminary work on new subtypes, which integrates copy number alterations with gene expression, will allow researchers in the future to focus their characterization efforts on representatives of these clusters and eventually pave the way to more personalized treatments for breast cancer.

## Molecular characterization of specific populations

### Invasive lobular breast carcinoma

The genomic landscape of ILC has been studied by several research groups, whose combined work represents the analysis of 757 cases of this disease. Two groups identified *CDH1* (as high as 65%) and *FOXA1* as the most frequent alterations. More than half the cases harbored alterations in genes of the PI3K pathway (*PIK3CA, PTEN, AKT1*). Alterations of the HER/ERBB pathway were enriched in comparison with invasive ductal carcinoma: *ERBB2* (5.1%) and *ERBB3* (3.6%). These alterations represent potential therapeutic targets and are currently being studied in basket clinical trials. Other frequent alterations were *ESR1* gains in 25% of the cases. With regard to prognosis, *ERBB2* and *AKT1* mutations were associated with increased risk of early relapse. Proliferation and immune-related signatures determined three ILC transcriptional subtypes associated with survival differences. Mixed IDC/ILC cases were molecularly classified as ILC-like and IDC-like revealing no true hybrid features. A third study identified two main subtypes of ILCs: (1) an immune-related subtype with mRNA upregulation of PD-L1, PD-1, and CTLA-4, and greater sensitivity to DNA-damaging agents in representative cell line models, and (2) a hormone-related subtype, associated with epithelial-to-mesenchymal transition, the gain of chromosomes 1q and 8q, and the loss of chromosome 11q. These three studies will provide guidance for future therapeutic strategies targeting ILC. This is important given that ILC today is treated in a similar manner as invasive ductal carcinoma.

### Male breast cancer

Male breast cancer is an uncommon disease accounting for <1% of all invasive breast cancers. Thus, all current treatment strategies for men with breast cancer are extrapolated from clinical trials enrolling female participants. The majority of male breast cancers are ER+/HER2 − invasive ductal carcinomas of NST. ERBB2 amplifications or the TNBC phenotypes are rare.

In one study, 59 male breast cancers underwent gene sequencing targeting the full exons of 241 recurrently mutated genes in breast cancer, as well as genes involved in DNA repair pathways. Classic clinicopathological features were also studied and, as expected, 91% were invasive ductal carcinoma NST. Of these, 21% were luminal A-like and 71% were luminal B-like. The most frequently mutated genes were *PIK3CA* (20%) and *GATA3* (15%), which was associated with worse progression-free survival. However, ER-positive/HER2-negative male breast cancers less frequently harbored 16q losses, or *PIK3CA* and *TP53* mutations than ER-positive/HER2-negative female breast cancers. In addition, male breast cancers were found to be significantly enriched for mutations affecting DNA repair−related genes (10 patients had *BRCA2* mutations). The authors concluded that, given these genomic differences, caution should be exercised when applying the biological and therapeutic findings from studies of female breast cancers to male breast cancers.

## Emerging targets

### ESR1 alterations

Despite continuous expression of the ER, many ER+ breast cancers in the metastatic setting become refractory to inhibition of estrogen action. Thirty-one biopsy samples collected after progression on hormonal therapy from patients with metastatic ER+ breast cancer were subjected to targeted DNA sequencing in a study. *ESR1* mutations (a total of 9/36 cases) were detected at much higher rates than in those reported by TCGA. Mutations in *ESR1* clustered in the ligand-binding domain. Furthermore, analysis of the available paired primary samples showed that these mutations were an acquired event. The mutations clearly occurred in a population of patients with hormone refractory disease and the biochemical and structural data demonstrated that these mutations promoted the agonist conformation of ER in the absence of ligand. Furthermore, the mutant ER isoforms were only partially inhibited by direct antagonists of the receptor such as tamoxifen or fulvestrant and needed higher doses than what is used in the clinic to achieve full inhibition. Thus, *ESR1* mutations have emerged as predictors of acquired resistance to endocrine therapy and should help select patients who are potential candidates for chemotherapy or other targeted agents.

*ERBB2 mutations*

*ERBB2* somatic mutations have been found in breast cancers classified as HER2-amplification negative. Mutations clustered in two functional regions: the extracellular domain and the kinase domain. RNA sequencing demonstrated that these mutations were expressed, and *in vitro* kinase assays were conducted on the ERBB2 kinase domain mutations, showing increased kinase activity for some variants when compared with wild type. Cell line experiments showed that while cell lines with *ERBB2*-activating mutations were resistant to lapatinib, some were sensitive to neratinib. Thus, despite a rare incidence, HER2-nonamplified breast cancers with HER2-activating mutations can be treated with pan-HER inhibitors such as neratinib. This is particularly true for ILCs that harbor HER2-activating mutations at a higher frequency.

## Conclusions

Breast cancer is a clinically heterogeneous disease and, for many years, this heterogeneity was explained by the differing histopathological characteristics identified mainly under the microscope. Guidelines based on histopathology have served as important risk stratification tools for therapy selection, but these guidelines are limited with respect to treatment tailoring for individual patients. Indeed, it has long been observed that among patients with similar anatomic and pathological risk profiles, there can be substantial variability in both the natural disease history and response to treatment.

Novel molecular technologies and a better understanding of the tumor biology of breast cancer have resulted in significant advances in recent years. Previous emphasis on refining the traditional histopathologic criteria, on developing indices of proliferation, on understanding genomic instability, and on analyzing single gene expression profiles has shifted dramatically to molecular profiling, reflecting the expression of many thousands of genes. This is particularly true for gene expression signatures that are already playing a role in defining treatment for early disease, knowing that pivotal prospective trials have reported results. Immune markers such as TILs and immune signatures might also help refine prognostic and predictive tools.

The advent of NGS is now playing a role in the treatment of metastatic breast cancer with the emergence of targeted agents matched to oncogenes or used in combination with endocrine therapy to overcome resistance. However, all of these molecular studies have been performed on primary cancer samples, and much remains to be learned about metastatic disease. AURORA, a large, multinational, collaborative metastatic breast cancer molecular screening program launched by the Breast International Group, will help fill this void, as 1000 patients will be recruited prospectively and will provide paired primary cancer and metastasis samples, as well as plasma DNA.

Another major technological advance that will play an important role in the future is the advent of technologies that can detect circulating tumor DNA in plasma samples from patients with cancer. Circulating DNA fragments carrying cancer-specific sequence alterations (circulating tumor DNA) are found in the cell-free fractions of blood, representing a variable and generally small fraction of the total circulating DNA. An interesting proof-of-principle study based on a case study of a patient with metastatic disease showed that there was significant heterogeneity between the mutational profiles of that patient's primary cancer and the metastatic tumors. Analysis of circulating tumor DNA in the same case study captured all the mutations present in both cancer samples. With new knowledge, new challenges emerge. In this chapter we have tried to summarize some of the most important developments related to the study of breast cancer at the molecular level in recent years, and how they have been or in the future may be implemented to potentially change the way we understand and treat the various forms of the disease. We hope that what we have described will over time help advance the concept of precision oncology, ultimately shedding light on the secrets of metastatic breast cancer and improving cure rates of early disease.

## Key concepts

1. Breast cancer is a biologically heterogeneous disease, with distinct molecular subtypes associated with recurrence risk and patterns of relapse.
2. The estrogen receptor (ER) and human epidermal growth factor-2 (HER2) receptor are predictive biomarkers that are routinely used in clinical practice to select patients for anti-hormonal and anti-HER2 therapy.
3. There is considerable inter-laboratory variation in ER and HER2 testing, highlighting the importance of compliance with standardized testing criteria.

4. There are several commercially available gene expression profiles for early breast cancer that provide additional prognostic information beyond traditional clinical and pathological characteristics.
5. The prognostic performance of these prognostic gene expression profiles is similar, driven largely by improved quantification of proliferation in the ER+/HER2- subtype.

## Suggested readings

[1] Lakhani SR, Ellis IO, Schnitt SJ, Tan PH, van de Vijver MJ. World Health Organization classification of tumours of the breast. Lyon: IARC; 2012.

[2] Viale G, Regan MM, Maiorano E, et al. Prognostic and predictive value of centrally reviewed expression of estrogen and progesterone receptors in a randomized trial comparing letrozole and tamoxifen adjuvant therapy for postmenopausal early breast cancer: BIG 1–98. J Clin Oncol 2007;25:3846–52.

[3] Hammond ME, Hayes DF, Dowsett M, et al. American Society of Clinical Oncology/College of American Pathologists guideline recommendations for immunohistochemical testing of estrogen and progesterone receptors in breast cancer. J Clin Oncol 2010;28:2784–95.

[4] Wolff AC, Hammond MEH, Allison KH, et al. Human epidermal growth factor receptor 2 testing in breast cancer: American Society of Clinical Oncology/College of American Pathologists Clinical Practice Guideline Focused Update. J Clin Oncol 2018;36:2105–22.

[5] Harris LN, Ismaila N, McShane LM, et al. Use of biomarkers to guide decisions on adjuvant systemic therapy for women with early-stage invasive breast cancer: American Society of Clinical Oncology practice guideline. J Clin Oncol 2016;34(10):1134–5011.

[6] Piccart-Gebhart MJ, Procter M, Leyland-Jones B, et al. Herceptin adjuvant (HERA) trial study Team trastuzumab after adjuvant chemotherapy HER2-positive breast cancer. N Engl J Med 2005;353:1659–72.

[7] Lehmann BD, Jovanović B, Chen X, et al. Refinement of triple-negative breast cancer molecular subtypes: implications for neoadjuvant chemotherapy selection. PLoS One 2016;11:e0157368.

[8] Denkert C, von Minckwitz G, Darb-Esfahani S, et al. Tumour-infiltrating lymphocytes and prognosis in different subtypes of breast cancer: a pooled analysis of 3771 patients treated with neoadjuvant therapy. Lancet Oncol 2018;19:40–50.

[9] Salgado R, Denkert C, Demaria S, et al. The evaluation of tumor-infiltrating lymphocytes (TILs) in breast cancer: recommendations by an international TILs working group 2014. Ann Oncol 2015;26:259–71.

[10] Gingras I, Gebhart G, de Azambuja E, et al. HER2-positive breast cancer is lost in translation: time for patient-centered research. Nat Rev Clin Oncol 2017;14:669–81.

[11] Perou CM, Sorlie T, Eisen MB, et al. Molecular portraits of human breast tumours. Nature 2000;406:747–52.

[12] Stephens PJ, Tarpey PS, Davies H, et al. The landscape of cancer genes and mutational processes in breast cancer. Nature 2012;486:400–4.

[13] Curtis C, Shah SP, Chin SF, et al. METABRIC group, the genomic and transcriptomic architecture of 2000 breast tumours reveals novel subgroups. Nature 2012;486:346–52.

[14] Sotiriou C, Pusztai L. Gene-expression signatures in breast cancer. N Engl J Med 2009;360:790–800.

[15] Cardoso F, van't Veer LJ, Bogaerts J, et al. 70-Gene signature as an aid to treatment decisions in early-stage breast cancer. N Engl J Med 2016;375:717–29.

[16] Sparano JA, Gray RJ, Makower DF, et al. Adjuvant chemotherapy guided by a 21-gene expression assay in breast cancer. N Engl J Med 2018;379:111–21.

[17] Sestak I, Buus R, Cuzick J, et al. Comparison of the performance of 6 prognostic signatures for estrogen receptor-positive breast cancer: a secondary analysis of a randomized clinical trial. JAMA Oncol 2018;4:545–53.

[18] Desmedt C, Haibe-Kains B, Wirapati P, et al. Biological processes associated with breast cancer clinical outcome depend on the molecular subtypes. Clin Cancer Res 2008;14:5158–65.

[19] Desmedt C, Zoppoli G, Gundem G, et al. Genomic characterization of primary invasive lobular breast cancer. J Clin Oncol 2016;34:1872–81.

[20] Dawson SJ, Tsui DW, Murtaza M, et al. Analysis of circulating tumor DNA to monitor metastatic breast cancer. N Engl J Med 2013;368:1199–209.

# Chapter 27

# Molecular basis of skin disease

**Vesarat Wessagowit**

*The Institute of Dermatology, Rajvithi Phyathai, Bangkok, Thailand*

**Summary**

The skin is a stratified epithelium with an inner layer of proliferating cells that can also give rise to multiple layers of terminally differentiating cells. It protects the body by providing a mechanical barrier against the external environment, a chemical barrier, cutaneous immune surveillance, and myriad proteins acting as antimicrobial peptides or signaling host responses through chemotactic, angiogenic, growth factor, and immunosuppressive activity. The structural integrity of the skin depends on several proteins. Mutations in genes encoding these proteins disrupt skin architecture, giving rise to genetic skin diseases. Mutations in hemidesmosome-related genes give rise to skin fragility syndromes called epidermolysis bullosa, whereas aberration in desmosomal genes give rise to variable abnormalities in skin, hair, and heart. Even common inflammatory skin dermatoses were revealed to be associated with genetic mutations, such as atopic dermatitis and its association with mutations in filaggrin gene (*FLG*), coding for filaggrin. Several hemidesmosomal and desmosomal proteins may serve as target antigens for both inherited and acquired vesiculobullous disorders. Skin cancers can also be associated with inherited and acquired mutations. Understanding the molecular pathology has also changed clinical practice. It provides insight into the clinical features in affected individuals and also allows for accurate diagnosis, improved genetic counseling, prenatal genetic diagnosis, and can also be used to diagnose cutaneous infections with pathogens difficult to identify using standard culture techniques, such as mycobacterial or human papilloma infections. New molecular data also provide a platform to develop new drugs as well as drug repositioning to treat genetic skin diseases. Thus, understanding the molecular pathology of skin diseases can have important implications that benefit patients.

## Molecular basis of healthy skin

A key role of skin is to provide a mechanical barrier against the external environment (Fig. 27.1). The cornified cell envelope and the stratum corneum restrict water loss from the skin, while keratinocyte-derived endogenous antibiotics provide innate immune defense against bacteria, viruses, and fungi. Normal skin has been shown to have a very effective defense system against microbes (Fig. 27.2). In the stratum corneum there is an effective chemical barrier maintained by the expression of S100A7 (psoriasin). This antimicrobial substance is very effective at killing *Escherichia coli*. Subjacent to this in the skin there is another class of antimicrobial peptides such as RNASE7, which is effective against a broad spectrum of microorganisms, especially *Enterococci*. RNASE7 serves as a protective minefield in the superficial skin layers and helps destroy invading organisms. Below this in the living layers of the skin are other antimicrobial peptides such as a β-defensins. The antimicrobial activity of most peptides occurs as a result of unique structural properties to enable them to destroy the microbial membrane while leaving human cell membranes intact. Some may play a specific role against certain microbes in normal skin, whereas others act only when the skin is injured and the physical barrier is disrupted.

Certain antimicrobial peptides influence host cell responses in specific ways. The human cathelicidin peptide LL-37 can activate mitogen-activated protein kinase (MAPK), an extracellular signal-related kinase in epithelial cells, and blocking antibodies to LL-37 hinder wound repair in human skin equivalents. Defensins and cathelicidins have immunostimulatory and immunomodulatory capacities as catalysts for secondary host defense mechanisms and can be chemotactic for distinct subpopulations of leukocytes as well as other inflammatory cells. Human β-defensins (hBDs) 1−3 are chemotactic for memory T-cells and immature dendritic cells—hBD2 attracts mast cells and activated neutrophils, whereas hBD3−4 is chemotactic for monocytes and macrophages. Cathelicidins are chemotactic for neutrophils, monocytes/macrophages, and CD4 T-lymphocytes.

Essential Concepts in Molecular Pathology. DOI: https://doi.org/10.1016/B978-0-12-813257-9.00027-9

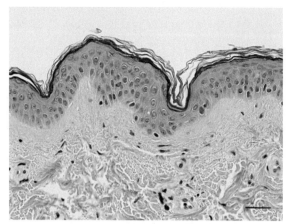

FIGURE 27.1    Light microscopic appearance of normal human skin. Hematoxylin and eosin stain. Bar=50 μm.

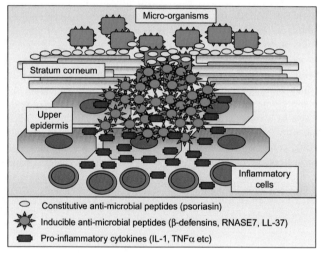

FIGURE 27.2    Immune defense system in human skin. Microorganisms that breach the human epidermis are faced with a constitutive antimicrobial system, for example, psoriasin. Further protection is provided by inducible antimicrobial peptides, such as the β-defensins RNASE7 and LL-37. Microorganisms may be targeted by pro-inflammatory cytokines.

    Skin immunity can be pathogenic, such as in allergic contact dermatitis (ACD), a major occupational disease affecting many people. ACD, or contact hypersensitivity in animal models, is comprised of two phases, the priming, or sensitization phase, when the skin is exposed to the antigen the first time, and elicitation phase in which the same antigen is re-exposed to the skin leading to skin inflammation that peaks after 24—48 h. ACD is generally induced by small haptens. They themselves must bind to self-carrier protein in the skin to be allergenic. When haptens get into the skin, they induce keratinocyte stress/injury, leading to ATP release and reactive oxygen species (ROS) production. ATP leads to NLRP3 activation through P2X7 signaling, whereas low-molecular-weight hyaluronic acid from ROS production leads to MAPK/NFκB activation through TLR2/4 signaling. This process results in the production of many mediators from keratinocytes and mast cells leading to inflammation and subsequent hapten-activated dendritic cell migration to the lymph nodes. In the paracortical region of lymph nodes, the dendritic cells express protein on its surface to present to naïve T-lymphocytes that can then undergo clonal proliferation to helper T-cells (Th) or cytotoxic T-cells (Tc). This priming phase is called sensitization phase of contact hypersensitivity (Fig. 27.3). The elicitation phase starts when the skin is reexposed to the same hapten, leading to keratinocytes producing interleukin 1 alpha (IL-1α). IL-1α stimulates M2-type macrophages around post-capillary venules to release CXCL2, which attracts dermal dendritic cells (dDC). In addition to CXCL2, LTB4 increases dDC motility. Together, this results in clustering of M2-type macrophage—dendritic cell—effector T-cells around post-capillary venules, so called inducible skin-associated lymphoid tissue (iSALT). Cytokines, such as IFN-γ released from iSALT, lead to skin inflammation (Fig. 27.4).

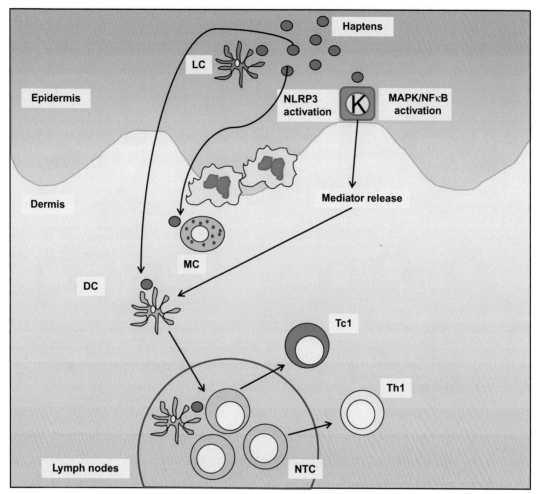

FIGURE 27.3 Contact hypersensitivity, sensitization phase. Haptens induce keratinocyte damage, leading to NLRP3 and MAPK/NFkB activation and release of mediators such as IL-1α. Antigen-captured activated DC travel to distant lymph nodes, presenting antigens to naïve T-cells, which eventually differentiate to Th-1 and Tc1 cells. *DC*, dendritic cells; *K*, keratinocytes; *LC*, Langerhans cells; *MC*, mast cells; *NTC*, naïve T-cells. *Adapted from Honda T, Kabashima K. Novel concept of iSALT (inducible skin-associated lymphoid tissue) in the elicitation of allergic contact dermatitis. Proc Jpn Acad Ser B Phys Biol Sci 2016;92:20—8.*

Langerhans cells, the most prevalent dendritic cells in the skin, were previously thought to be the main antigen-presenting cells in contact hypersensitivity. This theory was challenged as depletion of Langerhans cells did not abolish contact hypersensitivity. Recent studies show that Langerhans cells work as regulators in contact hypersensitivity, inducing tolerance in dinitrothiocyanobenzene exposure through Treg upregulation. Langerhans cells contribute to several skin pathologies, including infections, inflammation, and cancer. Thus, these cells play a pivotal role in regulating the balance between immunity and peripheral tolerance. Langerhans cells have characteristics that are different from dendritic cells, in that they are more likely to induce Th-2 responses than the Th-1 responses that are usually necessary for cellular immune responses against pathogens. Langerhans cells, or a subset thereof, may have immunoregulatory properties that counteract the pro-inflammatory activity of surrounding keratinocytes.

Besides the antigen detection and processing role of dendritic cells, cutaneous immune surveillance is carried out in the dermis by an array of macrophages, T-cells, and dendritic cells (Fig. 27.5). These immune sentinel and effector cells can provide rapid and efficient immunologic backup to restore tissue homeostasis if the epidermis is breached. The dermis contains a very large number of resident T-cells. Indeed, there are approximately $2 \times 10^{10}$ resident T-cells, which is twice the number of T-cells in the circulating blood. Dermal dendritic cells vary in their functionality. Some have potent antigen-presenting capacities, whereas others have potential to develop into CD1a-positive and Langerin-positive cells, while some are pro-inflammatory. A recent addition to the family of skin immune sentinels is type 1 interferon-producing plasmacytoid predendritic cells, which are rare in normal skin but which can accumulate in

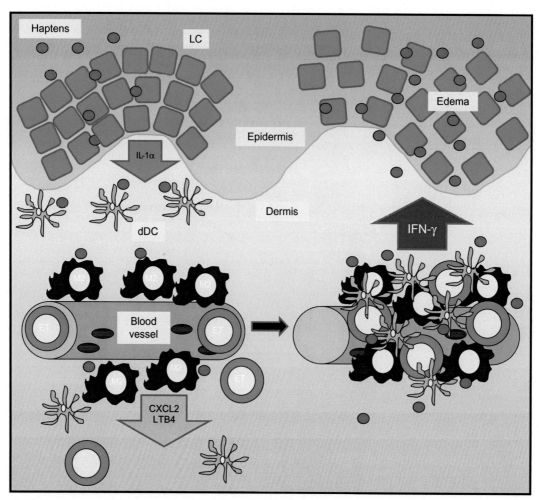

**FIGURE 27.4** Contact hypersensitivity, elicitation phase. IL-1α produced from keratinocytes stimulates M2 macrophages around post-capillary venules. These M2 macrophages release CXCL2, which attracts dDC, and LTB4, which increases dDC motility. ET cells are activated within these cell clusters, so-called iSALT, and release cytokines. *dDC*, dermal dendritic cells; *ET*, effector T cells; *iSALT*, inducible skin-associated lymphoid tissue; *M2*, M2-type macrophages. *Adapted from Honda T, Kabashima K. Novel concept of iSALT (inducible skin-associated lymphoid tissue) in the elicitation of allergic contact dermatitis. Proc Jpn Acad Ser B Phys Biol Sci 2016;92:20—8.*

inflamed skin. A further component of the dermal immune system is the dermal macrophage. Dermal immune sentinels exhibit flexibility or plasticity in function. Depending on microenvironmental factors and cues, they may acquire an antigen-presenting mode, a migratory mode, or a tissue-resident phagocytic mode.

## Skin development and maintenance provide new insight into the molecular mechanisms of disease

The development of a stratified epithelium such as the skin requires a detailed architecture which maintains an inner layer of proliferating cells, but can give rise to multiple layers of terminally differentiating cells that extend to the body surface and which are subsequently shed. A detailed understanding of this process that generates a self-perpetuating barrier to keep microbes out and essential body fluids in is becoming clearer, and this improved understanding is providing new insights into skin maintenance as well as the pathogenesis and molecular mechanisms underlying certain developmental disorders.

One fundamental issue has been trying to provide an explanation for how epithelial progenitor cells retain a self-renewing capacity. In 1999, it was shown that mice lacking the transcription factor p63 had thin skin and abnormal skin renewal. p63 is an evolutionary predecessor to the p53 protein, part of a family of transcriptional regulators of cell growth differentiation and apoptosis. While p53 is a major player in tumorigenesis, p63 and another family member,

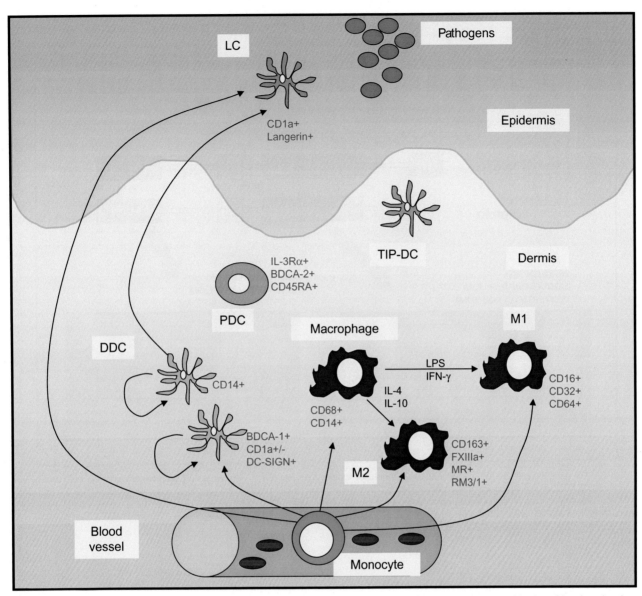

**FIGURE 27.5** Diversity of immune sentinels in human skin. These include CD1a+ Langerin+ Langerhans cells located in the epidermis and various subtypes of dendritic cells and macrophages in the dermis. This figure illustrates some of the recent immunophenotypic and functional findings of these immune sentinels. The macrophage population expressing CD68 and CD14 can be further subdivided into classically activated macrophages (M1) and alternatively activated macrophages (M2), which develop under the influence of IL-4 and IL-10. Several cells have self-renewing potential under conditions of tissue homeostasis. Under inflammatory conditions, circulating blood-derived monocytes are potential precursors of Langerhans cells, dermal dendritic cells, and macrophages. *Adapted from Nestle FO, Nickoloff BJ. Deepening our understanding of immune sentinels in the skin. J Clin Invest 2007;117:2382–5.*

p73, appear to have pivotal roles in embryonic development (Fig. 27.6). p73-deficient mice have neurological and inflammatory pathology, whereas p63-knockout mice have major defects in epithelial limb and craniofacial development. These observations suggest that p63 plays a crucial role in tissue morphogenesis and maintenance of epithelial stem cell compartments. Lack of p63 compromises skin formation either by creating an absence of lineage commitment and an early block in epithelial differentiation, or by causing skin failure through a defect in epithelial stem cell renewal. p63 has been linked to several important signaling pathways such as epidermal growth factor, fibroblast growth factor, bone morphogenic protein (BMP), and Notch/Wnt/hedgehog signaling. p63 directly regulates expression of extracellular matrix adhesion molecules, including the $\alpha_6\beta_4$ basal integrins and desmosomal proteins such as PERP, all of which are essential for epithelial integrity.

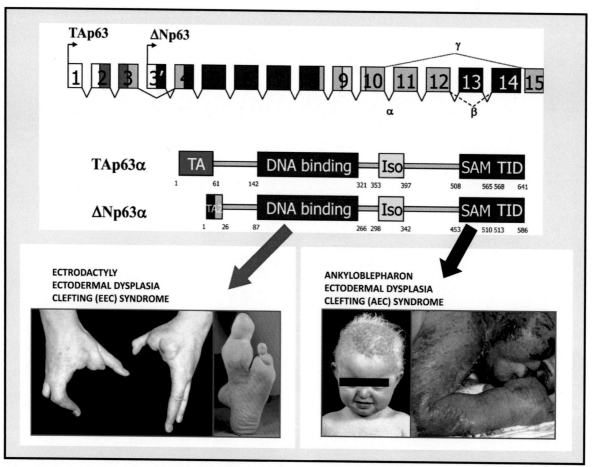

**FIGURE 27.6**  Genomic and functional domain organization of the transcription factor *p63*. At least six different isoforms can be generated by use of alternative translation initiation sites or alternative splicing. The main isoform expressed in human skin is ΔNp63α. Autosomal dominant mutations in the DNA-binding domain of the *p63* gene lead to ectrodactyly, ectodermal dysplasia, and clefting (EEC) syndrome. In contrast, autosomal dominant mutations in the SAM domain result in ankyloblepharon, ectodermal dysplasia, and clefting (AEC) syndrome. A number of other ectodermal dysplasia syndromes may result from mutations in the *p63* gene.

The human *p63* gene consists of 16 exons, located on chromosome 3q28. There are two different promoter sites and three different splicing routes, which create at least six different protein isoforms. Several functional domains have been identified (Fig. 27.6).

A further understanding of the role of p63 in skin development has been gleaned from studies on naturally occurring mutations in this gene. Heterozygous mutations cause developmental disorders, displaying various combinations of ectodermal dysplasia, limb malformations, and orofacial clefting. Thus far, seven different disorders have been linked to mutations in the *p63* gene. These conditions may have overlapping genotypic features, but there are some distinct genotype/phenotype correlations. The most common p63-associated ectodermal dysplasia is ectrodactyly, ectodermal dysplasia, and cleft/lip palate (EEC) syndrome (OMIM604292). It is characterized by three major clinical signs of cleft lip and/or palate, ectodermal dysplasia (abnormal teeth, skin, hair, nails, and sweat glands, or combinations of these), and limb malformations in the form of split hand/foot (ectrodactyly), and/or fusion of fingers/toes (syndactyly). Another group of *p63*-linked patients are those with Rapp-Hodgkin syndrome (OMIM129400) or AEC/Hay–Wells syndrome (OMIM106260). These syndromes fulfill the criteria of ectodermal dysplasia and orofacial clefting, but do not have the severe limb malformation (s) seen in EEC syndrome. The features of these syndromes may include eyelid fusion (ankyloblepharon filiforme adnatum), severe erosions at birth, and abnormal hair with pili torti or pili canaliculi. The *p63* mutations in EEC syndrome are clustered in the DNA-binding domain and most likely alter the DNA-binding properties of the protein. By contrast, mutations in Rapp-Hodgkin or AEC syndromes are clustered in the SAM and TI domains in the carboxy-terminus of p63α. The SAM domain is involved in protein-protein interactions, whereas the TI domain is combined intramolecularly to the

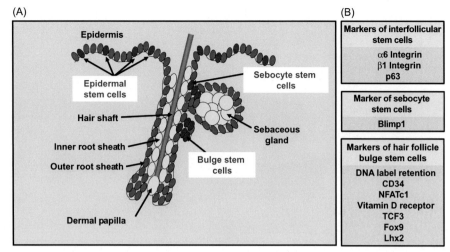

**FIGURE 27.7** Stem cells of the skin. (A) Localization of stem cells in human epidermis. Stem cells are located within the basal layer of interfollicular epidermis, as well as at the base of sebocytes and in the bulge area of hair follicles. (B) These epidermal stem cells are associated with a number of cellular markers.

TA domain, thereby inhibiting transcription activation. All *p63*-associated disorders are inherited in an autosomal dominant manner, and mutations are thought to have either dominant-negative or gain-of-function effects.

The epidermis contains a population of epidermal stem cells that reside in the basal layer, although it is not clear how many cells within the basal layer have a stem cell capacity (Fig. 27.7). Stem cells are proposed to express elevated levels of $\beta_1$ and $\alpha_6$ integrins and differentiate by delamination and upward movement to form the spinous layer, a granular layer, and the stratum corneum. The proliferation of epidermal stem cells is regulated positively by $\beta_1$ integrin and transforming growth factor $\alpha$ (TGF$\alpha$), and negatively by TGF$\beta$ signaling. Hair follicle stem cells reside in the bulge compartment below the sebaceous gland. These stem cells are slow cycling and express the cell surface molecules CD34 and VdR as well as the transcription factors TCF3, Sox9, Lhx2, and NFATc1. These bulge area stem cells generate cells of the outer root sheath, which drive the highly proliferative matrix cells next to the mesenchymal papillae. After proliferating, matrix cells differentiate to form the hair channel, the inner root sheath, and the hair shaft. The mechanisms that control the stem cell proliferation and differentiation in the skin provide new insights into skin homeostasis.

## Molecular pathology of mendelian genetic skin disorders

There are approximately 5000 single gene disorders, of which nearly 600 have a distinct skin phenotype. Many of these disorders have been characterized at a molecular level. Most inherited skin disorders are transmitted either by autosomal dominant, autosomal recessive, X-linked dominant, or X-linked recessive modes of inheritance. However, understanding the precise pattern of inheritance is essential for accurate genetic counseling. For example, the connective tissue disorder pseudoxanthoma elasticum (OMIM264800) was for many years thought to reflect a mixture of autosomal dominant and autosomal recessive genotypes. However, it has been shown that all forms of pseudoxanthoma elasticum are in fact autosomal recessive and that the disease is caused by mutations in the *ABCC6* gene, as the earlier reports of autosomal dominant inheritance were actually pseudodominant inheritance in consanguineous pedigrees. Understanding the molecular pathology of pseudoxanthoma elasticum identified similar autosomal recessive inherited skin diseases that have an overlapping phenotype, but in which there are subtle clinical differences. For example, a pseudoxanthoma elasticum-like disease with cutis laxa skin changes and coagulopathy (OMIM610842) has been shown to result from mutations in the *GGCX* gene. In both conditions, there is progressive accumulation of calcium phosphate in tissue, resulting in ocular and cardiovascular complications. The knowledge that these conditions are autosomal recessive helps with genetic counseling, prognostication, and management of families with affected members. Moreover, understanding the precise molecular pathology allows for more careful clinical monitoring and patient follow-up.

One of the principal functions of human skin is to provide a mechanical barrier against the external environment. The structural integrity of the skin depends on several proteins. These include intermediate filaments inside keratinocytes, intercellular junctional proteins between keratinocytes, and a network of adhesive macromolecules at the dermal-epidermal junction (Fig. 27.8). Since the late 1980s, several Mendelian genetic disorders resulting from autosomal

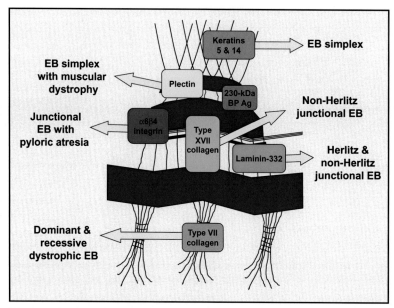

**FIGURE 27.8** Illustration of the integral structural macromolecules present within hemidesmosome-anchoring filament complexes and the associated forms of clinical epidermolysis bullosa that result from autosomal dominant or autosomal recessive mutations in the genes encoding these proteins.

dominant or autosomal recessive mutations in structural proteins in the skin have provided fascinating insights into skin structure and function. In addition, determining the key roles of particular proteins has provided a plethora of new clinically and biologically relevant data.

One of the best characterized groups of disorders is epidermolysis bullosa (EB), a group of skin fragility disorders associated with blister formation of the skin and mucous membranes that occurs following mild trauma (Table 27.1). EB simplex is the most common form of inherited EB and affects approximately 40,000 people in the United States. Transmission is mainly autosomal dominant, but recessive patients (OMIM601001) have been reported. Ultrastructurally, the level of split is through the cytoplasm of basal cells, often close to the inner hemidesmosomal plaque. In dominant forms of EB simplex, there may be disruption of keratin tonofilaments or aggregation of keratin filaments into bundles (Fig. 27.9). However, transmission electron microscopy may show only very subtle or indiscernible morphological changes in intermediate filaments. The molecular defects that cause EB simplex affect either the keratin 5 gene (*KRT5*) or the keratin 14 gene (*KRT14*), or in cases of the autosomal recessive EB simplex-muscular dystrophy, in the plectin gene (*PLEC1*). The molecular pathology of keratin gene mutations provides some insight into genotype/phenotype correlation. Notably mutations in the helix initiation/helix termination motifs in helices 1A and 2B of the keratin genes result in more severe subtypes. Clinically, EB simplex patients have the mildest skin lesions and scarring is not frequent. Extracutaneous involvement is rare, apart from cases with plectin pathology. The mildest clinical subtype of EB simplex is the localized EB simplex (Weber-Cockayne variant) (OMIM131800) in which blistering occurs mainly on the palms and soles (Fig. 27.10A). The molecular pathology involves autosomal dominant mutations that occur mainly outside the critical helix boundary motifs. In some forms of EB simplex, such as the generalized intermediate EB simplex (Köbner subtype), the disease is more generalized than localized EB simplex variant, although there is considerable overlap. The most severe form of EB simplex is the severe, generalized EB simplex (Dowling-Meara type) (OMIM131760). This often presents with generalized blister formation shortly after birth, and it can be fatal in neonates. A characteristic of this condition in later childhood is the grouping of lesions in a herpetiform clustering arrangement (Fig. 27.9). This form of EB simplex is associated with the greatest disruption of keratin tonofilaments. Another autosomal dominant form of EB simplex is one associated with mottled skin pigmentation (OMIM131960) (Fig. 27.10B). Apart from blisters, there is diffuse speckled hyperpigmentation as well as keratoderma of the palms and soles. The hypermelanotic macules are most evident in the axillae, limbs, and lower abdomen. The underlying molecular pathology in all reports is the same heterozygous proline to leucine substitution at codon 25 in the non-helical V1 domain of the *KRT5* gene. This proline residue is expressed on the outer part of polymerized keratin filaments and when mutated may result in abnormal interactions with melanosomes or other keratinocyte organelles.

**TABLE 27.1** Epidermolysis Bullosa classification.

| Level of skin cleavage | Major EB type | Major EB subtypes | Phenotype | Targeted proteins (genes) |
|---|---|---|---|---|
| Intraepidermal | EBS | Suprabasal EBS | Acral peeling skin syndrome | Transglutaminase 5 (*TGM5*) |
| | | | Skin fragility syndromes | |
| | | | • Desmoplakin deficiency<br>  1. EBS-desmoplakin<br>  2. kin fragility-woolly hair syndrome | Desmoplakin (*DSP*) |
| | | | • Plakoglobin deficiency<br>  1. EBS plakoglobin<br>  2. Skin fragility-plakoglobin deficiency | Plakoglobin (*JUP*) |
| | | | • Plakophilin deficiency<br>  1. EBS-plakophilin<br>  2. Skin fragility-ectodermal dysplasia syndrome | Plakophilin 1 (*PKP1*) |
| | | | Acantholytic EBS | Desmoplakin/plakoglobin (*DSP/JUP*) |
| | | | EBS, superficialis | Unknown |
| | | Basal EBS | EBS, localized | K5/K14 (*KRT5/KRT14*) |
| | | | EBS, generalized, severe | |
| | | | EBS, generalized intermediate | |
| | | | EBS, with mottled pigmentation | K5 (*KRT5*) |
| | | | EBS, migratory circinate | |
| | | | EBS, autosomal recessive K14 | K14 (*KRT14*) |
| | | | EBS with muscular dystrophy | Plectin (*PLEC*) |
| | | | EBS with pyloric atresia | Plectin/α6β4 integrin (*PLEC/ITGA6/ITGB4*) |
| | | | EBS Ogna | Plectin (*PLEC*) |
| | | | EBS, autosomal recessive— BP230 deficiency | BPAG1 (*DST*) |
| | | | EBS, autosomal recessive— exophilin5 deficiency | Exophilin5 (*EXPH5*) |
| Intralamina lucida | JEB | JEB, generalized | JEB, generalized severe | Laminin-332 (*LAMA3/LAMB3/LAMC2*) |
| | | | JEB, generalized intermediate | Laminin-332/type XVII collagen (*LAMA3/LAMB3/LAMC2/COL17A1*) |
| | | | JEB with pyloric atresia | α6β4 integrin (*ITGA6/ITGB4*) |
| | | | JEB, late onset | Type XVII collagen (*COL17A1*) |
| | | | JEB with respiratory and renal involvement | α3 integrin (*ITGA3*) |
| | | JEB, localized | JEB, localized | Type XVII collagen/α6β4 integrin/laminin-332 (*COL17A1/ITGB4/LAMA3/LAMB3/LAMC2*) |
| | | | JEB, inversa | Laminin-332 (*LAMA3/LAMB3/LAMC2*) |
| | | | JEB, Laryngo-onycho-cutaneous syndrome | Laminin-332 (*LAMA3*) |

*(Continued)*

**TABLE 27.1** (Continued)

| Level of skin cleavage | Major EB type | Major EB subtypes | Phenotype | Targeted proteins (genes) |
|---|---|---|---|---|
| Sublamina densa | DEB | DDEB | DDEB, generalized | Type VII collagen (*COL7A1*) |
| | | | DDEB, acral | |
| | | | DDEB, pretibial | |
| | | | DDEB, pruriginosa | |
| | | | DDEB, nail only | |
| | | | DDEB, bullous dermolysis of the newborn | |
| | | RDEB | RDEB, generalized severe | |
| | | | RDEB, generalized intermediate | |
| | | | RDEB, inversa | |
| | | | RDEB, localized | |
| | | | RDEB, pretibial | |
| | | | RDEB, pruriginosa | |
| | | | RDEB, centripetalis | |
| | | | RDEB, bullous dermolysis of the newborn | |
| Mixed | Kindler syndrome | — | Kindler syndrome | Kindlin-1(*FERMT1*) |

*DDEB*, dominant dystrophic epidermolysis bullosa; *DEB*, dystrophic epidermolysis bullosa; *EBS*, epidermolysis bullosa simplex; *JEB*, junctional epidermolysis bullosa; *RDEB*, recessive dystrophic epidermolysis bullosa.
Adapted from Fine JD, Bruckner-Tuderman L, Eady RA, et al. Inherited epidermolysis bullosa: updated recommendations on diagnosis and classification. J Am Acad Dermatol 2014;70:1103−26.

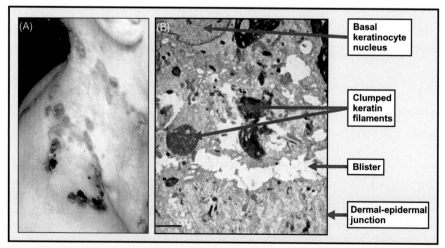

**FIGURE 27.9** Clinicopathological consequences of mutations in the gene encoding keratin 14 (*KRT14*), the major intermediate filament protein in basal keratinocytes. (A) The clinical picture shows autosomal dominant Dowling-Meara epidermolysis bullosa simplex. (B) The electron micrograph shows keratin filament clumping and basal keratinocyte cytolysis (bar=1 μm).

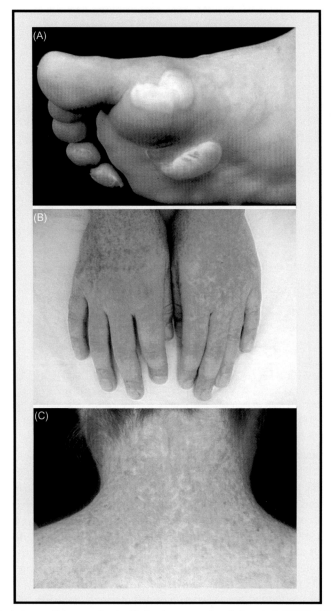

**FIGURE 27.10** Spectrum of clinical abnormalities associated with dominant mutations in keratin 5 (*KRT5*). (A) Missense mutations in the nonhelical end domains result in the most common form of EB simplex, which is localized to the hands and feet (localized EB simplex). (B) A specific mutation in keratin 5, p.P25L, is the molecular cause of epidermolysis bullosa simplex associated with mottled pigmentation. (C) Heterozygous nonsense or frameshift mutations in the *KRT5* gene lead to Dowling-Degos disease.

Molecular analysis of other Mendelian disorders has demonstrated that these skin structural proteins may be mutated in other inherited skin diseases, known as allelic heterogeneity. For example, heterozygous nonsense mutations in the *KRT14* gene have been shown to result in the Naegeli-Franceschetti-Jadassohn form of ectodermal dysplasia (OMIM161000). In addition, heterozygous loss-of-function mutations in the *KRT5* gene have been shown to underlie the autosomal dominant disorder Dowling-Degos disease (OMIM179850). This is characterized by clustered skin papules, the histology of which shows seborrheic keratosis-like morphology (Fig. 27.10C).

Junctional EB is an autosomal recessive condition in which the molecular pathology involves loss-of-function mutations in any one of at least six different genes encoding structural proteins within the hemidesmosome or lamina lucida at the cutaneous basement membrane zone. Clinical features include blistering, atrophic scarring, nail dystrophy, and defective dental enamel as well as other abnormalities affecting the hair, eyes, and genitourinary tract. Ultrastructurally,

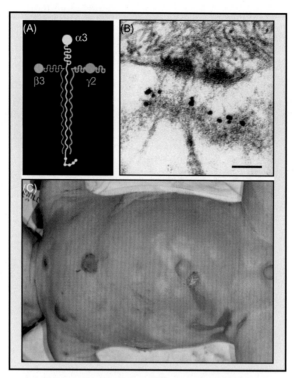

**FIGURE 27.11** Laminin-332 mutations result in junctional epidermolysis bullosa. (A) Laminin-332 consists of three polypeptide chains: α3, β3, and γ2. (B) Immunogold electron microscopy shows laminin-332 staining at the interface between the lamina lucida and lamina densa subjacent to a hemidesmosome (bar=50 nm). (C) Loss-of-function mutations in any one of these genes encoding these polypeptide chains results in severe generalized junctional epidermolysis bullosa, which is associated with a poor prognosis, usually with death in early infancy.

the level of split is mainly through the lamina lucida of the basement membrane zone, but blistering may occur just above the basal keratinocyte plasma membrane as a focal observation in biopsies from patients with junctional EB-associated with pyloric atresia. The most severe type of junctional EB is the generalized severe junctional EB (Herlitz subtype) (OMIM226700). There is typically widespread blistering with mucosal involvement in the mouth and upper respiratory tract. Many affected infants die from overwhelming secondary infection. In later infancy, patients may develop wounds with exuberant granulation tissue, particularly around the mouth, nose, and nails. Dystrophic nails with paronychia and swollen fingertips are frequent findings, and death in early infancy occurs in most cases. The underlying molecular pathology involves homozygous or compound heterozygous loss-of-function mutations in any of the three genes that encode the laminin-332 polypeptide: *LAMA3* or *LAMB3* or *LAMC2* (Fig. 27.11).

Some forms of junctional EB are termed generalized intermediate junctional EB (non-Herlitz) (OMIM226650). In these cases there may be extensive blisters at birth, but the disease typically lessens in severity with time, although atrophic wounds, abnormal dentition, and nail dystrophy persist, and alopecia is very common. This type of junctional EB is genetically heterogeneous. It may result from mutations in either the *LAMA3*, *LAMB3*, or *LAMC2* genes (the subcomponents of laminin-332) or alternatively due to loss-of-function mutations on both alleles of the gene-encoding type XVII collagen, *COL17A1*. The range of *COL17A1* gene mutations includes missense, nonsense, frameshift, or splice-site mutations, but usually there is total ablation of type XVII collagen protein. Another subtype of junctional EB is associated with a further extracutaneous abnormality, namely pyloric atresia. Affected pregnancies are usually complicated by polyhydramnios, as fetuses with pyloric atresia cannot swallow amniotic fluid. Postnatally, feeding results in non-bilious vomiting. The severity of blistering in neonates is variable, but all cases involve molecular pathology in the $\alpha_6\beta_4$ integrin complex. Most patients have mutations in the gene encoding $\beta_4$ integrin, *ITGB4*, with the most severe cases usually having nonsense or frame-shift mutations. Mutations in the *ITGA6* gene are seen less frequently.

Dystrophic EB represents a third type of inherited skin blistering (Fig. 27.12). Ultrastructurally, the level of split is below the lamina densa, and the underlying molecular defect in all types of dystrophic EB involves mutations in the type VII collagen gene (*COL7A1*). Transmission electron microscopy reveals abnormalities in the number and/or morphology of anchoring fibrils, which are principally composed of type VII collagen. Dominant forms of dystrophic EB are usually caused by heterozygous glycine substitution mutations within the type VII collagen triple helix. These result

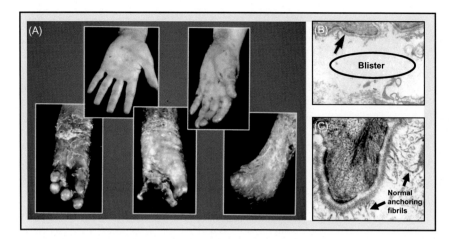

FIGURE 27.12 Clinicopathological abnormalities in the dystrophic forms of epidermolysis bullosa. (A) This form of epidermolysis bullosa is associated with variable blistering and flexion contraction deformities, here illustrated in the hands. (B) The disorder results from mutations in type VII collagen (*COL7A1* gene), the major component of anchoring fibrils at the dermal-epidermal junction. This leads to blister formation below the lamina densa (lamina densa indicated by arrow). (C) In contrast, in normal human skin there is no blistering, and the sublamina densa region is characterized by a network of anchoring fibrils.

in dominant-negative interference and disruption of anchoring fibril formation. Affected individuals have mild trauma-induced blisters, mainly on skin overlying bony prominences such as the knees, ankles, or fingers. Blistering is followed by scarring and milia formation. Nail dystrophy is common.

It is clear, however, that the *COL7A1* pathology alone cannot account for the phenotypic variability of dystrophic EB and that other modifying genes or environmental factors may play a role. The most severe form of dystrophic EB is the generalized severe recessive dystrophic EB (Hallopeau-Siemens subtype) (OMIM226600). Blister formation in affected individuals starts from birth or early infancy, and the skin is very fragile. Wound healing is often poor, leading to chronic ulcer formation with exuberant granulation tissue formation, repeated secondary infection, and frequent scar formation. Mucous membranes are extensively affected, and esophageal involvement causes dysphagia and obstruction due to stricture. Affected individuals have an increased risk of squamous cell carcinoma (Fig. 27.13). The molecular pathology involves loss-of-function mutations on both alleles of the *COL7A1* gene, leading to markedly reduced or completely absent type VII collagen expression at the dermal-epidermal junction. Overall, genotype/phenotype correlation suggests that in recessive dystrophic EB the amount of type VII collagen that is expressed at the dermal-epidermal junction is inversely proportional to clinical severity in terms of scarring and extent of blistering.

To maintain the structural function of the epidermis, a number of intercellular junctions exist, including desmosomes, tight junctions, gap junctions, and adherens junctions. Mutations in these junctional complexes result in several Mendelian inherited skin diseases. Desmosomes are important cell-cell adhesion junctions found predominantly in the epidermis and the heart. They consist of three families of proteins: the armadillo proteins, cadherins, and plakins (Fig. 27.14). Mutations in desmosomal proteins result in skin, hair, and heart phenotypes (Fig. 27.15). Armadillo proteins contain several 42 amino acid repeat domains and are homologous to the drosophila armadillo protein. They bind to other proteins through their armadillo domains and play a variety of roles in the cell, including signal transduction, regulation of desmosome assembly, and cell adhesion. Mutations in the plakophilin 1 gene (*PKP1*) result in autosomal recessive ectodermal dysplasia-skin fragility syndrome (OMIM604536). Affected individuals have a combination of skin fragility and inflammation and abnormalities of ectodermal development, such as scanty hair, keratoderma, and nail dystrophy. Pathogenic *PKP1* mutations are typically splice-site or nonsense mutations. Autosomal dominant and recessive mutations have been described in the plakoglobin gene (*JUB*). A recessive mutation results in Naxos disease (OMIM601214), a genodermatosis frequently seen on Naxos Island in the Mediterranean where approximately 1 in 1000 individuals is affected with clinical features of arrhythmogenic right ventricular dysplasia, diffuse palmar keratoderma, and woolly hair. Autosomal dominant mutations in plakoglobin can result in cardiomyopathy.

The cadherins comprise a group of desmogleins and desmocollins, transmembranous glycoproteins that are present between keratinocytes. Heterozygous mutations in the desmoglein 1 gene (*DSG1*) result in autosomal dominant striate palmoplantar keratoderma (OMIM148700). The molecular pathology in this disorder results from desmoglein 1 haploinsufficiency. Mutations in the desmoglein 4 gene (*DSG4*) result in localized autosomal recessive hypotrichosis (OMIM607903) in which affected individuals have hypotrichosis restricted to the scalp, chest, arms, and legs, but sparing of axillary or pubic hair. Papules on the scalp show atrophic curled-up hair follicles and shafts with marked swelling of the precortical region. Recessive mutations in desmoglein 4 may underlie some cases of autosomal recessive monilethrix. Plakins comprise a family of proteins that cross-link the cytoskeleton to desmosomes. They include desmoplakin, envoplakin, periplakin, plectin, bullous pemphigoid antigen 1, corneodesmosin, and microtubule actin cross-linking

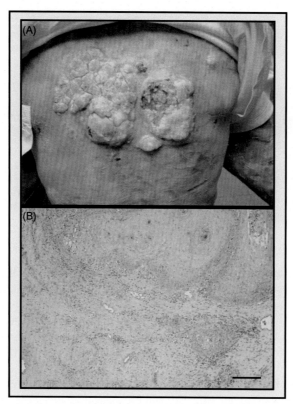

**FIGURE 27.13** Squamous cell carcinoma (SCC) in severe recessive dystrophic epidermolysis bullosa. (A) Affected individuals have a 70-fold increased risk of developing SCC, here illustrated on the mid-back. (B) Light microscopy reveals a moderately differentiated SCC.

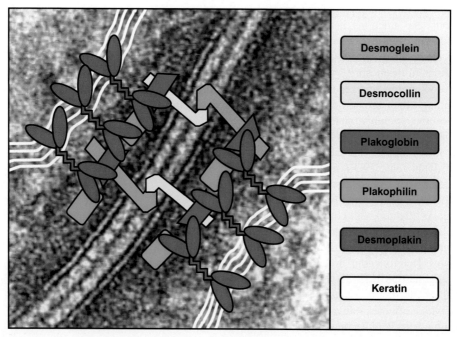

**FIGURE 27.14** Protein composition of the desmosome linking two adjacent keratinocytes a moderately differentiated SCC. The major transmembranous proteins are the desmogleins and the desmocollins. Several desmosomal plaque proteins, including desmoplakin, plakophilin, and plakoglobin provide a bridge that links binding between the transmembranous cadherins and the keratin filament network within keratinocytes.

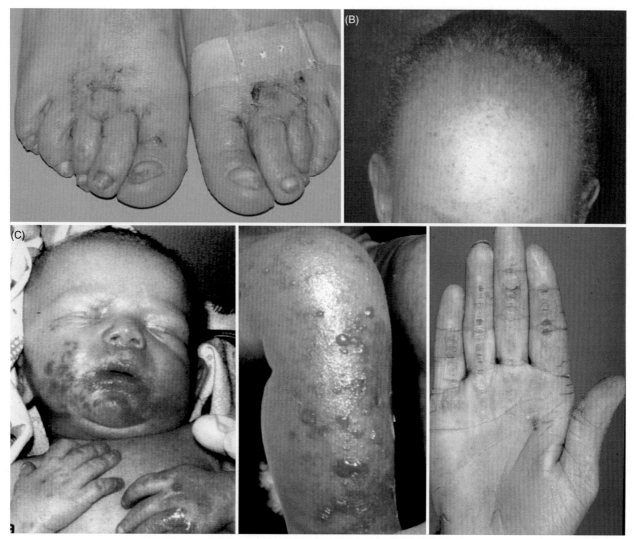

**FIGURE 27.15** Clinical abnormalities associated with inherited gene mutations in desmosome proteins. (A) Recessive mutations in plakophilin 1 result in nail dystrophy and skin erosions. (B) Woolly hair is associated with several desmosomal gene abnormalities, particularly mutations in desmoplakin. (C) Recessive mutations in plakophilin 1 can result in extensive neonatal skin erosions, particularly on the lower face. (D) Recessive mutations in desmoplakin can lead to skin blistering. (E) Autosomal dominant mutations in desmoplakin do not result in blistering but can lead to striate palmoplantar keratoderma.

factor. Mutations in the desmoplakin gene (*DSP*) result in a variable combination of skin, hair, and cardiac abnormalities. These can be autosomal dominant or recessive. The phenotype in autosomal dominant cases ranges from cutaneous striate palmoplantar keratoderma to cardiac arrhythmogenic ventricular dysplasia 8 (OMIM607450). The phenotype in autosomal recessive patients ranges from skin fragility-woolly hair syndrome (OMIM607655) to other syndromes in which the skin and heart may be abnormal, such as dilated cardiomyopathy with woolly hair and keratoderma (OMIM605676). Compound heterozygous mutations that almost completely ablate the desmoplakin tail have recently been demonstrated to produce the most severe clinical subtype known as lethal acantholytic epidermolysis bullosa. Autosomal dominant mutations in the corneodesmosin gene (*CDSN*) result in hypotrichosis of the scalp (OMIM146520).

## Molecular pathology of common inflammatory skin diseases

Atopic dermatitis and psoriasis represent two of the most common inflammatory skin dermatoses. Recent molecular insights are now providing new ideas about disease susceptibility and pathogenesis.

Atopic dermatitis is a chronic itching skin disease that results from a complex interplay between strong genetic and environmental factors. There are two forms of atopic dermatitis: the extrinsic and intrinsic forms. The former is associated with IgE-mediated sensitization, whereas the latter is characterized by a normal total serum IgE level and the absence of specific IgE responses to aeroallergens and food-derived allergens.

To dermatologists, the association between atopic dermatitis and the monogenic disorder ichthyosis vulgaris (OMIM146700) has been evident for many years, given that several patients with ichthyosis vulgaris also have atopic dermatitis. Histopathologically, many cases of ichthyosis vulgaris are associated with abnormal (diminished) keratohyalin granules within the granular layer, and there is reduced immunohistochemical labeling for filaggrin in ichthyosis vulgaris skin. Filaggrin is the major component of keratohyalin granules. *Filaggrin* is a composite phrase for <u>fila</u>ment <u>aggr</u>egating prote<u>ins</u>, repeat units of complex polypeptides derived from profilaggrin that help aggregate keratin filaments in the formation of the epidermal barrier. It was shown that ichthyosis vulgaris results from loss-of-function mutations in the *FLG* gene. Ichthyosis vulgaris is a semidominant condition with heterozygotes displaying no phenotype or just mild ichthyosis, whereas homozygotes or compound heterozygotes have a more severe form of ichthyosis vulgaris with skin barrier defects. Filaggrin mutations are very common in the general population, occurring in approximately 10% of Europeans. Subsequently, it has been shown that filaggrin gene mutations are a major primary predisposing risk factor for atopic dermatitis (Fig. 27.16). It is evident that approximately 50% of all cases of severe atopic dermatitis harbor mutations in the filaggrin gene. The presence of filaggrin mutations is also a risk factor for asthma, but only for asthma in combination with atopic dermatitis, and not for asthma alone. This finding suggests that asthma in individuals with atopic dermatitis is secondary to allergic sensitization, which develops because of the defective epidermal barrier that allows allergens to penetrate the skin to make contact with antigen-presenting cells. Filaggrin is not expressed in respiratory epithelium, and therefore the new data on filaggrin mutations offer an intriguing new concept that atopic asthma may be initiated as a result of a primary cutaneous (rather than respiratory) abnormality in some individuals. The hypothesis of a defective epidermal barrier underlying asthma, and indeed allergic sensitization, has been verified in several studies which have shown an association between filaggrin gene mutations and extrinsic atopic dermatitis associated with high total serum IgE levels and concomitant allergic sensitizations (Fig. 27.17).

Primary defects in filaggrin are not the entire basis of the molecular pathology of atopic dermatitis. It is likely that genes in several other factors, particularly in milder cases of atopic dermatitis, are also important.

Recent research focus on abnormalities of the epidermal barrier is providing fascinating new insights into understanding the nature and etiology of atopic dermatitis and perhaps novel treatments. The presence of filaggrin gene mutations is set to influence and accelerate the design of new treatments that restore filaggrin expression and skin barrier function, given the new evidence that restoration of an intact epidermis may prevent both atopic dermatitis and cases of atopic dermatitis–associated asthma as well as systemic allergies. One of the common mutations in filaggrin is a nonsense mutation (p.R501X), which may represent an attractive drug target for small molecule approaches that modify post-transcriptional mechanisms designed to increase read-through of nonsense mutations and thereby stabilize mRNA expression. Other approaches that involve drug library screening or *in silico* methods to identify compounds capable of

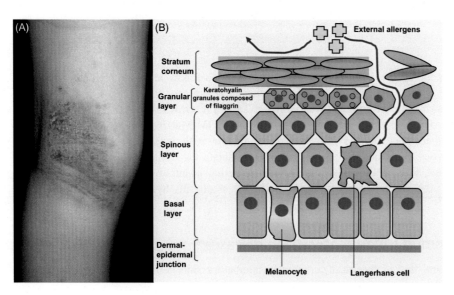

**FIGURE 27.16** Clinicopathological abnormalities in atopic dermatitis. (A) Clinically, there is inflammation in the antecubital fossa with erythema erosions and lichenification. (B) Genetic or acquired abnormalities that lead to reduction in filaggrin expression in the granular layer disrupt the skin barrier permeability, which allows penetration of external allergens and presentation to Langerhans cells. Reduced filaggrin in skin may be a major risk factor for atopic dermatitis and increases susceptibility to atopic asthma and systemic allergies.

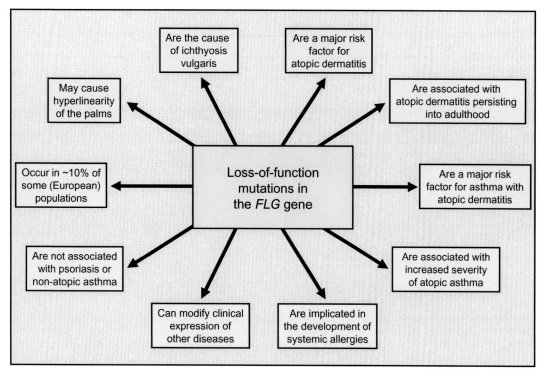

**FIGURE 27.17** Loss-of-function mutations in the filaggrin gene result in several common disease associations or susceptibilities.

increasing filaggrin expression in the epidermis are likely to lead to new evidence-based topical preparations suitable for the treatment of atopic dermatitis and ichthyosis vulgaris. Psoriasis is a common and complex disease. It may manifest with inflammation in the skin as well as the nails and joints of patients. In recent years, considerable evidence has emerged for specific patterns of immunologic abnormalities and for the critical role of certain cytokines and chemokine networks in psoriasis, findings that may be relevant to the design of new molecular therapies (Fig. 27.18).

There is overwhelming evidence that psoriasis has an important genetic component, in that there is a higher incidence among first- and second-degree relatives of patients than unaffected control subjects and there is a higher concordance in monozygotic compared to dizygotic twins. The disease has a bimodal distribution of age of onset, with an early peak between 16 and 22 years and a later one between 57 and 60 years. Linkage studies have identified several genetic loci. Aside from genetic risk factors, environmental risk factors include trauma (the Köbner phenomenon at sites of injury), infection (including *Streptococcal* bacteria and human immunodeficiency virus infections), drugs (lithium, anti-malarials, β-blockers, and angiotensin-converting enzymes), sunlight (in a minority), and metabolic factors (such as high-dose estrogen therapy and hypocalcemia). Other factors including psychogenic stress, alcohol, and smoking have also been implicated.

Psoriasis has traditionally been considered as a Th-1 disease, based on the identification of IFN-γ and TNFα in the lesions with little or no detection of IL-4, IL-5, or IL-10. More recently, other cytokines including IL-18, IL-19, IL-22, and IL-23 have been identified as being upregulated in psoriatic lesions. With relevance to the molecular pathology of psoriasis, one of the first clues came from the observation of clearing of psoriasis in renal transplant patients given the immunosuppressant cyclosporin A. In psoriatic skin that resolved in these individuals, there were reductions in levels of several cytokines, indicating a connection between immunocytes, cytokines, and maintenance of psoriatic plaques. The effects of neutralizing a single cytokine can be extremely helpful, particularly with the use of anti-TNFα therapy, a treatment that was initially developed for patients with sepsis. TNFα blockers or receptor blockers are beneficial in patients with psoriasis, but it is unclear if the improvement of psoriatic plaques and arthritis with these agents is due to local or systemic effects.

Intriguing new ideas, however, have recently been gleaned from specific mouse models, in which prepsoriatic skin is grafted onto a special type of immunodeficient mouse known as an AGR mouse (which lacks type I or IFN-α receptors and lacks type II or IFN-γ receptors, and is RAG-deficient). In such mice, psoriatic plaques develop spontaneously. However, if these mice receive injections of anti-TNFα agents, psoriasis does not develop. This demonstrates that

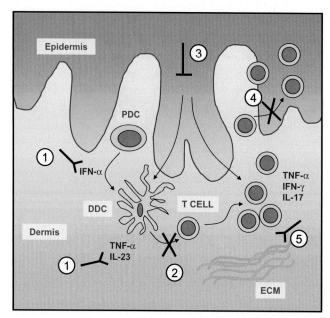

**FIGURE 27.18** Abnormalities and therapeutic potential for inflamed skin in psoriasis. There is increasing evidence for a role of tissue-resident immune cells in the immunopathology of psoriasis. New therapies may be developed by (1) antagonizing local cytokines and chemokines, such as IFN-α, (2) blocking of adhesion molecules (e.g., integrins) and costimulatory molecules within the tissue, (3) modification of keratinocyte proliferation and differentiation (e.g., use of corticosteroids or vitamin D preparations), (4) blocking of entry of dermal T-cells into the epidermis, and (5) modification of the microenvironment, including the extracellular matrix. *Adapted from Boyman O, Conrad C, Tonel G, Gilliet M, Nestle FO. The pathogenic role of tissue-resident immune cells in psoriasis. Trends Immunol 2007;28:51–7.*

resident immunocytes, contained within prepsoriatic skin, are necessary and sufficient to trigger psoriasis and that the local production of TNFα is critically important in the generation of skin lesions.

A new area of investigation in the field of chronic inflammation centers around a possible inflammatory axis in which IL-12/IL-23 influence levels of a cytokine known as IL-17. This has led to a new paradigm through which Th-17 type T-cells contribute to autoimmunity and chronic inflammation. Thus, besides a Th-1 or Th-2 type immune system, there is also a cytokine network dominated by a Th-17 type response. IL-17 is important in psoriasis because it can promote accumulation of neutrophils and can affect skin barrier function by inducing release of pro-inflammatory mediators by keratinocytes. Understanding the molecular role of IL-17 in psoriatic skin is providing new insights for developing novel therapies.

## Skin proteins as targets for inherited and acquired disorders

The integrity of the skin as a mechanical barrier depends, in part, on adhesive complexes that link cell-to-cell and cell-to-basement membrane. Two key junctional complexes in this task are the hemidesmosomes and the desmosomes. In addition to genetic diseases, however, further clues to the function of hemidesmosomal and desmosomal proteins have been derived from animal models or human diseases, in which the same structural proteins can be targeted and disrupted by autoantibodies (Fig. 27.19). Thus, several hemidesmosomal and desmosomal proteins may serve as target antigens for both inherited and acquired disorders (Fig. 27.20).

One of the main transmembranous hemidesmosomal proteins is type XVII collagen, known as the 180 kDa bullous pemphigoid antigen. Autoantibodies against this protein typically result in bullous pemphigoid, a chronic vesiculobullous disease that usually affects the elderly. Histology shows subepidermal blisters with eosinophils, and the pathogenic antibodies are usually directed against a particular epitope (within the NC16A domain) in the first non-collagenous extracellular part of type XVII collagen. IgG1 subclass autoantibodies are found in patients with active skin lesions, whereas IgG4 autoantibodies are found in patients in remission. Besides bullous pemphigoid, other blistering conditions may have autoantibodies against type XVII collagen. For example, pemphigoid gestationis, which is an acute, pruritic vesicular-bullous eruption, is seen in pregnant women. Patients produce autoantibodies against the same epitope of type XVII collagen. IgA autoantibodies against the NC16A domain of type XVII collagen give rise to two different skin diseases: (1) chronic bullous disease of childhood and (2) linear IgA bullous dermatosis. Chronic bullous disease of

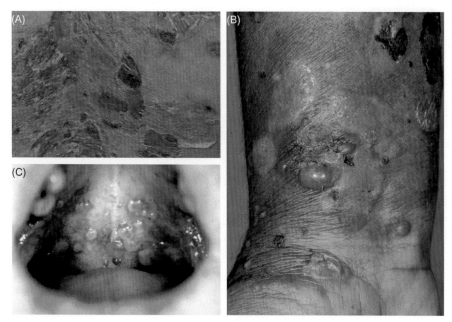

**FIGURE 27.19**   Clinical pathology resulting from autoantibodies against desmosomes or hemidesmosomes. (A) Pemphigus vulgaris resulting from antibodies against desmoglein 3. (B) Bullous pemphigoid associated with antibodies against type XVII collagen. (C) Mucous membrane pemphigoid associated with antibodies to laminin-332.

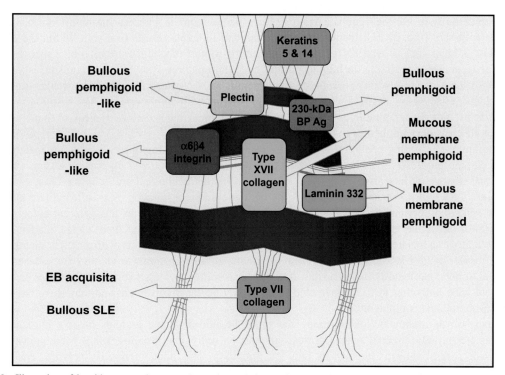

**FIGURE 27.20**   Illustration of hemidesmosomal structural proteins and the autoimmune diseases associated with antibodies directed against these individual protein components.

childhood usually presents in young children with clustered tense bullae, often in the perioral or perineal regions. Clinical manifestations can mimic bullous pemphigoid or dermatitis herpetiformis. A further disease with autoantibodies against type XVII collagen is lichen planus pemphigoides. Patients typically have lichen planus lesions mixed with tense blisters, either on lichen planus lesions or on normal skin. Autoantibodies react with epitopes within the NC16A domain, but the precise epitope is different from bullous pemphigoid.

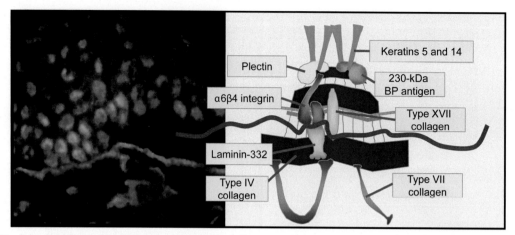

**FIGURE 27.21**    Salt-split skin technique to diagnose immunobullous disease. Incubation of normal human skin in 1 M NaCl overnight at 4 °C results in cleavage through the lamina lucida. This results in separation of some proteins to the roof of the split and some to the base (above and below *pink line* on the schematic). In the skin labeling shown, immunoglobulin from a patient's serum binds to the base of salt-split skin. Further analysis revealed that the antibodies were directed against type VII collagen. This technique is useful in delineating bullous pemphigoid from epidermolysis bullosa acquisita, both of which are associated with linear IgG at the dermal-epidermal junction in intact skin.

Autoantibodies to type VII collagen, the major component of anchoring fibrils, give rise to two different conditions: (1) epidermolysis bullosa acquisita and (2) bullous systemic lupus erythematosus. Patients with epidermolysis bullous acquisita have subepidermal blisters. In both epidermolysis bullosa acquisita and bullous systemic lupus erythematosus, there is linear deposition of IgG at the dermal-epidermal junction. However, a useful technique in immunodermatology to distinguish epidermolysis bullosa acquisita from bullous pemphigoid is indirect immunofluorescence using 1 M sodium chloride split-skin (Fig. 27.21). Incubation of normal skin in saline results in a split within the lamina lucida. Thus, certain skin antigens such as type XVII collagen map to the roof of the split, whereas others such as type VII collagen map to the base. This means that in bullous pemphigoid and epidermolysis bullosa acquisita, labeling with sera from patients with these conditions on salt-split skin can permit an accurate diagnosis to be made. Bullous systemic lupus erythematosus reflects non-specific bullous changes in patients with active disease. The histology resembles dermatitis herpetiformis, but the immunofluorescence and immunological analyzes are typical of lupus erythematosus patients, and not all cases have antibodies to type VII collagen.

Another autoimmune blistering condition that targets hemidesmosomal components is mucous membrane pemphigoid. This is a chronic progressive autoimmune subepithelial disease characterized by erosive lesions of the skin and mucous membranes that result in scarring. Lesions commonly affect the ocular and oral mucosa. Direct immunofluorescence studies show IgG and/or IgA autoantibodies at the dermal-epidermal junction. Salt-split indirect immunofluorescence can show epidermal, dermal, or roof and base labeling patterns, which reflects the different autoantigens seen in this disease. In most cases, the target epitope is part of the extracellular domain of type XVII collagen, although this differs from the epitope associated with bullous pemphigoid. A minority of patients with mucous membrane pemphigoid display antibodies against laminin-332. This is an important subset of patients to identify since there is association with certain solid tumors, particularly malignancies of the upper aerodigestive tract (Fig. 27.22). Cases of mucous membrane pemphigoid with antibodies to $\beta_4$ integrin tend to have only ocular involvement and may therefore be referred to as ocular mucous membrane pemphigoid.

Although many of the immunobullous diseases that target hemidesmosomal proteins involve immunopathology in which antibodies are initially directed against critical epitopes on individual proteins, it has been shown that in many cases there may be several antigens targeted by humoral immunity. This phenomenon is known as epitope spreading and reflects chronic inflammation that may alter the immunogenicity of other neighboring proteins involved in adhesion. In some patients, epitope spreading can lead to transition from one autoimmune disease to another. Published examples include transition from mucosal dominant pemphigus vulgaris to mucocutaneous pemphigus vulgaris, pemphigus foliaceus to pemphigus vulgaris, bullous pemphigoid to epidermolysis bullosa acquisita, dermatitis herpetiformis to bullous pemphigoid, pemphigus foliaceus to bullous pemphigoid, and concurrent bullous pemphigoid with pemphigus foliaceus. Antibodies to the plakin protein, desmoplakin, which are characteristically found in paraneoplastic pemphigus, have been detected in pemphigus vulgaris as well as oral and genital lichenoid reactions. In epidermolysis bullosa acquisita, antibodies to laminin-332 may be found in addition to the typical type VII autoantibodies. In bullous systemic

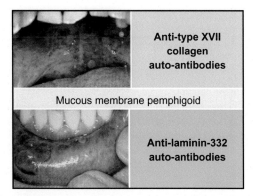

**FIGURE 27.22** Mucous membrane pemphigoid may be associated with autoantibodies against either type XVII collagen or laminin-332. Distinction between the two may have clinical relevance since anti-laminin-332 antibodies in mucous membrane pemphigoid can be associated with malignancy (especially of the upper aerodigestive tract) in some patients.

lupus erythematosus, multiple autoantibodies against type VII collagen, bullous pemphigoid antigen 1, laminin-332, and laminin-311 have been reported. The phenomenon of epitope spreading is perhaps best exemplified in paraneoplastic pemphigus, which is characteristically associated with autoantibodies against many desmosomal and hemidesmosomal proteins, including desmogleins, plakins, bullous pemphigoid antigen 1, plectin, and plakoglobin.

In addition to hemidesmosomal proteins, the structural components of desmosomes may be targets for autoimmune diseases. The skin blistering disease pemphigus is associated with autoantibodies against desmosomal cadherins, principally desmoglein 3 and desmoglein 1. However, the intraepithelial expression patterns of desmoglein 1 and desmoglein 3 differ between the skin and mucous membranes, giving rise to different clinical manifestations when patients have antibodies against different desmogleins. In the skin, desmoglein 1 is expressed throughout the epidermis but more so in superficial parts, as opposed to desmoglein 3, which is predominantly expressed in the basal epidermis. In the oral mucosa, both types of desmogleins are expressed throughout the epithelium, but desmoglein 1 is expressed at a much lower level than desmoglein 3. When there is coexpression of desmoglein 1 and desmoglein 3, these proteins may compensate for each other, but when expressed in isolation, specific pathology can arise when those proteins are targeted by autoantibodies. For example, when there are only desmoglein 1 antibodies in the skin, blisters appear in the superficial epidermis, a site where there is no coexpression of desmoglein 3, whereas in the basal epidermis, the presence of desmoglein 3 can compensate for the loss-of-function of desmoglein 1. In the oral epithelium, keratinocytes express desmoglein 3 at a much higher level than desmoglein 1, and despite the presence of desmoglein 1 antibodies, no blisters form. Therefore, sera containing only desmoglein 1 antibodies cause superficial blisters in the skin without mucosal involvement, and the clinical consequences are pemphigus foliaceus and its localized form, pemphigus erythematosus. Desmoglein 1 is the target antigen in endemic pemphigus, known as fogo selvagem. Desmoglein autoantibodies result in skin with very superficial blisters, such that crust and scale are the predominant clinical features. Desmoglein 1 may be targeted in a different manner—not by autoantibodies, but by bacterial toxins (Fig. 27.23). Specifically, exfoliative toxins A-D, which are produced by *Staphylococcus aureus*, specifically cleave the extracellular part of desmoglein 1. This leads to bullous impetigo and Staphylococcal scalded skin syndrome with superficial blistering in the epidermis, findings that are histologically similar to pemphigus foliaceus associated with autoantibodies to the extracellular part of desmoglein 1.

When the sera of a patient contains only desmoglein 3 autoantibodies, coexpressed desmoglein 1 can compensate for the impaired function of desmoglein 3, resulting in no or only limited skin lesions. However, in the mucous membranes, oral erosions predominate, as desmoglein 1 cannot compensate for the impaired desmoglein 3 function because of its low expression. Therefore, the patient typically has painful oral ulcers without much initial skin involvement. This accounts for the clinical phenotype in patients with the mucosal dominant type of pemphigus vulgaris. However, when sera contain antibodies to both desmoglein 1 and desmoglein 3, as seen in the mucocutaneous type of pemphigus vulgaris, patients have extensive blisters, mucosal erosions, and skin blisters, because the function of both desmogleins is disrupted.

Intercellular autoantibodies in pemphigus are typically composed of IgG isotypes, although some patients with superficial blistering may have intercellular autoantibodies of the IgA subclass. There are two clinical subtypes of IgA pemphigus. One is a subcorneal pustular dermatosis subtype, and this is associated with flaccid vesicopustules with clear fluid and pus arranged in an annular/polycyclic configuration. Indirect immunofluorescence studies show IgA autoantibodies reacting against the superficial epidermis. These antibodies usually recognize epitopes within the

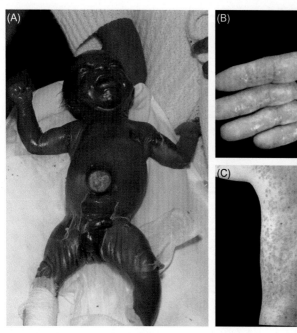

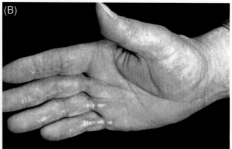

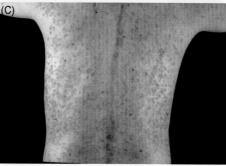

**FIGURE 27.23** Clinical consequences of disruption of desmoglein 1 in human skin. (A) Staphylococcal toxins cleave the extracellular part of desmoglein 1 and result in staphylococcal scalded skin syndrome. (B) Inherited autosomal dominant mutations in desmoglein 1 can result in striate palmoplantar keratoderma. (C) Autoantibodies against desmoglein 1 result in pemphigus foliaceus, which is associated with superficial blistering and crusting in human skin.

desmosomal cadherin, desmocollin 1. In contrast, IgA pemphigus patients may have a different phenotype known as the intraepidermal neutrophilic subtype that presents with a sunflowerlike configuration and pustules and vesicles with indirect immunofluorescence showing IgA intercellular autoantibodies throughout the entire thickness of the epidermis. As well as spontaneous onset cases of pemphigus, certain drugs such as D-penicillamine can lead to drug-induced pemphigus. Most patients have autoantibodies against the same target epitopes as pemphigus patients. In paraneoplastic pemphigus, the clinical presentation involves progressive blistering and erosions on the upper trunk, head, neck, and proximal limbs with an intractable stomatitis. Erythema multiforme-like lesions on the palms and soles can help distinguish the condition from ordinary pemphigus. Indirect immunofluorescent studies on rat bladder, an organ rich in transitional epithelial cells, demonstrate IgG intercellular autoantibodies consistent with autoimmunity against plakin proteins although numerous antigenic targets are usually present.

## Molecular pathology of skin cancer

Skin cancer is very common. Non-melanoma skin cancers, which include basal cell carcinomas (BCCs) and squamous cell carcinomas, are the most common forms of human neoplasia.

The main risk factor for skin neoplasia is environmental exposure to ultraviolet (UV) irradiation. The UV component of sunlight can be divided into three energy subtypes: UVC (100−280 nm), UVB (280−315 nm), and UVA (315−400 nm). The action spectra for UVB and UVC closely match the absorption spectrum of DNA, resulting in the formation of pyrimidine dimers, involving nucleotides thiamine (T) and/or cytosine (C). In contrast, UVA is absorbed by other cellular chromophores, thereby generating oxygen-reactive species such as hydroxyl radicals. These can result in DNA strand breaks and chromosome translocations. Of the UV light that reaches the earth, UVA accounts for more than 90%, and UVB approximately 10%. Most UVB-induced mutations are located almost exclusively at dipyrimidine nucleotide sites (TT, CC, CT, and TC). Approximately 70% of observed mutations are C-to-T transitions and 10% are CC-to-TT, the latter representing an UV signature mutation. Cellular mechanisms of DNA repair are not always effective, and these UVB-induced mutations can proceed unchecked, becoming permanent and subsequently inherited by all the progeny of the mutated cell, thereby allowing expression of the aberrant gene/protein function. Aside from UV radiation inducing DNA point mutations and small deletions, it may also result in gross chromosomal changes.

Collectively, these genetic/chromosomal changes initiate and promote cancer formation as well as the increased genomic instability and loss of heterozygosity (LOH) frequently observed in non-melanoma skin cancer. Cytogenic analysis has enabled the identification of a number of chromosomal abnormalities associated with non-melanoma skin cancer and has implicated certain regions containing oncogenes and tumor suppressor genes that may be involved in their development. For example, in BCCs, early LOH studies identified regions on chromosome 9q22 as a common observation specific to these tumors. These regions harbor the Patched tumor suppressor gene (*PTCH*), which is a transmembrane receptor involved in the regulation of hedgehog signaling (Fig. 27.24). Subsequently, the discovery of mutations that are known to activate hedgehog signaling pathways, including *PTCH*, Sonic hedgehog (*Shh*), and Smoothened (*Smo*), implicates hedgehog signaling as a fundamental transduction pathway in skin tumor development.

In the skin, the Shh pathway is crucial for maintaining the stem cell population and regulating the development of hair follicles and sebaceous glands. Although key embryonic developmental signaling pathways may be switched off during adulthood, aberrant activation of these pathways in adult tissue is often oncogenic. The Shh pathway may be activated in many neoplasms, including BCCs, medulloblastoma, and rhabdomyosarcoma, and abnormalities in Shh signaling pathway components, such as Shh, PTCH1, Smo, GLI1, and GLI2, are major contributing factors in the development of BCCs. The function of PTCH1 is to repress Smo signaling. This function is impaired when PTCH1 is mutationally inactivated or when stimulated by Shh binding, both of these lead to uncontrolled Smo signaling. Downstream of Smo are the GLI transcription factors. Overexpression of GLI1 or GLI2 can lead to BCC development, and GLI1 can activate Platelet-Derived Growth Factor receptor-$\alpha$, the expression of which may be increased in BCCs. The roles of other Smo target molecules such as the suppressor of fused Su(Fu) and protein kinase A (PKA) in the development of BCC are not fully understood. Other components of the pathway include a putative antagonist of Smo signaling, known as hedgehog interacting protein, and an actin-binding protein, missing in metastases (MIM), which is an Shh-responsive gene. MIM is a part of the GLI/Su(Fu) complex and potentiates GLI-dependent transcription using domains distinct from those used for monomeric actin binding. Alterations in Shh regulate cell proliferation and associated cell cycle events. Shh overexpression leads to epidermal hyperplasia, accompanied by the proliferation of normally growth-arrested cells. Shh-expressing cells fail to exit the S and G2/M phases in response to calcium-induced differentiation signals and are unable to block the p21CIP1/WAF1/induced growth arrest in skin keratinocytes. Furthermore, PTCH1 protein interacts with phosphorylated cyclin B1 and blocks its translocation to the nucleus. The Shh/GLI pathway upregulates expression of the phosphatase CDC25b, which is involved in G2/M-transition. Thus, the loss of regulation of cell cyclin control is associated with the development of epithelial cancers, including BCCs.

Patients with the autosomal dominant disorder known as nevoid basal cell carcinoma syndrome or Gorlin syndrome (OMIM109400) have substantially increased susceptibility to BCCs and other tumors, including medulloblastomas, meningiomas, fibromas, rhabdomyomas, and rhabdomyosarcomas. They may also manifest jaw cysts and ectopic calcification, spina bifida, rib defects, palate abnormalities, coarse facies, hypertelorism, microcephaly, and skeletal abnormalities with bony and soft tissue overgrowth. Affected individuals develop BCCs on any part of the body, particularly in sun-exposed skin. These patients have heterozygous germline mutations in the *PTCH* gene. The BCCs in these individuals retain a mutant germline and lose the wild-type allele as a second hit. Mutations in the *PTCH* gene have been demonstrated in up to 40% of sporadic BCCs, emphasizing that *PTCH* gene mutations are important in the development of BCCs. Other sporadic BCCs that do not have mutations in *PTCH* may carry mutations in the *Smo* gene. An understanding of the mutations that lead to activation of hedgehog signaling has thus expanded our knowledge of the genetic basis of BCCs.

Insight into the molecular pathology of BCC has led to development of a number of animal models that may be useful in developing chemoprevention strategies, as well as confirming specific contributions from hedgehog signaling pathway components. For example, drugs such as cyclopamine are known to be a specific inhibitor of Shh signaling. Other possible chemoprevention strategies might involve small molecule hedgehog signaling inhibitors or immunosuppressive agents such as rapamycin, which is an inhibitor of GLI1.

Like BCC, melanoma is another tumor that may be entirely curable by early recognition and therapeutic intervention. Nevertheless for patients with advanced metastatic melanoma, the 5-year survival is currently estimated at only 6% with a median survival time of 6 months. It is important, therefore, to try to understand the molecular pathology of melanoma to determine which patients are most at-risk and which tumors will exhibit the most aggressive biology.

Dissection of the melanoma genome has revealed a number of driver mutations (those that probably influence the growth of tumors) and passenger mutations (those that do not specifically promote tumor growth). One important pathway is the RAS/ MAPK pathway, which regulates cell proliferation and survival in several cell types (Fig. 27.25). Activating cell mutations in *NRAS* and *BRAF* have a combined prevalence of approximately 90% in melanoma and in benign melanocytic lesions, suggesting that activation of the MAPK pathway is an early essential step in melanocytic proliferation. Activating somatic *NRAS* mutations occur in 10%−20% of melanomas. Of note, three highly recurrent *NRAS* missense changes represent over 80% of all mutations in this gene. *NRAS* mutations are more common in chronically sun-exposed sites and appear to occur early in tumorigenesis as well as being common in congenital nevi. Downstream of *NRAS*, activating mutations in *BRAF* have been identified, including a common missense mutation at valine 600. This mutation is equally prevalent in benign nevi and in melanoma, suggesting that BRAF activation is

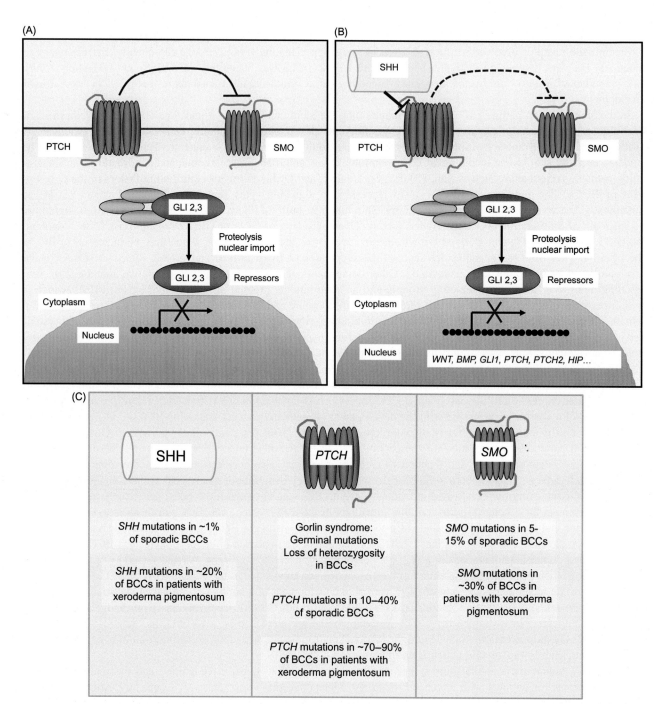

FIGURE 27.24    The SHH signaling pathway. (A) In the absence of SHH, PATCHED constitutively represses smoothened, a transducer of the SHH signal. (B) Binding of the ligand SHH to PTCH relieves its inhibition of SMO and transcriptional activation occurs through the GLI family of proteins, resulting in activation of target genes. (C) Mutations in *SHH* or *PTCH* or *SMO* may be associated with basal cell carcinomas, both in sporadic tumors as well as in certain genodermatoses, such as xeroderma pigmentosum, that are associated with increased risk of BCC. Germinal mutations in the *PTCH* gene underlie Gorlin syndrome. *Adapted from Daya-Grosjean L, Couve-Privat S. Sonic hedgehog signaling in basal cell carcinomas. Cancer Lett 2005;225:181–92.*

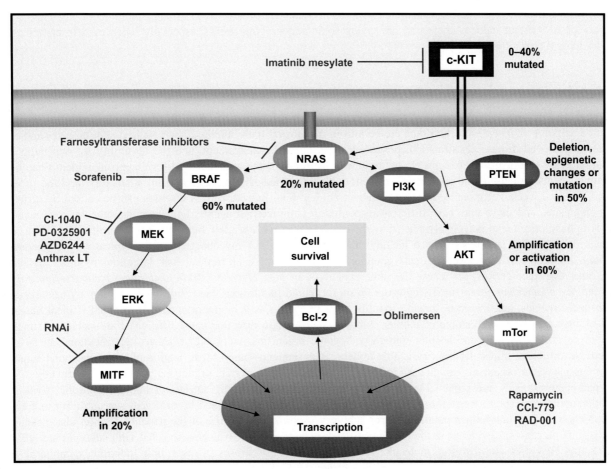

FIGURE 27.25 Potential for targeted therapies in melanoma. Recent improvement in defining the genetics of melanoma has led to the development of targeted therapeutic agents that are directed at specific molecular aberrations involved in tumor proliferation and resistance to chemotherapy. *Adapted from Singh M, Lin J, Hocker TL, Tsao H. Genetics of melanoma tumorigenesis. Br J Dermatol 2008;158:15–21.*

necessary for melanocytic proliferation, but not for tumorigenesis. To date, there has been no correlation between *BRAF* mutations, disease progression, and clinical outcome.

The *KIT* proto-oncogene encodes the stem cell factor receptor tyrosine kinase found on numerous cell types, including melanocytes. c-KIT signals via the MAPK pathway, and a downstream target is microphthalmia transcription factor (MITF), which is a critical regulator of melanocyte function. MITF regulates the development and differentiation of melanocytes and maintains melanocyte progenitor cells. MITF has a clear role in melanocyte survival, and one of its transcriptional targets is the apoptosis antagonist and proto-oncogene BCL-2.

Apart from MAPK signaling, another pathway, the V-RAS/phosphatidylinositol-3 kinase (PIK3) pathway, may be activated in certain melanomas. Although PIK3 itself is not mutated in melanoma, several downstream components may have a role in melanoma tumorigenesis, including PTEN and Akt. PTEN has a tumor suppressor role in melanomas, and its expression has been shown to be reduced or lost in some primary or metastatic melanomas. Restoration of PTEN in PTEN-deficient melanoma can reduce melanoma tumorigenicity and metastases. Changes in PTEN expression are not usually due to mutations, but involve epigenetic inactivation of both functional alleles. Akt is an important kinase in melanoma survival and progression. Akt3 is the main Akt isoform that is activated in melanomas. Strong expression of Phospho-Akt is found in several melanomas, suggesting a direct role of Akt in tumor progression. Phosphorylation of Akt is associated with increased PIK3 activity. Knowledge of these pathways is important if new treatments are to be developed for advanced melanoma, since it is clear that alterations in survival and growth signaling pathways in melanoma tumor cells can lead to increased tumorigenesis and resistance to chemotherapy.

A detailed knowledge of the molecular pathology of melanomas may have other therapeutic relevance. For example, it may be possible to render melanoma cells more sensitive to existing forms of chemotherapy, to enhance apoptosis, or to restrict the proliferation of the oncogene-driven cells by targeting these aberrant pathways. New drugs such as

Oblimersen (antisense oligonucleotide targeted to the anti-apoptotic protein BCL-2) and Sorafenib (a small molecule inhibitor of BRAF that induces apoptosis) are giving some encouraging results, especially when used in combination with traditional chemotherapy.

## Molecular diagnosis of skin disease

Understanding the molecular pathology of skin diseases has the potential to bring about several clinical and translational benefits for patients. Molecular data are helpful in diagnosing both inherited and acquired skin diseases and contribute to improved disease classification, prognostication, clinical management, and the feasibility for designing and developing new treatments for patients. In many infectious diseases, the gold standard for diagnosis is identification of infectious agents by culture and species identification. However, material yield can be unsatisfactory, and the process can take several weeks, or months, before a pathogen is identified. Indeed, certain diseases can be caused by different microorganisms, and these may have different responses to antimicrobial agents. For example, subcutaneous mycosis with lymphangiotic spread is typically caused by the mold *Sporothrix schenckii*, but lymphatic sporotrichoid lesions can be caused by diverse pathogens. These include other non-tuberculous mycobacteria, most typically *Mycobacterium marinum*, a common pathogen for fish tank granuloma in people who keep tropical fish. Infections by other mycobacteria, other than tuberculosis, that cause the same phenotype include *Mycobacterium chelonae*, *Mycobacterium fortuitum*, and *Mycobacterium abscessus*. Lymphangiotic sporotrichoid-like lesions can result from infection by bacteria such as *Nocardia* species, and therefore, precise identification of species is very important in the optimal clinical management of such infections. Mycetoma is another chronic localized skin infection caused by different species of fungi or actinomycetes. Grains from the lesions contain pathogenic organisms, but final organism identification can be protracted. Nevertheless, molecular biology is able to help since sequence-based identification of large ribosomal subunits specific to particular organisms can be used to identify and characterize certain organisms that contribute directly to disease pathogenesis. PCR and sequencing approaches are useful in identifying subsets of human papilloma virus. For example, epidermodysplasia verruciformis is a rare genodermatosis characterized by profound susceptibility to cutaneous infection with certain human papilloma virus subtypes. Approximately 50% of the patients develop non-melanoma skin cancers on sun-exposed areas in the fourth or fifth decade of life. The cancer-associated viral subtypes are HPV-5, HPV-8, and HPV-14d, and therefore molecular identification of these viruses in susceptible individuals can be helpful in promoting more vigorous surveillance of at-risk patients.

Molecular profiling is useful in the identification and classification of cutaneous malignancies. Notably, T-cell receptor gene rearrangements, which arise as a result of variability in the V-J segment, provide important information relevant to the diagnosis and prognosis of cutaneous T-cell lymphoma (Fig. 27.26). The presence of clonality usually (but not always) is associated with the presence of malignancy. Thus, in many cases, a T-cell receptor gene rearrangement can be helpful in distinguishing a malignant from a benign lymphocytic infiltrate, especially when clinical and histologic features are inconclusive. Clonality can influence prognosis. For example, Sézary syndrome patients with T-cell clonality are more likely to die from lymphoma/leukemia than their non-clonal counterparts. DNA microarray profiling has been used in

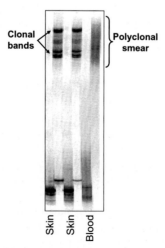

**FIGURE 27.26**   Clonal T-cell expansion in a patient with mycosis fungoides (cutaneous T-cell lymphoma). The clinical stage of the patient is stage 1b. This figure shows single-strand conformational polymorphism (SSCP) analysis and demonstrates an identical clonal T-cell receptor gene rearrangement in two lesional skin biopsies. The matched blood sample is polyclonal.

patients with diffuse large B-cell lymphoma. Specific gene signatures have been linked to certain subtypes (such as prominent germinal center B-cell profile or activated B-cell profile), which can be directly associated with different prognoses. The recent characterization of the molecular basis of the acquired immunobullous diseases has led to new diagnostic tests. For example, antigen-specific ELISA kits for desmogleins 1 and 3 are now commercially available for the assessment of serum samples from patients with pemphigus. These ELISA tests are more sensitive and specific than immunofluorescence microscopy when used in diagnosing pemphigus. More importantly, the titers have been shown to correlate with disease activity. ELISA tests for the NC16 A domain of type XVII collagen are useful in the diagnosis of bullous pemphigoid. In mucous membrane pemphigoid, antibodies can be directed against either type XVII collagen or laminin-332. Molecular approaches to determining the target antigen can be important since there is an increased risk of certain malignancies in mucous membrane pemphigoid associated with laminin-332 targeted auto-antibodies.

For inherited skin blistering diseases, several clinical subtypes can have similar phenotypes, and routine light microscopy is not usually able to assist diagnosis or help determine prognosis. Nevertheless, antibodies to structural components at the dermal-epidermal junction are now commercially available, and these can be helpful in diagnosing recessive forms of epidermolysis bullosa. Testing may involve labeling of skin sections with antibodies to type VII collagen, laminin-332, integrin-$\alpha_6$ and $\beta_4$, plectin, or type XVII collagen. In many recessive forms of epidermolysis bullosa, immunostaining with one of these antibodies is reduced or undetectable. This immunohistochemical approach is therefore a useful prelude to determining the candidate gene that can then be sequenced and used to identify the pathogenic mutations.

Molecular pathology can provide insight into other genodermatoses that conventional microscopy is unable to provide. For example, in some cases of hereditary leiomyomas, lesions may be multiple, and there can be an autosomal dominant mode of inheritance. In such families, skin leiomyomas usually appear in adolescence or early adulthood (Fig. 27.27). However, some cases may be complicated by the subsequent development of uterine leiomyomas, usually when the patient is in their early 20s. The molecular pathology may involve mutations in the fumarase (*FH*) gene, and identification of such cases through molecular diagnosis can have important clinical implications. For example, if an *FH* gene mutation is identified in a woman with hereditary multiple leiomyomas who plans to have children at some

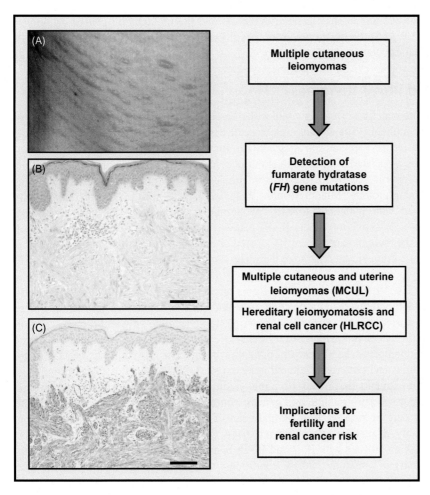

FIGURE 27.27 Impact of molecular diagnostics on clinical management. (A) Clinical appearances of multiple cutaneous leiomyomas. (B) Light microscopic appearance shows a spindle cell tumor within the dermis (bar=100 μm). (C) Immunostaining with smooth muscle actin identifies the dermal tumor as a leiomyoma (bar=100 μm). In patients with multiple cutaneous leiomyomas and an autosomal dominant family history, detection of fumarate hydratase (*FH* gene mutations) may indicate a diagnosis of specific syndromes that can have implications for fertility as well as the risk of developing rare forms of renal cancer.

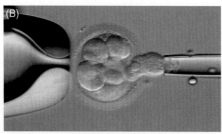

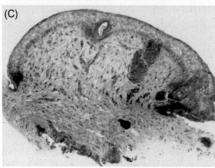

FIGURE 27.28 Options for prenatal testing for severe inherited skin diseases. (A) Chorionic villus samples taken at 10−12 weeks; (B) Preimplantation genetic diagnosis, here illustrating single cell extraction from a 72-h-old embryo; (C) Fetal skin biopsy performed at 16−22 weeks gestation, here showing the appearances of normal human fetal skin at 18 weeks (bar=25 μm).

stage, it may be prudent to advise her to consider having her children at an early age before the onset of uterine fibroids. Diagnosis of *FH* gene mutations in hereditary multiple leiomyomas may have other prognostic significance, in that a subset of patients may be at-risk from developing aggressive type II papillary renal cell carcinomas.

Understanding the molecular pathology of inherited skin diseases has changed clinical practice by allowing the development of newer techniques for prenatal diagnosis for certain disorders (Fig. 27.28). Advances in fetal medicine have led to further advances in which the molecular pathology of inherited skin diseases can be used to develop new techniques for prenatal testing. These include preimplantation genetic diagnosis and preimplantation genetic haplotyping. Further technical advances are likely to lead to the development of less invasive forms of fetal screening, such as the analysis of free fetal DNA in the maternal circulation.

## New molecular mechanisms and novel therapies

Two groups of dermatological diseases that are benefiting from new molecular-based therapies are the chronic autoinflammatory diseases as well as cutaneous T-cell lymphoma. The autoinflammatory diseases are complex and clinically difficult to manage. Nevertheless, new insights into molecular mechanisms are leading to immediate therapeutic benefits. The cryopyrinopathies represent a spectrum of diseases associated with mutations in the cold-induced autoinflammatory syndrome 1 (*CIAS1*) gene that encodes cryopyrin. Cryopyrin and pyrin (the protein implicated in familial Mediterranean fever) belong to the family of pyrin domain−containing proteins. Mutations in the gene that encodes for the CD2-binding protein 1 (*CD2BP1*) that binds pyrin are associated with pyogenic arthritis, pyodermic gangrenosum, and acne syndrome (OMIM604416). Common to all these diseases is activation of the IL-1β pathway, and this finding has been translated into direct benefits for patients. Specifically, the recombinant human IL-1 receptor antagonist, Anakinra, has proved to be helpful and in some cases results in a quick and dramatic treatment for these chronic autoinflammatory diseases.

Management of the cutaneous T-cell lymphomas will improve following a more detailed understanding of the molecular pathology of these disorders (Fig. 27.29). The most common forms of cutaneous T-cell lymphoma are mycosis fungoides and Sézary syndrome. Mycosis fungoides usually presents in a skin with erythematous patches, plaques, and sometimes tumors. Sézary syndrome is a triad of generalized erythroderma, lymphadenopathy, and the presence of circulating malignant T-cells with cerebriform nuclei (known as Sézary cells). Recent studies have identified a number of changes in various tumor suppressor and apoptosis-related genes in patients with mycosis fungoides and Sézary syndrome. Discovery of specific gene abnormalities and patterns of altered gene expression have diagnostic and prognostic value and offer new insight into developing treatments that transform a malignant disease into a less aggressive chronic illness. There are accumulating data regarding specific immune and genetic abnormalities in these diseases that might lead to new therapies that block the trafficking or proliferation of malignant T-cells. Of note, immunophenotyping suggests that mycosis fungoides and possibly Sézary cells are derived from CLA-positive effector memory cells. The malignant cells demonstrate altered cytokine profiles with IL-7 and IL-18 being upregulated in the plasma and skin of affected individuals. Loss of T-cell diversity is a feature of these conditions, and antigen-presenting dendritic cells

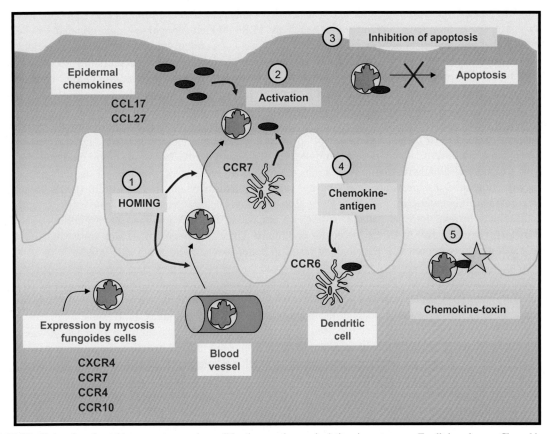

**FIGURE 27.29** Roles for chemokine receptors and possible therapeutic manipulation in cutaneous T-cell lymphoma. Chemokine receptors may have important roles in enabling malignant T-cells to enter and survive in the skin. (1) Homing: activation of T-cell integrins permits T-cell adhesion to endothelial cells in the skin and subsequent binding to extracellular matrix proteins. T-cells can then migrate along a gradient of chemokines, e.g., CCL17 and CCL27 to the epidermis. (2) Activation: chemokine receptors allow T-cells to interact with dendritic cells such as Langerhans cells, leading to T-cell activation and release of inflammatory cytokines. (3) Inhibition of apoptosis: chemokine receptor engagement can lead to upregulation of PI3K and AKT, which are prosurvival kinases. T-cells survive and proliferate in the skin. (4) Chemokine-antigen fusion proteins can be used to target tumor antigens from cutaneous T-cell lymphoma cells to CCR6+ presenting dendritic cells that can stimulate host anti-tumor immunity. (5) Chemokine toxin molecules target specific chemokine receptors found on cutaneous T-cell lymphoma cells to mediate direct killing.

can have an important role in the pathogenesis of the disorders, especially in maintaining the survival of proliferating malignant T-cells. Specific chemokine receptors are associated with mycosis fungoides and Sézary syndrome, and these are directly relevant to skin tropism of malignant T-cells.

Given these abnormalities, new treatments are being developed that go beyond current combinations of nitrogen mustard, corticosteroids, and radiotherapy. The overproduction of Th-2 cytokines suggests that cytokines that promote a Th-1 phenotype might be clinically useful. Indeed, IL-12 has shown clinical benefit. IL-12 is a Th-1-promoting cytokine synthesized by phagocytic cells and antigen-presenting cells. It enhances cytolytic T-cell and natural killer cell functions and is necessary for IFN-γ production by activated T-cells. Recombinant IFN-α and IFN-γ can shift the balance from a Th-2 toward a Th-1 phenotype and may have clinical relevance.

Other approaches using antibodies to CD4 (Zanolimumab) and CD52 (Alemtuzumab) broadly target T-cells and may help in relieving certain symptoms such as erythroderma and pruritus in Sézary syndrome although the immunosuppression may result in complications such as *Mycobacterium* and herpes simplex infections. Vaccine therapy for mycosis fungoides and Sézary syndrome is another intriguing but somewhat preliminary approach. This is due to the scarcity of target antigens, but identifying T-cell receptor sequences expressed by malignant lymphocytes should be technically feasible and may provide targets for immunotherapy. In similar fashion, loading autologous dendritic cells with tumor cells treated with Th-1-priming cytokines is another approach. Vaccination of patients with mimotopes (such as synthetic peptides that stimulate anti-tumor CD8-positive T-cells) is a promising clinical option. Specific molecular targeting can be attempted using histone deacetylase inhibitors such as Depsipeptide and Vorinostat. These drugs induce growth arrest in conjunction with cell differentiation and apoptosis and have been shown to improve erythroderma and pruritus and to reduce the number of circulating Sézary cells in some patients with Sézary syndrome.

Drugs such as Imiquimod, a toll-like receptor agonist, benefits cutaneous T-cell lymphoma by inducing IFN-α. Other toll-like receptor agonists may have clinical utility, such as TLR9, which recognizes unmethylated CpG-containing nucleotide motifs that are present in most bacteria and DNA viruses. Treatment with CpG oligodeoxynucleotides leads to plasmacytoid dendritic cell upregulation of costimulatory molecules and migratory receptors that subsequently generate type 1 interferons and promote a strong Th-1 immune response and enhance cellular immunity. These drugs may therefore represent a useful adjuvant in immunotherapy or addition to cytotoxic drugs.

Another approved drug for use in cutaneous T-cell lymphoma is Bexarotene. This is a retinoid that modulates gene expression through selective binding to retinoid receptors. These receptors form either homodimers or heterodimers with other nuclear receptors that then act as transcription factors. Bexarotene therapy may be combined with new treatments such as Denileukin diftitox, which is a fusion molecule containing the IL-2 receptor binding domain and the catalytically active fragment of diphtheria toxin. This targets the high-affinity IL-2 receptor present on activated T-cells and B-cells. A newer compound that probably works by the same mechanism is anti-Tac(Fv)-PE38 (LMB-2), which is an anti-CD25 recombinant immunotoxin.

Overall, these new insights into the pathophysiology of cutaneous T-cell lymphoma are providing opportunities to therapeutically target specific aspects of the molecular pathology. These approaches, in which more selective therapy is the goal, are starting to have benefits for patients that result in better survival and fewer side effects.

Understanding metabolic pathways of genodermatoses lead to novel therapies. For inborn errors of metabolism, enzyme replacement therapy can improve quality of life. Fabry disease, an X-linked recessive condition, is caused by mutations in *GLA*, coding for α-galactosidase A (GL-3). Deficiency of this enzyme leads to the accumulation of globotriaosylceramide in blood vessel walls throughout the body, leading to angiokeratomas, limb pains, and heart and kidney diseases. Replacement therapy with agalsidase has been commercially available for more than a decade. It was shown that agalsidase, given early and in conjunction with standard therapy, brings out clinical benefit to Fabry patients.

Mutant enzyme proteins, such as GL-3, are unstable in the cell, and unable to express catalytic activities. Chemical chaperones are a class of small molecules that function to enhance the folding and/or stability of proteins. It was shown that the mutant enzyme and chemical chaperone form a stable complex to be transported to the lysosome, to restore the catalytic activity of mutant enzyme after spontaneous dissociation under the acidic condition. 1-deoxygalactonojirimycin, a chemical chaperone, has been shown to help mutant GL-3 enzyme to fold properly and transport successfully to the lysosomes in end organs such as the heart or kidneys (Fig. 27.30).

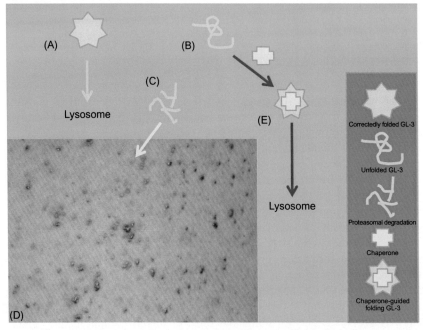

FIGURE 27.30   Chaperone therapy in Fabry disease. α-galactosidase is an enzyme responsible for the catabolism of glycosphingolipids, especially globotriaosylceramideGlobotriaosylceramide (Gb3). (A) Correctly folded α-galactosidase is transported to lysosomes for its function. Fabry disease (angiocorporis diffusum, OMIM301500) is frequently caused by missense mutation in *GLA*, which encodes for α-galactosidase. (B) Missense mutations in GLA cause misfolding of α-galactosidase enzyme, resulting in its retention in the endoplasmic reticulum and eventually (C) degradation. (D) The accumulation of Gb3 leads to vasculopathy in many organs, including the skin (multiple angiomata) and other tissues. (E) 1-deoxygalactonojirimycin is a pharmacological chaperone, which facilitates proper folding of the mutant enzyme, improving its stability and trafficking.

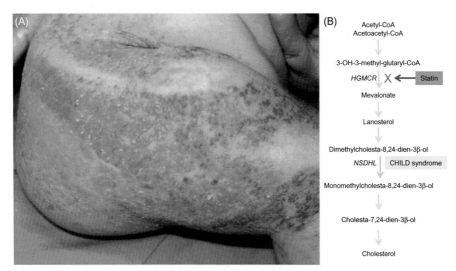

**FIGURE 27.31** Pathogenesis-based therapy for congenital Hemidysplasia with Ichthyosiform erythroderma and limb dystrophy (CHILD) syndrome. (A) CHILD syndrome patients have poor development of one side of the body, together with ipsilateral limb dystrophy and characteristic greasy, yellowish scaxles on the same side of the body. (B) Cholesterol pathway (abbreviated). CHILD syndrome is caused by mutations in *NSDHL* gene, encoding 3 β-hydroxy sterol dehydrogenase, resulting in the lack of cholesterol and accumulation of toxic metabolic intermediates. HMG-CoA reductase inhibitors such as statin block cholesterol pathway upstream to the mutation. *Adapted from Paller AS, van Steensel MA, Rodriguez-Martin M, et al. Pathogenesis-based therapy reverses cutaneous abnormalities in an inherited disorder of distal cholesterol metabolism. J Invest Dermatol 2011;131:2242—8.*

Understanding the molecular underpinning of the involved pathway can lead to novel therapy. Mutations in protein-coding enzymes can lead to accumulation of toxic by-products in the body's tissues. Mutations in *NSDHL*, encoding the enzyme 3 β-hydroxy sterol dehydrogenase, disrupt cholesterol biosynthetic pathway and lead to the buildup of cholesterol precursors and cholesterol deficiency in end organs (Fig. 27.31). Females with Congenital Hemidysplasia with Ichthyosiform erythroderma and Limb Dystrophy (CHILD) syndrome have poor development of one side of the body, together with ipsilateral limb dystrophy and characteristic greasy, yellowish scales on the same side of the body. Topical treatments with HMG-CoA reductase inhibitors together with cholesterol prevent accumulation of toxic metabolites and provide the deficient end-product, leading to normalized skin within 3 months.

New techniques such as high-throughput screening can lead to drug repositioning for genetic skin diseases. Pachyonychia congenita is a keratinizing haploinsufficiency disorder caused by mutations in *KRT6A* (50% of the cases), *KRT6B*, *KRT16*, or *KRT17*. Automated chemical library screening revealed that statins inhibit K6a promoter activity and K6a protein expression through cholesterol/mevalonate pathway inhibition. This leads to clinical trials for statins in pachyonychia congenita patients with *KRT6A* mutations.

## Key concepts

- A key role of skin is to protect the body against the external environment by providing a mechanical barrier and also a defense immune system.
- Integrity of this mechanical barrier property is provided by structural proteins in the hemidesmosomes and desmosomes.
- These structural proteins are targets for inherited and acquired skin disorders, many of which are mechanobullous.
- Defects in filaggrin cause ichthyosis vulgaris and is a major predisposing factor for its related condition, atopic dermatitis.
- Pathology of psoriasis involves tissue-resident immune cells.
- Sonic hedgehog signaling is a fundamental transduction pathway in skin tumor development.
- Molecular techniques are now being used in diagnosing inherited and acquired skin diseases, infectious diseases, identification and classification of cutaneous malignancies, and prenatal diagnosis.
- Understanding of the molecular mechanisms of skin pathology has led to the development of targeted molecular therapeutic agents.

# Suggested readings

[1] Smith FJ, Irvine AD, Terron-Kwiatkowski A, et al. Loss-of-function mutations in the gene encoding filaggrin cause ichthyosis vulgaris. Nat Genet 2006;38:337—42.

[2] Honda T, Kabashima K. Novel concept of iSALT (inducible skin-associated lymphoid tissue) in the elicitation of allergic contact dermatitis. Proc Jpn Acad Ser B Phys Biol Sci 2016;92:20—8.

[3] Rinne T, Hamel B, van Bokhoven H, Brunner HG. Pattern of p63 mutations and their phenotypes—update. Am J Med Genet A 2006;140:1396—406.

[4] Fine JD, Bruckner-Tuderman L, Eady RA, et al. Inherited epidermolysis bullosa: updated recommendations on diagnosis and classification. J Am Acad Dermatol 2014;70:1103—26.

[5] Chan YM, Yu QC, Fine JD, Fuchs E. The genetic basis of Weber-Cockayne epidermolysis bullosa simplex. Proc Natl Acad Sci USA 1993;90:7414—18.

[6] Pulkkinen L, Christiano AM, Gerecke D, et al. A homozygous nonsense mutation in the beta 3 chain gene of laminin 5 (LAMB3) in Herlitz junctional epidermolysis bullosa. Genomics 1994;24:357—60.

[7] Christiano AM, Greenspan DS, Hoffman GG, et al. A missense mutation in type VII collagen in two affected siblings with recessive dystrophic epidermolysis bullosa. Nat Genet 1993;4:62—6.

[8] McGrath JA, McMillan JR, Shemanko CS, et al. Mutations in the plakophilin 1 gene result in ectodermal dysplasia/skin fragility syndrome. Nat Genet 1997;17:240—4.

[9] Lowes MA, Bowcock AM, Krueger JG. Pathogenesis and therapy of psoriasis. Nature 2007;445:866—73.

[10] Hashimoto T. Skin diseases related to abnormality in desmosomes and hemidesmosomes—editorial review. J Dermatol Sci 1999;20:81—4.

[11] Fairley JA, Woodley DT, Chen M, Giudice GJ, Lin MS. A patient with both bullous pemphigoid and epidermolysis bullosa acquisita: an example of intermolecular epitope spreading. J Am Acad Dermatol 2004;51:118—22.

[12] Hanakawa Y, Schechter NM, Lin C, et al. Molecular mechanisms of blister formation in bullous impetigo and staphylococcal scalded skin syndrome. J Clin Invest 2002;110:53—60.

[13] Borman AM, Linton CJ, Miles SJ, Johnson EM. Molecular identification of pathogenic fungi. J Antimicrob Chemother 2008;61 i7—12.

[14] Alizadeh AA, Eisen MB, Davis RE, et al. Distinct types of diffuse large B-cell lymphoma identified by gene expression profiling. Nature 2000;403:503—11.

[15] Amagai M, Komai A, Hashimoto T, et al. Usefulness of enzyme-linked immunosorbent assay using recombinant desmogleins 1 and 3 for serodiagnosis of pemphigus. Br J Dermatol 1999;140:351—7.

[16] Fine JD, Eady RA, Bauer EA, et al. Revised classification system for inherited epidermolysis bullosa: report of the Second International Consensus Meeting on diagnosis and classification of epidermolysis bullosa. J Am Acad Dermatol 2000;42:1051—66.

[17] Fassihi H, Renwick PJ, Black C, McGrath JA. Single cell PCR amplification of microsatellites flanking the COL7A1 gene and suitability for preimplantation genetic diagnosis of Hallopeau-Siemens recessive dystrophic epidermolysis bullosa. J Dermatol Sci 2006;42:241—8.

[18] Hoffman HM, Rosengren S, Boyle DL, et al. Prevention of cold-associated acute inflammation in familial cold autoinflammatory syndrome by interleukin-1 receptor antagonist. Lancet 2004;364:1779—85.

[19] Hwang ST, Janik JE, Jaffe ES, Wilson WH. Mycosis fungoides and Sezary syndrome. Lancet 2008;371:945—57.

[20] Fan JQ, Ishii S, Asano N, Suzuki Y. Accelerated transport and maturation of lysosomal alpha-galactosidase A in Fabry lymphoblasts by an enzyme inhibitor. Nat Med 1999;5:112—15.

[21] Paller AS, van Steensel MA, Rodriguez-Martin M, et al. Pathogenesis-based therapy reverses cutaneous abnormalities in an inherited disorder of distal cholesterol metabolism. J Invest Dermatol 2011;131:2242—8.

[22] Zhao Y, Gartner U, Smith FJ, McLean WH. Statins downregulate K6a promoter activity: a possible therapeutic avenue for pachyonychia congenita. J Invest Dermatol 2011;131:1045—52.

[23] Boyman O, Conrad C, Tonel G, Gilliet M, Nestle FO. The pathogenic role of tissue-resident immune cells in psoriasis. Trends Immunol 2007;28:51—7.

[24] Daya-Grosjean L, Couve-Privat S. Sonic hedgehog signaling in basal cell carcinomas. Cancer Lett 2005;225:181—92.

[25] Singh M, Lin J, Hocker TL, Tsao H. Genetics of melanoma tumorigenesis. Br J Dermatol 2008;158:15—21.

# Chapter 28

# Molecular basis of bone diseases

**Emanuela Galliera[1,2] and Massimiliano M. Corsi Romanelli[3,4]**

[1]*Department of Biomedical, Surgical and Oral Sciences, Università degli Studi di Milano, Milan, Italy.,* [2]*IRCCS Galeazzi Orthopedic Institute, Milan, Italy.,* [3]*Department of Biomedical Sciences for Health, Università degli Studi di Milano, Milan, Italy.,* [4]*U.O.C SMEL-1 Patologia Clinica IRCCS Policlinico San Donato, Milan, Italy.*

## Summary

Bone is not an inert structure but a living and metabolic active tissue, which undergoes continuous remodeling throughout life, to allow bone architecture to respond to mechanical stress, consisting in bone tissue renewal to replace old bone tissue with new bone matrix. The equilibrium between osteoblasts and osteoclasts activity is the key of bone homeostasis, and any perturbation of this balance leads to the development of skeletal pathological condition characterized by decrease or an excess of bone mass. Bone is a living tissue undergoing a continuous remodeling, in order to respond to a variety of different stimuli, ranging from mechanical loading, to hormonal factors and calcium homeostasis. For this reason, bone is a metabolically active tissue, and bone metabolism is a complex series of event under the control of different genetic, hormonal, and biochemical factors. This chapter will describe the molecular basis of the main bone disease characterized by or an excess of bone mass, as well as metabolic bone disease. In addition, due to its structure, bone is usually protected from infections, but it can undergo infection. Moreover, bone can also be affected by primary tumor or metastatic invasion. This chapter also provides the most recent insight concerning bone infections and bone tumors and metastases.

## Introduction

Bone is a living tissue composed by an organic matrix and inorganic components, mainly represented by calcium hydroxyapatite. The inorganic part is responsible for calcium depots, and the strength and resistance properties of bone, while the inorganic component account for bone metabolism and turnover. The organic part of bone tissue is composed by bone cells and matrix proteins. The main bone cell types include (i) osteoblast, (ii) osteoclasts, and (iii) osteocytes. Osteoblasts are located on bone surface are responsible for new bone matrix deposition. They are under the control of various growth and differentiation factors. These cells express cell surface membrane receptors for parathyroid hormone (PTH), estrogen, cytokines and chemokines, and matrix metalloproteinases (MMPs). Once new bone matrix deposition is complete, these cells become defined as osteocytes. Osteocytes are the most abundant cell type in bone, and they are found throughout bone tissue and surrounded by bone matrix. Osteocytes are essential for immediate responses to alterations of calcium and phosphorus concentrations in the blood, and function to modulate the extracellular concentration of these two minerals. Osteocytes also have mechanical receptors, which are responsible for bone tissue response to mechanical stress mediated by activating *cyclic* adenosine monophosphate-mediated signals. In contrast, osteoclasts are responsible for bone matrix erosion and the process of bone resorption. Osteoclasts differentiate from a myeloid progenitor, and share several characteristics with monocytes and macrophages. Like monocytes and macrophages, osteoclasts express receptors for several cytokines (Interleukin 1 or IL-1, IL-6, IL-11) and tumor necrosis factor receptor (TNFR), granulocyte-macrophages colony stimulating factor (GM-CSF), and macrophage colony stimulating factor (M-CSF). Osteoclasts make and secrete digestive enzymes that break up or dissolve the bone tissue. Subsequently, osteoclasts absorb the bone debris and further break it down inside the cell. Collagen is broken down into amino acids, which are recycled to build other proteins, while the calcium and phosphate are released to be utilized elsewhere in the body. Osteoclasts are found on top of or next to existing bone tissue, sometimes in close proximity to osteoblasts. There is a balance between the two cell types: osteoblasts make bone tissue while osteoclasts reabsorb bone tissue.

Essential Concepts in Molecular Pathology. DOI: https://doi.org/10.1016/B978-0-12-813257-9.00028-0

# Molecular basis of bone modeling and remodeling

Bone is not an inert structure but a living and metabolic active tissue. Bone undergoes continuous remodeling throughout life to allow bone architecture to respond to mechanical stress. Bone remodeling consists of bone tissue renewal to replace old bone tissue with new bone matrix. The process of bone remodeling is based on the balanced action of the two cell types present in bone: osteoclasts remove old, mineralized bone, and osteoblasts form new bone matrix that becomes mineralized. The mechanism of this process consist in a cycle of four phases: the remodeling is initiated by the *activation* of quiescent bone cells, covered with bone lining cells, followed by the recruitment of osteoclasts precursors, which fuse together to form mature bone resorbing osteoclasts. This phase, called *resorption*, consist in the action of osteoclast, which create an acidic microenvironment to dissolve inorganic matrix and degrade the organic one by with specific enzymes. The next step, called *reversal*, is characterized by the recruitment of mononuclear cells to the bone surface to prepare it to the final step, or bone *formation*. Bone formation occurs when osteoblasts synthesize and deposit new bone matrix, which is covered by bone lining cells to keep the material quiescent until the next cycle. The equilibrium between osteoblasts and osteoclasts activity is the key of bone homeostasis. Perturbation of this balance leads to the development of pathological conditions of the skeleton characterized by decrease of bone mass (osteoporosis) or an excess of bone mass (osteopetrosis).

## Interplay between the immune system and bone: osteoimmunology and RANK-RANKL-OPG system

Cytokines produced by bone cells and immune system cells play a crucial role in regulation of bone remodeling. Over the past decades, several immune cell-derived bone remodeling modulators have been described. The importance of the interplay between the skeletal system and the immune system is reflected by the emerging interdisciplinary research field of *osteoimmunology*, which is focused on common aspects of osteology and immunology. The tight relationship between these systems is reflected by their common site of cellular origin in the bone marrow. In the bone marrow, immune cells and bone cells influence each other and share the common progenitors. Macrophages and osteoclasts develop from the same monocyte lineage progenitor cells and share the same degradation properties. The two major immune cell types (B-cells and T-cells) affect bone metabolism. T-lymphocytes stimulate or inhibit osteoclastogenesis depending on the activation stage and the pattern of cytokines produced, while B-lymphocytes are involved osteoclast development, sharing common regulators of differentiation such as steroids, in particular estrogens. These molecules, produced by immune cells affect bone cell function and vice versa, mediated by the receptor activator of nuclear factor κ B (RANK)/RANK ligand (RANKL)/OPG system, which was independently discovered in both the bone system and immune system. Osteoprotegerin (OPG) belongs to the TNFR superfamily and has bone-protective properties, being able to inhibit osteoclast development and activation. RANK belongs to the TNF superfamily and is mainly expressed by preosteoblast/stromal cells, as well as activated T-cells. The expression of both these molecules is strictly regulated by systemic and local factors: estrogens induce OPG expression, while RANKL expression is increased by prostaglandin and cytokines (including IL-6, IL-8, and IL-11). RANKL acts through its receptor, RANK, a transmembrane receptor belonging to the OPG family and expressed mainly by osteoclasts and dendritic cells. After RANKL binding, RANK activates a complex signaling cascade via TRAF6, AP-kt, and NF-kB, which lead to activation, differentiation, and survival of osteoclasts. This system is negatively modulated by OPG, which is able to compete with RANK for RANKL binding. However, OPG lacks the transmembrane domain and acts as a decoy receptor for RANKL, interfering with RANK-mediated signaling.

# Molecular basis of bone disease associated to bone matrix

Most bone matrix components have been characterized only recently. Emerging evidence underscores the importance of bone matrix regulation in the development of skeletal disease associated with bone mineral defects, including osteoporosis and osteogenesis imperfecta (OI).

## Osteogenesis imperfecta

Diseases in which collagen structure is abnormal produce brittle bones and are classified under the category of OI. OI, or brittle bone disease, is a heritable skeletal disease that results from various alterations of connective tissue resulting in skeletal abnormalities, bone fragility, and deformity. The clinical presentation is heterogeneous and characterized by low bone mass, reduced mineral strength, bone fragility, deformity, and skeletal growth deficiencies. According to

**TABLE 28.1** Osteogenesis Imperfecta subtypes.

| OI subtype | Genetic trasmission | Collagen defects | Clinical presentation |
|---|---|---|---|
| I | Autosomic dominant | Decreaded synthesis of pro-collagen I | Bone fragility, joint weakness blue sclera |
| II | Mostly Autosomic dominant | Structural defects in pro-collagen I and II, abnormal collagen helix | Perinatal lethal Bone fragility multiple fractures Blue sclera |
| III | 75% Autosomic dominant 25% Autosomic recessive | Structural defects in pro-collagen II and helix | Multiple fractures Blue sclera |
| IV | Autosomic dominant | Structural defects in helix, shorter pro-collagen II molecule | Moderate bone fragility susceptibility to fractures |

clinical presentation, aggressiveness, and genetic determinants, OI has been classified in four types (OI I-IV), as shown in Table 28.1.

The molecular basis of OI is a defect in the structure or processing of collagen type I, the main protein of the extracellular matrix (ECM) of bone tissue. This disease was first identified as an autosomal dominant bone disorder caused by mutation in the gene for type 1 collagen I. However, recent evidence identified several collagen-related recessive mutations as causative for OI. Hence, new insights into the development of this disease are continuing to emerge.

## Autosomal dominant osteogenesis imperfecta

The major mutation resulting in autosomal dominant OI are defects in the *COL1A* and *COL1A2* genes, encoding the α1 (I) and α2(I) chains if type I collagen. These mutations result in alteration of the quantity and structure of type I collagen leading to a variety of phenotypes ranging from subclinical to lethal, classified in the Sillence Classification. The molecular defect in OI type I is a null *CLO1A1* allele due to frame-shift or splice site defects, both resulting in reduced synthesis of structurally normal collagen. OI type II—IV result primarily from glycine substitutions and splice site mutations, leading to alteration in collagen I structure. Glycine substitutions alter helix folding and collagen conformation, inducing post-translational modifications, most of which have a lethal outcome. The observation of differences in lethality between the two collagen alpha chains suggests that they play different roles in collagen matrix composition, leading to different clinical outcomes. Procollagen can also be involved in OI development. A rare mutation affecting procollagen structure or processing results in different forms of OI.

## Recessive osteogenesis imperfecta

The recessive form of OI occurs in 2%—5% cases of North America and Europe and results from mutations in genes encoding the components of the collagen 3-hydroxylation complex. The collagen 3-hydroxylation complex is composed by Prolyl 3 —hydroxylase (P3HI), cartilage associated protein (CRTAP) and cyclophilin b (CyPB) and is responsible for the post translational modification of specific proline residues of unfolded collagen α chains. Each component of this complex can also act as a single multifunctional protein, exerting a specific function. The majority of CRTAP alterations result from frame shift mutations that impair CRTAP protein production, resulting in a loss of α1(I) collagen 3-hydroxylation and a consequent impaired collagen folding.

P3HI is the enzymatic component of the complex that is expressed in tissues rich in fibrillary collagen and abundant during skeletal development. Molecular defects in the P3HI encoding gene (*LEPREI*) results in type VII and VIII OI, characterized by severe to lethal osteogenesis, undertubulation of long bone, and extremely low bone mineral density (BMD).

CyBP is a peptidyl-proline cis-trans isomerase. The isomerization of the cis proline to the trans conformation is essential for correct collagen folding. Thus, mutation occurring at the CyBP encoding gene (*PPIB*) results in a misfolded protein and leads to severe to lethal OI.

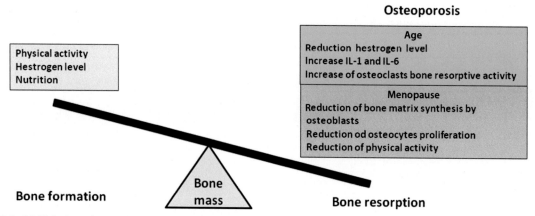

**FIGURE 28.1** Multiple determinants of osteoporosis. Osteoporosis is a heterogeneous disorder, because it results from alteration in the equilibrium between bone forming factors (young age, physical exercises) and bone erosive factors (age, hormones, menopause).

## Osteoporosis

Osteoporosis is characterized by low BMD and alterations in bone microstructure resulting in increased susceptibility to fracture. Osteoporotic-related fractures are the major cause of morbidity for this disease, in particular hip and vertebral fractures. Therefore, a large number of studied focused on the pathogenesis of osteoporosis in order to improve therapies and prevent fracture risk. As described in Fig. 28.1, A large number of studies have focused on the pathogenesis of osteoporosis in order to improve therapies and reduce fracture risk in affected patients. As described in Fig. 28.1, osteoporosis is a heterogeneous disorder, because it results from the complex interaction between local and systemic regulation of bone cell function, changes in signal receptors transduction mechanisms, nuclear transcription factors, and several enzymes that induce or inhibit local regulators. Therefore, the study of the molecular basis of osteoporosis must consider the various aspects of regulation of bone turnover, ranging from hormonal regulation to genetic and cell signaling regulation.

## Estrogen regulation

Estrogen deficiency is critical in the pathogenesis of osteoporosis, increasing the rate of bone remodeling and resorption. Studies in animal models have shown that estrogen regulates bone resorption at multiple levels. Estrogen Receptor α (ERα) is primarily involved in bone remodeling, being expressed on osteoblasts, and studies on ERα single nucleotide polymorphism (SNPs) suggest that different SNPs for ERα are associated with increased of fracture risk, independent of BMD. ERα (as well as ERβ) regulates bone resorption through the orphan receptor estrogen-related receptor α (ERRα), which is expressed by bone cells. ERRα is unable to bind estrogen and acts a negative regulator by interacting with ERα and ERβ to influence bone cell function. A regulatory variant of the ERRα-encoding gene was recently described to be associated with significant alteration of BMD in postmenopausal women. The bioavailability of sex hormone is regulated by sex hormone-binding protein (SHBP). Epidemiologic studies suggest that sex hormone−binding globulin (SHBG) influences bone fracture risk and bone loss.

Estrogen can act not only on cells of the osteoblastic lineage, but can also affect bone remodeling through osteoclast precursors, mature osteoclasts, and lymphocytes. In particular, estrogen affects local production of cytokines, such as IL-1, TNF-α, transforming growth factor beta (TGF-β), and IL-6 is mediated by action of T-cells.

## Genetic regulation

The pathogenesis of osteoporosis is influenced by genetic factors, and various recent studies have focused on the genetic determinants of osteoporosis using different investigational approaches. The most comprehensive approach is the genome-wide association study (GWAS), genotyping 300,000−1,000,000 polymorphic markers spread across the genome looking for as association with osteoporosis phenotypes. More than 50 candidate genes were found to be associated with BMD in the lumbar and femoral neck. In particular, recent GWAS study identified loci containing candidate genes involved in Wnt and RANKL pathways, confirming the role of these two pathways in the pathogenesis of osteoporosis.

## Regulation of gene expression by microRNA

MicroRNAs (miRNA) are small noncoding RNA molecules that acts as RNA silencer and post-transcriptional regulators of gene expression. miRNA binds target mRNAs with similar sequence, inhibiting the translation of different genes. Therefore, miRNAs play crucial regulatory roles in myriad biological processes, including bone mass determination. Recent evidence suggests that different miRNAs are involved in various osteoporosis phenotypes and osteoporotic fractures. The study of the expression profile of serum miRNA in postmenopausal women revealed that miR-422a and miR-133a are differentially expressed in postmenopausal women with high BMD. In addition, osteoporosis patients exhibit higher expression of miR-21. Another study on the expression profile of miRNA in patients with osteoporotic fractures detected a critical role for miR122-5p, miR125b-5p, and miR-21-5p. In addition, a recent study focused on vertebral fractures in postmenopausal osteoporosis patients revealed significantly different expression levels of miR-106b and miR-19b in osteoporosis patients with or without vertebral fractures.

## Regulation of signaling pathways

Different signal transduction pathways are involved in the pathogenesis of osteoporosis. These signal transduction pathways are critical in the regulation of osteoblast differentiation and function. Genetic studies show that downregulation of Runt-related transcription factor 2 (Runx2) or its downstream factor Osterix significantly influence osteoblast regulation. The Wnt signaling pathway is also important in the regulation of osteoblast function, and mutations affecting this pathway are critical for bone mass regulation. In the Wnt pathway, low-density lipoprotein (LDL) receptor—related protein 5 (LRP5) interacts with Frizzled receptor to transduce signals from Wnt ligands. Constitutive activation of LRP5 results in increased bone density, while deletion of the LRP5 encoding gene results in a severe osteoporotic phenotype. In vivo models show that activating mutations of LRP5 result in an increased response to mechanical loading and a consequent increased BMD. The Wnt signaling pathway ultimately activates β-catenin through interaction with bone morphogenetic protein 2 (BMP2). The Wnt inhibitor Sclerostin, which is encoded by the SOST gene, regulates this interaction. Inactivating mutations in *SOST* result in disorders characterized by high bone mass.

## Regulation of local and systemic growth factors

The pathogenesis of osteoporosis depends not only on bone cell—specific factors, but also on the production and activity of local and systemic regulatory factors. These include BMPs, TNF family factors, insulin-like growth factoe (IGF), and TGF-β. In particular, some association have been found between polymorphic forms of IGF-1 and TGF-β and osteoporotic phenotypes and fractures

# Molecular basis of bone disease associated with bone resorption

Skeletal diseases displaying increased bone density comprise a group of different disorders, characterized by an excess of bone matrix deposition, with variable distribution across the skeletal system. Dysplasia reflects different defects in osteoclastogenesis, mature osteoclast activity, and through osteoclast-osteoblast interaction.

## Osteopetrosis

Human osteopetrosis is a rare genetic disorder characterized by osteoclast failure and impaired bone resorption, resulting in a variety of heterogeneous symptoms, ranging from a neonatal fatal phenotype to an asymptomatic form in adult. The pathogenesis of osteopetrosis is based on heterogeneous molecular defects resulting in decreased bone resorption and increased bone density. Osteopetrosis is classified in three forms: autosomal dominant osteopetrosis (ADO, the most common), autosomal recessive osteopetrosis (ARO), and X-linked osteopetrosis. Given the variety of genetic defects resulting in osteopetrotic phenotypes, the etiology of osteopetrosis can be better understood by analyzing the genetic basis of the disease.

The most common and heterogeneous form of ADO is Albers—Shonberg disease. Approximately 70% of Albers-Shonberg patients present with dominant-negative missense mutations in the gene *ClCN7*, encoding for a chloride channel that transports Cl-in the resorbing lacuna and is required for correct bone resorption. So far 25 different mutations have been described in the *ClCN7* gene, most of which are missense mutation or frame shift mutations. Analysis of the G215R mutation (the most common) revealed reduced acid secretion into lysosomes due to retention of mutated protein in the endoplasmic reticulum.

In ADO patients, as well as in patients with other osteoclast-rich forms of osteopetrosis, bone resorption is impaired but bone formation is ongoing. Osteoclasts (independent of their bone resorptive activities) produce anabolic signals for osteoblasts, such as IGF-1, BMP6, and Wnt10b, that stimulate bone formation and increase formation of additional bone matrix.

The ARO, also called malignant infantile osteopetrosis, is rarer than the autosomal dominant form. The genetic basis of this disease is still poorly understood, but the main mutations found in this disorder can be divided into groups according to the abundance of osteoclasts. The osteoclast-rich ARO, characterized by high number of osteoclasts but severely impaired bone resorption, results from mutation of the genes *TCIRG*, *CLCN7*, *SNX10*, and *PLEKHM1*, while the osteoclast-poor ARO is due to mutations in the genes *TNFSF11* and *TNFRSF11A*.

## Paget disease of bone

Paget disease of bone (PDB) is a common disorder characterized by increased bone turnover within lesions sites (pagetic lesions) spread throughout the skeleton. Clinical features include bone hypertrophy, deformities of long bone, and presentation in middle age or old age (typically). From the cellular point of view, PDB is characterized by higher numbers of osteoclasts with larger size and an increased number of nuclei compared to normal osteoclasts. The majority of PDB mutations are found in a single gene *SQSTM1*, encoding for the p62 protein, but these mutations demonstrate incomplete penetrance, confirming an important role for environmental factors influencing PDB etiology. Mutations in *SQSTM1* can be divided into two groups according to the domain of SQSTM1 affected. One group of *SQSTM1* mutations affect the ubiquitin-associated domain (UBA) domain of p62. The majority of PDB-associated *SQSTM1* mutations are missense or truncation mutations in the C-terminal region of p62 where UBA is located. SQSTM1-p62 mutated protein is unable to bind ubiquitin and fails to be degraded by the proteasome. This results in an aberrant RANKL signaling via NFkB in osteoclasts, promoting an increase of bone resorption. Specifically, within RANKL pathway, p62 interacts with TRAF6 and PKC proteins, regulating the signal-dependent ubiquitination of TRAF6. Negative regulation of this process is mediated by the deubiquitinating enzyme CLYD, which removes the ubiquitin modification from TRAF6. CLYD is recruited trough the interaction with the p62 UBA domain, and the mutation P392L in the UBA domain of p62 abolished this interaction. In addition to osteoclasts, recent profiling studies investigated the role of osteoblasts in PDB. In particular the circulating level of DKK-1 (Dikkopf-1), an inhibitor of Wnt signaling pathway expressed by osteoblasts, are increased in PDB patients compared to healthy controls, suggesting DKK-1 as a potential serum biomarker for PDB.

A second group of *SQSTM1* mutations that do not affect the UBA domain of p62 have been identified. The D335E missense mutation is localized in the LC3 binding region of p62 involved in the regulation of autophagy. Autophagy is a regulated catabolic process whereby cytoplasmic proteins and damaged organelles are degraded. Cytosolic components are encapsulated in a double membrane−bound autophagosome, which fuses with lysosome to generate an autophagolysosome, and hydrolases are released to complete degradation. Ubiquitination, in addition to degradation via 26S proteasome, mediates degradation by autophagy. In this process, p62 targets polyubiquitinated protein to the autophagic machinery. Hence, p62 acts as an adaptor to drive specific molecules to undergo selective degradation by autophagy.

Incomplete penetrance of *SQSTM1* mutations suggests the importance of interaction between genetic factors and environmental factors in the determination of PDB phenotype. It has been proposed that *SQSTM1* mutation leads to increased risk factor for PDB, with manifestation of PDB after a second insult. An example of a secondary insult is viral infection, in particular infection by paramyxoviruses. Accordingly, recent evidence suggests that a viral survival strategy is to impair autophagy, where p62 is involved. In addition, another environmental factor affecting *SQSTM1* is oxygen radical stress, a condition that stimulates p62 expression. Oxidative stress could unmask susceptibility due to *SQSTM1* mutations leading to development of PDB. This could explain the prevalence of PDB in patients older than 50 years of age, when oxidative stress is increased compared to younger individuals.

## Molecular basis of metabolic bone disease

The molecular basis of bone metabolism disorders can be explored at different levels, involving genetic, hormonal, and end-organ mechanisms.

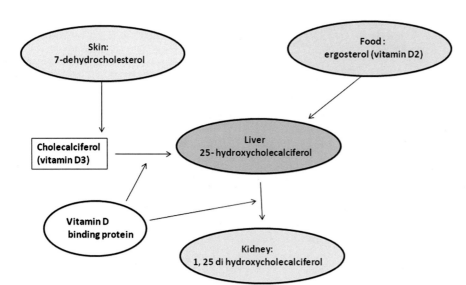

FIGURE 28.2 Vitamin D metabolism. In the skin, 7-dehydrocholesterol is converted into pre-vitamin D3 by UV light and then modified into vitamin D3 (cholecalciferol). The dietary source of vitamin D is in the form of ergocalciferol (vitamin D2). Vitamin D2 and D3 are prohormones transported into the blood by means of specific binding proteins and are hydroxylated in the liver into 25-hydroxyvitamin D (calcidiol). Calciferol is further hydroxylated in the renal tubule into 1,25 dihydroxyvitamin D (calcitrol), which is the active form of the hormone. The production of calcitrol is stimulated by increased levels of parathyroid hormone (PTH) and decreased level of phosphate (PO4).

## Vitamin D metabolism

Vitamin D metabolism is illustrated in Fig. 28.2. The major sources of vitamin D are synthesis in the skin under the influence of UV, and a small portion is derived from the diet. Calcitrol binds specific vitamin D receptors (VDR) at the target organ, activating transcription factors that regulate calcium metabolism and bone metabolism. VDRs are expressed by all three major bone cell types: osteoclasts, osteoblasts, and osteocytes. VDRs are essential regulators of vitamin D metabolism. Hence, defects in the VDR gene results in hereditary vitamin D-resistant rickets, characterized by an impairment of a bone matrix mineralization. In addition to the catabolic effect of vitamin D in bone, VDRs mediate anabolic function in osteoblasts. The overexpression of VDR in mature osteoblast cell lines result in increased mineral formation and decreased bone resorption.

Vitamin D deficiencies lead to impaired mineralization of bone mass and present with disease in the form of rickets during growth and osteomalacia in the adult age. Beyond insufficient dietary vitamin D intake, vitamin D-deficiency results from malabsorption in gastrointestinal, pancreatic, and hepatobiliary diseases. In addition, some drugs and toxins, such as anticonvulsants and bisphosphonates, can impair vitamin D metabolism.

Several genetic disorders cause heritable forms of rickets and osteomalacia. X-linked hypophosphatemia, also called vitamin D-resistant rickets, is the most common form of hereditary rickets and osteomalacia resulting from inactivating mutations of the *PHEX* genes that encode endopeptidases. Autosomal dominant hypophosphatemic rickets is a rare form of renal phosphate loss caused by a gain-of-function mutation of the gene encoding FGF23 (fibroblast growth factor 23), which has recently described as a marker of fracture risk. Autosomal recessive hypophosphatemic rickets is a rare form caused by mutation is the *DMP1*, *ENPP1*, and *FAM20C* genes, encoding for different proteins involved in vitamin D metabolism. Additional forms of hereditary rickets or osteomalacia are vitamin D-dependent rickets type I and II, characterized by reduced vitamin D biosynthesis, and hypophophatasia which is rare and due to deficient activity of tissue nonspecific enzyme alkaline phosphatase.

## Parathyroid hormone

PTH is the main regulator of serum calcium levels. The effect of PTH on bone metabolism is described in Fig. 28.3. PTH acts on kidney to increase reabsorption of calcium and decrease renal reabsorption of PO4, and on bone to induce bone resorption for the release of calcium from bone matrix and to decrease the release of PO4 from bone. At physiological concentrations, PTH exerts an anabolic effect of bone, while sustained elevation of PTH is catabolic, stimulating bone turnover and resorption. The effects of PTH on bone depends on the duration of low levels of calcium. PTH is secreted very shortly after reduction of serum calcium levels. Within the first hours, active degradation of PTH by parathyroid cells is decreased, followed by increased PTH expression along with the increase of parathyroid cell number. Conditions of excessive production of PTH are defined as hyperparathyroidism, which can be primary (when it is associated with increased serum calcium levels) or secondary (when the parathyroid gland response to low serum calcium

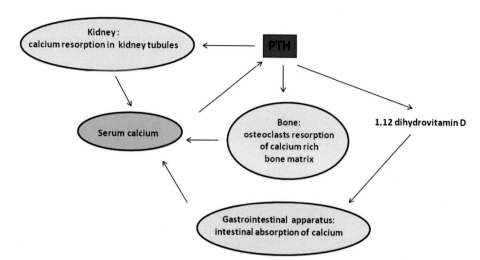

**FIGURE 28.3** Effect of PTH on calcium metabolism and bone turnover. PTH acts on kidney increasing reabsorption of calcium and decreasing renal reabsorption of phosphate (PO4), and on bone by inducing bone resorption for the release of calcium from bone matrix and decreasing release of PO4 from bone. At physiological concentration, PTH exerts an anabolic effect of bone, while sustained elevation of PTH is catabolic, stimulating bone turnover and resorption.

levels is adequate, but vitamin D is deficient or inactive). In both cases, chronic overproduction of PTH results in increased bone resorption.

## Renal osteodystrophy

Sustained loss of renal function results in a form of renal-derived hyperparathyroidism, characterized by a high bone turnover leading to metabolic bone disease and alteration of calcium and phosphorus metabolism, which are regulated by PTH and FGF23, respectively. Reduction of glomerular filtration rate (GFR) leads to a progressive retention of phosphate and a consequent deficit in vitamin D activation. Vitamin D acts as negative regulator of PTH. Cells of the parathyroid gland express VDR, and vitamin D binding exerts a negative effect on PTH gene expression. As a consequence, lack of vitamin D activation stimulates the production of PTH by the parathyroid gland. Moreover, parathyroid cells express calcium-sensing receptor, a G-protein–coupled receptor that mediates quick responses to alteration in extracellular calcium concentration.

An important role in the balance of bone resorption and formation is played by FGF23 (phosphatonin), a phosphaturic hormone regulating phosphate concentration and negatively regulating the production of active vitamin D. FGF23 (like PTH) is elevated in chronic kidney disease. However, while PTH increases in response to decreased serum calcium level and increased serum phosphate, FGF23 is upregulated in response to increased serum phosphate concentration.

## Molecular basis of bone infection and inflammation

Due to its structure, bone is usually protected from infections. However, bone can become infected via three main mechanisms: (1) by infection, driven by the bloodstream and originating in another site of the body; (2) by infection of nearby structures, such as natural or artificial joint and soft tissue; or (3) by direct invasion through open fractures or surgery procedures. When bone becomes infected, bone marrow begins to swell and press against the rigid bone, resulting in compression of blood vessels in the bone marrow.

The main causative agent of osteomyelitis that spreads through the blood are *Staphylococcus aureus* and *Mycobacterium tuberculosis*. However, other circulating pathogens can cause osteomyelitis (such as fungi), particularly in subjects with a weak or defective immune system due to comorbidity with HIV infection, autoimmune disease, cancer, or treatment with immunosuppressive drugs.

## Prosthetic joint infection

One of the most frequent causes of osteomyelitis is prosthetic implant, which becomes colonized by adherent bacteria, mainly *S. aureus* and *Staphylococcus epidermis*, forming a growing structure called biofilm. The establishment of the biofilm starts with the adhesion of bacteria to the prosthesis surface, followed by bacterial proliferation and formation of microcolonies. These colonies are able to produce an organized biofilm matrix, composed of various bacterial products (proteins and exopolysaccarides), in a bounded volume resulting in a high microbial density. At the early stage of formation, biofilms are quite unstable and are susceptible to host defenses. However, maturation of the biofilm is associated with a high density of microorganisms and establishment of a microbial communication system. This leads to development of resistance to host defenses and high-dose antibiotic drugs (particularly cell-wall active drugs). Microbial colonization of the prosthesis can occur at the time of implantation or subsequent to surgery from hematogenous seeding. For this reason, prosthetic infections manifest at different times after surgery and are classified according to time of appearance: early prosthetic joint infection (within 2 months), delayed prosthetic joint infection (within 24 months), and late prosthetic joint infection (after 24 months). Typically, early prosthetic joint infection and delayed prosthetic joint infection are endogenously acquired from *S. aureus* or *Escherichia coli* (early prosthetic joint infection), or *Propiobacteium acnes* (delayed prosthetic joint infection). Late prosthetic joint infection results from hematogenous seeding, as well as gastrointestinal, urinary, or even dental infections, and appear after a symptom-free postoperative period. The molecular basis of biofilm antibiotic resistance is multiple. First, biofilm matrix physically limits the diffusion of antibiotic drugs. Second, nutrient and oxygen depletion within the biofilm induces bacteria to enter into a nongrowing state, rendering the population nonsusceptible to growth-dependent antibacterial agents. In addition, biofilm matrix itself can bind antibiotics, further decreasing the antimicrobial activity and acting as an extra compartment in the tissue. As a consequence, biofilm represents a third gradient of antibiotic concentration from the blood, beyond the concentration in the bloodstream and tissue. For this reason, if an antibiotic has a short half-life and limiting ability to penetrate the biofilm, antibiotic concentration (even with high doses) never reaches sufficient local levels to eradicate the bacterial infection at the prosthesis site.

The molecular basis of prosthetic joint infection pathogenesis involves the immune response at different levels. The presence of a foreign body impairs host defense response. Granulocytes accumulate around the implant, but are defective in phagocytosis, leading to impaired ingestion and superoxide production, allowing the selective seeding of bacteria on the implant. Prosthetic joint infection typically evolves as a chronic infection, activating both innate and adaptive immunity. The immune response is mediated by polymorphonucleated granulocytes (PMN) recruited to the infection site by the production of IL-8, IL-1β, IL-6, and TNFα, and by the activation of the complement system by the alternative or lectin-binding pathway. The initial phase of the PMN-mediated inflammatory response is followed by the action of monocytes. Monocytes recruited in response to chemokine CCL2 amplify the inflammatory response by releasing different inflammatory mediators, such as the soluble urokinase plasminogen activator receptor (SuPAR). Due to their specific involvement in prosthetic joint infection inflammatory responses, both SuPAR and CCL2 have been suggested as potential serum diagnostic markers for clinical detection of prosthetic joint infection. The adaptive immune immunity further amplifies the response with the production of antibodies and the activation of the complement system through the classical pathway.

Prosthetic joint infection are mainly caused by Gram-positive bacteria, such as *S. aureus*. In the innate immune response, bacteria are recognized by TLR2 receptor, which drives the response of the innate immune system towards different bacteria. In particular, TLR2 specifically recognizes Gram-positive bacteria and TLR4 mediates the response directed to Gram-negative bacteria. Most prosthetic joint infections (>80%) are caused by Gram-positive bacteria, involving TLR2-mediated response, and a recent study indicated that the soluble form of TLR2 can be considered a serum diagnostic marker for the early detection of prosthetic joint infection.

## Molecular basis of bone neoplasms and bone metastasis

Bone neoplasms present with a variety of phenotypes, affecting bone formation, bone erosion, cartilage formation, and even neurological disorders. Bone tumors are broadly classified according to their metastatic potential: benign tumors (that do not metastasize) and malignant tumors (that are invasive and spread to distant sites).

### Benign tumors

Benign tumors of bone are not invasive and do not metastasize to other regions of the body. However, they can be dangerous because they can grow in any part of the skeleton, thereby compressing healthy bone tissue, causing pain and skeletal dysfunction. Osteochondroma is a cartilaginous tumor representing more than 30% of benign bone tumors. It arises mainly in the femur and tibia, at the metaphysis and diametaphysis, projecting out from the underlying bone. Giant cell bone tumor (GCT) represents 20% of benign bone tumors, arising between the age of 20 years and 40 years in long bone, especially in the knee. Osteoblastoma is a rare benign bone tumor accounting for 14% of bone tumors, arising usually in the second and third decades and with preferential site in the axial skeleton with spinal lesion. Osteoid osteoma is a benign bone tumor, accounting for 12% of bone tumors, composed of osteoid and woven bone affecting the cortex of long bone, in particular femur or tibia. Aneurysmal bone cyst (ABC) is a rare benign cystic lesion, accounting for 9% of bone tumors. The cysts contain a mix of osteoclasts, giant cells, and reactive woven bone. Fibrous dysplasia is a benign bone tumor existing in two forms: monostotic (affecting only one bone) or polystotic (affecting several bones). It consists of fibrous stroma with cellular component, mainly fibroblasts and osteoclasts displaying altered activity, resulting the production of aberrant trabeculae and woven bone. Enchondroma accounts for nearly 2% of benign bone tumors. This tumor consists of a mass of hyaline cartilage in a lobular formation, occurring in long tubular bone (hands and feet).

## Malignant bone tumors

### Osteosarcoma

Among bone cancers, the most commonly diagnosed primary malignant tumor is osteosarcoma (OSA). OSA represents more than 50% of all bone tumors, occurring preferentially in the young age. OSA is a very aggressive cancer, with high tendency to metastatic spread and resistance to conventional chemotherapy.

Different signaling pathways are involved in OSA pathogenesis. OSA is associated with aberrant activation of Wnt signaling, resulting in an accumulation of β-catenin in the cytoplasm and in the nucleus. Wnt/β-catenin signaling is responsible for FoxO3-mediated expression of syndecan 2, a crucial modulator of apoptosis and a regulator of chemosensitivity. Therefore, Wnt signaling is involved not only in the pathogenesis of OSA, but also in its mechanism of chemoresistance. Wnt signaling is subject to complex regulation, mediated by different Wnt antagonists (Fig. 28.4), which are also involved in OSA pathogenesis.

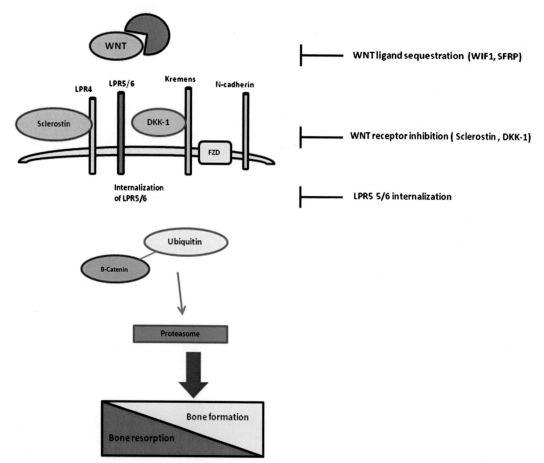

FIGURE 28.4 WNT signaling inhibition. The Wnt pathway can be negatively regulated at multiple levels: (1) sequestration of WNT ligand by soluble receptor antagonist WIF1 and SFRP, (2) blocking of WNT receptor by sclerostin (blocking Wnt receptor LPR4) and DKK-1 (blocking Wnt receptor Kremens), receptor internalization (LPR5/6). These events lead to β-catenin ubiquitination and following degradation into proteasome, leading to a downregulation of gene encoding for bone formation and an upregulation of bone resorption.

GSK3β is a serine/threonine protein kinase that exerts opposite functions (as tumor suppressor or promoter) depending upon tumor type. In OSA cells expression of active GSK3β results in a monogenic effect, through the hyperactivation of NFkB transcription factor. p53 is an extremely important oncosuppressor protein, being involved in a variety of tumors, including OSA. In OSA cells, p53 expression is impaired and murine models expressing null-p53 in osteoblasts develop OSA. Abnormal function of PI3K/AKT resulting from *PI3KCA* gene mutation is observed in OSA. The resulting phosphorylation of AKT and its substrates (FOXO transcription factor and GSK3) lead to the stimulation of proliferation and suppression of mitochondria and caspase-dependent apoptosis. Other signaling pathways involved in OSA pathogenesis include GLI2 transcription factor, the Hedgehog pathway, SPHK1/ASK1/JNK/CHK1 signaling, and pigmented epithelium derived factor (PEDF) signaling.

OSA is associated to an aberrant activation of Wnt signaling, resulting in an accumulation of β-catenin in the cytoplasm and in the nucleus. In particular in the nucleus β-catenin forms a complex with lymphocyte enhancer factor /T-cell factor family of transcription factor (LEF/TCF) to induce the expression of numerous oncoproteins, including c-Myc, cyclin D1, metalloproteinases, and others. In addition, Wnt/β-catenin signaling is responsible for FoxO3-mediated expression of syndecan 2, a crucial modulator of apoptosis and a regulator of chemosensitivity. Therefore, Wnt signaling is involved not only in the pathogenesis of OSA, but also in its mechanism of chemoresistance. Wnt signaling is subject to complex regulation, mediated by different Wnt antagonist (Fig. 28.4), which are also involved in OSA pathogenesis. Wnt inibitory factor (WIF) is a Wnt antagonist containing a Wnt-binding domain and an epidermal growth factor (EGF)−binding domain. WIF is downregulated in the majority of OSA cell lines and tissues secondary to methylation of the *WIF-1* promoter. Downregulation of WIF results in the activation of metalloproteinases MMP-9 and MMP-14. Both MMP-9 and MMP-14 are final targets of Wnt signaling and are associated with poor prognosis. High levels of MMP-9 expression was associated with OSA metastatic spread, in particular in the formation of micrometastasis typical of OSA.

## Ewing sarcoma

Sarcomas are bone tumors that can be divided according to their genetic determinants: (1) sarcomas characterized by a complex karyotype resulting from different mutations, severe genetic and chromosomal instability (osteosarcoma and chondrosarcoma) and (2) sarcomas characterized by recurring tumor-specific mutations, such as Ewing sarcoma (ES). ES is the result of a specific chromosomal translocation between the 5'-terminus of the *EWSR1* gene on chromosome 22 and the 3'-terminus of the gene encoding the FL11 transcription factor family. The result of this fusion is a chimeric protein (EWS-FL11) that acts as a potent transcriptional activator, thereby resulting in an oncogenic function. The pathogenesis of this gene rearrangement is still unknown, but recent studies suggest a genetic susceptibility associated with an *ESWR1* polymorphism affecting the intron 6 region. The effect of EWS-FL11 fusion protein is cell-type dependent. In undifferentiated rhabdomyosarcoma cells, EWS-FL11 induces an ES phenotype, while it blocks the differentiation of pluripotent mesenchymal stromal cells, suggesting that in addition to conferring genetic susceptibility, cell context affects the protumorigenic or antitumorigenic activity of EWS-FL11. EWS-FL11 behaves differently from typical oncogenic fusion proteins (like BCR-ABL) that have unique molecular targets.

## Chordoma

Chordomas are rare malignant tumors developing in the spheno-occipital, sacrum, or vertebrae, and cause neurological deficits. Chordoma appears as a multiloculated cystic gelatinous tumor that progressively invades trabecular space, causing bone erosion and compression of adjacent structures, leading to neurological deficit and pain symptoms. Chordoma recurrence is associated with expression of proangiogenic factors, including FGF2 and TGF-α. In addition, >50% of primary chordomas exhibit deletions in the 6p12 locus that contains the *VEGF* gene. Deletion mutations are a common feature of chordoma, resulting in enhanced DNA replication, aberrant cell cycle regulation, and increased cell division. Among genetic lesions found in chordoma, the three major forms include 1p36 loss, 9p21 loss, and 7q gain. Different signaling pathways are dysregulated in chordoma. Upregulation of genes involved in the synthesis of ECM, such as those encoding metalloproteinases MMP-1, MMP-2, MMP-9, and MMP-19, has been observed. Chordoma cells also express several tyrosine kinase receptors, including PDGFR-α and PDGFR-β, HER2, KIT, and C-Met. Activated downstream signaling has been observed for MAP kinase, Akt, and STAT3 pathways. Finally, a greater level of phosphorylated STAT3 has been described in primary chordoma and metastasis. Consistently, small molecule inhibitor of Stat3 and Jak2 inhibits chordoma cell proliferation and expression of antiapoptotic proteins Bcl-xL and MCL1, enabling the cytotoxic effects of doxorubicin and cisplatin treatment.

# Molecular basis of bone metastasis

Skeletal metastases are a frequent complication of several types of cancers, in particular in patients with advanced breast or prostate cancer, but also in carcinoma of the lung, colon, stomach, bladder, uterus, rectum, thyroid, kidney, and multiple myeloma. Skeletal metastasis is a major problem in the management of advanced cancer patients, because they result in a variety of adverse skeletal events, such as bone pain, hypercalcemia, spinal cord compression, and fractures. Moreover, once the cancer spreads to the bone tissue, it is difficult or impossible to cure, because bone metastases are highly resistant to therapy and account for significant morbidity and mortality. The process of metastasis to bone is facilitated by the fenestrated structure of bone tissue, high blood vasculature, and by the presence of bone marrow niches where tumor cells can remain in a dormant state for many years until they progress to micrometastatic lesions.

## Osteolytic and osteoblastic bone metastasis

Bone homeostasis is maintained by a balanced activity of osteoclasts and osteoblasts. The dysregulation of bone turnover induced by metastatic spread can lead to two different types of metastases: (1) the osteoblastic metastases, typical of prostate cancer, where the disruption of bone is due to an increase in both bone formation and resorption and (2) the most common osteolytic metastases, typical of breast and multiple myeloma, where there is a prevalence of bone erosion, leading to local bone destruction and fracture.

Osteolytic metastases are based on the aberrant activation of osteoclasts regulated by a balance between RANKL and OPG production. RANKL induces other mediators of bone turnover, such as MMP-1 and MMP9, as well as the MMP inducer CD147, the inflammatory cytokines IL-6 and IL-8, and the angiogenetic factor VEGF. Osteolytic metastatic lesions are maintained by a vicious cycle that is sustained by several tumor-derived growth factors, such as TGF-β, IGF-1, FGFs, and PDGFs. The main cancers giving rise to osteolytic metastases are breast and lung cancer. In breast cancer, osteoclast survival is maintained by IL-1, IL-6, PGE2, colony-stimulating factor (CSF-1), and TNF-α. Moreover, a breast cancer−specific metastatic pathway was described involving DLC1 (Deleted in Liver Cancer 1). DCL1 acts as a Rho GTPase-activating protein (RhoGAP), which inhibits RhoA, RhoB, RhoC, and cell division cycle 42 (cdc42), resulting in a block of TGF-β−induced PTHrP (PTH-related peptide), a mediator of osteoclast maturation. In lung cancer, a specific expression profile, including IL-1, IL-7, and RANKL is associated with development of osteolytic metastasis. Conversely, osteoblastic metastasis is typical of prostate cancer and is characterized by the accumulation of immature mineralized bone mass. Several mediators have been identified as promoter of osteoblastic metastasis, including TGF-β, VEGF, bone morphogenetic protein BMP, endothelin 1, as well as Wnt signaling. Survival and proliferation of osteoblasts in prostate cancer is promoted by FGFs, IGF, and TGF-β. Each of these factors originates from the tumor microenvironment.

## Genes involved in bone metastasis

The genes responsible for bone metastasis in breast cancer were identified using a model system based on MDA-MB-435 cells. Among the overexpressed genes, the most highly overexpressed genes included CXC4, FGF5, MMP-1, as well as IL-11 and connective tissue factor induced by TGF-β. Bone metastasis formation is also promoted by high expression of integrin β-like 1, which is highly homologous to EGF, is a target of transcription factor Runx2, and induces the expression of TGF-β1 and TGF-β3 as mediators of bone metastasis. In prostate cancer, TGF-β is also regulated by prostate transmembrane protein androgen induced 1 (PMEPA1), which exerts antimetastatic function by interfering with prometastatic TGF-β signaling by interaction with the receptor-regulated SMADs and ubiquitin ligase. PMEPA1 was reported to be decreased in metastatic prostate patients, and low PMEPA is a marker of decreased metastasis-free survival.

A key role in metastatic process is the one of CXC12/CXCR4 axis. Stromal cells secreting the chemokine CXC12 are able to attract CXCR4 overexpressing tumor cells, thereby promoting the mobilization of hematopoietic stem cells and the creation of the metastatic niche, as well as the recruitment of disseminating tumor cells into the niche. The CXC12-CXC4 signaling in metastatic invasion is supported by the induction of MMP9 and a decrease of the MMP tissue inhibitor 2, thereby promoting the ECM remodeling required for metastatic invasion.

## miRNA and bone metastasis

In the context of bone metastasis, miRNA can trigger the transition from epithelial to mesenchymal and vice-versa, osteomimicry, and osteoblasts/osteoclasts interplay. The expression of selected bone metastasis miRNA in tumor and in corresponding normal tissue was evaluated by The Cancer Genome Atlas. MiR203 was identified as a suppressor of bone metastasis. Low expression of miR203 correlates with prostate cancer progression, while miR218 was described as prometastatic, by upregulating BSP, CXCR4, and OPN in breast cancer cells. MiR30 family members are mediators of osteomimicry. Low levels of miR33a correlated with cancer-mediated bone erosion, while restoration of mir33 expression in lung cells decreased RANKL production and PTHrP expression, thereby reducing bone resorption. Similarly, miR30a resulted overexpression of nonstratified prostate cancer and downregulation in breast cancer, suggesting a role in osteolytic rather than osteoblastic bone metastases. Based on their ability to modulate bone metastasis, miRNA have recently become targets for antimetastatic therapy. For example, a current study is evaluating the replacement of miR34, a suppressor miRNA commonly downregulated in cancer.

## Key concepts

- Bone is not an inert structure, but a living and metabolically active tissue. Bone undergoes continuous remodeling throughout life, enabling bone architecture to respond to mechanical stress. Bone remodeling consists of bone tissue renewal to replace old bone tissue with new bone matrix. This process is based on the balanced action of the two cell types present in bone: osteoclasts remove old mineralized bone, and osteoblasts form new bone matrix which becomes mineralized.
- The interplay of the skeletal and immune systems is reflected by the emerging interdisciplinary research field of *osteoimmunology*. Osteoimmunology is based on bone and immune cell communication through the Rank/RANKL/OPG axis and the Wnt pathway, as well as via the action of cytokines and growth factors.
- Recent evidence highlights the growing importance of bone matrix regulation in the development of skeletal disease associated with bone mineral defects, such as osteoporosis and osteogenesis imperfecta.
- Skeletal diseases displaying increased bone density comprise a group of disorders including osteopetrosis and Paget disease of bone (PDB). These disorders are characterized by an excess of bone matrix deposition, with variable distribution all over the skeleton. Bone dysplasia occurs in response to defects in osteoclastogenesis, mature osteoclast activity, and through osteoclast-osteoblasts interaction.
- Bone can be the site of infection. The main causative agents of osteomyelitis that spread through the blood are *Staphylococcus Aureus* and *Mycobacterium Tubercolosis*. However, other circulating pathogens can cause osteomyelitis, including fungi, particularly in patients a weak or defective immune systems as a comorbidity with HIV infection, autoimmune disease, cancer, or treatment with immunosuppressive drugs.
- Bone tumors present with a variety of phenotypes, affecting bone formation, bone erosion, cartilage formation, and even neurological disorders. They can be divided in benign tumors (without the capability to metastasize) and malignant tumors (which display the ability to invade and spread to distant sites).
- Skeletal metastasis represent a frequent complication of several types of cancers, including advanced breast or prostate cancer, as well as cancers of the lung, colon, stomach, bladder, uterus, rectum, thyroid, kidney, and multiple myeloma.

## Suggested readings

[1]  Galliera E, Locati M, Mantovani A, Corsi MM. Chemokines and bone remodeling. Int J Immunopathol Pharmacol 2008;21:485−91.
[2]  Takayanagi H. Inflammatory bone destruction and osteoimmunology. J Periodontal Res 2005;40:287−93.
[3]  Takayanagi H. Introduction to osteoimmunology. Nippon Rinsho. 2005;63:1505−9.
[4]  Hadjidakis DJ, Androulakis II. Bone remodeling. Ann NY Acad Sci 2006;1092:385−96.
[5]  Rauner M, Sipos W, Pietschmann P. Osteoimmunology. Int Arch Allergy Immunol 2007;143:31−48.
[6]  Martin TJ, Sims NA. Osteoclast-derived activity in the coupling of bone formation to resorption. Trends Mol Med 2005;11:76−81.
[7]  Theoleyre S, Wittrant Y, Tat SK, Fortun Y, Redini F, Heymann D. The molecular triad OPG/RANK/RANKL: involvement in the orchestration of pathophysiological bone remodeling. Cytokine Growth Factor Rev 2004;15:457−75.
[8]  Wittrant Y, Theoleyre S, Chipoy C, et al. RANKL/RANK/OPG: new therapeutic targets in bone tumours and associated osteolysis. Biochim Biophys Acta 2004;1704:49−57.
[9]  Simonet WS, Lacey DL, Dunstan CR, et al. Osteoprotegerin: a novel secreted protein involved in the regulation of bone density. Cell 1997;89:309−19.

[10] Gingery A, Bradley E, Shaw A, Oursler MJ. Phosphatidylinositol 3-kinase coordinately activates the MEK/ERK and AKT/NFkappaB pathways to maintain osteoclast survival. J Cell Biochem 2003;89:165–79.

[11] Forlino A, Marini JC. Osteogenesis imperfecta. Lancet 2016;387:1657–71.

[12] Pagliari D, Ciro Tamburrelli F, Zirio G, Newton EE, Cianci R. The role of "bone immunological niche" for a new pathogenetic paradigm of osteoporosis. Anal Cell Pathol (Amst) 2015;2015:434389.

[13] Del Fattore A, Cappariello A, Teti A. Genetics, pathogenesis and complications of osteopetrosis. Bone 2008;42:19–29.

[14] Balemans W, Van Wesenbeeck L, Van Hul W. A clinical and molecular overview of the human osteopetroses. Calcif Tissue Int 2005;77:263–74.

[15] Sobacchi C, Schulz A, Coxon FP, Villa A, Helfrich MH. Osteopetrosis: genetics, treatment and new insights into osteoclast function. Nat Rev Endocrinol 2013;9:522–36.

[16] Michou L, Brown JP. Genetics of bone diseases: paget's disease, fibrous dysplasia, osteopetrosis, and osteogenesis imperfecta. Joint Bone Spine 2011;78:252–8.

[17] Ralston SH. Pathogenesis of Paget's disease of bone. Bone 2008;43:819–25.

[18] Gardiner EM, Baldock PA, Thomas GP, et al. Increased formation and decreased resorption of bone in mice with elevated vitamin D receptor in mature cells of the osteoblastic lineage. FASEB J 2000;14:1908–16.

[19] Nagata Y, Imanishi Y, Ishii A, et al. Evaluation of bone markers in hypophosphatemic rickets/osteomalacia. Endocrine 2011;40:315–17.

[20] Ryan JW, Anderson PH, Turner AG, Morris HA. Vitamin D activities and metabolic bone disease. Clin Chim Acta 2013;425:148–52.

[21] Costa-Reis P, Sullivan KE. Chronic recurrent multifocal osteomyelitis. J Clin Immunol 2013;33:1043–56.

[22] Esposito S, Leone S. Prosthetic joint infections: microbiology, diagnosis, management and prevention. Int J Antimicrob Agents 2008;32:287–93.

[23] Galliera E, Drago L, Marazzi MG, Romano C, Vassena C, Corsi Romanelli MM. Soluble urokinase-type plasminogen activator receptor (suPAR) as new biomarker of the prosthetic joint infection: correlation with inflammatory cytokines. Clin Chim Acta 2015;441:23–8.

[24] Galliera E, Drago L, Vassena C, et al. Toll-like receptor 2 in serum: a potential diagnostic marker of prosthetic joint infection? J Clin Microbiol 2014;52:620–3.

[25] Hakim DN, Pelly T, Kulendran M, Caris JA. Benign tumours of the bone: a review. J Bone Oncol 2015;4:37–41.

[26] Roodman GD. Mechanisms of bone metastasis. N Engl J Med 2004;350:1655–64.

[27] Fohr B, Dunstan CR, Seibel MJ. Clinical review 165: markers of bone remodeling in metastatic bone disease. J Clin Endocrinol Metab 2003;88:5059–75.

[28] Dougall WC, Chaisson M. The RANK/RANKL/OPG triad in cancer-induced bone diseases. Cancer Metastasis Rev 2006;25:541–9.

[29] Pectasides D, Farmakis D, Nikolaou M, et al. Diagnostic value of bone remodeling markers in the diagnosis of bone metastases in patients with breast cancer. J Pharm Biomed Anal 2005;37:171–6.

[30] Guise TA, Mohammad KS, Clines G, et al. Basic mechanisms responsible for osteolytic and osteoblastic bone metastases. Clin Cancer Res 2006;12:6213–6216ss.

# Chapter 29

# Molecular basis of diseases of the nervous system

Margaret Flanagan[1], Joshua A. Sonnen[2], Christopher Dirk Keene[3], Robert F. Hevner[4] and Thomas J. Montine[1]

[1]Department of Pathology, Stanford University, Palo Alto, CA, United States, [2]Department of Pathology, University of Utah, Salt Lake City, UT, United States, [3]Department of Pathology, University of Washington, Seattle, WA, United States, [4]Department of Neurological Surgery, Seattle Children's Hospital Research Institute, Seattle, WA, United States

## Abstract

The central nervous system (CNS) is composed of cellular components organized in a complex structure that is unlike other organ systems. At the macroscopic level, the parenchyma of the CNS can be categorized into two structurally and functionally unique components: gray and white matter. The CNS can be divided into a number of anatomic regions, each with specific neurologic or cognitive functions. Disease or damage to these regions produces neurologic or cognitive deficits that correlate with the anatomic location and extent of disease. There are several ways to divide these structures. The main divisions are the cerebrum, cerebellum, brainstem, and spinal cord. Developmental neuropathology encompasses a broad variety of cerebral malformations and functional impairments caused by disturbances of brain development, manifesting during ages from the embryonic period through adolescence and young adulthood. The neurologic and psychiatric manifestations of neurodevelopmental disorders range widely depending on the affected neural systems and include such diverse manifestations as epilepsy, mental retardation, cerebral palsy, breathing disorders, ataxia, autism, and schizophrenia. In terms of morbidity and mortality, the spectrum is extremely broad: the mildest neurodevelopmental disorders can be asymptomatic, whereas the worst malformations often lead to intrauterine or neonatal demise. This chapter focuses on the genetic disorders of brain development, caused by mutations of gene loci or chromosomal regions with important neurodevelopmental functions.

## Anatomy of the central nervous system

The central nervous system (CNS) is composed of cellular components organized in a complex structure that is unlike other organ systems. Broadly, its cellular components can be divided into neuroepithelial and mesenchymal elements. Neuroepithelial elements derive from the primitive neural tube and include neurons and glia. The mesenchymal elements include blood vessels and microglia. Microglia are a population of resident myeloid cells that play a central role in coordination and actualization of CNS inflammation. At the macroscopic level, the parenchyma of the CNS can be categorized into two structurally and functionally unique components: (1) gray matter, and (2) white matter. Gray matter is the location of most neurons and is the site of the integration of neural impulses by neurotransmitters across synapses between neurons. White matter functions to conduct these impulses efficiently and quickly between neurons in different gray matter regions.

## Microscopic anatomy

### Gray matter

Macroscopically, gray matter forms a ribbon of cortex in the human cerebrum and cerebellum, as well as the mass of the deep nuclei. Microscopically, it is composed of cells forming and embedded in a finely interdigitating network of cellular processes. This network, referred to as neuropil, is sufficiently dense that individual cellular processes cannot be distinguished. Indeed, under the microscope, neuropil appears as a homogenous matrix surrounding neurons and

Essential Concepts in Molecular Pathology. DOI: https://doi.org/10.1016/B978-0-12-813257-9.00029-2

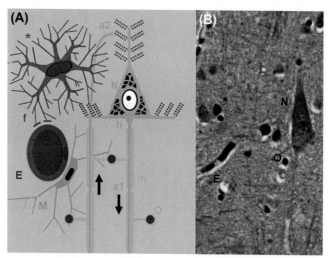

FIGURE 29.1    Microanatomy of gray matter. (A) Schematic of the microanatomy of gray matter. A neuron (N) contains a prominent nucleus with open chromatin, a conspicuous nucleolus and cytoplasmic Nissl substance that is composed of abundant rough endoplasmic reticulum. The neuron elaborates numerous apical and lateral dendrites that are decorated with many receptors. Electrochemical impulses are generated at dendrites and integrated across the neuron's body at the axonal hillock (h) and transmitted along the axon (a1). Oligodendrocytes (O) envelop the axon within a segmented myelin (m) sheath allowing more rapid and efficient conduction of impulses. An endothelial cell (E) forms a small capillary space surrounded by a resting microglial cell (M). A nearby astrocyte (*) has numerous cytoplasmic processes, some of which rest foot processes (f) on the vessel, helping to maintain the blood—brain barrier. An axon (a2) from a distant neuron forms an axonal terminal (t) on a dendrite of the pictured neuron, forming a synapse and releasing neurotransmitters to the receptors on the dendrite. (B) Photomicrograph of the microanatomy of gray matter. Section of frontal cortex stained with hematoxylin and eosin (H&E) and Luxol fast blue (LFB). Labels are as defined for (A).

other cells. Because of this, it was not until 1889 and the work of Santiago Ramón-y-Cajal that the neuron theory was fully applied to the brain, 50 years after cellular theory was proposed by Theodore Schwann.

The cellular constituents of gray matter include neurons, glia, endothelium, and perivascular cells of blood vessels and microglia (Fig. 29.1). Neurons are the electrically active cells of the brain. A neuron is composed of slender branching dendrites on which axons from other neurons synapse and propagate action potentials, a body or soma where the metabolic and synthetic processes of the neuron are orchestrated, an axonal hillock where electrochemical impulses are integrated, an axon along which the integrated electrochemical impulse is conducted, and an axonal terminal where the electrochemical signal is passed to another neuron's dendrites or an effector cell across a synapse. The soma of neurons is easily identifiable by routine histologic stains. Neurons have generally round nuclei with a prominent nucleolus and open chromatin. The cytoplasm of neurons is remarkable in that it generally contains abundant rough endoplasmic reticulum (called Nissl substance) that is demonstrable by numerous histologic techniques. Loss of Nissl substance is a sign of early neuronal injury and is seen in a variety of conditions including axonal transection and hypoxia/ischemia. The axons and dendrites of neurons are major components of neuropil.

Neurons require a constant supply of oxygen and glucose, and even short interruptions can cause neuronal death. Neurons are the most susceptible to diverse forms of CNS injury and are the first cells lost to necrosis or apoptosis under stressful conditions. Furthermore, for reasons that are not fully understood, populations of neurons have differential susceptibilities to different types of stress. For example, neurons in one region of the hippocampus may become necrotic in response to hypoxia, whereas neurons in adjacent regions are spared.

Glia (from Latin for glue) form the bulk of the CNS parenchyma and outnumber neurons on the order of 1000 to 1. The primary glial cell of gray matter is the star-shaped astrocyte. Astrocytes maintain a variety of supportive functions including structure, metabolic support for neurons, management of cellular waste products, uptake and release of neurotransmitters, regulation of extracellular ion concentration, and interactions with the vasculature including helping to maintain the blood—brain barrier (BBB) and responding to injury. Normally, astrocytes have irregular, potato-shaped nuclei and numerous fine cellular processes that are indistinct within the neuropil. In response to noxious stimuli, astrocytes increase production of their characteristic intermediate filament, glial fibrillary acidic protein (GFAP), as their soma swells and becomes prominent. This process is known as astrogliosis (or simply gliosis) and is a nonspecific reaction to injury in the CNS. A second population of glia in gray matter is the oligodendrocytes.

Microglia are a population of myeloid lineage resident cells in cerebral, cerebellar, brainstem, and spinal cord parenchyma. Microglia serve diverse functions but primarily surveil the CNS and coordinate and effectuate the immune response to internal and external insults. Usually unobtrusive, quiescent microglia have small rod-shaped nuclei and

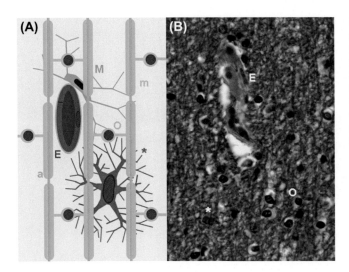

**FIGURE 29.2** Microanatomy of white matter. (A) Schematic of the microanatomy of white matter. (B) Photomicrograph of the microanatomy of white matter (hematoxylin and eosin (H&E)/Luxol fast blue (LFB)). Numerous axons (a) from distant neurons pass through white matter conducting nerve impulses. Oligodendroglia (O) are the most common cells and myelinate many adjacent segments of passing axons. The insulating myelin (m) allows rapid and efficient conduction of nerve impulses. Oligodendrocytes have high metabolic needs, and endothelium (E) forms numerous capillaries. In the absence of disease, astrocytes (*) and microglia (M) are inconspicuous.

inapparent cytoplasm with multiple delicate processes surveilling the local environment. These ramified (quiescent) microglia do not express major histocompatibility complex (MHC) I/II antigens, unlike other scavenger cells. However, in response to injury, microglia migrate to the site of injury and undergo morphologic and functional changes to coordinate the response to injury and elimination of toxins or microorganisms. In aggregate, this is termed microglial activation and typically occurs in concert with astrogliosis. Activated microglia are now known to exhibit a spectrum of behaviors that include MHC class I/II expression, inflammatory and cytotoxic signaling, and phagocytosis of material for antigen presentation to T-cells.

## White matter

White matter is composed of numerous axonal processes (from neurons whose bodies reside in gray matter), glia, blood vessels, and microglia (Fig. 29.2). The axons are insulated in segments by layers of myelin, and this insulation allows more rapid and efficient conduction of electrochemical signals along the axon. Myelin is composed of concentric proteolipid membranes that are extensions of cytoplasmic processes of the primary glia of white matter, the oligodendrocyte. Oligodendrocytes have small round nuclei with condensed chromatin and indistinct cytoplasm. A single oligodendrocyte myelinates numerous passing axons. Because oligodendrocytes are responsible for the maintenance of a large amount of myelin, they have relatively high metabolic demands and are consequently relatively sensitive to injury compared with other white matter elements. Injury to oligodendrocytes causes local loss of myelin. Astrocytes and microglia are minority populations within white matter and are less sensitive to injury than oligodendrocytes. Tissue response to noxious stimuli is mediated through astrocytic gliosis and microglial activation as in gray matter.

## Gross anatomy

The CNS can be divided into a number of anatomic regions, each with specific neurologic or cognitive functions. Regional damage due to disease or injury produces neurologic or cognitive deficits that correlate with the anatomic location and extent of disease. The main divisions are the cerebrum, cerebellum, brainstem, and spinal cord. The cerebrum is covered by a folded layer of gray matter called the cortex. The folding allows greater surface area to fit within the confines of the skull. The folds themselves are called gyri and are characteristics of normal cortical development in humans. Lesions of the cerebral cortex cause deficits of cognition and conscious movement or sensation. The cortex can be further divided into lobes that subserve different cognitive domains or neurologic functions (Fig. 29.3). The frontal lobes anteriorly are involved in executive function (self-control, planning) and personality. Damage to this region produces personality changes and socially inappropriate behavior. Posteriorly, the frontal lobe houses the primary motor cortex that controls voluntary movement for the opposite side (contralateral) of the body. Damage causes contralateral weakness or paralysis. The parietal lobe contains the primary sensory cortex for the contralateral half of the body, and damage causes anesthesia and neglect. The temporal lobe is not only involved in the conscious processing of sound but also contains the hippocampus, which is integral to the formation of new memories. Damage to the hippocampus produces memory dysfunction and, in cases where both hippocampi are damaged, produces severe anterograde amnesia wherein the unfortunate individual is incapable of forming new memories. The occipital lobe contains the primary

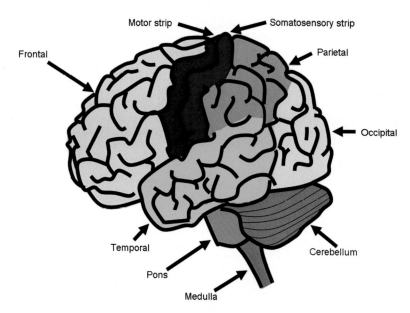

**FIGURE 29.3** Lateral view of the central nervous system.

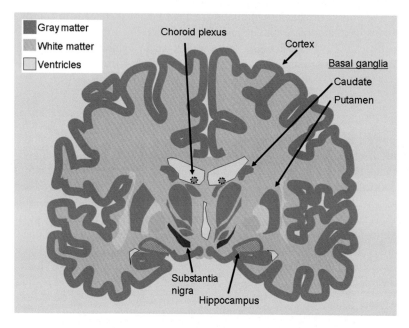

**FIGURE 29.4** Coronal section of cerebrum through the deep gray matter nuclei.

visual cortex for the field of vision on the opposite side of the body, and damage causes partial or complete loss of conscious perception of visual stimuli. The white matter underlying and connecting these cortical regions may also be damaged and produces deficits related to the regions connected. The cerebrum also contains deeply situated gray matter nuclei. The most prominent of these are the basal ganglia and the thalamus (Fig. 29.4). The basal ganglia are formed by a group of nuclei that compose important regulatory circuits with the cerebrum, and diseases that afflict the basal ganglia cause movement disorders and well as disorders of cognition. The thalamus is a complex structure whose primary function is a sensory and motor relay.

## Neurodevelopmental disorders

Developmental neuropathology encompasses a broad variety of malformations and functional impairments caused by disturbances of brain development, manifesting during ages from the embryonic period through adolescence and young adulthood. The neurologic and psychiatric manifestations of neurodevelopmental disorders range widely depending on the affected neural systems and include such diverse manifestations as epilepsy, mental retardation, cerebral palsy, breathing disorders, ataxia, autism, and schizophrenia. In terms of morbidity and mortality, the spectrum is extremely broad. The mildest neurodevelopmental disorders can be asymptomatic, whereas the worst malformations often lead to intrauterine or neonatal demise. This section will focus on genetic disorders of brain development, caused by mutations of gene loci or chromosomal regions with important neurodevelopmental functions. Excluded from consideration here are other categories of developmental neuropathology caused by general metabolic disorders (such as storage diseases and amino acidopathies), disruptions (amniotic web disruption sequence), environmental insults (hypoxia—ischemia or toxic exposure), infection, and neoplasia. More extensive coverage of these categories can be found in other specialized texts.

## Neural tube defects

Development of the brain and spinal cord begins with gastrulation, when the neural plate forms and differentiates into distinct regional subdivisions along the rostrocaudal and mediolateral axes. As morphogenesis continues, the processes of neurogenesis, gliogenesis, synaptogenesis, and myelination are initiated sequentially, and these processes then continue concurrently with broad overlap. Brain development is mostly complete by the end of adolescence; however, neurogenesis continues at decreasing levels in the dentate gyrus of the hippocampus throughout adulthood. In this sense, brain development continues until death.

### Organizing the central nervous system: signaling centers and regional patterning

The central nervous system begins as the neural plate, which folds along the midline and closes dorsally to form the neural tube. From its formation, the neural plate is patterned along the rostrocaudal and mediolateral axes in a gridlike fashion by gradients and boundaries of gene expression. HOX genes, for example, are differentially expressed along the rostrocaudal axis and play an important role in specifying segmental organization of the spinal cord and hindbrain. Such differences of gene expression are programmed by both the intrinsic developmental history of each region and by extracellular factors produced in signaling centers that define positional information through interactions with developing neural tissue. Gene expression gradients, compartments, boundaries, and signaling centers continue to be important throughout the embryonic period until each brain subdivision has been generated and acquired its specific identity.

### Neural tube closure and Wnt-PCP signaling

Neural tube closure is a key early event in brain and spinal cord development, in which the planar epithelium of the neural plate folds at the midline along the anteroposterior axis and the lateral edges of the neural plate move dorsally, contact each other, and fuse dorsally to form the neural tube (Fig. 29.5). Dorsal fusion first occurs in the region of the cervical spinal cord primordium, followed by separate closure events at midbrain and rostral telencephalic points. The exact location and number of dorsal fusion events vary between and within species. From each of these sites, fusion proceeds rostrally and caudally in a zipperlike mode until closure is complete (primary neurulation) from the rostral end (anterior neuropore) to the caudal (posterior neuropore). Interestingly, this mode of primary neural tube closure does not extend the full caudal length of the spinal cord but ends around the lumbosacral region. More caudal regions (mainly sacral) appear to develop by a distinct mechanism involving cavitation of the caudal eminence (tail bud) mesenchyme, known as secondary neurulation. A schematic of this process is shown in Fig. 29.6.

Importantly, the process of neurulation transforms the axes of the developing CNS neuroepithelium from a planar to a tubular coordinate system. The planar (two-dimensional) neural plate is defined by rostrocaudal and mediolateral axes, whereas the tubular (three-dimensional) neural tube is defined by rostrocaudal, mediolateral, and dorsoventral axes. During this reorganization process, the lateral edge of the neural plate becomes the dorsal midline (roof plate) of the neural tube, and the medial edge (midline) of the neural plate becomes the ventral midline (floor plate) of the neural tube (Fig. 29.5). The new axes become the substrate for further patterning and remain important throughout later development; however, additional axial transformations occur locally in the developing brain.

Finally, neural tube closure is essential for subsequent development of posterior tissues including the vertebral arches and cranial vault, paraspinal muscles, and posterior skin of the head and back. In some cases, neural tube closure proceeds normally, but the overlying skin and mesodermal structures fail to cover the posterior neural tube. It is unknown whether these malformations (such as encephalocele, myelocele, and sacral agenesis) are caused by primary defects of neural tube closure (as frequently assumed in the neuropathology literature) or mesodermal and ectodermal development.

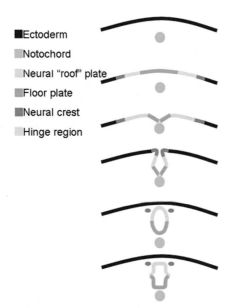

■Ectoderm
Notochord
Neural "roof" plate
■Floor plate
■Neural crest
Hinge region

**FIGURE 29.5** Neural tube formation. In cross section, the neural tube first derives from a planar (two-dimensional) epithelium with mediolateral organization and transforms to a tubular (three-dimensional) with both mediolateral and dorsoventral axes.

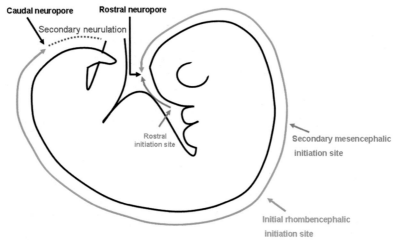

**FIGURE 29.6** Simplified schematic of the sequence of neural tube fusion. After the neural tube begins closure from rostral and posterior cerebral sites, closure propagates anteriorly and posteriorly. The locations of the neuropores are potential sites for defects of neural tube closures. Although two initiation sites of primary neural tube fusion are shown here, this has not been rigorously proven in humans, and other mammals may have more initiation sites. *Adapted from Padmanabhan R. Etiology, pathogenesis and prevention of neural tube defects. Congenit Anom (Kyoto) 2006;46:55−67.*

## Human neural tube defects

The most severe NTD, characterized by the complete failure of neural tube closure along the entire craniospinal axis, is called craniorachischisis (Fig. 29.7A). This lethal malformation corresponds to the severe form of NTD found in Lp mice and is probably caused by defects of Wnt-PCP signaling during convergent extension. Closure defects limited to the cranial region result in anencephaly, a severe, lethal defect (Fig. 29.7B). In anencephaly, extensive destruction of the brain tissue is secondary to direct exposure to amniotic fluid. Initially, the cerebral tissue proliferates and grows outside the surrounding skull base, a transient stage known as exencephaly. The most common human NTDs are limited to the lumbosacral region, sites of posterior neuropore closure and secondary neurulation. Typically, the spinal tissue has a ragged interface with mesodermal and ectodermal derivatives and is exposed externally. This appearance is called myelomeningocele (Fig. 29.7C), somewhat erroneously because there is usually no -cele or closed sac. Less often, malformations with a closed skin surface and sac are also seen. In the great majority of cases, lumbosacral myelomeningocele

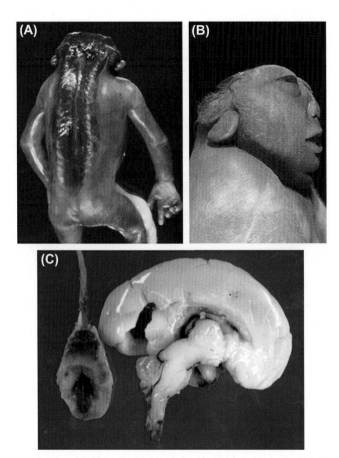

**FIGURE 29.7**   Neural tube defects. (A) Craniorachischisis. (B) Anencephaly. (C) Myelomeningocele with Chiari type II malformation. The spinal cord with open lumbosacral myelomeningocele is shown at left; the medial view of brain with hydrocephalus, tectal beaking, and herniation of the medulla over the spinal cord is shown at right.

is accompanied by a malformation of the midbrain, hindbrain, and skull base known as Chiari type II malformation, historically called the Arnold—Chiari malformation (Fig. 29.7C).

## Rostrocaudal and dorsoventral patterning of the neural tube: holoprosencephaly

A program of segmental, compartmentalized gene expression begins to define rostrocaudal subdivisions of the CNS during neural plate stages, and this process accelerates with neurulation. Morphologically, the spinal cord is partitioned into cervical, thoracic, lumbar, and sacral segments that are associated with mesodermal somites. Morphological development of the brain is more complicated (Fig. 29.8), as it is initially divided into three vesicles (prosencephalon/forebrain, mesencephalon/midbrain, and rhombencephalon/hindbrain) and then further subdivided into five vesicles (telencephalon, diencephalon, mesencephalon, metencephalon, and myelencephalon).

Concurrent with rostrocaudal segmentation, the neural tube is patterned along the dorsoventral axis by a different set of molecules and signaling centers. Important ventral signaling centers include (1) the notochord (primordium of vertebral body nucleus pulposus), a mesodermal structure located ventral to the spinal cord and hindbrain, (2) the prechordal plate mesendoderm, located ventral to the forebrain and midbrain, and (3) the floor plate of the neural tube, a specialized neuroepithelial structure in the ventral midline of spinal cord and brain. All three of these ventral signaling centers produce Sonic hedgehog (Shh), a small, posttranslationally modified, secreted protein with potent ventralizing activity on neural structures. The key dorsal signaling center is the roof plate, a specialized neuroepithelial structure in the dorsal midline that ultimately develops into choroid plexus. The roof plate, characterized by high-level expression of ZIC genes, produces secreted morphogens belonging to the bone morphogenetic protein (BMP) and Wnt families, with potent dorsalizing activity. The balance of antagonistic ventralizing and dorsalizing signals patterns the neural tube into basal and alar subdivisions associated with motor and sensory pathways, respectively. Accordingly, the ventral

**(A)**

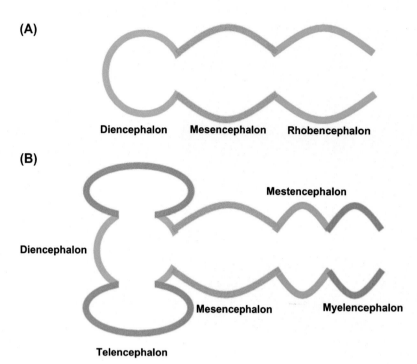

Diencephalon    Mesencephalon    Rhobencephalon

**(B)**

Mestencephalon

Diencephalon

Mesencephalon    Myelencephalon

Telencephalon

FIGURE 29.8 Segmentation and cleavage of the developing brain. (A) After fusion of the neural tube, the anterior-most portion forms swellings that will become the cerebrum (diencephalon), the midbrain (mesencephalon), and hindbrain (rhombencephalon). (B) Following this, proliferation of the cortical neuroepithelium forms telencephalic vesicles and will become the majority of the brain's paired hemispheres, and the rhombencephalon divides into the metencephalon and myelencephalon, which will become the pons and cerebellum and medulla, respectively.

spinal primordium generates motor neurons, whereas the dorsal spinal primordium generates sensory relay neurons. Sensory and motor distinctions become more subtle in the midbrain and forebrain. For example, most of the forebrain is composed of alar (sensory) plate neuroepithelium, including such ventral (anatomically speaking) structures as the neural retina and optic pathways. In contrast, the neurohypophysis (ventral hypothalamus) may be considered a motor pathway, inasmuch as hormone-producing neurons generate somatic responses and behaviors.

## Holoprosencephaly

The best-characterized and most common defect of dorsoventral patterning in humans is holoprosencephaly, defined as complete or partial failure of cerebral hemispheric separation. Holoprosencephaly exhibits a spectrum of severity. The mildest forms (lobar holoprosencephaly) show some development of the interhemispheric fissure with continuity of cortex across the midline. Forms with intermediate severity (semilobar holoprosencephaly) show partial development of the interhemispheric fissure, and severe forms (alobar holoprosencephaly) show complete absence of the interhemispheric fissure, with a single forebrain ventricle (Fig. 29.9). The defective midline structures in holoprosencephaly result from abnormalities of ventral or dorsal patterning molecules, including Shh and Zic2. Dorsoventral patterning is essential for differential proliferation and apoptosis of dorsal midline and ventrolateral forebrain structures, the essential underlying mechanisms of hemispheric separation. Overall proliferation is severely reduced in holoprosencephaly, and the brain is invariably small.

## Signaling centers and transcription factor gradients in the cerebral cortex: thanatophoric dysplasia

The cerebral cortex develops as a dorsal outpouching of the alar plate, with midline separation into right and left hemispheres. Early in cortical development, the cleavage of telencephalon into right and left hemispheres depends on dorsoventral patterning, such that proliferation is initially enhanced in lateral cortical regions (away from the dorsal and ventral midline), whereas apoptosis is enhanced in the dorsal midline (roof plate). Differential growth of the telencephalic roof plate (slow) and lateral telencephalon (rapid) causes growth of the hemispheric neuroepithelium accompanied by relative reduction of the dorsal midline, leading to formation of the interhemispheric fissure. Accordingly, defective dorsal or ventral patterning (due to mutations in SHH, ZIC2, and other genes) results in a failure of separation between the hemispheres.

Even as the cerebral hemispheres begin to separate, patterning of the cortical areas within each hemisphere is already being initiated along the rostrocaudal and mediolateral axes, through the elaboration of additional signaling

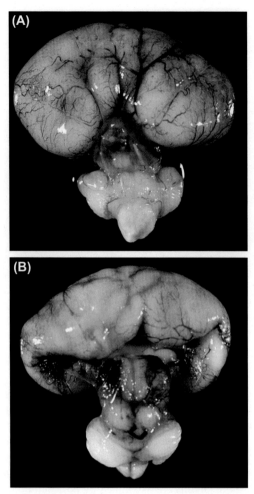

**FIGURE 29.9** Alobar holoprosencephaly in a 23-week gestational age fetus. (A) Anterior view: the single holosphere exhibits shallow sulci but no deep interhemispheric fissure. (B) Posterior view: cortex is continuous across the midline and the posteriorly open (due to disrupted delicate membranes) single forebrain ventricle is visible.

centers and their production of secreted morphogens. Principle signaling centers that pattern the cerebral cortex include the commissural plate (rostral signaling center), cortical hem (dorsomedial signaling center), and cortical antihem (ventrolateral signaling center). These signaling centers, located at edges of the cortical neuroepithelium, secrete morphogens that regulate regional growth and identity along diffusion gradients that, by differential activation of receptors, encode positional information in the cortical neuroepithelium. The commissural plate, for example, produces fibroblast growth factor-8 (FGF-8), which specifies frontal lobe identity near the FGF-8 source and parietal lobe more distally. Remarkably, ectopic expression of FGF-8 in caudal regions of embryonic mouse cortex induces the formation of a duplicate, mirror-image frontoparietal cortex (as indicated by molecular and cytoarchitectonic markers) at the occipital pole. The cortical hem, located at the medial edge of embryonic cortex, produces Wnt and BMP family molecules that specify hippocampus, medial temporal, and medial parietal cortex.

## Thanatophoric dysplasia

FGF signaling plays a major role in regulating cortical arealization and proliferation. FGF receptor 3 (FGFR3) is one of the four tyrosine kinase receptors for FGF signaling and is highly expressed in the embryonic cerebral cortex in a high caudal—low rostral gradient. Constitutive activating mutations of FGFR3 have been linked to thanatophoric dysplasia, a severe form of achondroplastic dwarfism in which the brain is enlarged and exhibits aberrant, prematurely forming sulci in temporal and occipital lobes (Fig. 29.10). The cortical surface area and overall mass are also markedly increased in thanatophoric dysplasia, illustrating the coordinate regulation of cortical surface area growth and areal patterning even in a pathological condition.

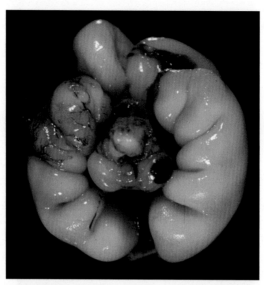

**FIGURE 29.10** Thanatophoric dysplasia in a midgestational fetal brain (inferior view). Aberrant, prematurely formed sulci are present in the medial temporal and occipital lobes, radiating from the hindbrain and cerebellum (with small subarachnoid hemorrhage) in a "bear claw" configuration.

## Compartmentalization of embryonic neurogenesis

Mature brain and spinal regions contain a mixture of excitatory and inhibitory cell types that utilize different neurotransmitters. The principle excitatory neurotransmitter at central synapses is L-glutamate, and the principle inhibitory neurotransmitter is γ-aminobutyric acid (GABA). Recently, it has been shown that developmental patterning of the brain and spinal cord plays a fundamental role in defining neurotransmitter types at very early stages, before the genesis of post-mitotic neurons. Moreover, it has been shown that many distinct regions of the developing brain are patterned to produce either excitatory (glutamatergic) neurons or inhibitory (GABAergic) neurons. Thus, the production of different types of neurons is compartmentalized, and mature circuits are assembled by post-mitotic neuronal migrations.

Important examples of compartmentalized neurogenesis have been discovered in development of the cerebral cortex and cerebellum, the two largest brain structures in humans. In the cerebral cortex, it has been shown that glutamatergic (pyramidal) cells, which account for 75%−80% of cortical neurons, are produced locally within the cortical neuroepithelium. In contrast, GABAergic interneurons are produced in subcortical progenitor compartments called the medial, caudal, and lateral ganglionic eminences that have long been known as basal ganglia progenitor regions. The interneurons then migrate long distances tangentially into the developing cortex and integrate into the circuitry along with the locally generated glutamatergic neurons. The compartmentalization of neurogenesis implies that different genes may control the development of pyramidal cells and interneurons, and indeed this is the case. In mice, the cortical and subcortical compartments express different transcription factors, including EMX family homeobox genes and TBR family T-box genes in the cortex, and DLX family homeobox genes in the subcortical regions. Mutations of these transcription factors selectively interfere with the development of the predicted set of neurons in mice; however, this remains to be confirmed in humans.

In the cerebellum, glutamatergic and GABAergic neurogeneses are likewise compartmentalized, but in contrast to cortex, glutamatergic neurons migrate tangentially and GABAergic neurons migrate radially. Cerebellar GABAergic neurons (Purkinje cells and inhibitory interneurons) are produced in the cerebellar ventricular neuroepithelium, whereas glutamatergic neurons (deep nuclei projection neurons, granule cells, and excitatory interneurons) are produced in a specialized compartment known as the rhombic lip, which surrounds the roof of the fourth ventricle and, its derivative, the external granular layer, produced by progenitor migration from the rhombic lip over the entire surface of the cerebellar primordium. Transcription factors play a prominent role in defining the neurogenic regions, such as Ptf1a (also known as cerebelless), a basic helix−loop−helix (bHLH) transcription factor, in GABAergic progenitors, and Math1, another bHLH family transcription factor, in rhombic lip glutamatergic progenitors. This view of cerebellar neurogenesis has been established in mice but remains to be confirmed in human brain. No malformations ascribed specifically to one type of neurotransmitter differentiation have been described as yet.

## Micrencephaly and megalencephaly

The overall and relative sizes of the brain and spinal cord, and their component regions, are determined in large part by the balance of proliferation by progenitor cells, and apoptosis by progenitors and postmitotic cells. Cell size also plays a significant role in determining nervous system size. In recent years, there has been tremendous progress in identifying various types of progenitor cells that differ in morphology, proliferative potential, neuronal or glial commitment, and maturation sequences.

Both neurogenesis and gliogenesis are functions of neural progenitor cells that are heterogeneous and, as a group, undergo progressive maturation and specialization from early to late stages. Timing has great significance, as different types of neurons and glia are produced sequentially according to consistent timetables. In the cortex, for example, deep-layer neurons are produced first, followed by superficial-layer neurons, and then glial cells.

### Neuroepithelial cells and radial glia

The nervous system begins as a true epithelium, with apical and basal surfaces that correspond to the ventricular and pial surfaces, respectively. Accordingly, the earliest progenitor cells, which have elongated morphology with processes that span the epithelium, are designated as neuroepithelial cells. Generally, neuroepithelial cells are capable of producing neurons, astrocytes, and oligodendrocytes and have virtually unlimited self-renewal proliferative potential, and thus qualify as a type of neural stem cells. Neuroepithelial cells and other neural stem cells express characteristic markers, such as Nestin (an intermediate filament); however, none of these markers are completely specific. With further maturation, the progenitors show evidence of astrocytic differentiation while maintaining their elongated (radial) morphology. These astrocyte-like cells are known as radial glia, and they serve both as progenitors and as a scaffold for the developing nervous system. Radial glia generally have high proliferative potential but heterogeneous properties of molecular expression and fate commitment (neurogenic or gliogenic). Neuroepithelial cells and radial glia both divide at the apical (ventricular) surface, which may serve to limit their proliferation, as well as maintain apical–basal polarity of the pseudostratified neuroepithelium.

### Intermediate neuronal progenitors

Intermediate neuronal progenitors, also known as transient amplifying cells, are produced from neuroepithelial cells and radial glia, but unlike them, do not contact the ventricular or pial surfaces. Instead, they have short or no processes and divide away from the apical surface. For this reason, they have also been called basal progenitors. Intermediate neuronal progenitors have a very limited proliferative potential, amounting to no more than one to three mitotic cycles, perhaps because they are detached from the ventricular surface and lack the constraints posed by attachment. Intermediate neuronal progenitors appear to produce only neurons (not glia), and their molecular and morphologic characteristics vary among brain regions. In the cerebral cortex, intermediate progenitors specifically express Tbr2/Eomes, a T-box transcription factor. The role of intermediate neuronal progenitors appears to be expansion of neuronal subsets from limited numbers of neuroepithelial cells or radial glia.

### Adult neurogenesis

It was long believed that neurogenesis did not occur in mature animals, whereas studies in experimental animals and humans have demonstrated that, in fact, neurogenesis continues at robust levels in two brain regions: (1) the subgranular zone in the dentate gyrus of hippocampus, and (2) the subventricular zone adjacent to striatum. The hippocampal subgranular zone contains Nestin-positive, neural stem-like cells (stem-like because they may not have unlimited self-renewal capacity) that produce new glutamatergic neurons that incorporate anatomically and functionally into the dentate gyrus, presumably to subserve new memory formation; however, the functional impact of adult hippocampal neurogenesis is still not clear. The striatal subventricular zone likewise contains Nestin-positive, neural stemlike cells that produce new GABAergic (and some glutamatergic) neurons, which migrate along the rostral migratory stream into the olfactory bulb where they incorporate into its circuitry. Neurogenesis in the subventricular zone and rostral migratory stream continues throughout adulthood in rodents but is restricted to the first year or two of life in humans. Interestingly, adult neurogenesis involves stages of progenitor maturation and molecular expression that highly resemble those in embryonic development of the hippocampus and olfactory bulb.

## Gliogenesis

Glial cells include astrocytes, oligodendrocytes, microglia, and NG2-positive oligodendrocyte precursor cells. Generally, gliogenesis follows neurogenesis with relatively little temporal overlap. Astrocytes appear to be produced directly from radial glia and by continued division of astrocytes and glial progenitors that maintain the potential to proliferate throughout life. In terms of progenitor lineages and molecular expression, oligodendroglia in most regions are more closely related to neurons than to astrocytes. NG2-positive cells are a more recently identified type of glia that serve as oligodendrocyte precursors and may have special progenitor properties, such as the ability to generate neurons under certain conditions.

## Apoptosis

Programmed cell death, or apoptosis, occurs as a normal and essential part of brain development that functions to eliminate unnecessary, excess progenitors and neurons. In fact, apoptosis is most active and important in progenitor compartments. Gene targeting to inactivate (or knockout) genes that are required for apoptosis in mice causes hyperplasia of some neuroepithelial regions, especially the developing cerebral cortex. In mice, the result is a dramatic increase of cortical surface area with aberrant formation of gyri. Unlike humans, mice have smooth-surfaced or lissencephalic brains that normally lack gyri. Apoptosis of postmitotic neurons does occur, but at lower levels. Here apoptosis may cull excess neurons produced at the end of neurogenesis that fail to integrate into the CNS circuitry.

## Micrencephaly

Small brain size (micrencephaly) is caused by defective proliferation of cortical progenitors. Commonly, it is associated with other malformations, such as holoprosencephaly and lissencephaly. Less often, it occurs in the absence of other malformations. This form of micrencephaly has been linked in some cases to mutations in ASPM, a gene encoding an essential protein for the mitotic spindle. Micrencephaly usually results in neurodevelopmental delay and other neurological problems.

## Megalencephaly

Abnormally large brain size (megalencephaly) may involve the entire brain, one cerebral hemisphere (hemimegalencephaly), or more focal regions. In most cases, brain overgrowth is accompanied by cytologic dysplasia, such as neuronal and glial cell hypertrophy. Such cases have been linked to mutations in genes that encode components of the phosphatidylinositol-3-kinase (PI3K)-AKT-mTOR signaling pathway (for example, AKT3), a key regulator of cell growth and proliferation. Limited forms of megalencephaly, such as hemimegalencephaly, are caused by somatic mosaicism with mutations affecting the involved brain regions. The most devastating consequences of brain overgrowth disorders are intractable seizures, often requiring resection of epileptogenic foci, although developmental delay, cerebral palsy, autism, and other disorders can also result from megalencephaly.

# Neuronal migration and differentiation: lissencephaly

## Postmitotic events and integration into neural circuits

Much evidence suggests that fundamental aspects of neuronal fate (including positional or laminar fate, neurotransmitter type, axonal connections, and molecular expression) are determined during the last mitotic cycle, when the neuron is produced or born. Thus, newly generated neurons are born with their mature phenotype already defined. To achieve this, the new neuron must (1) differentiate by expression of general neuronal genes and neuronal subtype-specific genes, (2) migrate to the correct location, (3) grow axons and dendrites and make appropriate synaptic connections, (4) refine connections with other neural elements according to critical environmental stimuli, and (5) acquire glial contacts (myelination and astrocytic contacts) that enhance neuronal function.

## Neuronal differentiation

The vast numbers of distinct neuron types in the brain (numbering in the hundreds or more) express different combinations of general neuronal genes, as well as neuronal subtype-specific genes. In the cerebral cortex, for example, each cortical layer contains multiple distinct types of glutamatergic (pyramidal) neurons and multiple distinct types of GABAergic (inhibitory) neurons. In general, the specific properties of each neuron type are largely specified by the combinatorial and sequential expression of transcription factors. Initial specification of neuronal fates begins in progenitors by the expression of neurogenic genes, such as Neurogenin1, a bHLH transcription factor. Following mitosis, a different set of bHLH transcription factors is activated, belonging to the neuronal differentiation group (such as Neurod1). Further differentiation into neuronal subtypes involves other transcription factors belonging to a great variety of families.

## Neuronal migration

Virtually all new neurons migrate from their sites of production in progenitor zones to sites of integration in postmitotic neural structures. Migrations are remarkably diverse among various brain regions. They may traverse relatively short or long distances, may be mainly radial or tangential or both, and may involve multiple sequential phases (for instance, tangential followed by radial migration). One example of a long-range, mainly tangential, migration involves movement of interneurons from subcortical forebrain into the developing cortex. Indeed, migrations are particularly robust in the cerebral cortex and in the cerebellum, and these regions are hardest hit by mutations that perturb cell migration.

Neuronal migration requires the coordinated activity of numerous molecules belonging to several categories. First, the direction of migration is selected by a leading process with a distal specialization or growth cone where guidance receptors from a variety of families may be expressed (such as Eph family tyrosine kinase receptors). Second, the leading process must grow in the indicated direction, a process that requires actin and microtubule cytoskeletal dynamics, as well as membrane turnover (mediated by endocytic vesicles). Third, the nucleus must be pulled toward the leading process by a network of microtubules oriented from the centrosome. Using these mechanisms, neurons migrate in a phasic rather than continuous manner, with repeated cycles of leading process growth and nuclear translocation. Whereas neuronal migration appears highly sensitive to disturbances in the underlying cellular machinery, it is also clear that the mechanisms of cell migration have a great deal of overlap with axon guidance (likewise mediated by a growth cone), membrane recycling, and cellular proliferation. Accordingly, neuronal migration disorders in humans often display associated abnormalities of axon pathways and brain growth.

Studies of the genetics of human and murine neuronal migration disorders have led to the identification of many key molecules guiding or promoting neuronal migration. In mice, one of the prototypical neuronal migration disorders was found in the spontaneous mutant strain reeler, in which the cerebral cortex and cerebellum show severe abnormalities of cellular organization. Further studies revealed that the disorder resulted from deficiency of a large secreted protein encoded by the reeler gene. The cognate protein, Reelin, was found to be highly localized in the cortical marginal zone, and in specific locations within the developing cerebellum, suggesting that it might function as a guidance molecule for migrating neurons. Further studies have supported this interpretation, as receptors for Reelin and downstream intracellular signaling molecules have been identified. A human counterpart of the mouse disorder has been identified in patients with lissencephaly and cerebellar hypoplasia, who are found to have mutations in the human RELN gene. Other genetic studies of human patients with lissencephaly or periventricular heterotopia have identified several additional critical genes, notably including DCX (Doublecortin), LIS1 (Lissencephaly-1), FLNA (Filamin A), and ARFGEF2 (a vesicular trafficking molecule). These molecules are important for microtubule dynamics or membrane dynamics and critical for cellular mechanisms of either nuclear translocation within migrating cells (microtubule dynamics) or addition of membrane to the leading process (membrane trafficking).

## Lissencephaly, periventricular heterotopia, and polymicrogyria

Lissencephaly, periventricular heterotopia, and polymicrogyria are the classic disorders of neuronal migration described in human neuropathology. Lissencephaly is an abnormality of cortical development which, in humans, produces a thickened cortical gray matter layer without normal folding (gyri). Lissencephaly may be divided into type I (smooth surface) and type II (cobblestone surface), which have distinct mechanisms. Type I lissencephaly is a primary defect of neuronal proliferation and migration (Fig. 29.11A), whereas type II lissencephaly is a primary defect of basal lamina integrity at the pial surface, in which basal lamina breakdown leads to neuronal migration through the basal lamina defects, and disorganization of the cortical structure. Mutations of cell migration molecules such as LIS1 and DCX are the cause of type I lissencephaly, whereas mutations in glycosylation enzyme genes (essential for basal lamina integrity) are the cause of type II lissencephaly. Periventricular heterotopia, consisting of abnormally localized gray matter islands abutting the ventricular surfaces (Fig. 29.11B), can be caused by FLNA mutations, chromosome abnormalities, and other unknown molecular or tissue abnormalities. Polymicrogyria, abnormally organized cortex with numerous small gyri (Fig. 29.11C), is a heterogeneous disorder with variable lobar, unilateral, or bilateral distribution, and genetic as well as nongenetic causes (such as fetal hypoxia—ischemia). Different histological varieties of polymicrogyria have been described, including four-layered and unlayered types, but the common feature in all forms of polymicrogyria is redundancy of cortical layer 1 (the marginal zone). Lissencephaly and periventricular heterotopia involve defective migration from progenitor compartments (ventricular zone and subventricular zone) to the cortical plate, whereas polymicrogyria involves defective laminar organization of neurons within the cortical plate. Hereditary forms of polymicrogyria have been linked to mutations in GPR56, a G protein—coupled receptor of unknown function. On the basis of the genes identified so far, polymicrogyria may be caused by defects in distinct classes of molecules from lissencephaly and periventricular heterotopia.

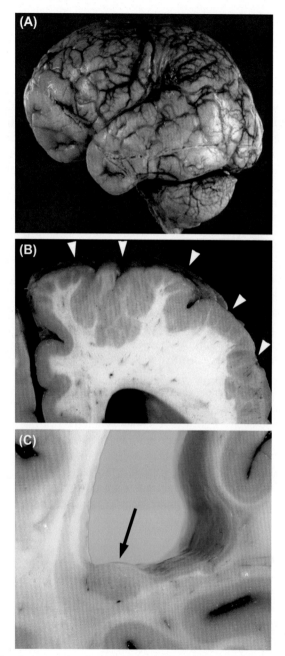

**FIGURE 29.11** Defects of cortical differentiation or migration. (A) Lissencephaly (type I) in Miller–Dieker syndrome (lateral view). The sulci are extraordinarily shallow, lending the brain surface a smooth surface appearance. (B) Polymicrogyria affecting lateral frontal cortex. (C) Periventricular heterotopia in the occipital horn of the lateral ventricle.

## Axon growth and guidance: agenesis of the corpus callosum

The cellular complexity of the human brain is exceeded only by its extraordinary connectional complexity. Axonal pathways develop concurrently with neuronal migrations and are mediated by numerous large families of axon guidance molecules, among them the Eph receptors, ephrin ligands, Semaphorins, neuropilins, plexins, Robo receptors, Slit ligands, diverse cell adhesion molecules, secreted morphogens, and extracellular matrix molecules and their receptors. Studies in experimental animals have shown that, to effectively navigate long distances, axon growth and guidance typically involve stepwise progression from one intermediate target to another. Hundreds of axon guidance phenotypes

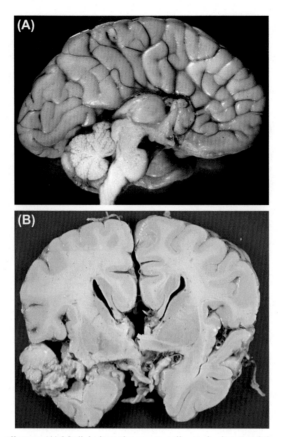

**FIGURE 29.12**  Agenesis of the corpus callosum. (A) Medial view: the corpus callosum is absent and the cingulate gyrus and sulcus are malformed. (B) Coronal slice: probst bundles are visible dangling below the malformed cingulate cortex, indicating growth toward the midline but failure to cross.

have been identified in mice and other experimental animals (fruit flies, worms, zebrafish) with mutations in axon guidance molecules, and this is an area of intense research. Somewhat surprisingly, only a few specific axon guidance disorders have been found in humans. This likely reflects the difficulty of tracing axon pathways in the human brain, where it is obviously impossible to inject actively transported axon tracers during life.

## Agenesis of the corpus callosum

Agenesis of the corpus callosum, a relatively common disorder ($\sim$1 per 1000), can be asymptomatic and is believed to be due to a failure of axon guidance. It has been linked in X-linked pedigrees to mutations of L1 cell adhesion molecule. Often, the callosal axons that fail to cross the midline instead form an aberrant tangle of fibers adjacent to the midline, called a Probst bundle (Fig. 29.12).

## Synaptogenesis, refinement, and plasticity: autism and schizophrenia

Synaptogenesis, refinement, and plasticity, which embody relatively late stages of neuronal development, are essential for proper CNS functioning, but not morphogenesis. Thus, perturbations of critical molecules for these processes cause complex cognitive, sensory, or motor disorders without obvious neuroanatomic malformations. Human diseases related to these processes are thought to include forms of autism, schizophrenia, and mental retardation (intellectual disability) and have many genetic as well as environmental etiologies. Many cases of autism, intellectual disability, and schizophrenia are caused by de novo (noninherited) mutations during germ cell or brain development. However, these disorders are genetically and phenotypically complex and remain poorly understood in relation to developmental neurobiology.

## Summary

Molecular expression changes due to genetic mutations are a frequent cause of cerebral and spinal malformations. Progress in developmental neurogenetics has transformed developmental neuropathology by providing a mechanistic understanding of malformations, by improving prognostic accuracy, and by providing insights that may one day allow for more effective treatments of these often devastating conditions.

## Neurological injury: stroke, neurodegeneration, and toxicants

Several mechanisms are not exclusive to any specific neurologic disease but are commonly proposed to contribute to varying degrees to the pathogenesis of stroke, neurodegeneration, and neurotoxicant injury. These include excitotoxicity (Fig. 29.13), mitochondrial dysfunction, free radical stress, lipid peroxidation, carbonyl stress, innate immune activation, and catechol metabolites. The adult CNS has limited regenerative ability. The natural history of CNS damage is

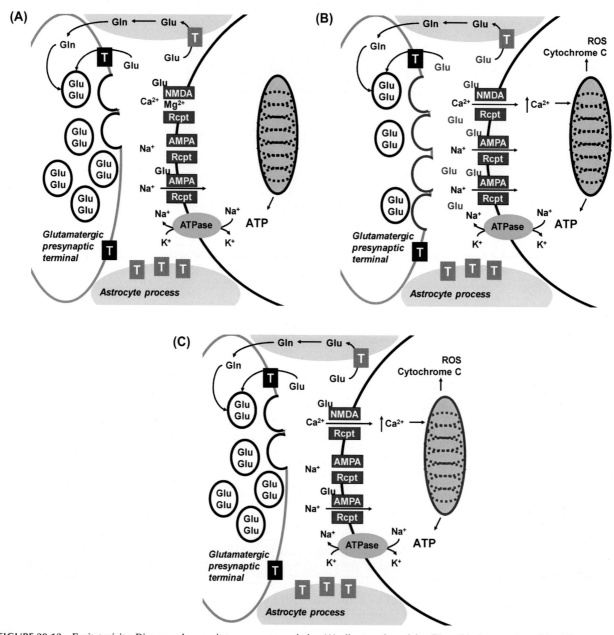

**FIGURE 29.13** Excitotoxicity. Diagrams show excitatory neurotransmission (A), direct excitotoxicity (B), and indirect excitotoxicity (C).

that cells vulnerable to injury become necrotic, apoptotic, or autophagocytic and their debris removed. The overall mass of the affected region is reduced and shrinks in a state called atrophy. The residual neuroepithelial cells, usually astrocytes and microglia, respond by gliosis.

## Vascular disease and injury

Disease in the cerebrovasculature can lead to compromised perfusion of regions of the brain or spinal cord, a process called vascular brain injury. The most common types of vascular diseases are atherosclerosis and arteriolosclerosis. The most common forms of vascular brain injury are ischemia and hemorrhage.

### Ischemia

The clinical consequences of acute ischemic stroke are dramatic and critically dependent on the precise regions of the CNS involved. The pathologic consequences of ischemia assume one of three general forms depending on the severity of the insult (Fig. 29.14). The first is a complete infarct with necrosis of all parenchymal elements. The second form of injury, an incomplete infarct, occurs when ischemia is less severe, producing necrosis of some cells, but not all tissue elements. Incomplete infarcts demonstrate that neurons, and to a lesser degree, oligodendroglia, are more vulnerable than other cells in the CNS to ischemia. The ultimate tissue manifestation of an incomplete infarct resembles the edge (penumbra) of a complete infarct: neuronal and oligodendroglial depopulation, myelin pallor, astrogliosis, and capillary prominence. The final, mildest form of ischemic injury results in damage and dysfunction, but without death of parenchymal elements.

Although necrotic brain is essentially irretrievably lost, it is important to realize that the zones with incomplete infarction or damage without necrosis hang in the balance between vulnerability to further injury, preservation, or perhaps even regeneration. Indeed, some functional recovery typically occurs in the days, weeks, and even months following an ischemic stroke, in part from resolution of reversible stressors, post-stroke neurogenesis in at least some regions of CNS, and functional reorganization of surviving elements. The balance of deleterious versus beneficial glial and immune responses in these surviving but damaged regions is thought to be key to optimal clinical outcome and is an area of intense investigation.

Regardless of the cause, CNS infarcts share a common evolution of coagulative necrosis, leading to liquefaction that culminates in cavity formation. Although death of cells occurs in CNS tissue within minutes of sufficient ischemia, the earliest structural sign of damage is not apparent until 12−24 h following the ictus, when it takes the form of coagulative necrosis of neurons, or red neuron formation (Fig. 29.15). The histologic hallmarks of this process are neuronal karyolysis and cytoplasmic hypereosinophilia. Infarcts become macroscopically apparent about 1 day after onset as a poorly delineated edematous lesion often with vascular congestion (Fig. 29.16). Subsequent evolution of infarcts is dictated by the inflammatory cell infiltrate. Around 1−2 days after infarction, the lesion is characterized by disintegration

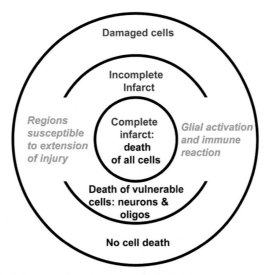

FIGURE 29.14  Diagram of varying levels of damage and reaction in vascular brain injury.

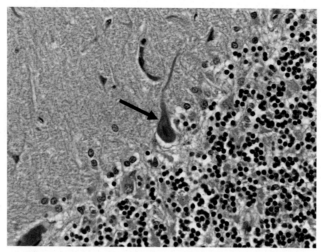

FIGURE 29.15 Red neuron. Photomicrograph of H&E (hematoxylin and eosin)-stained section of cerebellum shows Purkinje neuron with changes of coagulative necrosis, aka "red neuron" (black arrow).

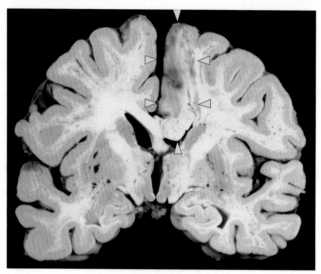

FIGURE 29.16 Coronal section of cerebrum shows acute infarct in the territory supplied by the left anterior cerebral artery (blue arrow heads).

of necrotic neurons, capillary prominence, endothelial cell hypertrophy, and a short-lived influx of neutrophils that are gradually replaced by macrophages starting within approximately 2−3 days after the insult. Initially, macrophages may be difficult to discern in the inflamed and necrotic tissue, but ultimately they accumulate to very high density and become enlarged with phagocytized material. This is the phase of liquefaction. These debris-laden macrophages eventually return to the bloodstream, in conjunction with increasing astrogliosis at the periphery of the lesion. When phagocytosis is completed, the solid mass of necrotic tissue that characterized the acute infarct has been transformed into a contracted, fluid-filled cavity that is traversed by a fine mesh of atretic vessels but does not develop a collagenous scar that is typical of other organs (Fig. 29.17). Usually, the cavitated infarct is surrounded by a narrow zone of astrogliosis and neuron loss that abuts with histologically normal brain parenchyma. Distal degeneration of fiber tracts is also a late manifestation of infarction in the CNS.

The type of vessel occluded leads to characteristic patterns of regional ischemic damage to CNS (Table 29.1). Territorial infarcts describe regions of necrotic tissue secondary to occlusion of an artery, such as the anterior cerebral artery (Fig. 29.16). A common mechanism for obstruction of large arteries is complicated atherosclerosis with thrombus formation (Fig. 29.18A,B). Some will progress to hemorrhage, as the obstruction is lysed and the necrotic regions are reperfused. The intraparenchymal large arterioles that arise from arteries at the base of the brain are vulnerable to processes that produce arteriolosclerosis such as hypertension and diabetes mellitus (Fig. 29.19). The two major

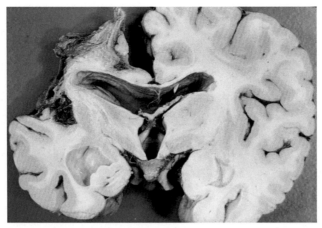

FIGURE 29.17    Coronal section of cerebrum shows remote infarct in the territory supplied by the left middle cerebral artery.

**TABLE 29.1** Characteristic central nervous system injuries related to size of vessel occluded.

| Vessel | Type of lesion | Examples of diseases |
| --- | --- | --- |
| Arteries | Territorial infarct, sometimes hemorrhagic | Atherosclerosis, vasculitis, vasospasm, dissection |
| Large arterioles | Lacunar infarct microaneurysm | Arteriolosclerosis, diabetes, hypertension |
| Microvessels | Widespread injury of varying severity | Arteriolosclerosis, CADASIL, fat or air emboli |
| Veins | Hemorrhagic infarct, often bilateral | Thrombosis |

CADASIL, cerebral autosomal dominant arteriopathy with subcortical infarcts and leukoencephalopathy.

complications of arteriolosclerosis in brain are smaller infarcts (<1 cm), called lacunar infarcts (Fig. 29.20), and formation of microaneurysms that may rupture and lead to intracerebral hemorrhage.

Rather than focal arteriolar disease leading to discrete lacunar infarcts, widespread disease of small arterioles and capillaries (also known as microvessels) can produce more pervasive ischemic damage to brain that often presents clinically with global neurologic dysfunction. There are several causes of this form of ischemic injury to brain, which typically vary in severity from diffuse degeneration with astrogliosis with or without microinfarcts (discernible only by microscopy) that are sometimes hemorrhagic. This variation in damage likely reflects a gradient of insults from oligemia to ischemia. One form of this type of injury thought to be secondary to widespread arteriolosclerosis is called subcortical arteriosclerotic encephalopathy or Binswanger disease that is associated with hypertension. Another cause of this type of ischemic injury is cerebral autosomal dominant arteriopathy with subcortical infarcts and leukoencephalopathy (CADASIL). CADASIL is associated with missense mutations in the NOTCH3 gene in the vast majority of patients and is characterized by a microangiopathy with degeneration of vascular smooth muscle cells, which leads to discontinuous immunoreactivity for smooth muscle actin in arterioles (Fig. 29.21). Although CADASIL is relatively rare, knowledge gained from investigation of this disease has spurred the search for other genetic causes and risk factors for ischemic brain injury. Other causes of widespread small arteriole/capillary occlusion are thrombotic thrombocytopenic, rickettsial infections, and fat or air emboli.

Thrombosis of veins leads to stagnation of blood flow in the territories being drained and is observed most frequently as a complication of systemic dehydration, hypercoagulable states, or phlebitis. Edema and extravasation of erythrocytes are observed initially. However, venous thrombosis can lead to infarcts that are commonly bilateral and conspicuously hemorrhagic.

In contrast to regional ischemic damage from diseases that affect blood vessels, global ischemic damage is produced by profound reductions in cerebral perfusion pressure or blood oxygenation. There are different forms of global ischemic damage that varies with the severity of the insult. In its most extreme form, global ischemia culminates in total cerebral necrosis or brain death. Although the patient's vital functions may be maintained artificially for some time, cerebral perfusion typically becomes blocked by the massively increased intracranial pressure, leading to necrosis of

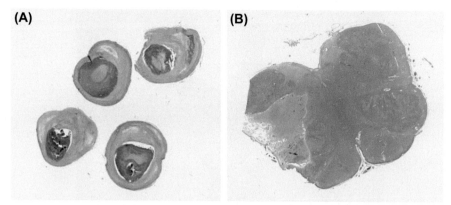

**FIGURE 29.18** Thrombosis. Photomicrographs of H&E (hematoxylin and eosin)/LFB (Luxol fast blue)-stained sections show thrombus occluding the basilar artery that also has complicated atherosclerosis (A) and the corresponding subacute infarct of the lateral medullary plate (B).

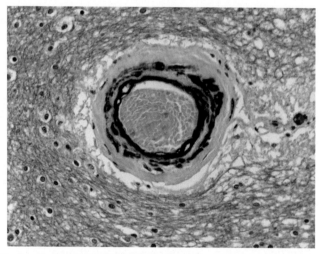

**FIGURE 29.19** Arteriolosclerosis. Section of subcortical white matter demonstrates an arteriole with a thickened vascular wall with immunohistochemical reaction for antimuscle-specific actin demonstrating continuous hypertrophy Gomori trichrome.

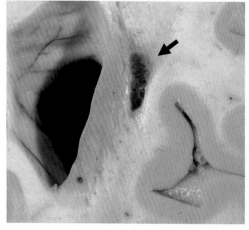

**FIGURE 29.20** Lacunar infarct. Coronal section of right cerebrum shows remote lacunar infarct in the internal capsule adjacent to caudate head (black arrow).

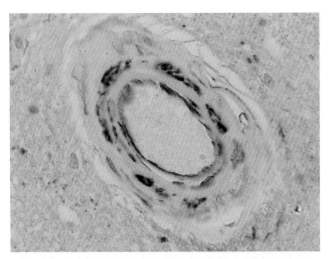

**FIGURE 29.21** Small vessel in CADASIL. Photomicrograph of immunohistochemical reaction for anti-muscle-specific actin shows discontinuous smooth muscle in a white matter arteriole from a patient with cerebral autosomal dominant arteriopathy with subcortical infarcts and leukoencephalopathy (CADASIL).

the entire brain. However, global ischemia need not be so severe as to produce total cerebral necrosis. These less severe instances are typified by sudden decrease in mean arterial pressure, such as shock or cardiac arrest, from which the patient is resuscitated in a relatively short period of time. The pattern of tissue damage reflects an exaggerated or selective vulnerability of some groups of neurons to ischemic injury. The most susceptible are the pyramidal neurons of the hippocampus and Purkinje cells of the cerebellum. In more severe cases, arterial border zone, or watershed, infarcts occur at the distal extreme of arterial territories, regions of the brain and spinal cord where distal vascular territories overlap. Although there are several arterial border zones in adults, incompletely developed anastomoses between the cerebral cortical long penetrating arteries and the basal penetrating arteries produce a periventricular arterial border zone that is inordinately sensitive to global ischemia in premature infants. The resulting lesion is called periventricular leukomalacia.

The pathophysiology of ischemic injury to the CNS is complex. Within seconds to minutes, there is failure of energy production, release of K+ and glutamate, and massive increase in intraneuronal calcium (direct excitotoxicity). What follows over hours to weeks is an intricate balance of further damage and response to injury that critically involves elements of immune activation. Enormous effort has been spent investigating potential therapeutic targets and experimental interventions focused on the acute phase of ischemic stroke. Sadly, so far none has translated to general patient care. Indeed, the current therapeutic approach to ischemic stroke is currently endovascular therapy or thrombolysis in eligible cases; however, this can be complicated by hemorrhage into injured brain.

## Hemorrhage

Intracranial hemorrhage occurs when a vessel ruptures and releases blood from the intravascular compartment into some other intracranial compartment. The key to appreciating the clinical significance of intracranial hemorrhages is an understanding of the type and location of the vessel involved (Table 29.2). Rupture of an artery leads to release of blood under high pressure into the spaces surrounding the brain that can rapidly lead to fatal compression, whereas rupture of an arteriole releases blood under high pressure into brain parenchyma with equally devastating effects. Rupture of microvessels leads to lesions called microhemorrhages that conspire with microinfarcts to produce apparently progressive widespread neurologic dysfunction. Finally, rupture or tearing of veins that bridge from the subarachnoid space to the dural sinuses releases blood under venous pressure that produces subdural hematomas.

Intraparenchymal hemorrhages from arteriolar disease, which can be large and life-threatening, deserve further discussion. As described in Table 29.1, aneurysm formation in cerebral arterioles is a consequence in at least some individuals with hypertension. Rupture of these weakened regions of vessels is a common cause of intraparenchymal hemorrhage that more commonly occur centrally or deep in the cerebral hemisphere (Fig. 29.22). Another common cause of intraparenchymal hemorrhage is rupture of arterioles sufficiently weakened by amyloid deposition, a condition called cerebral amyloid angiopathy (CAA). These more commonly occur in one of the cerebral lobes, such as the occipital lobe, and so they are termed lobar hemorrhages (Fig. 29.23). Amyloid means starchlike and describes a group of

**TABLE 29.2** Characteristic central nervous system injuries related to size of vessel ruptured.

| Vessel | Location of Ruptured Vessel | Outcome |
|---|---|---|
| Artery | Epidural | Medical emergency |
| | Subarachnoid | |
| Arteriole Microvessel | Intraparenchymal | Cumulative effects can contribute to widespread progressive neurologic deficits |
| Vein | Subdural | Varies, often slowly progressive |

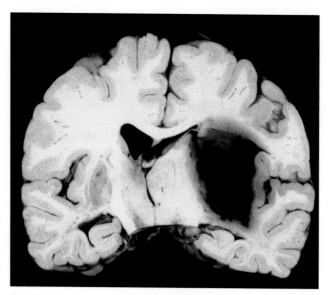

FIGURE 29.22 Hemorrhage. Coronal section of cerebrum shows intracerebral hemorrhage involving deep structures on the right.

proteins that share common interrelated features including high β sheet conformation, affinity for certain histochemical

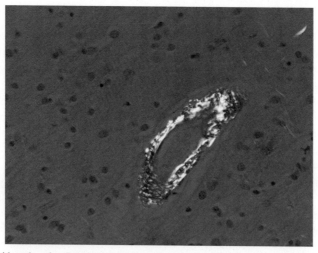

FIGURE 29.23 Congophilic amyloid angiopathy. Polarized light from Congo red–stained cerebral cortex shows birefringent parenchymal blood vessel. Note also adjacent parenchymal amyloid deposit from a patient with Alzheimer disease.

dyes, and the capacity to form fibrillar structures of limited solubility. CAA can result from the deposition of amyloid (A) β peptides, cystatin C, prion protein, ABri protein, transthyretin, or gelsolin. CAA with Aβ peptides can occur sporadically with advancing age and in patients with Alzheimer disease. Rarely, CAA is caused by autosomal dominant mutations and is called hereditary cerebral hemorrhage with amyloidosis (HCHWA). These include Dutch, Italian, and Flemish families with mutations in the amyloid precursor protein (APP) gene and the Icelandic type that is caused by mutations in the cystatin gene. It is important to note the other mutations in APP are rare autosomal dominant causes of Alzheimer disease. Furthermore, amyloid deposition in blood vessels is not always limited to the cerebrum; for example, HCHWA-Icelandic type typically shows more widespread involvement of vessels throughout the brain.

## Degenerative diseases

Neurodegenerative diseases range from very common illnesses such as Alzheimer disease to rare illnesses such as prion diseases. Function is localized within the CNS; therefore the clinical presentation of neurodegenerative illnesses is dictated by the regions of brain affected. For example, Alzheimer disease focuses initially in the hippocampus and closely related structures, and later involves primarily frontal, temporal, and parietal lobes. Clinically, Alzheimer disease is characterized by early impairment in declarative memory that is followed by impairments in other cognitive domains. Another example is amyotrophic lateral sclerosis (ALS) in which degeneration of Betz cells in the frontal lobe and anterior horn cells in the spinal cord produces a characteristic combination of weakness and paralysis. A fuller understanding of the human functional neuroanatomy is needed to appreciate the correlations between affected regions and clinical presentation. However, Table 29.3 presents a very broad overview for selected diseases.

### Etiology

The cause of each one of these illnesses is known to some extent. However, this mostly concerns forms that have patterns of highly penetrant autosomal dominant inheritance. For example, Huntington disease is caused by inheritance of an abnormally expanded trinucleotide repeat (CAG) in the HD gene that is translated into an expanded glutamine repeat in the Huntingtin protein. The situation is more complex for the other four neurodegenerative diseases that we are considering. Alzheimer disease, Parkinson disease, ALS, and Creutzfeldt—Jakob disease, the most common type of prion disease, all have an uncommon subset of patients with autosomal dominant forms of the disease, but also much more common sporadic forms that are not caused by inherited mutations; however, some have been associated with inherited risk factors (Table 29.4). It is key to understand that identification of genetic causes and risk factors defines relevance, but not mechanism. For example, the mutation that leads to Huntington disease has been known for over 20 years, but the pathophysiology of Huntington disease remains an area of intense investigation. Mechanism of disease is important because it forms the foundation for evidence-based therapeutic interventions.

### Pathogenesis of neurodegeneration

Although there is evidence for each of the basic mechanisms of neuronal injury contributing to neurodegenerative diseases, a relatively specific hallmark feature of several neurodegenerative diseases is the accumulation of aggregates of misfolded protein that in some instances form amyloid. Although protein misfolding is not limited to neurologic disease, within neurologic disease it seems especially important to neurodegenerative conditions. The focus on protein abnormalities in neurodegenerative diseases stems from the characteristic amyloid deposits known to neuropathologists for over a century: amyloid plaques in Creutzfeldt—Jakob disease that contain prion protein (PrP) amyloid, senile

**TABLE 29.3** Anatomic and clinical features of the neurodegenerative diseases.

| Disease | Affected region | Corresponding clinical features |
|---|---|---|
| Alzheimer disease | Hippocampus, regions of cerebral cortex | Dementia |
| Creutzfeldt—Jakob disease | Cerebral cortex, basal ganglia, cerebellum | Dementia, movement disorders |
| Parkinson disease | Midbrain and basal ganglia | Bradykinesia, rigidity, tremor |
| Huntington disease | Basal ganglia | Chorea |
| Amyotrophic lateral sclerosis | Primary motor neurons | Weakness and paralysis |

**TABLE 29.4** Autosomal dominant and sporadic forms of the neurodegenerative diseases.

| Disease | Prevalence | Autosomal dominant | | Sporadic | |
| --- | --- | --- | --- | --- | --- |
| | | Inherited cause | Frequency | Inherited risk factor | Frequency |
| Alzheimer disease | ~20% of people over 65 | Mutation in APP, PSEN1, or PSEN2 | Uncommon | APOE ε4 allele | Common |
| Parkinson disease | 3%–5% of people over 65 | Mutation in SNCA, PARK2, UCHL1, DJ1, or PINK1 | Uncommon | SNCA polymorphisms | Common |
| Amyotrophic lateral sclerosis | ~4 per million | Mutation in SOD1 | Uncommon | Not yet identified | Common |
| Creutzfeldt–Jakob disease | ~1 per million | Mutation in PRNP | Uncommon | PRNP polymorphisms | Common |
| Huntington disease | ~3 per 100,000 in Western Europeans | Expanded CAG repeat in HD | All | Not applicable | None |

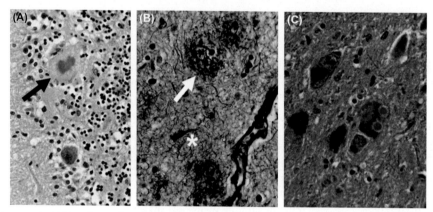

**FIGURE 29.24** Proteinaceous inclusions of neurodegenerative diseases. Photomicrographs show (A) plaque (black arrow) in cerebellum from patient with CJD (hematoxylin and eosin (H&E)), (B) senile plaque (white arrow) and neurofibrillary tangle (white asterisk) in a patient with Alzheimer disease (modified Bielschowsky), and (C) a pair of Lewy bodies, one among several pigmented (dopaminergic) neurons of the substantia nigra from a patient with Parkinson disease (hematoxylin and eosin (H&E)/Luxol fast blue (LFB)).

plaques of Alzheimer disease that contain amyloid β peptides, and α-synuclein-containing Lewy bodies in Parkinson disease (Fig. 29.24). Our understanding of the role of protein misfolding in neurodegenerative disease has advanced greatly from identification of amyloid. Current ideas about this aspect of neurodegeneration are best demonstrated by prion diseases such as Creutzfeldt–Jakob disease (Fig. 29.25).

The PRNP gene normally is transcribed (step 1) and translated (step 2) to PrP-C (green circle), a glycosyl phosphatidyl inositol–anchored protein that is expressed by many cells but at high levels by neurons. In some forms of prion diseases, PrP-C misfolds from its normal conformation to a pathogenic form (red square), PrP-Sc (step 3), which is high in β-sheet (Sc is an abbreviation for scrapie, a form of prion diseases in sheep). Some of the prion disease–causing mutations in PRNP encode for proteins with increased susceptibility to misfold into these pathogenic forms. These abnormal conformers of PrP-Sc are thought to seed or promote further subversion of PrP-C folding (step 5) leading to self-propagated generation of abnormal conformers that organize progressively into ordered complexes called fibrils (step 6) that can generate new seeds for further recruitment of PcP-C (step 7) and accumulate as amyloid in brain (step 8). Cellular defenses against protein misfolding, self-aggregation, and accumulation include chaperones, the ubiquitin–proteasome system, and autophagy, among others. Precisely how these abnormal conformers or higher ordered complexes lead to neuron death is not entirely clear but likely involves activation of at least some of the pathogenic processes described previously.

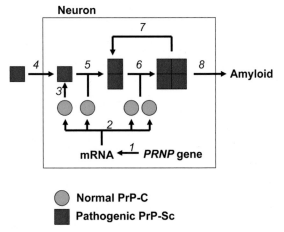

**Neuron**

○ **Normal PrP-C**
■ **Pathogenic PrP-Sc**

FIGURE 29.25   Diagram of prion disease pathogenesis. The PRNP gene is transcribed (1) and translated (2) to PrP-C (green circle). The pathogenic protein (red square) forms by misfolding of PrP-C (3) or can be transmitted (4). PrP-Sc promotes further recruitment of PrP-C (5) into fibrils (6) that can generate new seeds (7) and form amyloid (8).

Because similar pathologic features to those described for prion diseases are shared by several neurodegenerative diseases, many now propose that similar molecular mechanisms underlie the more common Alzheimer disease where the misfolded and accumulating proteins are Aβ peptides and pathologic forms of tau, a microtubule-binding protein. APP is the product of the APP gene that when mutated can cause a highly penetrant form of early-onset autosomal dominant Alzheimer disease. APP is a single membrane-spanning protein that is expressed at high levels by neurons and undergoes endoproteolytic cleavages by α-secretase to generate secreted and internalized segments, or by β-secretase and γ-secretase to generate secreted protein, amyloid β peptides, and internalized segments, all of which have biological activity. However, the major research focus has been on the neurotoxic properties of Aβ peptides. The identity of β-secretase is now known and is called BACE1 (β-site APP-cleaving enzyme), while part of the multicomponent γ-secretase appears to be the protein products of PSEN1 and PSEN2, mutations of which also cause highly penetrant forms of autosomal dominant Alzheimer disease. Indeed, the clustering of mutations that cause early-onset Alzheimer disease around the generation of Aβ peptides is the foundation of the amyloid hypothesis for this disease. Promiscuity in the cleavage site by γ-secretase is responsible for generating Aβ peptides of varying lengths. Of these, Aβ that are 40 (Aβ1−40) or 42 (Aβ1−42) amino acids in length are the most intensely studied with Aβ1−42 being more fibrillogenic and neurotoxic in model systems. Key to ultimately understanding Alzheimer disease will be unraveling the mechanistic connections between accumulation of Aβ peptides, not only in brain parenchyma, but also in cerebral blood vessels (Fig. 29.23), and the accumulation of pathologic forms of tau in structures called neurofibrillary tangles (Fig. 29.24B). Although this connection may involve some of the processes described, including seeding and prionlike spread, it currently remains enigmatic.

It is critically important to emphasize that a unique feature of prion diseases is their transmissibility (step 4 in Fig. 29.25). This is a real but rare clinical issue. However, it is achieved routinely in laboratory animals. Indeed, protease-resistant fragments of PrP-Sc are now widely viewed as the transmissible agent in prion diseases. This is not the case for Aβ peptides or aggregated proteins characteristic of other neurodegenerative diseases. Indeed, no other neurodegenerative disease has been shown to be transmissible between humans. Perhaps this simply reflects varying potency for transmission. Alternatively, this may point to a fundamental difference in mechanism among these diseases.

Before leaving the topic of neurodegenerative diseases, it is important to emphasize the immense looming public health challenge posed by late-onset Alzheimer disease (LOAD). LOAD describes those patients with Alzheimer disease who have onset in later adult life, typically older than 65 years, and who have not inherited a causative mutation. Sometimes this condition is referred to as sporadic Alzheimer disease. Although identification and investigation of autosomal dominant forms have provided invaluable insight into etiology and pathogenesis of Alzheimer disease, sporadic Alzheimer disease presents additional facets for investigation. Indeed, recent work has identified about 25 genetic risk loci for sporadic Alzheimer disease, the strongest being inheritance of the ε4 allele of the apolipoprotein E gene, APOE. Unlike other mammals, humans have three common alleles of APOE and inheritance of the ε4 allele increases the risk of Alzheimer disease in a gene dose−dependent manner. The mechanism(s) that underlie this increased risk caused by a particular isoform of apolipoprotein E are not entirely clear but include maintenance of synaptic

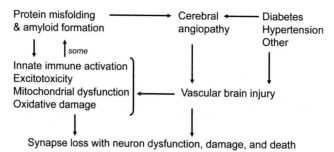

FIGURE 29.26    Diagram of potential interactions between Alzheimer disease and vascular brain injury in dementia.

physiology, trafficking of Aβ peptides, and regulation of immune response in brain. Another important facet of sporadic Alzheimer disease is that it frequently occurs along with other common age-related diseases of brain, mostly vascular brain injury as diagrammed in Fig. 29.26.

## Neurotoxicants

Neurotoxicology has significance that reaches beyond the identification of xenobiotics (neurotoxicants) or endogenous agents (neurotoxins) that are deleterious to the nervous system. Although considered separate fields of study, neurotoxicology, neurodegeneration, stroke, trauma, and metabolic diseases of the nervous system inform each other about mechanisms of neuronal dysfunction and death, and response to injury in the nervous system. Indeed, there are several examples of compounds first identified as neurotoxicants that subsequently came into use as models of human neurodegenerative disease. Perhaps the most striking example is MPTP as a model for Parkinson disease. Moreover, many compounds initially identified as neurotoxicants have become fundamental tools used by neuroscientists, such as tetrodotoxin, curare, kainic acid, and 6-hydroxydopamine.

### Neuropathological changes caused by neurotoxicants

For the most part, pathologic changes in brain following neurotoxicant exposure are nonspecific. Edema is the most striking macroscopic finding in acute toxic encephalopathies. The brain can be heavy and swollen, even after fixation, with broadened cortical gyri and obliterated sulci. In severe cases, transtentorial and cerebellar tonsillar herniations may occur. Cerebral edema secondary to vascular damage, as in lead encephalopathy, or direct damage to CNS myelin, as in triethyltin encephalopathy, is largely confined to the white matter. In contrast, edema resulting from diffuse cytotoxic damage, as in thallium intoxication, affects both gray matter and white matter, and has been observed to impart a moth-eaten appearance to the cerebral cortex. The histological manifestations of cerebral edema can be slight: myelin pallor and mild gliosis may be all that is observed by standard histochemical stains.

Neuronal degeneration following exposure to toxicants may be diffuse, for instance, in thallium intoxication. In contrast, a number of neurotoxicants are associated with damage to specific structures, including methylmercury-induced or dimethylmercury-induced degeneration of the calcarine and cerebellar cortices. Whether the primary lesion is focal or generalized, secondary degeneration of fiber pathways may be observed. Intraneuronal inclusions are uncommon in neurotoxicant-induced disease of the CNS. However, they have been reported in neurons following intoxication with heavy metals.

Glial response to injury includes astrocytic gliosis with strong immunoreactivity for GFAP. In many instances of systemic exposure to toxicants, GFAP immunoreactivity is most prominent around blood vessels. Another response to injury by astrocytes is formation of Alzheimer type II astrocytes that is associated with encephalopathy from hyperammonemic and other pathologic metabolic states. Diffuse microgliomatosis and myelin pallor have been reported following neurotoxicant exposure. Myelin pallor is more commonly due to a combination of edema and gliosis; however, examples of demyelination exist, including glue sniffer encephalopathy. Intramyelinic edema follows hexachlorophene or triethyltin intoxication.

Axonal degeneration is a common finding in toxicant-induced peripheral neuropathies. An optimal technique for viewing axons along several internodes is teased-fiber preparations showing linear collections of phagocytic cells that are digesting myelin and axonal debris, the appearance of axonal degeneration. Segmental demyelination is another common pathologic lesion in the peripheral nervous system following toxicant exposure. Inappropriately thin myelin sheaths, shortening of internodal distances, and variation of myelin thickness among internodal segments of the same axon are observed during the remyelination that follows demyelination. Recurrent episodes of demyelination with remyelination may lead to Schwann cell hyperplasia. Some neurotoxicant-induced diseases are characterized by giant axonal swellings. These large eosinophilic collections within axons are composed mostly of massive accumulations of neurofilaments. Similar neurofilament-filled axonal swellings also are seen in a rare familial neuropathy termed giant axonal neuropathy. This is in distinction to the morphologically similar axonal spheroids that contain tubulovesicular material and that have been observed in characteristic locations in older individuals as well as in patients with neuroaxonal dystrophies.

## Biochemical mechanisms of selected neurotoxicants

### Domoic acid

A large number of experiments and trials have provided indirect or pharmacological support for a role for excitotoxicity in neurological injury. Humans accidentally exposed to high doses of EAA receptor agonists and who subsequently developed neurologic disease underscore the importance of EAAs in disease. Perhaps the most striking example is the domoic acid intoxications that occurred in the Maritime Provinces of Canada in late 1987 (Fig. 29.27). A total of 107 patients were identified who suffered an acute illness that most commonly presented as gastrointestinal disturbance, severe headache, and short-term memory loss within 24—48 h after ingesting mussels. A subset of the more severely afflicted patients was subsequently shown to have chronic memory deficits, motor neuropathy, and decreased medial temporal lobe glucose metabolism by positron emission tomography. Neuropathological investigation of patients who died within 4 months of intoxication not only revealed neuronal loss with reactive gliosis that was most prominent in the hippocampus and amygdala but also affected regions of the thalamus and cerebral cortex. The responsible agent was identified as domoic acid, a potent structural analogue of L-glutamate that had been concentrated in cultivated mussels.

### MPTP

The history of MPTP-induced parkinsonism in young adults who inadvertently injected themselves with this compound is well known. MPTP is a protoxicant that, after crossing the BBB, is metabolized by glial MAO-B to a pyridinium intermediate (MPDP+) that undergoes further two-electron oxidation to yield the toxic metabolite methylphenyltetrahydropyridinium (MPP+) that is then selectively transported into nigral neurons via the mesencephalic dopamine transporter (DAT) (Fig. 29.28). Once inside these neurons, MPP+ is thought to act primarily as a mitochondrial toxin by inhibiting complex I activity in the mitochondrial electron transport chain, thereby reducing ATP production and increasing ROS generation. Indeed, MPTP-induced dopaminergic neurodegeneration can be diminished by free radical scavengers, inhibitors of the inducible form of nitric oxide synthase, and EAA receptor antagonists. Alternatively, transgenic mice lacking some elements of antioxidant defenses are significantly more vulnerable to MPTP-induced dopaminergic neurodegeneration. So far, the search for xenobiotics that may act similarly to MPTP and could be potential environmental toxicants that promote Parkinson disease has not yielded clear candidates. For this reason, MPTP

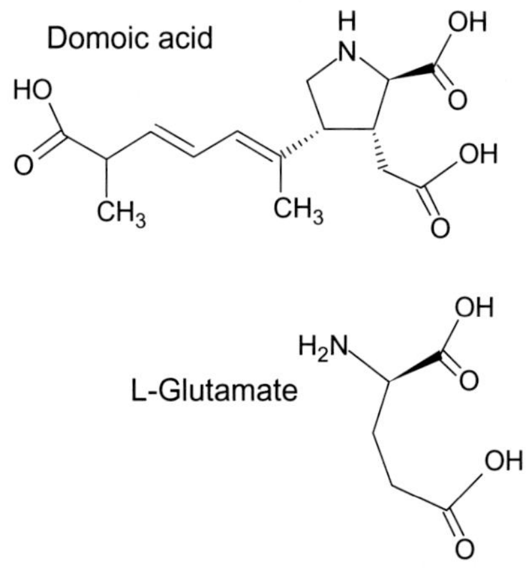

**FIGURE 29.27** Structures for domoic acid and L-glutamate.

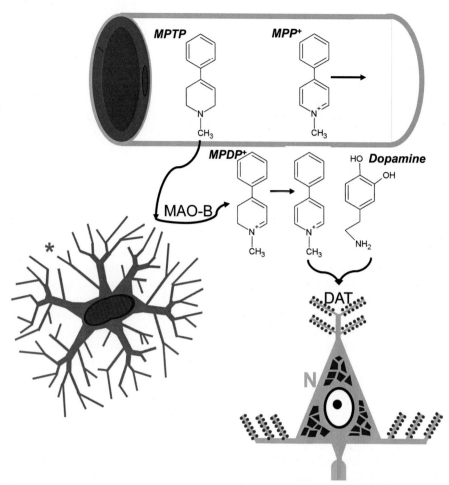

**FIGURE 29.28**   Diagram of steps in dopaminergic toxicity from 1-methyl-4-phenyl-1,2,3,6-tetrahydropyridine (MPTP) exposure. MAO-B in astrocytes (*) catalyzes oxidation of MPTP to ultimately form MPP+, which is selectively taken up by dopaminergic neurons (N) expressing the dopamine transporter (DAT).

might not be an accurate model of the etiology of Parkinson disease, but it does replicate relatively selective injury to mesencephalic dopaminergic neurons.

## Axonotoxicants

Distal sensorimotor polyneuropathy is probably the most common clinical manifestation of neurotoxicant exposure in humans. A variety of toxicants, including hexane, methyl n-butylketone (2-hexanone), carbon disulfide (CS2), acrylamide, and organophosphorus esters, result in degeneration of the distal portions of the longest, largest myelinated axons in the peripheral and central nervous systems, an observation encapsulated in the term central—peripheral distal axonopathy.

Two of these toxicants, n-hexane and CS2, are especially interesting because despite their very different chemical structures, chronic exposure to either can produce unusual pathological changes in nerve. The characteristic lesion produced by both n-hexane and CS2 is multifocal fusiform axonal swelling at the proximal side of nodes of Ranvier at distal but preterminal sites that consist of massive accumulations of disorganized 10-nm neurofilaments, decreased numbers of microtubules, thin myelin, and segregation of axoplasmic organelles and cytoskeletal components. Distal to swellings, axons may become shrunken and then degenerate. With continued exposure, more proximal swellings occur with subsequent degeneration. Investigations of n-hexane and CS2 have determined that the key to their shared clinical and pathological profiles appears to be the ability of each compound to generate protein-bound electrophilic species that can covalently cross-link proteins.

# Neoplasia

Neoplasms of the CNS can be divided into primary and metastatic. Metastatic neoplasms arise from cancer cells derived from non-CNS organs, such as lung (carcinoma), skin (melanoma), or blood (lymphoma), which are transported to the CNS via hematogenous routes or other means. Primary CNS neoplasms can be defined as uncontrolled growth of cells derived from normal CNS tissues. The most common type resembles glia histologically and is termed gliomas. Although it was previously believed that gliomas were derived from mature glial elements, such as astrocytes or oligodendrocytes, it is more likely that these tumors arise from glioneuronal progenitor cells, so-called cancer stem cells. Cancer stem cells represent a minority population of tumor cells that derive large numbers of variably differentiated glial elements that make up the bulk of the tumor and thus represent an attractive therapeutic target.

Gliomagenesis results from a number of known genetic defects associated with control of cell cycle and proliferation, apoptosis pathways, cell motility, and invasive potential. New discoveries continue to be made. Knowledge of the typical genetic alterations is being exploited in animal models of CNS neoplasia as potential targets for therapy and increasingly for diagnostic purposes. Not surprisingly, some mutations underlie the relationship between histologic appearance and biological behavior. Indeed, although assignment of tumor diagnosis and grade based on morphologic criteria has been the mainstay of diagnosis in the preceding 100 years, the next World Health Organization (WHO) classification scheme for tumors of the central nervous system implements recommendations for an integrated diagnosis that incorporates histological classification, molecular information, and grade into a single diagnostic scheme.

Unlike other organ systems that use the TNM (local Tumor growth; regional lymph Node spread; and distant Metastasis) staging system for establishing prognosis of tumors, primary CNS neoplasms are assigned a histologic grade that correlates with their predicted behaviors. The WHO classifies CNS neoplasms by grade, with grade I assigned to those potentially curable by resection alone with >10 year median survival, grade II if they are unlikely curable by resection alone and have 5−10 years' median survival, grade III for 2−3 years' median survival, and grade IV if they have <2 years' median survival. The histologic features underlying WHO grading include nuclear atypia, mitoses, vascular proliferation, and necrosis.

## Diffuse gliomas

Mutations in isocitrate dehydrogenase (IDH) 1 and 2 genes are present in the majority of low-grade diffuse gliomas. These mutations change enzyme function resulting in the production of 2-hydroxyglutarate, a possible oncometabolite. IDH1/2 mutations are oncogenic; however, whether the mechanism is through alterations in hydroxylases, redox potential, cellular metabolism, or gene expression is not clear. Gliomas with IDH1 and IDH2 mutations have a better prognosis compared with gliomas with wild-type IDH.

A defining characteristic of the diffuse gliomas is their ability to infiltrate widely throughout the CNS parenchyma, causing them to have no clearly recognizable border with normal tissue (Fig. 29.29) and making them impossible to

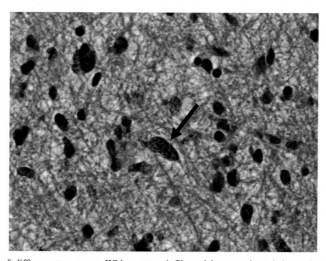

**FIGURE 29.29**  Photomicrograph of diffuse astrocytoma. White matter infiltrated by an enlarged, hyperchromatic (darkly stained), pleomorphic (irregularly shaped) neoplastic glial cell (arrow) in a diffuse astrocytoma (hematoxylin and eosin (H&E) 600× magnification).

cure by resection. The natural history of diffuse gliomas is a tendency toward progression from low-grade to high-grade by the accumulation of additional genetic defects. De novo or IDH wild-type, grade IV gliomas also occur but are usually associated with a different set of genetic defects. Broadly, there are four types of genetic alterations involved in gliomagenesis: those that affect (1) cell survival, (2) cell proliferation (cell cycle regulators), (3) brain invasion, and (4) neovascularization. Mutations in genes encoding Rb, p53, receptor tyrosine kinase, integrin, and other signaling pathways critical to cell cycle regulation are important, as are mutations in genes encoding proteins that constitute cell death pathways (TNFR, TRAIL, CD95, Bcl-2, and others). Transition from low-grade to high-grade is heralded by tumor enhancement by neuroimaging, which correlates with microvascular proliferation (angiogenesis), in which pathways regulated by vascular endothelial growth factor (VEGF), hypoxia-inducible factor (HIF), and other molecules are critical. Finally, brain invasion, which is probably the least understood of all the gliomagenic processes and the most critical to toxicity, is likely mediated by metalloproteinases and integrins, among others. For a comprehensive review of astrocytoma genetics, please see Refs.

## Diffuse astrocytoma-glioblastoma sequence

Diffuse astrocytomas (WHO grade II−IV) are gliomas in which the neoplastic cells resemble microscopically normal or reactive astrocytes and are often characterized by their cytoplasmic expression of GFAP. Histologically, grade II astrocytomas are typified by mild nuclear atypia and hypercellularity and an absence of mitotic activity, vascular proliferation, or necrosis. Anaplastic astrocytomas (WHO grade III) are histologically characterized by increased hypercellularity with nuclear atypia and mitotic activity in the absence of vascular proliferation and necrosis (Fig. 29.30).

Glioblastoma (WHO grade IV) is the most common primary glioma with an incidence of approximately 3/100,000 population per year. Glioblastomas are characterized by rapid progression and poor survival. They can be classified, based on clinical and molecular characteristics, into those that arise de novo and those that progress from lower-grade astrocytomas (Fig. 29.31).

Primary or de novo glioblastomas (mostly IDH wild-type) account for 90% of cases, generally arise in older patients, and are characterized by diverse histologic features that include cytologic atypia, mitotic activity, and microvascular proliferation and/or necrosis. The characteristic abnormalities of signal transduction in primary glioblastoma include overexpression or signal amplification of the epidermal growth factor receptor (EGFR) or loss-of-function mutations of PTEN (phosphatase and tensin homology). Deletion of EGFR exons 2−7 (EGFRvIII), the most common EGFR mutation (found in 20%−30% of glioblastoma multiform and 50%−60% of glioblastoma with EGFR amplification), results in constitutive EGFR activation and insensitivity to EGF. EGFRvIII stimulates a different signal transduction pathway than full-length EGFR. EGFR activation operates primarily through the Ras-Raf-MAPK and PI3K-Akt pathways. Constitutive activation of EGFRvIII mutants increases proliferation and survival by preferentially activating PI3K/Akt pathway and possibly through stimulation of a second messenger system not available to nonmutant EGFR.

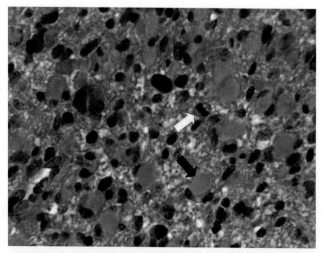

**FIGURE 29.30** Photomicrograph of an anaplastic astrocytoma. Numerous enlarged, hyperchromatic (darkly stained) neoplastic glial nuclei have infiltrated brain parenchyma in this anaplastic astrocytoma. A mitotic figure (white arrow) is present. A globule of eosinophilic cytoplasm is seen within neoplastic astrocytes (black arrow) likely representing glial fibrillary acidic protein, the primary intermediate filament of astrocytes (hematoxylin and eosin (H&E)).

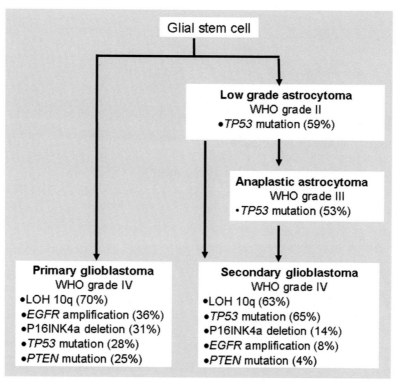

**FIGURE 29.31** Genetic alterations associated with primary and secondary glioblastoma. *Adapted from Ohgaki H, Kleihues P. Genetic pathways to primary and secondary glioblastoma. Am J Pathol 2007;170:1445—53.*

The characteristic genetic alterations involving cell cycle control in primary glioblastoma multiform include overexpression of MDM2 (murine double minute 2 protein), which suppresses p53, and deletions of CDKN2A, which encodes the tumor suppressor p16INK4A, a potent regulator of retinoblastoma (RB) tumor suppressor gene, or through an alternate reading frame (ARF) of p14ARF, an important accessory to p53 activation. Other common findings include loss of heterozygosity (LOH) on chromosome 10p and overexpression of Bcl2-like-12 protein, a potent anti-apoptotic molecule. The genetic abnormalities occur together in a random distribution and are not progressive. Homozygous deletion of CDKN2A is associated with EGFR overexpression and higher proliferative activity and may account for poorer overall survival of primary glioblastomas. It is interesting to note that the small cell phenotype in primary glioblastomas appears to be most closely associated with EGFR overexpression, suggesting a molecular-histologic link.

Secondary glioblastomas progress from lower-grade astrocytomas (mostly IDH mutant) with stepwise accumulation of additional genetic defects and generally occur in younger individuals. In a typical sequence of genetic alterations, first an IDH1 mutation occurs followed by common early genetic abnormalities that include direct mutations of the cell cycle suppressor genes TP53 and RB1, or overexpression of platelet-derived growth factor ligand and/or receptors. LOH of chromosomes 11p and 19q are common in progression of these low-grade astrocytomas to anaplastic astrocytomas. The transition from anaplastic astrocytoma to glioblastoma is not well characterized but may involve LOH of chromosome 10q, mutations indirectly affecting EGFR/PTEN pathways, and alterations to the cell cycle inhibitory p16INK4a/RB1 pathway. These pathways are summarized in Fig. 29.32.

By definition, all glioblastomas are prone to spontaneous necrosis and bizarre microvascular proliferation (Fig. 29.33). Aberrations in the genetic control of growth and cell cycle give rise to the hypoxia, necrosis, and angiogenesis that underlie these characteristic histologic features. A model of this process (Fig. 29.34) begins with genetic alterations in tumor cells, which cause a variety of downstream effects that alter the BBB, damage endothelium, or directly promote intravascular thrombosis (such as secreting tissue factor). Small vessels in the tumor become occluded, and the resulting damaged tissue releases thrombin as part of the coagulative cascade, which in turn binds to its ligand the protease-activated receptor 1 (PAR-1). Local hypoxia leads to focal necrosis. Tumor cells react to PAR-1 by forming a wave of migration away from the area of hypoxia, the histologic counterpart of which is a pseudopalisade. These migrating cells are severely hypoxic and express high levels of HIF, which activates hypoxia-responsive element

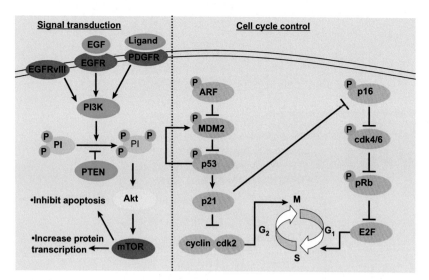

FIGURE 29.32 Molecular pathways implicated in gliomagenesis. Multiple molecules can be involved in the pathway to glioma. *Adapted from Fuller CE, Perry A. Molecular diagnostics in central nervous system tumors. Adv Anat Pathol 2005;12:180−94 and Behin A, Hoang-Xuan K, Carpentier AF, et al. Primary brain tumors in adults. Lancet 2003;361:323−31.*

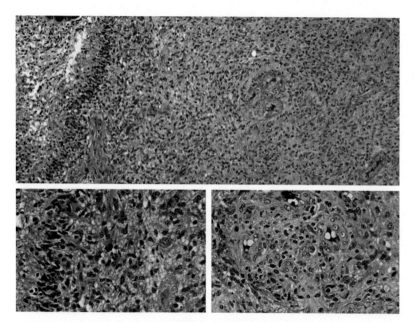

FIGURE 29.33 Photomicrograph of glioblastoma. (A) Sections from a brain infiltrated by glioblastoma stained with hematoxylin and eosin (H&E) at 100 × magnification and (B) 600 × magnification (below left). A line of parallel nuclei forming a pseudopalisade can be seen extending from the panel above through the left lower panel. The pseudopalisade is adjacent to necrosis. A neuron can be seen surrounded by neoplastic cells (arrow). (C) The collagen network of a proliferative vessel is highlighted blue-green by a trichrome stain.

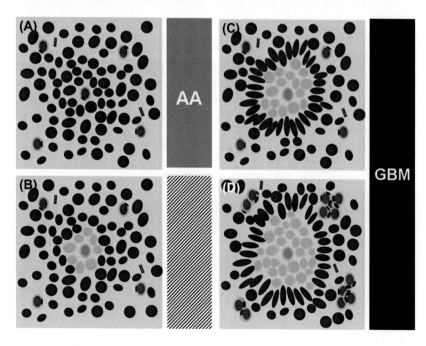

FIGURE 29.34 Progression from anaplastic astrocytoma (AA) to glioblastoma (GBM). (A) AA is mitotically active producing an area of hypercellularity but no necrosis or vascular cell hyperplasia. (B) Neoplastic glia produce thrombophilic compounds such as thrombin and tissue factor causing thrombosis of a microvessel, focal ischemia and hypoxia, and local necrosis. This change is the hallmark of the transition from AA to GBM. (C) Neoplastic glia respond to hypoxia by migrating away, forming a "pseudopalisade" of nuclei surrounding the necrosis. (D) Migrating neoplastic glia also produce angiogenic factors such as vascular endothelial growth factor and hypoxia-inducible factor, causing vascular cell hyperplasia and abnormal vessel formation. *Adapted from Rong Y, Durden DL, Van Meir EG, et al. 'Pseudopalisading' necrosis in glioblastoma: a familiar morphologic feature that links vascular pathology, hypoxia, and angiogenesis. J Neuropathol Exp Neurol 2006;65:529−39.*

domains, of which VEGF is the most characteristic in glioblastoma. High levels of VEGF cause the vascular proliferation characteristic of this tumor.

An epigenetic feature of glioblastoma that has therapeutic and prognostic implications is the methylation status of the DNA repair gene MGMT (O6-methylguanine-DNA methyltransferase). MGMT is a DNA repair protein that removes alkyl groups from the O6 position of guanine. Epigenetic silencing of this gene is identifiable in a little less than half of glioblastomas and predicts response to alkylating chemotherapy.

## Oligodendroglioma

Oligodendrogliomas are diffusely infiltrating gliomas and are the second most common parenchymal neoplasm of the adult CNS. The neoplastic cells in oligodendrogliomas histologically resemble mature oligodendrocytes. They generally have round nuclei with condensed chromatin and inconspicuous cytoplasm and few cytoplasmic processes. A classical appearance of the neoplastic cells in oligodendroglioma is a round hyperchromatic nucleus with a perinuclear clearing or halo giving the cells the classic appearance of fried eggs. They may express cytoplasmic GFAP, but generally less than astrocytic neoplasms. Histologically, they form monotonous proliferations of infiltrating neoplastic cells with occasional hypercellular nodules. They have characteristic vasculature composed of a network of fine branching capillaries resembling chicken wire (Fig. 29.35A). The neoplastic cells also have a tendency to cluster, or satellite, around parenchymal neurons and vessels (Fig. 29.35B).

The majority of oligodendrogliomas in adults contain an IDH mutation, and those with IDH1 and IDH2 mutations have a better prognosis compared with wild-type IDH. Oligodendrogliomas in adults usually also have a chromosomal translocation-mediated loss of the short arm of chromosome 1 (1p) and the long arm of chromosome 19 (19q), commonly detected using in situ hybridization or DNA amplification techniques. Loss of 1p/19q has been associated with both classic histomorphology and better clinical behavior. Patients whose tumors have this relative codeletion demonstrate improved disease-free survival, median survival, and typically respond better to alkylating chemotherapeutics. Loss of 1p/19q is inversely correlated with mutations in TP53, loss of chromosomal arms 9p and 10q, and amplification of EGFR. Codeletion of 1p/19q is also associated with lower MGMT expression and MGMT promoter hypermethylation and may explain the greater chemosensitivity of these tumors. Because of the robust correlation of this specific genetic aberration to both histomorphology and prognosis, oligodendrogliomas exist at the forefront of molecular diagnostics in gliomas. Molecular testing of infiltrating gliomas for IDH mutations and loss of 1p/19q for diagnostic and prognostic purposes is routinely performed. Loss of 1p/19q has proven useful in differentiating oligodendrogliomas from histologic mimics and in cases of ambiguous histomorphology. The molecular mechanisms by which loss of 1p and 19q mediates their effects have not been elucidated. The chromosomal translocation that underlies most losses of 1p/19q occurs at pericentromeric sites, excluding the possibility of a transgene product driving neoplastic

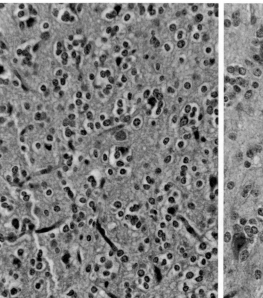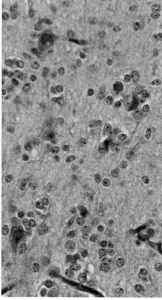

FIGURE 29.35 Photomicrograph of oligodendroglioma. Left panel: oligodendrocytes-like neoplastic cells with round nuclei and perinuclear clearing give this oligodendroglioma a fried egg appearance. Right panel: nonneoplastic neurons are clustered about by neoplastic cells. Fine capillaries are seen in the background of both panels (hematoxylin and eosin (H&E)).

transformation. Attempts to identify tumor suppressor genes on 1p and 19q have been unsuccessful. Proteomic analysis of tumors with and without the paired deletion have demonstrated increased expression of proteins associated with malignant behavior in tumors without the deletion, but definitively decreased expression of proteins coded on the lost arms has not been demonstrated.

Oligodendrogliomas are classified in the WHO system into two grades. WHO grade II oligodendrogliomas have the same histomorphology as described previously (Fig. 29.35). Grade III anaplastic oligodendrogliomas are histologically similar but display additional nuclear atypia and mitotic activity, and in addition to the classic features they often have vascular proliferation reminiscent of that seen in glioblastoma. Anaplastic oligodendrogliomas generally have a more favorable prognosis than glioblastoma.

## Circumscribed gliomas

Circumscribed gliomas are generally low-grade neoplasms that commonly occur in children and young adults and include many varieties including pilocytic astrocytoma (PA) most commonly as well as pleomorphic xanthoastrocytoma (PXA). Circumscribed gliomas are different from the diffuse gliomas in that they can be cured by resection alone. The histopathologic and genetic alterations seen in adult diffuse gliomas are uncommon in the circumscribed gliomas and, when present, may suggest misdiagnosis or malignant transformation, and certainly portend more malignant behavior. There are molecular alterations in circumscribed gliomas. The mitogen-activated protein kinase (MAPK) pathway signaling has a key role in the development and behavior of PA. Several mechanisms lead to activation of this pathway in PA, usually in a mutually exclusive manner, and most commonly with constitutive BRAF kinase activation subsequent to gene fusion between KIAA1549 and the BRAF oncogene. The high specificity of this fusion in PA when compared with other CNS neoplasm has diagnostic utility in clinical practice. Another alteration in BRAF is the V600E mutation that is thought to mimic phosphorylation of the activating amino acids T599 and S602, thereby leading to constitutive activation of the protein. Such activation affects cell proliferation, differentiation, and survival. Approximately two-thirds of PXAs harbor a mutation in BRAF V600E. A much smaller percentage of PAs harbor BRAF V600E mutations and these are strongly associated with extracerebellar localization. Other cytogenetic abnormalities in these entities have been reported, including gains on chromosomes 5 through 9 in Pas and in PXAs gains on chromosome 7, loss on 8p, and p53 mutations. Progression of PAs to higher grade gliomas through the accumulation of additional genetic defects is uncommon.

## Medulloblastoma

Medulloblastoma is the most common CNS malignancy of childhood. They tend to arise in the midline at the cerebellar vermis and are highly malignant, aggressive neoplasms. They tend to present with either cerebellar signs (truncal ataxia) or with signs of hydrocephalus. Unlike the diffuse gliomas, medulloblastomas do not generally diffusely infiltrate CNS parenchyma but rather form large necrotic masses that displace surrounding brain tissue. Moreover, they have a disturbing tendency to metastasize through the cerebrospinal fluid. Histologically, medulloblastomas form broad sheets of tightly packed cells with large nuclei, finely distributed chromatin, and little cytoplasm, an appearance that resembles the primitive neuroglial germinal matrix or external granular layer of cerebellum seen in development. Medulloblastomas often have brisk mitotic rates and abundant apoptosis, and spontaneous tumor necrosis is exceptionally common. Occasionally, the neoplastic cells form foci with more mature neuroglial elements such as ganglion cells or neuropil.

The molecular mechanisms of tumorigenesis in medulloblastoma have been extensively characterized because of its association with several genetic cancer syndromes (Fig. 29.36). The best characterized of these is overactivation of the Sonic hedgehog (Shh) pathway. The Shh pathway was first identified as a pathway in medulloblastoma in Gorlin syndrome, a genetic disorder in which individuals have multiple basal cell carcinomas of skin and an increased rate of medulloblastoma. Shh is a glycoprotein that normally activates neuronal precursor proliferation in the primitive external granular layer of the perinatal cerebellum by two primary mechanisms. In the absence of secreted Shh, Patched 1 (PTCH1) inhibits Smoothened (SMO) activation of the Gli family of cell fate and proliferation-related transcription factors. Individuals with Gorlin syndrome carry an autosomal mutation in PTCH1. Shh overactivation also drives proliferation by activation of MYCN, which activates D-type cyclins and inhibits cyclin-dependent kinase inhibitors. As long as MYCN remains active, the proliferating cells cannot exit the cell cycle. In normal development, MYCN is phosphorylated by glycogen synthase kinase-3 beta (GSK3β) and degraded.

A second partially convergent pathway involved in medulloblastoma is the Wnt signaling pathway. This process may be inhibited by the activity of insulin-like growth factor (IGF) in medulloblastoma. Wnt normally binds to its

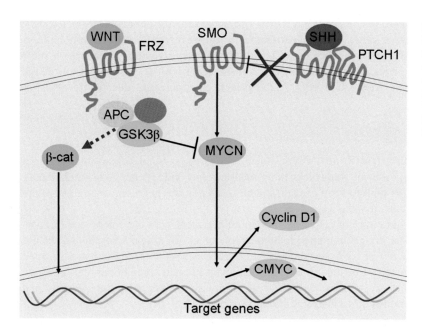

**FIGURE 29.36** Two pathways implicated in medulloblastoma. Defects in the Shh and Wnt signaling release beta catenin (β-cat) and MYCN to activate cell cycle and transcription factors. β-cat, β-catenin; APC, adenomatous polyposis coli protein; CMYC, Myc transcription factor; FRZ, frizzled receptor; GSK3β, glycogen synthase kinase 3 beta; MYCN, N-myc oncogene protein; PTCH1, patched 1 receptor; SMO, smoothened receptor.

receptor Frizzled and destabilizes a complex containing adenomatous polyposis coli (APC) gene product and the serine—threonine kinase glycogen synthase kinase 3 beta (GSK3β). This action stabilizes β-catenin, which then translocates to the nucleus and positively regulates transcription of genes that ultimately drive cellular proliferation. GSK3β was first implicated in medulloblastoma in a variant of Turcot syndrome, a familial tumor syndrome characterized by colon cancer and malignant brain tumors. Those individuals with Turcot syndrome who are predisposed to develop medulloblastomas carry a loss-of-function mutation of APC. Less commonly implicated pathways in medulloblastoma include the NOTCH and ErbB signaling pathways or defective DNA repair mechanisms.

Medulloblastomas now have distinct genetically defined categories that include WNT-activated, Shh-activated (TP53 mutated), Shh-activated (TP53 wild-type), and non-WNT/non-Shh, which has two subcategories (group 3 and 4). WNT-activated medulloblastomas are thought to arise from lower rhombic lip progenitor cells, usually occur in children, display classic histologic features, and have monosomy 6. Frequent genetic alterations in WNT-activated medulloblastomas include CTNNB1 mutation, DDX3X mutation, TP53 mutation, and APC germ-line mutation. Shh-activated (TP53 wild-type and mutant) medulloblastomas are thought to arise from cerebellar granule neuron cell precursors of the external granule cell later and cochlear nucleus. Shh-activated (TP53 wild-type) medulloblastomas may occur in infancy or adulthood and are associated with desmoplastic/nodular pattern, PTCH1 deletion/mutation, TERT promoter mutation, and 10q loss. SMO mutations may occur in adults, whereas SUFU mutations may occur in infants. Shh-activated (TP53 mutant) medulloblastomas present in childhood and have large cells with anaplastic features. Frequent copy number alterations include MYCN amplification, GLI2 amplification, 17p loss, and germ-line mutation in TP53. Non-WNT/non-Shh Group 3 medulloblastomas are thought to arise from CD133+/lineage neural stem cells. These present in infancy or childhood and have classic, large cell, or anaplastic features. These neoplasms often have MYC amplification and isocentric 17q, as well as genetic alterations in PVT-1MYC and GF11/GF11B structural variants. The cell of origin in non-WNT/non-Shh group 4 medulloblastomas is unknown. These neoplasms present in all ages and have classic medulloblastoma histologic features. Frequent copy number alterations include MYCN amplification and isocentric 17q. KDM6A genetic alterations and GF11/GF11B structural variants are also common. The modern approach to managing patients with medulloblastomas involves therapeutic stratification through a combination of histologic and molecular characteristics and provides a paradigm for integrated diagnostics. For example, targeted therapies are available for certain SHH and classic WNT tumors, which improves outcomes in these patients. Additionally, some TP53 mutations among SHH tumors are present in the germ line, which has implications for the patient's family members.

## Other neuroepithelial tumors

There are a number of other neoplasms that arise in the central nervous system, which are beyond the scope of this chapter. The most common of these are the ependymomas. These tumors have both glial and epithelial differentiation and occur most commonly around the brainstem and spinal cord. Supratentorial ependymomas, which tend to be more clinically aggressive, are associated with oncogenic fusions between RELA (principal effector of canonical nuclear factor-κB signaling) and C11orf95 (an uncharacterized gene). The nuclear factor κB (NF-κB) family of transcriptional regulators is a central mediator of the cellular inflammatory response. Constitutive NF-κB signaling is present in most human neoplasms, but mutations in pathway members are rare, which has made it difficult to understand and block aberrant NF-κB activity in cancer. The resulting RELA-C11orf95 fusion protein is a potential therapeutic target in supratentorial ependymoma

### Metastasis and lymphoma

The most common intracranial neoplasms are metastases from a distant organ. The most common primary malignancies that metastasize to the brain are, in descending order of frequency: lung, breast, colon, kidney, and skin (melanoma). Unlike diffusely infiltrating gliomas, these tumors tend to form discrete nodules, often at the junction of gray and white matter in cerebral cortex where arteriolar diameters are smallest.

Primary lymphomas of the CNS were once quite rare but have become more prevalent with the emergence of HIV-AIDS and an increasing numbers of individuals taking immune-modulating therapies for organ transplants. Diffuse large B cell lymphoma represents the vast majority of CNS lymphomas and expresses Epstein−Barr virus−associated genes and proteins, hinting at the oncogenic properties of this virus. Like malignant gliomas, lymphomas can infiltrate the CNS and have imaging characteristics that are similar to high-grade gliomas, including central necrosis and enhancement with contrast agents, making them difficult to differentiate radiographically.

## Disorders of myelin

Myelin is composed of concentric layers of a complex proteolipid membrane that allows rapid and efficient conduction of action potentials along the axons of neurons. Loss or dysfunction of the normal myelin sheath causes abnormalities of this normal electrical signaling. There are two broad categories that describe ways myelin can be damaged: dysmyelination and demyelination. Dysmyelination is the loss of abnormally formed myelin or loss of the oligodendrocytes that produce and maintain myelin and is due to biochemically deranged myelin processing. Demyelination is the loss of structurally and biochemically normal myelin either by immune-mediated attack against myelin or oligodendrocytes or due to metabolic derangement of oligodendrocytes.

### Leukodystrophies

Leukodystrophies are genetic disorders that cause damage to white matter (leuko—Greek: white). The leukodystrophies are autosomal recessive (AR) or X-linked disorders that result from loss-of-function mutations in enzymes involved in either production of myelin or normal myelin turnover and degradation (Fig. 29.37). These enzymatic defects are not necessarily specific only to the CNS, and other organ systems may be affected. Clinically, these diseases typically present in early childhood with progressive loss of motor control, cognitive function, seizures, and eventual death.

Metachromatic leukodystrophy (MLD) is the most common leukodystrophy, is AR, and is caused by deficiency of the arylsulfatase A enzyme, which breaks down galactosyl-3-sulfatide to galactocerebroside in lysosomes. Galactosyl-3-sulfatide accumulates in many tissues, but the symptoms are related to destruction of myelin. MLD is so named because the accumulated material will change the color (metachromasia) of acidified cresyl violet stain from violet to brown in tissue. Krabbe globoid cell leukodystrophy is another AR leukodystrophy caused by an enzyme deficiency in the sulfatide breakdown pathway. Here, there is a defect in the enzyme galactocerebroside β-galactocerebrosidase, which breaks down galactocerebroside to galactose and ceramide.

Adrenoleukodystrophy (ALD) affects both the CNS myelin and adrenal glands. ALD is usually X-linked and is unusual among the leukodystrophies in that it is due to a defect in a peroxisomal transporter protein. ALD is most often caused by a defect in ABCD1, a member of the ATP-binding cassette transporter family. ABCD1 encodes a transmembrane protein, which is half of a heterodimeric transporter that transfers the enzyme peroxisomal acyl coenzyme A into peroxisomes, where it β-oxidizes long-chain fatty acids. Deficiencies in this enzyme cause a buildup of very long-chain fatty acids, especially hexacosanoic (C26:0) and tetracosanoic (C24:0) acid. This form of ALD is X-linked. A more severe, autosomal recessive form of the disease is due to absent or reduced peroxisome receptor-1, which causes more widespread peroxisomal dysfunction.

Numerous other leukodystrophies exist, and the molecular defects are known in some. Pelizaeus−Merzbacher leukodystrophy is X-linked and caused by abnormalities in a highly abundant myelin structural protein, proteolipid protein

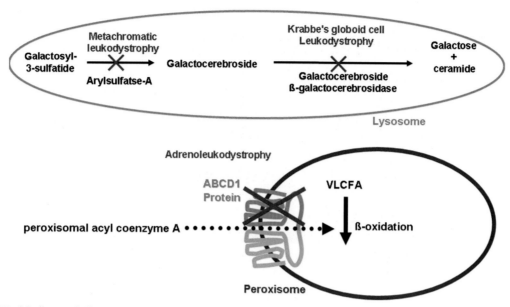

**FIGURE 29.37** Myelin metabolic pathway and loss-of-function mutations associated with dysmyelinating disease. VLCFA, very long-chain fatty acid.

(PLP), which interferes with normal myelin formation. Interestingly, complete loss of the protein causes a relatively mild form of the disease, while missense mutations confer more severe changes, probably because the abnormal protein product interferes conformationally with the compact layering of normal myelin. Alexander disease is a leukodystrophy in which the pathology is primary to astrocytes, not oligodendrocytes. It is caused by a dominant gain of toxic function mutation in the GFAP gene in which the protein forms abnormal aggregates. Large amounts of GFAP and other proteins accumulate in astrocyte processes. The mechanism of damage to oligodendrocytes or to myelin is not known.

## Demyelination

### Idiopathic demyelinating disease

MS describes a variety of related disorders characterized by relapsing and remitting multifocal immune-mediated damage to oligodendrocytes and myelin that are separated by time and anatomic location. These diseases are further clinically subclassified by the rate of progression of disability and the presence or absence of partial recovery between attacks. Histologically, there is loss of myelin with relative preservation of axons (Fig. 29.38). Although the patterns and mechanisms of injury in MS are increasingly well characterized, the ultimate cause(s) of MS is unknown. The incidence of the disease is characterized by a markedly uneven geographic distribution generally with higher latitudes having higher risk, perhaps suggesting a role for environmental factors. Genetic factors also clearly have a role, as the incidence of disease is elevated in family members of an affected individual, with highest risk in monozygotic twins, which reaches a probability of almost 1/3. There is a prevailing multistep hypothesis of MS (Fig. 29.39). Inflammatory cells outside the CNS are activated and upregulate adhesion molecules that guide them to the BBB. In this activated state, they are also more reactive to local chemokines and secrete matrix proteases to gain entry into the CNS, where they mediate damage. The disease mechanism of MS is generally understood to be T-cell—mediated autoimmune processes, which requires a yet-to-be-characterized trigger. T-cells outside the CNS are activated via the T-cell receptor (TCR) binding with antigen presented on the MHC molecule and additional signals by antigen-presenting cells. Additional signaling pathways also play a role in immune activation. One of these pathways that has been characterized involves the B7 signaling molecule on antigen-presenting cells. B7 interacts with signaling molecules on different T-cell subsets (most notably CD28 and CTLA-4) that modify the TCR response, which can either cause activation and development of T-cell function, or unresponsiveness and apoptosis. Antigen-presenting cells also release chemokines, such as interleukin (IL)-12 and IL-23, which generate CD4-positive T helper type 1 cells (Th1). Th1 cells are proinflammatory, interferon $\gamma$-producing cells that are part of the normal anti-viral response. These and other proinflammatory T-cell subsets have been implicated in demyelinating disease.

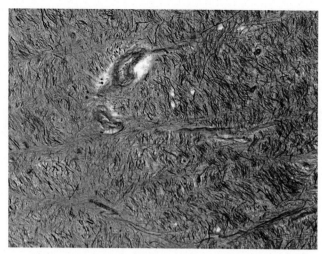

**FIGURE 29.38** Photomicrograph of old demyelinating lesion in multiple sclerosis. Sections of cerebral white matter from the edge of demyelinating lesion in multiple sclerosis. A Luxol fast blue (LFB) demonstrates loss of myelin (blue) in the left half of the figure, whereas a Holmes silver stain shows preserved axons in black.

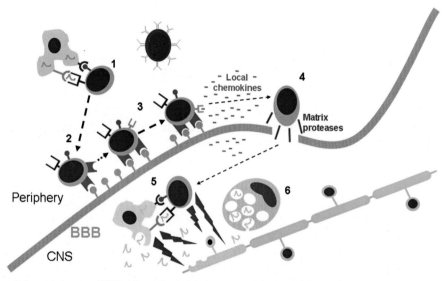

**FIGURE 29.39** Central nervous system (CNS)—directed autoimmunity in demyelinating disease. T-cells are activated in the periphery by antigen-presenting cells through the interaction of the T-cell receptor (TCR) and antigen bound to major histocompatibility complex (MHC) class II along with coregulatory signals (1). Activated T-cells express molecules and surface receptors for adhesion molecules on the luminal surface of the endothelium of the blood—brain barrier (BBB) (2). Activated T-cells also respond to local chemokines (3). The activated T-cells, under the influence of local chemokines, release matrix proteases that allow them to traverse the BBB (4). In the CNS, T-cells interact with microglia-derived or CNS infiltrating antigen-presenting cells through the interaction of the TCR and antigen-MHC and costimulatory pathways, thus mediating damage to myelin and oligodendrocytes (5). Macrophages remove debris (6). *Adapted from Bar-Or A. The immunology of multiple sclerosis. Semin Neurol 2008;28:29—45.*

Neuromyelitis optica (NMO), also known as Devic disease, is a variant of MS for which an etiology has been elucidated. Clinically, NMO is characterized by generally monophasic demyelination of the optic nerve, brainstem, and spinal cord without demyelination of the intervening cerebrum. Pathologically, the disease is characterized by a circulating autoantibody called the NMO-immunoglobulin (Ig)G. The NMO-IgG is directed against aquaporin 4, a water transporter in the foot processes of astrocytes where they abut CNS microvessels. The mechanism by which oligodendrocytes and glia are damaged is not fully understood.

## Animal model of demyelination

Experimental allergic (autoimmune) encephalomyelitis (EAE) is an animal model that was developed to investigate immune-mediated demyelinating disease and, although developed in nonhuman primates and guinea pigs, is now used in inbred mice. Animals are sensitized against brain antigens, myelin basic protein or PLP, which are usually not available to immune surveillance. Commonly, myelin basic protein and PLP are used. Adjuvants designed to open the BBB are infused to cause an acute, monophasic T-cell–mediated attack on myelin and demyelination. In its acute, monophasic course, EAE resembles acute disseminated encephalomyelitis (ADEM). ADEM is a human disease characterized by acute, monophasic demyelination without a chronic phase. It is usually a post-viral complication but has been known to follow vaccination, bacterial or parasitic infections, or without known history. Chronic demyelinating diseases such as MS can be modeled in EAE by concomitant use of immune-modulating agents. Table 29.5 shows the relationship between human demyelinating disease and EAE models.

## Acquired metabolic demyelination

There are two primary osmotic demyelination syndromes. Central pontine myelinolysis (CPM) is a dire complication associated with rapid correction of systemic hyponatremia. Rapid replacement of sodium intravenously in persons with low serum sodium concentrations causes an acute loss of myelin that is usually confined to the middle of the base of the pons. The disease usually occurs in the setting of chronic alcoholism, liver or renal disease, post-transplant, or malnourishment. Although the disease mechanism is not well understood, hypotheses include rapid shifts of water in and out of glia, glial dehydration, myelin degradation, or apoptosis of oligodendrocytes.

Marchiafava–Bignami disease is a similar disease entity in which the white matter of the corpus callosum, the major tract that connects the left and right cerebral hemispheres, undergoes a similar demyelination to that seen in CPM, albeit with more necrosis. Initially described in poorly nourished Italian men who consumed large quantities of crude red wine, it is now known to be associated with alcoholism and to have a worldwide distribution.

**TABLE 29.5** Relationship between human demyelinating disease and experimental allergic encephalomyelitis (EAE).

| Time course | Human | Experimental |
|---|---|---|
| Acute | Acute disseminated encephalomyelitis | EAE |
| | Neuromyelitis optica | |
| Chronic | Progressive multiple sclerosis | Treated EAE |
| | Relapsing/remitting multiple sclerosis | |

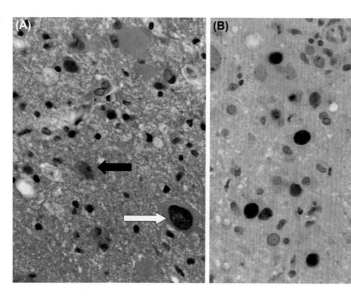

FIGURE 29.40 Photomicrograph of progressive multifocal leukoencephalopathy (PML). Sections of cerebral white matter from a demyelinating lesion of PML. (A) Hematoxylin and eosin (H&E) stains demonstrate enlarged oligodendrocyte nuclei with homogenous ground glass viral inclusions (black arrow) and bizarre enlarged astrocytes (white arrow) simulating a neoplasm. (B) Immunoperoxidase staining for antibodies against Simian virus 40, which cross-react with JC virus nuclear inclusions in oligodendroglia (brown).

## Infectious demyelination

Progressive multifocal leukoencephalopathy (PML) is an acute, progressive demyelinating disorder caused by a polyomavirus. JC virus is a small (50 nm), double-stranded, icosahedral DNA virus lacking a membranous envelope. Approximately 40% of individuals in the developed world are seropositive for the virus but are asymptomatic. Recrudescence of the JC virus causes disease in immunocompromised individuals, such as in AIDS, of which it is a defining illness. JC virus preferentially infects oligodendrocytes in the CNS. Histologically, PML is characterized by demyelination, lymphocytic inflammation, oligodendrocytes with enlarged nuclei with a ground glass appearance, and bizarre reactive astrocytes (Fig. 29.40). Electron microscopy shows oligodendrocyte nuclei packed with viral particles. Of note, the JC virus is oncogenic in rodents producing astrocytomas.

## Key concepts

- The central nervous system (CNS) is organized into a complex structure that is unlike other organ systems. Its cellular components can be divided into neuroepithelial (neurons and glia) and mesenchymal (blood vessels and microglia) elements. CNS parenchyma comprises two structurally and functionally unique components: gray matter, and white matter.

- Neurodevelopmental disorders can manifest from the embryonic period through adolescence and young adulthood and range from epilepsy, mental retardation, cerebral palsy, breathing disorders, ataxia, autism, and schizophrenia. The mildest neurodevelopmental disorders can be asymptomatic, whereas the most severe malformations lead to intrauterine or neonatal demise.

- Adult CNS has limited regenerative ability. CNS cells vulnerable to injury from factors such as exocytosis, mitochondrial dysfunction, or free radical stress become necrotic, apoptotic, or autophagocytic and their debris removed. The residual neuroepithelial cells, usually astrocytes and microglia, respond by gliosis.

- Neurodegenerative diseases range from very common illnesses such as Alzheimer disease to rare illnesses such as prion diseases. Function is localized within the CNS; therefore the clinical presentation of neurodegenerative illnesses is dictated by the regions of brain affected. A hallmark of several of these diseases is the accumulation of aggregates of misfolded protein.

- Primary CNS neoplasms arise from cells derived from normal CNS tissues, in contrast to metastatic neoplasms. The most common type are gliomas. Unlike other organ systems that use the TNM (local Tumor growth; regional lymph Node spread; and distant Metastasis) staging system for establishing prognosis of tumors, primary CNS neoplasms are assigned a histologic grade that correlates with their predicted behaviors.

- Myelin damage may arise from the loss of abnormally formed myelin or of the oligodendrocytes that produce and maintain myelin (dysmyelination), as well as the loss of structurally and biochemically normal myelin (demyelination), either by immune-mediated attack against myelin or oligodendrocytes or due to metabolic derangement of oligodendrocytes.

## Suggested readings

[1] Lopez-Munoz F, Boya J, Alamo C. Neuron theory, the cornerstone of neuroscience, on the centenary of the Nobel Prize award to Santiago Ramon y Cajal. Brain Res Bull 2006;70:391–405.

[2] Aloisi F. Immune function of microglia. Glia 2001;36:165–79.

[3] Baumann N, Pham-Dinh D. Biology of oligodendrocyte and myelin in the mammalian central nervous system. Physiol Rev 2001;81:871–927.

[4] Golden JA, Harding BN, editors. Developemental neuropathology: pathology and genetics. Basel: ISN Neuropath Press; 2004.

[5] Padmanabhan R. Etiology, pathogenesis and prevention of neural tube defects. Congenit Anom (Kyoto) 2006;46:55–67.

[6] Sadler TW. Embryology of neural tube development. Am J Med Genet C Semin Med Genet 2005;135C:2–8.

[7] Hoshino M. Molecular machinery governing GABAergic neuron specification in the cerebellum. Cerebellum 2006;5:193–8.

[8] Langston JW. The etiology of Parkinson's disease with emphasis on the MPTP story. Neurology 1996;47:S153–60.

[9] Prusiner SB. Shattuck lecture–neurodegenerative diseases and prions. N Engl J Med 2001;344:1516–26.

[10] Louis DN, Perry A, Reifenberger G, et al. The 2016 World Health Organization classification of tumors of the central nervous system: a summary et al Acta Neuropathol 2016;131:803–20.

[11] van den Bent MJ, Kros JM. Predictive and prognostic markers in neuro-oncology. J Neuropathol Exp Neurol 2007;66:1074–81.

[12] Furnari FB, Fenton T, Bachoo RM, et al. Malignant astrocytic glioma: genetics, biology, and paths to treatment. Genes Dev 2007;21:2683–710.

[13] Fuller CE, Perry A. Molecular diagnostics in central nervous system tumors. Adv Anat Pathol 2005;12:180–94.

[14] Bansal K, Liang ML, Rutka JT. Molecular biology of human gliomas. Technol Cancer Res Treat 2006;5:185–94.

# Chapter 30

# Molecular diagnosis of human disease

**Eli S. Williams and Lawrence M. Silverman**

*Department of Pathology, University of Virginia Health System, Charlottesville, VA, United States*

**Summary**

Molecular diagnostics are increasingly used to guide patient management, from diagnosis to treatment, particularly in the fields of cancer, infectious disease, and congenital abnormalities. The increased demand for genetic and genomic information has led to the rapid expansion of molecular techniques within clinical laboratories. Critical to the success of clinical genomics is the maintenance of good laboratory practices and regulatory adherence, which can be challenging in the face of rapid growth, emerging technologies, and an evolving regulatory landscape. This chapter addresses the regulations put forth by CLIA and other regulatory agencies that have shaped the implementation of molecular diagnostics. We also provide pre- and postanalytic considerations for common applications of molecular techniques. As molecular diagnostics continues to expand, we discuss its utility in emerging areas including applications of cell-free DNA, gene expression profiling, and pharmacogenomics.

## Introduction

Molecular techniques are now commonplace in most clinical laboratories for (1) disease diagnosis, (2) optimizing therapeutic choices, and (3) monitoring response. Frequently, molecular analysis (particularly in oncology) leads to individualized treatment, referred to as personalized medicine. Concurrently, changes in the regulatory environment have been proposed for *laboratory developed tests* (LDTs) to maintain good laboratory practices without restraining growth of this exciting new area. In this chapter, we will provide an overview of molecular diagnostics and describe the impact of regulatory agencies, professional organizations, and ad hoc committees on the analytical and clinical applications of LDTs as they are applied in infectious diseases, oncology, and diagnosing and treating heritable disorders. We provide specific examples from these clinical areas to illustrate the pre-analytical, analytical, and post-analytical considerations that comprise good laboratory practice. These examples also serve to present new concepts or mechanisms regarding disease pathogenesis and/or treatment. In particular, the application of molecular diagnostics in (1) clinical diagnosis, (2) population screening, (3) selected screening, and (4) tests associated with disease prediction and risk assessment. Finally, future growth areas in molecular diagnostics are discussed, including multivariate analyzes, personalized medicine, and miniaturization.

## Regulatory agencies and CLIA

The major driving force in clinical molecular diagnostics is the original *Clinical Laboratory Improvement Amendments* (CLIA). Since the original legislation (1988) and publication of the final rule (1992), only CLIA-certified laboratories are allowed to provide testing of human specimens that will result in clinical decision making (other than for forensic and research purposes). While most molecular testing is well covered by existing CLIA mandates, genetic testing (heritable disorders) represents a challenge, particularly regarding pre-analytical and post-analytical areas.

It was regarding the pre-analytical and post-analytical areas of genetic testing that most public and media concern has focused. As early as 1994, the *Institute of Medicine of the National Academy of Sciences* convened a committee to deal with issues regarding genetic testing. Among the many concerns were issues surrounding the use of genetic testing to predict future outcomes of individuals and their families. It is this unique aspect of genetic testing that differs from most other clinical laboratory tests and demonstrates the need for special attention in pre-analytical and post-analytical areas. In 1995, a second committee was commissioned to deal with these issues. The *Department of Energy* convened a

*Essential Concepts in Molecular Pathology.* DOI: https://doi.org/10.1016/B978-0-12-813257-9.00030-9

*Task Force on Genetic Testing* and emphasized the importance of restricting genetic testing to CLIA-certified clinical laboratories. Finally, in 1998 the *Department of Health and Human Services* Secretary Donna Shalala established the *Secretary's Advisory Committee on Genetic Testing* (SACGT)—later renamed as SACGHS (*Secretary's Advisory Committee on Genetics Health and Society*)—to take a more global view of genetic testing and the regulatory agencies who oversaw various aspects described by the previous committees. Over the past 20 years, this committee has been very active in assessing genetic testing and its wider implications, including genetic discrimination, patents, and the roles of various regulatory agencies regarding genetic testing. However, molecular technology and applications continue to develop in previously unanticipated areas.

Because of the nature of molecular genetic testing, many analytes were developed by laboratories for use only in their laboratory. These LDTs presented unusual challenges to traditional regulatory agencies, such as the *Food and Drug Administration* (FDA). To this day, LDTs are the lifeblood of molecular diagnostics, and they differentiate this area from most other areas of the clinical laboratory, which depend more on kit and instrument manufacturers. However, regulatory oversight for LDTs is more stringent, by necessity, than those having FDA approval. Thus, we enter the universe of ASR (analyte-specific reagents manufactured for LDTs), RUO (research use only kits), and IVDMIA (in vitro diagnostic multivariate index assays). Each of these designations imparts specific FDA guidelines and additional regulatory oversight. Typically, a clinical molecular laboratory uses a mixture of FDA-approved kits and LDTs, which include many ASRs. Occasionally, an RUO kit may be in use for research projects, and, perhaps, in the future labs may incorporate IVDMIAs, although currently these are primarily found in commercial laboratories.

In 2015, the FDA announced a draft guidance on LDTs, which would dramatically increase the cost and complexity of developing this lifeblood of laboratory medicine, and specifically clinical genomics. Various professional organizations have rendered assessments regarding the effects of such regulations, such as the following from the *American College of Medical Genetics and Genomics* (ACMG). "...*Genetic and genomic tests are highly complex tests based on recently acquired and rapidly evolving knowledge; they are not tests that produce individualized results on their own but require expert interpretation informed by medical and family histories to ensure their safe and effective use by providers...*" In addition, the ACMG suggests that changes be made to strengthen CLIA regulations to cover many of the concerns expressed by the FDA.

To maintain CLIA certification, clinical laboratories must undergo inspections, usually every 2 years, by accrediting organizations recognized by the *Centers for Medicare and Medicaid Services* (CMS), such as the *College of American Pathologists* (CAP). A CAP inspection, which is unannounced, uses checklists to monitor laboratory performance and quality, including personnel qualifications, and maintains an ongoing oversight program for molecular testing. However, CAP inspectors must have extensive inspection experience and training to effectively evaluate the variety of molecular tests performed for a range of clinical applications.

## Quality assurance, quality control, and external proficiency testing

An additional regulatory aspect involves quality assurance (QA) programs consisting of both external proficiency testing (PT) and the use of quality control (QC) materials. While QA is a part of all laboratory testing areas, unique problems surround the dearth of appropriate control materials and external PT programs. Molecular testing is confounded by many factors, including (1) the rare nature of many heritable conditions, (2) the multiplicity of mutations associated with a single disorder (for example, cystic fibrosis), and (3) the level of tissue heterogeneity, particularly when dealing with malignant conditions. For molecular infectious disease testing, particular care must be given to ensure that lack of an amplified product signifies no measurable target as opposed to amplification failure. In this latter case, use of internal controls added to the amplification reaction assures the laboratory that a negative result is a true negative. Also, because nucleic acid is found in living and dead organisms, the presence of a positive signal may not represent active infectious agents.

## Method validation

Adding to this complexity are the difficulties associated with validating molecular tests. When validation is applied to LDTs, the FDA (via the ASR rules) states "...*clinical laboratories that develop tests are acting as manufacturers of medical devices and are subject to FDA jurisdiction...*" To underscore this, CLIA requirements for test validation vary by test complexity and whether the test is FDA-approved, FDA-cleared, or laboratory developed. CLIA requires that each lab establishes or verifies the performance specifications of moderate-complexity and high-complexity test systems that are introduced for clinical use. For an FDA-approved test system without any modifications, the manufacturer validates the system. Therefore, the laboratory needs only to verify (confirm) the manufacturer's performance specifications. In contrast, for a modified FDA-approved test or an in-house developed test, the laboratory must establish the performance specifications for the test, including accuracy, precision, reportable range, and reference range. The laboratory must also develop and plan procedures for calibration and control of the test system. In addition, for a modified FDA-approved test or a laboratory-developed test, the laboratory must establish analytic sensitivity and specificity, and any other relevant indicators of test performance. How and to what extent performance specifications should be verified or established is ultimately under the purview of medical laboratory professionals and is overseen by the CLIA certification process, including laboratory inspectors.

## Clinical utility

In 1997, the *Department of Energy Task Force on Genetic Testing* proposed three criteria for the evaluation of genetic tests: (1) analytic validity, (2) clinical validity, and (3) clinical utility. By clinical utility, the report referred to "...*the balance of benefits to risks*..." However, more frequently, especially with new molecular genetic tests, clinical utility may not be readily available when a test is put into clinical use. In fact, it may take years to accumulate sufficient data on clinical utility. For example, when the molecular test for hereditary hemochromatosis was first introduced, there were limited data on penetrance, or on the association of specific phenotype with specific genotype. At this time, it was assumed that penetrance was close to complete because almost all patients demonstrating the phenotype had two disease-causing mutations. When population studies were performed for hereditary hemochromatosis, penetrance was found to be incomplete, with estimates ranging from 3% to 30% of individuals who carried two disease-causing mutations demonstrating the classical phenotype. Now we know that homozygous (two disease-causing mutations) individuals frequently do not demonstrate the phenotype, at least at the time of testing. Clinical utility is based on positive and negative results defining the disease phenotype.

## Molecular laboratory subspecialties

In this section, we describe examples of clinical molecular tests representing (1) heritable disorders, (2) infectious disease, and (3) oncology. In each case, we address (1) method validation, (2) pre-analytical and post-analytical concerns, and (3) clinical applications with the benefits and risks associated with both positive and negative results.

### Heritable disorders

#### Fragile X syndrome

The most common form of heritable intellectual disability is fragile X syndrome (FXS), so-called because of the cytogenetic appearance of increased fragility at a locus on the long arm of the X chromosome, under specific cell culture conditions. This observation is frequently associated with characteristic clinical features, primarily in males, including mild to moderate intellectual disability, a prominent facial appearance with large low-set ears, and enlarged testes in males, postpuberty. Affected females have a less well-defined phenotype, due to the variable effect of the second X chromosome. However, the inherent difficulties of this cytogenetic test led to the development of a molecular test once the disease gene was identified. The first molecular test was based on Southern blot analysis, coupled with specific restriction enzymes, which could identify the characteristic molecular defect.

The *FMR1* gene contains a CGG repeat region in the 5' untranslated region (Fig. 30.1). The number of CGG repeats is variable in the general population. When increased beyond 200 CGG repeats, the gene is hypermethylated, leading to absence of expression of the gene product, FMR1. In the normal population, the number of repeats varies from 5 to 44 repeats, with a median of 29. In individuals with FXS, the number of repeats usually exceeds 200. Pre-mutation carriers have between 55 and 200 repeats, which are associated with (1) increased risk of full expansion during female (but not male) meiosis leading to FXS in her offspring, (2) increased risk of ovarian dysfunction (premature ovarian failure), and (3) increased risk of FXTAS (fragile x-associated tremor/ataxia syndrome). Clearly, an accurate and precise method for determining the number of CGG repeats, and the associated hypermethylation status, has significant clinical implications. However, the Southern blot technique was able to assess only hypermethylation, but not accurately assess the number of CGG repeats (Fig. 30.2).

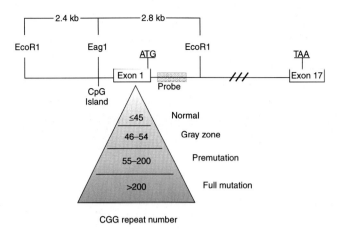

**FIGURE 30.1** Schematic representation of the CGG repeat in exon 1 of *FMR1* and associated alleles. A CGG repeat number less than or equal to 45 is normal. A CGG repeat number of 46–54 is in the *gray zone* and has been reported to expand to a full mutation in some families. A CGG repeat number of 55–200 is considered a pre-mutation allele and is prone to expansion to a full mutation during female meiosis. A CGG repeat number in excess of 200 is considered a full mutation and is diagnostic of fragile X syndrome.

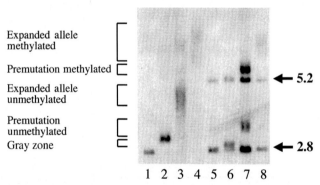

**FIGURE 30.2** Southern blot analysis for the diagnosis of fragile X syndrome. Patient DNA is simultaneously digested with restriction endonucleases *EcoR*1 and *Eag*1, blotted to a nylon membrane, and hybridized with a $^{32}$P-labeled probe adjacent to exon 1 of *FMR1* (see Fig. 30.1). *Eag*1 is a methylation-sensitive restriction endonuclease that will not cleave the recognition sequence if the cytosine in the sequence is methylated. Normal male control DNA with a CGG repeat number of 22 on his single X chromosome (*lane 1*) generates a band about 2.8 kb in length corresponding to *Eag*1-*EcoR*1 fragments (see Fig. 30.1). Normal female control DNA with a CGG repeat number of 20 on one X chromosome and a CGG repeat number of 25 on her second X chromosome (*lane 5*) generates two bands, one at about 2.8 kb and a second at 5.2 kb *EcoR*1–*EcoR*1 fragments approximately 5.2 kb in length represent methylated DNA sequences characteristic of the lyonized chromosome in each cell that is not digested with restriction endonuclease *Eag*1. DNA in *lane 2* contains an *FMR1* CGG repeat number of 90 and is characteristic of a normal transmitting male. The banding pattern observed in *lane 3* is representative of a mosaic male with a single X chromosome with a full mutation (>200 repeats). However, the full mutation in some cells is unmethylated; in other cells, the full mutation is fully methylated, hence the term *mosaic*. In those cells in which the full mutation is unmethylated, digestion by both *Eag*1 and *EcoR*1 occurs, and in those cells in which the full mutation is fully methylated, digestion of the DNA by *Eag*1 is inhibited. The banding pattern observed in *lane 4* is diagnostic of a male with fragile X syndrome, illustrating the typical expanded allele fully methylated in all cells. *Lane 6* is characteristic of a female with a normal allele and a CGG repeat number of 29 and a larger gray zone allele with a CGG repeat number of 54. *Lane 7* is the banding pattern observed from a pre-mutation carrier female with one normal allele having a CGG repeat number of 23 (band at about 2.8 kb) and a second pre-mutation allele with CGG repeats of 120 to about 200 (band at about 3.1 kb). In pre-mutation carrier females, in cells in which the X chromosome with the pre-mutation allele is lyonized, the normal 5.2 kb *EcoR*1–*EcoR*1 band is larger because of the increased CGG repeat number and is about 5.5 kb in length. *Lane 8* is diagnostic of a female with fragile X syndrome with one full expansion mutation allele that is completely methylated and transcriptionally silenced on one X chromosome but with a second normal allele with a CGG repeat number of 33.

## Validation of molecular sizing of FMR1

The Southern blot procedure for FXS is well documented. The focus here is on validation of a sizing assay for the CGG region of *FMR1* performed on a capillary electrophoresis platform (ABI 3100, Applied Biosystems, Foster City, California). The LDT is based on the report by Fu, while an RUO assay is commercially available from Celera Diagnostics (Alameda, California). Inherent to any method validation, well-characterized samples must be available. These can be patient samples analyzed by an independent laboratory, immortalized cell lines (Coriell Cell Repositories), or artificially prepared standards (*National Institute of Standards and Technology*). Each of these materials has advantages and disadvantages. Ideally, all three sources support complete assessment of method validation. Conditions for both the LDT and the RUO are available from the literature.

## PCR and capillary electrophoresis

Sizing of the *FMR1* 5′ UTR is performed on an automated capillary electrophoresis platform following PCR amplification of the critical region. Choice of the forward and reverse primers and PCR amplification details are discussed elsewhere, as are the capillary electrophoresis conditions.

## Statistical analysis

Validation studies should determine reproducibility (imprecision) and accuracy, allowing the laboratory to report values with confidence intervals (45 ± 2 repeats, for example). These estimates of precision can be obtained by multiple determinations of the same sample during a single run of samples and by analyzing the same sample on multiple runs, followed by determining the mean and the standard deviation. These terms are referred to as within-run and between-run precision.

## Post-analytical considerations

Interpretation of *FMR1* sizing analysis depends on the clinical considerations for ordering the test. These may be (1) diagnosis, (2) screening, (3) prediction within a family with affected members, or (4) risk assessment. For example, a diagnostic application would include determining whether an individual with the characteristic phenotype has FXS and depends on the sex of the patient. A male, having a single X chromosome, should have a single band following PCR amplification, with a normal number of CGG repeats (5–44 repeats). One of the limitations of the PCR procedure is the size of the CGG repeat region. Depending on the PCR primers and amplification conditions, the maximum number of CGG repeats that can be amplified is approximately 100 repeats. Thus, a fully amplified CGG region of >200 repeats would not produce a discernable PCR product. However, a female, with two X chromosomes, with a full expansion would still have an unaffected X chromosome, resulting in a single band following PCR amplification, representing the normal-sized allele. Another limitation with females is that both alleles may be the same size, resulting in a single band following amplification. Any analysis that yields a single band following PCR in females, or with the absence of a band in males, would necessitate Southern blot analysis, as this analysis is not limited by the constraints of PCR amplification. Thus, Southern blot analysis should detect fully expanded alleles. Screening applications may involve newborn screening in the future. However, methodological constraints are considerable at this time.

If an affected family member has been identified, carrier determination can be performed, exemplifying a predictive use of this testing. However, another level of complexity involves risk assessment of female pre-mutation carriers. These individuals have repeats between 55 and 200, not large enough for full manifestation of disease, but sufficient for the associated conditions (premature ovarian failure, FXTAS, and others), in addition to increased risk of passing on a fully expanded allele to offspring (the risk is proportional to the number of CGG repeats). Note that individual possessing alleles with 45–54 repeats are considered intermediate and may carry an increased risk for being pre-mutation carriers, but not for passing on full mutations. No data exist on the risk of these individuals for ovarian dysfunction or FXTAS. Obviously, knowing the precision and accuracy of the assay is paramount in interpreting the number of CGG repeats.

## Benefit/risk assessment (clinical utility)

Because defining clinical utility takes considerable clinical experience with newer assays, it is difficult, if not impossible, to assess clinical utility before setting up an LDT, as advocated by some of the regulatory agencies. For example, *FMR1* sizing assays demonstrate the evolving nature of clinical utility. When Southern blot analysis was the only molecular method available, the clinical utility was based on the contribution of the *FMR1* assay to the diagnosis of FXS. However, by this method, limited information was available for assessing pre-mutation carrier status. With the development of *FMR1* sizing assays, more accurate assessment of the pre-mutation allele could be made and the clinical utility for ovarian dysfunction and FXTAS could be assessed. Thus, improvements in methodology can affect the clinical utility of an analyte.

## Infectious diseases

### Hepatitis C viral genotype

Approximately 4 million people in the United States are chronically infected with hepatitis C virus (HCV), and many are unaware that they are infected. Recent studies estimate that, in the United States, 30,000 individuals per year are newly infected with HCV, and approximately 10,000 people will die annually from the sequelae of hepatitis C. HCV is the most frequent indication for liver transplant in adults in the United States. Since the outcome of HCV infection depends on both the viral load and HCV genotype, the determination of HCV genotype provides important clinical information regarding the duration and type of antiviral therapy used for a given HCV-infected patient. In addition, the genotype is an independent predictor of the likelihood of sustained HCV clearance after therapy. There are more than 11 genotypes and more than 50 subtypes of HCV. In the United States, genotypes 1a, 1b, 2a/c, 2b, and 3a are most common. Genotypes 4, 5, 6, and 7 are endemic to Egypt, South Africa, China, and Thailand, respectively, and are rarely found in the United States. Because patients infected with HCV genotype 1 may benefit from a longer course of therapy, and genotypes 2 and 3 are more likely to respond to combination interferon-ribavirin therapy, the clinically relevant distinction among the affected population of the United States is between genotype 1 and non–type 1. Not enough data currently exists to determine the likelihood of therapeutic response for HCV genotypes 4, 5, 6, and 7. Therefore, it has been recommended that infections with genotypes 4, 5, 6, or 7 be treated similar to patients infected with HCV type 1. Line-probe and sequencing-based methods are available for HCV genotyping, but are expensive, time-consuming, and provide more information than is currently needed by clinicians for patient management decisions. With this in mind, we developed a rapid HCV genotyping assay for the LightCycler to differentiate HCV type 1 from non–type 1 infections.

### Validation of hepatitis C virus genotyping

The determination of HCV genotype is based on isolation of HCV viral RNA and cDNA synthesis followed by PCR amplification of the 5′ untranslated region (5′ UTR). Viral RNA is isolated from infected serum after HCV viral load determination using the MagNA Pure Total Nucleic Acids Extraction Kit (Roche Diagnostics) or the Qiagen EZ1 BioRobot. The extracted nucleic acids are subjected to RT-PCR using primers specific for 5′ UTR sequence that is conserved among all known HCV genotypes. The RT-PCR step is performed using an Applied Biosystems GeneAmp 9600 PCR System (block cycler). A second PCR amplification using a seminested pair of primers is accomplished in the

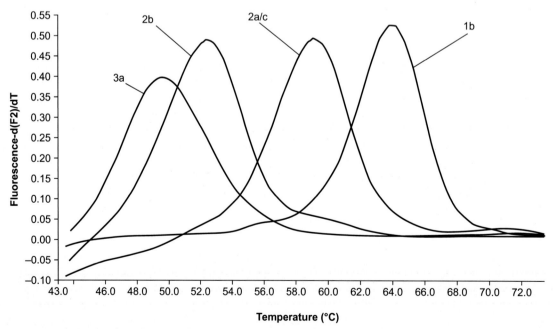

**FIGURE 30.3** HCV genotyping assay. Comparison of samples from patients infected with HCV genotypes 1b, 2a/c, 2b, and 3a. Shown are genotype-specific melting transitions for four samples in 2 mM $MgCl_2$. Data were obtained by monitoring the fluorescence of the LCRed640-labeled FRET sensor probe during heating from 40 to 80 °C at a temperature transition rate of 0.1 °C/s.

LightCycler in the presence of a pair of FRET oligonucleotide probes. The FRET detection probe allows the discrimination of HCV genotypes 1a/b, 2a/c, 2b, 3a, and 4 during melting curve analysis (Fig. 30.3) because it hybridizes with differential affinity to a region of the 5′ UTR that varies among the different HCV genotypes. Magnesium concentration is critical for optimal genotype discrimination. Primer and probe design and PCR details have been described. For method validation, patient samples analyzed by an independent reference laboratory were used. Positive QC materials consisted of pooled patient serum samples of previously determined genotype (genotypes 1, 2a/c, 2b, and 3a). Negative QC material consisted of pooled patient serum samples previously determined to be HCV negative. A water blank is also always included with each run. A serial limiting dilution of positive patient samples with known viral titers was used to determine the limits of detection. The seminested method described will fail when viral titers are less than 12,600 IU per mL. A more sensitive fully nested PCR method was developed and is able to genotype specimens with 130 IU per mL. The fully nested PCR method uses the freshly extracted RNA and the same RT and the first PCR steps used for the seminested approach. The second PCR step uses a new set of nested primers and is done in the block cycler. The products of this reaction are subjected to melting curve analysis in the LightCycler to determine genotype using the same set of FRET probes described for the semi-nested genotyping method. The fully nested approach is rarely needed because patients usually have high viral loads on initial presentation, and this is the ideal time to assess the HCV genotype.

## Statistical analysis

In addition to determination of the accuracy of the genotyping assay, melting point precision was determined within run and between runs using QC materials. An analysis of assay performance after 18 months showed no assay failures across 70 runs with 411 patient samples using three different LightCycler instruments operated by six different technicians. The genotype-specific melting temperatures remained nonoverlapping with coefficients of variation that were less than 1% for all genotypes. Melting temperatures are continuously monitored to assure assay performance.

## Pre-analytical considerations

The seminested and nested RT-PCR methods are especially susceptible to contamination by RNases, previously amplified DNA, and unrelated nucleic acids. Unidirectional flow of samples through physically separated pre-amplification and post-amplification areas of the laboratory should be maintained to prevent contamination of reagents or specimens with amplified DNA. Equipment, reagents, plasticware, and specimens should be handled with gloves to prevent contamination with fingertip RNases or cross-contamination of specimens. The use of plastic microfuge tube cap openers instead of gloved fingertips is recommended to prevent cross-contamination. Samples should be added one at a time, followed by positive QC, negative QC, and finally the water blank (in that order). The only known interference to this procedure is heparin, which may inhibit the RT-PCR reaction. Evaluation of isolation of nucleic acid by the MagNA Pure system has shown that heparin is removed during the isolation of nucleic acid by this method. A minimum of 500 μL of serum is the required sample for this method. Whole blood samples should be centrifuged, and serum should be separated within 6 h of collection. HCV is an enveloped RNA virus and is labile to repeated freeze-thaw cycles. Viral titers will drop upon storage at −20 °C. Therefore, fresh samples are preferred. Because aqueous solutions of RNA are unstable and susceptible to spontaneous hydrolysis even at −20 °C, it is best to perform the RT-PCR step on the same day as RNA extraction. MagNA Pure-extracted nucleic acid samples should be stored at −80 °C if the RT-PCR step cannot be performed on the day of isolation.

## Post-analytical considerations

Interpretation of HCV genotype is reported as type 1, type 2, type 3, type 4, or none detected if negative. Both early viral load measurements and viral genotype are used to determine likelihood of clinical response and influence management decisions during combination antiviral therapy. Therefore, these tests should be ordered together during the initial workup of a patient with viral hepatitis. Reflexive genotyping of patients who are positive for HCV with sufficiently high viral loads and no prior genotype should be considered. Once the genotype is determined, there is no need for additional genotyping unless there is clinical suspicion of a second infection with a different HCV genotype. The HCV LC-PCR assay is able to detect two genotypes in one patient sample. One limitation with this assay is the inability to detect some rare genotypes that may occur in the United States, such as genotypes 5 or 6. Thus, the detection of a novel melting temperature could indicate an infection with a non−type 1, non−type 2, non−type 3, or non−type 4 HCV genotype. This specimen would be sent out for sequencing by a reference laboratory. Similarly, a specimen with a high viral load but no detectable genotype indicates a technical failure or a novel genotype and would also be sent to a reference laboratory for sequencing.

### Benefit/risk assessment (clinical utility)

HCV genotyping assays are an essential component of an evolving clinical decision tree used to determine the duration of therapy and the likelihood of sustained virological response (SVR) to this therapy. Absolute viral load, log decline in viral load from baseline, and viral genotype are clinically useful in predicting SVR and non−sustained virological response (NSVR) to treatment. The *NIH Consensus Development Conference Statement, Management of Hepatitis C: 2002*, stresses the importance of early prediction of SVR and NSVR in patient management in the first 12 weeks of therapy. Combination interferon plus ribavirin therapy is given for 24 or 48 weeks, depending on viral genotype and viral load. The therapy is expensive and difficult for patients to tolerate due to side effects. HCV genotype and viral loads are used to predict the likelihood of SVR or NSVR. The likelihood of SVR and NSVR is useful for clinicians and patients when making decisions about the duration of therapy. Non−type 1 genotypes have a much greater rate of SVR than type 1 genotype−infected patients. HCV type 1 with a high viral load requires a longer course of therapy than non−type 1 infections. HCV genotype determination is clinically useful in risk assessment and making therapeutic decisions instead of making a diagnosis.

## Oncology

### B-cell and T-cell clonality

B-cell and T-cell clonality assays are based on the detection of a predominant antigen receptor gene rearrangement that represents the outgrowth of a predominant lymphocyte clone. In the correct clinical and pathologic context, the presence of a predominant clonal lymphocyte population is consistent with lymphoma or lymphoid leukemia. Southern blots were traditionally the method used to detect a predominant antigen receptor gene rearrangement. Multiplex PCR-based B-cell and T-cell clonality assays are replacing Southern blot−based assays because they are robust, faster to perform, and easier to interpret. Unlike Southern blots, PCR-based assays are amenable to formalin-fixed, paraffin-embedded (FFPE) tissues. PCR-based assays amplify the DNA between primers that target the conserved framework (FR) and joining (J) sequence regions within the lymphoid antigen receptor genes. These conserved regions flank the V−J gene segments that randomly rearrange during the maturation of B lymphocytes and T lymphocytes.

Random genetic rearrangement of the V−J segments occurs in the immunoglobulin heavy chain gene (*IGH*) and the kappa and lambda immunoglobulin light chain genes in prefollicular center B-cells. Random genetic rearrangement of the V−J segments also occurs in the alpha (*TCRA*), beta (*TCRB*), gamma (*TCRG*), and delta (*TCRD*) T-cell receptor genes during thymic T-cell development. Antigen receptor gene rearrangement occurs in a sequential fashion one allele at a time. If the first allele fails to successfully recombine (produces a nonfunctional protein product), then the other allele will undergo rearrangement. It is important to remember that the nonfunctional rearrangement can still be detected by the clonality assay. Because of the sequential manner in which these genes rearrange, a mature lymphocyte may carry one (monoallelic clone) or two (biallelic clone) rearranged antigen receptor genes of a given type. Each maturing B-cell or T-cell has either one or two gene-specific V−J rearrangements unique in both sequence and length. Therefore, when the V−J region is amplified, one or two unique amplicons will be produced per cell. If the population of lymphocytes being examined is a lymphoma, then one or two unique amplicons will dominate the population. In contrast, if the population is a normal polyclonal lymphoid infiltrate, then there will be a Gaussian distribution of many different-sized amplicons centered on a statistically favored, average-sized rearrangement (Fig. 30.4).

Because the antigen receptors are polymorphic and subject to mutagenesis as part of the normal rearrangement process, multiple primer sets are used to increase the ability of the assay to detect the majority of possible V−J rearrangements. Each set of multiplex primers has a defined valid amplicon size range. After antigen exposure, germinal center B-cells undergo somatic hypermutation of the V regions, and this may affect the binding of the V-region specific (FR) primers and prevent the detection of lymphoma clones arising from the germinal center or post-germinal center B-cell compartments. Primer sets that target *IGH* D−J rearrangements are less susceptible to somatic hypermutation and allow the detection of more terminally differentiated B-cell malignancies.

### Validation of B-cell and T-cell clonality assays

PCR-based clonality assays are routinely used by many laboratories. However, prior to 2004, referral lab PCR−based clonality assays used non-standardized, laboratory-specific PCR primer sets and methods, making proficiency testing

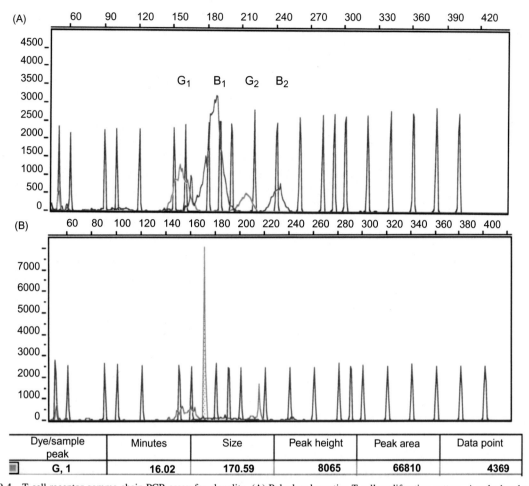

| Dye/sample peak | Minutes | Size | Peak height | Peak area | Data point |
|---|---|---|---|---|---|
| G, 1 | 16.02 | 170.59 | 8065 | 66810 | 4369 |

FIGURE 30.4  T-cell receptor gamma chain PCR assay for clonality. (A) Polyclonal reactive T-cell proliferation pattern. A polyclonal population of T-cells with randomly rearranged T-cell receptor gamma chain genes produces a normal or Gaussian distribution of fluorescently labeled PCR products from each primer pair in the multiplex reaction. This produces four bell-shaped curves that represent the valid size range for an individual primer pair. Two of the valid size ranges are green and two are blue; $G_1$, $B_1$, $G_2$, $B_2$. The red peaks represent size standards. (B) Clonal T-cell proliferation pattern. A clonal T-cell proliferation results in a relative dominance of a single T-cell receptor gamma chain gene producing a predominant spike of a discrete size on the corresponding electropherogram. Data were obtained using an ABI 3100 capillary electrophoresis system and ABI prism software.

and lab performance comparisons difficult. In addition, many of these institution-specific B-cell and T-cell clonality assays are not comprehensive and targeted a limited number of possible V−J rearrangements. To address this concern, we used the optimized InVivoScribe (IVS) multiplex PCR primer master mixes. The IVS master mixes have been standardized and extensively validated by testing more than 400 clinical samples using the *World Health Organization* (WHO) classification scheme. Validation was performed at >30 prominent independent testing centers throughout Europe in a collaborative study known as the BIOMED-2 Concerted Action.

The *TCRG* locus is on chromosome 7 (7q14) and includes 14 V gene segments divided into four subgroups. Six of the 14 V segments are functional, three are open reading frames, and five are pseudogenes. The 200 kilobase pair *TCRG* region also contains five J segments and two C genes. Although most mature T-cells express the alpha/beta TCR, the gamma/delta TCR genes rearrange first and are carried by almost all alpha/beta T-cells. The IVS *TCRG* assay uses only two master mixes and is able to detect 89% of all T-cell lymphomas tested in the BIOMED-2 Concerted Action. Master mix tube A targets the V gamma 1−8+10 and most J gamma gene segments. Master mix tube B targets V gamma 9+11 and most J gamma segments. The included positive and negative kit controls are used to assess each of the master mixes. The specimen control size standard is a separate master mix included with the kit that evaluates each patient DNA sample by targeting six different housekeeping genes potentially producing amplicons of 84, 96, 200, 300, 400, and 600 bp. Under ideal conditions and using capillary electrophoresis to analyze the amplicons, the *TCRG* assay can resolve a 1% clonal population in a polyclonal background.

The *IGH* locus is on chromosome 14 (14q32.3) and includes more than 66 V, 27 D, and 6 J gene segments. The IVS *IGH* assay uses five master mixes and is able to detect 92% of all B-cell lymphomas tested in the BIOMED-2 Concerted Action validation study. Master mixes A, B, and C target the framework (FR) 1, 2, and 3 regions of the V gene segments, respectively, and all J gene segments. Master mixes D and E target the D and J regions. The *IGH* locus rearranges the D and J segments first during maturation, followed by the addition of a random V segment to the previously rearranged D–J fusion product. In addition, the D segments are less likely to be mutated by somatic mutation during postgerminal center development. Because the primers in tubes D and E target D–J regions, these tubes are more likely to detect clonal *IGH* rearrangements in lymphoma cells that arose either very early or late in B-cell ontogeny. Positive, negative, and specimen extraction controls are included and run on each master mix and specimen, respectively. The specimen control size standard is used to assess the integrity of the DNA extracted from the specimen and is identical to the specimen control size standard described previously for the *TCRG* kit. The specimen control master mix is useful in determining the amount of DNA target molecule degradation that has occurred during fixation and DNA extraction. Capillary electrophoresis (CE) using the ABI 3100 Avant Genetic Analyzer was better at determining the size of predominant amplicons than both agarose and polyacrylamide gel systems during method development. Under ideal conditions and using CE to analyze the amplicons, the IGH assay can resolve a 1% clonal population in a polyclonal background.

Genomic DNA is isolated from whole blood, body fluids, fresh tissues, and formalin-fixed paraffin-embedded tissue blocks, using the appropriate Qiagen DNA isolation protocol. A slightly modified Qiagen DNA isolation protocol for FFPE samples was selected after a comparison to three other DNA isolation protocols during validation. After DNA isolation, multiple dilutions are tested for beta-globin amplification by PCR to assess DNA quality prior to analysis using the IVS system. Samples that fail to amplify beta-globin at any concentration are quantified by UV spectrophotometry. Approximately 100–200 ng of input DNA is ideal for *TCRG* or *IGH* PCR reactions. Samples that are undetectable by UV spectrophotometry and fail beta-globin PCR should be reported as quantity insufficient and retested by increasing the amount of starting material by two- to threefold, if possible. Samples that fail beta-globin PCR but contain sufficient quantities of DNA are most likely not amplifiable due to inhibitors and should be reported as non-amplifiable due to PCR inhibitors in the sample. Additional dilutions could be retested by beta-globin PCR if desired. The least dilute DNA that gives a strong beta-globin signal should be used for the InVivoScribe B-cell or T-cell clonality assay. Each of the products of the IVS PCR is combined with 0.5 mL ROX400HD size standards, denatured at 95 °C in formamide and electrophoresed through a capillary that contains the POP4 matrix. The amplified fragments are detected by the 6FAM, HEX, or NED fluorescent labels on the conserved, downstream *IGH* or *TCRG* J-region PCR primers. The size standards are detected by the ROX label. After assay validity is confirmed by checking all controls, the results are analyzed and a report is generated.

After the method was established, a variety of clinically and histopathologically diagnostic cases of B-cell and T-cell lymphomas were used to evaluate the performance of the method. This laboratory-modified method was able to detect 92% of B-cell lymphomas and 85% of T-cell lymphomas used in the validation set. Ideal QC control material is not available for this assay. The ideal QC material would consist of a high and low positive FFPE QC and a negative FFPE QC. FFPE QC material would control for the DNA extraction in addition to the PCR and interpretation phases of this method. The currently available QC and PT control materials do not assess the DNA extraction phase of this test.

## Pre-analytical considerations

Special considerations regarding specimen type and handling are important for this method. One of the most common specimens submitted for *IGH* or *TCRG* gene rearrangements is FFPE tissue blocks. PCR inhibitors are commonly found with this specimen type. Xylene, heme, and excessive divalent cation chelators are also inhibitory and need to be removed during DNA extraction procedures. The fixative used prior to paraffin embedding is critical for optimum results because formalin of inferior quality can introduce PCR inhibitors or prevent the extraction of amplifiable DNA. Fresh, high-quality, 10% neutral-buffered formalin is acceptable, and this should be communicated to the pathologists who will be submitting specimens for gene rearrangement studies. Fixatives that contain heavy metals such as Hg (Zenkers and B5) should not be used for PCR. Decalcified bone marrow biopsy specimens are unacceptable as are specimens that were fixed with strong acid–based fixatives such as Bouin's solution. Tissues should be fixed at room temperature using proper histologic standards. Fixation depends on formaldehyde-mediated cross-linking of proteins and nucleic acids. Excessive cross-linking will inhibit DNA extraction and shorten the average size of the recovered DNA molecules. Therefore, over-fixation should be avoided. Fixation using the approved fixatives at room temperature progresses at a rate of approximately 1 mm/h. Typical 0.5 mm thick tissue sections should fix for a minimum of 6 h and a maximum of 48 h. After 48 h in fixative, tissues can be stored in 50% ethanol for up to 2 weeks prior to processing and embedding. The volume of fixative used should be at least 10 times the volume of tissue to achieve optimal results. Some laboratories use razor blades to gouge tissue out of the block, but this destroys the block and prevents any additional studies that may be necessary in the future. The best way to obtain FFPE samples for PCR is to cut 4-micron thick sections on the microtome. Histotechnologists should cut the tissue sections for PCR first thing in the morning with a fresh microtome blade and new water. Several initial sections are discarded to remove any possible contaminating DNA on the surface of the block. One to five tissue sections are obtained, depending on the size and type of tissue sample and the relative proportion of the tissue that is involved by the suspicious lymphoid infiltrate. PCR of lymphoid infiltrates in FFPE skin or brain tissue blocks are somewhat refractory to PCR analysis and require more sections for analysis. Skin tissues are more likely to be over-fixed and yield highly fragmented DNA. Brain tissues could be problematic due to the higher lipid content than most tissues submitted for lymphocyte gene rearrangement studies.

To snap freeze fresh tissues, one should cut them into 1- to 3-mm$^3$ pieces, place them into an appropriately labeled cryovial, snap freeze them in liquid nitrogen, and then store them frozen at $-70\,°C$ until processed. Specimens snap frozen at a different location should be shipped on dry ice to arrive during normal business hours and be stored at $-70\,°C$ until processed. Whole blood specimens should be at least 200 μL EDTA or citrated blood, stored at $4\,°C$ for up to 3 days before DNA extraction. Bone marrow aspirates should be at least 200 μL EDTA or citrated bone marrow, stored at $4\,°C$ for up to 3 days before DNA extraction.

CSF specimens should be at least 200 μL depending on cellularity. All CSF samples should be collected in an appropriate sterile container and stored at $4\,°C$ for up to 24 h or at $-20\,°C$ for long-term storage. Other sample types, such as stained or unstained tissue on slides, may be accepted with the approval of the laboratory director. Extracted DNA samples should be stored at $-20\,°C$ for a minimum of 2 months after analysis is complete.

### Post-analytical considerations

The results of this test must be interpreted in the context of the other clinicopathologic data for the specimen. The detection of a clonal lymphoid proliferation in a tissue sample does not necessarily indicate a diagnosis of lymphoma or leukemia. On the other hand, the failure to detect a clonal lymphoid proliferation does not necessarily rule out a lymphoma/leukemia. The test is one puzzle piece that must be integrated with other clinicopathologic data before rendering a diagnosis. The molecular B-cell and T-cell clonality assay is an adjunct ordered by pathologists for cases that are suspicious, but not obviously lymphoma/leukemia, and the results of this test do not stand alone.

Although helpful guidelines for the interpretation of results are provided by the manufacturer, this assay is not an out-of-the-box assay. After method development, an important part of validation is the development of interpretive guidelines that are appropriate for the established method, specimen types, and the patient population. An important consideration is how the results will be interpreted by the ordering group of pathologists and oncologists. Specimen handling, DNA extraction procedures, and fragment analysis procedures will affect the final results from this assay, and this will affect the ability to detect a clonal proliferation in a polyclonal background. Our laboratory has established a set of interpretive guidelines that include a required minimum peak height for positive peaks and positive peak to background peak ratio values that are appropriate for our method and reduce the number of false positives. The details of our interpretive guidelines are beyond the scope of this chapter and have been described.

Results are interpreted and reported to the ordering pathologist in an interpretive report format. The results are interpreted as (1) positive for a clonal lymphoid proliferation, (2) suspicious but not diagnostic for a clonal lymphoid proliferation (with a recommendation to repeat the assay if clinical concerns persist), (3) oligoclonal reactive pattern (commonly seen in the peripheral blood of elderly patients with rheumatologic disease or viral infections), (4) polyclonal lymphoid population, or (5) insufficient material/inadequate quality. The reports also contain information on the pertinent clinical and laboratory history, the specimen submitted for PCR, and the relevant surgical pathology case if appropriate. Suspicious results are usually repeated for confirmation, especially *TCRG* studies.

### Benefit/risk assessment (clinical utility)

In many cases of atypical lymphoid proliferations suspicious for lymphoma, the morphologic, immunohistochemical, flow cytometric, or clinical features are equivocal or contradictory. Detection of a predominant antigen receptor gene rearrangement (detection of a T-cell or B-cell clone) can support a diagnosis of lymphoma. Early diagnosis and appropriate treatment of lymphoma can prolong patient survival and in many cases result in a complete remission or cure. Because the morphology of many lymphomas is indistinguishable from benign lymphoid hyperplasia, ancillary tests that demonstrate an atypical clonal lymphocyte proliferation are essential in early diagnosis and subsequent treatment of many lymphoma cases. Multiplex PCR assays for clonality have become the standard of practice in lymphoma diagnosis and may eventually be used on all cases of atypical lymphoid proliferations to either rule in or rule out lymphoma earlier. For those cases that are obviously lymphoma, molecular characterization of the specific lymphoma cell clone may also prove useful in patient follow-up after treatment to screen for residual or recurrent disease. In addition, many of the morphologic subtypes of lymphoma have not been extensively characterized from the molecular diagnostics standpoint. For example, for biopsy cases that are suspicious but not diagnostic for lymphoma, detection of a clonal B-cell or T-cell population predicts lymphoma behavior or patient outcome. Diagnostic tools that allow these types of correlative studies have only recently become available, and this assay represents one of the assays that will allow these kinds of questions to be considered. In the future, B-cell and T-cell clonality assays will become a standard part of a lymphoma workup and may alter clinical outcomes by providing a more comprehensive earlier diagnosis of lymphoma and preventing unnecessary treatment of non-lymphomas.

Another important consideration is the improved performance and diagnostic utility of the newer generation of multiplex PCR assays since 2004. Many descriptions on the utility of PCR-based clonality assays refer to multiplex PCR assays that cannot detect all possible *IGH* or *TCRG* gene rearrangements because of suboptimal primer sets. This can increase the number of false negatives because the PCR primers that could recognize the clonal gene rearrangement are missing. In addition, oligoclonal reactive proliferations may appear to be clonal because a limiting number of primers may not detect all of the members of the oligoclonal proliferation. To address this issue, we have initiated an investigation of the diagnostic utility of the IVS assay at our institution over the past 4 years.

## Future directions

### Liquid biopsies

New applications of next generation sequencing technology have emerged that may improve cancer diagnosis, treatment, and management. The detection of circulating tumor DNA (ctDNA) in a patient's blood (or other bodily fluids) offers oncologists a chance to monitor the tumor without the need for more invasive tissue biopsies. Circulating tumor DNA are short fragments of cell-free DNA, typically near 100 bp in size that arise through the death and degradation of cancer cells. Liquid biopsies are being used to detect clinically actionable mutations found in the ctDNA collected from a tube of blood. Another promising approach of NGS technology on ctDNA is to monitor for either minimal residual disease (MRD) or disease recurrence. For these applications, a tissue biopsy is used to determine the molecular profile of the tumor prior to treatment, and then a liquid biopsy may be used to detect MRD or recurrence of that abnormal clone. These approaches have been applied more readily to leukemias and lymphomas, and it has been demonstrated that the ability to detect the abnormal clone via a liquid biopsy increases the risk of a relapse. It is hoped that liquid biopsies may eventually be able to detect cancer before other symptoms appear. Liquid biopsies provide an exciting new avenue of cancer monitoring, but knowledge related to several areas of the test must be developed before liquid biopsies can become part of standard of care. Cancer genomes are complex and heterogeneous, and it is unknown whether ctDNA offers a complete (or sufficiently complete) genomic picture of the tumor. Some cancer researchers are exploring the use of liquid biopsies to collect circulating tumor cells (CTCs), which are far less abundant than ctDNA, but offer the possibility of an intact tumor genome, as well as RNA for detection of fusion events and gene expression, and possibly other clinically useful biomarkers. Regardless of the approach of liquid biopsies (ctDNA vs. CTCs), the reliable isolation of these entities from an individual's blood remains challenging. This is important to consider given the rarity of the event that is to be detected, and more work must be done to establish a reliable limit of detection. Perhaps most importantly, there is very limited data on what clinical steps should be taken in the case of a positive liquid biopsy in the setting of minimal residual disease or disease recurrence.

### Gene expression profiling

Another emerging area of molecular diagnostics is the use of gene expression profiling (GEP) for risk stratification, prediction of treatment response, and determination of the cell or tissue of origin. RNA expression continues to be evaluated primarily by microarray-based techniques, but increasingly NGS-based RNA sequencing approaches are being employed. Both approaches yield expression data on a large number of genes, and sophisticated bioinformatics approaches are needed to integrate this data to provide clinically actionable results. One of the most mature clinical applications of GEP focuses on stratifying breast cancer patients into risk groups. This stratification is based on the expression profile of approximately 50 genes and clinical data, such as tumor size and lymph node status. The results of this testing can provide the oncologist with information that will improve the treatment and management of the cancer.

The use of GEP in leukemias has identified potential therapeutic options that were not previously considered. Philadelphia chromosome-positive acute lymphoblastic leukemia (Ph+ALL) has a reproducible gene expression profile and is known to respond well to treatment with the tyrosine kinase inhibitor Imatinib. Researchers have shown that ALL lacking the Philadelphia chromosome, but with a GEP largely overlapping Ph+ALL, also responded to Imatinib. These studies suggest that the GEP may be a better predictor of response than the presence of the Philadelphia chromosome. Work with GEP in multiple myeloma has provided a risk stratification system that is independent of the traditional system based on cytogenetic abnormalities. The use of GEP to determine the cell of origin in lymphoma is beginning to be investigated as a prognostic indicator. These applications of GEP have shown promise in the clinical setting and will likely lead to the broader application of GEP in genomic testing.

### Pharmacogenomics

Pharmacogenomics refers to using specific genomic data to predict response to therapeutic drugs. This concept is well established in oncology, where the presence or absence of a particular variant will determine which treatment modality to use. A well-established example can be found in the management of colorectal cancer, where multiple studies have demonstrated that patients with mutations in the *KRAS* oncogene are unlikely to benefit from anti-EGFR antibody therapy. However, the increased access to genomic sequencing is likely to identify new links between genetic variants and drug response. The cytochrome P450 (CYP family) family of enzymes plays a particularly important role in drug metabolism, activating nearly 60% of all drugs. Variations in the genes that encode the CYP family of enzymes have been shown to affect the pharmacodynamics of numerous drugs in multiple clinical areas. Variation in the CYP2C19 gene plays an important role in cardiovascular medicine, as the presence of a particular variant predicts the decreased efficacy of the platelet inhibitor, clopidogrel. These poor responders have an increased risk of major adverse cardiovascular events, and molecular testing prior to drug administration allows the care team to consider other platelet inhibition regimens not involving CYP2C19. Despite the promise of personalized treatment plans based on a patient's genetic profile, pharmacogenomics has not seen wide clinical adoption. Genetic testing typically takes days to weeks from sample collection to final report, a time frame that is unmanageable for physicians who need to begin treatment immediately. Further slowing wide-spread implementation is the relatively small impact that genetic variants typically have on pharmacogenomics. For instance, the CYP2C19 genotype is the strongest individual factor determining platelet aggregation but accounts for only approximately 10% of variability.

## Key concepts

- The Clinical Laboratory Improvement Amendments (CLIA) is the primary regulatory legislation for clinical molecular diagnostics laboratories. However, the unique ability of genetic testing to predict future outcomes of individuals and families continues to raise concerns surrounding the use of such testing.
- Many molecular diagnostic laboratories rely heavily on laboratory developed tests (LDTs), differentiating genetic testing from most other areas of the clinical laboratory. Regulatory agencies, such as the FDA, apply more stringent oversight to LDTs than to FDA-approved tests, and the future of such oversight remains uncertain.
- Clinical utility, i.e. the balance of benefits to risk, has been proposed as a core criteria for the evaluation of genetic testing. However, this information may not be readily available when a test is put into clinical use.
- Thorough method validation and consideration of pre-analytical and post-analytical factors are essential to the successful implementation of clinical molecular diagnostic tests.

## Suggested readings

[1] Jennings L, Van Deerlin VM, Gulley ML. Principles and practice for validation of clinical molecular pathology tests. Arch Pathol Lab Med 2009;133:743—55.

[2] Wilson JA, et al. Consensus characterization of 16 FMR1 reference materials: a consortium study. J Mol Diag 2008;10:2—12.

[3] Germer JJ, Zein NN. Advances in the molecular diagnosis of hepatitis C and their clinical implications. Mayo Clin Proc 2001;76:911—20.

[4] Terrault NA, Pawlotsky JM, McHutchison J, et al. Clinical utility of viral load measurements in individuals with chronic hepatitis C infection on antiviral therapy. J Viral Hepat 2005;12:465—72.

[5] National Institutes of Health. National Institutes of Health Consensus Development conference statement: management of hepatitis C 2002 (June 10—12, 2002). Gastroenterology 2002;123:2082—99.

[6] Chute DJ, Cousar JB, Mahadevan MS, et al. Detection of immunoglobulin heavy chain gene rearrangements in classic Hodgkin lymphoma using commercially available BIOMED-2 primers. Diagn Mol Pathol 2008;17:65—72.

[7] Srinivasan M, Sedmak D, Jewell S. Effects of fixatives and tissue processing on the content and integrity of nucleic acids. Am J Pathol 2002;161:1961−71.

[8] Lanman RB, Mortimer SA, Zill OA, et al. Analytical and clinical validation of a digital sequencing panel for quantitative, highly accurate evaluation of cell-free circulating tumor DNA. PLoS One 2015;10:e0140712.

[9] Roschewski M, Dunleavy K, Pittaluga S, et al. Circulating tumour DNA and CT monitoring in patients with untreated diffuse large B-cell lymphoma: a correlative biomarker study. Lancet Oncol 2015;16:541−9.

[10] van't Veer LJ, Dai H, van de Vijver MJ, et al. Gene expression profiling predicts clinical outcome of breast cancer. Nature 2002;415:530−6.

[11] Roberts KG, Li Y, Payne-Turner D, et al. Targetable kinase-activating lesions in Ph-like acute lymphoblastic leukemia. N Engl J Med 2014;371:1005−15.

[12] Scott DW, Wright GW, Williams PM, et al. Determining cell-of-origin subtypes of diffuse large B-cell lymphoma using gene expression in formalin-fixed paraffin-embedded tissue. Blood 2014;123:1214−17.

[13] Scott DW, Mottok A, Ennishi D, et al. Prognostic significance of diffuse large B-cell lymphoma cell of origin determined by digital gene expression in formalin-fixed paraffin-embedded tissue biopsies. J Clin Oncol 2015;33:2848−56.

[14] Simon T, Varstuyft C, Mary-Krause M, et al. Genetic determinants of response to clopidogrel and cardiovascular events. N Engl J Med 2009;360:363−75.

[15] Chan NC, Eikelboom JW, Ginsberg JS, et al. Role of phenotypic and genetic testing in managing clopidogrel therapy. Blood 2014;124:689−99.

# Chapter 31

# Molecular assessment of human diseases in the clinical laboratory

M. Rabie Al-Turkmani, Sophie J. Deharvengt and Joel A. Lefferts

*Laboratory for Clinical Genomics and Advanced Technology (CGAT), Department of Pathology and Laboratory Medicine, Dartmouth Hitchcock Medical Center, Lebanon, NH, United States*

## Summary

The past decades have seen significant advances in the field of genomics. Perhaps most notably, major advances in nucleic acid sequence analysis and computing technology enabled the successful completion of the Human Genome Project. This new knowledge at the genetic level has been followed at the transcriptomic (RNA) and proteomic (protein) levels. These advances supported the development of targeted molecular diagnostics and therapeutics, facilitating a more personalized approach to health care, reducing unnecessary treatments and improving health outcomes. For example, infectious diseases and cancer are detected earlier and more accurately. Molecular testing is now an integral part of disease management and therapy and spans the entire spectrum of applications that are performed on a routine basis in most hospital clinical laboratories. While the ability of clinical laboratories to detect human genetic variation has historically been limited to a rather small number of traditional genetic diseases, the current understanding of many diseases at a molecular level has expanded testing capabilities to both human and nonhuman applications. The identification of numerous new genes and disease-causing mutations, as well as benign polymorphisms that may influence a specific phenotype, has created a rapidly growing demand for molecular diagnostic testing.

## Introduction

The past decades have seen significant advances pertaining to the field of genomics. Perhaps most notable are the major advances in sequence analysis and computing technology that led to the successful completion of the Human Genome Project. This new knowledge at the genetic level has been followed at the transcriptomic (RNA) and proteomic (protein) levels. These advances have led to the development of targeted diagnostics and therapeutics allowing for a more personalized approach to healthcare, thus reducing unnecessary treatments and improving health outcomes. While the ability of clinical laboratories to detect human genetic variation has historically been limited to a rather small number of traditional genetic diseases, our current understanding of many diseases at a molecular level has expanded our testing capabilities to both human and nonhuman applications. The identification of numerous new genes and disease-causing DNA variants, as well as variants that may influence drug metabolism, has created a rapidly growing demand for molecular diagnostic testing.

In order to understand the capabilities and limitations of molecular testing, we begin with a review of several major technologies used in clinical laboratories. We will then discuss their applications in the field of oncology, infectious diseases and genetic diseases. Finally, we present the liquid biopsy application that promises to revolutionize the practice of molecular pathology.

Essential Concepts in Molecular Pathology. DOI: https://doi.org/10.1016/B978-0-12-813257-9.00031-0

# Molecular pathology techniques

## PCR-based methods

### The polymerase chain reaction

The polymerase chain reaction (PCR) has had an unprecedented impact in the field of molecular diagnostics. The amplification of a target sequence uses a DNA polymerase that synthesizes new DNA molecules by adding free nucleotides to an existing strand of DNA. Primers are used to produce a starting point for DNA synthesis. A primer is a short single-stranded nucleic acid sequence (generally about 16—30 base pairs) that is complementary to the desired target sequence to amplify. The polymerase starts replication at the 3'-end of the primer, synthesizing a sequence of DNA complementary to the target strand. PCR represents a rapid, sensitive, and specific method for amplification of specific nucleic acid sequences, allowing for detection and analysis of specific gene sequences starting from minimal patient sample. By eliminating the need to extract large amounts of sample, PCR has revolutionized the field of molecular diagnostics. Numerous developments related to methodological modifications and new forms of instrumentation have greatly helped in enhancing this technology.

### Reverse transcriptase-PCR

In order to expand the power of PCR amplification to the transcriptome level, RNA must first be reverse-transcribed into complementary DNA (cDNA). Reverse transcriptase PCR (RT-PCR) is an excellent method for analysis of RNA transcripts, especially when working with limited amounts of starting material. Classic blotting hybridization assays require much more RNA for analysis and they lack the speed and ease offered by PCR-based applications. Other advantages of RT-PCR include versatility, sensitivity, rapid turnaround time, and the ability to compare multiple samples simultaneously.

### Real-time PCR (qPCR)

A significant advancement in the development of diagnostic applications of PCR was the development of real-time or quantitative PCR (qPCR) assays. Various detection chemistries for real-time PCR were rapidly introduced, and many of the older detection methods (including gel electrophoresis, allele-specific oligonucleotide blots, and others) started to vanish from the laboratory.

qPCR combines the amplification steps of traditional PCR with simultaneous detection steps that do not require post-PCR manipulation, as the PCR is monitored directly within the reaction vessel. In qPCR, the exponential phase of PCR is monitored as it occurs, using fluorescently labeled molecules (Fig. 31.1). During the exponential phase, the amount of PCR product present in the reaction vessel is directly proportional to the amount of emitted fluorescence and

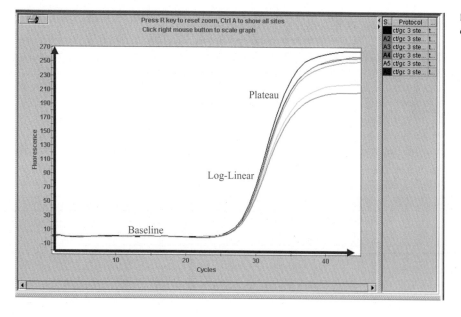

FIGURE 31.1 Real-time polymerase chain reaction amplification curve.

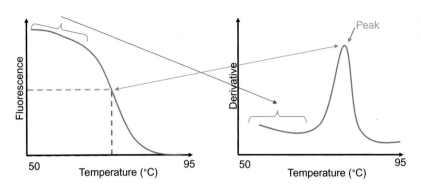

**FIGURE 31.2** Melt curve from real-time polymerase chain reaction.

the amount of initial target present. Thus, these reactions can also be quantitative. There are two main types of detection chemistries for qPCR: (a) those that use intercalating DNA binding dyes such as SYBR® Green I and (b) those that use various types of fluorescently labeled probes such as TaqMan. The main advantages of qPCR is the speed with which samples can be analyzed as there are no post-PCR processing steps required, the closed-tube nature of the technology that helps to prevent contamination, and its quantitative nature.

There are two main post-qPCR analyzes that are not exclusive: one using the amplification curve and another using the melt curve. Amplification curve analysis allows quantification of the amplicon amount by comparison to a known standard. Melt curve analysis is based on the fundamental property of double strand DNA (dsDNA) to denature with heat. This separation can be monitored with the use of fluorescent dyes that specifically fluoresce when bound to dsDNA. As the temperature is raised, the double strand begins to dissociate releasing the dye and thus inducing a decrease in the fluorescence intensity. The fluorescence data generated are represented by a curve of fluorescence intensity versus temperature. The point at which 50% of the DNA is in the double stranded state is called the melting temperature (Tm) and it represents the peak of the melting curve derivative (Fig. 31.2). In the presence of saturating concentrations of DNA binding dyes, a specific amplicon sequence will present with specific Tm and melting curve shape. This distinctive melting curve can be used to detect DNA sequence variations in the amplicon without the need for any post-PCR processing. The development of both high-resolution qPCR instruments and new saturating DNA dyes has allowed for a more precise assessment of sequence variations based on melting analysis. High resolution melting analysis (HRMA) can discriminate DNA sequences based on their composition, length, GC content, or strand complementarity. It can be used for mutation scanning experiments, methylation studies and genotyping. The advantages of this method are its ease of use, high sensitivity and specificity, low-cost and rapid turn-around-time. These characteristics make HRMA ideal for use in routine diagnostic settings. However, the accuracy of the technique depends on appropriate instrumentation, saturation dyes, and the software used for analysis, thus inducing potential variability depending on the specific clinical diagnostic setting it is used for.

## Multiplex PCR

Another important milestone in the evolution of PCR involves the expansion of assay multiplexing capabilities. Multiplex PCR is a powerful technique that enables amplification of two or more products in parallel in a single reaction tube. It typically employs different primer or probe pairs in the same reaction for simultaneous amplification of multiple targets. Multiplex PCR is used widely in genotyping applications and multiple areas of DNA testing in research, forensic, and diagnostic laboratories. Applications include gene expression and deletion analysis, SNP genotyping, forensic identity testing (e.g., STR typing), and pathogen detection. Quantification of multiple genes in a single reaction also reduces reagent costs, conserves sample material, and allows increased throughput.

## Digital PCR

qPCR enables relative quantification of target sequences and not absolute quantification, due to the issue of amplification bias and the requirement for standard references that can differ among laboratories. A recent advancement in PCR technology, digital PCR (dPCR), considerably improves the performance of PCR by eliminating the need for standard curves and bypassing the limitation related to reaction efficiency.

Digital PCR is derived from a digital readout that utilizes numerical digits to identify presence or absence of a specific target in single reactions. dPCR is performed by diluting target nucleic acid across a large number of reactions (termed partitions or fractions), which statistically results in zero, one or more target molecules per reaction prior to amplification. A strong sample partitioning enables reliable measurement of small fold differences in target DNA sequence or copy numbers among samples. Additionally, the creation of individual reaction compartments avoids the effects of cross-contamination and achieves absolute target quantification in each sample. However there is always the possibility that more than one molecule will be present per reaction, thus requiring the need to use statistical methods, such as the Poisson distribution or Negative Binomial distributions.

There are four main ways to partition samples: droplet based sample dispersion (oil-water emulsions), microwell based sample dispersion, channel based sample dispersion, and printing based sample dispersion. The next step consists of thermal cycling amplification or isothermal amplification. After amplification, each compartment is evaluated individually. This eliminates PCR bias and decreases error rates, thus enabling the detection of small differences. In addition, evaluation of an individual compartment in dPCR reduces the background signal by increasing the signal-to-noise ratio, thus significantly improving the detection sensitivity. Fluorescent probes are generally used to detect the amplification products. A single template reaction yields a positive signal, which can be differentiated from fluorescence intensity generated by multiple target reactions. Depending on the sample dispersion strategies, the quantification processes in dPCR mainly relies on two methods. The first is droplet-based fluorescence signal counting, a method that determines the number of droplets that have different fluorescence intensities, similar to the principle of flow cytometry. The other is chip-based fluorescent dot image processing, which involves acquiring, and analyzing the two-dimensional (2D) fluorescence bitmap images.

dPCR offers significant benefits over other PCR methods by providing an absolute count of target DNA copies per input sample without the need for standard curves, making this technique ideal for nucleic acid quantification, rare mutation detection, pathogen detection (viral load analysis, and microbial quantification), copy number variation (CNV) detection, microRNA (miRNA) expression analysis, single-cell gene expression analysis, chromosome abnormality detection and other potential applications.

However, dPCR encounters the same challenges as PCR and qPCR in terms of experimental design and the need for internal controls. Though dPCR offers the potential to perform precise low abundance quantification, it is imperative to measure sources of false-positive signal due to low-level contamination and/or non-target amplification (caused by mispriming and/or primer dimers) and may lead to overestimation of the target. Conversely, PCR inhibition can lead to underestimation due to the fact that dPCR is only able to count positive partitions. This problem is also encountered when detecting RNA by reverse transcriptase dPCR. Consequently, the use of internal positive and negative controls is highly recommended when analyzing clinical samples by dPCR.

## Array-based nucleic acid analysis

Available microarray techniques allows for numerous clinical applications, depending on the probe used. They can be used to determine a gene expression signature based on a multitude of genes. They are also used in the determination of copy number variation and the detection of single-nucleotide polymorphisms (SNPs) and mutations. Several microarray-based assays have been FDA approved (Table 31.1) for different applications in oncology, microbiology, and genetics.

Microarray technology generally involves probes (e.g. DNA sequence) that are usually synthesized and immobilized to a solid surface (i.e. membrane or beads) at specific spots. The precise knowledge of the location of each individual spot is necessary for the analysis. The target of interest is usually fluorescently labeled and hybridization between the probe and the labeled target is detected and quantified using fluorescence-based, silver-based, or chemiluminescence-based methods.

**TABLE 31.1 FDA-approved microarray assay.**

| Name | Use |
| --- | --- |
| Agendia's MammaPrint | Breast cancer gene signature |
| Pathwork tissue of origin test kit | Tissue of origin |
| Roche AmpliChip CYP450 microarray | Drug metabolizing enzymes |
| Luminex xTAGTM respiratory viral panel (RVP) | Detection of 12 respiratory viruses and subtypes |
| Affymetrix CytoScan Dx assay | Chromosome abnormalities |

Due to technological advances and the variable experimental details (including length and density of probes and types of solid surface), the spectrum of microarrays is very large. In addition to DNA based microarrays that are widely used, protein microarrays using antibodies have real potential for use in the clinical laboratory.

## Mass spectrometry

Mass spectrometry (MS) is used to identify the chemical constitution or molecular structure of an analyte by separating ionic chemical species according to their differing mass-to-charge (m/z) ratio. MS instruments usually consist of an ionization chamber, a mass analyzer, and an ion detector. The sample is pulsed with an energy source that ionizes the individual molecules and desorbs the solid or liquid phase samples into gas phase. The vaporized samples subsequently travel to the mass analyzer, allowing the separation of ions based on their $m/z$ ratio. When ionized particles collide with the ion detector, the $m$ and $z$ of each molecule are measured as derived from their individual force and time of impact. The results are represented in a mass spectrum, which plots the relative abundance on the y-axis versus the $m/z$ ratio on the x-axis.

Several MS instruments have been developed for specific applications. In order to be used on complex biomolecules, the method of ionization had to evolve to avoid degradation by using a low energy source.

### MALDI-TOF MS

One of these low energy methods is the Matrix-Assisted Laser Desorption/Ionization-Time-Of-Flight Mass Spectrometry (MALDI-TOF). This method uses a matrix, which embeds the sample and when it desorbs into the gas phase it absorbs the majority of the pulsed ionizing energy, thus protecting the sample from degradation. After irradiation, the matrix is positively charged and it then transfers its positive charge to the sample by collision. Once ionized, the sample is directed into the time of flight mass analyzer, which consists of an empty, pressurized tube where the ions travel toward the ion detector. The velocity of travel is dependent on the $m/z$ ratio, allowing lighter samples to travel faster. At the end of their travel, the ions hit a sensor, which detects ions and creates a mass spectrum. The mass spectrum obtained from a particular molecule is unique to the molecule and can be compared to a stored database for identification.

MALDI-TOF-MS is a versatile method that can be used to analyze a wide range of molecules in the clinical laboratory. Its high accuracy and sensitivity, combined with the high throughput and its wide mass range (1–300 kDa) make MALDI-TOF-MS a promising technique that can be used routinely to study nucleic acids, proteins and polysaccharides and has applications in clinical chemistry, microbiology, and molecular laboratories.

Nucleic acid detection by MS enables readout by molecular mass as a highly accurate alternative to fluorescent-based methods for quantitative and qualitative genomic analysis. Single-base and multiple-base primer extension methods in combination with MALDI-TOF MS analysis of primer extension products have been used routinely for the analysis of SNPs and mutations. In addition, by including bisulfite treatment step, MALDI-TOF MS can be used to screen and analyze the methylation status of CpG islands. It can also be used for targeted genetic analysis, copy number quantification, comparative sequence, and somatic mutation analysis. An example of commercially available MALDI-TOF MS systems that can be used for these applications is the MassARRAY® system (Agena Bioscience, San Diego, CA).

## Nucleic acid sequencing

While qPCR and dPCR assays are highly sensitive and specific, they require a prior knowledge of the target sequence. In contrast, sequencing methods have an unbiased approach to nucleic acid detection.

### Sanger sequencing

The Sanger method utilizes the natural properties of DNA polymerase as well as terminator dideoxy-nucleotides (ddNTP) tagged with specific fluorescent dyes to generate sequence data using capillary electrophoresis.

After amplification and purification of the target DNA, a primer anneals adjacent to the sequence of interest and is extended by DNA polymerase. During the extension reaction, the nascent chain is terminated by the random incorporation of fluorescently labeled ddNTPs complementary to the base on the opposite strand. The different size products are then resolved by capillary electrophoresis, and the resultant pattern of fluorescent peaks determines the DNA sequence. The technique is rapid, robust, accurate, and can typically achieve read lengths of up to 1 kb with relatively low cost. The entire Human Genome Project was completed using Sanger Sequencing, and it remains the gold standard of DNA sequence analysis.

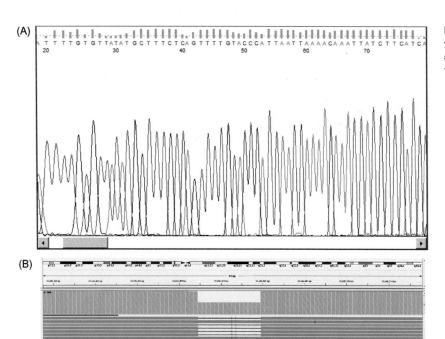

**FIGURE 31.3** Comparison of data obtained with Sanger sequencing (electropherogram) (A) and next-generation sequencing (alignment with integrative genomics viewer) (B).

## Next generation sequencing or massively parallel sequencing

Next generation sequencing (NGS) involves massively parallel sequencing of millions of DNA templates, rapidly producing an enormous volume of data at a relatively lower cost than Sanger sequencing per target (Fig. 31.3). With NGS, DNA sequencing libraries are clonally amplified *in vitro*, eliminating the need to clone the DNA library into bacteria. DNA is sequenced by synthesis, (addition of complementary nucleotides) and the use of molecular barcodes allows the DNA templates to be sequenced simultaneously in a parallel fashion.

NGS workflow consists of four main steps: library preparation, template amplification, sequencing, and data analysis (Fig. 31.4). Extensive reviews have been published on various NGS platforms. Due to the rapidly evolving technologies, we will focus on the basic principles used for sequencing and encourage readers to visit the websites of commercially available platforms for the latest information.

### Library preparation

DNA library preparation usually involves fragmentation of isolated DNA into smaller, random, over-lapping fragments followed by end repair and adapter ligation. DNA is fragmented usually by physical/mechanical (i.e., ultrasonication) or enzymatic methods into a range of 150−800 bp fragments, depending on the platform being used.

While direct sequencing of RNA molecules is possible, RNA sequencing is made possible on instruments that sequence DNA by using complimentary DNA (cDNA). cDNA libraries are prepared by capturing mRNA, random priming, and cDNA synthesis followed by the end polishing and platform specific adapter sequences ligation. Since the average size of RNA is smaller than DNA, no fragmentation is usually required. The cDNA library preparation method varies depending on the RNA species under investigation, which can differ in size, sequence, structural features and abundance. Custom gene panels can be performed using libraries enriched with the targeted sequences.

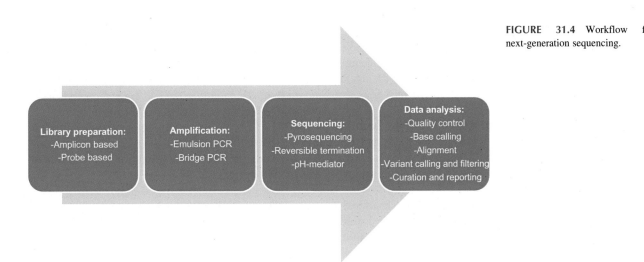

FIGURE 31.4 Workflow for next-generation sequencing.

TABLE 31.2 Amplicon-based versus hybridization-based enrichment methods.

|  | Amplicon based | Hybridization based |
|---|---|---|
| Enrichment | PCR primers | Probe |
| Turnaround time | Shorter | Longer |
| Allele dropout (false negative) | Sensitive | Resistant |
| DNA input | Lower input | Higher input |
| Capturing neighboring region from the regions of interest | No | Yes |
| Library complexity | Lower | Higher |
| Uniformity | Lower | Higher |
| False-positive and negative SNVs | Vulnerable | Less likely |
| Workflow | Simpler | Longer |
| Repetitive and GC-rich regions |  | Can be difficult |

***Amplification-based enrichment methods*** Amplicon-based methods use PCR to generate the sequencing libraries and are ideal for sequencing specific regions of interest (ROI). They are ideal when analyzing archived tissues that have limited material and rely on carefully designed sets of primers to avoid or minimize the amplification of pseudogenes, which are genomic regions with high sequence homology to the ROI. Amplicon-based enrichment strategies can achieve 100% coverage of a region with little off-target sequence as the start and stop coordinates of each amplicon are predetermined. One limitation of amplification-based target enrichment is its susceptibility to allele dropout, which occurs when a variant is located in the primer-binding site, leading to failed amplification and allele bias.

***Hybrid capture enrichment methods*** Specific nucleotide probes are designed to pull out or capture specific regions of interest by hybridization. These probes are much longer than typical PCR primers and therefore not as affected by variants in the probe-binding site, thus avoiding dropout issues. One disadvantage of hybrid capture enrichment is that probes can tolerate mismatches, thus reducing specificity and increasing off-target reads. It also requires more starting material than amplification-based enrichment, posing a problem if the amount of clinical sample is limited (Table 31.2).

## Template amplification

After library preparation, selected DNA fragments are subjected to clonal amplification. Clonal amplification involves solid phase amplification of the template to reach detectable signal during sequencing. Single DNA fragments to be sequenced are either bound to beads or flow cells. Depending upon the sequencing platform, emulsion PCR (emPCR) or bridge PCR is used to amplify the anchored DNA fragments into millions of spatially separated template fragments.

***Emulsion PCR*** DNA template, through the adapter sequences, hybridizes to complementary capture oligonucleotides covalently linked to beads. The template molecules and beads are then mixed to achieve an average of one template molecule per bead in an individual compartment created by the emulsion in an oil and water mixture. During emulsion PCR, the surface of the bead becomes coated with clonal copies of the DNA template.

***Isothermal bridge amplification (bridge PCR)*** Denatured template DNA fragments, through the adapter sequences, hybridize to complementary capture oligonucleotides covalently linked to the surface of the flow cell. These captured oligonucleotides are used as primers for the PCR amplification. After a second denaturation, the newly synthesized strand can then bend to hybridize with an adjacent capture oligonucleotide primer, and the cycle can be repeated to generate a cluster of identical template molecules. Bridge PCR allows paired-end sequencing (sequencing the DNA fragment from both ends), resulting in high coverage, high numbers of reads, and more data as compared to single end sequencing systems. This facilitates the detection of genomic rearrangements (insertions, deletions, and inversions), repetitive sequence elements, gene fusions and novel transcripts.

## Sequencing

The majority of NGS chemistries use sequencing by synthesis including the following categories.

***Sequencing by reversible termination*** This technique uses reversible terminator (RT) nucleotides that are fluorescently labeled and modified at the 3' end with a cleavable terminator moiety to ensure that only a single nucleotide incorporation event occurs with each sequencing cycle. Sequencing by reversible termination starts with the annealing of sequencing primer to the template molecules. Then each cycle consists of three steps: (1) incorporation of the complementary RT nucleotide by DNA polymerase to the DNA strand, followed by (2) detection of the different fluorescence signal for each cluster, and (3) finally both the fluorescent label and 3' terminator moiety are cleaved and removed, regenerating the growing strand for another sequencing cycle. Repetition of these cycles leads to sequencing of the template.

***pH-mediated sequencing*** This method is based on detection of hydrogen ions liberated upon incorporation of each nucleotide and is not dependent upon altered nucleotides or optical detection. Template DNA obtained after library preparation and clonal amplification is bound to ion sphere particles. Ideally, each bounded sphere is separated in a unique microwell of a chip. Subsequently, a single type of dNTP is flowed on the chip at a time and if it is incorporated during DNA synthesis, the release of a hydrogen ion induces a change in pH, which is detected by ion-sensitive sensors. This technique is rapid, with read length of up to 400 nucleotides and it operates at relatively low cost since it does not involve modified bases. With this sequencing chemistry, homopolymer sequencing is error-prone as repeats will be incorporated in one cycle, which leads to a proportionally higher electronic signal due to corresponding number of released hydrogen ion. Life Technologies has introduced the Hi-Q polymerase that has improved sequencing accuracy, decreased indel error rates, and reduced GC-bias.

## Data analysis

Data analysis consists of three main steps: base calling and quality score computation, assembly and alignment, and variant calling and annotation. The data output from an NGS instrument essentially consists of a text file of millions of raw sequence reads. The massive amount of sequence data generated by NGS creates new challenges for the laboratory, requiring significant investment in bioinformatics infrastructure and personnel. The accurate interpretation of the data generated by NGS is labor intensive, requiring expertize in genetics, pathology, clinical medicine and informatics. The establishment of well-curated genomic databases with phenotypic information is a critical step for the widespread clinical implementation of NGS-based tests.

## Methylation detection techniques

Analysis of CpG island methylation requires some method of discrimination between the methylated and the unmethylated DNA sequence. This can be achieved using non-bisulfite or bisulfite methods.

The non-bisulfite methods use methylation-sensitive restriction endonuclease digestion analysis. These methods are not the first choice for clinical laboratories, due to the limited availability of restriction sites, the occurrence of false positive results due to incomplete digestion, and the requirement of high quality DNA.

The majority of methods for methylation analysis use bisulfite-modified DNA. Bisulfite converts unmethylated cytosine to uracil (and then to thymine following *in vitro* DNA synthesis), while methylated cytosine is not affected, allowing the differentiation between methylated and unmethylated DNA. As this conversion happens only on denatured DNA, partial DNA denaturation can cause artifacts and impair the total conversion of cytosines to uracils. This conversion is critical to the success of the analysis and is necessary to evaluate its efficiency.

The benefit of bisulfite-based assays is that they require very small amounts of DNA and are compatible with DNA derived from formalin-fixed paraffin-embedded (FFPE) tissue samples. The products of the bisulfite conversion can be then analyzed either by direct sequencing or by various PCR based assays.

# Clinical applications

## Oncology

Molecular pathology has evolved to play a substantial role in oncology due to the genetic complexity of cancer etiology and the introduction of targeted therapies. Genetic alterations in cancer can be found at different levels: Genomics (change in the DNA sequence), Epigenomics (DNA methylation and histone modification), and Transcriptomics (RNA level). The molecular pathology laboratory generates critical diagnostic, prognostic and therapeutic information to guide treatment of cancer patients.

### Genomics

Different somatic mutational events are implicated in the tumor evolution and they range from single nucleotide variants (e.g., *BRAF* V600E, *KRAS* G12D) to gene fusions (e.g., *BCR−ABL, ELM4−ALK*), gene amplifications (e.g., *HER2, MET*), and global aneuploidy.

### Somatic Mutation Profiling

One important role for molecular diagnostics in oncology is to provide a somatic mutation profile using either archival tumor tissues or fresh/frozen tissue biopsies. Presence or absence of somatic variants can offer prognostic and diagnostic information to clinicians and impact treatment options. Several oncology drugs have been approved with a companion diagnostic test, mainly tyrosine kinase inhibitors. Companion diagnostic tests provide essential information for the safe and effective use of a corresponding therapeutic agent. The molecular laboratory can measure the status of the drug target driver mutation (such *EGFR, KRAS, KIT*), by using single-analyte tests, such as the Cobas EGFR mutation Test v2.

Numerous methods are available for tumor characterization, including immunohistochemistry (IHC), *in situ* hybridization (ISH), quantitative PCR (qPCR), digital PCR (dPCR), microarray, Sanger sequencing and Next Generation Sequencing (NGS). The choice of the method is dependent upon the need. Some methods are faster, but are limited to detect only known hotspot mutations in oncogenes. In contrast, sequencing provides more information without needing pre-existing knowledge.

Somatic mutation detection in solid tumors by Sanger sequencing presents several limitations. These include the limited number of targets per reaction, low throughput, and poor sensitivity that becomes more problematic if there is significant contamination with normal tissue. When several genes are tested for alterations, NGS can overcome these limitations with a diversity of tumor profiling approaches. Indeed, NGS allows Whole Genome Sequencing (WGS), which provides extensive genomic information ranging from single nucleotide variation to deletions, insertions, copy number variation, and structural variants. Whole exome sequencing (WES) can also be performed on tumor tissue and it is less comprehensive and costly compared to WGS. However, WES is still not cost effective in many cancer cases due to lengthy analysis and the large amount of non-actionable data obtained. Targeted gene panels have become widely adopted by clinical laboratories. Due to the rapid reduction in sequencing costs, gene panels have become more affordable for clinical applications. For example, the Ion Torrent AmpliSeq Cancer Hotspot Panel v2 is capable of identifying multiple somatic mutations in 50 genes in a single assay.

### Copy Number Variation

Copy Number corresponds to the number of copies of a particular gene present in the genome. The list of clinically relevant and actionable copy number variations (CNVs) is growing and along with it, a recognition of the importance of genome-wide copy number and loss of heterozygosity (LOH) analysis for solid tumors. The number and complexity of CNVs have also been shown to be an indicator of patient prognosis.

Genome-wide analysis of solid tumor samples is technically challenging due to limited amounts of highly modified and degraded DNA extracted from heterogeneous FFPE samples. In addition, the yield of extracted DNA may be low due to the amount of starting material available, especially in biopsy specimens. Traditional FFPE sample analysis techniques such as fluorescence *in situ* hybridization (FISH) are limited to locus-specific, low-resolution copy number information. Digital PCR has also been used for the detection of CNVs with detection limit as low as $1-100$ copies, but still only allows the detection of a limited number of genes.

Chromosomal Microarray Analysis (CMA) enables simultaneous detection of variations in cancers at a genome-wide level. CMA is used to look for chromosomal duplications and deletions ranging in size from several kb to entire chromosomes (hundreds of Mb). A robust microarray assay should be applicable to low quantity and quality DNA and have analytical tools to discern information from cancer cells versus that of normal cells. In order to generate high quality microarray data from FFPE specimens, microarray protocols have been modified by the introduction of molecular inversion probes (MIPs) into the sample preparation process. Unlike other array-based CNV analysis methods, MIP-based protocols only use the specimen DNA in the initial probe hybridization. As a result, the assay is not as sensitive to DNA quality as other array-based methods that use labeled specimen DNA to probe the array. The OncoScan® FFPE Assay Kit (Affymetrix, a Thermo Fisher Scientific company) is MIP-based and it is the first microarray designed specifically for use with degraded DNA found in FFPE tissue. The Oncoscan array utilizes SNP probes to provide copy number as well as allele frequency information.

Microarrays are cost effective and have a significantly higher resolution than G-banded chromosome analysis. However, they cannot detect balanced translocations and inversions. CNV determination by NGS has challenges related to the need for deep coverage to provide accurate copy number information from heterogeneous FFPE samples. Advances in computational tools have increased the accuracy of CNV determination by NGS.

### *Epigenomics*

Epigenetic modifications have been implicated in tumorigenesis. The two predominant epigenetic mechanisms are DNA methylation and histone modification. Cancer specific DNA methylations have been shown to contribute to disease initiation and progression. Epigenetic analyzes bring information for diagnosis, prognosis, outcome prediction, and/or treatment selection. CpG methylation analysis is clinically relevant as many cancer-associated promoter hypermethylation patterns have been identified and used as cancer biomarkers. CpG island promoter hypermethylation can transcriptionally silence tumor suppressor or mismatch repair genes. The epigenetic silencing of these genes is often specific to the tissue type and several clinical tests have been developed to analyze their methylation status. Tests used in the clinical laboratory can be FDA-approved (for colorectal cancer Cologuard® and Epi proColon®) or can be validated as laboratory developed tests (MGMT MethyLight assay). These tests can be performed not only on tumor tissue but also on blood or body fluids.

### *Transcriptomics*

A hallmark of cancer is aberrant expressions of certain genes, such as those implicated in cell cycle, invasion and angiogenesis. Oncogenes can be overexpressed and tumor-suppressor genes can be silenced. Gene-expression profiling assesses the relative expression of these genes by using differential gene expression (experimental samples versus controls), and the data obtained can then be used for cancer classification.

There are three main techniques for tackling the transcriptome: quantitative reverse transcriptase PCR (RT-qPCR), microarrays and NGS (RNA-Seq). The RT-qPCR is quantitative and sensitive but limited to a relatively small number of transcripts. Microarrays and RNA-Seq offer genome-wide surveys of the transcriptome. An example of the microarrays used in the clinical laboratory is the FDA approved 70-gene MammaPrint™ from Agilent. Microarrays are relatively inexpensive, widely available and easily analyzed. However they can only detect specific sequences homologous to the ones on the array. Increasingly, RNA-Seq is becoming the method of choice for studying gene expression and identifying novel RNA species, but it remains costly and requires an extensive analytical infrastructure.

## Gene Fusion

Created by structural chromosomal rearrangements, gene fusions are the result of the exchange of coding DNA sequences between genes. Some gene fusions are strong driver mutations in cancer, and their identification can be used for diagnostic purposes and treatment management, as some fusions are specific drug targets.

Fusion events can be detected by cytogenetic methods such as fluorescence *in situ* hybridization (FISH). Genome-wide screening approaches have also been developed for detection of breakpoints associated with gene fusions. More recently, RNA-Seq data along with specific bioinformatics methods have been used for fusion detection. Detection of a fusion event is typically revealed by reads containing fusion junctions or by differences in expression between the 5' and 3' ends of genes that are fused. Commercial kits are available to detect fusion events, such the ones by NuGene (San Carlos, CA) and ArcherDX (Boulder, CO).

## miRNA

MicroRNAs are a growing class of small, noncoding RNAs (17−27 nucleotides) that regulate gene expression by targeting mRNAs for translational repression, degradation, or both. These molecules are emerging as important modulators in cellular pathways such as growth and proliferation, apoptosis, and developmental timing. Since their discovery, microRNAs (miRNAs) have shown great promise in a wide array of clinical applications, such as using them as new diagnostic markers and targets for novel therapies that could affect an entire pathway. Expression profiling of miRNA is still in the preclinical phase and is characterized mainly by RT-qPCR, microarrays and NGS.

# Infectious diseases

Compared to traditional microbial techniques, molecular techniques offer higher sensitivity and specificity with shorter turnaround times. They can also make identifications at the species/strain level, help characterize resistance markers, and improve biosafety. Techniques such as those that are PCR-based, NGS, arrays and MALDI-TOF mass spectrometry are changing the way clinical microbiology is practiced.

## Qualitative analysis

### Pathogen Detection and Identification

Qualitative testing determines the presence or absence of a particular genetic sequence. Prompt detection and characterization of microorganisms that cause infection allow providers to institute adequate measures to interrupt transmission and/or to begin proper therapeutic management of the patient. The main techniques used for pathogen detection are PCR-based due to their multiplexing ability and high throughput. However, it is important to note that the main limitation for pathogen detection by molecular techniques is their inherent inability to distinguish between live and dead cells.

The application of nucleic acid amplification technology, particularly multiplex PCR, coupled with fluidic or fixed microarrays has created an important new approach for the detection of multiple microorganisms in a single test. These multiplex amplification tests are sensitive enough to detect infections at an early stage in patients. The U.S. Food and Drug Administration (FDA) has approved several multiplex PCR assays for the detection of microorganisms including the ePlex Respiratory Pathogen Panel (GenMark Diagnostics, Carlsbad, CA) and the FilmArray® Respiratory panel (BioFire Diagnostics, Salt Lake City, UT).

Another technique for molecular identification of bacteria in the clinical laboratory is MALDI-TOF-MS, with a short turnaround time, low cost and precision at the species level. An advantage of MALDI-TOF MS is the ability to work directly with blood cultures. However, MALDI-TOF MS cannot identify species among mixed populations or directly from solid specimens.

The development of the metagenomics field (the study of microbial communities sampled directly from their natural environment, without prior culturing) is entering the clinical world and could be the ultimate approach in detecting all microorganisms (e.g. bacteria, viruses, fungi) in a clinical sample. Whole Genome Sequencing (WGS) provides access to the whole genetic content of a strain, and can identify and characterize several pathogens directly from a patient specimen without the need for culture. For bacterial species, sequencing can target the 16S rRNA gene sequence, as it is ubiquitous but contains nine hyper-variable regions allowing discrimination among different bacteria. However, due to the high sequence similarities in 16S rRNA genes between certain bacterial species, this method does not always provide an unambiguous identification.

In addition to pathogen identification, various PCR- and sequencing-based molecular approaches have been introduced to provide rapid and reliable screening and detection of common antibiotic resistance genes.

### Genotypic Characterization

Typing methods can help to discriminate between different bacterial or viral isolates of the same species. The importance of genotyping is well-represented with the example of the Human Papilloma Virus (HPV). HPV infection has been correlated with the development of several tumor types. HPV is very heterogeneous, with over 100 distinct HPV genotypes. Some genital mucosal HPV types are mainly found in low-grade cervical lesions and are thus designated low risk. Certain genotypes are considered high-risk (hr)-HPV genotypes (i.e. HPV16 and HPV18) and are found regularly in high-grade dysplasia and cervical cancers.

A cost-effective method of detecting HPV is the evaluation of p16 expression by IHC. Unlike DNA-based methods, p16 staining distinguishes transcriptionally active forms of HPV from inactive HPV, but it is not specific to the presence of HPV. Several studies advise complementing p16 staining with HPV DNA testing to obtain reliable identification of the HPV infection. Several sensitive, DNA-based HPV genotyping methods are currently available including the Roche Linear Array® HPV genotyping test and the Roche COBAS® HPV test (Roche Molecular Diagnostic, Branchburg, NJ), which is a qPCR based qualitative *in vitro* test that is FDA-approved for the detection of HPV in liquid cytology specimens.

## Quantitative analysis

Quantitative testing of organisms helps in monitoring therapeutic efficacy or distinguishing a clinically significant infection from a latent or resolving infection. Quantification of viral load often correlates with the virus activity and outcome in most acute and chronic infections. Quantitative qPCR-based assays are currently used routinely in the diagnosis and management of infections by HIV, Hepatitis C Virus (HCV), Hepatitis B Virus (HBV), and Cytomegalovirus (CMV) using plasma or serum samples. Several automated viral load assays have been cleared by the FDA, including multiple assays on the cobas® (Roche, Pleasanton, CA) and m2000 (Abbott, Abbott Park, IL) analyzers. There is an ongoing effort to standardize and decrease variability among assays by different platforms.

## Genetic disorders

Recent advancement in molecular technologies has allowed direct mutation detection for diagnostic testing, newborn and carrier screening, prenatal diagnosis, and pharmacogenomics. In this section, we highlight few examples of these applications.

### Single gene disorders- cystic fibrosis (CF)

Single gene disorders are known to be caused by variants in a specific gene. A catalogue of human genes and diseases can be found at the Online Mendelian Inheritance in Man website (OMIM, http://omim.org/). Human diseases such as cystic fibrosis, sickle cell anemia, Fragile X syndrome, spinal muscular dystrophy, and alpha-1-antitrypisn deficiency are all examples of single gene disorders. The inheritance patterns of these disorders include autosomal dominant, autosomal recessive, X-linked dominant, X-linked recessive, and mitochondrial.

CF is a common autosomal recessive disorder caused by mutations in the *CFTR* gene, which codes for the cystic fibrosis transmembrane conductance regulator (CFTR), a member of the ATP binding cassette sub-family C. Over 1900 different variants have been identified in CF patients. The prevalence of CF is highest in the Caucasian population (1 in 2500) and Ashkenazi Jewish population (1 in 2300), with carrier frequencies of 1 in 29 and 1 in 27, respectively. In 2001, CF carrier screening became the first recommended national genetic disease screening program. The American College of Medical Genetics (ACMG) and the American College of Obstetricians and Gynecologists identified a pan-ethnic screening panel consisting of 23 *CFTR* variants with an allele frequency of at least 0.1% in the US general population as a standard of care for CF carrier testing. Several technologies are available for CF carrier screening including line probe assay (LiPA), Oligonucleotide Ligation Assay (OLA), Invader, and microbead arrays. Most recently, the FDA has cleared two next generation sequencing assays for CF testing, the Illumina MiSeqDx Cystic Fibrosis

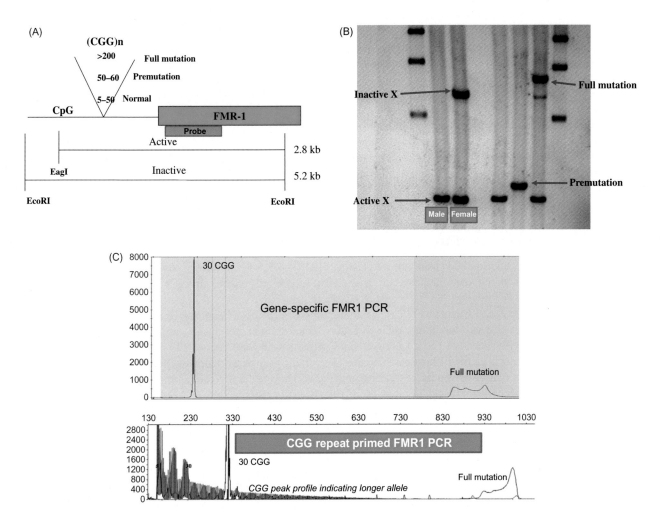

**FIGURE 31.5** (A) Schematic diagram of restriction digest sites and fragment sizes detected by southern blot for fragile X syndrome. (B) Fragile X syndrome southern blot analysis. (C) Fragile X syndrome testing by polymerase chain reaction and capillary electrophoresis.

139-Variant Assay and the MiSeqDx Cystic Fibrosis Clinical Sequencing Assay. Each assay has specific indications for carrier screening and diagnostics, respectively.

## Trinucleotide repeat disorders – fragile X syndrome (FXS)

FXS is the most common form of inherited mental retardation and is associated with expansion of a trinucleotide repeat, (CGG)n, in the 5' untranslated region of the *FMR1* gene which is located on Xq27.3. Individuals with 5-44 CGG repeats are considered to have a normal gene, while those with 45-54 are in an intermediate or gray zone. Premutation carriers have been described as those having 55-200 CGG repeats and at risk of passing an expanded gene to their offspring. The fully expanded repeat size of >200 is associated with FXS and results from an increased CGG copy number as well as abnormal methylation of the *FMR1* gene. Abnormal CGG repeat size has also been shown to occur in individuals with fragile X tremor ataxia syndrome and premature ovarian insufficiency.

Diagnostically, the fragile X chromosome was initially detected on karyograms. Once the gene region was identified, laboratories began to use Southern blotting for direct identification of both the CGG repeat size and presence or absence of methylation. This was performed by digesting DNA with routine and methylation sensitive restriction endonucleases, gel electrophoresis, transfer to membrane and hybridization with an *FMR1* gene-specific probe. With advances in PCR technology, many laboratories have been using PCR amplification methods followed by capillary electrophoresis for more accurate sizing of repeat sizes (Fig. 31.5).

*Germline copy number variants (CNVs) or contiguous gene deletions/duplications*

Intellectual disabilities are known to affect a significant percentage of the population and oftentimes the causative underlying genetic abnormality is not identified. Included in this category of diseases are the developmental delay disorders, autism spectrum disorders and psychiatric diseases for which a variety of genomic studies have been performed. These studies include karyotyping studies, mutation testing of certain genes, and sequencing of gene panels, clinical exome, whole exome or whole genome.

Chromosomal microarray analysis (CMA) techniques including array comparative genomic hybridization (aCGH) and SNP arrays are routinely used to screen the entire genome for copy number variations. In 2010, the American College of Medical genetics recommended chromosomal microarray (CMA) as the first line of diagnostic testing for patients with developmental delay and intellectual disabilities.

## Emerging technologies

### Liquid biopsies for molecular diagnosis

Recent advances in molecular methods have allowed testing of DNA/RNA derived from circulating tumor cells (CTCs) and cell-free nucleic acids (cfNAs: include cell free DNA (cfDNA), cell-free mRNA (cfRNA), and circulating miRNA).

cfNAs are small nucleic acid fragments released into the bloodstream from apoptotic or necrotic cells. cfDNA concentrations raises in physiological conditions such as cancer, acute trauma, pregnancy, transplantation, and infection. The developments of highly sensitive and accurate technologies such as NGS opened the possibility of using cfDNA for clinical applications such as liquid biopsies. Compared to traditional tissue biopsies, liquid biopsies have several advantages including minimal invasiveness, lower cost, and avoiding procedural complications and sampling biases that arise from genetic heterogeneity. With liquid biopsies, it is possible to get sequential samples over the course of the disease, allowing for close monitoring of disease progression and therapeutic response. The applications of cfDNA testing and liquid biopsies are vast and powerful, particularly for oncology and prenatal screening.

*Oncology*

The clinical utility of cfDNA is growing in oncology. Tumor derived cfDNA (circulating tumor DNA; ctDNA) is released from both primary and metastatic cancer lesions. Levels of ctDNA are highly variable, ranging from <0.1% to >50% of the total cfDNA.

Circulating tumor cells (CTCs) can also be used as source of DNA for molecular testing. CTCs are tumor cells shed into the bloodstream from both primary and metastatic tumor sites. However, CTCs are extremely rare in the blood, particularly at early stages, making their detection a challenge. By contrast, cfDNA is significantly more abundant and easier to purify than CTCs, making it a preferred source for molecular diagnosis, particularly in early-stage cancers. Liquid biopsies could allow early and serial assessment of metastatic disease, including follow-up during remission, characterization of treatment effects, and clonal evolution. Isolation and characterization of ctDNA are likely to improve cancer diagnosis, treatment, and minimal residual disease monitoring.

### Cancer diagnosis and screening

Early-stage cancers are highly curable by surgical resection but usually hard to detect because they manifest few to no symptoms. Liquid biopsies can potentially be used for cancer diagnosis and screening, as the ctDNA concentration correlates with the cancer stage and ctDNA can be detectable in early-stage cancer patients.

### Cancer molecular profiling

Somatic mutation profiling, copy number alterations and chromosomal rearrangements can be detected in cfDNA using both PCR-based (e.g., quantitative PCR, digital PCR) and NGS-based technologies. Liquid biopsies could also help to overcome the challenge of genetic heterogeneity of tumor, as the ctDNA is released from multiple tumor regions or different tumor foci. Liquid biopsies could also help in treatment selection of targeted therapy. FDA approval of the cobas® EGFR mutation Test V2 for cfDNA mutation detection as a companion diagnostic in selecting patients for anti-EGFR targeted therapies is a promise of liquid biopsy implementation in personalized oncology. The biggest technical challenge to analyzing cfDNA compared to tissue biopsies is its low mutant allele fraction and highly variable range of ctDNA levels. Therefore, the analytical sensitivity and dynamic range of the assay are crucial for maximizing the clinical utility of cfDNA.

### Prognosis and residual disease monitoring

The correlation between ctDNA levels and cancer stages suggests prognostic utility of ctDNA. In addition to ctDNA levels, mutational patterns can help to group patients into molecular subtypes with different prognoses.

### Monitoring of disease burden and treatment response

Due to the short half-life of ctDNA and its non-invasive nature, liquid biopsies can be used for real-time monitoring of disease progression and treatment response. Despite the highly-variable ctDNA level across different cancer types, the level within the same patient correlates well with the disease progression. The detection of ctDNA, even in the absence of clinical symptoms, could indicate the presence of minimal residual disease (MRD) and/or potential relapse due to treatment resistance.

Finally, elucidating the molecular changes involved in drug resistance is crucially important because a large portion of patients treated with matched targeted therapies ultimately develop drug resistance and disease progression.

## Obstetrics

Non-invasive prenatal screening (NIPS), which tests cfDNA released from the placenta during pregnancy to detect aneuploidy and other fetal chromosomal abnormalities, has already transitioned into clinical practice. NIPS has demonstrated superior sensitivity and specificity for aneuploidy screening compared to the standard prenatal screening approaches.

## Key concepts

- The discovery of molecular biomarkers and their clinical application to human disease has evolved along with advances in molecular techniques.
- The polymerase chain reaction (PCR) in its various forms propelled the field of molecular diagnostics forward.
- Real-time PCR made accurate quantification of nucleic acid targets in the fields of infectious diseases and oncology; other techniques including digital PCR are beginning to impact these fields as well.
- The impact of Next Generation Sequencing (NGS or Massively Parallel Sequencing) is starting to surpass the important role PCR has maintained for years in molecular diagnostics.
- The clinical fields of oncology, genetics and infectious diseases continue to evolve due to the applications of improved molecular techniques.
- Cell-free DNA is a biomarker that will continue to grow in importance within molecular diagnostics.

## Suggested readings

[1] Montgomery JL, Sanford LN, Wittwer CT. High-resolution DNA melting analysis in clinical research and diagnostics. Expert Rev Mol Diagn 2010;10:219−40.
[2] Tellinghuisen J, Spiess AN. Bias and imprecision in analysis of real-time quantitative polymerase chain reaction data. Anal Chem 2015;87:8925−31.
[3] Whale AS, Huggett JF, Cowen S, Speirs V, Shaw J, Ellison S, et al. Comparison of microfluidic digital PCR and conventional quantitative PCR for measuring copy number variation. Nucleic Acids Res 2012;40:e82.
[4] Schaaf CP, Wiszniewska J, Beaudet AL. Copy number and SNP arrays in clinical diagnostics. Annu Rev Genomics Hum Genet 2011;12:25−51.
[5] Tost J, Gut IG. Genotyping single nucleotide polymorphisms by MALDI mass spectrometry in clinical applications. Clin Biochem 2005;38:335−50.
[6] Buermans HP, den Dunnen JT. Next generation sequencing technology: advances and applications. Biochim Biophys Acta 2014;1842:1932−41.
[7] Chang F, Li MM. Clinical application of amplicon-based next-generation sequencing in cancer. Cancer Genet 2013;206:413−19.
[8] Zheng Z, Liebers M, Zhelyazkova B, Cao Y, Panditi D, Lynch KD, et al. Anchored multiplex PCR for targeted next-generation sequencing. Nat Med 2014;20:1479−84.
[9] Ambardar S, Gupta R, Trakroo D, Lal R, Vakhlu J. High throughput sequencing: an overview of sequencing chemistry. Indian J Microbiol 2016;56:394−404.

[10] Rothberg JM, Hinz W, Rearick TM, et al. An integrated semiconductor device enabling non-optical genome sequencing. Nature 2011;475:348−52.

[11] Dracopoli NC, Boguski MS. The evolution of oncology companion diagnostics from signal transduction to immuno-oncology. Trends Pharmacol Sci 2017;38:41−54.

[12] Gray PN, Dunlop CL, Elliott AM. Not all next generation sequencing diagnostics are created equal: understanding the nuances of solid tumor assay design for somatic mutation detection. Cancers (Basel) 2015;7:1313−32.

[13] Jung HS, Lefferts JA, Tsongalis GJ. Utilization of the oncoscan microarray assay in cancer diagnostics. Appl Cancer Res 2017;37:1.

[14] Sepulveda AR, Jones D, Ogino S, et al. CpG methylation analysis--current status of clinical assays and potential applications in molecular diagnostics: a report of the association for molecular pathology. J Mol Diagn 2009;11:266−78.

[15] Bartels CL, Tsongalis GJ. MicroRNAs: novel biomarkers for human cancer. Clin Chem 2009;55:623−31.

[16] Ruggiero P, McMillen T, Tang YW, Babady NE. Evaluation of the BioFire FilmArray respiratory panel and the GenMark eSensor respiratory viral panel on lower respiratory tract specimens. J Clin Microbiol 2014;52:288−90.

[17] Seng P, Rolain JM, Fournier PE, La Scola B, Drancourt M, Raoult D. MALDI-TOF-mass spectrometry applications in clinical microbiology. Future Microbiol 2010;5:1733−54.

[18] Frickmann H, Masanta WO, Zautner AE. Emerging rapid resistance testing methods for clinical microbiology laboratories and their potential impact on patient management. Biomed Res Int 2014;2014:375681.

[19] Brianti P, De Flammineis E, Mercuri SR. Review of HPV-related diseases and cancers. New Microbiol 2017;40:80−5.

[20] Strom CM, Crossley B, Buller-Buerkle A, et al. Cystic fibrosis testing 8 years on: lessons learned from carrier screening and sequencing analysis. Genet Med 2011;13:166−72.

[21] Grigsby J. The fragile X mental retardation 1 gene (FMR1): historical perspective, phenotypes, mechanism, pathology, and epidemiology. Clin Neuropsychol 2016;30:815−33.

[22] Rosenfeld JA, Patel A. Chromosomal microarrays: understanding genetics of neurodevelopmental disorders and congenital anomalies. J Pediatr Genet 2017;6:42−50.

[23] De Vlaminck I, Martin L, Kertesz M, et al. Noninvasive monitoring of infection and rejection after lung transplantation. Proc Natl Acad Sci USA 2015;112:13336−41.

[24] Wan JC, Massie C, Garcia-Corbacho J, et al. Liquid biopsies come of age: towards implementation of circulating tumour DNA. Nat Rev Cancer 2017;17:223−38.

[25] Skrzypek H, Hui L. Noninvasive prenatal testing for fetal aneuploidy and single gene disorders. Best Pract Res Clin Obstet Gynaecol 2017;42:26−38.

Chapter 32

# Pharmacogenomics and personalized medicine in the treatment of human diseases

Robert D. Nerenz

*Assistant Professor of Pathology and Laboratory Medicine, Dartmouth-Hitchcock Medical Center, Lebanon, NH, United States*

**Summary**

It is clear that patients respond to therapeutic drugs differently, and much of this variation can be attributed to differences in gene sequence or copy number, resulting in proteins with altered stability, cellular concentration, metabolic activity, or signaling properties. The application of pharmacogenetics, the study of how inherited variants affect drug response, is expected to reduce the number of adverse drug reactions by facilitating the selection of an appropriate drug and dose more efficiently than using a trial-and-error approach. Similarly, detection of somatic variants in cancer cells can help select an effective chemotherapeutic agent and identify appropriate targeted therapies for the individual patient's unique tumor phenotype. This chapter will discuss the anticipated benefits and potential challenges of implementing pharmacogenetic testing to guide patient care decisions.

## Introduction

In 2001, the first draft of the Human Genome Project was simultaneously published by two groups. This marked the beginning of an era that promised better diagnostic and prognostic testing that would lead to preventive medicine and more personalized therapy. Because of our greater understanding of the human genome, clinicians can select and dose therapeutic drugs using information gained from genetic testing. Although other environmental factors also play a role in this selection, an individual's genetic makeup is an important determinant of drug metabolism and may be used to guide therapeutic decisions. The need to incorporate this knowledge into routine clinical practice is highlighted by the more than 2 million annual hospitalizations and greater than 100,000 annual deaths in the United States due to adverse drug reactions (ADRs).

Current medical practice often utilizes a trial-and-error approach to select the proper medication and dosage for a given patient (Fig. 32.1). As an attempt to improve on this trial-and-error approach, pharmacogenetics (PGx) comprises the study of inherited genetic variants that affect drug response. When combined with pharmacokinetics (effect of body on the drug) and pharmacodynamics (effect of drug on the body), PGx can help clinicians tailor their therapeutic decisions according to the individual needs of each patient (Fig. 32.2). This application of human genetics associates genomic sequence variations with alterations in absorption, distribution, metabolism, and excretion of therapeutic drugs at a systemic level. The overall aim of PGx testing is to decrease adverse responses to therapy and increase efficacy by ensuring the appropriate drug and dose selection. A second component of this application of genomics to personalized medicine is in the identification of gene variants that may be targeted by novel therapies.

Numerous genetic variants or polymorphisms in genes that code for drug metabolizing enzymes, drug transporters, and drug targets that alter response to therapeutics have been identified (www.pharmgkb.org/) (Table 32.1). Classification of these enzymes follows a specific pattern of nomenclature. For example, CYP2D6*1 includes the name of the enzyme (cytochrome P450, CYP), followed by family (2), subfamily (D), and gene (6). Allelic variants are indicated by an asterisk followed by a number (*1). The technologies that are currently available in clinical laboratories to routinely test for these variants are shown in Table 32.2.

**Essential Concepts in Molecular Pathology. DOI: https://doi.org/10.1016/B978-0-12-813257-9.00032-2**

## Current reactive medical care model

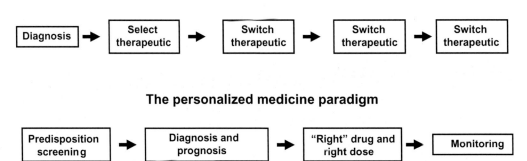

```
Diagnosis  →  Select        →  Switch       →  Switch       →  Switch
              therapeutic      therapeutic     therapeutic     therapeutic
```

## The personalized medicine paradigm

```
Predisposition  →  Diagnosis and  →  "Right" drug and  →  Monitoring
screening          prognosis         right dose
```

**FIGURE 32.1** Schematic diagram illustrating current medical practice and the changes that occur in a personalized medicine model.

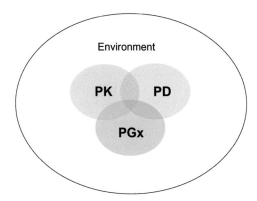

**FIGURE 32.2** Venn diagram of pharmacogenomics showing the interactions of pharmacogenetics (PGx), pharmacokinetics (PK), and pharmacodynamics (PD), along with other environmental factors that affect drug response.

**TABLE 32.1 Examples of enzymes and their designated polymorphic alleles.**

| Enzyme and alleles |
| --- |
| Cytochrome P450 2D6 (CYP2D6) <br> *2, *3, *4, *5, *6, *7, *8, *11, *12, *13, *14, *15, *16, *18, *19, *20, *21, *38, *40, *42, *10, *17, *36, *41 |
| Cytochrome P450 2C9 (CYP2C9) <br> *1, *2, *3, *4, *5, *6 |
| Cytochrome P450 2C19 (CYP2C19) <br> *2, *3 |
| UGT1A1 <br> *6, *28, *37, *60 |

Single nucleotide polymorphisms (SNPs) in these genes may have no significant phenotypic effect. On the other hand, SNPs may increase or decrease the rate of drug metabolism by >10,000-fold or alter protein binding by >20-fold. More than 3.5 million SNPs have been identified in the human genome (www.hapmap.org), making analysis quite challenging. As shown in Table 32.1, different genes can contain various numbers of SNPs that alter the function of the encoded protein. It is important to clearly differentiate benign polymorphisms (present in greater than 1% of the general population) from rare disease-causing mutations that are present in less than 1% of the general population. The SNPs relevant to the practice of PGx do not typically result in any form of genetic disease. However, we do know that these

**TABLE 32.2 Genotyping methodologies currently in use in the clinical laboratory.**

| Technology | Methodology[a] | Company[a] |
|---|---|---|
| Gel electrophoresis | Southern blot, PCR-RFLP[b] | Generic equipment |
| Real-time PCR | Allele-specific PCR | Applied Biosystems |
| | | Biorad |
| | | Cepheid |
| | | Roche molecular |
| | High-resolution melt analysis | Canon bioMedical |
| DNA sequencing | Sanger—fluorescent detection | Applied biosystems |
| | | Beckman Coulter |
| | Next-generation sequencing | Admera, Archer, SwiftBio, Illumina, Thermofisher |
| Microarrays | Hybridization/fluorescent detection | Roche molecular |
| | | Autogenomics |
| Liquid beads | Hybridization/fluorescent detection | Luminex |
| Mass spectrometry | MALDI-TOF[c] | Agena |

[a]Representative examples.
[b]PCR-RFLP, polymerase chain reaction–restriction fragment length polymorphism analysis.
[c]Matrix-assisted laser desorption/ionization time of flight.

SNPs may be used to evaluate individual risk for ADRs. Incorporating this information into clinical decision making is expected to decrease ADRs, facilitate selection of optimal therapy, increase patient compliance, promote development of safer and more effective drugs, revive withdrawn drugs, and reduce time and cost of clinical trials.

## Genotyping technologies

We currently have the means to rapidly and accurately identify variant sequences in genes that play important roles in drug metabolism (Table 32.2). Genotyping involves the identification of defined genetic variants that give rise to the specific drug response phenotypes. Genotyping methods are easier to perform and more cost effective than the more traditional phenotyping methods that assess enzymatic activity or rate of drug removal from circulation. Because an individual's genotype is not expected to change over time, most genotyping applications need to be performed only once in an individual's lifetime. The exception to this is testing for the acquired genetic variants that occur in cancer where tumor cell subclones with different mutations may emerge during the course of the disease. Although in most noncancer cases, a single genotype can determine responses to numerous therapeutic drugs, it is worth noting that patients and providers will need to refer back to genotype information on numerous occasions.

Early genotyping efforts utilized conventional procedures such as Southern blotting and the polymerase chain reaction (PCR) followed by restriction endonuclease digestion to interrogate human gene sequences for polymorphisms. Direct sequencing reactions have been developed on automated capillary electrophoresis instruments to detect specific base changes that result in a variant allele. The introduction of real-time PCR to clinical laboratories has resulted in assays that can be performed much faster and in a multiplexed fashion so that more variants are evaluated in a single assay (Fig. 32.3). More recently, platforms utilizing various array-based technologies have been introduced into the clinical laboratory for genotyping purposes.

## PGx and drug metabolism

Most of the enzymes involved in drug metabolism are members of the cytochrome P450 (CYP450) superfamily. CYP450 enzymes are mainly located in the liver and gastrointestinal tract and include greater than 30 isoforms. The most polymorphic of these enzymes responsible for the majority of biotransformations are the CYP3A subfamily,

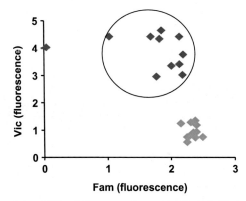

**FIGURE 32.3** Allelic discrimination using real-time PCR and Taqman probes to identify individuals who are homozygous or heterozygous (*circle*) for a particular single nucleotide polymorphism.

**TABLE 32.3 Some therapeutic drugs metabolized by CYP2D6 (cytochrome P450 2D6).**

| | |
|---|---|
| Amitriptyline | Lidocaine |
| Aripiprazole | Metoclopramide |
| Atomoxetine | Metoprolol |
| Carvedilol | Nortriptyline |
| Chlorpromazine | Oxycodone |
| Clomipramine | Paroxetine |
| Codeine | Propafenone |
| Desipramine | Propranolol |
| Dextromethorphan | Risperidone |
| Duloxetine | Tamoxifen |
| Flecainide | Thioridazine |
| Fluoxetine | Timolol |
| Fluvoxamine | Tramadol |
| Haloperidol | Venlafaxine |
| Imipramine | Zuclopenthixol |

CYP2D6, CYP2C19, and CYP2C9. Benign genetic variants or polymorphisms in these genes can lead to the following phenotypes: poor, intermediate, extensive, and ultrarapid metabolizers. Poor metabolizers (PM) have no detectable enzymatic activity, intermediate metabolizers (IM) have decreased enzymatic activity, extensive metabolizers (EM) demonstrate typical enzymatic activity and have at least one copy of an active gene, and ultrarapid metabolizers (UM) contain duplicated or amplified gene copies that result in increased drug metabolism. The following are several examples of polymorphic drug metabolizing enzymes that can affect response to therapy.

## CYP2D6

CYP2D6 is an example of one of the most widely studied members of this enzyme family. It is highly polymorphic and contains 497 amino acids. More than 50 alleles of *CYP2D6* have been described, of which alleles *3, *4, *5, *6, *7, *8, *11, *12, *13, *14, *15, *16, *18, *19, *20, *21, *38, *40, *42, and *44 were classified as nonfunctioning and alleles *9, *10, *17, *36, and *41 were reported to have substrate-dependent decreased activity. CYP2D6 alone is responsible for the metabolism of 20%−25% of prescribed drugs (Table 32.3).

An estimated 5%−10% of patients are considered poor metabolizers and have absent or very low CYP2D6 enzymatic activity. Screening for *CYP2D6\*3*, *CYP2D6\*4*, and *CYP2D6\*5* alleles identifies at least 95% of poor metabolizers in the Caucasian population. An assessment of CYP2D6 activity is particularly important in patients receiving medication for pain as the commonly prescribed opioids, codeine and hydrocodone, are partially metabolized by CYP2D6 to the more potent morphine and hydromorphone, respectively. Individuals with two inactive copies of *CYP2D6* are unable to convert enough of the parent compound to the more potent metabolite to achieve adequate pain relief. In these patients, alternate analgesic options are recommended.

Conversely, an estimated 1%−2% of patients are considered ultra-rapid metabolizers and have increased CYP2D6 enzymatic activity. These patients demonstrate increased conversion of codeine to morphine and are at risk for morphine overdose if a standard dose of codeine is administered. As in the case of poor metabolizers, alternate analgesic options are recommended in ultra-rapid metabolizers.

## CYP2C9

Another example of CYP450 enzyme polymorphisms causing alterations in drug metabolism is demonstrated by *CYP2C9* and warfarin. *CYP2C9* polymorphisms (*CYP2C9\*2* and *CYP2C9\*3*) were first found to reduce the metabolism of S-warfarin, which can lead to supratherapeutic concentrations when a standard dose is administered. Soon after, the identification of a polymorphism in the vitamin K epoxide reductase (*VKORC1*) gene proved that drug target polymorphisms must also be taken into account when selecting the appropriate warfarin dose. Since then, several warfarin dosing algorithms that incorporate patient demographics, *CYP2C9* and *VKORC1* polymorphisms, and concurrent drug use have been developed. In August 2007, the FDA announced an update to the warfarin package insert to include information on *CYP2C9* and *VKORC1* testing.

However, in 2015, the Centers for Medicare and Medicaid Services (CMS) amended their policy regarding *CYP2C9* and *VKORC1* PGx testing to limit reimbursement for Medicare beneficiaries to scenarios in which the following criteria are met: (1) previous *CYP2C9* and *VKORC1* testing has not been performed, (2) patient has received fewer than 5 days of warfarin treatment, and (3) patient is enrolled in a CMS-approved prospective, randomized, controlled clinical study evaluating PGx testing in the prediction of warfarin responsiveness. Similarly, Aetna, a major health insurance provider, does not cover PGx testing in the prediction of warfarin response because it considers the clinical value of this testing to be unproven.

## Drug−drug interactions

In addition to the genotype predicting a phenotype, drug metabolizing enzyme activity can be induced or inhibited by various drugs. Induction leads to the production of more enzymes within three or more days of exposure to inducers. Conversely, enzyme inhibition is usually the result of competition between two drugs for metabolism by the same enzyme. Given that many patients are prescribed multiple medications, drug−drug interactions can complicate dosing decisions. A better understanding of drug metabolism has led to a number of amended labels for commonly prescribed drugs (Table 32.4).

**TABLE 32.4 Examples of therapeutic drugs that include PGx information or have had labels amended due to new knowledge of metabolism.**

| | |
|---|---|
| Atomoxetine | Strattera |
| Thioridazine | Mellaril |
| Voriconazole | Vfend |
| 6-Mercaptopurine | Purinethol |
| Azathioprine | Imuran |
| Irinotecan | Camptosar |
| Warfarin | Coumadin |

## PGx and drug transporters

Although the genes that code for drug metabolizing enzymes have received more attention in recent years as markers for PGx testing, the genes that code for proteins used to transport drugs across membranes also need to be considered when discussing PGx. These drug transporter proteins move substrates across cell membranes, bringing them into cells or removing them from cells. These proteins are essential in the absorption, distribution, and elimination of various endogenous and exogenous substances including pharmaceutical agents.

### ABCB1

ABCB1 is a member of the ATP-binding cassette (ABC) superfamily of proteins. Also known as *P*-glycoprotein (*P*-gp) or MDR1, it is a 170 kDa glycosylated membrane protein expressed in various locations including the liver, intestines, kidney, brain, and testis. Generally, ABCB1 is located on the membrane of cells in these locations and serves to eliminate metabolites and a wide range of hydrophobic foreign substances from cells by acting as an efflux transporter. Due to the localization of ABCB1 on specific cells, ABCB1 aids in eliminating drugs into the urine or bile and helps maintain the blood—brain barrier.

ABCB1 was first identified in cancer cells that had developed a resistance to several anticancer drugs because of an overexpression of the transporter. When expressed at typical levels in noncancerous cells, ABCB1 has been shown to transport other classes of drugs out of cells including cardiac drugs (digoxin), antibiotics, steroids, HIV protease inhibitors, and immunosuppressants (cyclosporin A).

Genetic variations in the ABCB1 gene expressed in normal cells have been shown to have a role in inter-individual variability in drug response. Although many polymorphisms have been detected in the *ABCB1* gene, correlations between genotype and either protein expression or function have been described for only a few of the genetic variants. Most notable among these is the 3435 C>T polymorphism, found in exon 26, which has been found to result in decreased expression of ABCB1 in individuals homozygous for the T allele. However, these results have been questioned by subsequent studies. The possible importance of the 3435 C>T is particularly interesting considering the mRNA levels are not affected by this polymorphism, which is found within a coding exon. Although the correlation between the 3435 C>T polymorphism and ABCB1 protein levels may be due to linkage of this polymorphism to others in the *ABCB1* gene, a recent study showed that this polymorphism alone does not affect *ABCB1* mRNA or protein concentration, but does result in ABCB1 protein with an altered structure. The altered configuration due to 3435 C>T is hypothesized to be caused by the usage of a rare codon, which may affect proper folding or insertion of the ABCB1 into the membrane, affecting the function, but not the concentration. Genetic variations affecting the expression or function of drug transporter proteins, such as ABCB1, could drastically alter the pharmacokinetics and pharmacodynamics of a given drug.

## PGx and drug targets

Most therapeutic drugs act on specific targets, including receptors, enzymes, or proteins involved in various cellular events such as signal transduction and cell replication. Classical PGx evaluates metabolic enzyme polymorphisms that can be associated with differences in drug absorption, biotransformation, and toxicity. Recently, the field of PGx has expanded to include physiogenetics, the evaluation of how polymorphisms in drug targets render the patient more resistant or sensitive to the particular therapeutic agent. Without affecting the rate of drug metabolism, these drug target polymorphisms can have important effects on endocrine regulatory networks, signal transduction pathways, and neurochemistry, and can ultimately be associated with differences in patient response to drug. The human μ-opioid receptor gene *OPRM1* and the dopamine receptor genes *DRD1* and *DRD2* represent examples of the effects of drug target polymorphisms.

### OPRM1

The human μ-opioid receptor, coded by the *OPRM1* gene, is the major site of action for endogenous β-endorphin and most exogenous opioids. Forty-three SNPs have been discovered in the *OPRM1* gene, but the 118A>G SNP is the most well studied and has an allele frequency of 10%—30% in Caucasians. This SNP causes substantial decreases in *OPRM1* mRNA concentration and a corresponding decrease in μ-opioid receptor concentration, as well as an increased affinity of the receptor for β-endorphin, which improves pain tolerance in humans. However, the effect of this SNP is controversial in people taking exogenous opioids. Studies showed that 118G allele carriers required an increased dose of morphine in patients with acute and chronic pain, but a decreased dose of fentanyl in patients requiring epidural or intravenous fentanyl for pain control during labor. Subsequent work has demonstrated an association of the *OPRM1* 118A > G SNP with an increased risk of opiate addiction, but *OPRM1* testing is not routinely incorporated into therapeutic decisions in either the pain management or substance abuse settings. However, multiple companies offering PGx testing are beginning to include *OPRM1* and other drug receptors in their substance abuse panels.

## DRD1/2

The dopamine receptors 1 and 2, encoded by the *DRD1* and *DRD2* genes, respectively, are involved in several neurological processes including pleasure, memory, learning, fine motor control, and motivation, and several SNPs responsible for atypical dopamine receptor signaling have been linked with various neuropsychiatric disorders. The *DRD1* SNP rs4532 (minor allele frequency 25%−30%) has been associated with resistance to antipsychotic medication as schizophrenic patients homozygous for the atypical G allele have a fivefold greater risk of treatment resistance than patients homozygous for the wild-type A allele. This same SNP was associated with an increased risk of impulse control behaviors in patients with Parkinson disease treated with dopaminergic medications. Lastly, several SNPs in *DRD1* and *DRD2* have been associated with an increased risk of opioid addiction. As in the case of *OPRM1*, detection of SNPs in *DRD1* and *DRD2* has not been incorporated into routine clinical practice, but it is conceivable that this testing may guide therapeutic decisions in the future.

# PGx applied to oncology

Cancer represents a complex set of deregulated cellular processes that are often the result of alterations in underlying molecular mechanisms. Although there is recognized inter-individual variability with respect to an observed chemotherapeutic response, it is also apparent that there is considerable cellular heterogeneity within a single tumor. It is becoming clear that molecular genetic variants in genes coding for drug-metabolizing enzymes contribute significantly to an individual's response to a particular therapy. As an additional level of complexity, in the cancer patient the acquired genetics of the tumor cell must also be taken into account. Unlike traditional PGx testing that focuses on the underlying genome of the individual, PGx testing for cancer patients must also identify acquired somatic variants in the tumor cells.

Cancer patients exhibit a heterogeneous response to chemotherapy with only 25%−30% efficacy. PGx can improve on chemotherapeutic and targeted therapy responses by providing a more informative evaluation of the multiple underlying molecular determinants associated with tumor cell heterogeneity. In the cancer patient, PGx can be used to integrate information related to the patient's inherited variants that may predict chemotherapeutic responsiveness with somatically acquired alterations in molecular biomarkers. Thus, as with previous examples of PGx testing, therapeutic management of the cancer patient can be tailored to the individual patient or tumor phenotype.

## UGT1A1

The uridine diphosphate glucuronosyltransferase (UGT) superfamily of endoplasmic reticulum−bound enzymes is responsible for conjugating a glucuronic acid moiety to a variety of compounds, rendering these compounds more water-soluble and facilitating excretion. It is a member of this family (UGT1A1) that catalyzes the glucuronidation of bilirubin, allowing it to be excreted in the bile. As irinotecan therapy for advanced colorectal cancers became more widely used, it was observed that patients who had Gilbert syndrome (decreased *UGT1A1* expression, leading to mild hyperbilirubinemia) suffered severe toxicity. The observation made in Gilbert syndrome patients revealed that the irinotecan metabolite SN-38 (Fig. 32.4) shares a glucuronidation pathway with bilirubin. The decreased glucuronidation of bilirubin and SN-38 can be attributed to polymorphisms in the *UGT1A1* gene. The wild-type allele of this gene, *UGT1A1*1*, has six tandem TA repeats in the regulatory TATA box of the *UGT1A1* promoter. The most common polymorphism associated with low activity of *UGT1A1* is the *28 variant, which has seven TA repeats. In August 2005, the FDA amended the irinotecan package insert to recommend genotyping for the *UGT1A1* polymorphism and suggested a dose reduction in patients homozygous for the *28 allele, estimated to constitute approximately 10% of the North American population.

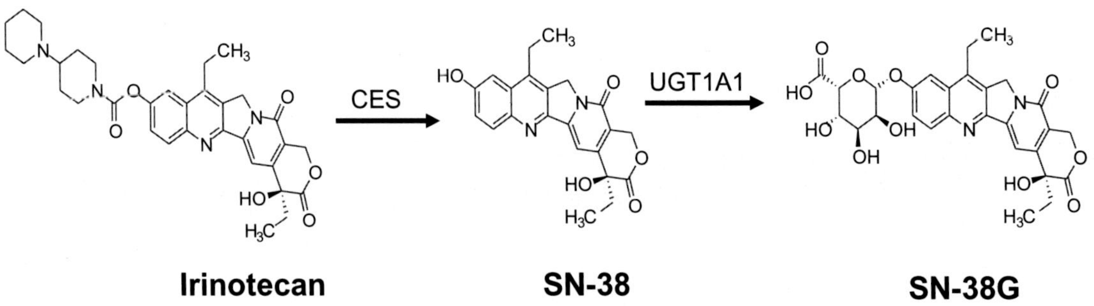

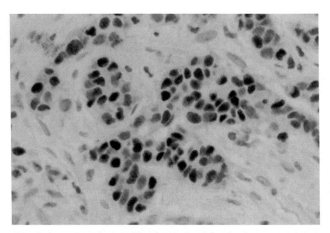

**FIGURE 32.4** Schematic diagram of irinotecan metabolism.

**FIGURE 32.5** Estrogen receptor—positive breast cancer. Immunohistochemical staining for the estrogen receptor in breast cancer.

## TPMT

Thiopurine *S*-methyltransferase (TPMT) is a cytosolic enzyme that inactivates thiopurine drugs such as 6-mercaptopurine and azathioprine through methylation. There is substantial variability in TPMT enzyme activity between individuals, and it has been found that this variability can be attributed to polymorphisms in the *TPMT* gene. The most common variant alleles, *TPMT*2*, *TPMT*3A*, and *TPMT*3C*, account for 95% of TPMT deficiency. Molecular testing is a relatively convenient method for assessment of TPMT enzyme function in patients before treatment with thiopurines. Insufficient TPMT activity could put a patient at risk for developing hematopoietic toxicity, because too much drug would be converted to 6-TGNs. In these patients, a reduced thiopurine dose is recommended. On the other hand, a patient with high TPMT activity would need higher than standard doses of a thiopurine drug to elicit a therapeutic response.

## Targeted therapies in oncology

### Estrogen receptor

One of the first examples of targeted therapeutics in oncology is the treatment of estrogen receptor (ER)-positive (Fig. 32.5) breast cancers with hormonal therapies that inhibit the pro-proliferative effects of estrogens in breast tissue, but demonstrate partial agonist activity in other estrogen-sensitive tissues. One of these estrogen analogs is tamoxifen (TAM), which itself has now been shown to have altered metabolism due to CYP450 genetic polymorphisms. Conversion of TAM to its active metabolites occurs predominantly through the CYP450 system (Fig. 32.6). Conversion of TAM to primary and secondary metabolites is important because these metabolites can have a greater affinity for the ER than TAM itself. For example, 4-OH *N*-desmethyl-TAM (endoxifen) has approximately 100-fold greater affinity for the ER than tamoxifen. Activation of tamoxifen to endoxifen is primarily due to the action of CYP2D6. Therefore, patients with defective *CYP2D6* alleles derive less benefit from tamoxifen therapy than patients with functional copies of *CYP2D6*.

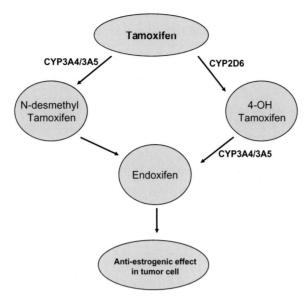

**FIGURE 32.6**    Metabolism of tamoxifen. Simplified schematic diagram showing metabolism of tamoxifen by CYP450 enzymes with the resulting production of the active metabolite endoxifen.

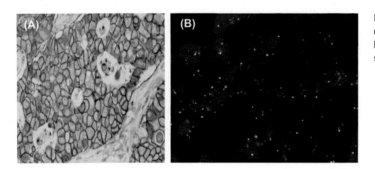

**FIGURE 32.7**    Determination of *HER2* status in breast cancer. (A) HER2 protein overexpression can be shown using immunohistochemistry. (B) *HER2* gene amplification can be demonstrated using fluorescence in situ hybridization.

## HER2

The human epidermal growth factor receptor 2 (*ERBB2* or *HER2*) gene is amplified in up to 25% of all breast cancers. While this gene can be expressed in low levels in a variety of normal epithelia, amplification of the *HER2* gene is the primary mechanism of HER2 overexpression that results in increased receptor tyrosine kinase activity. Assessment of HER2 status has been implemented as an indicator of both prognosis as well as a predictive marker for response to therapy such that HER2-positive breast cancers have a worse prognosis and are often resistant to hormonal therapies and other chemotherapeutic agents. As a predictive marker, HER2 status is utilized to determine sensitivity to anthracycline-based chemotherapy regimens. (Fig. 32.7). Several FDA-cleared tests for immunohistochemical detection of HER2 protein and fluorescence in situ hybridization (FISH) detection of gene amplification are commercially available and approved as companion diagnostics for Herceptin therapy.

## EGFR

The epidermal growth factor receptor (*EGFR* or *HER1*) is a member of the ErbB family of tyrosine kinases that also includes HER2. On binding ligand, the receptor undergoes homodimerization or heterodimerization. This dimerization induces phosphorylation, which initiates signaling to downstream pathways such as PI3K/AKT and RAS/RAF/MAPK, which in turn regulates cell proliferation and apoptosis.

EGFR is expressed in some lung cancers and is the target of newly developed small molecule drugs, including afatinib, gefitinib, and erlotinib. It has been shown that tumors that respond to these tyrosine kinase (TK) inhibitors (TKIs) contain somatic mutations in the EGFR TK domain. Mutation analysis in the *EGFR* represents a new application for molecular diagnostics as some mutations confer a favorable response, while others are not so favorable. It becomes critical to the management of the NSCLC patient that the clinical molecular diagnostics laboratory be able to identify such mutations on a routine basis.

### FLT3

FLT3 is a receptor tyrosine kinase expressed and activated in most cases of acute myeloid leukemia (AML), which has a relatively high relapse rate due to acquired resistance to traditional chemotherapies. An internal tandem duplication (ITD) mutation in the *FLT3* gene is found in up to 30% of AML patients, whereas point mutations have been shown to account for approximately 5% of refractory AML. The FLT3-ITD induces activation of this receptor and results in downstream constitutive phosphorylation in STAT5, AKT, and ERK pathways. This mutation in *FLT3* is a negative prognostic factor in AML. Recently, a novel multi-targeted receptor tyrosine kinase inhibitor, ABT-869, has been developed to suppress signal transduction from constitutively expressed kinases such as a mutated FLT3.

### BCR-ABL

Bcr-Abl is a chimeric oncoprotein formed through the fusion of the *ABL1* gene on chromosome 9 and the breakpoint cluster gene (*BCR*) on chromosome 22. *ABL1* encodes a tyrosine kinase involved in cellular differentiation, division, and adhesion that typically requires activation by cytokines to initiate signal transduction. However, the Bcr-Abl fusion generates a constitutively active tyrosine kinase that drives uncontrolled cellular growth and differentiation and is responsible for more than 90% of cases of chronic myelogenous leukemia (CML). Prior to 2001, CML was treated with nonspecific, cytotoxic chemotherapeutics that were poorly tolerated and relatively ineffective. In 2001, Imatinib (Gleevec), a tyrosine kinase inhibitor was introduced and significantly improved the survival rate in CML patients (4-year survival rate of 86% with Imatinib vs. 43% with other therapies). Detection of the *BCR-ABL1* fusion gene is now routinely incorporated into the initial diagnosis and monitoring of CML patients.

## Challenges encountered

Despite the promise of improved therapeutic outcomes, several challenges must be overcome before PGx testing becomes a routine part of clinical care. First, although CMS has determined that in some disease states, genetic testing facilitates the selection of specific, targeted therapies and clearly improves clinical outcomes, CMS argues that a conclusive benefit has yet to be demonstrated for many other gene-drug interactions. As a result, medical centers that wish to offer PGx testing must do so with the understanding that testing will be reimbursed only for certain genes and under well-defined clinical circumstances. Second, clinicians are often uncomfortable interpreting and acting on PGx test results, in many cases leading to underutilization and misinterpretation. Robust decision support tools and extensive provider education will be required to ensure correct utilization and interpretation of PGx testing to ultimately improve patient care.

## Conclusion

Incorporation of PGx into routine clinical practice offers the promise of fewer ADRs, greater patient compliance, and ultimately better therapeutic outcomes. New technological advances facilitate the routine performance of these tests in the clinical laboratory, and PGx principles continue to be applied in novel clinical areas. Currently, PGx testing can be performed to detect polymorphisms in the genes encoding metabolizing enzymes, transporter proteins, and some drug targets. One growing application in cancer patients is to detect polymorphisms and mutations associated with responses to newly developed small molecule–targeted therapies. The ultimate goal of these efforts is to truly provide a *personalized medicine* approach to patient management for the purpose of selecting more efficacious therapeutics and improving the overall well-being of the patient.

## Key concepts

- Sequence variants in genes that encode drug metabolizing enzymes, drug transporters and drug targets determine an individual's response to treatment
- Detection of these variants can facilitate selection of the appropriate therapeutic compound and dose to improve efficacy and avoid adverse drug reactions

- Traditional pharmacogenomic testing evaluates inherited sequence variants and only needs to be performed once in a person's lifetime
- Pharmacogenomic testing for cancer patients should include an assessment of both inherited variants and somatic mutations that may only be present in a subset of tumor cells
- Widespread adoption of pharmacogenomic testing in routine clinical use will require robust decision support tools and demonstrated improvement in clinical outcomes

## Suggested readings

[1]   Lander ES, Linton LM, Birren B, et al. Initial sequencing and analysis of the human genome. Nature 2001;409:860−921.

[2]   Venter JC, Adams MD, Myers EW, et al. The sequence of the human genome. Science 2001;291:1304−51.

[3]   Evans WE, McLeod HL. Pharmacogenomics − drug disposition, drug targets, and side effects. N Engl J Med 2003;348:538−49.

[4]   Kirchheiner J, Fuhr U, Brockmöller J. Pharmacogenetics-based therapeutic recommendations - ready for clinical practice? Nat Rev Drug Discov 2005;4:639−47.

[5]   Shastry BS. Pharmacogenetics and the concept of individualized medicine. Pharmacogenomics J 2006;6:16−21.

[6]   Nebert DW, Ingelman-Sundberg M, Daly AK. Genetic epidemiology of environmental toxicity and cancer susceptibility: human allelic polymorphisms in drug-metabolizing enzyme genes, their functional importance, and nomenclature issues. Drug Metab Rev 1999;31:467−87.

[7]   Lazarou J, Pomeranz BH, Corey PN. Incidence of adverse drug reactions in hospitalized patients: a meta-analysis of prospective studies. JAMA 1998;279:1200−5.

[8]   Evans WE, Relling MV. Pharmacogenomics: translating functional genomics into rational therapeutics. Science 1999;286:487−91.

[9]   Linder MW, Prough RA, Valdes R. Pharmacogenetics: a laboratory tool for optimizing therapeutic efficiency. Clin Chem 1997;43:254−66.

[10]  Ingelman-Sundberg M. Genetic polymorphisms of cytochrome P450 2D6 (CYP2D6): clinical consequences, evolutionary aspects and functional diversity. Pharmacogenomics J 2005;5:6−13.

[11]  Crews KR, Gaedigk A, Dunnenberger HM, et al. Clinical Pharmacogenetics implementation consortium guidelines for cytochrome P450 2D6 genotype and codeine therapy: 2014 update. Clin Pharmacol Ther 2014;95:376−82.

[12]  Maitland ML, Vasisht K, Ratain MJ. TPMT, UGT1A1 and DPYD: genotyping to ensure safer cancer therapy? Trends Pharmacol Sci 2006;27:432−7.

[13]  Early Breast Cancer Trialists' Collaborative Group. Tamoxifen for early breast cancer: an overview of the randomised trials. Lancet 1998;351:1451−67.

[14]  Slamon DJ, Godolphin W, Jones LA, et al. Studies of the HER-2/neu proto-oncogene in human breast and ovarian cancer. Science 1989;244:707−12.

[15]  Lynch TJ, Bell DW, Sordella R, et al. Activating mutations in the epidermal growth factor receptor underlying responsiveness of non-small-cell lung cancer to gefitinib. N Engl J Med 2004;350:2129−39.

# Index